HARMONIC VECTOR FIELDS

T0348847

HARMONIC VECTOR FIELDS

VARIATIONAL PRINCIPLES AND DIFFERENTIAL GEOMETRY

SORIN DRAGOMIR

Università degli Studi della Basilicata

DOMENICO PERRONE

Università del Salento

ELSEVIER

Amsterdam • Boston • Heidelberg • London
New York • Oxford • Paris • San Diego
San Francisco • Singapore • Sydney • Tokyo

Elsevier
225 Wyman Street, Waltham, MA 02451, USA
The Boulevard, Langford Lane, Kidlington, Oxford, OX5 1GB, UK

© 2012 Elsevier Inc. All rights reserved.

No part of this publication may be reproduced or transmitted in any form or by any means, electronic or mechanical, including photocopying, recording, or any information storage and retrieval system, without permission in writing from the publisher. Details on how to seek permission, further information about the Publisher's permissions policies and our arrangements with organizations such as the Copyright Clearance Center and the Copyright Licensing Agency, can be found at our website: *www.elsevier.com/permissions*.

This book and the individual contributions contained in it are protected under copyright by the Publisher (other than as may be noted herein).

Notices

Knowledge and best practice in this field are constantly changing. As new research and experience broaden our understanding, changes in research methods, professional practices, or medical treatment may become necessary.

Practitioners and researchers must always rely on their own experience and knowledge in evaluating and using any information, methods, compounds, or experiments described herein. In using such information or methods they should be mindful of their own safety and the safety of others, including parties for whom they have a professional responsibility.

To the fullest extent of the law, neither the Publisher nor the authors, contributors, or editors, assume any liability for any injury and/or damage to persons or property as a matter of products liability, negligence or otherwise, or from any use or operation of any methods, products, instructions, or ideas contained in the material herein.

Library of Congress Cataloging-in-Publication Data

Dragomir, Sorin, 1955–
Harmonic vector fields : variational principles and differential geometry / Sorin Dragomir and Domenico Perrone.
p. cm.
Includes bibliographical references and index.
ISBN 978-0-12-415826-9 (alk. paper)
1. Vector fields. 2. Geometry, Differential. I. Perrone, Domenico, 1949– II. Title.
QA613.619.D73 2011
514'.72–dc23

2011030072

British Library Cataloguing-in-Publication Data

A catalogue record for this book is available from the British Library.

For information on all Elsevier publications
visit our website at *www.elsevierdirect.com*

Printed and bound by CPI Group (UK) Ltd, Croydon, CR0 4YY

Working together to grow
libraries in developing countries

www.elsevier.com | www.bookaid.org | www.sabre.org

ELSEVIER BOOK AID International Sabre Foundation

CONTENTS

PREFACE

The main object of study in this monograph are *harmonic vector fields* on Riemannian manifolds. Let (M,g) be a real n-dimensional Riemannian manifold and $S^{n-1} \to S(M) \to M$ its tangent sphere bundle. It is a classical fact that $S(M)$ may be endowed with a Riemannian metric $\tilde{G}_s$ (the *Sasaki metric*) naturally associated to the Riemannian metric g on the base manifold (cf. e.g., D.E. Blair, [42]). Therefore any smooth unit tangent vector field $X : M \to S(M)$ may be looked at as a smooth map among the Riemannian manifolds (M,g) and $(S(M), \tilde{G}_s)$ so that one may consider (by assuming that M is compact and orientable) the ordinary Dirichlet energy

$$E(X) = \frac{1}{2} \int_M \|dX\|^2 d\,\mathrm{vol}(g) \tag{P.1}$$

familiar in the theory of harmonic maps (cf. e.g., H. Urakawa, [292]). A *harmonic vector field* is then a critical point $X \in \Gamma^\infty(S(M))$ of $E : \Gamma^\infty(S(M)) \to \mathbb{R}$ i.e., for any smooth 1-parameter variation $\{X_t\}_{|t|<\epsilon}$ of X (with $X_0 = X$) through unit tangent vector fields $X_t \in \Gamma^\infty(S(M))$, $|t| < \epsilon$, one has $\{dE(X_t)/dt\}_{t=0} = 0$. Any harmonic vector field is a smooth solution to the nonlinear elliptic PDE system

$$\Delta_g X - \|\nabla X\|^2 X = 0. \tag{P.2}$$

These are precisely the Euler-Lagrange equations associated to the constrained variational principle $\delta E(X) = 0$ and $g(X,X) = 1$ ([309], [316]). Here Δ_g is a second order elliptic operator acting on vector fields, e.g., if X is a C^2 vector field on M then locally

$$\Delta_g X = -\sum_{i=1}^{n} \left\{ \nabla_{E_i} \nabla_{E_i} X - \nabla_{\nabla_{E_i} E_i} X \right\}$$

with respect to a local orthonormal frame $\{E_i : 1 \leq i \leq n\}$ of $T(M)$. One may think of the covariant derivative ∇X as a section in the vector bundle $T^*(M) \otimes T(M) \to M$ hence consider the map $\nabla : \mathfrak{X}(M) \to \Omega^0(T^*(M) \otimes T(M))$. If ∇^* is the formal adjoint of ∇ i.e., $(\nabla^* \varphi, X) = (\varphi, \nabla X)$ for any $\varphi \in \Omega^0(T^*(M) \otimes T(M))$ and any $X \in \mathfrak{X}_0^\infty(M)$, then $\Delta_g = \nabla^* \nabla$ and although (P.2) is nonlinear, an obvious notion of weak solution to (P.2) may be introduced. Here we made use of the L^2 inner products $(X, Y) = \int_M$

$g(X,Y)d\text{vol}(g)$ and $(\varphi,\psi) = \int_M g^*(\varphi,\psi)d\text{vol}(g)$. A systematic study of weak solutions (e.g., existence and local properties) to (P.2) (the *harmonic vector field* system) is missing from the current mathematical literature. Indeed this book is mostly confined to the study of differential geometric properties of harmonic vector fields and of the geometric background (mainly within contact Riemannian and pseudohermitian geometry) supporting such vector fields. One notable exception is Theorem 2.53 in Chapter 2 showing the existence of minimizers for the total bending functional $\mathcal{B} : \mathcal{H}^1_g(S(M)) \to \mathbb{R}$ by a standard mix of functional analysis and calculus of variations (e.g., the Eberlein-Smulian theorem on the characterization of reflexive Banach spaces, Kondrakov's compact embedding theorem, and existence theorems for minimizers of lower semicontinuous functionals). Here

$$\mathcal{H}^1_g(S(M)) = \left\{ X \in \mathcal{H}^{1,2}_g(T(M)) : g(X,X) = 1 \text{ a.e. in } M \right\}$$

and the Sobolev type spaces of vector fields $\mathcal{H}^{k,p}_g(T(M))$ are described in Section 2.11 of Chapter 2. Other important exceptions are Theorem 3.43 (due to E. Boeckx & L. Vanhecke, [51]) and Theorem 3.44 (due to G. Nunes & J. Ripoll, [224]). Both results furnish examples of weak solutions to the harmonic vector fields system (cf. Definition 2.52 in Chapter 2) within interesting geometric contexts [E. Boeckx & L. Vanhecke's result is that radial vector fields on harmonic manifolds are weakly harmonic while G. Nunes & J. Ripoll's result is that normal vector fields to the principal orbits of a cohomogeneity one action (of a compact Lie group of isometries of a compact orientable Riemannian manifold) are weakly harmonic].

Harmonic vector fields aren't harmonic maps unless the additional curvature condition

$$\text{trace}_g \{R(\nabla.X, X)\cdot\} = 0 \tag{P.3}$$

is satisfied. The resulting theory of harmonic vector fields is similar in many respects to the more consolidated theory of harmonic maps yet presents new and intriguing aspects captured in a rapidly growing specific literature cf. e.g., E. Boeckx & L. Vanhecke, [51]–[54], E. Boeckx & J.C. Gonzales-Davila & L. Vanhecke, [56]–[57], V. Borrelli, [61], V. Borrelli & F. Brito & O. Gil-Medrano, [62], V. Borrelli & O. Gil-Medrano, [63], F. Brito, [71], F. Brito & P.G. Walczak, [72], G. Calvaruso & D. Perrone, [77], P.B. Chacon & A.M. Naveira & J.M. Westonn, [81], B-Y. Choi & J.W. Yim, [89], O. Gil-Medrano, [126]–[127], O. Gil-Medrano

& J.C. Gonzales-Davila & L. Vanhecke, [128], O. Gil-Medrano & A. Hurtado, [130], O. Gil-Medrano & E. Llinares-Fuster, [131]–[132], H. Gluck & W. Ziller, [133], J.C. Gonzàles-Dàvila & L. Vanhecke, [144]–[146], S.D. Han & J.W. Yim, [157], K. Hasegawa, [159], D. Perrone, [241]–[243]. The first basic results (e.g., the first and second variation formulae and applications) are due to C.M. Wood, [316] (cf. also [317]–[318]). A slightly different approach was undertaken by G. Wiegmink, [309–310], who introduced the functional $\mathcal{B} : \Gamma^\infty(S(M)) \to \mathbb{R}$ given by

$$\mathcal{B}(X) = \int_M \|\nabla X\|^2 d\,\text{vol}(g).$$

This is the *total bending* functional, a measure of the failure of $X \in \Gamma^\infty(S(M))$ to be parallel. A closer look at the properties of the Sasaki metric G_s on $S(M)$ shows however that the Dirichlet and total bending functionals are related

$$E(X) = \frac{n}{2}\,\text{Vol}(M) + \frac{1}{2}\,\mathcal{B}(X) \tag{P.4}$$

so that the theories in [316] and [309] are identical. The relation (P.4) may be used to show that the search for vector fields which are critical points of $E : \Gamma^\infty(S(M)) \to \mathbb{R}$, rather than critical points of $E : C^\infty(M, T(M)) \to \mathbb{R}$ or $E : \mathfrak{X}(M) \to \mathbb{R}$, is the only appropriate choice. Indeed the only smooth vector fields which are critical points of $E : C^\infty(M, T(M)) \to \mathbb{R}$ (respectively of $E : \mathfrak{X}(M) \to \mathbb{R}$) are the parallel vector fields.

The authors' interest in the theory of harmonic vector fields arose in relationship to the study of the geometry of contact Riemannian manifolds (cf. [237]–[247]) and of nonlinear subelliptic systems of variational origin appearing in the theory of Hörmander systems of vector fields (cf. e.g., J. Jost & C-J. Xu, [180]) and CR geometry (cf. e.g., E. Barletta et al., [25]).

The exposition of the material collected in this book is organized as follows. Chapter 1 is devoted to the basic geometric properties of the tangent bundle over a Riemannian manifold. A description of the tangent sphere bundle $S^1 \to S(T^2) \to T^2$ over a torus T^2 and a classification of its smooth sections (up to homotopy) complete Chapter 1 (and prepare several instances where the general theory may be applied, cf. Sections 2.6 and 2.10 in Chapter 2 and Section 3.9 in Chapter 3 of this book).

Chapter 2 presents the basic theoretic material while the remaining chapters deal mainly with applications and generalizations. From a technical point of view, the main achievement of Chapter 2 is perhaps the explicit expression (2.21) of the tension field

$$\tau(X) = \left\{ \left(\mathrm{trace}_g R(\nabla.X, X)\cdot \right)^H - \left(\Delta_g X \right)^V \right\} \circ X \tag{P.5}$$

of a C^∞ vector field $X : M \to T(M)$. Of course X is thought of as a map of the Riemannian manifolds (M,g) and $(T(M), G_s)$ where G_s is the Sasaki metric on $T(M)$ (so that $\tilde{G}_s = \iota^* G_s$ is the first fundamental form of $\iota : S(M) \hookrightarrow T(M)$). The proof of (P.5) is a rather involved calculation exploiting the relationship among the Levi-Civita connections of $(T(M), G_s)$ and (M,g) (related through the formalism of vertical and horizontal lifting and including curvature calculations, cf. e.g., [42], p. 139–141). A consequence of (P.5) (that is (2.27) in Theorem 2.19) may then be used to characterize unit vector fields which are harmonic maps (as the smooth unit vector fields satisfying (P.2)–(P.3)). Given a unit tangent vector field $X \in \Gamma^\infty(S(M))$ and a smooth 1-parameter variation $X_t \in \Gamma^\infty(S(M))$ of X, one derives (cf. (2.32) and (2.36) in Chapter 2) the first and second variation formulae

$$\frac{dE(X_t)}{dt}(0) = \int_M g(\Delta_g X, V)\, d\,\mathrm{vol}(g), \tag{P.6}$$

$$\frac{d^2E(X_t)}{dt^2}(0) = \frac{1}{2}\int_M \{\|\nabla V\|^2 - \|\nabla X\|^2 \|V\|^2\} d\,\mathrm{vol}(g), \tag{P.7}$$

where $V = \{dU_t/dt\}_{t=0}$. Related to (P.7) one discusses stability results for critical points of $E : \Gamma^\infty(S(M)) \to \mathbb{R}$. The Dirichlet problem

$$\Delta_g X - \|\nabla X\|^2 X = 0 \quad \text{in } \Omega, \tag{P.8}$$

$$X = X_0 \quad \text{on } \partial\Omega, \tag{P.9}$$

is considered in Section 2.9. Here Ω is a smoothly bounded strictly pseudoconvex domain in $\mathbb{C}^n$ (with $n \geq 2$) endowed with the Bergman metric g. We report on a recent result by E. Barletta, [22], dealing with C^2 regularity up to the boundary of solutions $X \in \mathfrak{X}(M)$ to (P.8)–(P.9). The existence problem is open so far.

Section 2.10 in Chapter 2 reports on a result by G. Wiegmink, [309], on the behavior of the total bending functional under conformal changes $\tilde{g} = e^{2u} g$ of the metric on the base manifold. As argued in Section 2.10 the study is confined to the 2-dimensional case (and if M is 2-dimensional, compact, orientable and admits globally defined nowhere zero vector fields then M must be the torus T^2) because the term $(\text{div}(X)X - \nabla_X X)u$ in the identity

$$\tilde{\mathcal{B}}(\tilde{X}) = \mathcal{B}(X) + \int_{T^2} \left\{ \|\nabla u\|_g^2 + 2(\text{div}(X)X - \nabla_X X)u \right\} d\,\text{vol}(g).$$

(cf. (2.110) in Chapter 2) may be calculated in terms of the Gaussian curvature of M in dimension $n = 2$, a calculation which appears to admit no obvious analog in higher dimension. The main result is G. Wiegmink's (cf. *op. cit.*) Theorem 2.43 implying that $\inf_{X \in \mathcal{E}} \mathcal{B}(X)$ is achieved in $\mathcal{E} = \Gamma^\infty(S(T^2))$.

Chapter 2 ends up with the construction of Sobolev type spaces of vector fields (appropriate for the study of weak solutions to (P.2)).

The harmonicity and stability of Hopf and Killing vector fields is discussed in Chapter 3. There we introduce (following F. Brito, [71]) the functional

$$\tilde{E}(X) = E(X) + \frac{(n-1)(n-3)}{2} \int_M \|H_X\|^2 d\,\text{vol}(g) \qquad \text{(P.10)}$$

where H_X is the mean curvature vector of the distribution $(\mathbb{R}X)^\perp$. This is referred to as *Brito's functional*. Brito's functional is an attempt to avoid the difficulties arising from the fact that Hopf vector fields on a sphere S^{2m+1} are (by a result of C.M. Wood, [316]) unstable critical points of $E : \Gamma^\infty(S(S^{2m+1})) \to \mathbb{R}$ yet they are absolute minima of $\tilde{E} : \Gamma^\infty(S(S^{2m+1})) \to \mathbb{R}$. Section 3.8 furnishes a detailed proof of the result by G. Nunes & J. Ripoll (cf. [224]) mentioned earlier in the preface. Section 3.9 of Chapter 3 reports on a beautiful result by G. Wiegmink, [309], giving a complete description of harmonic vector fields on a Riemannian torus.

Harmonicity and stability of special vector fields appearing on contact Riemannian manifolds (such as the Reeb vector field underlying a contact Riemannian structure) are studied in Chapter 4. One of the main notions in Chapter 4 is that of an *H-contact manifold* (a contact metric manifold whose Reeb vector is a harmonic vector field). In a long series of papers one of the authors of this monograph has emphasized (cf. [237]–[247]) that H-contact metric manifolds possess special features and may be quite explicitly described, especially in real dimension 3.

For each Riemannian manifold (M,g), its Sasaki metric G_s belongs to a large family of Riemannian metrics on $T(M)$, the family of Riemannian g-natural metrics, parametrized by elements of $C^\infty(\mathbb{R}_0^+, \mathbb{R}^6)$ (cf. [6]). In

Chapter 5, we endow $T(M)$ with an arbitrary Riemannian g-natural metric G and study smooth vector fields V on M thought of as maps of (M,g) into $(T(M), G)$ (cf. [2]) i.e., we look at harmonicity of V as a map of (M,g) into $(S(M), \tilde{G})$, where $\tilde{G}$ is the metric induced by G on $S(M)$ (cf. [3], [250]). One decomposes the tension field into vertical and horizontal components and derives two equations describing the harmonicity of $V : (M,g) \to (S(M), \tilde{G})$. As it turns out, one of these equations does not depend upon the choice of $\tilde{G}$. In particular the equation (P.2) is invariant under a 4-parameter deformation of the Sasaki metric. The remaining equation is a natural generalization of (P.3).

Many of the results in Chapters 2 to 4 admit versions holding for sections in Riemannan vector bundles (cf. e.g., J.J. Konderak, [194]). The main findings in this direction are presented in Chapter 6.

Chapter 7 is devoted to generalizations of the notion of a harmonic vector field within CR and pseudohermitian geometry (cf. e.g., G. Tomasini et al., [110]). Sections 7.1 to 7.3 are an attempt (based on the results in D. Perrone et al., [107]) to relate harmonicity of vector fields to the geometry of the Fefferman metric (a Lorentz metric on the total space of the canonical circle bundle over a strictly pseudoconvex CR manifold). Sections 7.4 to 7.7 rely on results by Y. Kamishima et al., [103], and generalize harmonic vector fields in the spirit of the work by J. Jost & C-J. Xu, [180], and E. Barletta et al., [25] (dealing with generalizations of harmonic maps within the theory of Hörmander systems of vector fields and CR geometry). The treatment in Chapter 7 leads naturally to nonlinear subelliptic systems of variational origin and exhibits a nontrivial link among the differential geometry of harmonic vector fields and the analysis of subelliptic partial differential equations.

In Chapter 8 we discuss, within the framework of Lorentz geometry, a version of the energy functional (the *spacelike energy*) due to O. Gil-Medrano & A. Hurtado, [130], defined on reference frames (unit timelike vector fields) and the corresponding critical points (the *spatially harmonic* vector fields). These are ordinary harmonic vector fields when geodesic. The treatment is tentative, as the study of harmonic vector fields on a Lorentz manifold is still in its infancy. Nevertheless the subject looks promising, especially in its potential applications to the general relativity theory.

The Authors,
Potenza-Lecce, June 9, 2011.

CHAPTER ONE

Geometry of the Tangent Bundle

Contents

The scope of this chapter is to briefly review the basic facts in the geometry of the tangent bundle $T(M)$ over a Riemannian manifold (M, g), such as nonlinear connections, the Dombrowski map and the Sasaki metric G_s. Remarkably $T(M)$ also carries a natural almost complex structure J (arising from g) compatible to G_s and such that $(T(M), J, G_s)$ is an almost Kähler manifold. The almost complex structure J (discovered by P. Dombrowski, [99]) is rarely integrable (in fact only when the base Riemannian manifold is locally Euclidean) yet J appears to be but one of the many *isotropic* almost complex structures $J_{\delta,\sigma}$ built by R.M. Aguilar, [11]. On the other hand the existence of an integrable isotropic almost complex structure only requires that (M, g) has constant sectional curvature and, if this is the case, the family $J_{\delta,\sigma}$ contains a large subfamily of complex structures (among which the *invariant* ones may be completely determined, cf. Theorem 1.20). When an almost complex structure $J_{\delta,\sigma}$ is non integrable the geometry of $(T(M), J_{\delta,\sigma})$ is related to the properties of the *twisted* Dolbeau complex (a description of which is given in Appendix A of this book). Further information on the geometry of $T(M)$ (over a semi-Riemannian manifold M) is furnished in Chapter 7. Chapter 1 also contains the calculation (due to G. Wiegmink, [309]) of the Bruschlinsky group of a torus T^2 endowed with an arbitrary Riemannian metric leading to the classification up to homotopy of the unit tangent vector fields on T^2. For the classical aspects of the geometry of the tangent bundle over a Riemannian manifold

Harmonic Vector Fields
© 2012 Elsevier Inc. All rights reserved.

(the Sasaki metric, the almost contact metric structure of the unit tangent bundle, etc.) the reader may also consult the books by K. Yano & S. Ishihara, [324], and D.E. Blair, [42].

1.1. THE TANGENT BUNDLE

Let M be a real n-dimensional C^∞ manifold and $\pi : T(M) \to M$ its tangent bundle. If $(U, \tilde{x}^1, \ldots, \tilde{x}^n)$ is a local coordinate system on M then let $(\pi^{-1}(U), x^i, y^i)$ be the naturally induced local coordinates on $T(M)$ i.e.,

$$x^i(v) = \tilde{x}^i(\pi(v)), \quad v = y^i(v) \frac{\partial}{\partial \tilde{x}^i}\Big|_{\pi(v)} \in \pi^{-1}(U).$$

Hence $T(M)$ is a real $2n$-dimensional C^∞ manifold. We set

$$\partial_i = \frac{\partial}{\partial x^i}, \quad \dot{\partial}_i = \frac{\partial}{\partial y^i}, \quad 1 \leq i \leq n,$$

for the sake of simplicity. For didactic reasons, and only through this section, we distinguish notationally between the local coordinates $\tilde{x}^i$ (defined on U) and x^i (defined on $\pi^{-1}(U)$). Let

$$\sigma_0 : M \to T(M), \quad \sigma_0(x) = 0_x \in T_x(M), \quad x \in M,$$

be the zero section. Then $\sigma_0 : M \to T(M)$ is an embedding of M in the (total space of its) tangent bundle. For each tangent vector $v \in T(M)$ the subspace $\mathcal{V}_v \equiv \mathrm{Ker}(d_v\pi) \subset T_v(T(M))$ is the *vertical space* at v. A tangent vector $X \in \mathcal{V}_v$ is a *vertical vector*. The assignment

$$\mathcal{V} : v \in T(M) \mapsto \mathcal{V}_v \subset T_v(T(M))$$

is a C^∞ distribution of rank n on $T(M)$ and $\{\dot{\partial}_i : 1 \leq i \leq n\}$ is a local frame of $\mathcal{V}$ defined on the open subset $\pi^{-1}(U)$.

Definition 1.1 $\mathcal{V}$ is called the *vertical distribution* on $T(M)$. A vector field $X \in \mathcal{V}$ (i.e., $X_v \in \mathcal{V}_v$ for any $v \in T(M)$) is a *vertical vector field* on $T(M)$. ■

The vertical distribution is involutive (as it may be easily seen by using the local frame $\{\dot{\partial}_i : 1 \leq i \leq n\}$). Therefore, by the classical Frobenius theorem, $\mathcal{V}$ is completely integrable and its maximal integral manifold passing through $v \in T(M)$ is the tangent space $T_x(M)$ where $x = \pi(v) \in M$.

Let $\pi^{-1}T(M) \to T(M)$ be the pullback by π of the tangent bundle $T(M) \to M$. Its total space $\pi^{-1}TM$ is a submanifold of the product manifold $T(M) \times T(M)$. The fibre over $v \in T(M)$ is

$$\left(\pi^{-1}T(M)\right)_v = \{v\} \times T_{\pi(v)}(M).$$

Alternatively $\pi^{-1}T(M)$ is the largest subset of $T(M) \times T(M)$ such that the diagram

$$\begin{array}{cccc} T(M) \times T(M) \supset & \pi^{-1}TM & \stackrel{\tilde{\pi}}{\longrightarrow} & T(M) \\ & p \downarrow & & \downarrow \pi \\ & T(M) & \stackrel{\pi}{\longrightarrow} & M \end{array}$$

is commutative. Here $\tilde{\pi}$ and p are the (restrictions to $\pi^{-1}TM$ of the) first and second canonical projections of the product manifold $T(M) \times T(M)$.

Definition 1.2 Let $X : M \to T(M)$ be a tangent vector field on M. The cross-section $\hat{X} : T(M) \to \pi^{-1}T(M)$ defined by $\hat{X}(v) = (v, X_{\pi(v)})$, for any $v \in T(M)$, is called the *natural lift* of X. ■

Let $X_i : \pi^{-1}(U) \to \pi^{-1}T(M)$ be the natural lift of the (local) tangent vector field $\partial/\partial\tilde{x}^i : U \to T(M)$. Then $\{X_i : 1 \le i \le n\}$ is a local frame of the pullback bundle $\pi^{-1}T(M) \to T(M)$ defined on the open set $\pi^{-1}(U)$.

By a customary language abuse, one may identify the tangent bundle $T(M)$ and the vertical bundle. Of course $T(M) \to M$ and $\mathcal{V} \to T(M)$ both have rank n yet different base manifolds and the precise statement is that there is a natural vector bundle isomorphism $\pi^{-1}T(M) \approx \mathcal{V}$. Indeed for any $v \in T(M)$ and any $X \in T_{\pi(v)}(M)$ let $C : (-\epsilon, \epsilon) \to T(M)$ be the curve given by

$$C(t) = v + tX, \quad |t| < \epsilon \quad (\epsilon > 0).$$

For each $v \in T(M)$ let $\gamma_v : \left(\pi^{-1}T(M)\right)_v \to T_v(T(M))$ be the map given by

$$\gamma_v(v, X) = \frac{dC}{dt}(0), \quad (v, X) \in \left(\pi^{-1}T(M)\right)_v.$$

Let $(U, \tilde{x}^i)$ be a local coordinate system on M such that $x = \pi(v) \in U$. Then

$$\gamma_v X_i(v) = \left.\frac{\partial}{\partial y^i}\right|_v \in \mathcal{V}_v, \quad 1 \le i \le n.$$

In particular, γ_v is an $\mathbb{R}$-linear isomorphism $\left(\pi^{-1}T(M)\right)_v \approx \mathcal{V}_v$.

Definition 1.3 The vector bundle isomorphism $\gamma : \pi^{-1}T(M) \to \mathcal{V} = \mathrm{Ker}(d\pi)$ is called the *vertical lift*. ■

For each section $s : T(M) \to \pi^{-1}T(M)$ the vertical vector field γs is the *vertical lift* of s. Note that in general s may fail to be the natural lift of a vector field on M. The *vertical lift* $X^V : T(M) \to T(T(M))$ of a tangent vector field $X : M \to T(M)$ is the vertical lift $X^V \equiv \gamma \hat{X}$ of the natural lift $\hat{X}$ of X.

1.2. CONNECTIONS AND HORIZONTAL VECTOR FIELDS

Let ∇ be a linear connection on M. For any local coordinate system (U, x^i) on M let $\Gamma^i_{jk} : U \to \mathbb{R}$ be the local connection coefficients i.e.,

$$\nabla_{\partial/\partial x^j} \frac{\partial}{\partial x^k} = \Gamma^i_{jk} \frac{\partial}{\partial x^i}, \quad 1 \le j, k \le n.$$

Any linear connection ∇ on M induces a connection $\hat{\nabla}$ in the vector bundle $\pi^{-1}T(M) \to T(M)$. This is easiest to describe locally, as follows. Let (U, x^i) be a local coordinate system on M and let $(\pi^{-1}(U), x^i, y^i)$ be the induced local coordinates on $T(M)$. We set by definition

$$\hat{\nabla}_{\partial_j} X_k = \left(\Gamma^i_{jk} \circ \pi\right) X_i, \quad \hat{\nabla}_{\dot{\partial}_j} X_k = 0. \tag{1.1}$$

It may be easily checked that the definition doesn't depend upon the choice of local coordinates on M and that (1.1) gives rise to a (globally defined) connection in the vector bundle $\pi^{-1}T(M) \to T(M)$.

Definition 1.4 A C^∞ distribution $\mathcal{H}$ on $T(M)$ is called a *nonlinear connection* on $T(M)$ if

$$T_v(T(M)) = \mathcal{H}_v \oplus \mathcal{V}_v, \quad v \in T(M), \tag{1.2}$$

where $\mathcal{V} \equiv \mathrm{Ker}(d\pi)$ is the vertical distribution. ■

A nonlinear connection $\mathcal{H}$ on $T(M)$ is also referred to as a *horizontal distribution* on $T(M)$ while $\mathcal{H} \to T(M)$ is the corresponding *horizontal bundle*. By (1.2) any horizontal distribution on $T(M)$ has rank n.

Let us consider the bundle morphism $L : T(T(M)) \to \pi^{-1}T(M)$ given by

$$L_v A = (v, (d_v\pi)A), \quad A \in T_v(T(M)), \quad v \in T(M).$$

Locally

$$L(\partial_i) = X_i, \quad L(\dot{\partial}_i) = 0, \quad 1 \le i \le n.$$

Moreover the following sequence of vector bundles and vector bundle morphisms

$$0 \to \pi^{-1}T(M) \xrightarrow{\gamma} T(T(M)) \xrightarrow{L} \pi^{-1}T(M) \to 0$$

is exact. In particular given a nonlinear connection $\mathcal{H}$ on $T(M)$ the restriction of L to $\mathcal{H}$ gives a vector bundle isomorphism $L: \mathcal{H} \to \pi^{-1}T(M)$ whose inverse is denoted by $\beta : \pi^{-1}T(M) \to \mathcal{H}$.

Definition 1.5 Let $\mathcal{H}$ be a nonlinear connection on $T(M)$. The bundle isomorphism $\beta : \pi^{-1}T(M) \to \mathcal{H}$ is called the *horizontal lift* (associated to $\mathcal{H}$). ■

Let (U, x^i) be a local coordinate system on M and let $\mathcal{H}$ be a nonlinear connection on $T(M)$. Then

$$\beta X_i = M_i^j \partial_j - N_i^j \dot{\partial}_j$$

for some (uniquely defined) C^∞ functions $M_i^j, N_i^j : \pi^{-1}(U) \to \mathbb{R}$. Applying L to both sides of this identity gives $M_i^j = \delta_i^j$. Let us set $\delta_i = \delta/\delta x^i = \beta X_i$. Then $\{\delta_i : 1 \le i \le n\}$ is a local frame of $\mathcal{H}$ defined on the open set $\pi^{-1}(U)$. The functions N_j^i in $\delta_j = \partial_j - N_j^i \dot{\partial}_i$ are the *local coefficients* of the nonlinear connection $\mathcal{H}$.

Definition 1.6 The cross-section $\mathcal{L}: T(M) \to \pi^{-1}T(M)$ in the vector bundle $\pi^{-1}T(M) \to T(M)$ defined by

$$\mathcal{L}(v) = (v, v) \in \left(\pi^{-1}T(M)\right)_v, \quad v \in T(M),$$

is called the *Liouville vector.* ■

Let $\hat{\nabla}$ be a connection in the vector bundle $\pi^{-1}T(M) \to T(M)$ and let us define $\mathcal{H} \equiv \mathcal{H}_{\hat{\nabla}}$ by setting

$$\mathcal{H}_v = \left\{ A \in T_v(T(M)) : \left(\hat{\nabla}_{\tilde{A}} \mathcal{L}\right)_v = 0 \right\}, \quad v \in T(M),$$

where $\tilde{A}$ is any smooth extension of A to $T(M)$ i.e., $\tilde{A}$ is a C^∞ vector field on $T(M)$ such that $\tilde{A}_v = A$. Clearly the definition of $\mathcal{H}_v$ doesn't depend upon the choice of smooth extension of $A \in T_v(T(M))$.

Definition 1.7 A connection $\hat{\nabla}$ in $\pi^{-1}T(M) \to V(M)$ is said to be *regular* if $\mathcal{H}_{\hat{\nabla}}$ is a nonlinear connection on $T(M)$. ■

Proposition 1.8 *For any linear connection ∇ on M the induced connection $\hat{\nabla}$ in $\pi^{-1}T(M) \to T(M)$ is regular.*

Proof. Let A be a tangent vector field on $T(M)$ locally described as $A = A^i\partial_i + B^i\dot{\partial}_i$. Then $A \in \mathcal{H}_{\hat{\nabla}}$ if and only if

$$B^k = -A^i y^j \left(\Gamma^k_{ij} \circ \pi\right). \tag{1.3}$$

Thus $\mathcal{H}_{\hat{\nabla}}$ is locally the span of

$$\left\{\partial_i - y^j \Gamma^k_{ij}\dot{\partial}_k : 1 \le i \le n\right\} \tag{1.4}$$

so that $\mathcal{H}_{\hat{\nabla}}$ is a C^∞ distribution of rank n on $T(M)$. Therefore to check (1.2) it suffices to show that the sum $\mathcal{H}_{\hat{\nabla},v} + \mathcal{V}_v$ is direct for any $v \in T(M)$. Indeed let $A \in \mathcal{H}_{\hat{\nabla}} \cap \mathcal{V}$. As $A \in \mathcal{V}$ one has $LA = 0$ hence $A^i = 0$ and then (by (1.3)) $B^k = 0$ as well. Thus $\mathcal{H} \cap \mathcal{V} = (0)$. ■

An inspection of (1.4) also shows that

Corollary 1.9 *The local coefficients of the nonlinear $\mathcal{H}_{\hat{\nabla}}$ connection on $T(M)$ determined by a linear connection ∇ on M are given by $N^i_j = \left(\Gamma^i_{jk} \circ \pi\right) y^k$.*

Let $\mathcal{H}$ be a nonlinear connection on $T(M)$. A tangent vector $A \in \mathcal{H}_v$ is a *horizontal vector*. A tangent vector field $A \in \mathcal{H}$ is a *horizontal vector field* on $T(M)$. For any section s in $\pi^{-1}T(M) \to T(M)$ the vector field βs is the *horizontal lift* of s. Given a vector field X on M its *horizontal lift* is the vector field $X^H \equiv \beta\hat{X}$ i.e., the horizontal lift of the natural lift $\hat{X}$ of X.

1.3. THE DOMBROWSKI MAP AND THE SASAKI METRIC

Definition 1.10 Let $\mathcal{H}$ be a nonlinear connection on $T(M)$. Bundle morphism $K : T(T(M)) \to \pi^{-1}TM$ defined by

$$K_v A = \gamma_v^{-1} Q_v A, \quad A \in T_v(T(M)), \quad v \in T(M),$$

is called the *Dombrowski map* associated to $\mathcal{H}$. Here $Q : T(T(M)) \to \mathcal{V}$ is the natural projection associated to the decomposition $T(T(M)) = \mathcal{H} \oplus \mathcal{V}$. ■

Let $\mathcal{H}$ be a nonlinear connection on $T(M)$ and $\beta : \pi^{-1}TM \to \mathcal{H}$ and $K: T(T(M)) \to \pi^{-1}TM$ respectively the horizontal lift and the Dombowski map associated to $\mathcal{H}$. It is an easy consequence of definitions that the sequence of vector bundles and vector bundle morphisms

$$0 \to \pi^{-1}TM \xrightarrow{\beta} T(T(M)) \xrightarrow{K} \pi^{-1}TM \to 0$$

is exact. Locally

$$K(\delta_i) = 0, \quad K(\dot{\partial}_i) = X_i, \quad 1 \leq i \leq n.$$

In particular $K(\partial_i) = N_i^j X_j$.

Let g be a Riemannian metric on M. It induces a Riemannian bundle metric $\hat{g}$ in the vector bundle $\pi^{-1}TM \to T(M)$ as follows. Let $v \in T(M)$ and $X, Y \in \left(\pi^{-1}TM\right)_v$. Let (U, x^i) be a local coordinate system on M such that $\pi(v) \in U$. Let X_i be the natural lift of $\partial/\partial x^i$. We set

$$\hat{g}_v(X, Y) = g_{ij}(\pi(v)) X^i Y^j$$

where $X = X^i X_i(v)$ and $Y = Y^i X_i(v)$. The definition of $\hat{g}_v(X, Y)$ doesn't depend upon the choice of local coordinates at $\pi(v)$ so that $\hat{g}$ is a globally defined Riemannian bundle metric in $\pi^{-1}TM$. In particular $\hat{g}(X_i, X_j) = g_{ij} \circ \pi$.

Definition 1.11 Let $\mathcal{H}$ be a nonlinear connection on $T(M)$ and g a Riemannian metric on M. The Riemannian metric G_s on $T(M)$ defined by

$$G_s(A, B) = \hat{g}(LA, LB) + \hat{g}(KA, KB), \quad A, B \in \mathfrak{X}(T(M)), \tag{1.5}$$

is called the *Sasaki metric* on $T(M)$ associated to the pair $(\mathcal{H}, g)$. ■

Of course the Sasaki metric G_s may be thought of as induced by the given Riemannian metric g on M while the given nonlinear connection $\mathcal{H}$ is but the lifting tool (from M to $T(M)$).

Let $\mathcal{H}$ be a nonlinear connection on $T(M)$. Let $J: T(T(M)) \to T(T(M))$ be the $(1,1)$-tensor field on $T(M)$ defined by

$$J\beta X = \gamma X, \quad J\gamma X = -\beta X, \quad X \in \Gamma^\infty(\pi^{-1}TM). \tag{1.6}$$

It is then immediate that

Proposition 1.12 *Let (M, g) be a Riemannian manifold. For any nonlinear connection $\mathcal{H}$ on $T(M)$ the synthetic object $(T(M), J, G_s)$ is an almost Hermitian manifold.*

Proof. One has $J^2 = -I$ as a consequence of (1.6), where I is the identical transformation of $T(T(M))$, so that J is an almost complex structure on $T(M)$. Moreover (1.5)–(1.6) imply $G_s(JA, JB) = G_s(A,B)$ for any $A, B \in \mathfrak{X}(T(M))$. ■

It may be easily shown that

Proposition 1.13 (P. Dombrowski, [99]) *J is integrable if and only if (M,g) is locally Euclidean.*

Our convention for the sign of the curvature tensor field R of a connection $\nabla : \Gamma^\infty(E) \to \Gamma^\infty(T^*(M) \otimes E)$ in a vector bundle $E \to M$ is

$$R(X,Y)s = -\nabla_X \nabla_Y s + \nabla_Y \nabla_X s + \nabla_{[X,Y]} s$$

for any $X, Y \in \mathfrak{X}(M)$ and any $s \in \Gamma^\infty(E)$.

Proposition 1.14 *Let (M,g) be a Riemannian manifold and $\hat{\nabla}$ a regular connection in $\pi^{-1}TM \to T(M)$. Let $\mathcal{H} = \mathcal{H}_{\hat{\nabla}}$ be the nonlinear connection on $T(M)$ associated to $\hat{\nabla}$ and $\beta : \pi^{-1}TM \to \mathcal{H}$ the corresponding horizontal lift. Let G_s be the Sasaki metric associated to the pair $(\mathcal{H}, g)$. Then the Levi-Civita connection D of $(T(M), G_s)$ is expressed by*

$$D_{\beta X} \beta Y = \beta \hat{\nabla}_{\beta X} Y + \frac{1}{2} \gamma \hat{R}(\beta X, \beta Y)\mathcal{L},$$

$$D_{\beta X} \gamma Y = \gamma \hat{\nabla}_{\beta X} Y - \frac{1}{2} \beta \hat{R}(\beta \mathcal{L}, \beta Y)X,$$

$$D_{\gamma X} \beta Y = \beta \hat{\nabla}_{\gamma X} Y - \frac{1}{2} \beta \hat{R}(\beta \mathcal{L}, \beta X)Y,$$

$$D_{\gamma X} \gamma Y = \gamma \hat{\nabla}_{\gamma X} Y,$$

for any $X, Y \in \Gamma^\infty(\pi^{-1}TM)$, where $\hat{R}$ is the curvature tensor field of the connection $\hat{\nabla}$.

For a proof of Proposition 1.14 one may see [42], p. 139–140.

The almost complex structure $J = J_{1,0}$ given by (1.6) was introduced by P. Dombrowski, [99]. It is but one of a larger family of *isotropic* almost complex structures $J_{\delta,\sigma}$ due to R.M. Aguilar, [11]. Any isotropic almost complex structure determines an almost Kähler metric whose Kähler 2-form is the pullback to $T(M)$ via $\flat : T(M) \to T^*(M)$ of the canonical symplectic form on $T^*(M)$. Moreover isotropic *complex* structures exist precisely when M has constant sectional curvature (cf. Theorem 1 in [11], p. 431). The remainder of this section is devoted to a brief presentation of the results in [11], which are believed (as we argue later on) to offer the possibility of further development for the harmonic vector field theory.

1.3.1. Preliminaries on Local Calculations

Let (M,g) be a Riemannian manifold and $\{X_i : 1 \leq i \leq n\}$ be a local orthonormal frame of $T(M)$, defined on the open set $U \subseteq M$. Let $\{\theta^i : 1 \leq i \leq n\}$ be the dual coframe on U i.e., $\theta^i(X_j) = \delta^i_j$ for any $1 \leq i,j \leq n$. The metric may be locally written as $g = \sum_{i=1}^n \theta^i \otimes \theta^i$. Also $\{X_i^H, X_i^V : 1 \leq i \leq n\}$ is a local frame of $T(T(M))$ defined on the open set $\pi^{-1}(U)$ and $\{\pi^*\theta^i, \theta^i \circ p \circ K : 1 \leq i \leq n\}$ is the corresponding dual coframe. We set

$$\eta^i = \pi^*\theta^i, \quad \xi^i = \theta^i \circ p \circ K, \quad 1 \leq i \leq n,$$

for the sake of simplicity. Let $\Theta \in \Omega^1(T(M))$ be the 1-form on $T(M)$ defined by

$$\Theta_v(A) = g_{\pi(v)}((d_v\pi)A, v), \quad A \in T_v(T(M)), \quad v \in T(M).$$

Then

$$\Theta = \sum_{i=1}^n f^i \eta^i \tag{1.7}$$

where the functions $f^i : \pi^{-1}(U) \to \mathbb{R}$ are given by

$$f^i(v) = \theta^i_{\pi(v)}(v), \quad v \in \pi^{-1}(U), \quad 1 \leq i \leq n.$$

It may be easily shown that

$$d\Theta = \sum_{i=1}^n \xi^i \wedge \eta^i. \tag{1.8}$$

Let $E : T(M) \to \mathbb{R}$ be given by $E(v) = \frac{1}{2} g_{\pi(v)}(v,v)$ for any $v \in T(M)$. Then

$$dE = \sum_{i=1}^n f^i \xi^i. \tag{1.9}$$

Let $\omega^i_j \in \Omega^1(U)$ be the connection 1-forms of ∇ (the Levi-Civita connection of (M,g)) associated to the local frame $\{X_i : 1 \leq i \leq n\}$ i.e., $\nabla X_j = \omega^i_j \otimes X_i$ and $\Omega^i_j \in \Omega^2(U)$ the curvature 1-forms i.e.,

$$\Omega^i_j = d\omega^i_j + \omega^i_k \wedge \omega^k_j = \frac{1}{2} R^i_{jk\ell} \theta^k \wedge \theta^\ell.$$

Then (by the structure equations for ∇)

$$d\eta^i = \eta^k \wedge \pi^* \omega_k^i, \tag{1.10}$$

$$d\xi^i = \xi^k \wedge \pi^* \omega_k^i + f^k \pi^* \Omega_k^i. \tag{1.11}$$

Next we establish the following

Lemma 1.15 (R.M. Aguilar, [11]) *With the notations above*

$$df^i = \xi^i - f^k \pi^* \omega_k^i \tag{1.12}$$

on $\pi^{-1}(U)$ *for any* $1 \leq i \leq n$.

Proof. Let us take the exterior derivative of (1.7)

$$d\Theta = \sum_i \left\{ df^i \wedge \eta^i + f^i \, d\eta^i \right\}$$

and substitute from (1.8) to get

$$\sum_i \left(\xi^i - df^i\right) \wedge \eta^i = \sum_i f^i \, d\eta_i.$$

Next we substitute $d\eta^i$ from (1.10)

$$\sum_i \left(\xi^i - df^i\right) \wedge \eta^i = \sum_i f^i \eta^k \wedge \pi^* \omega_k^i$$

and profit from $\omega_j^i = -\omega_i^j$ to write

$$\sum_i \left(df^i - \xi^i + f^k \pi^* \omega_k^i \right) \wedge \eta^i = 0$$

so that, by Cartan's lemma

$$df^i = \xi^i - f^k \pi^* \omega_k^i + C_j^i \eta^j \tag{1.13}$$

for some $C_j^i \in C^\infty(\pi^{-1}(U))$ such that $C_j^i = C_i^j$. On the other hand we may differentiate (1.9)

$$0 = \sum_i \left(df^i \wedge \xi^i + f^i \, d\xi^i \right)$$

and substitute df^i from (1.13)

$$\sum_i \left\{ -f^k \pi^* \omega_k^i \wedge \xi^i + C_j^i \eta^j \wedge \xi^i + f^i \, d\xi^i \right\} = 0.$$

In the last identity we replace $d\xi^i$ from (1.11)

$$\sum_i \left\{ C^i_j \eta^j \wedge \xi^i + f^i f^k \pi^* \Omega^i_k \right\} = 0$$

and observe that $\sum_{ij} f^i f^j \pi^* \Omega^i_j = 0$ because of $\Omega^i_j = -\Omega^j_i$. It remains that $\sum_i C^i_j \eta^j \wedge \xi^i = 0$ hence $C^i_j = 0$ and (1.13) implies (1.12) as desired. ∎

1.3.2. Isotropic Almost Complex Structures

We adopt the following

Definition 1.16 Let $\mathcal{A} \subseteq T(M)$ be an open subset. An almost complex structure J on $\mathcal{A}$ is said to be *isotropic* with respect to the Riemannian metric on M if there are smooth functions $\alpha, \delta, \sigma : \mathcal{A} \to \mathbb{R}$ such that $\alpha\delta - \sigma^2 = 1$ and

$$JX^H = \alpha X^V + \sigma X^H, \quad JX^V = -\sigma X^V - \delta X^H, \tag{1.14}$$

for any $X \in \mathfrak{X}(\mathcal{A})$. ∎

Given an isotropic almost complex structure J on $\mathcal{A}$ defined by (1.14) we write $J = J_{\delta,\sigma}$ to capture the dependence of J on the parameters $\delta, \sigma \in C^\infty(\mathcal{A})$. For $\alpha = \delta = 1$ and $\sigma = 0$ one obtains the almost complex structure $J_{1,0}$ due to P. Dombrowski (cf. *op. cit.*).

Theorem 1.17 (R.M. Aguilar, [11]) *Let (M, g) be a Riemannian manifold. There is an integrable isotropic almost complex structure $J_{\delta,\sigma}$ on some open subset $\mathcal{A} \subseteq T(M)$ if and only if $\pi(\mathcal{A})$ has constant sectional curvature.*

Remark 1.18

i. If M is real analytic there is a unique germ $\mathcal{M}$ of complex manifold, of complex dimension n, such that M is embedded in $\mathcal{M}$ as a totally real submanifold. Then we may argue, together with R.M. Aguilar, [11], that the result in Theorem 1.17 is about putting such complex structure into "isotropic" form (similar to the classical Beltrami theorem on geodesic maps onto the Euclidean space).

ii. Let $J_{\delta,\sigma}$ be an isotropic almost complex structure defined on some open subset $\mathcal{A} \subseteq T(M)$. Then

$$g_{\delta,\sigma}(A, B) = (d\Theta)(J_{\delta,\sigma} A, B), \qquad A, B \in \mathfrak{X}(\mathcal{A}),$$

is a Riemannian metric on $\mathcal{A}$ provided that $\alpha > 0$. It should be observed that $g_{1,0}$ is the Sasaki metric G_s and $\sigma_0^* g_{\delta,\sigma} = \alpha g$ (where

$\sigma_0 : \pi(\mathcal{A}) \to \mathcal{A}$ is the zero section). Also $(J_{\delta,\sigma}, g_{\delta,\sigma})$ is an almost Kähler structure on $\mathcal{A}$. It is an open problem to study harmonic (unit) vector fields $X \in \Gamma^\infty(S(M, g_{\delta,\sigma}))$ as in Section 2.6 of this book (by replacing the Sasaki metric G_s with one of the metrics $g_{\delta,\sigma}$ associated to a given isotropic almost complex structure). Are $g_{\delta,\sigma}$ among the g-natural metrics in Section 4.4 of this book?

iii. As $J_{\delta,\sigma}$ are in general non-integrable almost complex structures $(\Omega^{0,\bullet}(T(M)), \overline{\partial})$ is in general only a pseudocomplex (in the sense of [297]). The *Dolbeau cohomology* of $(T(M), J_{\delta,\sigma})$ is then the cohomology of the complex

$$\overline{D} : \Omega^{0,q}(T(M)) \times \overline{\partial}^2\, \Omega^{0,q-1}(T(M)) \to \Omega^{0,q+1}(T(M)) \times \overline{\partial}^2\, \Omega^{0,q}(T(M)),$$

$$\overline{D}(\lambda, \mu) = \left(\overline{\partial}\lambda - \mu, \overline{\partial}\left(\overline{\partial}\lambda - \mu\right)\right),$$

$$\lambda \in \Omega^{0,q}(T(M)), \quad \mu \in \overline{\partial}^2\, \Omega^{0,q-1}(T(M)).$$

The Dolbeau cohomology groups of $(T(M), J_{\delta,\sigma})$ turn out to be isomorphic to the cohomology groups of the pseudocomplex $(\Omega^{0,\bullet}(T(M)), \overline{\partial})$. The calculation of the Dolbeau cohomology is an open problem even in the simplest instances (cf. Appendix A of this book for a preliminary discussion of $(T(M), J_{1,0})$ together with a brief recollection of the needed notions of homological algebra). ■

Let M and N be two manifolds and $f : M \to N$ a smooth map. We define $f_* : T(M) \to T(N)$ by setting $f_*(v) = (d_{\pi(v)} f) v$ for any $v \in T(M)$. In particular $f_{**} : T(T(M)) \to T(T(N))$ is given by

$$f_{**}(V) = (d_{\Pi(V)} f_*) V, \quad V \in T(T(M)),$$

where $\Pi : T(T(M)) \to T(M)$ is the projection.

Definition 1.19 An isotropic almost complex structure $J = J_{\delta,\sigma}$ on $\mathcal{A} \subseteq T(M)$ is said to be *invariant* if for any isometry $F \in \mathrm{Isom}(M, g)$

$$J_{F_*(v)} F_{**} A_v = F_{**}(JA)_v$$

for any $A \in \mathfrak{X}(\mathcal{A})$ and any $v \in \mathcal{A}$. ■

Theorem 1.20 (R.M. Aguilar, [11]) *Let $M^n(\kappa) = (M, g)$ be a real space form of sectional curvature $\kappa \in \mathbb{R}$. Let us consider the set*

$$\mathcal{I} = \left\{ \left(\frac{1}{\sqrt{2\kappa E + b}}, 0 \right) : b \in \mathbb{R} \right\} \cup \left\{ \left(\delta, a\kappa\delta^2\right) : a \in \mathbb{R} \setminus \{0\},\ b \in \mathbb{R} \right\}$$

where $\delta^{-2} = \frac{1}{2}\left[2\kappa E + b + \sqrt{(2\kappa E + b)^2 + 4a^2\kappa^2}\right]$. *Then*

$$\{J_{\delta,\sigma} : (\delta, \sigma) \in \mathcal{I}\}$$

is the set of all invariant isotropic complex structures on some open subset of $T(M)$.

Let us prove Theorem 1.17. Let $J_{\delta,\sigma}$ be an isotropic almost complex structure on some open set $\mathcal{A} \subseteq T(M)$. The holomorphic tangent bundle $T^{1,0}(\mathcal{A}) = \{X - iJ_{\delta,\sigma}X : X \in T(\mathcal{A})\}$ is locally the span of

$$Z_j = (\sigma + i)\frac{\delta}{\delta x^j} + \alpha \frac{\partial}{\partial y^j}, \quad 1 \leq j \leq n.$$

The dual complex 1-forms ω^j are determined by

$$\omega^j(Z_k) = \delta^j_k, \quad \omega^j(\overline{Z}_k) = 0,$$

hence

$$\omega^j = -\frac{i}{2}\, dx^j + f\, \delta y^j, \quad f = \frac{1 + i\sigma}{2\alpha}\, \delta y^j.$$

Then

$$dx^j = (\sigma + i)\omega^j + (\sigma - i)\omega^{\bar{j}}, \quad \delta y^j = \alpha\left(\omega^j + \omega^{\bar{j}}\right),$$

where $\omega^{\bar{j}} = \overline{\omega^j}$. Consequently

$$dx^j \wedge dx^k = (\sigma + i)^2\, \omega^j \wedge \omega^k + \alpha\delta\Big(\omega^j \wedge \omega^{\bar{k}} + \omega^{\bar{j}} \wedge \omega^k\Big) + (\sigma - i)^2\, \omega^{\bar{j}} \wedge \omega^{\bar{k}},$$

$$\frac{1}{\alpha}\, dx^j \wedge \delta y^k = (\sigma + i)\left[\omega^j \wedge \omega^k + \omega^j \wedge \omega^{\bar{k}}\right] + (\sigma - i)\left[\omega^{\bar{j}} \wedge \omega^k + \omega^{\bar{j}} \wedge \omega^{\bar{k}}\right],$$

$$\delta y^j \wedge \delta y^k = \alpha^2\left(\omega^j \wedge \omega^k + \omega^j \wedge \omega^{\bar{k}} + \omega^{\bar{j}} \wedge \omega^k + \omega^{\bar{j}} \wedge \omega^{\bar{k}}\right).$$

Let us compute $d\omega^k$. Using

$$d\omega^j = df \wedge \delta y^k + f\, d\delta y^k,$$

$$df = \frac{\delta f}{\delta x^i}\, dx^i + \frac{\partial f}{\partial y^i}\, \delta y^i,$$

one finds

$$
\begin{aligned}
d\omega^k &= \partial\omega^k + \overline{\partial}\omega^k + \overline{\mathcal{N}}\omega^k,\\
\partial\omega^k &= \left[\alpha(Z_j f)\delta^k_\ell + \frac{f}{2} R^k_{j\ell}\right] \omega^j \wedge \omega^\ell,\\
\overline{\partial}\omega^k &= \left[\alpha(Z_j f)\delta^k_\ell - \alpha(\overline{Z}_\ell f)\delta^k_j - f\left(R^k_{j\ell} + 2i\Gamma^k_{j\ell}\right)\right] \omega^j \wedge \omega^{\overline{\ell}},\\
\overline{\mathcal{N}}\omega^k &= \left[\alpha\delta^k_\ell \overline{Z}_j f + \frac{f}{2} R^k_{j\ell}\right] \omega^{\overline{j}} \wedge \omega^{\overline{\ell}}.
\end{aligned}
$$

Here one exploited

$$
\begin{aligned}
d\Omega^{1,0}(T(M)) &\subset \Omega^2(T(M)),\\
\Omega^2(T(M)) &= \Omega^{2,0}(T(M)) \oplus \Omega^{1,1}(T(M)) \oplus \Omega^{0,2}(T(M)),
\end{aligned}
$$

where $\Omega^{p,q}(T(M))$ is the bundle of complex forms of type (p,q) on $T(M)$ i.e., locally sums of monomials of the form

$$
\omega^{i_1} \wedge \cdots \wedge \omega^{i_p} \wedge \omega^{\overline{j}_1} \wedge \cdots \wedge \omega^{\overline{j}_q}
$$

with complex $C^\infty(\mathcal{A} \cap \pi^{-1}(U))$-coefficients. See also (A.6) in Appendix A to this book. By a result in [297], p. 363, the almost complex structure $J_{\delta,\sigma}$ is integrable if and only if $\overline{\mathcal{N}} = 0$ i.e.,

$$
f R^k_{j\ell} = \alpha\left(\delta^k_j \overline{Z}_\ell f - \delta^k_\ell \overline{Z}_j f\right). \tag{1.15}
$$

Let $R_{ij} = R^k_{jki}$ be the Ricci tensor of (M,g). Contraction of k and j in (1.15) and $R^k_{j\ell} = R^k_{ji\ell} Y^i$ furnishes

$$
\overline{Z}_j f = \frac{f}{(n-1)\alpha} R_{jk} y^k. \tag{1.16}
$$

Substitution from (1.16) into (1.15) leads to

$$
R^k_{j\ell r} = \frac{1}{n-1}\left(\delta^k_\ell \delta^s_j - \delta^k_j \delta^s_\ell\right) R_{sr}. \tag{1.17}
$$

Let $x \in M$ and let $Y \in T_x(M)$ be a unit tangent vector. Let $p \subset T_x(M)$ be a 2-plane tangent to M at x such that $Y \in p$. Let us complete Y to an orthonormal basis $\{X, Y\}$ of p. Then the sectional curvature $k(p)$ is given by

$$
\begin{aligned}
k(p) &= R_x(Y,X,Y,X) = \langle R_x(Y,X)Y, X\rangle\\
&= Y^i Y^j X^k X^\ell g(R(\partial_i, \partial_k)\partial_j, \partial_\ell)_x
\end{aligned}
$$

where $\langle , \rangle = g_x$. Yet (by (1.17))

$$g(R(\partial_i, \partial_k)\partial_j, \partial_\ell) = R^s_{ikj} g_{\ell s} = \frac{1}{n-1}\left(g_{k\ell} R_{ij} - g_{\ell i} R_{kj}\right)$$

so that

$$k(p) = \frac{1}{n-1}\left[\|X\|^2 \mathrm{Ric}_x(Y,Y) - \langle X, Y\rangle \mathrm{Ric}_x(X,Y)\right]$$

or

$$k(p) = \frac{1}{n-1}\mathrm{Ric}_x(Y,Y). \tag{1.18}$$

Let now p, q be two 2-planes tangent to M at x. If $p \cap q \neq (0)$ then we may choose $Y \in p \cap q$ such that $\|Y\| = 1$ so that (by (1.18)) $k(p) = k(q)$. When $p \cap q = (0)$ we may consider orthonormal basis $\{X, Y\} \subset p$ and $\{A, B\} \subset q$. Let then $r \subset T_x(M)$ be the 2-plane spanned by $\{Y, A\}$. As $p \cap r \neq (0)$ and $r \cap q \neq (0)$ we may conclude that $k(p) = k(r) = k(q)$ that is k is constant on $G_2(M)_x$ (the fibre over x in the Grassmann bundle $G_2(\mathbb{R}^n) \to G_2(M) \longrightarrow M$ of all 2-planes tangent to M). That is to say there is a smooth function $\kappa : M \to \mathbb{R}$ such that $k = \kappa \circ \Pi$, where $\Pi : G_2(M) \to M$ is the natural projection. Then

$$R(X,Y)Z = \kappa\{g(X,Z)Y - g(Y,Z)X\}, \quad X, Y, Z \in \mathfrak{X}(M), \tag{1.19}$$

and by the classical Schur theorem κ is a constant provided that $n \geq 3$. To include the case $n = 2$ we need a more detailed analysis. However we impose no restriction on the dimension. Let $\{X_i : 1 \leq i \leq n\}$ be a local orthonormal frame of $T(M)$ and let us adopt the notations and conventions in Section 1.3.1. Then

$$W_j = (\sigma + i)X_j^H + \alpha X_j^V, \quad 1 \leq j \leq n,$$

is a local frame of $T^{1,0}(\mathcal{A})$. Consequently the local complex 1-forms

$$u^j = (1 - i\sigma)\eta^j + i\delta\xi^j, \quad 1 \leq j \leq n, \tag{1.20}$$

are of type $(1,0)$. We may take the exterior derivative of (1.20) and use the structure equations (1.10)–(1.11) to obtain

$$du^j = u^k \wedge \pi^*\omega^i_k + i\delta f^k \pi^*\Omega^i_k + i\left(d\delta \wedge \xi^j - d\sigma \wedge \eta^j\right). \tag{1.21}$$

Let us substitute from

$$\eta^j = \frac{1}{2}\left(u^j + u^{\bar{j}}\right), \quad \xi^j = \frac{1}{2\delta}\left[(\sigma - i)u^j + (\sigma + i)u^{\bar{j}}\right],$$

into (1.21) to get

$$du^j = u^k \wedge \pi^*\omega^i_k + i\delta f^k \pi^*\Omega^j_k + \frac{1}{2}\left(\frac{1+i\sigma}{\delta}\, d\delta - i d\sigma\right) \wedge u^j - \frac{1}{2}\left(\frac{1-i\sigma}{\delta}\, d\delta + i d\sigma\right) \wedge u^{\bar{j}}. \tag{1.22}$$

Let us assume that $J_{\delta,\sigma}$ is integrable. Then the identity (1.19) holds good. It may be easily rewritten as

$$\Omega^i_j = \kappa\, \theta^i \wedge \theta^j,$$

where for the time being $\kappa \in C^\infty(M)$. Then (by (1.7))

$$f^k \pi^*\Omega^j_k = \frac{\kappa \circ \pi}{2}\left(u^j \wedge u^{\bar{j}}\right) \wedge \Theta$$

and one may substitute into (1.22)

$$du^j = u^k \wedge \left[\pi^*\omega^j_k + \frac{\delta\, i\, \kappa \circ \pi}{2}\, \delta^j_k \Theta + \left(\frac{i}{2}\, d\sigma - \frac{1+i\sigma}{2\delta}\, d\delta\right)\delta^j_k\right] + u^{\bar{j}} \wedge \left(\frac{\delta\, i\, \kappa \circ \pi}{2}\, \Theta + \frac{i}{2}\, d\sigma + \frac{1-i\sigma}{2\delta}\, d\delta\right). \tag{1.23}$$

As $J_{\delta,\sigma}$ is assumed to be integrable the $(0,2)$ component of the 2-form du^j must vanish i.e.,

$$i\delta(\kappa \circ \pi)\Theta^{0,1} + i\bar{\partial}\sigma + \frac{1-i\sigma}{\delta}\,\bar{\partial}\delta = 0. \tag{1.24}$$

Here $\Theta^{0,1}$ is the $(0,1)$ component of

$$\Theta = \sum_j f^j \eta^j = \frac{1}{2}\sum_j f^j\left(u^j + u^{\bar{j}}\right)$$

i.e.,

$$\Theta^{0,1} = \frac{1}{2}\sum_j f^j u^{\bar{j}}. \tag{1.25}$$

Let us prove (1.24). As a consequence of

$$u^j(W_k) = 2(\sigma + i)\delta^j_k$$

for each smooth function $f : \pi^{-1}(U) \to \mathbb{R}$

$$df = f_j u^j + f_{\bar{j}} u^{\bar{j}},$$

$$f_j = \frac{1}{2(\sigma+i)} W_j(f), \quad f_{\bar{j}} = \overline{f_j}, \quad 1 \le j \le n.$$

Therefore

$$i\delta(\kappa\circ\pi)\Theta + i\,d\sigma + \frac{1-i\sigma}{\delta}\,d\delta = \lambda_k u^k + \mu_{\overline{k}} u^{\overline{k}},$$

$$\lambda_k = \frac{i\delta\,(\kappa\circ\pi)}{2} f^k + i\sigma_k + \frac{1-i\sigma}{\delta}\delta_k,$$

$$\mu_{\overline{k}} = \frac{i\delta\,(\kappa\circ\pi)}{2} f^k + i\sigma_{\overline{k}} + \frac{1-i\sigma}{\delta}\delta_{\overline{k}}.$$

Then the (0,2)-component of du^j is $\overline{\mathcal{N}}u^j = u^{\overline{j}} \wedge \mu_{\overline{k}} u^{\overline{k}}$. As argued above this must vanish

$$\mu_{\overline{k}}\, u^{\overline{j}} \wedge u^{\overline{k}} = 0$$

and then (by applying the last identity to the pair $(W_{\overline{r}}, W_{\overline{s}})$)

$$\mu_{\overline{k}}\left(\delta_r^j \delta_s^k - \delta_s^j \delta_r^k\right) = 0.$$

Then we may contract the indices j and r to yield $(n-1)\mu_{\overline{s}} = 0$ which is equivalent to (1.24). We need the following

Lemma 1.21 $\overline{\partial}\Theta^{1,0} = 0$.

Proof. The identity (1.9) may be written

$$dE = \frac{1}{2\delta}\sum_j f^j\left[(\sigma - i)u^j + (\sigma + i)u^{\overline{j}}\right]$$

hence

$$\overline{\partial}E = \frac{\sigma + i}{2\delta}\sum_j f^j u^{\overline{j}}.$$

Then (by taking into account (1.25))

$$\Theta^{0,1} = \frac{\delta}{\sigma + i}\,\overline{\partial}E.$$

Next

$$\overline{\partial}\Theta^{0,1} = \left[\frac{\overline{\partial}\delta}{\sigma + i} - \frac{\delta\,\overline{\partial}\sigma}{(\sigma + i)^2}\right] \wedge \overline{\partial}E. \tag{1.26}$$

Let us express $\overline{\partial}\sigma$ from (1.24)

$$\overline{\partial}\sigma = -\delta(\kappa\circ\pi)\,\Theta^{0,1} + (\sigma + i)\,\overline{\partial}\log|\delta|$$

to compute

$$(\sigma + i)\overline{\partial}\delta - \delta\,\overline{\partial}\sigma = \delta(\kappa\circ\pi)\,\Theta^{0,1} = \frac{\delta^2\,(\kappa\circ\pi)}{\sigma + i}\,\overline{\partial}E$$

and then (1.26) yields $\overline{\partial}\Theta^{0,1} = 0$. Lemma 1.21 is proved. ■

Let us go back to the proof of Theorem 1.17. To this end we apply $\overline{\partial}$ to (1.24) and use Lemma 1.21

$$(\kappa\circ\pi)\,\overline{\partial}\delta\wedge\Theta^{0,1} + \delta\,\overline{\partial}(\kappa\circ\pi)\wedge\Theta^{0,1} - \overline{\partial}\sigma\wedge\overline{\partial}\log|\delta| = 0.$$

Once again we replace $\overline{\partial}\sigma$ from (1.24) and obtain

$$\overline{\partial}(\kappa\circ\pi)\wedge\Theta^{0,1} = 0. \tag{1.27}$$

Then

$$W_{\bar{j}}(\kappa\circ\pi)f^{\ell} - W_{\bar{\ell}}(\kappa\circ\pi)f^{j} = 0 \tag{1.28}$$

is a local version of (1.27) (got by applying (1.27) to the pair $(W_{\bar{j}}, W_{\bar{\ell}})$). On the other hand

$$W_{\bar{j}}(\kappa\circ\pi) = (\sigma - i)X_j(\sigma)\circ\pi,$$

and

$$f^j(v) = \theta^j_{\pi(v)}(v) = v^j, \quad v = v^j X_j(\pi(v)).$$

In particular $f^j = \left(Y^j_k\circ\pi\right)y^k$ where $\left[Y^j_k\right] = \left[X^j_k\right]^{-1}$ and the smooth function $\left[X^j_k\right] : \pi^{-1}(U)\to \mathrm{GL}(n,\mathbb{R})$ is given by

$$X_k = X^j_k\,\frac{\partial}{\partial x^j}, \quad 1\le k\le n.$$

Then (1.28) may be written

$$(X_j(\kappa)\circ\pi)\left(Y^{\ell}_k\circ\pi\right)y^k - (X_{\ell}(\kappa)\circ\pi)\left(Y^j_k\circ\pi\right)y^k = 0$$

or (by differentiating with respect to y^k)

$$X_j(\kappa)Y^{\ell}_k - X_{\ell}(\kappa)Y^j_k = 0.$$

Let us contract with X^k_s and then contract the indices ℓ and s to obtain $(n-1)X_j(\kappa) = 0$. As M is tacitly assumed to be connected it follows that κ is a constant.

Vice versa, let us assume that M is a real space form $M^n(\kappa)$ of (constant) sectional curvature κ and let us look for smooth functions $\alpha,\sigma,\delta:\mathcal{A}\to\mathbb{R}$, defined on some open subset $\mathcal{A}\subseteq T(M)$, such that $\alpha\delta-\sigma^2=1$ and the corresponding isotropic almost complex structure $J_{\delta,\sigma}$ is integrable. By the considerations above such $J_{\delta,\sigma}$ is integrable if and only if the functions δ and σ satisfy (1.24) i.e.,

$$\overline{\partial}\sigma+\delta\kappa\Theta^{1,0}-\delta^{-1}(\sigma+i)\overline{\partial}\delta=0. \tag{1.29}$$

Therefore it suffices to determine particular solutions δ,σ to (1.29). Let us look for solutions with spherical symmetry i.e.,

$$\sigma=f\circ E,\quad \delta=g\circ E,$$

for some C^∞ functions $f(t)$ and $g(t)$ with $t\in I$ for some open interval $I\subseteq\mathbb{R}$. Then

$$\overline{\partial}\sigma=(f'\circ E)\,\overline{\partial}E,\quad \overline{\partial}\delta=(g'\circ E)\,\overline{\partial}E,$$

and (by taking into account (1.29) and the previously established identity $\Theta^{0,1}=[\delta/(\sigma+i)]\overline{\partial}E$) it suffices to solve the first order ODE system

$$f'(t)+\frac{\kappa g(t)^2}{f(t)+i}-\frac{f(t)+i}{g(t)}g'(t)=0$$

or

$$f'(t)-2f(t)g(t)^{-1}g'(t)=0, \tag{1.30}$$

$$f(t)f'(t)+\kappa g(t)^2-(f(t)^2-1)g(t)^{-1}g'(t)=0. \tag{1.31}$$

At this point the ODE system (1.30)–(1.31) may be solved to yield the existence of an isotropic complex structure $J_{\delta,\sigma}$ as desired. The solutions δ,σ are in general defined only on some open subset of $T(M)$. The proof of Theorem 1.17 is complete.

If $f=0$ then (1.31) becomes $g'+\kappa g^3=0$ hence $g^{-1}=\sqrt{2\kappa t+b}$ where $b\in\mathbb{R}$ is a constant of integration.

Let us look for a solution (f,g) with $f\neq 0$ on some open interval $I\subseteq\mathbb{R}$. The equation (1.30) may be written $f'/f=2g'/g$ hence $f=a\kappa g^2$ where $a\in\mathbb{R}\setminus\{0\}$ is a constant (the form of the constant is chosen to obtain a simple expression of g^{-2}). Then (1.31) becomes

$$\frac{1+a^2\kappa^2g^4}{g^3}\,g'+\kappa=0$$

and we may integrate to obtain

$$-\frac{1}{2g^2}+a^2\kappa^2\frac{g^2}{2}+\kappa+\frac{b}{2}=0. \qquad (1.32)$$

Again the form of the constant of integration is chosen in such a way to get an elegant form of g^{-2}. Indeed we may solve for g^{-2} in (1.32) to obtain

$$g^{-2}=\frac{1}{2}\left[2\kappa t+b+\sqrt{(2\kappa t)^2+4a^2\kappa^2}\right].$$

Summarizing the information obtained so far we have obtained the following spherically symmetric solutions to (1.29)

$$\delta^{-1}=\sqrt{2\kappa E+b},\quad \sigma=0,\quad b\in\mathbb{R}, \qquad (1.33)$$

$$\delta^{-2}=\frac{1}{2}\left[2\kappa E+b+\sqrt{(2\kappa E+b)+4a^2\kappa^2}\right], \qquad (1.34)$$

$$\sigma=a\kappa\delta^2,\quad a\in\mathbb{R}\setminus\{0\},\quad b\in\mathbb{R}.$$

It is a beautiful result (cf. Theorem 1.20 above) of R.M. Aguilar, [11] that the complex structures $J_{\delta,\sigma}$ corresponding to the functions δ,σ given by (1.33)–(1.34) are precisely the *invariant* (in the sense of Definition 1.19 above) isotropic complex structures on (some open subset of) $T(M)$.

1.3.3. Invariant Isotropic Complex Structures

Let $f: M\to M$ be a smooth map. Then

$$\begin{array}{ccccccc} T(M) & \xrightarrow{f_*} & T(M) & \qquad & T(T(M)) & \xrightarrow{f_{**}} & T(T(M)) \\ \pi\downarrow & & \downarrow\pi & & \pi_*\downarrow & & \downarrow\pi_* \\ M & \xrightarrow{f} & M & & T(M) & \xrightarrow{f_*} & T(M) \end{array}$$

are commutative diagrams for any smooth map $f: M\to M$. Indeed let $\Pi: T(T(M))\to T(M)$ be the projection and $V\in T(T(M))$. Then

$$\Pi(f_{**}(V))=\Pi((d_{\Pi(V)}f_*)V)=f_*(\Pi(V)),$$

$$\begin{aligned}\pi_*(f_{**}(V))&=(d_{\Pi(f_{**}V)}\pi)(d_{\Pi(V)}f_*)V=(d_{f_*(\Pi(V))}\pi)(d_{\Pi(V)}f_*)V\\&=(d_{\Pi(V)}(\pi\circ f_*))V=(d_{\Pi(V)}(f\circ\pi))V=f_*(\pi_*(V)).\end{aligned}$$

Lemma 1.22 *If $f \in \mathrm{Isom}(M,g)$ then the diagram*

$$\begin{array}{ccccc} T(T(M)) & \stackrel{K}{\longrightarrow} & \pi^{-1}TM & \stackrel{p}{\longrightarrow} & T(M) \\ f_{**}\downarrow & & & & \downarrow f_* \\ T(T(M)) & \stackrel{K}{\longrightarrow} & \pi^{-1}TM & \stackrel{p}{\longrightarrow} & T(M) \end{array}$$

is commutative i.e.,

$$f_* \circ p \circ K = p \circ K \circ f_{**} \tag{1.35}$$

where $K : T(T(M)) \to \pi^{-1}TM$ is the Dombrowski map and $p : \pi^{-1}TM \to T(M)$ is the restriction to $\pi^{-1}TM$ of the projection $p : T(M) \times T(M) \to T(M)$ given by $p(v,\xi) = \xi$ for any $v, \xi \in T(M)$.

Proof. As $f : M \to M$ is an isometry of (M,g), it preserves the Levi-Civita connection i.e.,

$$\frac{\partial^2 f^i}{\partial x^j \partial x^k} - \Gamma^{\ell}_{jm} \frac{\partial f^m}{\partial x^k} \frac{\partial f^i}{\partial x^\ell} + \left(\Gamma^i_{\ell m} \circ f\right) \frac{\partial f^\ell}{\partial x^j} \frac{\partial f^m}{\partial x^k} = 0. \tag{1.36}$$

At this point we may check (1.35). First

$$(f_* \circ p \circ K)\delta_{j,v} = 0$$

hence $f_* \circ p \circ K$ vanishes on $\mathcal{H}_v$ for any $v \in T(M)$. Next

$$(p \circ K \circ f_{**})\delta_{j,v} = pK\left(d_{\Pi(\delta_{j,v})} f_*\right)\delta_{j,v} = pK(d_v f_*)\delta_{j,v}.$$

The map $f_* : T(M) \to T(M)$ has the local components

$$f^i_* = f^i \circ \pi, \quad f^{i+n}_* = \left(\frac{\partial f^i}{\partial x^j} \circ \pi\right) y^j,$$

so that

$$(d_v f_*)\delta_{j,v} = \frac{\delta f^i_*}{\delta x^j}(v)\partial_i\Big|_v + \frac{\delta f^{i+n}_*}{\delta x^j}(v)\dot{\partial}_i\Big|_v$$

$$= \frac{\partial f^i}{\partial x^j}(x)\left(\delta_i + N^k_i \dot{\partial}_k\right)_{f_*(v)} + \left[\frac{\partial^2 f^i}{\partial x^j \partial x^k}(x) - N^k_j(v)\frac{\partial f^i}{\partial y^k}(x)\dot{\partial}_k\right]\dot{\partial}_i\Big|_{f_*(v)}$$

where $x = \pi(v) \in M$. As $K\delta_j = 0$ and $K\dot{\partial}_j = X_j$ it follows that

$$K(d_v f_*)\delta_{j,v} = \left(\frac{\partial^2 f^i}{\partial x^j \partial x^k}(x) - \Gamma^{\ell}_{jm}(x)\frac{\partial f^m}{\partial x^k}(x)\frac{\partial f^i}{\partial x^\ell}(x)\right.$$

$$\left. + \left(\Gamma^i_{\ell m}(f(x))\right)\frac{\partial f^\ell}{\partial x^j}(x)\frac{\partial f^m}{\partial x^k}(x)\right) v^k \delta_{j,v} = 0$$

by $N^i_j = \Gamma^i_{jk} y^k$ and (1.36). Hence $p \circ K \circ f_{**}$ vanishes on $\mathcal{H}_v$ as well. Moreover

$$(f_* \circ p \circ K)\dot{\partial}_{j,v} = (f_* \circ p)X_{j,v} = f_* \left.\frac{\partial}{\partial x^j}\right|_x = \frac{\partial f^i}{\partial x^j}(x) \left.\frac{\partial}{\partial x^i}\right|_{f(x)}.$$

Similarly

$$(p \circ K \circ f_{**})\dot{\partial}_{j,v} = pK(d_v f_*)\dot{\partial}_{j,v} = pK\left(\frac{\partial f^i_*}{\partial y^j}(v)\, \partial_i|_{f_*(v)} + \frac{\partial f^{i+n}_*}{\partial y^j}(v)\, \dot{\partial}_i|_{f_*(v)}\right)$$

$$= p\frac{\partial f^i}{\partial x^j}(x) X_i(f_*(v)) = \frac{\partial f^i}{\partial x^j}(x) \left.\frac{\partial}{\partial x^i}\right|_{f(x)}.$$

Therefore $f_* \circ p \circ K$ and $p \circ K \circ f_{**}$ coincide on $\mathcal{V}_v$ for any $v \in T(M)$. Lemma 1.22 is proved. ∎

Lemma 1.23 *Let $f \in \mathrm{Isom}(M,g)$. For each $X \in \mathfrak{X}(M)$*

$$f_{**}X^H_v = (f_*X)^H_{f_*v}, \quad f_{**}X^V_v = (f_*X)^V_{f_*v}, \tag{1.37}$$

for any $v \in T(M)$.

Proof. Let $v \in T(M)$ be an arbitrary tangent vector and $x = \pi(v) \in M$. Let $y = f(x) \in M$. Note that

$$f_{**}X^H_v = \left(d_{\Pi(X^H_v)} f_*\right) X^H_v = (d_v f_*)\, X^H_v$$

hence $\Pi\left(f_{**}X^H_v\right) = f_*(v)$. Then $f_*X \in \mathfrak{X}(M)$ is given by

$$\begin{aligned}(f_*X)_y &= \left(d_{f^{-1}(y)}f\right) X_{f^{-1}(y)} = \left(d_{\pi(v)}f\right) X_{\pi(v)} = f_* X_{\pi(v)} \\ &= f_*(d_v\pi)\, X^H_v = f_*\left(d_{\Pi(X^H_v)}\pi\right) X^H_v = f_*\pi_* X^H_v = \pi_* f_{**} X^H_v \\ &= \left(d_{\Pi(f_{**}X^H_v)}\pi\right) f_{**}X^H_v = \left(d_{f_*(v)}\pi\right) f_{**}X^H_v.\end{aligned}$$

Consequently

$$\left(d_{f_*(v)}\pi\right) f_{**}X^H_v = (f_*X)_y = \left(d_{f_*(v)}\pi\right)(f_*X)^H_{f_*(v)}$$

and then

$$f_{**}X^H_v - (f_*X)^H_{f_*(v)} \in \mathrm{Ker}\left(d_{f_*(v)}\pi\right) = \mathcal{V}_{f_*(v)}.$$

Equivalently

$$f_{**}X^H_v = (f_*X)^H_{f_*(v)} + V \tag{1.38}$$

for some $V \in \mathcal{V}_{f_*(v)}$. We ought to show that $V = 0$. Let us apply $p \circ K$ to both sides of (1.38). We get

$$pK_{f_*(v)} V = pK_{f_*(v)} f_{**} X_v^H = f_* p K X_v^H = 0$$

that is

$$K_{f_*(v)} V = (f_*(v), 0) = 0$$

and then $V = 0$ as the restriction of $K_{f_*(v)}$ to $\mathcal{V}_{f_*(v)}$ is a $\mathbb{R}$-linear isomorphism $\mathcal{V}_{f_*(v)} \approx \left(\pi^{-1} TM\right)_{f_*(v)}$. Therefore (1.38) yields the first of the identities (1.37). To prove the second identity in (1.37) we conduct the following calculation

$$\begin{aligned} pKf_{**} X_v^V &= f_* pKX_v^V = f_*(p\hat{X}_v) = f_*(X_{\pi(v)}) = (d_x f) X_x \\ &= (d_{f^{-1}(y)} f) X_{f^{-1}(y)} = (f_* X)_y. \end{aligned}$$

Note that $y = f(x) = f(\pi(v)) = \pi(f_*(v))$ so that

$$\begin{aligned} K_{f_*(v)} f_{**} X_v^V &= \left(f_*(v), (f_* X)_y\right) = \left(f_*(v), (f_* X)_{\pi(f_*(v))}\right) \\ &= \widehat{f_* X}(f_* v) = \left(K \gamma \widehat{f_* X}\right)_{f_*(v)} = K_{f_*(v)} \left(f_* X\right)_{f_*(v)}^V \end{aligned}$$

implying the second of the identities (1.37). Lemma 1.23 is proved. ■

A piece of Theorem 1.20 follows from

Proposition 1.24 *Let $\mathcal{A} \subseteq T(M)$ be an open set and $\sigma, \delta : \mathcal{A} \to \mathbb{R}$ two smooth functions depending only on E i.e., $\sigma = f \circ E$ and $\delta = g \circ E$ for some smooth functions $f(t), g(t)$, $t \geq 0$. Let $\alpha : \mathcal{A} \to \mathbb{R}$ be determined by $\alpha\delta - \sigma^2 = 1$. Then the isotropic almost complex structure $J = J_{\delta,\sigma}$ is invariant.*

Proof. Let $F \in \mathrm{Isom}(M, g)$. Then

$$E(F_* v) = (1/2) g_{\pi(F_* v)}(F_* v, F_* v) = (1/2) g_{\pi(v)}(v, v) = E(v)$$

for any $v \in T(M)$. Therefore $\sigma(F_* v) = f(E(F_* v)) = f(E(v)) = \sigma(v)$ and a similar property holds for the functions α, δ as well. Consequently (by Lemma 1.23)

$$\begin{aligned} J_{F_*(v)} F_{**} X_v^H &= J_{F_*(v)} (F_* X)_{F_*(v)}^H \\ &= \alpha(F_* v)(F_* X)_{F_*(v)}^V + \sigma(F_* v)(F_* X)_{F_*(v)}^H \\ &= \alpha(v) F_{**} X_v^V + \sigma(v) F_{**} X_v^H = F_{**} \left(\alpha X^V + \sigma X^H\right)_v \end{aligned}$$

i.e.,

$$J_{F_*(v)} F_{**} X_v^H = F_{**}(JX^H)_v.$$

Of course the identity

$$J_{F_*(v)} F_{**} X_v^V = F_{**}(JX^V)_v$$

may be proved in a similar manner. We may conclude that J is invariant. Proposition 1.24 is proved. ■

To complete the proof of Theorem 1.20, one should consider an invariant isotropic complex structure $J_{\delta,\beta}$ defined on some open subset $\mathcal{A} \subseteq T(M)$ and show that the functions $\delta, \sigma : \mathcal{A} \to \mathbb{R}$ are necessarily given by the formulae (1.33)–(1.34). Let $v \in \mathcal{A}$ and $x = \pi(v) \in M$. Let $(U, x^1, \ldots, x^n)$ be a normal coordinate neighborhood with origin at x and s_x the local symmetry defined by $(x^1, \ldots, x^n) \mapsto (-x^1, \ldots, -x^n)$. Then $(s_x)_* X = -X$ for any $X \in \mathfrak{X}(U)$ and consequently

$$(s_x)_{**} X_v^H = -X_{-v}^H, \quad (s_x)_{**} X_v^V = -X_{-v}^V,$$

for any $X \in \mathfrak{X}(U)$ and any $v \in \pi^{-1}(U)$. Let us assume that $J_{\delta,\sigma}$ is an invariant complex structure and apply the formula in Definition 1.19 for $F = s_x$. Then

$$\alpha(v) = \alpha(-v), \quad \sigma(v) = \sigma(-v), \quad \delta(v) = \delta(-v), \tag{1.39}$$

for any $v \in \pi^{-1}(U)$. We claim that $J_{\delta,-\sigma}$ is integrable, as well. To prove the claim we consider the $(1,0)$-forms

$$v^j = (1 + i\sigma)\eta^j + i\delta\xi^j,$$

i.e., a local frame of $T^{1,0}(T(M), J_{\delta,-\sigma})^*$. Let N_{-1} denote fibrewise multiplication by -1. A simple calculation shows that

$$(N_{-1})^* \eta^j = \eta^j, \quad (N_{-1})^* \xi^j = -\xi^j. \tag{1.40}$$

Next we consider the $(0,1)$-forms

$$u^{\bar{j}} = \overline{u^j} = (1 + i\sigma)\eta^j - i\delta\xi^j$$

locally spanning $T^{0,1}(T(M), J_{\sigma,\delta})$. Then (by (1.39)–(1.40))

$$(N_{-1})^* u^{\bar{j}} = v^j, \quad 1 \leq j \leq n.$$

In particular N_{-1} is an anti-holomorphic map of $(T(M), J_{\delta,\sigma})$ into $(T(M), J_{\delta,-\sigma})$. As shown earlier, the integrability of $J_{\delta,\sigma}$ is equivalent to $du^j \equiv 0$, mod u^k hence

$$dv^j = d(N_{-1})^* u^{\bar{j}} = (N_{-1})^* du^{\bar{j}} \equiv 0, \quad \text{mod } (N_{-1})^* u^{\bar{k}},$$

i.e., $dv^j \equiv 0$, mod v^k, a fact implying the integrability of $J_{\delta,-\sigma}$. Let $\overline{\partial}_\sigma$ be the $\overline{\partial}$-operator associated to the complex structure $J_{\delta,\sigma}$. Then (by replacing σ with $-\sigma$ in (1.24))

$$i\delta(\kappa \circ \pi)\,\Pi^{0,1}_{-\sigma}\Theta - i\overline{\partial}_{-\sigma}\sigma + \frac{1+i\sigma}{\delta}\overline{\partial}_{-\sigma}\delta = 0, \tag{1.41}$$

where, once again to distinguish among the complex structures $J_{\delta,\sigma}$ and $J_{\delta,-\sigma}$, we denote by $\Pi^{0,1}_\sigma : \Omega^1(\mathcal{A}, J_{\delta,\sigma}) \to \Omega^{0,1}(\mathcal{A}, J_{\delta,\sigma})$ the natural projection. The formulae (1.24) and (1.41) may be thought of as a linear system with the unknowns

$$X_j^H(\delta), \quad X_j^V(\delta), \quad X_j^H(\sigma), \quad X_j^V(\sigma). \tag{1.42}$$

By solving for the unknowns (1.42) in the equations (1.24) and (1.41) one may easily see that δ and σ depend only on E (one also exploits the identity

$$\Pi^{0,1}_\sigma \Theta = \frac{\delta}{\sigma + i}\overline{\partial}E,$$

cf. the proof of Lemma 1.21 above).

Remark 1.25

i. Let $\sigma_0 : M \to T(M)$ be the zero section. Then $(d_x\sigma_0)X_x = X^H_{0_x}$ for any $X \in \mathfrak{X}(M)$ and any $x \in M$. Consequently $\sigma_0(M)$ is a totally real submanifold of $(T(M), J_{\delta,\sigma})$ i.e.,

$$\left(J_{\delta,\sigma}T(\sigma_0(M))\right) \cap T(\sigma_0(M)) = (0).$$

Thus any isotropic complex structure is a "thickening" of $\sigma_0(M) \approx M$ (if defined there) and the resulting complex manifold is locally (i.e., in a neighborhood of M) biholomorphic to any other standard thickening of M.

ii. Let X be a complex manifold, $M \subset X$ a smooth real submanifold with $\dim_{\mathbb{R}} M = \dim_{\mathbb{C}} X = n$, and u a nonnegative bounded or unbounded exhaustion function of X such that $M = \{z \in X : u(z) = 0\}$. Let us assume that u^2 is smooth and strictly plurisubharmonic and satisfies the complex Monge-Ampère equation

$$\left(\partial\overline{\partial}u\right)^n = 0 \quad \text{on} \quad X \setminus M. \tag{1.43}$$

The synthetic object (X, M, u) is commonly referred to as a *Monge-Ampère model* (of *bounded type* when u is bounded). A program of classifying Monge-Ampère models with a given *center* M was started by L. Lempert & R. Szöke, [206]. One of their main results is

Theorem 1.26 (L. Lempert & R. Szöke, [206]) *Let (X, M, u) and (X', M', u') be two Monge-Ampère models, not necessarily of bounded type. Let g and g' be the Riemannian metrics on M and M' associated respectively to u and u'. Let us assume that (M, g) and (M', g') are isometric and $\sup u = \sup u'$. Then there is a biholomorphism $F : X \to X'$ such that $u' \circ F = u$.*

Theorem 1.26 generalizes a result by G. Patrizio & P-M. Wong, [236], where the Monge-Ampère model (X, M, u) is constructed from a compact symmetric space of rank 1 (cf. [236], p. 356). When $J_{\delta,\sigma}$ is one of the invariant isotropic complex structures in Theorem 1.20 the equation (1.43) with $u^{-1}(0) = \sigma_0(M)$ reduces to an ODE in one real variable. Let $S^n = \{x = (x_1, \ldots, x_{n+1}) \in \mathbb{R}^{n+1} : \sum_{j=1}^{n+1} x_j^2 = 1\}$ and let $Q^n = \{z = (z_1, \ldots, z_{n+1}) \in \mathbb{C}^{n+1} : \sum_{j=1}^{n+1} z_j^2 = 1\}$ be the natural "thickening" of S^n (cf. [236], p. 357–360). As for any invariant isotropic thickening of S^n one has $\sup u = \infty$, any of these is globally biholomorphic to Q^n. If $J_{\delta,\sigma}$ is singled by $a = 0$ and $b = 1$ then one obtains the biholomorphism $T(S^n) \approx Q$ expressed by $(X, Y) \mapsto \sqrt{1 + \sum_{j=1}^{n+1} y_j^2}\, X + iY \in Q^n$ where $z_j = x_j + iy_j$. A deep circle of ideas relates the geometry of the tangent bundle over a Riemannian manifold to the study of the global solutions to the homogeneous complex Monge-Ampère equation (1.43). A brief introduction to the subject is given in Appendix C of this book. ∎

1.4. THE TANGENT SPHERE BUNDLE

Let (M, g) be a real n-dimensional Riemannian manifold and $S^{n-1} \to S(M) \to M$ its *tangent sphere bundle* i.e., the fibre in $S(M)$ over each $x \in M$ is given by

$$S(M)_x = \{v \in T_x(M) : g_x(v, v) = 1\}.$$

The total space $S(M)$ of the tangent sphere bundle over M is a real hypersurface in $T(M)$. If (U, x^i) is a local coordinate system on M and $(\pi^{-1}(U), x^i, y^i)$ are the induced local coordinates on $T(M)$ then $S(M)$ is locally given by

$$S(M) \cap \pi^{-1}(U) = \{v \in \pi^{-1}(U) : g_{ij}(\pi(v))v^i v^j = 1\}$$

where $v^i = y^i(v)$, $1 \le i \le n$. Therefore the local equation of $S(M)$ in $T(M)$ is $g_{ij}(x)y^i y^j = 1$. Consequently a tangent vector field $A \in \mathfrak{X}(T(M))$ locally represented as

$$A = A^i \partial_i + A^{i+n}\dot{\partial}_i$$

is tangent to $S(M) \cap \pi^{-1}(U)$ if and only if

$$A^i \frac{\partial g_{jk}}{\partial x^i} y^j y^k + 2A^{i+n} g_{ij} y^j = 0 \tag{1.44}$$

on $\pi^{-1}(U)$. Let $\mathcal{H}$ be the nonlinear connection associated to the Levi-Civita connection of (M,g) so that $N^i_j = \Gamma^i_{jk} y^k$ and

$$\Gamma^i_{jk} = g^{i\ell}\Gamma_{jk\ell}, \quad \Gamma_{ijk} = \frac{1}{2}\left(\frac{\partial g_{ik}}{\partial x^j} + \frac{\partial g_{jk}}{\partial x^i} - \frac{\partial g_{ij}}{\partial x^k}\right).$$

When $A = \delta_j$ for some $1 \le j \le n$ then $A^i = \delta^i_j$ and $A^{i+n} = -N^i_j$ hence

$$\begin{aligned} A^i \frac{\partial g_{k\ell}}{\partial x^i} y^k y^\ell + 2A^{i+n} g_{ik} y^k &= \frac{\partial g_{k\ell}}{\partial x^j} y^k y^\ell - 2\Gamma^i_{j\ell} y^\ell g_{ik} y^k \\ &= y^k y^\ell \left(\frac{\partial g_{k\ell}}{\partial x^j} - 2\Gamma_{j\ell k}\right) = 0. \end{aligned}$$

We proved the following result:

Proposition 1.27 *Let (M,g) be a Riemannian manifold and $\mathcal{H} = \mathcal{H}_{\hat{\nabla}}$ the nonlinear connection on $T(M)$ associated to $\hat{\nabla}$, where $\hat{\nabla}$ is the connection in $\pi^{-1}TM \to M$ induced by the Levi-Civita connection ∇ of (M,g). Then $\mathcal{H}_v \subset T_v(S(M))$ for any $v \in T(M)$ i.e., each horizontal vector is tangent to $S(M)$.*

Which vertical vectors are tangent to $S(M)$? To answer this question let $A = A^{i+n}\dot{\partial}_i$ be a vertical vector field and

$$\rho : \pi^{-1}(U) \to \mathbb{R}, \quad \rho(v) = g_{ij}(\pi(v))y^i(v)y^j(v) - 1, \quad v \in \pi^{-1}(U).$$

Then $A(\rho) = 0$ if and only if

$$A^{i+n} g_{ij} y^j = 0. \tag{1.45}$$

Therefore we may state the following

Proposition 1.28 *Let $X \in \Gamma^\infty(\pi^{-1}TM)$. Then the vertical tangent vector field γX is tangent to $S(M)$ if and only if $\hat{g}(X, \mathcal{L}) = 0$ where $\hat{g} = \pi^{-1}g$ is* *the Riemannian bundle metric induced by g in $\pi^{-1}TM \to T(M)$ and $\mathcal{L}$ is the Liouville vector field.*

Let $A = A^i \partial_i + A^{i+n}\dot{\partial}_i$ be an arbitrary tangent vector field on $S(M)$ i.e., the functions (A^i, A^{i+n}) satisfy (1.44). A vector field $B = B^i\partial_i + B^{i+n}\dot{\partial}_i$ is orthogonal to $S(M)$ if and only if

$$g_{ij}A^iB^j + g_{k\ell}(A^{k+n} + N_i^k A^i)(B^{\ell+n} + N_j^\ell B^j) = 0. \tag{1.46}$$

We shall show that

Theorem 1.29 *Let (M,g) be a Riemannian manifold and $S(M) \to M$ its tangent sphere bundle. Then $\nu = \gamma\,\mathcal{L}$ is a unit normal vector field on $S(M)$ and*

$$T(S(M)) = \mathcal{H} \oplus \gamma\,\mathrm{Ker}(\omega) \tag{1.47}$$

where $\omega_x : (\pi^{-1}TM)_x \to \mathbb{R}$ is defined by $\omega_x(X) = \hat{g}_x(X, \mathcal{L}_x)$ for any $X \in (\pi^{-1}TM)_x$ and any $x \in M$.

Proof. The first statement in Theorem 1.29 follows from (1.46). Indeed if $B = \gamma\,\mathcal{L}$ then $B^i = 0$ and $B^{i+n} = y^i$ hence

$$g_{ij}A^iB^j + g_{k\ell}\left(A^{k+n} + N_i^k A^i\right)\left(B^{\ell+n} + N_j^\ell B^j\right)$$
$$= g_{k\ell}\left(A^{k+n} + N_i^k A^i\right)y^\ell = 0$$

as a consequence of (1.44). The formula (1.47) follows from Propositions 1.27 and 1.28 by comparing dimensions. ∎

Let (M,g) be a Riemannian manifold. Let $J = J_{1,0}$ be the standard almost complex structure on $T(M)$. The *geodesic flow* is the horizontal vector field ξ' on $T(M)$ given by

$$\xi' = -J\nu = \beta\,\mathcal{L} = y^i \frac{\delta}{\delta x^i}.$$

By Proposition 1.27 the vector field ξ' is tangent to $S(M)$. Let $\tilde{G}_s$ be the metric on $S(M)$ induced by the Sasaki metric on $T(M)$. Let $\eta' \in \Omega^1(S(M))$ be given by $\eta'(V) = \tilde{G}_s(V, \xi')$ for any $V \in \mathfrak{X}(S(M))$. Also let ϕ' be the $(1,1)$-tensor field on $S(M)$ given by

$$\phi' V = JV - \eta'(V)\xi', \quad V \in \mathfrak{X}(S(M)).$$

We set

$$\phi = \phi', \quad \xi = 2\xi', \quad \eta = \frac{1}{2}\eta', \quad G_{cs} = \frac{1}{4}\tilde{G}_s,$$

so that $(\phi, \xi, \eta, G_{cs})$ turns out to be a contact metric structure (the *standard* contact metric structure) on $S(M)$. Let $H(S(M)) = \mathrm{Ker}(\eta) \subset T(S(M))$ and let us set

$$T_{1,0}(S(M)) = \{X - iJX : X \in H(S(M))\} \quad \left(i = \sqrt{-1}\right).$$

This is the *standard* almost CR structure on $S(M)$. It is discussed in some detail in Chapter 6 of this book. Cf. also [110], p. 321–329. By a result of S. Tanno, [282], $T_{1,0}(S(M))$ is integrable if and only if (M,g) (with $n > 2$) has constant sectional curvature. Similarly any isotropic almost complex structure $J_{\delta,\sigma}$, defined on an open subset $\mathcal{A} \subseteq T(M)$, induces an almost CR structure $T_{1,0}(T(M))_{\delta,\sigma}$ on $\mathcal{A} \cap S(M)$. It is noteworthy that $J_{\delta,\sigma}$ with

$$\delta = \frac{1}{\sqrt{\kappa(2E-1)+1}}, \qquad \sigma = 0,$$

induces the same almost CR structure as $J_{1,0}$ yet (unlike $J_{1,0}$) has the same integrability condition as $T_{1,0}(S(M))$.

1.5. THE TANGENT SPHERE BUNDLE OVER A TORUS

Let $d_1, d_2 \in \mathbb{R}^2$ be two linearly independent vectors and let $\Gamma \subset \mathbb{R}^2$ be the lattice given by

$$\Gamma = \{m\, d_1 + n\, d_2 : m, n \in \mathbb{Z}\}.$$

Let $T^2 = \mathbb{R}^2/\Gamma$ and let $\pi : \mathbb{R}^2 \to T^2$ be the natural projection. Then $\mathbb{R}^2$ is the universal covering space of T^2. We assume T^2 is oriented such that $\pi : \mathbb{R}^2 \to T^2$ is orientation preserving. Let J be the almost complex structure on T^2 induced by the fixed orientation. Next we assume that T^2 is equipped with an arbitrary Riemannian metric g and let $\{S, W\}$ be an orthonormal frame of $T(T^2)$ such that $W = JS$. Let $S^1 \to S(T^2,g) \to T^2$ be the tangent sphere bundle and let us set $\mathcal{E} = \Gamma^\infty(S(T^2,g))$.

Definition 1.30 Let $X \in \mathcal{E}$ be a unit tangent vector field on T^2. The functions $\varphi, \psi \in C^\infty(T^2, \mathbb{R})$ given by

$$\varphi = g(X, S), \quad \psi = g(X, W),$$

are called the (S, W)-*coordinates* of X. ■

If $X \in \mathcal{E}$ and $\varphi, \psi \in C^\infty(T^2, \mathbb{R})$ are its (S, W)-coordinates then

$$X = \varphi\, S + \psi\, W, \quad \varphi^2 + \psi^2 = 1.$$

Also the map $\mathcal{E} \to C^\infty(T^2, S^1)$ given by $X \mapsto \varphi + \sqrt{-1}\,\psi$ is a bijection.

Definition 1.31 Let $X \in \mathcal{E}$. A function $\alpha : \mathbb{R}^2 \to \mathbb{R}$ is called a (S, W)-*angle function* for X if

$$X \circ \pi = (\cos\alpha)\, S \circ \pi + (\sin\alpha)\, W \circ \pi \tag{1.48}$$

everywhere on $\mathbb{R}^2$. ■

Let $X \in \mathcal{E}$ and let $\alpha : \mathbb{R}^2 \to \mathbb{R}$ be a (S, W)-angle function for X. Then

$$\varphi \circ \pi = \cos\alpha, \quad \psi \circ \pi = \sin\alpha$$

i.e., the following diagram is commutative

$$\begin{array}{ccc} \mathbb{R}^2 & \xrightarrow{\ \alpha\ } & \mathbb{R} \\ {\scriptstyle 1_{\mathbb{R}^2}}\downarrow & & \downarrow{\scriptstyle (\cos,\sin)} \\ \mathbb{R}^2 & \xrightarrow[(\varphi,\psi)\circ\pi]{} & S^1 \end{array}$$

that is $\alpha : \mathbb{R}^2 \to \mathbb{R}$ is a lift of $(\varphi + \sqrt{-1}\,\psi) \circ \pi : \mathbb{R}^2 \to S^1$ (and clearly such a lift always exists).

If $I \subseteq \mathbb{R}$ is an open interval such that $0 \in I$ we set

$$\mathcal{E}_I = \{(X_t)_{t\in I} \in \mathcal{E}^I : (X_t)_{t\in I} \text{ of class } C^\infty\}.$$

Here $(X_t)_{t\in I}$ is said to be of class C^∞ if the map $X : T^2 \times I \to \mathbb{R}$ given by $X(p, t) = X_t(p)$, for any $p \in T^2$ and any $t \in I$, is of class C^∞.

Given $(X_t)_{t\in I} \in \mathcal{E}_I$ we set

$$\varphi_t(p) = g(X_t, S)_p, \quad \psi_t(p) = g(X_t, W)_p,$$

so that $\varphi_t, \psi_t \in C^\infty(T^2, \mathbb{R})$ are the (S, W)-coordinates of X_t for each $t \in I$. Let $\alpha_t : \mathbb{R}^2 \to \mathbb{R}$ be a (S, W)-angle function for X_t so that

$$\cos\alpha_t(\xi) = \varphi_t(p), \quad \sin\alpha_t(\xi) = \psi_t(p), \tag{1.49}$$

for any $\xi \in \mathbb{R}^2$ where $p = \pi(\xi) \in T^2$. We shall need the following elementary fact

Lemma 1.32 *There is a unique* $(m, n) \in \mathbb{Z}^2$ *such that*

$$\alpha_t(\xi + d_1) - \alpha_t(\xi) = 2m\pi, \quad \alpha_t(\xi + d_2) - \alpha_t(\xi) = 2n\pi, \tag{1.50}$$

for any $\xi \in \mathbb{R}^2$ *and any* $t \in I$.

Proof. As $\pi(\xi + d_1) = \pi(\xi)$ for any $\xi \in \mathbb{R}^2$ one may write

$$\cos\alpha_t(\xi + d_1) = \varphi_t(p), \quad \sin\alpha_t(\xi + d_1) = \psi_t(p). \tag{1.51}$$

Then (by comparing (1.49) and (1.51))

$$\sin\frac{\alpha_t(\xi + d_1) + \alpha_t(\xi)}{2} \sin\frac{\alpha_t(\xi + d_1) - \alpha_t(\xi)}{2} = 0, \tag{1.52}$$

$$\sin\frac{\alpha_t(\xi + d_1) - \alpha_t(\xi)}{2} \cos\frac{\alpha_t(\xi + d_1) + \alpha_t(\xi)}{2} = 0, \tag{1.53}$$

As a consequence of (1.52) we have I) $\alpha_t(\xi + d_1) - \alpha_t(\xi) = 2\pi\, m(\xi, t)$ for some $m(\xi, t) \in \mathbb{Z}$ or II) $\alpha_t(\xi + d_1) + \alpha_t(\xi) = 2\pi\, k(\xi, t)$ for some $k(\xi, t) \in \mathbb{Z}$. In the first case, the continuity of the function $(\xi, t) \mapsto \alpha_t(\xi)$ implies that $(\xi, t) \mapsto m(\xi, t)$ is a continuous map of $\mathbb{R}^2 \times I$ into $\mathbb{Z}$ hence $m(\xi, t) =$ constant, thus proving the first identity in (1.50). In the second case we need to exploit (1.53) so that to conclude that III) $\alpha_t(\xi + d_1) - \alpha_t(\xi) = 2\pi\,\ell(\xi, t)$ for some $\ell(\xi, t) \in \mathbb{Z}$ or IV) $\alpha_t(\xi + d_1) + \alpha_t(\xi) = (2p(\xi, t) + 1)\pi$ for some $p(\xi, t) \in \mathbb{Z}$. Yet case (II) rules out case (IV) hence case (III) must occur and once again $\ell(\xi, t)$ is the constant function by continuity. ■

Definition 1.33 Let $(m, n) \in \mathbb{Z}^2$ and $I \subseteq \mathbb{R}$ an open interval containing the origin. We set

$$\mathrm{Per}_I(m, n) = \{\alpha \in C^\infty(\mathbb{R}^2 \times I) : \alpha(\xi + d_1, t) - \alpha(\xi, t) = 2m\pi,$$
$$\alpha(\xi + d_2, t) - \alpha(\xi, t) = 2n\pi, \quad \forall \xi \in \mathbb{R}^2, \quad \forall t \in I\}.$$

An element $\alpha \in \mathrm{Per}_I(m, n)$ is referred to as a *(m, n)-semiperiodic function.* ■

We also set for further use

$$\mathcal{W}_I = \bigcup_{(m,n)\in\mathbb{Z}^2} \mathrm{Per}_I(m, n).$$

We shall omit the index I when the elements α do not depend upon t. That is $\mathrm{Per}(m_1, m_2)$ consists of all C^∞ functions $\alpha : \mathbb{R}^2 \to \mathbb{R}$ such that $\alpha(\xi + d_i) - \alpha(\xi) = 2\pi\, m_i$ for all $\xi \in \mathbb{R}^2$ and $\mathcal{W} = \bigcup_{(m,n)\in\mathbb{Z}^2} \mathrm{Per}(m, n)$. The contents of Lemma 1.32 is that given $X = (X_t)_{t\in I}$ and a (S, W)-angle function $\alpha_t : \mathbb{R}^2 \to \mathbb{R}$ for each X_t the function $\alpha : \mathbb{R}^2 \times I \to \mathbb{R}$ given by $\alpha(\xi, t) = \alpha_t(\xi)$, for any $\xi \in \mathbb{R}^2$ and any $t \in I$, is (m, n)-semiperiodic for some $(m, n) \in \mathbb{Z}^2$.

Lemma 1.34 *Let $\{S, W\}$ be a fixed orthonormal frame of $(T(T^2), g)$ and let $\alpha \in C^\infty(\mathbb{R}^2 \times I)$. Then the following statements are equivalent*

i. *There is* $\overline{\alpha} \in C^\infty(T^2 \times I)$ *such that* $\alpha(\xi, t) = \overline{\alpha}(\pi(\xi), t)$ *for any* $\xi \in \mathbb{R}^2$ *and any* $t \in I$.
ii. $\alpha \in \mathrm{Per}_I(0,0)$.

The proof of Lemma 1.34 is elementary. If such $\overline{\alpha}$ exists then

$$\alpha(\xi + d_i, t) - \alpha(\xi, t) = \overline{\alpha}(\pi(\xi + d_i), t) - \overline{\alpha}(\pi(\xi), t) = 0.$$

Vice versa if $\alpha \in \mathrm{Per}_I(0,0)$ then we may set by definition

$$\overline{\alpha}(p,t) = \alpha(\xi, t), \quad \xi \in \pi^{-1}(p), \quad p \in T^2, \quad t \in I,$$

and the definition of $\overline{\alpha}(p,t)$ doesn't depend upon the choice of $\xi \in \pi^{-1}(p)$ precisely because α is $(0,0)$-semiperiodic.

Lemma 1.35 *For each* $(m,n) \in \mathbb{Z}^2$ *the set* $\mathrm{Per}_I(m,n)$ *is an affine subspace of* $C^\infty(\mathbb{R}^2 \times I)$ *and* $\mathrm{Per}_I(0,0)$ *is its associated vector space. In particular for any* $\alpha \in \mathrm{Per}_I(m,n)$ *and any* $Y \in \mathfrak{X}(\mathbb{R}^2)$ *one has*

$$Y(\alpha), \ \frac{\partial \alpha}{\partial t} \in Per_I(0,0).$$

Proof. Clearly $\mathrm{Per}_I(0,0)$ is a real vector space (with the usual operations with functions $\alpha : \mathbb{R}^2 \times I \to \mathbb{R}$). Moreover let us observe that for any pair (α, β) of (m,n)-semiperiodic functions their difference $\beta - \alpha$ is $(0,0)$-semiperiodic and consider the map

$$f : \mathrm{Per}_I(m,n) \times \mathrm{Per}_I(m,n) \to \mathrm{Per}_I(0,0),$$

$$f(\alpha, \beta) = \beta - \alpha, \quad (\alpha, \beta) \in \mathrm{Per}_I(m,n)^2.$$

Then a) $f(\alpha,\beta) + f(\beta,\gamma) = f(\alpha,\gamma)$ for any $\alpha, \beta, \gamma \in \mathrm{Per}_I(m,n)$ and b) if $\alpha_0 \in \mathrm{Per}_I(m,n)$ is fixed then for any $\beta \in \mathrm{Per}_I(0,0)$ there is $\alpha \in \mathrm{Per}_I(m,n)$ such that $f(\alpha_0, \alpha) = \beta$ (indeed one may set by definition $\alpha = \alpha_0 + \beta$). Therefore the synthetic object

$$\mathcal{A} = (\mathrm{Per}_I(m,n), \mathrm{Per}_I(0,0), f)$$

is an affine space (whose associated vector space is $\mathrm{Per}_I(0,0)$). The proof of the last statement in Lemma 1.35 is immediate (there Y acts in the ξ-variable). ∎

Lemma 1.36 a) *For each* $X \in \mathcal{E}_I$ *there is an angle function* $\alpha \in \mathcal{W}_I$. *If* $X \in \mathcal{E}_I$ *is fixed then its angle functions differ solely by integer multiples of* 2π *and lie in but one* $\mathrm{Per}_I(m,n)$ *for some* $(m,n) \in \mathbb{Z}^2$ *depending on* $\{S, W\}$. b) *Let*

$\alpha \in C^\infty(\mathbb{R}^2 \times I)$. *Then* α *is an angle function for some* $X \in \mathcal{E}_I$ *if and only if* $\alpha \in \mathcal{W}_I$.

As a consequence of Lemma 1.36 we may adopt

Definition 1.37 We define a function $\text{htp}^{(S,W)} : \mathcal{E} \to \mathbb{Z}^2$ as follows. Let $X \in \mathcal{E}$ be a unit vector field on the torus T^2. According to Lemma 1.36 there is a unique $(m,n) \in \mathbb{Z}^2$ such that all angle functions of X belong to $\text{Per}(m,n)$. Then we set by definition $\text{htp}^{(S,W)}(X) = (m,n)$. ■

The next lemma gives the homotopy classification of the unit vector fields tangent to T^2.

Lemma 1.38 *Two unit vector fields* $X, Y \in \mathcal{E}$ *are homotopic in* $\mathcal{E}$ *if and only if* $\text{htp}^{(S,W)}(X) = \text{htp}^{(S,W)}(Y)$. *Thus homotopy classes of elements of* $\mathcal{E}$ *are classified by the elements of* $\mathbb{Z}^2$. *For each* $X \in \mathcal{E}$ *its homotopy class* $[X] \in \pi(T^2; S(T^2))$ *doesn't depend upon the choice of* $\{S, W\}$ *yet the index* (m,n) *does. Precisely if* $\{\tilde{S}, \tilde{W}\}$ *is another orthonormal frame of* $(T(T^2), g)$ *such that* $J\tilde{S} = \tilde{W}$ *then there is a unique* $(k,\ell) \in \mathbb{Z}^2$ *such that*

$$\text{htp}^{(S,W)}(\tilde{S}) = \text{htp}^{(S,W)}(\tilde{W}) = (k,\ell),$$

and for any $X \in \mathcal{E}$

$$\text{htp}^{(S,W)}(X) = \text{htp}^{(\tilde{S},\tilde{W})}(X) + (k,\ell).$$

Let $\mathcal{E}^{(S,W)}_{(m,n)} \in \pi\left(T^2; S(T^2)\right)$ *denote the homotopy class of* $X \in \mathcal{E}$ *where* $(m,n) = \text{htp}^{(S,W)}(X)$. *Then the unit vector fields* S, W *lie in the homotopy class* $\mathcal{E}^{(S,W)}_{(0,0)}$.

As to the notations in Lemma 1.38 we adopt the conventions in S-T. Hu, [172].

Proof of Lemma 1.38. Let $X, Y \in \mathcal{E}$ such that $X \sim Y$ i.e., X and Y are homotopic. Then there is a path $C : [0,1] \to \mathcal{E} \subset S(T^2)^{T^2}$ connecting X and Y i.e., $C(0) = X$ and $C(1) = Y$. Let $H = (C(t))_{0 \le t \le 1} \in \mathcal{E}_{[0,1]}$ and let $\alpha \in \mathcal{W}_{[0,1]}$ be an angle function for H and $\alpha_t(\xi) = \alpha(\xi, t)$ for any $\xi \in \mathbb{R}^2$ and any $0 \le t \le 1$. Let $(m,n) \in \mathbb{Z}^2$ such that $\alpha \in \text{Per}_{[0,1]}(m,n)$. Then $\alpha_0, \alpha_1 \in \text{Per}(m,n)$ are angle functions for X and Y respectively so that

$$\text{htp}^{(S,W)}(X) = (m,n) = \text{htp}^{(S,W)}(Y).$$

Vice versa let us assume that $\text{htp}^{(S,W)}(X) = \text{htp}^{(S,W)}(Y)$ and let us denote their common value by (m,n) so that the angle functions of both X, Y

must lie in $\mathrm{Per}(m,n)$. Let $\alpha_0, \alpha_1 \in \mathrm{Per}(m,n)$ be angle functions for X and Y respectively and let us consider

$$\alpha : \mathbb{R}^2 \times [0,1] \to \mathbb{R},$$
$$\alpha(\xi,t) = (1-t)\alpha_0(\xi) + t\alpha_1(\xi), \quad \xi \in \mathbb{R}^2, \quad 0 \le t \le 1.$$

Clearly $\alpha \in \mathrm{Per}_{[0,1]}(m,n)$. Let us consider the path $C : [0,1] \to \mathcal{E}$ given by

$$C(t)_p = (\cos\alpha_t(\xi))\, S_p + (\sin\alpha_t(\xi))\, W_p \in T_p(T^2),$$
$$\xi \in \pi^{-1}(p), \quad p \in T^2, \quad 0 \le t \le 1.$$

The definition of the unit tangent vector $C(t)_p$ doesn't depend upon the choice of $\xi \in \pi^{-1}(p)$ due to the general observation that composition of elements in $\mathrm{Per}_I(m,n)$ with the trigonometric functions cos and sin induces natural maps

$$\cos, \sin : \mathrm{Per}_I(m,n) \to \mathrm{Per}_I(0,0).$$

Then C connects X and Y so that $X \sim Y$.

By the first part of Lemma 1.38 the map of

$$\{[X] \in \pi(T^2; S(T^2)) : X \in \mathcal{E}\}$$

into $\mathbb{Z}^2$ given by $[X] \mapsto \mathrm{htp}^{(S,W)}(X)$ is a well defined bijection hence the announced classification of the homotopy classes $\{[X] : X \in \mathcal{E}\}$.

Let us look at the dependence on the frame $\{S, W\}$. If $\{\tilde{S}, \tilde{W}\}$ is another orthonormal frame such that $J\tilde{S} = \tilde{W}$ then

$$(\tilde{S}, \tilde{W})^t = a(S, W)^t$$

for some $a \in \mathrm{O}(2)$ (where v^t is the transpose of $v \in \mathbb{R}^2$). If $a = [a_{ij}]$ then $JS = W$ and $J\tilde{S} = \tilde{W}$ imply $a_{11} = a_{22}$ and $a_{21} = -a_{12}$ hence given an angle function $\theta \in \mathrm{Per}(k,\ell)$ for $\tilde{S}$

$$\tilde{S} \circ \pi = (\cos\theta)\, S \circ \pi + (\sin\theta)\, W \circ \pi$$

it follows that

$$\tilde{W} \circ \pi = \cos\left(\theta + \frac{\pi}{2}\right) S \circ \pi + \sin\left(\theta + \frac{\pi}{2}\right) W \circ \pi$$

that is $\theta + \pi/2$ is an angle function for $\tilde{W}$ belonging to $\mathrm{Per}(k,\ell)$. Thus

$$\mathrm{htp}^{(S,W)}(\tilde{S}) = (k,\ell), \quad \mathrm{htp}^{(S,W)}(\tilde{W}) = (k,\ell).$$

Let $\alpha \in \text{Per}(m,n)$ be an angle function for $X \in \mathcal{E}$. Since

$$S \circ \pi = (\cos\theta)\ \tilde{S} \circ \pi - (\sin\theta)\ \tilde{W} \circ \pi,$$
$$W \circ \pi = (\sin\theta)\ \tilde{S} \circ \pi + (\cos\theta)\ \tilde{W} \circ \pi,$$

it follows that

$$X \circ \pi = \cos(\alpha - \theta)\ \tilde{S} \circ \pi + \sin(\alpha - \theta)\ \tilde{W} \circ \pi$$

i.e., $\alpha - \theta$ is an angle function for X with respect to the frame $\{\tilde{S}, \tilde{W}\}$. Note that

$$(\alpha - \theta)(\xi + d_1) - (\alpha - \theta)(\xi) = 2\pi(m - k), \quad \xi \in \mathbb{R}^2,$$

etc., hence $\alpha - \theta \in \text{Per}(m - k, n - \ell)$ and we may conclude that

$$\text{htp}^{(\tilde{S},\tilde{W})}(X) = (m - k, n - \ell) = \text{htp}^{(S,W)}(X) - (k, \ell). \qquad \blacksquare$$

Given a topological space $\mathfrak{X}$ its *Bruschlinsky group* is the abelian group $\pi^1(\mathfrak{X})$ of all homotopy classes of maps $f : \mathfrak{X} \to S^1$ (the additive structure of $\pi^1(\mathfrak{X})$ is naturally induced by the additive structure of $S^1 \subset \mathbb{C}$). Cf. [172], p. 47. Let $\{S, W\}$ be an orthonormal frame on T^2 such that $JS = W$. If $X \in \mathcal{E}$ is a given unit vector field on T^2 and $(m,n) = \text{htp}^{(S,W)}(X) \in \mathbb{Z}^2$ then let $\alpha \in \text{Per}(m,n)$ be an angle function for X. Next we consider the function $e^{i\alpha} : T^2 \to S^1$ given by

$$\left(e^{i\alpha}\right)(p) = e^{i\alpha(\xi)}, \quad \xi \in \pi^{-1}(p), \quad p \in T^2,$$

so that the map

$$\mathcal{E} \to C^\infty\left(T^2, S^1\right), \quad X \mapsto e^{i\alpha}, \tag{1.54}$$

is a bijection. The bijection (1.54) induces a group isomorphism $\{[X] : X \in \mathcal{E}\} \approx \pi^1(T^2)$. Therefore Lemma 1.38 (due to G. Wiegmink, cf. Lemma 3 in [309], p. 334–335) is nothing but the calculation of the Bruschlinsky group of the torus i.e., $\pi^1(T^2) = \mathbb{Z} \oplus \mathbb{Z}$.

The calculations in this section will be used in Section 2.6 (where we compute the total bending functional on (T^2,g) in terms of the angle functions of a unit vector field on T^2) and in Section 2.10 (where we examine the behavior of the total bending functional on (T^2,g) under conformal transformations $\tilde{g} = e^{2u}g$ of the metric). Conclusive and quite complete results (due to G. Wiegmink, [309]) on harmonic vector fields on Riemannian tori are presented in Section 3.8 of Chapter 3 in this book.

CHAPTER TWO

Harmonic Vector Fields

Contents

In the present chapter we discuss fundamental matters such as the Dirichlet energy and tension tensor of a unit tangent vector field X on a Riemannian manifold (M,g), first and second variation formulae and the harmonic vector fields system, to be applied in the remaining chapters of this book. Indeed one may look at a smooth vector field $X: M \to T(M)$ as a map of Riemannian manifolds (M,g) and $(T(M), G_s)$ where G_s is the Sasaki metric (cf. Chapter 1) and apply the ordinary results in variational calculus to the functional

$$E(X) = \frac{1}{2} \int_M \|dX\|^2 d\,\mathrm{vol}(g)$$

where $\|dX\|$ is the Hilbert-Schmidt norm of dX, very much in the spirit of the theory of harmonic maps (cf. e.g., J. Eells & J.H. Sampson, [113]). One of the main references here is the work of C.M. Wood, [316]. The Dirichlet

© 2012 Elsevier Inc. All rights reserved.

energy functional turns out to be related to the *total bending functional*

$$\mathcal{B}(X) = \int_M \|\nabla X\|^2 d\,\mathrm{vol}(g)$$

as introduced by G. Wiegmink, [309]. Precisely we find (cf. the identity (2.3) below) that $E(X) = \frac{1}{2}\left[n\,\mathrm{Vol}(M) + \mathcal{B}(X)\right]$. The search for critical points for $E : C^\infty(M, T(M)) \to [0, +\infty)$ or $E : \mathfrak{X}(M) \to [0, +\infty)$ shows that both domains $C^\infty(M, T(M))$ and $\mathfrak{X}(M)$ are inappropriate. Indeed these domains are too "large": if a smooth vector field $X : M \to T(M)$ is a harmonic map or a critical point of $E : \mathfrak{X}(M) \to [0, +\infty)$ (that is to say with respect to smooth 1-parameter variations of X within $\mathfrak{X}(M)$) then X must be parallel (cf. Corollary 2.14 and Theorem 2.17 below) and $E(X) = \frac{n}{2}\mathrm{Vol}(M)$. As it turns out, the appropriate functional to look at is $E : \Gamma^\infty(S(M)) \to [0, +\infty)$ where $S(M)$ is the total space of the tangent sphere bundle over (M, g). The search for critical points for this functional leads to a new and appealing theory of *harmonic vector fields* on M. These are smooth solutions to the PDE system $\Delta_g X - \|\nabla X\|^2 X = 0$ (cf. Theorem 2.23 below) where Δ_g is a second order elliptic operator acting on tangent vector fields (the rough Laplacian). A study of the weak solutions to this system (existence and local properties) is missing from the present day mathematical literature. As opposed to the fundamental matters discussed in this chapter, the remainder of this book is devoted to the investigation of various instances where harmonic vector fields occur (especially in contact and CR geometry, cf. Chapter 4 and Sections 7.1 to 7.3 in Chapter 7) and to generalizations (within subelliptic theory, cf. Sections 7.4 to 7.7 in Chapter 7).

2.1. VECTOR FIELDS AS ISOMETRIC IMMERSIONS

Let M be an n-dimensional C^∞ manifold. Let $X : M \to T(M)$ be a tangent vector field on M. We may show that

Proposition 2.1 1) *$X : M \to T(M)$ is an immersion.* 2) *If g is a Riemannian metric on M and $\mathcal{H}$ a nonlinear connection on $T(M)$ then $X : (M, g) \to (T(M), G_s)$ is an isometric immersion if and only if*

$$\left(\frac{\partial \lambda^j}{\partial x^i} \circ \pi + N_i^j\right)_{X_x} = 0, \quad 1 \le i, j \le n, \tag{2.1}$$

for any local coordinate system (U,x^i) *and any* $x \in U$, *where* $X = \lambda^i \partial/\partial x^i$ *on* U *with* $\lambda^i \in C^\infty(U)$. 3) *Let* ∇ *be a linear connection on* M *and* $\hat{\nabla}$ *the induced regular connection in* $\pi^{-1}TM \to T(M)$ *and* $\mathcal{H} = \mathcal{H}_{\hat{\nabla}}$ *the associated nonlinear connection. Then* X *is an isometric immersion of* (M,g) *into* $(T(M), G_s)$ *if and only if* X *is parallel with respect to* ∇.

Proof. Let us assume that $(d_x X)v = 0 \in T_{X(x)}(M)$. Applying $d_{X(x)}\pi$ one has (by the chain rule)

$$0 = (d_{X(x)}\pi) \circ (d_x X)v = d_x(\pi \circ X)v = (d_x 1_M)v$$

hence $v = 0$. Here 1_M denotes the identical transformation of M. Consequently $\mathrm{Ker}(d_x X) = (0)$ i.e., X is an immersion as stated above. Let $X^* G_s$ be the pullback by X of the Sasaki metric G_s (associated to the nonlinear connection $\mathcal{H}$). Then for any $x \in M$ and any $v, w \in T_x(M)$

$$\begin{aligned}(X^* G_s)_x(v,w) &= G_{s\,X(x)}((d_x X)v,\ (d_x X)w) \\ &= \hat{g}_{X(x)}(L_{X(x)}(d_x X)v,\ L_{X(x)}(d_x X)w) \\ &\quad + \hat{g}_{X(x)}(K_{X(x)}(d_x X)v,\ K_{X(x)}(d_x X)w).\end{aligned}$$

On the other hand

$$\begin{aligned}L_{X(x)}(d_x X)v &= \big(X(x),\ (d_{X(x)}\pi)(d_x X)v\big) \\ &= (X(x),\ d_x(\pi \circ X)v) = (X(x), v)\end{aligned}$$

hence locally $L_{X(x)}(d_x X)v = v^i X_i(X_x)$ where $v = v^i \left(\partial/\partial x^i\right)_x$. Moreover, in order to compute

$$K_{X(x)}(d_x X)v = \gamma_{X(x)}^{-1} Q_{X(x)}(d_x X)v$$

we need to calculate the differential of X at x i.e.,

$$(d_x X)\frac{\partial}{\partial x^i}\bigg|_x = \frac{\partial X^j}{\partial x^i}(x)\partial_j\bigg|_{X(x)} + \frac{\partial X^{j+n}}{\partial x^i}(x)\dot{\partial}_j\bigg|_{X(x)}$$

where X^j and X^{j+n} are the local components of X (with respect to the local coordinate systems (U,x^i) and $(\pi^{-1}(U), x^i, y^i)$ on M and $T(M)$ respectively) that is

$$X^j(x) = x^j(x), \quad X^{j+n}(x) = \lambda^j(x), \quad 1 \le j \le n,$$

where $X = \lambda^j \, \partial/\partial x^j$ on U. Thus

$$(d_x X)\frac{\partial}{\partial x^i}\bigg|_x = \left[\partial_i + \left(\frac{\partial \lambda^j}{\partial x^i} \circ \pi\right) \dot{\partial}_j\right]_{X(x)}$$

$$= \left\{\delta_i + \left[\left(\frac{\partial \lambda^j}{\partial x^i} + N_i^j\right) \circ \pi\right] \dot{\partial}_j\right\}_{X(x)}$$

so that

$$K_{X(x)}(d_x X)v = v^i \left(\frac{\partial \lambda^j}{\partial x^i}(x) + N_i^j(X_x)\right) X_j(X_x).$$

Consequently

$$\hat{g}_{X(x)}(L_{X(x)}(d_x X)v, \, L_{X(x)}(d_x X)w) = v^i w^j g_{ij}(x) = g_x(v, w),$$

$$\hat{g}_{X(x)}(K_{X(x)}(d_x X)v, \, K_{X(x)}(d_x X)w)$$

$$= v^i w^k g_{j\ell}(x)\left(\frac{\partial \lambda^j}{\partial x^i}(x) + N_i^j(X_x)\right)\left(\frac{\partial \lambda^\ell}{\partial x^k}(x) + N_k^\ell(X_x)\right),$$

hence $(X^* G_s)_x(v, w) = g_x(v, w)$ if and only if

$$v^i w^k g_{j\ell}(x)\left(\frac{\partial \lambda^j}{\partial x^i}(x) + N_i^j(X_x)\right)\left(\frac{\partial \lambda^\ell}{\partial x^k}(x) + N_k^\ell(X_x)\right) = 0$$

or (for $v = \partial/\partial x^r$ and $w = \partial/\partial x^s$)

$$g_{j\ell}(x)\left(\frac{\partial \lambda^j}{\partial x^r}(x) + N_r^j(X_x)\right)\left(\frac{\partial \lambda^\ell}{\partial x^s}(x) + N_s^\ell(X_x)\right) = 0. \tag{2.2}$$

We set

$$V_i = \left(\frac{\partial \lambda^j}{\partial x^i} \circ \pi + N_i^j\right) X_j \in \Gamma^\infty(\pi^{-1}(U), \, \pi^{-1}TM), \quad 1 \le i \le n,$$

so that (2.2) becomes $\hat{g}(V_i, V_j)_{X_x} = 0$ for any $1 \le i, j \le n$. Hence X is an isometry of (M, g) into $(T(M), G_s)$ if and only if $V_i(X_x) = 0$, $1 \le i \le n$. This proves the second statement in Proposition 2.1. To prove the last statement, let ∇ be a linear connection on M. Then the local coefficients of the nonlinear connection $\mathcal{H}_{\hat{\nabla}}$ are given by $N_j^i = \left(\Gamma_{jk}^i \circ \pi\right) y^k$ (cf. Corollary 1.9) where Γ_{jk}^i are the connection coefficients. Thus $N_j^i(X_x) = \Gamma_{jk}^i(x)\lambda^k(x)$. Taking into account (2.1) it follows that $X^* G_s = g$ if and only if

$$0 = \frac{\partial \lambda^j}{\partial x^i}(x) + \Gamma_{ik}^j(x)\lambda^k(x) = \left(\nabla_i \lambda^j\right)_x$$

i.e., if and only if $\nabla X = 0$. ■

2.2. THE ENERGY OF A VECTOR FIELD

Let (M,g) be an n-dimensional compact oriented Riemannian manifold and $X : M \to T(M)$ a tangent vector field on M. Let ∇ be the Levi-Civita connection of (M,g). Let $\hat{\nabla}$ be the regular connection in $\pi^{-1}TM \to T(M)$ induced by ∇ and $\mathcal{H}_{\hat{\nabla}}$ the corresponding nonlinear connection on $T(M)$. Let G_s be the Sasaki metric on $T(M)$ (associated to the pair $(g, \mathcal{H}_{\hat{\nabla}})$).

Definition 2.2 The *energy functional* is the map $E : \mathfrak{X}(M) = \Gamma^\infty(M, T(M)) \to [0,+\infty)$ given by

$$E(X) = \int_M e(X)\, d\,\mathrm{vol}(g), \quad e(X) = \frac{1}{2}\|dX\|^2,$$

for any $X \in \mathfrak{X}(M)$. Here $\|dX\|$ is the Hilbert-Schmitd norm of dX i.e.,

$$\|dX\|^2 = \mathrm{trace}_g(X^* G_s)$$

and $d\,\mathrm{vol}(g)$ is the Riemannian volume form on M. ∎

Given a local g-orthonormal frame $\{V_1, \ldots, V_n\}$ of $T(M)$ defined on the open set U

$$\|dX\|^2(x) = \sum_{j=1}^n (X^* G_s)(V_j, V_j)_x$$
$$= \sum_{j=1}^n G_{s\,X(x)}((d_x X) V_j(x), (d_x X) V_j(x))$$

for any $x \in U$.

For each Riemannian manifold (M,g), a pointwise inner product on $\Gamma^\infty(M, T^*(M) \otimes T(M))$ is defined as follows. Let $x \in M$ and $\{V_j : 1 \le j \le n\}$ be a g-orthonormal local frame of $T(M)$ defined on an open set $U \subseteq M$ such that $x \in U$. For any $S, R \in \Gamma^\infty(M, T^*(M) \otimes T(M))$ we define $g^*(S,R) : M \to [0,+\infty)$ by setting

$$g^*(S,R)_x = \sum_{j=1}^n g(SV_j, RV_j)_x.$$

The definition of $g^*(S,R)_x$ doesn't depend upon the choice of local orthonormal frame at x. Moreover, when either M is compact or one of the tensor fields S, R has compact support, the L^2 inner product of S and R is defined by

$$(S, R) = \int_M g^*(S, R) d\,\mathrm{vol}(g).$$

For any vector field $X \in \mathfrak{X}(M)$ its covariant derivative ∇X is a cross-section in the vector bundle $T^*(M) \otimes T(M) \to M$, hence its pointwise norm $\|\nabla X\| = g^*(\nabla X, \nabla X)^{1/2}$ is well defined.

Proposition 2.3 *Let (M, g) be an n-dimensional compact oriented Riemannian manifold. For any tangent vector field $X \in \mathfrak{X}(M)$, its energy is given by*

$$E(X) = \frac{n}{2} \mathrm{vol}(M) + \frac{1}{2} \int_M \|\nabla X\|^2 d\,\mathrm{vol}(g) \tag{2.3}$$

where $\mathrm{vol}(M) = \int_M d\,\mathrm{vol}(g)$ *is the volume of* M.

Proof. Let $\{V_i : 1 \leq i \leq n\}$ be a g-orthonormal local frame of $T(M)$. Then (by taking into account the proof of Proposition 2.1)

$$L_{X(x)}(d_x X) V_i(x) = V_i^j(x) X_j(X_x),$$

$$K_{X(x)}(d_x X) V_i(x) = V_i^j(x) \left(\frac{\partial \lambda^k}{\partial x^j}(x) + \Gamma^k_{j\ell}(x) \lambda^\ell(x) \right) X_k(X_x)$$

$$= V_i^j(x) \left(\nabla_j \lambda^k \right)_x X_k(X_x) = \widehat{\nabla_{V_i} X}\Big|_{X(x)}$$

where $V_i = V_i^j \partial / \partial x^j$ and a hat indicates as usual the natural lift of a vector field (cf. Definition 1.2). Thus (using the identity $\hat{g}(\hat{V}, \hat{W}) = g(V, W) \circ \pi$ for any $V, W \in \mathfrak{X}(M)$)

$$2e(X) = \sum_{i=1}^{n} G_{s\,X(x)}((d_x X) V_i(x), (d_x X) V_i(x))$$

$$= \sum_{i=1}^{n} \left\{ g(V_i, V_i)_x + g(\nabla_{V_i} X, \nabla_{V_i} X)_x \right\} = n + \|\nabla X\|_x^2$$

and Proposition 2.3 is proved. ∎

Remark 2.4 Under a homothetic transformation of the given metric the energy of a vector field behaves as follows. Let $c > 0$ and $\bar{g} = cg$.

If $\overline{X} = (1/\sqrt{c})\, X$ then

$$E_{\overline{g}}(\overline{X}) = c^{n/2-1}\left[\frac{n(c-1)}{2}\operatorname{Vol}(M,g) + E_g(X)\right] \tag{2.4}$$

where we adopt for a moment the more precise notation $E_g(X)$ for the energy of X relative to the metric g. ■

Remark 2.5 Let $(\nabla X)^t$ be the transpose of ∇X i.e.,

$$g((\nabla X)^t V, W) = g(V, \nabla_W X), \quad V, W \in \mathfrak{X}(M), \tag{2.5}$$

and let us consider the endomorphism of the tangent bundle $L_X \in \Gamma^\infty(M, T^*(M) \otimes T(M))$ defined by

$$L_X = I + (\nabla X)^t \circ (\nabla X) \tag{2.6}$$

where I is the identical transformation. Then

$$\begin{aligned}\operatorname{trace}(L_X) &= \sum_{i=1}^{n} g(L_X V_i, V_i)\\ &= n + \sum_{i=1}^{n} g((\nabla X)^t \circ (\nabla X) V_i, V_i) = n + \|\nabla X\|^2\end{aligned}$$

so that the energy of X may also be written as

$$E(X) = \frac{1}{2}\int_M \operatorname{trace}(L_X)\, d\operatorname{vol}(g). \tag{2.7}$$

■

Remark 2.6 The differential of a smooth map $\phi : M \to N$ may be thought of as a section in the vector bundle $T^*(M) \otimes \phi^{-1}T(N)$. In particular dX is a section in $T^*(M) \otimes X^{-1}T(T(M))$. Let $E \to M$ be the vector bundle whose fibres are $E_x = \{X(x)\} \times T_x(M)$ for any $x \in M$. We define sections $X_*^V, X_*^H \in \Gamma^\infty(T^*(M) \otimes E)$ by setting

$$\begin{gathered}\left(X_*^V\right)_x : T_x(M) \to \left(\pi^{-1}TM\right)_{X(x)} = \{X(x)\} \times T_x(M) = E_x,\\ \left(X_*^V\right)_x = K_{X(x)} \circ (d_x X), \quad x \in M,\\ \left(X_*^H\right)_x : T_x(M) \to E_x, \quad \left(X_*^H\right)_x = L_{X(x)} \circ (d_x X), \quad x \in M.\end{gathered}$$

Let (U, x^i) be a local coordinate system on M and $e_i \in \Gamma^\infty(U, E)$ defined by $e_i(x) = (X_x, (\partial/\partial x^i)_x)$ for any $x \in U$. Thus $\{e_i : 1 \le i \le n\}$ is a local frame in E. Also $E \to M$ inherits the Riemannian bundle metric g^E given by $g^E_x((X_x, v), (X_x, w)) = g_x(v, w)$ for any $v, w \in T_x(M)$ and any $x \in M$. An inner product on $T^*(M) \otimes E$ is then defined by

$$(dx^i \otimes e_k, dx^j \otimes e_\ell) = g^{ij} g_{k\ell} .$$

As X^V_* and X^H_* are sections in $T^*(M) \otimes E$, it makes sense to consider their norms $\|X^V_*\|$ and $\|X^H_*\|$. Note that $\|X^H_*\|^2 = n$. Indeed if $\{V_i : 1 \le i \le n\}$ is a g-orthonormal local frame of $T(M)$ on U, then for any $x \in U$

$$\left(X^H_*\right)_x w = (X(x), w), \quad w \in T_x(M),$$

$$\|X^H_*\|^2_x = \sum_{i=1}^n g^E_x(\left(X^H_*\right)_x V_{i,x}, \left(X^H_*\right)_x V_{i,x}) = \sum_{i=1}^n g_x(V_i, V_i)_x = n.$$

On the other hand for any $w \in T_x(M)$

$$\begin{aligned}(d_x X)w \in T_{X(x)}(T(M)) &= \mathcal{H}_{X(x)} \oplus \mathcal{V}_{X(x)} \\ &= \left(\beta \pi^{-1} TM\right)_{X(x)} \oplus \left(\gamma \pi^{-1} TM\right)_{X(x)}\end{aligned}$$

hence $(d_x X)w = \beta_{X(x)} V + \gamma_{X(x)} W$ for some $V, W \in \left(\pi^{-1} TM\right)_{X(x)}$. By applying $K_{X(x)}$, respectively $L_{X(x)}$, to the previous identity we obtain

$$V = \left(X^H_*\right)_x w, \quad W = \left(X^V_*\right)_x w,$$

hence

$$d_x X = \beta_{X(x)} \circ \left(X^H_*\right)_x + \gamma_{X(x)} \circ \left(X^V_*\right)_x . \tag{2.8}$$

By the very definition of the Sasaki metric $\beta_v : \left(\pi^{-1} TM\right)_v \to T_v(T(M))$ and $\gamma_v : \left(\pi^{-1} TM\right) \to T_v(T(M))$ are linear isometries of the inner product spaces $((\pi^{-1} TM)_v, \hat{g}_v)$ and $T_v(T(M)), G_{s\,v})$, for any $v \in T(M)$. In particular it follows that the maps $\beta_{X(x)} : E_x \to T_{X(x)}(T(M))$ and $\gamma_{X(x)} : E_x \to T_{X(x)}(T(M))$ are linear isometries of (E_x, g^E_x) into $(T_{X(x)}(T(M)), G_{s\,X(x)})$, for any $x \in M$. Hence (by (2.8))

$$\begin{aligned}\|dX\|^2_x &= \left(\mathrm{trace}_g X^* G_s\right)(x) = \sum_{i=1}^n G_{s\,X(x)}\left((d_x X) V_{i,x}, (d_x X) V_{i,x}\right) \\ &= \sum_i \{\| \left(X^H_*\right)_x V_{i,x}\|^2 + \| \left(X^V_*\right)_x V_{i,x}\|^2\} = \|X^H_*\|^2_x + \|X^V_*\|^2_x\end{aligned}$$

from which

$$E(X) = \frac{n}{2}\,\mathrm{vol}(M) + \frac{1}{2}\int_M \|X_*^V\|^2 d\,\mathrm{vol}(g).$$

■

Our calculations through Remark 2.6 are in the spirit of C.M. Wood, [316], where $\int_M \|X_*^V\|^2 d\,\mathrm{vol}(g)$ is referred to as the *vertical energy* of X. Of course this is precisely the integral $\int_M \|\nabla X\|^2 d\,\mathrm{vol}(g)$.

Definition 2.7 (G. Wiegmink, [309]) Let (M,g) be a compact orientable Riemannian manifold and $X \in \mathfrak{X}(M)$ a tangent vector field on M. The number

$$\mathcal{B}(X) = \int_M \|\nabla X\|^2\, d\,\mathrm{vol}(g)$$

is called the *total bending* (or *biegung*) of X. ■

Clearly the biegung of X is a measure of the failure of X to be parallel.

Corollary 2.8 *Let M be a compact oriented Riemannian manifold. For any tangent vector field X on M*

$$E(X) \geq \frac{n}{2}\,\mathrm{vol}(M) \tag{2.9}$$

with equality if and only if $\nabla X = 0$.

Remark 2.9 By a result in [258] (cf. Proposition 5.10, p. 170)

$$\int_M \|\nabla X\|^2 d\,\mathrm{vol}(g) = \int_M \left\{\mathrm{Ric}(X,X) + \frac{1}{2}\|\mathcal{L}_X g\|^2 - (\mathrm{div}\,X)^2\right\} d\,\mathrm{vol}(g).$$

Then on a compact n-dimensional Riemannian manifold (M,g), the energy of a unit Killing vector field X is given by

$$E(X) = \frac{n}{2}\,\mathrm{Vol}(M) + \frac{1}{2}\int_M \mathrm{Ric}(X,X)\, d\,\mathrm{vol}(g).$$

If additionally (M,g) is an Einstein manifold and $n \geq 3$, then all unit Killing vector fields have the same energy $(n^2 + \rho)\mathrm{Vol}(M)/(2n)$ where ρ is the scalar curvature. ■

2.3. VECTOR FIELDS WHICH ARE HARMONIC MAPS

Let (M,g) be a compact oriented n-dimensional Riemannian manifold and ∇ its Levi-Civita connection. Let $\mathcal{H} = \mathcal{H}_{\hat{\nabla}}$ be the associated nonlinear connection, with local coefficients $N^i_j(x,y) = \Gamma^i_{jk}(x)y^k$, relative to a local coordinate system (U,x^i) on M. Let G_s be the Sasaki metric associated to the pair $(\mathcal{H},g)$. Without further specifications, this will be our choice of Sasaki metric on $T(M)$ for the remainder of the present chapter.

Theorem 2.10 (T. Ishihara, [176], O. Nouhaud, [223]) *Let $X \in \mathfrak{X}(M)$ be a tangent vector field. The following statements are equivalent*

i. *X is a harmonic map of (M,g) into $(T(M), G_s)$.*

ii. *X is an absolute minimum of the energy functional*

$$E : \mathfrak{X}(M) \to [0,+\infty), \quad E(X) = \frac{1}{2}\int_M \|dX\|^2 \, d\,\mathrm{vol}(g), \quad X \in \mathfrak{X}(M).$$

iii. *X is parallel i.e., $\nabla X = 0$.*

See also J.J. Konderak, [194]. *Proof of Theorem 2.10.* (ii) $\Longleftrightarrow$ (iii) as a consequence of Corollary 2.8. The smooth map $X : M \to T(M)$ is harmonic if and only if X is a critical point of $E : C^\infty(M, T(M)) \to [0,+\infty)$ i.e., $\{dE(X_t)/dt\}_{t=0} = 0$ for any smooth 1-parameter variation $\mathcal{X} : M \times (-\epsilon, \epsilon) \to T(M)$ of X by smooth maps. Here we set

$$X_t : M \to T(M), \quad X_t(x) = \mathcal{X}(x,t), \quad x \in M, \ |t| < \epsilon,$$

and then $X_0 = X$. In particular we may consider the variation

$$\mathcal{X}(x,t) = (1-t)X_x, \quad x \in M, \ |t| < \epsilon,$$

hence

$$0 = \left.\frac{dE(X_t)}{dt}\right|_{t=0} = \frac{d}{dt}\left\{\frac{n}{2}\mathrm{vol}(M) + \frac{1}{2}\int_M \|\nabla X_t\|^2 d\,\mathrm{vol}(g)\right\}_{t=0}$$

$$= \frac{d}{dt}\left\{\frac{(1-t)^2}{2}\int_M \|\nabla X\|^2 d\,\mathrm{vol}(g)\right\}_{t=0} = -\int_M \|\nabla X\|^2 d\,\mathrm{vol}(g)$$

hence $\nabla X = 0$.

Vice versa, if $\nabla X = 0$ then (by Proposition 2.1) $X : M \to T(M)$ is an isometric immersion. We shall show that X is totally geodesic (and in particular minimal i.e., X is harmonic). Let $\gamma(t)$ be a geodesic of (M,g) and let us set $\tilde{\gamma}(t) = X(\gamma(t))$ for any value of the parameter t. Then

$$\frac{d\tilde{\gamma}}{dt}(t) = d_t(X \circ \gamma)\frac{d}{dt}\bigg|_t = (d_{\gamma(t)}X) \circ (d_t\gamma)\frac{d}{dt}\bigg|_t = (d_{\gamma(t)}X)\frac{d\gamma}{dt}(t)$$

$$= \frac{d\gamma^i}{dt}(t)\left[\delta_i + \left(\frac{\partial \lambda^j}{\partial x^i} \circ \pi + N_i^j\right)\dot{\partial}_j\right]_{X(\gamma(t))}.$$

The remainder of the proof requires the Levi-Civita connection D of $(T(M), G_s)$. It is locally given by

$$D_{\delta_i}\delta_j = \Gamma_{ij}^k \delta_k - \frac{1}{2}y^k R_{ijk}^\ell \dot{\partial}_\ell, \tag{2.10}$$

$$D_{\delta_i}\dot{\partial}_j = \frac{1}{2}y^k R_{kji}^\ell \delta_\ell + \Gamma_{ij}^k \dot{\partial}_k, \tag{2.11}$$

$$D_{\dot{\partial}_i}\delta_j = \frac{1}{2}y^k R_{kij}^\ell \delta_\ell, \tag{2.12}$$

$$D_{\dot{\partial}_i}\dot{\partial}_j = 0, \tag{2.13}$$

where $R_{jk\ell}^i$ is the curvature tensor field of ∇. The formulae (2.10)–(2.13) follow easily from Proposition 1.14. For simplicity we set

$$Z_i = \delta_i + \left(\frac{\partial \lambda^j}{\partial x^i} + N_i^j\right)\dot{\partial}_j$$

so that $(d\tilde{\gamma}/dt)(t) = (d\gamma^i/dt)(t) Z_{i,X(\gamma(t))}$. Using (2.10)–(2.13) we may compute

$$D_{Z_i}Z_j = D_{\delta_i}\delta_j + \delta_i\left(\frac{\partial \lambda^\ell}{\partial x^j} + N_j^\ell\right)\dot{\partial}_\ell + \left(\frac{\partial \lambda^\ell}{\partial x^j} + N_j^\ell\right)D_{\delta_i}\dot{\partial}_\ell$$

$$+ \left(\frac{\partial \lambda^k}{\partial x^i} + N_i^k\right)\Big\{D_{\dot{\partial}_k}\delta_j$$

$$+ \dot{\partial}_k\left(\frac{\partial \lambda^\ell}{\partial x^j} + N_j^\ell\right)\dot{\partial}_\ell + \left(\frac{\partial \lambda^\ell}{\partial x^j} + N_j^\ell\right)D_{\dot{\partial}_k}\dot{\partial}_\ell\Big\}$$

$$= \left\{\Gamma_{ij}^s + \frac{1}{2}y^k R_{k\ell i}^s\left(\frac{\partial \lambda^\ell}{\partial x^j} + N_j^\ell\right) + \frac{1}{2}y^\ell R_{\ell kj}^s\left(\frac{\partial \lambda^k}{\partial x^i} + N_i^k\right)\right\}\delta_s$$

$$+\left\{\frac{\partial^2\lambda^s}{\partial x^i\partial x^j}+\frac{\partial\Gamma^s_{jk}}{\partial x^i}\gamma^k-N^k_i\Gamma^s_{jk}\right.$$

$$\left.+\left(\frac{\partial\lambda^\ell}{\partial x^j}+N^\ell_j\right)\Gamma^s_{i\ell}+\left(\frac{\partial\lambda^k}{\partial x^i}+N^k_i\right)\Gamma^s_{jk}\right\}\dot{\partial}_s$$

hence the identity

$$\left(\frac{\partial\lambda^\ell}{\partial x^j}+N^\ell_j\right)_{X(\gamma(t))}=\left(\nabla_j\lambda^\ell\right)_{\gamma(t)}$$

yields

$$\begin{aligned}(D_{Z_i}Z_j)_{X(\gamma(t))}&=\left\{\Gamma^s_{ij}(\gamma(t))+\frac{1}{2}\lambda^k(\gamma(t))R^s_{k\ell i}(\gamma(t))\left(\nabla_j\lambda^\ell\right)_{\gamma(t)}\right.\\&\quad\left.+\frac{1}{2}\lambda^\ell(\gamma(t))R^s_{\ell kj}(\gamma(t))\left(\nabla_i\lambda^k\right)_{\gamma(t)}\right\}\delta_s\Big|_{X(\gamma(t))}\\&\quad+\left\{\left[\frac{\partial}{\partial x^i}\left(\nabla_j\lambda^s\right)-\Gamma^s_{jk}\nabla_i\lambda^k\right]_{\gamma(t)}\right.\\&\quad\left.+\left(\nabla_j\lambda^\ell\right)_{\gamma(t)}\Gamma^s_{i\ell}(\gamma(t))+\left(\nabla_i\lambda^k\right)_{\gamma(t)}\Gamma^s_{jk}(\gamma(t))\right\}\dot{\partial}_s\Big|_{X(\gamma(t))}\end{aligned}$$

due to the identity

$$\begin{aligned}&\frac{\partial^2\lambda^s}{\partial x^i\partial x^j}(\gamma(t))+\frac{\partial\Gamma^s_{jk}}{\partial x^i}(\gamma(t))\lambda^k(\gamma(t))-\Gamma^k_{i\ell}(\gamma(t))\Gamma^s_{jk}(\gamma(t))\lambda^\ell(\gamma(t))\\&\quad=\left[\frac{\partial}{\partial x^i}\left(\nabla_j\lambda^s\right)-\Gamma^s_{jk}\nabla_i\lambda^k\right]_{\gamma(t)}.\end{aligned}$$

Therefore (as $\nabla_i\lambda^j=0$)

$$(D_{Z_i}Z_j)_{X(\gamma(t))}=\Gamma^s_{ij}(\gamma(t))\delta_s\Big|_{X(\gamma(t))}. \tag{2.14}$$

Finally (by (2.14) and by the fact that $\gamma(t)$ is a geodesic of ∇)

$$\begin{aligned}\left(D_{d\tilde{\gamma}/dt}\frac{d\tilde{\gamma}}{dt}\right)_{\tilde{\gamma}(t)}&=\frac{d^2\gamma^j}{dt^2}(t)Z_{j,X(\gamma(t))}+\frac{d\gamma^i}{dt}(t)\frac{d\gamma^j}{dt}(t)\left(D_{Z_i}Z_j\right)_{X(\gamma(t))}\\&=\left(\frac{d^2\gamma^k}{dt^2}(t)+\Gamma^k_{ij}(\gamma(t))\frac{d\gamma^i}{dt}(t)\frac{d\gamma^j}{dt}(t)\right)\delta_k\Big|_{X(\gamma(t))}=0.\end{aligned}$$

What we just proved is that the isometric immersion $X : M \to T(M)$ maps geodesics of the submanifold (M, g) into geodesics of the ambient space $(T(M), G_s)$, hence X is totally geodesic. Theorem 2.10 is completely proved.

2.4. THE TENSION OF A VECTOR FIELD

Let $\phi : M \to N$ be a smooth map of Riemannian manifolds (M, g) and (N, h) and $\phi^{-1}TN \to M$ the pullback of $T(N) \to N$ via ϕ. Let ∇ and ∇^N be the Levi-Civita connections of g and h, respectively. Let $\phi^{-1}\nabla^N$ be the connection in $\phi^{-1}TN \to M$ induced by ∇^N. This is most easily described in local coordinates as follows. Let (U, x^i) and (V, y^α) be local coordinate systems on M and N respectively such that $\phi(U) \subseteq V$. Let us set $\phi^\alpha = y^\alpha \circ \phi$ for any $1 \le \alpha \le \nu = \dim(N)$. Let $Y_\alpha = (\partial/\partial y^\alpha)^\phi$ be the natural lift of the local tangent vector field $\partial/\partial y^\alpha$ i.e.,

$$Y_\alpha(x) = \left.\frac{\partial}{\partial y^\alpha}\right|_{\phi(x)}, \quad x \in U.$$

Let $\left(\Gamma^N\right)^\alpha_{\beta\gamma}$ be the Christoffel symbols of the second kind of $h_{\alpha\beta}$. We set by definition

$$\left(\phi^{-1}\nabla^N\right)_{\partial/\partial x^i} Y_\beta = \frac{\partial \phi^\alpha}{\partial x^i}\left[\left(\Gamma^N\right)^\gamma_{\alpha\beta} \circ \phi\right] Y_\gamma.$$

It is easily checked that the definition of $\phi^{-1}\nabla^N$ doesn't depend on the choices of local coordinates. Given a tangent vector field X on M we may consider the cross-section $\phi_* X \in \Gamma^\infty(\phi^{-1}TN)$ given by

$$(\phi_* X)(x) = (d_x\phi)X_x, \quad x \in M.$$

The following formula is also useful in calculations

$$\left(\phi^{-1}\nabla^N\right)_X \phi_* Y = [X(Y\phi^\alpha) \circ \phi] Y_\alpha + [X(\phi^\alpha) Y(\phi^\beta) Z_{\alpha\beta}] \circ \phi, \tag{2.15}$$

$$Z_{\alpha\beta} \equiv \nabla^N_{\partial/\partial y^\alpha} \frac{\partial}{\partial y^\beta},$$

for any $X, Y \in \mathfrak{X}(M)$. The *second fundamental form* of ϕ is

$$\beta_\phi(X, Y) = \left(\phi^{-1}\nabla^N\right)_X \phi_* Y - \phi_* \nabla_X Y, \quad X, Y \in \mathfrak{X}(M).$$

Then the *tension* tensor field $\tau_\phi \in \Gamma^\infty(\phi^{-1}TN)$ is given by

$$\tau(\phi) = \text{trace}_g \beta_\phi.$$

Locally, with respect to a g-orthonormal local frame $\{V_i : 1 \le i \le n\}$ of $T(M)$ defined on the open set $U \subseteq M$

$$\tau(\phi)_x = \sum_{i=1}^{n} \beta_\phi(V_i, V_i)_x, \quad x \in U.$$

Let us apply these notions to the case of a tangent vector field $X : M \to T(M)$ and compute the tension field $\tau(X) \in \Gamma^\infty(X^{-1}TT(M))$.

Let $X \in \mathfrak{X}(M)$ be locally written as $X = \lambda^i \partial/\partial x^i$. We need to work with natural lifts relative to the pullback bundle $X^{-1}TT(M) \to M$ i.e., given a tangent vector field $A : T(M) \to T(T(M))$ its natural lift is the cross-section $A \circ X : M \to X^{-1}TT(M)$. Then for any local coordinate system (U, x^i) on M the natural lifts

$$\{\delta_i \circ X,\ \dot{\partial}_i \circ X : 1 \le i \le n\}$$

of the local vector fields $\{\delta_i, \dot{\partial}_i : 1 \le i \le n\}$ form a local frame in $X^{-1}TT(M) \to M$ defined on the open set $X^{-1}(\pi^{-1}(U)) = U$. Let us endow $N = T(M)$ with the Sasaki metric $h = G_s$ associated to the pair $(g, \mathcal{H}_{\hat{\nabla}})$. Then for any $V \in \mathfrak{X}(M)$ written locally as $V = V^i \partial/\partial x^i$ with $V^i \in C^\infty(U)$

$$(X_* V)_x = V^i(x) \left\{ \delta_i + \left(\frac{\partial \lambda^j}{\partial x^i} \circ \pi + N_i^j \right) \dot{\partial}_j \right\}_{X(x)} \tag{2.16}$$

for any $x \in U$. A coordinate-free reformulation of (2.16) is

$$X_* V = \left\{ V^H + (\nabla_V X)^V \right\} \circ X. \tag{2.17}$$

We wish to compute

$$\beta_X(V, V) = \left(X^{-1} D \right)_V X_* V - X_* \nabla_V V.$$

We start by establishing the following

Lemma 2.11 *For any tangent vector fields $X, V \in \mathfrak{X}(M)$*

$$D_{V^H} V^H = V^i \frac{\partial V^j}{\partial x^i} \delta_j \tag{2.18}$$

$$+ V^i V^j \left\{ D_{\partial_i} \partial_j - N_j^\ell D_{\partial_i} \dot{\partial}_\ell - N_i^k D_{\dot{\partial}_k} \partial_j + \left(N_i^k \Gamma_{jk}^\ell - y^k \frac{\partial \Gamma_{jk}^\ell}{\partial x^i} \right) \dot{\partial}_\ell \right\},$$

$$D_{V^H}(\nabla_V X)^V = V^i \frac{\partial}{\partial x^i}(\nabla_V X)^j \dot{\partial}_j + V^i (\nabla_V X)^j D_{\partial_i}\dot{\partial}_j, \tag{2.19}$$

$$D_{(\nabla_V X)^V} V^H = (\nabla_V X)^i V^j \left\{ D_{\dot{\partial}_i}\partial_j - \Gamma^k_{ji}\dot{\partial}_k \right\}. \tag{2.20}$$

Throughout $V^H = \beta \hat{V}$ and $V^V = \gamma \hat{V}$ where natural lifting is relative to the pullback bundle $\pi^{-1}TM \to M$. The proof of Lemma 2.11 is a straightforward calculation. For instance if $V = V^i \partial/\partial x^i$ then

$$\begin{aligned} D_{V^H} V^H &= V^i D_{\delta_i}\left(V^j \delta_j\right) = V^i \left\{ \frac{\partial V^j}{\partial x^i}\delta_j + V^i V^j D_{\delta_i}\delta_j \right\} \\ &= V^i \frac{\partial V^j}{\partial x^i}\delta_j + V^i V^j D_{\partial_i - N_i^k \dot{\partial}_k}\left(\partial_j - N_j^\ell \dot{\partial}_\ell\right) = V^i \frac{\partial V^j}{\partial x^i}\delta_j \\ &\quad + V^i V^j \left\{ D_{\partial_i}\partial_j - \frac{\partial N_j^\ell}{\partial x^i}\dot{\partial}_\ell - N_j^\ell D_{\partial_i}\dot{\partial}_\ell - N_i^k \left(D_{\dot{\partial}_k}\partial_j - \frac{\partial N_j^\ell}{\partial y^k}\dot{\partial}_\ell \right) \right\} \end{aligned}$$

yielding (2.18). Similarly

$$\begin{aligned} D_{V^H}(\nabla_V X)^V &= V^i D_{\delta_i}(\nabla_V X)^j \dot{\partial}_j \\ &= V^i \left\{ \frac{\partial}{\partial x^i}(\nabla_V X)^j \, \dot{\partial}_j + (\nabla_V X)^j D_{\delta_i}\dot{\partial}_j \right\} \end{aligned}$$

yields (2.19) while (2.20) is a consequence of

$$D_{(\nabla_V X)^V} V^H = (\nabla_V X)^i D_{\dot{\partial}_i}\left(V^j \delta_j\right) = (\nabla_V X)^i V^j \left\{ D_{\dot{\partial}_i}\partial_j - \frac{\partial N_j^k}{\partial y^i}\dot{\partial}_k \right\}.$$

Our main purpose in this section is to establish the following

Proposition 2.12 *Let (M,g) be a Riemannian manifold, not necessarily compact, and $X \in \mathfrak{X}(M)$ a tangent vector field on M. Then the tension field $\tau(X) \in \Gamma^\infty(M, X^{-1}T\,T(M))$ is given by*

$$\tau(X) = \left\{ \left(\mathrm{trace}_g R(\nabla . X, X)\cdot\right)^H - \left(\Delta_g X\right)^V \right\} \circ X. \tag{2.21}$$

Therefore $X : (M,g) \to (T(M), G_s)$ is a harmonic map if and only if $\mathrm{trace}_g \{R(\nabla . X, X)\cdot\} = 0$ *and* $\Delta_g X = 0$.

Our convention for the sign in the definition of the curvature tensor field of a Riemannian manifold (M,g) is

$$R(X,Y)Z = -\nabla_X \nabla_Y Z + \nabla_Y \nabla_X Z + \nabla_{[X,Y]} Z, \quad X, Y, Z \in \mathfrak{X}(M).$$

The differential operator Δ_g appearing in Proposition 2.12 is defined as follows. Let $x \in M$ and let $\{V_i : 1 \le i \le n\}$ be a local orthonormal field on $T(M)$ defined on an open neighborhood $U \subseteq M$ of x. Then we set by definition

$$\left(\Delta_g X\right)(x) = -\sum_{i=1}^{n} \left\{\nabla_{V_i}\nabla_{V_i} X - \nabla_{\nabla_{V_i} V_i} X\right\}_x. \tag{2.22}$$

The reader will check easily that the right hand side of (2.22) is invariant under a transformation

$$V_i' = a_i^j V_j, \quad 1 \le i \le n,$$

for any C^∞ map $[a_i^j] : U \cap U' \to O(n)$, hence the definition of $(\Delta_g X)(x)$ doesn't depend upon the choice of local orthonormal frame at x. The choice of sign in the definition (2.22) makes Δ_g into a positive operator i.e., $(\Delta_g X, X) \ge 0$ for any $X \in \mathfrak{X}(M)$. The L^2 inner product on $\mathfrak{X}(M)$ is given by $(X, Y) = \int_M g(X, Y)\, d\,\mathrm{vol}(g)$, for any $X, Y \in \mathfrak{X}(M)$, at least one of compact support.

Definition 2.13 Let (M, g) be a Riemannian manifold. The differential operator $\Delta_g : \mathfrak{X}(M) \to \mathfrak{X}(M)$ is referred to as the (rough) *Laplacian* on vector fields. ■

Let us compute the symbol of Δ_g. We adopt the notations and conventions in [308], p. 114–115. Let $T'(M) = T^*(M) \setminus (0)$ and let $\pi : T'(M) \to M$ be the projection. The symbol $\sigma_2(\Delta_g)$ is a bundle endomorphism $\sigma_2(\Delta_g) : \pi^{-1}T(M) \to \pi^{-1}T(M)$. Let $\omega \in T'(M)$ and let $x = \pi(\omega) \in M$. To recall the definition of

$$\sigma_2(\Delta_g)_\omega : T_x(M) \to T_x(M)$$

let $v \in T_x(M)$ and let us choose $f \in C^\infty(M)$ and $X \in \mathfrak{X}(M)$ such that $(df)_x = \omega$ and $X_x = v$. Then

$$\sigma_2(\Delta_g)_\omega(v) = \Delta_g\left(\frac{i^2}{2!}(f - f(x))^2 X\right)(x) \in T_x(M)$$

(with $i = \sqrt{-1}$). Using the relations

$$\Delta_g(uX) = u\Delta_g X + (\Delta u)X - 2\nabla_{\nabla u} X,$$
$$\Delta(u^2) = 2u\Delta u - 2\|du\|^2,$$

one has

$$\Delta_g\left[(f-f(x))^2 X\right](x) = \left\{\Delta[(f-f(x))^2]\, X\right\}_x = -2\|\omega\|^2 v$$

that is $\sigma_2(\Delta_g)_\omega(v) = \|\omega\|^2 v$. In particular $\sigma_2(\Delta_g)_\omega$ is a linear isomorphism of $T_x(M)$ into itself hence Δ_g is an elliptic operator.

As a corollary of Proposition 2.12

Theorem 2.14 *Let* (M,g) *be an oriented Riemannian manifold and* $X \in \mathfrak{X}(M)$. *If* X *is parallel then* X *is a harmonic map of* (M,g) *into* $(T(M), G_s)$. *Vice versa if* $\tau(X) = 0$ *and either* i) M *is compact or* ii) M *is complete and* $*X^\flat \in L^1(\Omega^{n-1}(M))$, $d * X^\flat \in L^1(\Omega^n(M))$, *then* $\nabla X = 0$.

Here $X^\flat \in \Omega^1(M)$ is given by $g(X,Y) = X^\flat(Y)$, for any $Y \in \mathfrak{X}(M)$, and $* : \Omega^1(M) \to \Omega^{n-1}(M)$ is the Hodge operator. Also $L^1(\Omega^r(M))$ is the space of differential forms of degree r whose pointwise norm $|\omega|$ is integrable i.e., if $\omega \in L^1(\Omega^r(M))$ then $\int_M |\omega| * 1 < \infty$. The proof of Theorem 2.14 requires the following

Lemma 2.15 *For any tangent vector field* X *on* M

$$g(\Delta_g X, X) = \frac{1}{2}\Delta(\|X\|^2) + \|\nabla X\|^2 \tag{2.23}$$

where Δ *is the ordinary Laplace-Beltrami operator on functions.*

Proof. Let $\{V_i : 1 \le i \le n\}$ be a local orthonormal frame in $T(M)$. Then (due to $\nabla g = 0$)

$$\begin{aligned} g(\Delta_g X, X) &= -\sum_i \{g(\nabla_{V_i}\nabla_{V_i} X, X) - g(\nabla_{\nabla_{V_i} V_i} X, X)\} \\ &= -\sum_i \{V_i(g(\nabla_{V_i} X, X)) - \|\nabla_{V_i} X\|^2 - \frac{1}{2}(\nabla_{V_i} V_i)(\|X\|^2)\} \\ &= -\sum_i \{\frac{1}{2} V_i^2(\|X\|^2) - \|\nabla_{V_i} X\|^2 - \frac{1}{2}(\nabla_{V_i} V_i)(\|X\|^2)\} \\ &= \frac{1}{2}\Delta(\|X\|^2) + \|\nabla X\|^2. \end{aligned}$$

Here we made use of the local expression (with respect to the local orthonormal frame $\{V_i : 1 \le i \le n\}$) of the ordinary Laplacian on functions i.e.,

$$\Delta u = -\sum_{i=1}^n \{V_i(V_i u) - (\nabla_{V_i} V_i) u\}, \quad u \in C^2(M).$$

■

Proof of Theorem 2.14. If $\nabla X = 0$ then (by (2.21)–(2.22)) $\tau(X) = 0$. Vice versa, let us assume that X is a harmonic map. Then $\tau(X) = 0$ yields $\Delta_g X = 0$ (by Proposition 2.12). Then (by Lemma 2.15)

$$0 = \int_M g(\Delta_g X, X)\, d\,\mathrm{vol}(g) = \int_M \|\nabla X\|^2 d\,\mathrm{vol}(g)$$

as $\int_M \mathrm{div}(X)\, d\,\mathrm{vol}(g) = 0$ either by Green's lemma (under hypothesis (i) i.e., when M is is compact) or by the Stokes formula on a complete Riemannian manifold (cf. M.P. Gaffney, [121], p. 141, or Appendix B of this book). Hence $\nabla X = 0$.

Let us give now a proof of Proposition 2.12. For any $V \in \mathfrak{X}(M)$ locally written as $V = V^i \partial/\partial x^i$ we may use the identity

$$X_* V = V^i \left(\partial_i + \frac{\partial \lambda^j}{\partial x^i} \dot{\partial}_j \right) \circ X$$

to perform the following (rather involved) calculation

$$\begin{aligned}
\left(X^{-1} D\right)_V X_* V &= V(V^i) \left(\partial_i + \frac{\partial \lambda^j}{\partial x^i} \dot{\partial}_j \right) \circ X \\
&\quad + V^i V^j \left(X^{-1} D\right)_{\partial/\partial x^j} \left(\partial_i + \frac{\partial \lambda^k}{\partial x^i} \dot{\partial}_k \right) \circ X \\
&= V(V^i) \left\{ \delta_i + \left(\frac{\partial \lambda^j}{\partial x^i} + N_i^j \right) \dot{\partial}_j \right\} \circ X \\
&\quad + V^i V^j \left\{ D_{\partial_j} \partial_i + \frac{\partial \lambda^k}{\partial x^j} D_{\dot{\partial}_k} \partial_i \right. \\
&\quad \left. + \frac{\partial^2 \lambda^k}{\partial x^i \partial x^j} \dot{\partial}_k + \frac{\partial \lambda^k}{\partial x^i} \left(X^{-1} D\right)_{\partial/\partial x^j} \dot{\partial}_k \right\} \circ X \\
&= \left[V^i \frac{\partial V^j}{\partial x^i} \delta_j + V^k \frac{\partial V^i}{\partial x^k} \left(\frac{\partial \lambda^j}{\partial x^i} + N_i^j \right) \dot{\partial}_j + V^i V^j \frac{\partial^2 \lambda^k}{\partial x^i \partial x^j} \dot{\partial}_k \right. \\
&\quad \left. + V^i V^j \left(D_{\partial_j} \partial_i + \frac{\partial \lambda^k}{\partial x^j} D_{\dot{\partial}_k} \partial_i + \frac{\partial \lambda^k}{\partial x^i} D_{\partial_j} \dot{\partial}_k \right) \right] \circ X
\end{aligned}$$

(replacing partials by covariant derivatives with respect to ∇)

$$= \left[V(V^j) \delta_j + V^k \frac{\partial V^i}{\partial x^k} \left(\nabla_i \lambda^j\right) \dot{\partial}_j \right.$$

$$+ V^i V^j \frac{\partial}{\partial x^j}\left(\frac{\partial \lambda^k}{\partial x^i} + N_i^k\right)\dot{\partial}_k - V^i V^j \frac{\partial}{\partial x^j}\left(\Gamma_{i\ell}^k \lambda^\ell\right)\dot{\partial}_k$$

$$+ V^i V^j \left(D_{\partial_i}\partial_j + \left(\nabla_j \lambda^k\right) D_{\dot{\partial}_k}\partial_i - N_j^k D_{\dot{\partial}_k}\partial_i\right.$$

$$\left.\left. + \left(\nabla_i \lambda^k\right) D_{\partial_j}\dot{\partial}_k - N_i^k D_{\partial_j}\dot{\partial}_k\right)\right] \circ X$$

$$= \Big[V(V^j)\delta_j + V^i V^j \left\{D_{\partial_i}\partial_j - N_j^\ell D_{\partial_i}\dot{\partial}_\ell - N_i^k D_{\dot{\partial}_k}\partial_j\right\}$$

$$- V^i V^j \frac{\partial \Gamma_{i\ell}^k}{\partial x^j}\lambda^\ell \dot{\partial}_k + V^i V^j \Gamma_{i\ell}^k N_j^\ell \dot{\partial}_k$$

$$+ V^j \frac{\partial V^i}{\partial x^j}\left(\nabla_i \lambda^k\right)\dot{\partial}_k + V^i V^j \frac{\partial}{\partial x^j}\left(\nabla_i \lambda^k\right)\dot{\partial}_k - V^i V^j \Gamma_{i\ell}^k \left(\nabla_j \lambda^\ell\right)\dot{\partial}_k$$

$$+ V^i V^j \left\{\left(\nabla_j \lambda^k\right) D_{\dot{\partial}_k}\partial_i + \left(\nabla_i \lambda^k\right) D_{\partial_j}\dot{\partial}_k\right\}\Big] \circ X$$

(replacing the first two rows from (2.18))

$$= \Big[D_{V^H} V^H + V^j \frac{\partial}{\partial x^j}\left(V^i \nabla_i \lambda^k\right)\dot{\partial}_k + V^i V^j \left(\nabla_i \lambda^k\right) D_{\partial_j}\dot{\partial}_k$$

$$- V^i V^j \Gamma_{i\ell}^k \left(\nabla_j \lambda^\ell\right)\dot{\partial}_k + V^i V^j \left(\nabla_j \lambda^k\right) D_{\dot{\partial}_k}\partial_i\Big] \circ X$$

(replacing the second row from (2.19))

$$= \Big[D_{V^H} V^H + D_{V^H} (\nabla_V X)^V$$

$$- V^i (\nabla_V X)^\ell \, \Gamma_{i\ell}^k \dot{\partial}_k + V^i (\nabla_V X)^k D_{\dot{\partial}_k}\partial_i\Big] \circ X$$

or (replacing the last row from (2.20))

$$\left(X^{-1}D\right)_V X_* V$$
$$= \left(D_{V^H} V^H + D_{V^H} (\nabla_V X)^V + D_{(\nabla_V X)^V} V^H\right) \circ X. \tag{2.24}$$

On the other hand (by Proposition 1.14 in Chapter 1)

$$D_{V^H} V^H = (\nabla_V V)^H,$$

$$D_{V^H} (\nabla_V X)^H = (\nabla_V \nabla_V X)^V + \frac{1}{2}(R(\nabla_V X, X) V)^H,$$

$$D_{(\nabla_V X)^V} V^H = \frac{1}{2}(R(\nabla_V X, X) V)^H,$$

so that (2.24) becomes

$$\left(X^{-1}D\right)_V X_* V = \left\{(\nabla_V V + R(\nabla_V X, X)V)^H + (\nabla_V \nabla_V X)^V\right\} \circ X.$$

Taking into account that (by (2.17))

$$X_* \nabla_V V = \left\{(\nabla_V V)^H + \left(\nabla_{\nabla_V V} X\right)^V\right\} \circ X$$

we may conclude that the second fundamental form of X is given by

$$\begin{aligned}\beta_X(V,V) &= \left(X^{-1}D\right)_V X_* V - X_* \nabla_V V \\ &= \left\{(R(\nabla_V X, X)V)^H + \left(\nabla_V \nabla_V X - \nabla_{\nabla_V V} X\right)^V\right\} \circ X.\end{aligned}$$

We may now take traces to get the statement in Proposition 2.12. ■

Remark 2.16 Let g and $\overline{g}$ be two Riemannian metrics on M and G_s the Sasaki metric on $T(M)$ associated to (M,g). Let $\overline{\tau}(X) \in \Gamma^{-1}(X^{-1}T\,T(M))$ be the tension field of a tangent vector field $X \in \mathfrak{X}(M)$ thought of as a map of $(M,\overline{g})$ into $(T(M), G_s)$. By a result of O. Gil-Medrano (cf. [126])

$$\overline{\tau}(X) = \left(\text{trace}_g R(\nabla . X, X) \cdot + \tau_{\overline{g}}(1_M)\right)^H + \left(-\Delta_g X + (\nabla X)\tau_{\overline{g}}(1_M)\right)^V$$

along X. This clearly coincides with (2.21) in Proposition 2.12 when $g = \overline{g}$. Here $\tau_{\overline{g}}(1_M)$ denotes the tension field of the identical map $1_M : M \to M$ thought of as a map between the map between the Riemannian manifolds $(M,\overline{g})$ (the source) and (M,g) (the target). ■

2.5. VARIATIONS THROUGH VECTOR FIELDS

The results in the previous section show that there aren't actually any smooth vector fields $X : M \to T(M)$ which are harmonic maps (as maps between the Riemannian manifolds (M,g) and $(T(M), G_s)$) except for parallel vector fields. Of course these are but trivial examples (the integrand function in the total bending functional vanishes identically). The reason is that the domain $C^\infty(M, T(M))$ on which the energy functional is *a priori* defined is too large. Nevertheless the same phenomenon occurs if critical points of the energy functional are looked for in the smaller domain $\mathfrak{X}(M)$. Precisely

Theorem 2.17 (O. Gil-Medrano, [126]) *Let M be a compact oriented Riemannian manifold. Let $X \in \mathfrak{X}(M)$ be a tangent vector field on M. Then X is a critical point of the functional $E : \mathfrak{X}(M) \to [0, +\infty)$ if and only if $\nabla X = 0$.*

As we shall shortly see, nontrivial (i.e., not necessarily parallel) examples of critical points occur when the energy functional is restricted to the space of all *unit vector fields*.

Proof of Theorem 2.17. Let $X \in \mathfrak{X}(M)$ be a critical point of $E|_{\mathfrak{X}(M)}$. Then $\{dE(X_t)/dt\}_{t=0} = 0$ for any smooth 1-parameter variation $\{X_t\}_{|t|<\epsilon}$ of X *through vector fields*. Now the first part of the proof of Theorem 2.10 applies to conclude that $\nabla X = 0$ (as the 1-parameter variation $X_t = (1-t)X$ used there is already a variation through vector fields). Vice versa, if $\nabla X = 0$ then (by Theorem 2.10) X is a critical point of E thus $\{dE(\phi_t)/dt\}_{t=0} = 0$ for *any* smooth 1-parameter variation $\{\phi_t\}_{|t|<\epsilon}$ of X hence for smooth variations through vector fields, as well.

Theorem 2.18 *Let (M, g) be a compact oriented Riemannian manifold and $E : \mathfrak{X}(M) \to [0, +\infty)$ the energy functional restricted to the space of all vector fields. Then*

$$\frac{d}{dt}\{E(X_t)\}\bigg|_{t=0} = \int_M g(\Delta_g X, V)\, d\,\mathrm{vol}(g) \tag{2.25}$$

for any smooth 1-parameter variation $\mathfrak{X} : M \times (-\epsilon, \epsilon) \to T(M)$ of X through vector fields i.e., $X_t \in \mathfrak{X}(M)$ for any $|t| < \epsilon$. Here $X_t(x) = \mathfrak{X}(x, t)$ for any $x \in M$ and $|t| < \epsilon$. Also $V : M \to T(M)$ is the tangent vector field on M given by[1]

$$V(x) = \lim_{t \to 0} \frac{1}{t}\{X_t(x) - X(x)\} = \frac{d\mathcal{X}_x}{dt}(0), \quad x \in M,$$

where $\mathfrak{X}_x(t) = X_t(x)$, $(x, t) \in M \times (-\epsilon, \epsilon)$.

Proof. Let $\mathfrak{X} : M \times (-\epsilon, \epsilon) \to T(M)$ be a smooth 1-parameter variation of X (i.e., $\mathfrak{X}(x, 0) = X(x)$ for any $x \in M$) such that $X_t(x) = \mathfrak{X}(x, t) \in T_x(M)$ for any $x \in M$ and any $|t| < \epsilon$. Let us set

$$\mathcal{E}(t) = E(X_t) = \frac{1}{2}\int_M \|dX_t\|^2, \quad |t| < \epsilon.$$

Then, as well known from the theory of harmonic maps (cf. e.g., Theorem 11.1 in [109], p. 104)

$$\mathcal{E}'(0) = -\int_M G_s{}^X(\mathcal{V}, \tau(X))\, d\,\mathrm{vol}(g)$$

[1] The limit is taken in $(T_x(M), g_x)$ (a finite dimensional Banach space).

where $X^{-1}G_s = G_s^X$ is the Riemannian bundle metric induced by G_s in $X^{-1}T\,T(M) \to M$ and $\mathcal{V} \in \Gamma^\infty(X^{-1}T\,T(M))$ is the infinitesimal variation induced by $\mathfrak{X}$ i.e.,

$$\mathcal{V}(x) = (d_{(x,0)}\mathfrak{X})\frac{\partial}{\partial t}\bigg|_{(x,0)}, \quad x \in M.$$

Let us set

$$\mathcal{X}^i(x,t) = x^i(\mathcal{X}(x,t)), \quad \mathcal{X}^{i+n}(x,t) = y^i(\mathcal{X}(x,t)), \quad 1 \le i \le n.$$

Then, on one hand $\mathcal{X}^i(x,t) = \tilde{x}^i(x)$ hence

$$\frac{\partial \mathcal{X}^i}{\partial t}(x,0) = 0, \quad 1 \le i \le n.$$

On the other hand if $V = V^i\, \partial/\partial\tilde{x}^i$ then

$$\frac{\partial \mathcal{X}^{i+n}}{\partial t}(x,0) = \lim_{t\to 0}\frac{1}{t}\left\{\mathcal{X}^{i+n}(x,t) - \mathcal{X}^i(x,0)\right\} = V^i(x)$$

hence

$$\begin{aligned}\mathcal{V}_x &= \frac{\partial \mathcal{X}^i}{\partial t}(x,0)\partial_{i,X(x)} + \frac{\partial \mathcal{X}^{i+n}}{\partial t}(x,0)\dot{\partial}_{i,X(x)} \\ &= V^i(x)\dot{\partial}_{i,X(x)} = (\gamma\hat{V})_{X(x)} = V^V_{X(x)}\end{aligned}$$

and we may conclude that

$$\mathcal{V} = V^V \circ X. \tag{2.26}$$

Finally, by taking into account (2.21) and (2.26)

$$\begin{aligned}\mathcal{E}'(0) &= -\int_M G_s(V^V, (\mathrm{trace}_g R(\nabla.X, X)\cdot)^H - (\Delta_g X)^V)d\,\mathrm{vol}(g) \\ &= \int_M g(V, \Delta_g X)d\,\mathrm{vol}(g).\end{aligned}$$

Theorem 2.18 is proved. ∎

2.6. UNIT VECTOR FIELDS

Let (M,g) be an n-dimensional compact orientable Riemannian manifold and $S^{n-1} \to S(M) \to M$ the tangent sphere bundle over M i.e.,

$$S(M)_x = \{v \in T_x(M) : g_x(v,v) = 1\}, \quad x \in M.$$

Any smooth cross-section U in the tangent sphere bundle $S(M) \to M$ is a *unit vector field* on M i.e., $U \in \mathfrak{X}(M)$ and $g(U,U) = 1$ everywhere on M. As is well known, the existence of global nowhere vanishing smooth sections in $S(M) \to M$ is actually tied to the topological restriction $\chi(M) = 0$ (where $\chi(M)$ is the Euler-Poincaré characteristic of M).

Let U be a unit vector field thought of as a map of (M,g) into $(S(M), G_s)$ where G_s is the Riemannian metric induced on $S(M)$ (a smooth real hypersurface in $T(M)$) by the Sasaki metric. Then

Theorem 2.19 (S.D. Han & J.W. Yim, [157]) *Let $\tau_1(U)$ be the tension field (a smooth section in $U^{-1}T(S(M)) \to M$) of U as a map among the Riemannian manifolds (M,g) and $(S(M), G_s)$. Then*

$$\tau_1(U) = \left\{ \left(\mathrm{trace}_g \, \{R(\nabla.U, U)\cdot\} \right)^H - \tan\left(\Delta_g U \right)^V \right\} \circ U \tag{2.27}$$

where $\tan_v : T_v(T(M)) \to T_v(S(M))$ *is the natural projection associated to the direct sum decomposition* $T_v(T(M)) = T_v(S(M)) \oplus \mathbb{R}\nu_v$ *for any* $v \in S(M)$. *Also* $\nu = \gamma \mathcal{L}$. *Consequently U is a harmonic map of (M,g) into $(S(M), G_s)$ if and only if*

$$\Delta_g U - \|\nabla U\|^2 U = 0, \tag{2.28}$$

$$\mathrm{trace}_g \, \{R(\nabla.U, U)\cdot\} = 0. \tag{2.29}$$

Proof. Let $\tau(U)$ be the tension field of U as a map of (M,g) into $(T(M), G_s)$ (a smooth section in $U^{-1}T\,T(M) \to M$) and $\tau_1(U)$ the tension field of U as a map of (M,g) into $(U(M), G_s)$ (a smooth section in $U^{-1}TS(M) \to M$). The symbol G_s denotes both the Sasaki metric on $T(M)$ and the induced metric $i^* G_s$ on $S(M)$, where $i : S(M) \hookrightarrow T(M)$ is the canonical inclusion. It is a well-known fact in the theory of harmonic maps that $\tau_1(U)$ is the orthogonal projection of $\tau(U)$ on $T(S(M))$ i.e.,

$$\tau_1(U) = \tan \tau(U).$$

By (2.21)

$$\tau(U) = \left\{ \left(\mathrm{trace}_g R(\nabla.U, U)\cdot \right)^H - \left(\Delta_g U \right)^V \right\} \circ U.$$

On the other hand (by Theorem 1.29)

$$\tau(U)_x \in T_{U(x)}(T(M)) = T_{U(x)}(S(M)) \oplus \mathbb{R}\nu_{U(x)}, \quad x \in M.$$

By Proposition 1.27, the horizontal component $\left(\mathrm{trace}_g R(\nabla.U, U)\cdot \right)^H_{U(x)}$ of $\tau(U)_x$ is tangent to $S(M)$. It remains that we compute the tangential component of $\left(\Delta_g U\right)^V_{U(x)}$. As $\nu = \gamma\,\mathcal{L}$ is a unit normal on $S(M)$

$$\operatorname{nor}\left(\Delta_g U\right)^V = G_s(\left(\Delta_g U\right)^V, \nu)\nu$$

where $\operatorname{nor}_v : T_v(T(M)) \to \mathbb{R}\nu_v$ is the natural projection, $v \in S(M)$. Note that $\mathcal{L} \circ U = \hat{U} \circ U$ hence

$$G_s((\Delta_g U)^V, \nu) \circ U = \hat{g}(\widehat{\Delta_g U}, \mathcal{L}) \circ U = g(\Delta_g U, U) = \|\nabla U\|^2$$

as a consequence of (2.23) for $X = U$ (as $\|U\| = 1$). Here $\{E_j : 1 \le j \le n\}$ is a local g-orthonormal frame of $T(M)$. Note that this is the contents of the proof of Lemma 2.15 (the same result follows from (2.23) for $X = U$). Finally

$$\nu_{U(x)} = (\gamma\,\mathcal{L})_{U(x)} = \gamma_{U(x)}\hat{U}_{U(x)} = (\gamma\,\hat{U})_{U(x)} = U^V_{U(x)}$$

hence

$$\begin{aligned}\tan\left(\Delta_g U\right)^V_{U(x)} &= \left(\Delta_g U\right)^V_{U(x)} - \operatorname{nor}_{U(x)}\left(\Delta_g U\right)^V_{U(x)} \\ &= \left(\Delta_g U\right)^V_{U(x)} - \left(\|\nabla U\|^2 U\right)_x = \left(\Delta_g U - \|\nabla U\|^2 U\right)^V_{U(x)}\end{aligned}$$

for any $x \in M$. The identity (2.27) may be explicitly written as

$$\tau_1(U) = \left\{\left(\operatorname{trace}_g R(\nabla_{\cdot} U, U)\cdot\right)^H - \left(\Delta_g U - \|\nabla U\|^2 U\right)^V\right\} \circ U \qquad (2.30)$$

and we may conclude that $\tau_1(U) = 0$ if and only if (2.28)–(2.29) hold. Theorem 2.19 is proved. ■

Next we shall give a geometric interpretation of the condition (2.29) in Theorem 2.19.

Proposition 2.20 *Let (M,g) be a real space-form of (constant) sectional curvature $c \neq 0$. Let U be a unit vector field on M. Then* $\operatorname{trace}_g\{R(\nabla_{\cdot} U, U)\cdot\} = 0$ *if and only if U is geodesic (i.e., $\nabla_U U = 0$) and the distribution $(\mathbb{R}U)^\perp$ is minimal.*

Proof. The curvature tensor field R of a Riemannian manifold (M,g) of constant sectional curvature c is given by

$$R(X,Y)Z = c\{g(X,Z)Y - g(Y,Z)X\}$$

for any $X, Y, Z \in \mathfrak{X}(M)$. Then

$$\begin{aligned}
\text{trace}_g\{R(\nabla_{\cdot}U, U)\cdot\} &= \sum_{j=1}^{n} R(\nabla_{E_j}U, U)E_j \\
&= c\sum_{j=1}^{n}\{g(\nabla_{E_j}U, E_j)U - g(U, E_j)\nabla_{E_j}U\} \\
&= c\{\text{div}(U)U - \nabla_{\sum_j g(U,E_j)E_j}U\} = c\{\text{div}(U)U - \nabla_U U\}
\end{aligned}$$

hence $\text{trace}_g\{R(\nabla_{\cdot}U, U)\cdot\} = 0$ if and only if

$$\nabla_U U - \text{div}(U)U = 0. \tag{2.31}$$

Yet $g(\nabla_U U, U) = \frac{1}{2}U(\|U\|^2) = 0$ i.e., $\nabla_U U$ and U are orthogonal and then linearly independent at each point of M. Therefore (2.31) is equivalent to $\nabla_U U = 0$ and $\text{div}(U) = 0$. It remains to be shown that $\text{div}(U) = 0$ is equivalent to the minimality of the distribution $(\mathbb{R}U)^{\perp}$ (the orthogonal complement of $\mathbb{R}U$ in $(T(M), g)$).

Let $\mathcal{D}$ be a smooth distribution of rank m on M and $\mathcal{D}^{\perp}$ its orthogonal complement in $T(M)$ with respect to g. Let $\pi^{\perp} : T(M) \to \mathcal{D}^{\perp}$ be the natural projection with respect to the decomposition $T(M) = \mathcal{D} \oplus \mathcal{D}^{\perp}$. Let us consider the bilinear form

$$B_{\mathcal{D}}(X, Y) = \pi^{\perp}\nabla_X Y, \quad X, Y \in \mathcal{D}.$$

Let us set

$$H_{\mathcal{D}} = \frac{1}{m}\,\text{trace}_g\, B_{\mathcal{D}}.$$

If $\{X_a : 1 \leq a \leq m\}$ is a local g-orthonormal ($g(X_a, X_b) = \delta_{ab}$) frame of $\mathcal{D}$ defined on the open set $U \subseteq M$ then

$$H_{\mathcal{D}} = \frac{1}{m}\sum_{a=1}^{m} B_{\mathcal{D}}(X_a, X_a).$$

Clearly, if $\mathcal{D}$ is completely integrable then the (pointwise) restriction of $H_{\mathcal{D}}$ to a leaf of $\mathcal{D}$ is the mean curvature vector of the leaf (as a submanifold of (M, g)). It is customary to call $H_{\mathcal{D}}$ the *mean curvature* vector of $\mathcal{D}$. Also if $H_{\mathcal{D}} = 0$ then $\mathcal{D}$ is said to be *minimal*. Let us apply these concepts to the distribution $\mathcal{D} = (\mathbb{R}U)^{\perp}$. If this is the case then $\mathcal{D}^{\perp} = \mathbb{R}U$ hence

$$\pi^{\perp}X = g(X, U)U, \quad X \in T(M).$$

Let $\{E_a : 1 \le a \le n-1\}$ be a local g-orthonormal frame of $\mathcal{D} = (\mathbb{R}U)^\perp$ (i.e., $g(E_a, E_b) = \delta_{ab}$ and $g(E_a, U) = 0$). Let $E_n = U$ (so that $\{E_j : 1 \le j \le n\}$ is a local g-orthonormal frame of $T(M)$). Then

$$
\begin{aligned}
(n-1)H_{\mathcal{D}} &= \sum_{a=1}^{n-1} B_{\mathcal{D}}(E_a, E_a) = \sum_{a=1}^{n-1} \pi^\perp \nabla_{E_a} E_a \\
&= \sum_{a=1}^{n-1} g(\nabla_{E_a} E_a, U)U = \sum_{a=1}^{n-1} \{E_a(g(E_a, U)) - g(E_a, \nabla_{E_a} U)\}U \\
&= -\sum_{a=1}^{n-1} g(E_a, \nabla_{E_a} U)U = -\sum_{j=1}^{n} g(E_j, \nabla_{E_j} U)U
\end{aligned}
$$

because of $\nabla_U U = 0$. Finally

$$(n-1)H_{\mathcal{D}} = -\mathrm{div}(U)\, U$$

and Proposition 2.20 is proved. ■

The following concept is central to the present monograph.

Definition 2.21 Let M be an n-dimensional compact orientable Riemannian manifold. A unit vector field $U \in \Gamma^\infty(S(M))$ is called a *harmonic vector field* if U is a critical point of the energy functional

$$E : \Gamma^\infty(S(M)) \to [0, +\infty)$$

given by

$$E(X) = \frac{n}{2}\mathrm{Vol}(M) + \frac{1}{2}\int_M \|\nabla X\|^2 d\,\mathrm{vol}(g)$$

for any $X \in \Gamma^\infty(S(M))$. ■

Here E is the usual Dirichlet energy functional yet its domain consists solely of the unit vector fields (and 1-parameter variations of U are through unit vector fields alone). Obviously any unit vector field that is a harmonic map is also a harmonic vector field. As we shall shortly see the converse is false in general.

Proposition 2.22 *Let M be a compact orientable Riemannian manifold. Let U be a unit vector field. Let $\mathcal{U} : M \times (-\delta, \delta) \to S(M)$ be a smooth 1-parameter variation of U through unit vector fields i.e., $U_t \in \Gamma^\infty(S(M))$ for any $|t| < \delta$*

where $U_t(x) = \mathcal{U}(x,t)$, $x \in M$, $|t| < \delta$. *Let* $\mathcal{E}(t) = E(U_t)$. *Then* $g(U,V) = 0$ *and*

$$\mathcal{E}'(0) = \int_M g(\Delta_g U, V)\, d\,\mathrm{vol}(g) \tag{2.32}$$

where $V_x = \frac{d}{dt}\{t \mapsto U_t(x)\}_{t=0} \in T_x(M)$ *for any* $x \in M$.

Proof. For any smooth 1-parameter variation $\{U_t\}_{|t|<\delta}$ of U as in Proposition 2.22 one has

$$g(U_t, U_t) = 1, \quad |t| < \delta,$$

hence

$$\begin{aligned} 0 &= \frac{d}{dt}\{t \mapsto g(U_t, U_t)_x\}_{t=0} = \frac{d}{dt}\{t \mapsto U_t^i(x) U_t^j(x) g_{ij}(x)\}_{t=0} \\ &= \frac{d}{dt}\left\{\mathcal{U}^i(x,\cdot)\mathcal{U}^j(x,\cdot) g_{ij}(x)\right\}_{t=0} = 2\frac{\partial \mathcal{U}^i}{\partial t}(x,0)\mathcal{U}^j(x,0) g_{ij}(x) \\ &= 2V^i(x) U^j(x) g_{ij}(x) = g(V,U)_x = 0. \end{aligned}$$

Moreover (2.32) is an immediate consequence of Theorem 2.18. Proposition 2.22 is proved. ■

The proof of Proposition 2.22 as given above relies on the explicit calculation of the tension field $\tau(X)$ as the trace of the second fundamental form of X (cf. the identity (2.21) in Proposition 2.12) and the ordinary first variation formula in the theory of harmonic maps. One may give a direct proof of the first variation formula (2.32) as follows. Let $N = M \times (-\delta, \delta)$ and let $p: N \to M$ be the projection. Let $p^{-1}TM \to N$ be the pullback of the tangent bundle $T(M) \to M$ by p. Then $\mathcal{U}$ may be thought of as a C^∞ section in $p^{-1}TM \to N$. If Y is a tangent vector field on M we set $\hat{Y} = Y \circ p$. The Riemannian metric g induces a bundle metric $\hat{g}$ in $p^{-1}TM \to N$ uniquely determined by $\hat{g}(\hat{Y}, \hat{Z}) = g(Y,Z) \circ p$. Also let D be the connection in $p^{-1}TM \to N$ induced by the Levi-Civita connection ∇. Precisely let $\tilde{Y}$ be the tangent vector field on $T(M)$ given by

$$\tilde{Y}_{(x,t)} = (d_x i_t) Y_x, \quad x \in M, \quad |t| < \delta,$$

where $i_t : M \to N$, $i_t(x) = (x,t)$. Then D is determined by

$$D_{\tilde{Y}} \hat{Z} = \widehat{\nabla_Y Z}, \quad D_{\partial/\partial t} \hat{Z} = 0, \quad Y, Z \in T(M).$$

Moreover a simple calculation shows that $D\hat{g} = 0$ and

$$(D_{\tilde{Y}}\mathcal{X})_{(x,t)} = (\nabla_Y X_t)_x, \quad (x,t) \in N.$$

Let $\{E_i : 1 \le i \le n\}$ be a local orthonormal frame of $T(M)$. Then

$$\begin{aligned}\mathcal{B}(U_t) &= \frac{1}{2}\int_M \sum_{i=1}^n \hat{g}(\nabla_{E_i} U_t, \nabla_{E_i} U_t)_x (d\,\mathrm{vol}(g))_x \\ &= \frac{1}{2}\int_M \sum_i \hat{g}(D_{\tilde{E}_i}\mathcal{U}, D_{\tilde{E}_i}\mathcal{U})_{(x,t)} (d\,\mathrm{vol}(g))_x\end{aligned}$$

hence

$$\begin{aligned}\frac{d}{dt}\mathcal{B}(U_t) &= \int_M \sum_i \hat{g}(D_{\partial/\partial t} D_{\tilde{E}_i}\mathcal{U}, D_{\tilde{E}_i}\mathcal{U})_{(x,t)} (d\,\mathrm{vol}(g))_x \\ &= \int_M \sum_i \hat{g}(D_{\tilde{E}_i} D_{\partial/\partial t}\mathcal{U}, D_{\tilde{E}_i}\mathcal{U})_{(x,t)} (d, \mathrm{vol}(g))_x\end{aligned}$$

as $R^D(\partial/\partial t, \tilde{E}_i)\mathcal{U} = 0$ and $[\partial/\partial t, \tilde{E}_i] = 0$. Moreover (by $D\hat{g} = 0$)

$$\begin{aligned}\frac{d}{dt}\mathcal{B}(U_t) = \int_M \sum_i &\{\tilde{E}_i(\hat{g}(D_{\partial/\partial t}\mathcal{U}, D_{\tilde{E}_i}\mathcal{U})) \\ &- \hat{g}(D_{\partial/\partial t}\mathcal{U}, D_{\tilde{E}_i} D_{\tilde{E}_i}\mathcal{U})\}_{(x,t)} (d\,\mathrm{vol}(g))_x.\end{aligned}$$

For each fixed $|t| < \delta$ we define $Y_t \in T(M)$ by setting

$$g(Y_t, Y)_x = \hat{g}(D_{\partial/\partial t}\mathcal{U}, D_{\tilde{Y}}\mathcal{U})_{(x,t)}$$

for any $Y \in T(M)$ and any $x \in M$. Then (by $\nabla g = 0$)

$$\begin{aligned}\tilde{E}_i(\hat{g}(D_{\partial/\partial t}\mathcal{U}, D_{\tilde{E}_i}\mathcal{U})) &= E_i(g(Y_t, E_i)) \\ &= g(\nabla_{E_i} Y_t, E_i) + g(Y_t, \nabla_{E_i} E_i).\end{aligned}$$

As $\nabla d\,\mathrm{vol}(g) = 0$ the divergence operator is given by

$$\mathrm{div}(Y) = \mathrm{trace}\{Z \mapsto \nabla_Z Y\} = \sum_{j=1}^n g(\nabla_{E_j} Y, E_j).$$

Finally (by Green's lemma)

$$\frac{d}{dt}\{E(U_t)\}_{t=0} = -\int_M \hat{g}(D_{\partial/\partial t}\mathcal{U}, \sum_i \{D_{\tilde{E}_i} D_{\tilde{E}_i}\mathcal{U} - D_{\widetilde{\nabla_{E_i} E_i}}\mathcal{U}\})_{(x,0)} (d\,\mathrm{vol}(g))_x$$

$$= \int_M g(V, \Delta_g X)_x (d\,\mathrm{vol}(g))_x$$

and (2.32) is proved.

Theorem 2.23 (G. Wiegmink, [309]) *Let M be a compact orientable Riemannian manifold and U a unit vector field on M. Then U is a harmonic vector field if and only if $\Delta_g U - \|\nabla U\|^2 U = 0$.*

See also O. Gil-Medrano, [126], and C.M. Wood, [316].

Proof of Theorem 2.23. Let U be a harmonic vector field on M. Let us set $\mathcal{S} = \{V \in \mathfrak{X}(M) : g(U, V) = 0\}$. As U is a harmonic vector field (2.32) yields

$$\int_M g(\Delta_g U, V) d\,\mathrm{vol}(g) = 0 \tag{2.33}$$

for any $V \in \mathcal{S}$. Indeed let $V \in \mathcal{S}$ be an arbitrary element of $\mathcal{S}$ and let us set

$$W_t = U + tV, \quad U_t = \|W_t\|^{-1} W_t, \quad |t| < \epsilon. \tag{2.34}$$

Then $\{U_t\}_{|t|<\epsilon}$ is a smooth 1-parameter variation of U through unit vector fields. On the other hand if $f(t) = \|W_t\|$ then $f(0) = 1$ and

$$f(t)^2 = \|W_t\|^2 = 1 + 2tg(U, V) + t^2\|V\|^2 = 1 + O(t^2),$$
$$2f(t)f'(t) = O(t),$$

hence $f'(0) = 0$. Consequently

$$\left.\frac{dU_t}{dt}\right|_{t=0} = \left(-\frac{f'(t)}{f(t)^2} W_t + \frac{1}{f(t)}\frac{dW_t}{dt}\right)_{t=0} = V.$$

As U is a harmonic vector field $\mathcal{E}'(0) = 0$ for any smooth 1-parameter variation of U including the variation defined by (2.34). Then (2.32) yields (2.33) for any $V \in \mathcal{S}$ as announced.

At this point we may take into account the decomposition

$$T_x(M) = (\mathbb{R}U_x) \oplus (\mathbb{R}U_x)^\perp, \quad x \in M,$$

to conclude that

$$\Delta_g U = \lambda U + V$$

for some $\lambda \in C^\infty(M)$ and some $V \in \mathcal{S}$. Let us take the (pointwise) inner product with V to get

$$\|V\|^2 = g(\Delta_g U, V).$$

Let us integrate over M and use (2.33) so that

$$\int_M \|V\|^2 d\,\mathrm{vol}(g) = \int_M g(\Delta_g U, V)\, d\,\mathrm{vol}(g) = 0$$

hence $V = 0$ i.e., $\Delta_g U = \lambda U$. Finally we may take the (pointwise) inner product with U so that

$$\lambda = g(\Delta_g U, U) = \|\nabla U\|^2$$

i.e., $\Delta_g U = \|\nabla U\|^2 U$.

Vice versa if $\Delta_g U$ is proportional to U then (by (2.32)) $\mathcal{E}'(0) = 0$ for any smooth 1-parameter variation $\mathcal{U}$ of U through unit vector fields (indeed for such variations $g(U, V) = 0$ hence the integrand in the right hand side of (2.32) vanishes identically) so that U is a harmonic vector field. Theorem 2.23 is proved. ■

A reformulation of Theorem 2.19 is then

Corollary 2.24 *Let M be a compact orientable Riemannian manifold and U a unit vector field on M. Then U is a harmonic map of (M,g) into $(S(M), G_s)$ if and only if* i) *U is a harmonic vector field and* ii) $\mathrm{trace}_g\{R(\nabla.U, U)\cdot\} = 0$.

Remark 2.25

i. Let M be a compact orientable Riemannian manifold and $f : M \to S^m$ a smooth map into the standard sphere $S^m \subset \mathbb{R}^{m+1}$. Let us set $\Psi = i \circ f$ where $i : S^m \to \mathbb{R}^{m+1}$ is the inclusion. It may be shown (cf. J. Eells & A. Ratto, [114]) that f is a harmonic map if and only if $\Delta\Psi = \|d\Psi\|^2\Psi$ where Δ is the Laplace-Beltrami operator of M on functions (if $\Psi = (\Psi^1, \ldots, \Psi^{m+1})$ then $\Delta\Psi = (\Delta\Psi^1, \ldots, \Delta\Psi^{m+1})$). The reader should observe the formal analogy to the harmonic vector fields (equation 2.28). See also Theorem 10.2 in [109], p. 86.

ii. The Hodge-de Rham operator $\Delta_1 = dd^* + d^*d$ on differential 1-forms $\Omega^1(M)$ may be thought of as acting on smooth vector fields, up to the musical isomorphism $\sharp : \Omega^1(M) \to \mathfrak{X}(M)$ i.e., $g(\omega^\sharp, X) = \omega(X)$

for any $X \in \mathfrak{X}(M)$. It may be shown that Δ_1 is related to the rough Laplacian $\Delta_g : \mathfrak{X}(M) \to \mathfrak{X}(M)$ by $\Delta_1 = \Delta_g + Q$ where Q is the Ricci operator i.e., $g(QX, Y) = \text{Ric}(X, Y)$ for any $X, Y \in \mathfrak{X}(M)$. Hence in general a harmonic vector field, i.e., $U \in \Gamma^\infty(S(M))$ with $\Delta_g U - \|\nabla U\|^2 U = 0$, isn't necessarily Hodge-de Rham harmonic. Nevertheless one may prove that whenever M is an Einstein manifold and U is a unit vector field tangent to M such that $\Delta_1 U = 0$ then U is a harmonic vector field. See S.I. Goldberg, [138], p. 86–88.

iii. A *harmonic morphism* is a smooth mapping $\phi : M \to N$ among two Riemannian manifolds such that the pullback via ϕ of any local harmonic function on N is a (local) harmonic function on M. By a well-known characterization (cf. e.g., Proposition 4.3.10 in [19], p. 113) a smooth mapping $\phi : M \to N$ is a harmonic morphism if and only if $\phi^*\omega$ is a harmonic 1-form on M for any harmonic 1-form ω on N. In particular the induced map on de Rham cohomology $\phi^* : H^1(N, \mathbb{R}) \to H^1(M, \mathbb{R})$ is injective (cf. Proposition 4.3.11 in [19], p. 113). This suggests the following notion and (open) problem. Let $\phi : M \to N$ be a smooth mapping of Riemannian manifolds which is a morphism in the following sense. Let $Y \in \mathfrak{X}(N)$ be a harmonic vector field on N and $\omega = Y^\flat$ the corresponding 1-form i.e., $\omega(V) = h(V, Y)$ for any $V \in \mathfrak{X}(N)$, where h denotes the Riemannian metric on N. We request that the vector field $X = (\phi^*\omega)^\sharp \in \mathfrak{X}(M)$ be harmonic. The properties of such morphisms ϕ are unknown. For instance is ϕ a harmonic map?

iv. Let (M, J, g) be a Hermitian manifold of complex dimension n, where J is the complex structure and g the Hermitian metric. Let $\tilde{\nabla}$ be a holomorphic torsion-free connection on M i.e., $\tilde{\nabla} J = 0$ and $T_{\tilde{\nabla}} = 0$. Let us consider the second order differential operator

$$\tilde{\Delta} u = -\text{trace}_g \left(\tilde{\nabla} du \right), \quad u \in C^2(M). \tag{2.35}$$

Here the Hessian $\tilde{\nabla} du$ is defined by

$$\left(\tilde{\nabla} du \right)(X, Y) = X(Yu) - \left(\tilde{\nabla}_X Y \right) u, \quad X, Y \in \mathfrak{X}(M).$$

As $\tilde{\nabla}$ is torsion free, the Hessian $\tilde{\nabla} du$ is symmetric i.e.,

$$\left(\tilde{\nabla} du \right)(X, Y) = \left(\tilde{\nabla} du \right)(Y, X), \quad X, Y \in \mathfrak{X}(M).$$

Let $\{Z_\alpha : 1 \leq \alpha \leq n\}$ be a local frame of the holomorphic tangent bundle $T^{1,0}(M)$, defined on some open subset $U \subseteq M$, such that $g(Z_\alpha, \overline{Z}_\beta) = \delta_{\alpha\beta}$.

Let us set

$$E_\alpha = \frac{1}{\sqrt{2}}\left(Z_\alpha + \overline{Z}_\alpha\right), \quad E_{\alpha+n} = \frac{i}{\sqrt{2}}\left(Z_\alpha - \overline{Z}_\alpha\right),$$

so that $\{E_a : 1 \le a \le 2n\} = \{E_\alpha, E_{\alpha+n} : 1 \le \alpha \le n\}$ is a local orthonormal (i.e., $g(E_a, E_b) = \delta_{ab}$) frame of $T(M)$ defined on U. Then

$$\begin{aligned}\tilde{\Delta} u &= -\sum_{a=1}^{2n} (\tilde{\nabla} du)(E_a, E_a) \\ &= -\frac{1}{2}\sum_{\alpha=1}^{n} \{(\tilde{\nabla} du)(Z_\alpha, Z_{\overline{\alpha}}) + (\tilde{\nabla} du)(Z_{\overline{\alpha}}, Z_\alpha)\}\end{aligned}$$

where $Z_{\overline{\alpha}} = \overline{Z}_\alpha$. Using also the symmetry of the Hessian we may express $\tilde{\Delta}$ as

$$\tilde{\Delta} u = -\sum_{\alpha=1}^{n} (\tilde{\nabla} du)(Z_\alpha, Z_{\overline{\alpha}}) \tag{2.36}$$

on U. Let $(z^1, \ldots, z^n)$ be local complex coordinates on U. Then $Z_\alpha = U_\alpha^\beta \, \partial/\partial z^\beta$ for some C^∞ functions $U_\alpha^\beta : U \to \mathbb{C}$ such that $\sum_{\alpha=1}^n U_\alpha^\beta U_{\overline{\alpha}}^{\overline{\gamma}} = g^{\beta\overline{\gamma}}$. Here $U_{\overline{\alpha}}^{\overline{\beta}} = \overline{U_\alpha^\beta}$. Also $g_{\alpha\overline{\beta}} = g(\partial/\partial z^\alpha, \partial/\partial \overline{z}^\beta)$ and $g_{\alpha\overline{\beta}} g^{\overline{\beta}\gamma} = \delta_\alpha^\gamma$. As an elementary consequence (2.36) becomes

$$\tilde{\Delta} u = -g^{\alpha\overline{\beta}}\left(\tilde{\nabla} du\right)\left(\frac{\partial}{\partial z^\alpha}, \frac{\partial}{\partial \overline{z}^\beta}\right) = -g^{\alpha\overline{\beta}}\left\{\frac{\partial^2 u}{\partial z^\alpha \partial \overline{z}^\beta} - \Gamma_{\alpha\overline{\beta}}^A \frac{\partial u}{\partial z^A}\right\}$$

where

$$\tilde{\nabla}_{\partial_A} \partial_B = \Gamma_{AB}^C \partial_C, \quad \partial_A = \frac{\partial}{\partial z^A},$$

with the convention $A, B, C, \ldots \in \{1, \ldots, n, \overline{1}, \ldots, \overline{n}\}$ and $z^{\overline{\alpha}} = \overline{z}^\alpha$. On the other hand, as ∇ parallelizes J and is torsion free, all connection coefficients Γ_{BC}^A vanish except eventually for $\Gamma_{\beta\gamma}^\alpha$ (and of course $\Gamma_{\overline{\beta}\overline{\gamma}}^{\overline{\alpha}}$ as the complex conjugates of $\Gamma_{\beta\gamma}^\alpha$). We may end our local calculation of $\tilde{\Delta}$ by concluding that

$$\tilde{\Delta} u = -g^{\alpha\overline{\beta}} \frac{\partial^2 u}{\partial z^\alpha \partial \overline{z}^\beta}. \tag{2.37}$$

J. Jost & S-T. Yau, [179], considered the system

$$-\tilde{\Delta}\phi^i + g^{\alpha\overline{\beta}}\left(\left|\begin{matrix} i \\ jk \end{matrix}\right| \circ \phi\right) \frac{\partial \phi^j}{\partial z^\alpha} \frac{\partial \phi^k}{\partial \overline{z}^\beta} = 0. \tag{2.38}$$

A smooth map $\phi : M \to N$ into a Riemannian manifold (N, h) is called *Hermitian harmonic* if ϕ is a solution to (2.38). Here $\left|\begin{matrix} i \\ jk \end{matrix}\right|$ are the coefficients of the Levi-Civita connection of h. It should be observed that (2.38) is a nonlinear elliptic system possessing neither a divergence nor a variational structure. See also L. Ni, [222]. Let us set

$$\left(\nabla_{\partial/\partial z^j} J\right) \frac{\partial}{\partial \bar{z}^k} = J^A_{\bar{k},j}\, \partial_A,$$

for some C^∞ functions $J^A_{\bar{k},j} : U \to \mathbb{C}$. Exploiting

$$\Delta u = -\mathrm{trace}_g\,(\nabla du)$$

it may be easily seen that the Laplace-Beltrami operator Δ of (M, g) is related to $\tilde{\Delta}$ by

$$\Delta u = \tilde{\Delta} u + \frac{i}{2} g^{\alpha\overline{\beta}} \left[J^\ell_{\bar{k},j} \frac{\partial u}{\partial z^\ell} - J^{\bar{\ell}}_{j,\bar{k}} \frac{\partial u}{\partial \bar{z}^\ell} \right]. \tag{2.39}$$

Here $J^{\bar{\ell}}_{j,\bar{k}} = \overline{J^\ell_{\bar{j},k}}$. As a consequence of (2.39) when g is a Kählerian metric the system (2.38) is nothing but the ordinary harmonic map system on M. It is well known (cf. e.g., Theorem 13.1 in [109], p. 116) that a holomorphic map of Kählerian manifolds is harmonic. This simple result fails in general if the source manifold M is only Hermitian. For instance if H^n, $n \geq 2$, is the complex Hopf manifold (cf. [189], Vol. II, p. 137) endowed with the Boothby metric (a locally conformal Kähler metric on H^n, cf. [106], p. 22) then by a result of J. Jost & S-T. Yau, [179], there is no nonconstant Hermitian harmonic map $\phi : H^n \to S^1$. Due to this obstruction, it has been argued (cf. [179], p. 221–222) that the study of (2.38), rather than the harmonic map system, is more appropriate on a Hermitian manifold and that results about Hermitian harmonic maps could be useful in studying rigidity of complete Hermitian manifolds (cf. [222], p. 232). Going back to the mainstream of this book, one ought to mention that harmonic vector fields on Hermitian manifolds haven't been studied so far. A simple calculation shows that the rough Laplacian Δ_g on a Hermitian manifold may be locally expressed as

$$\Delta_g X = -g^{\alpha\overline{\beta}} \left\{ \nabla_{\partial/\partial z^\alpha} \nabla_{\partial/\partial \bar{z}^\beta} X - \nabla_{\nabla_{\partial/\partial z^\alpha} \partial/\partial \bar{z}^\beta} X \right\} - i \left(\Lambda_g R\right) X,$$

where the trace $\Lambda_g R$ of the curvature tensor field R of ∇ is defined by

$$\Lambda_g R = \frac{i}{2} \sum_{\alpha=1}^{n} R(Z_\alpha, Z_{\overline{\alpha}}).$$

Let us consider the first order differential operator L given by

$$(LX)^a_{bc} = 2 \left| \begin{matrix} a \\ bd \end{matrix} \right| \frac{\partial X^d}{\partial x^c} + \left(\frac{\partial}{\partial x^c} \left| \begin{matrix} a \\ bd \end{matrix} \right| + \left| \begin{matrix} e \\ bd \end{matrix} \right| \left| \begin{matrix} a \\ ce \end{matrix} \right| - \left| \begin{matrix} e \\ bc \end{matrix} \right| \left| \begin{matrix} a \\ ed \end{matrix} \right| \right) X^d$$

for any $X \in \mathfrak{X}(M)$ locally written as $X = X^a \, \partial / \partial x^a$. Here $\left| \begin{matrix} a \\ bc \end{matrix} \right|$ are the Christoffel symbols of the second kind of g and the range of the indices is $a, b, c, \ldots \in \{1, \ldots, 2n\}$. A calculation shows that

$$\left(\Delta_g X\right)^a = \Delta X^a - G^{bc}(LX)^a_{bc}$$

where $G_{ab}G^{bc} = \delta^c_a$ and $G_{ab} = g(\partial/\partial x^a, \partial/\partial x^b)$. Consequently the harmonic vector fields system may be locally written

$$\Delta X^a = G^{ab}(LX)^a_{bc} + \|\nabla X\|^2 X^a, \quad 1 \le a \le 2n. \tag{2.40}$$

Then the system

$$\tilde{\Delta} X^a = G^{ab}(LX)^a_{bc} + \|\nabla X\|^2 X^a \tag{2.41}$$

is a Hermitian analog to (2.40) and coincides with (2.40) when the metric g is Kählerian. The study of (2.41) is an open problem. ■

We end this section by computing the total bending of a unit vector field on a torus $M = T^2$ endowed with an arbitrary Riemannian metric g (see Section 1.5 in Chapter 1). Let $\{S, W\}$ be an orthonormal frame of $(T(T^2), g)$ such that $JS = W$. Let us consider the 1-forms $\Theta_S, \Theta_W \in \Omega^1(T^2)$ given by

$$\Theta_S = g(S, \cdot), \quad \Theta_W = g(W, \cdot).$$

Let us consider the volume form $\omega_T = \Theta_S \wedge \Theta_W \in \Omega^2(T^2)$. Also we shall need the 1-form $\omega_{SW} \in \Omega^1(T^2)$ given by

$$\omega_{SW} = g(\nabla S, W).$$

Moreover let $a, b \in C^\infty(T^2)$ be given by

$$\omega_{SW} = a\Theta_S + b\Theta_W.$$

If $Z \in \mathfrak{X}(T^2)$ is given by $Z = aS + bW$ then $\omega_{SW} = g(Z, \cdot)$. Let ∇ be the Levi-Civita connection of (T^2, g). A straightforward calculation shows that

$$\nabla_S S = aW, \quad \nabla_S W = -aS, \quad \nabla_W S = bW, \quad \nabla_W W = -bS,$$
$$\operatorname{div}(Z) = S(a) + W(b).$$

Now given $X \in \mathcal{E} = \Gamma^\infty(S(T^2))$ and $\varphi, \psi : T^2 \to \mathbb{R}$ are the (S, W)-coordinates of X (cf. Definition 1.30) then

$$\mathcal{B}(X) = \int_{T^2} \|\nabla X\|^2 \, d\operatorname{vol}(g) = \int_{T^2} \big\{(S\varphi - a\psi)^2 + (S\psi + a\varphi)^2$$
$$+ (W\varphi - b\psi)^2 + (W\psi + b\varphi)^2\big\} \, d\operatorname{vol}(g).$$

If $\pi : \mathbb{R}^2 \to T^2$ is the projection we set $\hat{g} = \pi^* g$. Also let $\hat{S}$, $\hat{W}$, $\hat{Z} \in \mathfrak{X}(\mathbb{R}^2)$ be π-related to $S, W, Z \in \mathfrak{X}(T^2)$. For instance $(d_\xi \pi)\hat{S}_\xi = S_{\pi(\xi)}$ for any $\xi \in \mathbb{R}^2$. Let

$$Q = \{sd_1 + td_2 \in \mathbb{R}^2 : (s, t) \in [0, 1]^2\}.$$

Then $\pi(Q) = T^2$ and $\pi : Q \setminus \partial Q \to T^2$ is injective, where $\partial Q = \{sd_1 + td_2 : (s, t) \in \partial[0, 1]^2\}$. Let (U, x^i) be a local coordinate system on T^2. We set

$$\frac{\partial}{\partial x^i} = \lambda_i S + \mu_i W, \quad i \in \{1, 2\},$$

and assume that the coordinates have been chosen such that $\lambda_1 \mu_2 - \mu_1 \lambda_2 > 0$. Moreover if we set

$$dx^i = \lambda^i \Theta_S + \mu^i \Theta_W$$

then $\lambda^i \lambda_j + \mu^i \mu_j = \delta^i_j$. If $G = g_{11} g_{22} - g_{12}^2$ then $\sqrt{G} = \lambda_1 \mu_2 - \mu_1 \lambda_2$ hence

$$d\operatorname{vol}(g) = \sqrt{G}\, dx^1 \wedge dx^2 = (\lambda_1 \mu_2 - \mu_1 \lambda_2)(\lambda^1 \mu^2 - \mu^1 \lambda^2)\, \Theta_S \wedge \Theta_W$$

so that $d\operatorname{vol}(g) = 2\omega_T$. Next

$$(S\varphi - a\psi)_{\pi(\xi)} = S_{\pi(\xi)}(\varphi) - a(\pi(\xi))\psi(\pi(\xi))$$
$$= \Big(\hat{S}(\varphi \circ \pi) - (a \circ \pi)(\psi \circ \pi)\Big)_\xi$$

for any $\xi \in \mathbb{R}^2$. Consequently if $\alpha \in \mathcal{W}$ is an angle function for X so that $\cos\alpha = \varphi \circ \pi$ and $\sin\alpha = \psi \circ \pi$ then

$$(S\varphi - a\psi) \circ \pi = \hat{S}(\cos\alpha) - (a \circ \pi)\sin\alpha = -\Big\{\hat{S}(\alpha) + a \circ \pi\Big\}\sin\alpha.$$

Similar calculations furnish the expressions of $(S\psi + a\varphi) \circ \pi$, $(W\varphi - b\psi) \circ \pi$ and $(W\psi + b\varphi) \circ \pi$ in terms of the angle function α and its derivatives. In the end

$$\pi^* \left\{ \left[(S\varphi - a\psi)^2 + (S\psi + a\varphi)^2 + (W\varphi - b\psi)^2 + (W\psi + b\varphi)^2 \right] d\,\mathrm{vol}(g) \right\}$$
$$= 2\left\{ \left(\hat{S}(\alpha) + a \circ \pi \right)^2 + \left(\hat{W}(\alpha) + b \circ \pi \right)^2 \right\} \pi^* \omega_T.$$

With a change of variables under the integral sign we have

$$\mathcal{B}(X) = 2 \int_Q \left\{ \left(\hat{S}(\alpha) + a \circ \pi \right)^2 + \left(\hat{W}(\alpha) + b \circ \pi \right)^2 \right\} \pi^* \omega_T. \tag{2.42}$$

Let us denote the right hand side of (2.42) by $\mathcal{B}(\alpha)$ to recast the total bending as a functional $\mathcal{B} : \mathcal{W} \to [0, +\infty)$ defined on the set $\mathcal{W}$ of all angle functions. On the other hand the identities

$$\hat{\nabla}\alpha = \hat{S}(\alpha)\,\hat{S} + \hat{W}(\alpha)\,\hat{W}, \quad \hat{Z} = (a \circ \pi)\hat{S} + (b \circ \pi)\,\hat{W},$$

lead to

$$\mathcal{B}(\alpha) = 2 \int_Q \|\hat{\nabla}\alpha + \hat{Z}\|_{\hat{g}}^2\, \pi^* \omega_T. \tag{2.43}$$

We may conclude that

Theorem 2.26 (G. Wiegmink, [309]) *There is an action of the additive reals $\mathbb{R}$ on $\mathcal{W}$ leaving the total bending functional $\mathcal{B}$ invariant and preserving each* $\mathrm{Per}(m, n) \subset \mathcal{W}$, $(m, n) \in \mathbb{Z}^2$. *Precisely*

$$\mathbb{R} \times \mathcal{W} \to \mathcal{W}, \quad (r, \alpha) \mapsto \alpha + r,$$

is an action with the required properties. At the level of unit vector fields on T^2 it corresponds to the action

$$\mathrm{SO}(2) \times \mathcal{E} \to \mathcal{E}, \tag{2.44}$$

$$\left(\begin{pmatrix} \cos r & -\sin r \\ \sin r & \cos r \end{pmatrix}, X \right) \mapsto (\cos r)X + (\sin r)JX,$$

for any $r \in \mathbb{R}$ and any $X \in \mathcal{E}$.

The proof of Theorem 2.26 follows easily from (2.43) (as the right hand side of (2.43) depends only on the first order derivatives of α). Of course Theorem 2.26 admits the following geometric interpretation: the total bending of a unit vector field X on the torus T^2 does not change when X is rotated simultaneously in all tangent spaces by a common angle.

2.7. THE SECOND VARIATION OF THE ENERGY FUNCTION

The purpose of the present section is to establish the second variation formula for $E : \Gamma^\infty(S(M)) \to [0, +\infty)$ in the neighborhood of a critical point (i.e., a harmonic vector field).

Theorem 2.27 (G. Wiegmink, [309]) *Let (M, g) be a compact orientable Riemannian manifold and U a tangent vector field on M. Let us consider a smooth 2-parameter variation of U*

$$\mathcal{V} : M \times I_\delta^2 \to T(M), \quad I_\delta = (-\delta, \delta), \quad \delta > 0,$$
$$U_{t,s} = \mathcal{V} \circ i_{t,s}, \quad t, s \in I_\delta, \quad U_{0,0} = U.$$

Here we set $N = M \times I_\delta^2$ and $i_{t,s} : M \to N$, $i_{t,s}(x) = (x, t, s)$ for any $x \in M$ and any $t, s \in I_\delta$. Let $V = (\partial U_{t,s}/\partial t)_{t=s=0}$ and $W = (\partial U_{t,s}/\partial s)_{t=s=0}$. Let us assume that $U_{t,s} \in \Gamma^\infty(S(M))$ for any $t, s \in I_\delta$. If $U \in \Gamma^\infty(S(M))$ is a harmonic vector field then

$$\frac{\partial^2}{\partial t \partial s} \{\mathcal{B}(U_{t,s})\}_{t=s=0} = \int_M g(V, \Delta_g W - \|\nabla U\|^2 W) d\,\mathrm{vol}(g). \tag{2.45}$$

In particular for any smooth 1-parameter variation $\mathcal{U}$ of U through unit vector fields $U_t = \mathcal{U}(\cdot, t) \in \Gamma^\infty(S(M))$, $|t| < \delta$

$$\frac{d^2}{dt^2} \{\mathcal{B}(U_t)\}_{t=0} = \int_M \{\|\nabla V\|^2 - \|\nabla U\|^2 \|V\|^2\} d\,\mathrm{vol}(g) \tag{2.46}$$

where $V = (dU_t/dt)_{t=0}$.

The identity (2.46) is the *second variation formula* (of the *biegung* functional). See also C.M. Wood, [316], and O. Gil-Medrano & E. Llinares-Fuster, [132].

Proof of Theorem 2.27. Let $p : N \to M$ be the projection and $p^{-1}TM \to N$ the pullback of $T(M)$ by p. Then $\mathcal{V}$ is a C^∞ section in $p^{-1}TM \to N$. Let $h = p^{-1}g$ and $D = p^{-1}\nabla$ be respectively the Riemannian bundle metric induced by g and the connection induced by the Levi-Civita connection ∇ in $p^{-1}TM \to N$. We set

$$\tilde{Y}_{(x,t,s)} = (d_x i_{t,s}) Y_x, \quad x \in M, \quad t, s \in I_\delta.$$

For simplicity we set $\mathbb{T} = \partial/\partial t$ and $\mathbb{S} = \partial/\partial s$ ($\mathbb{T}, \mathbb{S} \in \mathfrak{X}^\infty(N)$). Then (as $p^{-1}g$ is parallel with respect to D)

$$\frac{\partial}{\partial t}\mathcal{B}(U_{t,s}) = \int_M \sum_{i=1}^n h(D_{\mathbb{T}} D_{\tilde{E}_i}\mathcal{V}, D_{\tilde{E}_i}\mathcal{V})\, d\,\text{vol}(g)$$

$$= \int_M \sum_i h(D_{\tilde{E}_i} D_{\mathbb{T}}\mathcal{V}, D_{\tilde{E}_i}\mathcal{V})\, d\,\text{vol}(g)$$

due to

$$[\mathbb{T}, \tilde{E}_i] = 0, \quad R^D(\mathbb{T}, \tilde{E}_i)\mathcal{V} = 0.$$

Then

$$\frac{\partial^2}{\partial s \partial t}\mathcal{B}(U_{t,s}) = \int_M \sum_i \frac{\partial}{\partial s} h(D_{\tilde{E}_i} D_{\mathbb{T}}\mathcal{V}, D_{\tilde{E}_i}\mathcal{V})\, d\,\text{vol}(g) \tag{2.47}$$

$$= \int_M \sum_i \{h(D_{\mathbb{S}} D_{\tilde{E}_i} D_{\mathbb{T}}\mathcal{V}, D_{\tilde{E}_i}\mathcal{V}) + h(D_{\tilde{E}_i} D_{\mathbb{T}}\mathcal{V}, D_{\mathbb{S}} D_{\tilde{E}_i}\mathcal{V})\}\, d\,\text{vol}(g)$$

(as $[\mathbb{S}, \tilde{E}_i] = 0$ and $R^D(\mathbb{S}, \tilde{E}_i)\mathcal{V} = 0$)

$$= \int_M \sum_i \{h(D_{\tilde{E}_i} D_{\mathbb{S}} D_{\mathbb{T}}\mathcal{V}, D_{\tilde{E}_i}\mathcal{V}) + h(D_{\tilde{E}_i} D_{\mathbb{T}}\mathcal{V}, D_{\tilde{E}_i} D_{\mathbb{S}}\mathcal{V})\}\, d\,\text{vol}(g)$$

$$= \int_M \sum_i \{\tilde{E}_i(h(D_{\mathbb{S}} D_{\mathbb{T}}\mathcal{V}, D_{\tilde{E}_i}\mathcal{V})) - h(D_{\mathbb{S}} D_{\mathbb{T}}\mathcal{V}, D_{\tilde{E}_i} D_{\tilde{E}_i}\mathcal{V})$$

$$+ \tilde{E}_i(h(D_{\mathbb{T}}\mathcal{V}, D_{\tilde{E}_i} D_{\mathbb{S}}\mathcal{V})) - h(D_{\mathbb{T}}\mathcal{V}, D_{\tilde{E}_i} D_{\tilde{E}_i} D_{\mathbb{S}}\mathcal{V})\}\, d\,\text{vol}(g).$$

For each fixed $(t,s) \in I_\delta^2$ we define $Y_{t,s} \in T(M)$ by

$$g(Y_{t,s}, Z)_x = h(D_{\mathbb{S}} D_{\mathbb{T}}\mathcal{V}, D_{\tilde{Z}}\mathcal{V})_{(x,t,s)}, \quad Z \in T(M).$$

Then

$$\sum_i \tilde{E}_i(h(D_{\mathbb{S}} D_{\mathbb{T}}\mathcal{V}, D_{\tilde{E}_i}\mathcal{V})) = \sum_i E_i(g(Y_{t,s}, E_i)) \circ p$$

$$= \sum_i \{g(\nabla_{E_i} Y_{t,s}, E_i) + g(Y_{t,s}, \nabla_{E_i} E_i)\} \circ p$$

$$= \text{div}(Y_{t,s}) \circ p + h\left(D_{\mathbb{S}} D_{\mathbb{T}}\mathcal{V}, \sum_i D_{\widetilde{\nabla_{E_i} E_i}}\mathcal{V}\right).$$

Similarly, given $Z_{t,s} \in T(M)$ determined by

$$g(Z_{t,s}, Z)_x = h(D_{\mathbb{T}}\mathcal{V}, D_{\tilde{Z}} D_{\mathbb{S}}\mathcal{V})_{(x,t,s)}$$

one has

$$\sum_i \tilde{E}_i(h(D_{\mathbb{T}}\mathcal{V}, D_{\tilde{E}_i} D_{\mathcal{S}}\mathcal{V}))$$

$$= \operatorname{div}(Z_{t,s}) \circ p + h\left(D_{\mathbb{T}}\mathcal{V}, \sum_i D_{\widetilde{\nabla_{E_i} E_i}} D_{\mathbb{S}}\mathcal{V} \right).$$

Going back to (2.47) one has (by Green's lemma)

$$\frac{\partial^2}{\partial s \partial t} \left\{ \mathcal{B}(U_{t,s}) \right\}_{t=s=0}$$

$$= \int_M \left\{ h(D_{\mathbb{S}} D_{\mathbb{T}}\mathcal{V}, \sum_i \{ D_{\widetilde{\nabla_{E_i} E_i}}\mathcal{V} - D_{\tilde{E}_i} D_{\tilde{E}_i}\mathcal{V}\}) \right.$$

$$\left. + \, h(D_{\mathbb{T}}\mathcal{V}, \sum_i \{ D_{\widetilde{\nabla_{E_i} E_i}} D_{\mathbb{S}}\mathcal{V} - D_{\tilde{E}_i} D_{\tilde{E}_i} D_{\mathbb{S}}\mathcal{V}\}) \right\}_{t=s=0} d\operatorname{vol}(g)$$

$$= \int_M \{ g(V, \Delta_g U) + g(V, \Delta_g W)\} d\operatorname{vol}(g)$$

where we have set $V = (\partial^2 U_{t,s}/\partial t \partial s)_{t=s=0}$. Moreover (by differentiating $h(\mathcal{V}, \mathcal{V}) = 1$)

$$g(V, U)_x = h(D_{\mathbb{S}} D_{\mathbb{T}}\mathcal{V}, \mathcal{V})_{(x,0,0)}$$

$$= \{ \mathbb{S}(h(D_{\mathbb{T}}\mathcal{V}, \mathcal{V})) - h(D_{\mathbb{T}}\mathcal{V}, D_{\mathbb{S}}\mathcal{V})\}_{(x,0,0)} = -g(V, W)_x$$

and (as U is harmonic i.e., a smooth solution to (2.28))

$$\frac{\partial^2}{\partial t \partial s} \left\{ \mathcal{B}(U_{t,s}) \right\}_{t=s=0} = \int_M \{ \|\nabla U\|^2 g(V, U) + g(V, \Delta_g W)\} d\operatorname{vol}(g)$$

$$= \int_M g(V, \Delta_g W - \|\nabla U\|^2 W) d\operatorname{vol}(g)$$

and (2.45) is proved. Finally given an arbitrary smooth 1-parameter variation $\mathcal{U} : M \times I_\delta \to T(M)$ of X through unit vector fields the identity (2.46) follows from (2.45) for the particular 2-parameter variation $\mathcal{V} : M \times I_{\delta/2} \to$

$T(M)$ given by $\mathcal{V}(x,t,s) = \mathcal{U}(x,t+s)$ for any $x \in M$ and any $t, s \in I_{\delta/2}$. Indeed

$$\frac{d^2}{dt^2}\{\mathcal{B}(U_t)\}_{t=0} = \int_M g(V, \Delta_g V - \|\nabla U\|^2 V)\, d\,\mathrm{vol}(g). \tag{2.48}$$

On the other hand, for any smooth vector field V on M

$$\begin{aligned}\Delta_g \|V\|^2 &= -\sum_{i=1}^{n} \{E_i E_i \|V\|^2 - (\nabla_{E_i} E_i)\|V\|^2\} \\ &= -2\sum_i \{E_i(g(\nabla_{E_i} V, V)) - g(\nabla_{\nabla_{E_i} E_i} V, V)\} \\ &= -2\sum_i \{g(\nabla_{E_i}\nabla_{E_i} V, V) + g(\nabla_{E_i} V, \nabla_{E_i} V) - g(\nabla_{\nabla_{E_i} E_i} V, V)\}\end{aligned}$$

hence

$$\Delta_g \|V\|^2 = 2\{g(\Delta_g V, V) - \|\nabla V\|^2\}. \tag{2.49}$$

Now (2.46) follows from (2.48)–(2.49) and Green's lemma. ■

Definition 2.28 Let $U \in \Gamma^\infty(S(M))$ be a harmonic vector field. The *Jacobi operator* $J_U : \mathcal{S} \to \mathcal{S}$ is given by

$$J_U V = p_{\mathcal{S}}\, \Delta_g V - \|\nabla U\|^2 V, \quad V \in \mathcal{S},$$

where $\mathcal{S} = \{V \in \mathfrak{X}(M) : g(U, V) = 0\}$ and $(p_{\mathcal{S}})_x : T_x(M) \to (\mathbb{R}U_x)^\perp$ is the natural projection associated to the direct sum decomposition $T_x(M) = \mathbb{R}U_x \oplus (\mathbb{R}U_x)^\perp$ for any $x \in M$. ■

It may be shown that the Jacobi operator J_U is a self-adjoint elliptic second order differential operator.

Definition 2.29 A harmonic vector field $U \in \Gamma^\infty(S(M))$ is said to be *stable* if

$$\int_M \left(\|\nabla V\|^2 - \|V\|^2 \|\nabla U\|^2\right) d\,\mathrm{vol}(g) \geq 0$$

for any $V \in \mathcal{S}$. ■

Stability of U may be shown to be equivalent to the requirement that the eigenvalues of the Jacobi operator J_U be nonnegative.

Definition 2.30 Let U be a harmonic vector field on M. Let $(\text{Hess}\, E)_U$ be the bilinear form on $\mathcal{S} = \{V \in \mathfrak{X}(M) : g(U, V) = 0\}$ determined by

$$(\text{Hess}\, E)_U(V, V) = \int_M \left(\|\nabla V\|^2 - \|V\|^2 \|\nabla U\|^2\right) d\,\text{vol}(g), \quad V \in \mathcal{S},$$

(by polarization). The *index* and *nullity* of U are the index and nullity of $(\text{Hess}\, E)_U$. ■

Corollary 2.31 *Let U be a stable harmonic vector field. Then $E(U) \le E(V)$ for any $V \in \mathcal{S} \cap \Gamma^\infty(S(M))$.*

That is to say stable harmonic vector fields are absolute minima for the energy functional $E : \mathcal{S} \cap \Gamma^\infty(S(M)) \to [0, +\infty)$ given by (2.3).

Proof of Corollary 2.31. Let $V \in \mathcal{S}$. As previously shown, there is a smooth 1-parameter variation $\mathcal{U} : M \times (-\epsilon, \epsilon) \to T(M)$ of U by unit vector fields such that $(dU_t/dt)_{t=0} = V$, where $U_t(x) = \mathcal{U}(x, t)$ for any $x \in M$ and any $|t| < \epsilon$. If V is also a unit vector field then another construction of $\mathcal{U}$ is to set

$$U_t = (\cos t) U + (\sin t) V, \quad |t| < \epsilon.$$

Let $\mathcal{E}(t) = E(U_t)$ for any $|t| < \epsilon$. Then (by stability and by the second variation formula (2.46))

$$0 \le \mathcal{E}''(0) = \int_M \left(\|\nabla V\|^2 - \|\nabla U\|^2\right) d\,\text{vol}(g)$$

hence $\mathcal{B}(U) \le \mathcal{B}(V)$. Corollary 2.31 is proved. ■

Proposition 2.32 *Let M be an orientable 2-dimensional Riemannian manifold and U a unit vector field tangent to M. Let $\{E_1, E_2\}$ be a local orthonormal frame of $T(M)$ defined on the open set $\Omega \subseteq M$ such that $E_1 = U$. Let us consider the functions $a, b \in C^\infty(\Omega)$ given by*

$$a = g(\nabla_U U, E_2), \quad b = g(\nabla_{E_2} E_2, U).$$

Then

i. *U is a harmonic vector field if and only if $U(a) = E_2(b)$ for any local orthonormal frame as above.*

ii. *The following statements are equivalent* 1) *U is Killing,* 2) *U is parallel, and* 3) *U is geodesic and* $\text{div}(U) = 0$. *Moreover if statement* (1) *holds then g is flat and U is a harmonic map.*

Proof. **i.** As $\|U\| = 1$ one has $g(\nabla_U U, U) = g(\nabla_{E_2} U, U) = 0$ hence

$$\begin{aligned}\nabla_U U &= g(\nabla_U U, E_2)E_2 = aE_2,\\ \nabla_{E_2} U &= g(\nabla_{E_2} U, E_2)E_2\\ &= \{E_2(g(U,E_2)) - g(U, \nabla_{E_2} E_2)\}E_2 = -bE_2,\end{aligned}$$

that is we were able to determine the covariant derivative of U. Similarly we may calculate ∇E_2. Indeed $\|E_2\| = 1$ yields $g(\nabla_U E_2, E_2) = g(\nabla_{E_2} E_2, E_2) = 0$ hence

$$\begin{aligned}\nabla_U E_2 &= g(\nabla_U E_2, U)U = -aU,\\ \nabla_{E_2} E_2 &= g(\nabla_{E_2} E_2, U)U = bU.\end{aligned}$$

Gathering up the information obtained so far

$$\begin{aligned}\nabla_{E_1} U = aE_2, &\quad \nabla_{E_2} U = -bE_2,\\ \nabla_{E_1} E_2 = -aE_1, &\quad \nabla_{E_2} E_2 = bE_1,\end{aligned}$$

or (by identifying the endomorphisms ∇U, $\nabla E_2 \in \Gamma^\infty(T^*(M) \otimes T(M))$ with their matrices with respect to the frame $\{E_1, E_2\}$)

$$\nabla U = \begin{pmatrix} 0 & a \\ 0 & -b \end{pmatrix}, \quad \nabla E_2 = \begin{pmatrix} -a & 0 \\ b & 0 \end{pmatrix}.$$

On the other hand

$$\|\nabla U\|^2 = \sum_{i=1}^{2} g(\nabla_{E_i} U, \nabla_{E_i} U) = a^2 + b^2.$$

Hence

$$\begin{aligned}\Delta_g U &= -\sum_{j=1}^{2} \{\nabla_{E_j} \nabla_{E_j} U - \nabla_{\nabla_{E_j} E_j} U\}\\ &= -\nabla_{E_1} \nabla_{E_1} U + \nabla_{\nabla_{E_1} E_1} U - \nabla_{E_2} \nabla_{E_2} U + \nabla_{\nabla_{E_2} E_2} U\\ &= -\nabla_{E_1}(aE_2) + \nabla_{aE_2} U - \nabla_{E_2}(-bE_2) + \nabla_{bE_1} U\\ &= -E_1(a)E_2 + a\{\nabla_{E_1} E_2 + \nabla_{E_2} U\}\\ &\quad + E_2(b)E_2 + b\{\nabla_{E_2} E_2 + \nabla_{E_1} U\}\\ &= -U(a)E_2 + a(aE_1 - bE_2) + E_2(b)E_2 + b(bE_1 + aE_2)\\ &= (a^2 + b^2)E_1 - [U(a) - E_2(b)]E_2\end{aligned}$$

that is

$$\Delta_g U - \|\nabla U\|^2 U = -[U(a) - E_2(b)]E_2$$

and (i) in Proposition 2.32 is proved.

ii. Proof of (1) $\Longrightarrow$ (2). Let U be a unit Killing vector field on M i.e., $U \in \Gamma^\infty(S(M))$ and $\mathcal{L}_U g = 0$. Then

$$0 = (\mathcal{L}_U g)(X, Y) = g(\nabla_X U, Y) + g(\nabla_Y U, X)$$

for any $X, Y \in \mathfrak{X}(M)$, where $\mathcal{L}$ denotes the Lie derivative. Consequently $\nabla U \in \Gamma^\infty(T^*(M) \otimes T(M))$ is a skew-symmetric endomorphism so that $g(\nabla_X U, X) = 0$ for any $X \in \mathfrak{X}(M)$. Let $\{E_1, E_2\}$ be a local orthonormal frame on $T(M)$ such that $E_1 = U$. Thus

$$\nabla_U U = g(\nabla_U U, E_2)E_2 = -g(\nabla_{E_2} U, U) = 0,$$
$$\nabla_{E_2} U = g(\nabla_{E_2} U, E_2)E_2 = 0,$$

that is $\nabla U = 0$. In particular U is harmonic.

The proof of (2) $\Longrightarrow$ (3) is easy (hence omitted).

Proof of (3) $\Longrightarrow$ (1). As $0 = \nabla_U U = aE_2$ it follows that $a = 0$. Moreover $0 = \operatorname{div}(U) = -b$ yields $b = 0$. Hence U is parallel and in particular Killing. Finally

$$R(U, E_2)U = -\nabla_U \nabla_{E_2} U + \nabla_{E_2} \nabla_U U + \nabla_{[U,E_2]} U = 0,$$
$$g(R(U, E_2)E_2, U) = -g(R(U, E_2)U, E_2) = 0,$$
$$g(R(U, E_2)E_2, E_2) = 0,$$

so that $R = 0$.

■

Remark 2.33 Let M be a compact orientable Riemannian manifold. As already emphasized earlier in this chapter the existence of a globally defined unit vector field on M is equivalent to $\chi(M) = 0$. If additionally M is 2-dimensional then (by the classification of compact connected real surfaces) M is a torus. ■

Proposition 2.34 *Let M be a real 2-dimensional compact orientable Riemannian manifold. Then any harmonic vector field U is stable. Moreover for each $V \in \mathcal{S}$ one has* $(\operatorname{Hess} E)_U(V, V) = 0$ *if and only if* $\|V\| =$ *constant.*

Proof. Let $U \in \Gamma^\infty(S(M))$ and $V \in \mathcal{S}$ where $\mathcal{S} = \{V \in \Gamma^\infty(M) : g(U, V) = 0\}$. For each $x_0 \in M$ let E_2 be a tangent vector field defined on an open neighborhood $\Omega \subseteq M$ of x_0 such that $\{U, E_2\}$ is a (local) orthonormal frame of $T(M)$. As M is 2-dimensional, it follows that $\dim_{\mathbb{R}} (\mathbb{R}U_x)^\perp = 1$ for any $x \in \Omega$. Yet $V_x \in (\mathbb{R}U_x)^\perp$ for any $x \in \Omega$ hence $V = \lambda E_2$ for some $\lambda \in C^\infty(\Omega)$. Next $\|V\| = |\lambda| \|E_2\| = |\lambda|$ on Ω. Let us compute $\|\nabla V\|$. First, if $E_1 = U$ and (as in Proposition 2.32)

$$a = g(\nabla_U U, E_2), \quad b = g(\nabla_{E_2} E_2, U),$$

then

$$\begin{aligned}
\nabla_U V &= \sum_{i=1}^{2} g(\nabla_U V, E_i) E_i \\
&= \sum_i \{U(\lambda) g(E_2, E_i) + \lambda g(\nabla_U E_2, E_i)\} E_i \\
&= U(\lambda) \|E_2\|^2 E_2 + \lambda g(\nabla_U E_2, E_1) E_1 = U(\lambda) E_2 - \lambda g(E_2, \nabla_U U) E_1
\end{aligned}$$

or

$$\nabla_U V = U(\lambda) E_2 - a\lambda U. \tag{2.50}$$

Similarly

$$\begin{aligned}
\nabla_{E_2} V &= \sum_i g(\nabla_{E_2} V, E_i) E_i \\
&= \sum_i \{E_2(\lambda) g(E_2, E_i) + \lambda g(\nabla_{E_2} E_2, E_i)\} E_i \\
&= E_2(\lambda) \|E_2\|^2 E_2 + \lambda g(\nabla_{E_2} E_2, E_1) E_1
\end{aligned}$$

or

$$\nabla_{E_2} V = E_2(\lambda) E_2 + b\lambda U. \tag{2.51}$$

Next (by taking into account (2.50)–(2.51))

$$\begin{aligned}
\|\nabla V\|^2 &= \sum_i g(\nabla_{E_i} V, \nabla_{E_i} V) \\
&= g(\nabla_{E_1} V, \nabla_{E_1} V) + g(\nabla_{E_2} V, \nabla_{E_2} V) \\
&= U(\lambda)^2 + a^2 \lambda^2 + E_2(\lambda)^2 + b^2 \lambda^2
\end{aligned}$$

or

$$\|\nabla V\|^2 = \|d\lambda\|^2 + \|V\|^2 \|\nabla U\|^2. \tag{2.52}$$

Finally (by (2.52))

$$(\text{Hess}\, E)_U(V,V) = \int_M \|d\lambda\|^2 \, d\,\text{vol}(g) \geq 0$$

with equality if and only if $\lambda =$ constant on Ω. This amounts to saying that $(\text{Hess}\, E)_U(V,V) = 0$ if and only if $\|V\|$ is locally constant (and M is tacitly thought of as connected). ■

2.8. UNBOUNDEDNESS OF THE ENERGY FUNCTIONAL

Under the assumptions of Proposition 2.22 we may prove the following

Corollary 2.35 (G. Wiegmink, [309]) *For any nonempty open subset $\Omega \subseteq M$ and any unit vector field X on M there is a sequence $\{Y_\nu\}_{\nu \geq 1}$ of unit vector fields such that each Y_ν coincides with X outside Ω and $E(Y_\nu) \to \infty$ for $\nu \to \infty$. In particular the energy functional E is unbounded from above.*

Proof. Let $h = (x^1, \dots, x^m) : U \to \mathbb{R}^m$ be a local coordinate system on M such that $U \subseteq \Omega$, $h(U) \supset [-2\pi, 2\pi]^m$ and $X = \partial/\partial x^1$ on U (cf. the proof of the classical Frobenius theorem, e.g., [235], p. 91–92). Moreover let $\varphi \in C_0^\infty(M)$ be a test function such that i) $0 \leq \varphi(x) \leq 1$ for any $x \in M$, ii) $\varphi = 1$ in a neighborhood V of the compact set $K = h^{-1}([-\pi, \pi]^m)$ such that $\overline{V} \subset U$, and iii) $\varphi = 0$ outside $h^{-1}([-2\pi, 2\pi]^m)$. For each $\nu \in \mathbb{Z}$, $\nu \geq 1$, let f_ν be the C^∞ extension to M of the function $\sin(\nu x^1)$ (thought of as defined on the closed set $\overline{V}$) and let us set $\alpha_\nu = \varphi f_\nu$. Let us consider a unit vector field F on U such that X and F are mutually orthogonal. Next we set

$$Y_\nu = (\cos \alpha_\nu) X + (\sin \alpha_\nu) F, \quad \nu \geq 1.$$

Then Y_ν is a unit vector field coinciding with X outside Ω. As we may complete X to a local frame of $T(M)$

$$\begin{aligned} |\nabla Y_\nu|^2 &\geq g_\theta(\nabla_X Y_\nu, \nabla_X Y_\nu) \\ &= X(\alpha_\nu)^2 + \|\nabla_X X\|^2 \cos^2 \alpha_\nu + \|\nabla_X F\|^2 \sin^2 \alpha_\nu \\ &\quad + 2X(\alpha_\nu) g(\nabla_X X, F) + g(\nabla_X X, \nabla_X F) \sin(2\alpha_\nu) \end{aligned}$$

because of $g(X,F)=0$, $g(X,\nabla_X X)=0$ and $g(F,\nabla_X F)=0$. Next let us set

$$C_1=\sup_K \left|g(\nabla_X X,F)\right|, \quad C_2=\sup_K \left|g(\nabla_X X,\nabla_X F)\right|,$$

so that

$$|\nabla Y_\nu|^2 \geq X(\alpha_\nu)^2 - 2C_1\,|X(\alpha_\nu)| - C_2 \tag{2.53}$$

On the other hand $X(\alpha_\nu)=X(\varphi)f_\nu+\varphi\nu(\cos\nu x^1)$ on U so that $X(\alpha_\nu)=\nu\cos\nu x^1$ on $V\supset K$ and then $|X(\alpha_\nu)|\geq\nu$ on K. Hence (by (2.53))

$$\begin{aligned}
2E(Y_\nu) &\geq \int_K |\nabla Y_\nu|^2\, d\,\mathrm{vol}(g)\\
&\geq \int_K \left\{X(\alpha_\nu)^2-2C_1\nu-C_2\right\} d\,\mathrm{vol}(g)\\
&= \nu^2\int_K \cos^2(\nu x^1)\,d\,\mathrm{vol}(g) - (2C_1\nu+C_2)\mathrm{Vol}(K).
\end{aligned}$$

If $d\,\mathrm{vol}(g)=\sqrt{G(x)}\,dx^1\wedge\cdots\wedge dx^m$ is the Riemannian volume form of (M,g) (with $G(x)=\det[g_{ij}(x)]$) and we set $a=\inf_{x\in K}\sqrt{G(x)}$ then $a>0$ and

$$\int_K \cos^2(\nu x^1)\,d\,\mathrm{vol}(g) \geq a \int_{[-\pi,\pi]^m} \cos^2(\nu t^1)\,dt^1\cdots dt^m = (2\pi)^m\,\frac{a}{2}$$

Hence

$$E(Y_\nu)\geq (a/4)(2\pi)^m\nu^2-(2C_1\nu+C_2)\,\mathrm{Vol}(K)\to\infty$$

for $\nu\to\infty$. ■

2.9. THE DIRICHLET PROBLEM

Let $\Omega\subset\mathbb{C}^n$ be a smoothly bounded strictly pseudoconvex domain endowed with its Bergman metric g. Following E. Barletta, [22], we consider the Dirichlet problem for the harmonic vector fields equation

$$\Delta_g X - |\nabla^g X|^2\, X = 0 \quad \text{in } \Omega, \tag{2.54}$$

$$X = X_0 \quad \text{on } \partial\Omega, \tag{2.55}$$

for some continuous vector field X_0 tangent to $\partial\Omega$. Moreover, if Δ is the Bergman Laplacian then

$$(\Delta_g X)^i = \Delta X^i - 2g^{jk}\Gamma^i_{j\ell}\frac{\partial X^\ell}{\partial x^k} - g^{jk}\left(\frac{\partial \Gamma^i_{j\ell}}{\partial x^k} + \Gamma^s_{j\ell}\Gamma^i_{ks} - \Gamma^s_{jk}\Gamma^i_{s\ell}\right)X^\ell \quad (2.56)$$

where $X = X^i\,\partial/\partial x^i$ and Γ^i_{jk} are the Christoffel symbols of ∇^g (the Levi-Civita connection of (Ω, g)) with respect to the Cartesian coordinates (x^i) on $\mathbb{R}^{2n}$. Also $g_{ij} = g(\partial_i, \partial_j)$ with $\partial_i = \partial/\partial x^i$ and $[g^{ij}] = [g_{ij}]^{-1}$. Hence (2.54) is a nonlinear elliptic system similar to the harmonic map system (cf. e.g., (11.16) in [109], p. 107). We study arbitrary C^2 solutions to (2.54) (not necessarily *unit* vector fields). It is well known that the ellipticity of the Bergman Laplacian degenerates at the boundary e.g., the Bergman Laplacian on the unit ball $\mathbb{B}^n = \{z \in \mathbb{C}^n : |z| < 1\}$ is given by

$$\Delta u = -4(1 - |z|^2)\sum_{j,k=1}^{n}(\delta^{jk} - z^j\overline{z}^k)\frac{\partial^2 u}{\partial z^j \partial \overline{z}^k}$$

(here g is the Bergman metric on $\mathbb{B}^n$). By a result of C.R. Graham, [148], the Dirichlet problem $\Delta u = 0$ in $\mathbb{B}^n$ and $u|_{S^{2n-1}} = f$ may be solved for any $f \in C^0(S^{2n-1})$. Nevertheless in order that the solution be C^∞ up to the boundary it is necessary that f be the boundary value of a pluriharmonic function (see also [149]). On the same line of thought one may seek for the necessary conditions satisfied by the boundary data (2.55) of a solution $X^i \in C^2(\overline{\Omega})$ to (2.54). One may state

Theorem 2.36 (E. Barletta, [22]) *Let $\Omega \subset \mathbb{C}^n$ be a bounded strictly pseudoconvex domain and g its Bergman metric. Let us assume that a solution $X^i \in C^2(\overline{\Omega})$, with $1 \leq i \leq 2n$, to the Dirichlet problem (2.54)–(2.55) exists. If $X = X^i\,\partial/\partial x^i$ is tangent to the Levi distribution of each level hypersurface of $\varphi(z) = -K(z,z)^{-1/(n+1)}$, $z \in \Omega$, then $X^i = 0$ everywhere on $\partial\Omega$.*

Here $K(z,\zeta)$ is the Bergman kernel of Ω. The existence problem for (2.54)–(2.55) is open (even when $\Omega = \mathbb{B}^n$). The methods one adopts are similar to those in [21] i.e., one derives the equations induced by (2.54) on a level set $\varphi(z) = -\epsilon$ of a defining function φ of Ω (in a small one-sided neighborhood of $\partial\Omega$) and then shows that in the boundary limit (i.e., when $\epsilon \to 0$) these equations yield $X = 0$. The main technical ingredient is the Graham-Lee connection (cf. Proposition 1.1 in [151], p. 701) allowing us to express (2.54) in terms of functions (such as the *transverse curvature* of the foliation by level sets of φ, and its derivatives) which stay bounded as $z \to \partial\Omega$. Cf. also S. Dragomir & S. Nishikawa, [105], or Appendix A in [26], for a new axiomatic description of the Graham-Lee connection. The same approach was used in [26] to study the C^∞ regularity up to the boundary of Yang-Mills fields on (Ω, g). Similar to Theorem 2.36 (and with the conventions there) we obtain

Theorem 2.37 *Let $u \in C^2(\overline{\Omega})$ be a function such that uT is a harmonic vector field in Ω. Let $u_b = u|_{\partial\Omega}$ be the boundary values of u. Then*

$$u^3\left[n-1-\frac{4}{(1-r\varphi)^2}\right] = O(\varphi), \quad \varphi \to 0, \tag{2.57}$$

hence $u_b = 0$ provided that $n \neq 5$. Moreover if $n = 5$

$$2T(u_b) - T(r)u_b = 0, \quad \nabla^H u_b = 0 \quad on\ \partial\Omega, \tag{2.58}$$

where T is the characteristic direction of $i\partial\overline{\partial}\varphi$ and r is the transverse curvature of the foliation $\mathcal{F}$ by level sets of φ. Consequently $u_b(z) = c$ for some $c \in \mathbb{R}$ and any $z \in \partial\Omega$. Finally if $T(r)_z \neq 0$ for some $z \in \partial\Omega$ then again $u_b = 0$.

See the next section for the definition of the transverse curvature r of φ. By a result of J.M. Lee & R. Melrose, [202], r is C^∞ up to the boundary. A similar result (to that in Theorem 2.37) is expected to hold for solutions to (2.54) of the form uN with $u \in C^2(\overline{\Omega})$, where $N = -JT$ (and J is the complex structure on $\mathbb{C}^n$). This is claimed in [22] yet no proof is given.

2.9.1. The Graham-Lee Connection

Let $\Omega \subset \mathbb{C}^n$ be a bounded domain and $K(z, \zeta)$ its Bergman kernel. We endow Ω with the Bergman metric g given by

$$g_{j\overline{k}} = \frac{\partial^2 \log K(z,z)}{\partial z^j \partial \overline{z}^k}, \quad 1 \leq j, k \leq n.$$

Let us assume from now on that Ω is strictly pseudoconvex i.e., its boundary $\partial\Omega$ is a smooth real hypersurface in $\mathbb{C}^n$ whose induced CR structure $T_{1,0}(\partial\Omega)$ is strictly pseudoconvex (cf. [110], Chapter 1, for the notions of CR and pseudohermitian geometry needed through this section). Then $\varphi(z) = -K(z,z)^{-1/(n+1)}$, $z \in \Omega$, is a defining function for Ω (as a consequence of Fefferman's asymptotic development of the Bergman kernel, cf. [116]). For sufficiently small $\epsilon > 0$ the level sets $M_\epsilon = \{z \in \Omega : \varphi(z) = -\epsilon\}$ are strictly pseudoconvex CR manifolds and each M_ϵ is a leaf of a foliation $\mathcal{F}$ of an open neighborhood V of $\partial\Omega$ in Ω. Let us set $\theta = \frac{i}{2}(\overline{\partial} - \partial)\varphi(z)$ so that the pullback of θ to M_ϵ is a contact form on M_ϵ and the corresponding Levi form

$$G_\theta(X,Y) = (d\theta)(X,JY), \quad X,Y \in H(\mathcal{F}),$$

is positive definite. Here $H(\mathcal{F}) \to V$ is the subbundle of $T(\mathcal{F}) \to V$ determined by $H(\mathcal{F})|_{M_\epsilon} = H(M_\epsilon)$, $\epsilon > 0$, and $H(M_\epsilon)$ is the Levi, or maximally complex, distribution of M_ϵ. By a result of J.M. Lee & R. Melrose, [202], there is a unique complex vector field ξ on V, of type (1,0), such that $\partial\varphi(\xi) = 1$ and ξ is orthogonal to $T_{1,0}(\mathcal{F})$ with respect to $\partial\overline{\partial}\varphi$ i.e., $\partial\overline{\partial}\varphi(\xi,\overline{Z}) = 0$, for any $Z \in T_{1,0}(\mathcal{F})$. Here $T_{1,0}(\mathcal{F}) \to V$ is the complex subbundle of $T(\mathcal{F}) \otimes \mathbb{C} \to V$ determined by $T_{1,0}(\mathcal{F})\big|_{M_\epsilon} = T_{1,0}(M_\epsilon)$, $\epsilon > 0$, and $T_{1,0}(M_\epsilon)$ is the CR structure of M_ϵ. Moreover, we consider the function $r = 2\,\partial\overline{\partial}\varphi(\xi,\overline{\xi})$. r is the *transverse curvature* of φ. It should be observed that we work with the defining function $\varphi(z) = -K(z,z)^{-1/(n+1)}$ (as in A. Korányi & H.M. Reimann, [195]) yet the same constructions can be performed for any defining function $\varphi(z) < 0$ in Ω such that $\log(-1/\varphi)$ is strictly plurisubharmonic near $\partial\Omega$ (and then $\frac{i}{2}\partial\overline{\partial}\log(-1/\varphi)$ is the Kähler 2-form associated to a Kähler metric g on V). For instance, if $\Omega \subset \mathbb{C}^n$ is the Siegel domain with $\varphi(z) = \sum_{\alpha=1}^{n-1} |z_\alpha|^2 - \mathrm{Im}(z_n)$ then the transverse curvature is $r = 0$. Also if $\Omega = \mathbb{B}^n$ then $K(z,z) = C_n(1-|z|^2)^{-(n+1)}$ (with $C_n = n!/\pi^n$) and $r(z) = 1/(a_n|z|^2)$ (with $a_n = C_n^{-1/(n+1)}$) and the regularity of r up to $\partial\Omega$ can be directly read off from this simple explicit formula. In the general case of an arbitrary strictly pseudoconvex domain there is no such formula and we need the result in [202] that $r \in C^\infty(\overline{V})$. Let $\xi = \frac{1}{2}(N - iT)$ be the real and imaginary parts of ξ. Then

$$(d\varphi)(N) = 2, \quad (d\varphi)(T) = 0,$$
$$\theta(N) = 0, \quad \theta(T) = 1,$$
$$\partial\varphi(N) = 1, \quad \partial\varphi(T) = i.$$

In particular T is tangent to the leaves of $\mathcal{F}$. $\mathcal{F}$ carries the tangential Riemannian metric g_θ defined by

$$g_\theta(X,Y) = G_\theta(X,Y), \quad g_\theta(X,T) = 0, \quad g_\theta(T,T) = 1,$$

for any $X,Y \in H(\mathcal{F})$. The pullback of g_θ to each leaf M_ϵ of $\mathcal{F}$ is the Webster metric of M_ϵ (associated to the contact form $j_\epsilon^*\theta$, where $j_\epsilon : M_\epsilon \hookrightarrow V$).

By a result in [26]

$$g(X,Y) = -\frac{n+1}{\varphi} g_\theta(X,Y), \quad X,Y \in H(\mathcal{F}). \tag{2.59}$$

$$g(X,T) = 0, \quad g(X,N) = 0, \quad X \in H(\mathcal{F}), \tag{2.60}$$

$$g(T,N) = 0, \quad g(T,T) = g(N,N) = \frac{n+1}{\varphi}\left(\frac{1}{\varphi} - r\right). \tag{2.61}$$

In particular $1 - r\varphi > 0$ everywhere in Ω. We shall need the following basic result

Theorem 2.38 (C.R. Graham & J.M. Lee, [151]) *There is a unique linear connection ∇ on V such that*

i. *$T_{1,0}(\mathcal{F})$ is parallel with respect to ∇,*
ii. *$\nabla G_\theta = 0$, $\nabla T = 0$, $\nabla N = 0$, and*
iii. *the torsion T_∇ of ∇ is pure i.e*

$$T_\nabla(Z,W) = 0, \quad T_\nabla(Z,\overline{W}) = 2iL_\theta(Z,\overline{W})T, \tag{2.62}$$

$$T_\nabla(N,W) = rW + i\tau(W), \tag{2.63}$$

for any $Z, W \in T_{1,0}(\mathcal{F})$, and

$$\tau(T_{1,0}(\mathcal{F})) \subseteq T_{0,1}(\mathcal{F}), \tag{2.64}$$

$$\tau(N) = -J\nabla^H r - 2rT. \tag{2.65}$$

where $\tau(X) = T_\nabla(T,X)$ for any $X \in T(\mathcal{F})$. Also $\nabla^H r$ is defined by $\nabla^H r = \pi_H \nabla r$ and $g_\theta(\nabla r, X) = X(r)$ for any $X \in T(\mathcal{F})$, where $\pi_H : T(\mathcal{F}) \to H(\mathcal{F})$ is the projection associated to the direct sum decomposition $T(\mathcal{F}) = H(\mathcal{F}) \oplus \mathbb{R}T$.

The unique linear connection ∇ furnished by Theorem 2.38 is referred to as the *Graham-Lee connection* of (V,φ). While Theorem 2.38 belongs to [151] its statement in the form above (together with a coordinate-free proof) is given in Appendix A of [26]. ∇ has the remarkable property that its pointwise restriction to each leaf M_ϵ of $\mathcal{F}$ is precisely the Tanaka-Webster connection of the leaf.

2.9.2. The Levi-Civita Connection of the Bergman Metric

Using (2.59)–(2.61) one may relate the Levi-Civita connection ∇^g of (V,g) to the Graham-Lee connection ∇. Precisely (cf. [26])

Theorem 2.39 *Let $\Omega \subset \mathbb{C}^n$ be a smoothly bounded strictly pseudoconvex domain, $K(z,\zeta)$ its Bergman kernel, and $\varphi(z) = -K(z,z)^{-1/(n+1)}$. Then the*

Levi-Civita connection of the Bergman metric and the Graham-Lee connection of (Ω, φ) *are related by*

$$\nabla^g_X Y = \nabla_X Y + \left\{ \frac{\varphi}{1-\varphi r} g_\theta(\tau X, Y) + g_\theta(X, \phi Y) \right\} T$$

$$- \left\{ g_\theta(X, Y) + \frac{\varphi}{1-\varphi r} g_\theta(X, \phi \tau Y) \right\} N, \tag{2.66}$$

$$\nabla^g_X T = \tau X - \left(\frac{1}{\varphi} - r \right) \phi X - \frac{\varphi}{2(1-r\varphi)} \{X(r)T + (\phi X)(r)N\}, \tag{2.67}$$

$$\nabla^g_X N = -\left(\frac{1}{\varphi} - r \right) X + \tau \phi X + \frac{\varphi}{2(1-r\varphi)} \{(\phi X)(r)\, T - X(r)\, N\}, \tag{2.68}$$

$$\nabla^g_T X = \nabla_T X - \left(\frac{1}{\varphi} - r \right) \phi X - \frac{\varphi}{2(1-r\varphi)} \{X(r)T + (\phi X)(r)N\}, \tag{2.69}$$

$$\nabla^g_N X = \nabla_N X - \frac{1}{\varphi} X + \frac{\varphi}{2(1-r\varphi)} \{(\phi X)(r)T - X(r)N\}, \tag{2.70}$$

$$\nabla^g_N T = -\frac{1}{2} \phi \nabla^H r - \frac{\varphi}{2(1-r\varphi)} \left\{ \left(N(r) + \frac{4}{\varphi^2} - \frac{2r}{\varphi} \right) T + T(r) N \right\}, \tag{2.71}$$

$$\nabla^g_T N = \frac{1}{2} \phi \nabla^H r - \frac{\varphi}{2(1-r\varphi)} \left\{ \left(N(r) + \frac{4}{\varphi^2} - \frac{6r}{\varphi} + 4r^2 \right) T + T(r) N \right\}, \tag{2.72}$$

$$\nabla^g_T T = -\frac{1}{2} \nabla^H r - \frac{\varphi}{2(1-r\varphi)} \left\{ T(r) T - \left(N(r) + \frac{4}{\varphi^2} - \frac{6r}{\varphi} + 4r^2 \right) N \right\}, \tag{2.73}$$

$$\nabla^g_N N = -\frac{1}{2} \nabla^H r + \frac{\varphi}{2(1-r\varphi)} \left\{ T(r) T - \left(N(r) + \frac{4}{\varphi^2} - \frac{2r}{\varphi} \right) N \right\}, \tag{2.74}$$

for any $X, Y \in H(\mathcal{F})$. *Here* $\phi : T(\mathcal{F}) \to T(\mathcal{F})$ *is the bundle morphism given by* $\phi(X) = JX$ *for any* $X \in H(\mathcal{F})$ *and* $\phi(T) = 0$.

2.9.3. Proof of Theorem 2.36

Let $\{W_\alpha : 1 \le \alpha \le n-1\}$ be a local orthonormal frame of $T_{1,0}(\mathcal{F})$ i.e., $G_\theta(W_\alpha, W_{\overline{\beta}}) = \delta_{\alpha\beta}$. Here $W_{\overline{\beta}} = \overline{W_\beta}$. Let us set $f = \varphi/(1 - r\varphi)$ for simplicity. Then

$$E_\alpha = \sqrt{-\frac{\varphi}{n+1}}\, W_\alpha, \quad E_n = \sqrt{\frac{2f\varphi}{n+1}}\, \xi,$$

is a local orthonormal frame of $T^{1,0}(V)$ i.e., $g(E_j, E_k) = \delta_{jk}$. Then

$$\Delta_g X = -\sum_{j=1}^{n} \left\{ \nabla^g_{E_j} \nabla^g_{E_{\overline{j}}} X + \nabla^g_{E_{\overline{j}}} \nabla^g_{E_j} X - \nabla^g_{\nabla^g_{E_j} E_{\overline{j}}} X - \nabla^g_{\nabla^g_{E_{\overline{j}}} E_j} X \right\}$$

$$= \frac{\varphi}{n+1} \sum_{\alpha=1}^{n-1} \left\{ \nabla^g_{W_\alpha} \nabla^g_{W_{\overline{\alpha}}} X + \nabla^g_{W_{\overline{\alpha}}} \nabla^g_{W_\alpha} X - \nabla^g_{\nabla^g_{W_\alpha} W_{\overline{\alpha}}} X - \nabla^g_{\nabla^g_{W_{\overline{\alpha}}} W_\alpha} X \right\}$$

$$+ \frac{2f\varphi}{n+1} \left\{ \nabla^g_\xi \nabla^g_{\overline{\xi}} X + \nabla^g_{\overline{\xi}} \nabla^g_\xi X - \nabla^g_{\nabla^g_\xi \overline{\xi}} X - \nabla^g_{\nabla^g_{\overline{\xi}} \xi} X \right\}.$$

Let us assume from now on that $X \in H(\mathcal{F})$. Then (by (2.66))

$$\nabla^g_{W_\alpha} X = \nabla_{W_\alpha} X + \{f\, g_\theta(\tau W_\alpha, X) - i g_\theta(W_\alpha, X)\} T$$
$$- \{g_\theta(W_\alpha, X) - if\, g_\theta(W_\alpha, \tau X)\} N.$$

Consequently (again by (2.66))

$$\nabla^g_{W_{\overline{\alpha}}} \nabla^g_{W_\alpha} X = \nabla_{W_{\overline{\alpha}}} \nabla_{W_\alpha} X$$
$$+ \{f\, g_\theta(\tau W_\alpha, X) - i g_\theta(W_\alpha, X)\} \left\{ \tau W_{\overline{\alpha}} + \frac{i}{f} W_{\overline{\alpha}} \right\}$$
$$+ \{g_\theta(W_\alpha, X) - if\, g_\theta(W_\alpha, \tau X)\} \left\{ \frac{1}{f} W_{\overline{\alpha}} + i\tau W_{\overline{\alpha}} \right\}$$
$$+ \{f\, W_{\overline{\alpha}}(g_\theta(\tau W_\alpha, X)) + f\, g_\theta(\tau W_{\overline{\alpha}}, \nabla_{W_\alpha} X)$$
$$- i W_{\overline{\alpha}}(g_\theta(W_\alpha, X)) + i g_\theta(W_{\overline{\alpha}}, \nabla_{W_\alpha} X)$$
$$- \frac{f}{2} r_{\overline{\alpha}} [f\, g_\theta(\tau W_\alpha, X) - i g_\theta(W_\alpha, X)]$$
$$+ \frac{if}{2} r_{\overline{\alpha}} [g_\theta(W_\alpha, X) - if\, g_\theta(W_\alpha, \tau X)]\} T$$
$$+ \{if\, W_{\overline{\alpha}}(g_\theta(W_\alpha, \tau X) - if\, g_\theta(W_{\overline{\alpha}}, \tau \nabla_{W_\alpha} X)$$
$$- W_{\overline{\alpha}}(g_\theta(W_\alpha, X)) - g_\theta(W_{\overline{\alpha}}, \nabla_{W_\alpha} X)$$

$$+\frac{if}{2} r_{\overline{\alpha}} [f g_\theta(\tau W_\alpha, X) - i g_\theta(W_\alpha, X)]$$

$$+\frac{f}{2} r_{\overline{\alpha}} [g_\theta(W_\alpha, X) - if g_\theta(W_\alpha, \tau X)]\}N \tag{2.75}$$

where $r_\alpha = W_\alpha(r)$ and $r_{\overline{\alpha}} = \overline{r_\alpha}$. Moreover

$$\nabla^g_{W_{\overline{\alpha}}} W_\alpha = \nabla_{W_{\overline{\alpha}}} W_\alpha + iT - N \tag{2.76}$$

hence

$$\nabla^g_{\nabla^g_{W_{\overline{\alpha}}} W_\alpha} X = \nabla_{\nabla_{W_{\overline{\alpha}}} W_\alpha} X + i\nabla_T X - \frac{i}{f}\phi X - \nabla_N X + \frac{1}{\varphi} X \tag{2.77}$$

$$+ \{f g_\theta(\tau \nabla_{\overline{\alpha}} W_\alpha, X) - i g_\theta(\nabla_{W_{\overline{\alpha}}} W_\alpha, X) - \frac{if}{2} X(r) - \frac{f}{2}(\phi X)(r)\}T$$

$$- \{g_\theta(\nabla_{W_{\overline{\alpha}}} W_\alpha, X) - if g_\theta(\nabla_{W_{\overline{\alpha}}} W_\alpha, \tau X) + \frac{if}{2}(\phi X)(r) - \frac{f}{2} X(r)\}N.$$

On the other hand (by (2.69)–(2.70))

$$\nabla^g_\xi X = \frac{1}{2}\Big\{\nabla_N X - \frac{1}{\varphi} X - i\nabla_T X + \frac{i}{f}\phi X$$

$$+ \frac{f}{2}[(\phi X)(r) + iX(r)](T + iN)\Big\} \tag{2.78}$$

allowing one to compute $\nabla^g_N \nabla^g_\xi X$. First let us observe that

$$N\left(\frac{1}{\varphi}\right) = -\frac{2}{\varphi^2}, \quad N\left(\frac{1}{f}\right) = -\frac{2}{\varphi^2} - N(r),$$

$$N(f) = \frac{2}{(1-\varphi r)^2} + f^2 N(r).$$

Moreover the following identities

$$\nabla^g_N \nabla_N X = \nabla_N \nabla_N X - \frac{1}{\varphi}\nabla_N X + \frac{f}{2}\{(\phi \nabla_N X)(r) T - (\nabla_N X)(r) N\}$$

$$\nabla^g_N \nabla_T X = \nabla_N \nabla_T X - \frac{1}{\varphi}\nabla_T X + \frac{f}{2}\{(\phi \nabla_T X)(r) T - (\nabla_T X)(r) N\}$$

$$\nabla^g_N \phi X = \nabla_N \phi X - \frac{1}{\varphi}\phi X - \frac{f}{2}\{X(r) T + (\phi X)(r) N\}$$

yield

$$
\begin{aligned}
2\nabla^g_N \nabla^g_\xi X = {} & \nabla_N \nabla_N X - \frac{2}{\varphi}\nabla_N X + \frac{3}{\varphi^2}X - i\nabla_N \nabla_T X + \frac{i}{\varphi}\nabla_T X \\
& - i\left(N(r) + \frac{2}{\varphi^2} + \frac{1-\varphi r}{\varphi^2}\right)\phi X + \frac{i}{f}\nabla_N \phi X \\
& - \frac{f}{4}F\,\phi\nabla^H r - \frac{if}{4}F\,\nabla^H r \\
& + \left\{\frac{f}{2}(\phi\nabla_N X)(r) - \frac{1}{2(1-\varphi r)}(\phi X)(r) - \frac{if}{2}(\phi\nabla_T X)(r)\right. \\
& \left. - \frac{i}{2}X(r) + \frac{1}{2}N(fF) - \frac{f^2}{4}FH + \frac{if^2}{4}F\,T(r)\right\} T \\
& + \left\{-\frac{f}{2}(\nabla_N X)(r) + \frac{1}{2(1-\varphi r)}X(r) + \frac{if}{2}(\nabla_T X)(r)\right. \\
& \left. - \frac{i}{2}(\phi X)(r) + \frac{i}{2}N(fF) - \frac{f^2}{4}F\,T(r) - \frac{if^2}{4}FH\right\} N \qquad (2.79)
\end{aligned}
$$

where the functions $F, G, H \in C^\infty(V)$ are given by

$$
F = (\phi X)(r) + iX(r), \quad G = N(r) + \frac{4}{\varphi^2} - \frac{6r}{\varphi} + 4r^2, \quad H = N(r) + \frac{4}{\varphi^2} - \frac{2r}{\varphi}.
$$

Similarly, using

$$
T(f) = f^2 T(r), \quad T\left(\frac{1}{f}\right) = -T(r),
$$

and the identities

$$
\begin{aligned}
\nabla^g_T \nabla_N X &= \nabla_T \nabla_N X - \frac{1}{f}\phi\nabla_N X - \frac{f}{2}\{(\nabla_N X)(r)T + (\phi\nabla_N X)(r)N\}, \\
\nabla^g_T \nabla_T X &= \nabla_T \nabla_T X - \frac{1}{f}\phi\nabla_T X - \frac{f}{2}\{(\nabla_T X)(r)T + (\phi\nabla_T X)(r)N\}N, \\
\nabla^g_T \phi X &= \nabla_T \phi X + \frac{1}{f}X - \frac{f}{2}\{(\phi X)(r)T - X(r)N\}, \\
\nabla^g_T T &= -\frac{1}{2}\nabla^H r - \frac{f}{2}\{T(r)T - GN\}, \\
\nabla^g_T N &= \frac{1}{2}\phi\nabla^H r - \frac{f}{2}\{GT + T(r)N\},
\end{aligned}
$$

we find

$$
\begin{aligned}
2\nabla^g_T\nabla^g_\xi X = {} & \nabla_T\nabla_N X - \frac{1}{f}\phi\nabla_N X - \frac{1}{\varphi}\nabla_T X + \frac{1-\varphi r}{\varphi^2}\phi X \\
& - i\nabla_T\nabla_T X + \frac{i}{f}\phi\nabla_T X - iT(r)\phi X + \frac{i}{f}\nabla_T\phi X + \frac{i}{f^2}X \\
& - \frac{f}{4}F\nabla^H r + \frac{if}{4}F\phi\nabla^H r \\
& + \left\{ -\frac{f}{2}(\nabla_N X)(r) + \frac{1}{2(1-\varphi r)}X(r) + \frac{if}{2}(\nabla_T X)(r) \right. \\
& \left. - \frac{i}{2}(\phi X)(r) + \frac{1}{2}T(fF) - \frac{f^2}{4}F\,T(r) - \frac{if^2}{4}FG \right\} T \\
& + \left\{ -\frac{f}{2}(\phi\nabla_N X)(r) + \frac{1}{2(1-\varphi r)}(\phi X)(r) + \frac{if}{2}(\phi\nabla_T X)(r) \right. \\
& \left. + \frac{i}{2}X(r) + \frac{f^2}{4}FG + \frac{i}{2}T(fF) - \frac{if^2}{4}FT(r) \right\} N. \qquad (2.80)
\end{aligned}
$$

The identities (2.79)–(2.80) yield

$$
\begin{aligned}
4\nabla^g_{\bar\xi}\nabla^g_\xi X = {} & \nabla_N\nabla_N X - \frac{2}{\varphi}\nabla_N X + \frac{3}{\varphi^2}X - i\nabla_N\nabla_T X + \frac{i}{\varphi}\nabla_T X \\
& - i\left(N(r) + \frac{3}{\varphi^2} - \frac{r}{\varphi}\right)\phi X + \frac{i}{f}\nabla_N\phi X - \frac{f}{4}F\phi\nabla^H r - \frac{if}{4}F\nabla^H r \\
& + i\nabla_T\nabla_N X - \frac{i}{f}\phi\nabla_N X - \frac{i}{\varphi}\nabla_T X + \frac{i(1-\varphi r)}{\varphi^2}\phi X \\
& + \nabla_T\nabla_T X - \frac{1}{f}\phi\nabla_T X + T(r)\phi X - \frac{1}{f}\nabla_T\phi X \\
& - \frac{1}{f^2}X - \frac{if}{4}F\nabla^H r - \frac{f}{4}F\phi\nabla^H r \\
& + \left\{ \frac{f}{2}(\phi\nabla_N X)(r) - \frac{1}{2(1-\varphi r)}(\phi X)(r) - \frac{if}{2}(\phi\nabla_T X)(r) \right. \\
& - \frac{i}{2}X(r) + \frac{1}{2}N(fF) - \frac{f^2}{4}FH + \frac{if^2}{4}FT(r) \\
& - \frac{if}{2}(\nabla_N X)(r) + \frac{i}{2(1-\varphi r)}X(r) - \frac{f}{2}(\nabla_T X)(r)
\end{aligned}
$$

$$+\frac{1}{2}(\phi X)(r)+\frac{i}{2}T(fF)-\frac{if^2}{4}FT(r)+\frac{f^2}{4}FG\Big\}\,T$$
$$+\Big\{-\frac{f}{2}(\nabla_N X)(r)+\frac{1}{2(1-\varphi r)}X(r)+\frac{if}{2}(\nabla_T X)(r)$$
$$-\frac{i}{2}(\phi X)(r)+\frac{i}{2}N(fF)-\frac{f^2}{4}FT(r)-\frac{if^2}{4}FH$$
$$-\frac{if}{2}(\phi\nabla_N X)(r)+\frac{i}{2(1-\varphi r)}(\phi X)(r)-\frac{f}{2}(\phi\nabla_T X)(r)$$
$$-\frac{1}{2}X(r)+\frac{if^2}{4}FG-\frac{1}{2}T(fF)+\frac{f^2}{4}FT(r)\Big\}\,N. \tag{2.81}$$

A calculation based on (2.75) and (2.77) leads to

$$\sum_{\alpha=1}^{n-1}\Big\{\nabla^g_{W_{\overline{\alpha}}}\nabla^g_{W_\alpha}X+\nabla^g_{W_\alpha}\nabla^g_{W_{\overline{\alpha}}}X-\nabla^g_{\nabla^g_{W_{\overline{\alpha}}}W_\alpha}X-\nabla^g_{\nabla^g_{W_\alpha}W_{\overline{\alpha}}}X\Big\}$$
$$=-\Delta_b X+2f\tau^2X+\frac{2}{f}X+2(n-1)(\nabla_N X-\frac{1}{\varphi}X)$$
$$+\big\{2\,\mathrm{div}(\phi X)+2f\,\mathrm{div}(\tau X)$$
$$-f\,\mathrm{trace}[Y\mapsto(\nabla_Y\tau)X]+(n-2)f(\phi X)(r)\big\}\,T$$
$$-\big\{2\,\mathrm{div}(X)+2f\,\mathrm{div}(\phi\tau X)$$
$$+(n-2)f\,X(r)-f\,\mathrm{trace}[Y\mapsto\phi(\nabla_Y\tau)X]\big\}\,N \tag{2.82}$$

where Δ_b is the second order differential operator defined (locally) by

$$\Delta_b X=-\sum_{\alpha=1}^{n-1}\{\nabla_{W_\alpha}\nabla_{W_{\overline{\alpha}}}X+\nabla_{W_{\overline{\alpha}}}\nabla_{W_\alpha}X-\nabla_{\nabla_{W_\alpha}W_{\overline{\alpha}}}X-\nabla_{\nabla_{W_{\overline{\alpha}}}W_\alpha}X\}.$$

Also the divergence operator in (2.82) above is given by

$$\mathrm{div}(X)=\mathrm{trace}\{Y\in T(\mathcal{F})\mapsto\nabla_Y X\}.$$

Moreover the identity

$$\nabla^g_\xi\xi=-\frac{1}{4}\nabla^H r+\frac{r}{2}(iT-N),$$

yields

$$\nabla^g_{\nabla^g_\xi\xi}X=-\frac{1}{4}\nabla_{\nabla^H r}X-\frac{r}{2}(\nabla_N X-i\nabla_T X)+\frac{r}{2}\left(\frac{1}{\varphi}X-\frac{i}{f}\phi X\right)$$

$$-\left\{\frac{f}{4}(\tau X)(r) + \frac{1+rf}{4}(\phi X)(r) - \frac{ifr}{4}X(r)\right\} T$$
$$+\left\{\frac{f}{4}(\phi\tau X)(r) + \frac{1+rf}{4}X(r) - \frac{ifr}{4}(\phi X)(r)\right\} N$$

hence

$$\nabla^g_{\nabla^g_\xi \overline{\xi}} X + \nabla^g_{\nabla^g_{\overline{\xi}} \xi} X = -\frac{1}{2}\nabla_{\nabla^H r} X - r\nabla_N X + \frac{r}{\varphi}X \tag{2.83}$$
$$-\left\{\frac{f}{2}(\tau X)(r) + \frac{1+rf}{2}(\phi X)(r)\right\} T + \left\{\frac{f}{2}(\phi\tau X)(r) + \frac{1+rf}{2}X(r)\right\} N.$$

Using (2.81) and (2.83) and

$$N(f(F+\overline{F})) = 2fN((\phi X)(r)) + \left[\frac{4}{(1-\varphi r)^2} + 2f^2 N(r)\right](\phi X)(r),$$
$$iT(f(F-\overline{F})) = -2fT(X(r)),$$
$$iN(f(F-\overline{F})) = -2fN(X(r)) - \left[\frac{4}{(1-\varphi r)^2} + 2f^2 N(r)\right]X(r),$$

we derive

$$\nabla^g_\xi \nabla^g_{\overline{\xi}} X + \nabla^g_{\overline{\xi}} \nabla^g_\xi X - \nabla^g_{\nabla^g_\xi \overline{\xi}} X - \nabla^g_{\nabla^g_{\overline{\xi}} \xi} X = -\frac{1}{f}\phi\nabla_T X$$
$$+\frac{1}{2}(\nabla_T\nabla_T X + \nabla_N\nabla_N X) + \left(r - \frac{1}{\varphi}\right)\nabla_N X$$
$$+\left(\frac{3}{2\varphi^2} - \frac{1}{2f^2} - \frac{r}{\varphi}\right)X + \frac{1}{2}T(r)\phi X$$
$$+\frac{1}{2}\nabla_{\nabla^H r}X + \frac{f}{4}X(r)\nabla^H r - \frac{f}{4}(\phi X)(r)\phi\nabla^H r$$
$$+\left\{-\frac{f}{4}(\nabla_T X)(r) + \frac{f}{4}(\phi\nabla_N X)(r) - \frac{f}{4}T(X(r)) + \frac{f}{2}(\tau X)(r)\right.$$
$$\left.+\frac{f}{4}N((\phi X)(r)) + \left[\frac{1}{2(1-\varphi r)^2} + \frac{f^2}{4}N(r) + \frac{1-fr}{4}\right](\phi X)(r)\right\} T$$
$$+\left\{-\frac{f}{4}(\phi\nabla_T X)(r) - \frac{f}{4}(\nabla_N X)(r) - \frac{f}{4}T((\phi X)(r)) - \frac{f}{4}N(X(r))\right.$$
$$\left.-\frac{f}{2}(\phi\tau X)(r) - \left[\frac{1}{2(1-\varphi r)^2} + \frac{f^2}{4}N(r) + \frac{2-fr}{4}\right]X(r)\right\} N. \tag{2.84}$$

Next we compute

$$|\nabla^g X|^2 = 2\sum_{j=1}^{n} g\left(\nabla^g_{E_j} X, \nabla^g_{E_{\bar{j}}} X\right)$$
$$= -\frac{2\varphi}{n+1}\sum_{\alpha=1}^{n-1} g\left(\nabla_{W_\alpha} X, \nabla^g_{W_{\bar{\alpha}}} X\right) + \frac{4f\varphi}{n+1} g\left(\nabla^g_{\xi}, X, \nabla^g_{\bar{\xi}} X\right)$$

and obtain (by (2.59)–(2.61))

$$\begin{aligned}
|\nabla^g X|^2 = {} & |\nabla^H X|^2_\theta \\
& -\frac{2}{f}\left[f^2|\tau X|^2_\theta + |X|^2_\theta\right] + \frac{f^2}{2}\left[X(r)^2 + (\phi X)(r)^2\right] \\
& -f\left[\left(\frac{1}{\varphi^2} + \frac{1}{f^2}\right)|X|^2_\theta + |\nabla_T X|^2_\theta + |\nabla_N X|^2_\theta \right. \\
& \left. -\frac{2}{f} g_\theta(\nabla_T X, \phi X) - \frac{2}{\varphi} g_\theta(\nabla_N X, X)\right]
\end{aligned} \tag{2.85}$$

where $|X|_\theta = g_\theta(X,X)^{1/2}$. Finally we substitute from (2.82) and (2.84)–(2.85) into (2.54) and use the uniqueness in the direct sum decomposition $T(V) = H(\mathcal{F}) \oplus \mathbb{R}T \oplus \mathbb{R}N$. The $H(\mathcal{F})$-component of (2.54) reads

$$\begin{aligned}
& -\Delta_b X + 2f\tau^2 X + \frac{2}{f} X + 2(n-1)\left(\nabla_N X - \frac{1}{\varphi} X\right) \\
& -f\left\{\nabla_T \nabla_T X + \nabla_N \nabla_N X - \frac{2}{f}\phi \nabla_T X + \frac{2}{f}\nabla_N X \right. \\
& + \left(\frac{3}{\varphi^2} - \frac{1}{f^2} - \frac{2r}{\varphi}\right) X + T(r)\phi X \\
& \left. + \nabla_{\nabla^H r} X + \frac{f}{2} X(r)\nabla^H r - \frac{f}{2}(\phi X)(r)\phi \nabla^H r\right\} \\
& - \frac{n+1}{\varphi}\left\{|\nabla^H X|^2_\theta - \frac{2}{f}\left[f^2|\tau X|^2_\theta + |X|^2_\theta\right] + \frac{f^2}{2}\left[X(r)^2 + (\phi X)(r)^2\right] \right. \\
& -f\left[\left(\frac{1}{\varphi^2} + \frac{1}{f^2}\right)|X|^2_\theta + |\nabla_T X|^2_\theta + |\nabla_N X|^2_\theta \right. \\
& \left.\left. -\frac{2}{f} g_\theta(\nabla_T X, \phi X) - \frac{2}{\varphi} g_\theta(\nabla_N X, X)\right]\right\} X = 0.
\end{aligned} \tag{2.86}$$

Note that $f = O(\varphi)$ as $\varphi \to 0$. Multiplying (2.86) by φf gives

$$\left(3 + \frac{f^2}{\varphi^2}\right) |X|^2_\theta \, X = O(\varphi)$$

and then (by passing to the limit with $z \to \partial\Omega$) it follows that $X = 0$ on $\partial\Omega$. ∎

2.9.4. Proof of Theorem 2.37

With the notations in the previous section

$$\Delta(uT) = (\Delta_g u)T + u\Delta T - 2\nabla^g_{\nabla^g u} T. \tag{2.87}$$

Let us compute ΔT. First (by (2.67))

$$\nabla^g_{W_\alpha} T = \tau W_\alpha - \frac{i}{f} W_\alpha - \frac{f}{2} r_\alpha (T + iN).$$

Consequently (as $W_\alpha(f) = 0$)

$$\nabla^g_{W_{\overline{\alpha}}} \nabla^g_{W_\alpha} T = \nabla^g_{W_{\overline{\alpha}}}(\tau W_\alpha - \frac{i}{f} W_\alpha)$$
$$- \frac{f}{2} W_{\overline{\alpha}}(r_\alpha)(T + iN) - \frac{f}{2}\left(\nabla^g_{W_{\overline{\alpha}}} T + i\nabla^g_{W_{\overline{\alpha}}} N\right)$$

and the identities

$$\nabla^g_{W_{\overline{\alpha}}}(\tau W_\alpha - \frac{i}{f}\nabla_{W_{\overline{\alpha}}} W_\alpha) = \nabla_{W_{\overline{\alpha}}}\tau W_\alpha - \frac{i}{f}\nabla_{W_{\overline{\alpha}}} W_\alpha$$
$$+ \left\{ f g_\theta(\tau^2 W_{\overline{\alpha}}, W_\alpha) + \frac{1}{f} \right\} T - i \left\{ f g_\theta(W_{\overline{\alpha}}, \tau^2 W_\alpha) - \frac{1}{f} \right\} N,$$

$$\nabla^g_{W_\alpha} N = -\frac{1}{f} W_\alpha + i\tau W_\alpha + \frac{f}{2} r_\alpha (iT - N),$$

lead to

$$\sum_\alpha \nabla^g_{W_{\overline{\alpha}}} \nabla^g_{W_\alpha} T = \sum_\alpha \left\{ \nabla_{W_{\overline{\alpha}}}\tau W_\alpha - \frac{i}{f}\nabla_{W_{\overline{\alpha}}} W_\alpha - f r_\alpha \tau W_{\overline{\alpha}} \right\}$$
$$+ \left\{ \frac{f}{2}\operatorname{trace}(\tau^2) + \frac{n-1}{f} - \frac{f}{2}\sum_\alpha W_{\overline{\alpha}}(r_\alpha) \right\} T$$
$$- i \left\{ \frac{f}{2}\operatorname{trace}(\tau^2) - \frac{n-1}{f} + \frac{f}{2}\sum_\alpha W_{\overline{\alpha}}(r_\alpha) \right\} N$$

hence

$$\sum_\alpha \{\nabla^g_{W_\alpha}\nabla^g_{W_{\overline{\alpha}}} T + \nabla^g_{W_{\overline{\alpha}}}\nabla^g_{W_\alpha} T\} = \sum_\alpha \{\nabla_{W_\alpha}\tau W_{\overline{\alpha}} + \nabla_{W_{\overline{\alpha}}}\tau W_\alpha$$
$$+ \frac{i}{f}\left(\nabla_{W_\alpha} W_{\overline{\alpha}} - \nabla_{W_{\overline{\alpha}}} W_\alpha\right) - f\left(r_{\overline{\alpha}}\,\tau W_\alpha - r_\alpha\,\tau W_{\overline{\alpha}}\right)\}$$
$$+ \{f \operatorname{trace}(\tau^2) + \frac{2(n-1)}{f} - \frac{f}{2}\sum_\alpha (W_\alpha r_{\overline{\alpha}} + W_{\overline{\alpha}} r_\alpha)\} T$$
$$+ \frac{if}{2}\sum_\alpha (W_\alpha r_{\overline{\alpha}} - W_{\overline{\alpha}} r_\alpha) N. \tag{2.88}$$

A similar calculation (the details of which we omit) shows that

$$\nabla^g_{\nabla^g_{W_{\overline{\alpha}}} W_\alpha} T = \tau \nabla_{W_{\overline{\alpha}}} W_\alpha - \frac{i}{f}\nabla_{W_{\overline{\alpha}}} W_\alpha + \frac{1}{2}\phi \nabla^H r - \frac{i}{2}\nabla^H r$$
$$- \frac{f}{2}\left\{(\nabla_{W_{\overline{\alpha}}} W_\alpha)(r) - iT(r) - H\right\} T$$
$$- \frac{f}{2}\left\{i(\nabla_{W_{\overline{\alpha}}} W_\alpha)(r) - iG - T(r)\right\} N. \tag{2.89}$$

Note that

$$\sum_\alpha \left\{W_\alpha r_{\overline{\alpha}} - W_{\overline{\alpha}} r_\alpha - (\nabla_{W_\alpha} W_{\overline{\alpha}})(r) + (\nabla_{W_{\overline{\alpha}}} W_\alpha)(r)\right\} = -2i\,T(r).$$

Therefore (by (2.88)–(2.89))

$$\sum_\alpha \left\{\nabla^g_{W_\alpha}\nabla^g_{W_{\overline{\alpha}}} T + \nabla^g_{W_{\overline{\alpha}}}\nabla^g_{W_{\overline{\alpha}}}\nabla^g_{W_\alpha} T\right.$$
$$\left. - \nabla^g_{\nabla^g_{W_\alpha} W_{\overline{\alpha}}} T - \nabla^g_{\nabla^g_{W_{\overline{\alpha}}} W_\alpha} T\right\} = -(n-2) f T(r) N$$
$$+ \operatorname{trace}_{G_\theta}(\nabla\tau) - (n-1)\phi \nabla^H r - f\,\tau \nabla^H r$$
$$+ \left\{f \operatorname{trace}(\tau^2) + \frac{2(n-1)}{f} - f(n-1)H - \frac{f}{2}\Delta_b r\right\} T. \tag{2.90}$$

Next (by (2.71)–(2.73))

$$2\nabla^g_\xi T = -\frac{1}{2}\phi \nabla^H r + \frac{i}{2}\nabla^H r - \frac{f}{2}\{H - iT(r)\} T - \frac{f}{2}\{T(r) + iG\}$$

and we may compute $\nabla^g_{\overline{\xi}} \nabla^g_{\xi} T$. Indeed the identities

$$
\begin{aligned}
(\nabla^H r)(r) &= |\nabla^H r|^2_\theta, \quad (\phi \nabla^H r)(r) = 0, \\
\nabla^g_N \phi \nabla^H r &= \phi \nabla_N \nabla^H r - \frac{1}{\varphi} \phi \nabla^H r - \frac{f}{2} |\nabla^H r|^2_\theta T, \\
\nabla^g_N \nabla^H r &= \nabla_N \nabla^H r - \frac{1}{\varphi} \nabla^H r - \frac{f}{2} |\nabla^H r|^2_\theta N, \\
&\quad - \frac{f}{2}(H - iT(r)) \nabla^g_N T - \frac{f}{2}(T(r) + iG) \nabla^g_N N \\
&= \frac{f}{4}(H - iT(r)) \phi \nabla^H r + \frac{f}{4}(T(r) + iG) \nabla^H r \\
&\quad + \frac{f^2}{4} \{H(h - iT(r)) - T(r)(T(r) + iG)\} T \\
&\quad + \frac{f^2}{4} \{T(r)(H - iT(r)) + H(T(r) + iG)\} N,
\end{aligned}
$$

imply that

$$
\begin{aligned}
2\nabla^g_N \nabla^g_{\xi} T = &-\frac{1}{2} \phi \nabla_N \nabla^H r + \frac{i}{2} \nabla_N \nabla^H r \\
&+ \left[\frac{1}{2\varphi} + \frac{f}{4}(H - iT(r))\right] \phi \nabla^H r + \left[\frac{f}{4}(T(r) + iG) - \frac{i}{2\varphi}\right] \nabla^H r \\
&+ \left\{ \frac{f}{4} |\nabla^H r|^2_\theta - N\left(\frac{f}{2}\{H - iT(r)\}\right) \right. \\
&\left. + \frac{f^2}{4} [H(H - iT(r)) - T(r)(T(r) + iG)] \right\} T \\
&+ \left\{ -\frac{if}{4} |\nabla^H r|^2_\theta - N\left(\frac{f}{2}\{T(r) + iG\}\right) \right. \\
&\left. + \frac{f^2}{4} [T(r)(H - iT(r)) + H(T(r) + iG)] \right\} N.
\end{aligned} \tag{2.91}
$$

Similarly the identities

$$
\begin{aligned}
\nabla^g_T \phi \nabla^H r = \nabla_T \phi \nabla^H r + \frac{1}{f} \nabla^H r + \frac{f}{2} |\nabla^H r|^2_\theta N, \\
- \frac{if}{2}(H - iT(r)) \nabla^g_T T - \frac{if}{2}(T(r) + i) \nabla^g_T N
\end{aligned}
$$

$$= \frac{if}{4}(H - iT(r))\nabla^H r - \frac{if}{4}(T(r) + iG)\phi\nabla^H r$$
$$+ \frac{if^2}{4}[T(r)(H - iT(r)) + G(T(r) + iG)]\,T$$
$$- \frac{if^2}{4}[G(H - iT(r)) - T(r)(T(r) + iG)]\,N,$$

imply

$$2i\nabla^g_T\nabla^g_\xi T = -\frac{i}{2}\nabla_T\phi\nabla^H r - \frac{1}{2}\nabla_T\nabla^H r$$
$$+ \left[\frac{1}{2f} - \frac{if}{4}(T(r) + iG)\right]\phi\nabla^H r + \left[\frac{if}{4}(H - iT(r)) - \frac{i}{2f}\right]\nabla^H r$$
$$+ \left\{\frac{f}{4}|\nabla^H r|^2_\theta - iT\left(\frac{f}{2}\{H - iT(r)\}\right)\right.$$
$$\left. + \frac{if^2}{4}[T(r)(H - iT(r)) + G(T(r) + iG)]\right\}T$$
$$- \left\{\frac{if}{4}|\nabla^H r|^2_\theta + iT\left(\frac{f}{2}\{T(r) + iG\}\right)\right.$$
$$\left. + \frac{if^2}{4}[G(H - iT(r)) - T(r)(T(r) + iG)]\right\}N. \tag{2.92}$$

Then (2.91)–(2.84) yield

$$4\left(\nabla^g_\xi\nabla^g_{\overline{\xi}}T + \nabla^g_{\overline{\xi}}\nabla^g_\xi T\right)$$
$$= -\phi\nabla_N\nabla^H r - \nabla_T\nabla^H r + f\,T(r)\,\nabla^H r + \left\{\frac{1}{\varphi} + \frac{f}{2}H + \frac{1}{f} + \frac{f}{2}G\right\}\phi\nabla^H r$$
$$+ \left\{f|\nabla^H r|^2_\theta + \frac{f^2}{2}(H^2 - G^2) - N(fH) - T(fT(r))\right\}T$$
$$+ \left\{f^2HT(r) - f^2GT(r) - N(fT(r)) + T(fG)\right\}N. \tag{2.93}$$

Similar calculations that we omit lead to

$$\nabla^g_{\nabla^g_\xi\overline{\xi}}T + \nabla^g_{\nabla^g_{\overline{\xi}}\xi}T = -\frac{1}{2}\tau\nabla^H r + \frac{1}{2}\left(\frac{1}{f} + r\right)\phi\nabla^H r$$
$$+ \frac{f}{2}\left\{\frac{1}{2}|\nabla^H r|^2_\theta + rH\right\}T + \frac{rf}{2}T(r)N. \tag{2.94}$$

Summing up (by (2.90) and (2.93)–(2.94))

$$\Delta T = \frac{\varphi}{n+1}\left\{\mathrm{trace}_{G_\theta}(\nabla\tau) - (n-1)\phi\nabla^H r - f\tau\nabla^H r\right\}$$
$$+ \frac{f\varphi}{n+1}\left\{\frac{1}{2}\phi\nabla_N\nabla^H r + \frac{1}{2}\nabla_T\nabla^H r - \frac{f}{2}T(r)\nabla^H r - \tau\nabla^H r\right\}$$
$$+ \frac{\varphi}{n+1}\left\{\left[f\,\mathrm{trace}(\tau^2) + \frac{2(n-1)}{f} - f(n-1)H - \frac{f}{2}\Delta_b r\right]\right.$$
$$\left. + f\left[\frac{f^2}{4}(H^2 - G^2) - \frac{1}{2}N(fH) - \frac{1}{2}T(fT(r)) - fr H\right]\right\} T$$
$$+ \frac{\varphi}{n+1}\left\{(n-2)fT(r) - f\left[\frac{f^2}{2}T(r)(H-G)\right.\right.$$
$$\left.\left. - \frac{1}{2}N(fT(r)) + \frac{1}{2}T(fG) - rfT(r)\right]\right\} N. \tag{2.95}$$

Next the identities

$$\sum_\alpha g(\nabla^g_{W_\alpha}(uT), \nabla^g_{W_{\overline{\alpha}}}(uT)) = \frac{n+1}{2\varphi f}\left\{|\nabla^H u|^2_\theta\right.$$
$$\left. + u^2\left[\frac{f^2}{2}|\nabla^H r|^2_\theta - f\mathrm{trace}(\tau^2) - (n-1)\frac{2}{f}\right]\right\},$$
$$g(T, \nabla^g_\xi T) = -\frac{n+1}{4\varphi}(H - iT(r)),$$
$$g(\nabla^g_\xi(uT), \nabla^g_{\overline{\xi}}(uT)) = \frac{n+1}{\varphi f}|\xi(u)|^2$$
$$- \frac{n+1}{4\varphi}u\left[\xi(u)(H + iT(r)) - \overline{\xi}(u)(H - iT(r))\right]$$
$$+ \frac{(n+1)}{8\varphi}u^2\left[\frac{f}{2}\left(H^2 + G^2 + 2T(r)^2\right) - |\nabla^H r|^2_\theta\right],$$

furnish

$$|\nabla^g(uT)|^2 = \frac{1}{f}\left\{|\nabla^H u|^2_\theta\right.$$
$$\left. + u^2\left[\frac{f^2}{2}|\nabla^H r|^2_\theta - f\,\mathrm{trace}(\tau^2) - (n-1)\frac{2}{f}\right]\right\}$$

$$+f\left\{\frac{4(n+1)}{f}|\xi(u)|^2 - u(HN(u)+T(r)T(u))\right.$$
$$\left. + u^2\left[\frac{f}{4}\left(H^2+G^2+2T(r)^2\right) - \frac{1}{2}|\nabla^H r|_\theta^2\right]\right\}. \tag{2.96}$$

Similar straightforward calculations show that

$$\nabla^g_{\nabla^g u} T = -\frac{\varphi}{n+1}\left\{\tau\nabla^H u - \frac{1}{f}\phi\nabla^H u\right.$$
$$\left. +\frac{f}{2}\left[N(u)\phi\nabla^H r + T(u)\nabla^H r\right]\right\}$$
$$+\frac{\varphi f}{2(n+1)}\left\{(\nabla^H u)(r) - f[N(u)H + T(u)T(r)]\right\}T$$
$$+\frac{\varphi f}{2(n+1)}\left\{(\phi\nabla^H u)(r) - f[N(u)T(r) - T(u)G]\right\}N, \tag{2.97}$$

$$\Delta_g u = \frac{\varphi}{n+1}\{\Delta_b u + 2(n-1)N(u)\}$$
$$-\frac{\varphi f}{n+1}\left\{N^2(u) + T^2(u) + (\nabla^H u)(r) + 2rN(u)\right\}. \tag{2.98}$$

Finally we may use (2.87) and (2.95)–(2.98) to write the equation

$$\Delta(uT) - |\nabla^g(uT)|^2 uT = 0 \tag{2.99}$$

as

$$u\left\{\mathrm{trace}_{G_\theta}(\nabla\tau) - (n-1)\phi\nabla^H r - f\tau\nabla^H r\right. \tag{2.100}$$
$$+f\left[\frac{1}{2}\phi\nabla_N\nabla^H r + \frac{1}{2}\nabla_T\nabla^H r - \frac{f}{2}T(r)\nabla^H r - \tau\nabla^H r\right.$$
$$\left.\left. - \left(\frac{1}{2\varphi} + \frac{f}{4}(H+G) - \frac{1}{2f} - r\right)\phi\nabla^H r\right]\right\}$$
$$-\frac{2\varphi}{n+1}\left\{\tau\nabla^H u - \frac{1}{f}\phi\nabla^H u + \frac{f}{2}\left(N(u)\phi\nabla^H r + T(u)\nabla^H r\right)\right\} = 0$$

(the $H(\mathcal{F})$-component of (2.99)) and

$$-\Delta_b u + 2(n-1)N(u)$$
$$-f\left[N^2(u) + T^2(u) + (\nabla^H u)(r) + 2rN(u)\right]$$
$$+u\left\{f\,\mathrm{trace}(\tau^2) + \frac{2(n-1)}{f} - f(n-1)H\right.$$

$$+\frac{f}{2}\Delta_b r - \frac{f}{8}(H^2 - G^2) + \frac{1}{4}T(fT(r)) + \frac{fr}{2}H\Big\}$$
$$-f\left\{(\nabla^H u)(r) - f[N(u)H + T(u)T(r)]\right\}$$
$$+\frac{(n+1)u}{f\varphi}\Bigg\{|\nabla^H u|_\theta^2 + u^2\left[\frac{f^2}{2}|\nabla^H r|_\theta^2 + f\,\mathrm{trace}(\tau^2) - \frac{2(n-1)}{f}\right]$$
$$-f^2\Bigg[\frac{4(n+1)}{f}|\xi(u)|^2 - u(HN(u) + T(r)T(u))$$
$$+u^2\left(\frac{f}{4}(H^2 + G^2 + 2T(r)^2\right) - \frac{1}{2}|\nabla^H r|_\theta^2\Bigg]\Bigg\} = 0 \qquad (2.101)$$

(the T-component of (2.99)) and

$$u\Bigg\{(n-2)fT(r) + \frac{f^2}{4}T(r)(H - G)$$
$$-\frac{1}{4}N(fT(r)) + \frac{1}{4}T(fG) - \frac{rf}{2}T(r)\Bigg\}$$
$$+f\left\{(\phi\nabla^H u)(r) - f[N(u)T(r) - T(u)G]\right\} = 0 \qquad (2.102)$$

(the N-component of (2.99)). An elementary asymptotic analysis of (2.100)–(2.102) as $\varphi \to 0$ will now complete the proof of Theorem 2.37. Indeed

$$T(fG) = f\left[T(N(r)) - \frac{6}{\varphi}T(r) + 8rT(r)\right],$$
$$N(fT(r)) = fN(T(r)) + \left[\frac{2}{(1-\varphi r)^2} + f^2 N(r)\right]T(r),$$

so that (2.102) (as $\varphi \to 0$) implies $2T(u) - T(r)u = 0$ on $\partial\Omega$, which is the first statement in (2.58). Similarly, by taking into account that

$$\varphi N(fH) = \frac{8}{1-r\varphi}\left(\frac{1}{1-r\varphi} - 2\right)\frac{1}{\varphi} + O(\varphi),$$
$$\varphi fH = \frac{4}{1-r\varphi} + O(\varphi),$$
$$\varphi f(H^2 - G^2) = \frac{32r}{1-r\varphi}\left(\frac{1}{\varphi} - 2r\right) + O(\varphi),$$

(2.101) may also be written as (2.57). Finally

$$f^2(H+G) = \frac{8}{(1-r\varphi)^2} + O(\varphi)$$

hence (2.100) becomes $\nabla^H u = O(\varphi)$ as $\varphi \to 0$. In particular the boundary values u_b of u is a real valued CR function (of class C^2) on $\partial\Omega$ so that (as $\partial\Omega$ is nondegenerate) u_b must be constant.

As to the last statement in Theorem 2.37, a remark is in order. While in general $T(r) \neq 0$ at the boundary, there is a wealth of examples of strictly pseudoconvex domains $\Omega \subset \mathbb{C}^n$ such that $T(r) = 0$ on $\partial\Omega$. For instance if $\Omega = \mathbb{B}^n$ then $T(r) = 0$ and the phenomenon is related to the spherical symmetry of the example. Indeed, in general let $\Omega = \{z \in \mathbb{C}^n : \phi(|z_1|, \ldots, |z_n|) < 1\}$ be the domain considered by L. Hörmander, [170] (an example for which $\mathcal{J}(a_n K(z,z)^{-1/(n+1)}) = 1$ on $\partial\Omega$, where $\mathcal{J}$ is the complex Monge-Ampère operator, cf. (A) in [116], p. 395). We can show that

Proposition 2.40 (E. Barletta, [22]) *Let $\phi(t_1, \ldots, t_n)$ be a smooth function of $(t_1, \ldots, t_n) \in \mathbb{R}^n$ such that* i) *$\phi_{jk} = 0$ for any $j \neq k$ and that* ii) *$|z_j|\phi_{jj}(|z_1|, \ldots, |z_n|) + \phi_n(|z_1|, \ldots, |z_n|)$ is ≥ 0 for any $1 \leq j \leq n$ and $\neq 0$ for some $1 \leq j \leq n$. Then the transverse curvature of the foliation $\mathcal{F}$ by level sets of $\varphi(z) = \phi(|z_1|, \ldots, |z_n|) - 1$ is*

$$r = \frac{1}{f^0 + |\phi_n|\,|f^n|} \tag{2.103}$$

where

$$f^0 = \sum_{\alpha=1}^{n-1} \frac{|z_\alpha|\phi_\alpha^2}{|z_\alpha|\phi_{\alpha\alpha} + \phi_\alpha}, \quad f^n = \frac{z_n\phi_n}{|z_n|\phi_{nn} + \phi_n},$$

and $\phi_j = \partial\phi/\partial t_j$ and $\phi_{jk} = \partial^2\phi/\partial t_j \partial t_k$. Moreover $T(r) = 0$.

Proof. One has (with $z^j = z_j$)

$$4\partial\overline{\partial}\varphi = \sum_{j,k=1}^{n} \phi_{jk} \frac{\overline{z}^j}{|z^j|} \frac{z^k}{|z^k|} dz^j \wedge d\overline{z}^k + \sum_{k=1}^{n} \frac{1}{|z^k|} \phi_k \, dz^k \wedge d\overline{z}^k$$

so that (ii) in Proposition 2.40 implies strict pseudoconvexity. The CR structure $T_{1,0}(M_\epsilon)$ is locally spanned by

$$Z_\alpha = \frac{\partial}{\partial z^\alpha} - \frac{(\overline{z}^\alpha/|z^\alpha|)\phi_\alpha}{(\overline{z}_n/|z_n|)\phi_n} \frac{\partial}{\partial z^n}, \quad 1 \leq \alpha \leq n-1.$$

Then $\partial\overline{\partial}\varphi(\xi,\overline{Z}_\alpha)=0$ and $\partial\varphi(\xi)=1$ imply

$$\xi=\frac{2}{f^0+|\phi_n||f^n|}\left(f^\alpha\frac{\partial}{\partial z^\alpha}+f^n\frac{\partial}{\partial z^n}\right)$$

where $f^j=z_j\phi_j/\left(|z_j|\phi_{jj}+\phi_j\right)$ $(1\le j\le n)$ thus yielding (2.103). Finally a calculation shows that $T(r)=0$. ■

2.9.5. Final Comments

Let M be a $(2n-1)$-dimensional $(n\geq 2)$ compact strictly pseudoconvex CR manifold, of CR dimension $n-1$, with the CR structure $T_{1,0}(M)\subset T(M)\otimes\mathbb{C}$. Let θ be a contact form on M such that the Levi form

$$G_\theta(X,Y)=(d\theta)(X,JY),\quad X,Y\in H(M),$$

is positive definite. Here $H(M)=\mathrm{Re}\{T_{1,0}(M)\oplus T_{0,1}(M)\}$ and $J(Z+\overline{Z})=i(Z-\overline{z})$, $Z\in T_{1,0}(M)$, is the complex structure in $H(M)$. Let $\Psi=\theta\wedge(d\theta)^{n-1}$ be the corresponding volume form. A C^∞ map $\phi:M\to N$, where N is a given ν-dimensional Riemannian manifold, is a *subelliptic harmonic map* (cf. [25]) if it is a critical point of the energy functional

$$E(\phi)=\frac{1}{2}\int_M \mathrm{trace}_{G_\theta}(\pi_H\phi^*h)\,\Psi \tag{2.104}$$

where $\pi_H\phi^*h$ is the restriction of ϕ^*h to $H(M)$ and h is the Riemannian metric on N. The Euler-Lagrange equations of the variational principle $\delta E(X)=0$ are

$$\Delta_b\phi^A+\sum_{a=1}^{2(n-1)}(\Gamma^A_{BC}\circ\phi)X_a(\phi^B)X_a(\phi^C)=0,\quad 1\le A\le\nu, \tag{2.105}$$

where Δ_b (the *sublaplacian*) is the formally self-adjoint second order differential operator given by

$$\Delta_b u=\mathrm{div}(\nabla^H u),\quad u\in C^2(M).$$

The divergence operator is meant with respect to the volume form Ψ i.e., $\mathcal{L}_X\Psi=\mathrm{div}(X)\Psi$ where $\mathcal{L}_X$ is the Lie derivative. Also Γ^A_{BC} are the Christoffel symbols of the second kind of the Riemannian manifold (N,h) while $\{X_a:1\le a\le 2(n-1)\}$ is a local G_θ-orthonormal frame of $H(M)$. Let X_a^* be the formal adjoint of X_a (with respect to the L^2-inner product $(u,v)=\int_M u\overline{v}\Psi$). As $\{X_a:1\le a\le 2(n-1)\}$ is a Hörmander system and locally $\Delta_b u=\sum_{a=1}^{2(n-1)}X_a^*X_a u$ (the *Hörmander operator*) the notion of a subelliptic harmonic map admits a local reformulation in terms of an arbitrary Hörmander system of vector fields (on an open subset $U \subseteq \mathbb{R}^m$), which do not necessarily arise from a CR structure (and perhaps aren't even linearly independent at all points of U). This is the context in [180], where the notion of a subelliptic harmonic map was first introduced (there the adjoints of the X_a's are meant with respect to the Lebesgue measure on U). As shown by Y. Kamishima et al., [103], the boundary values of harmonic maps $\Phi : (\Omega, g) \to N$ (where g is the Bergman metric on Ω) are subelliptic harmonic maps, provided that the boundary values have vanishing normal derivatives (cf. Theorem 1 in [103]). In other words if $\Phi : \Omega \to N$ is a C^∞ solution to the Dirichlet problem

$$\Delta_g \Phi^A + g^{ij} \left(\Gamma^A_{BC} \circ \Phi \right) \frac{\partial \Phi^B}{\partial x^i} \frac{\partial \Phi^C}{\partial x^j} = 0 \quad \text{in} \quad \Omega, \quad 1 \leq A \leq \nu, \tag{2.106}$$

$$\Phi = \phi \quad \text{on} \quad \partial\Omega, \tag{2.107}$$

with $\phi \in C^\infty(\partial\Omega)$ then the boundary data ϕ must satisfy the compatibility relations (2.105) provided that $N(\phi^A) = 0$. A first attempt of generalizing the notion of a subeliptic harmonic map to the context of unit vector fields $X : M \to U(M)$ [thought of as smooth maps from a strictly pseudoconvex CR manifold M to the unit tangent bundle, endowed with the Sasaki metric associated to the Webster metric (cf. [110], Chapter 1)] was made by S. Dragomir & D. Perrone, [107] [there one requests that $X \in H(M)$ and that the horizontal lift $X^\uparrow$ to the total space of the canonical circle bundle be harmonic in the usual sense (cf. [316]) with respect to the Fefferman metric (cf. [110], Chapter 2)]. Another generalization is proposed in [103]. There one first produces an analog to Dombrowski's horizontal lifting technique [by replacing the Levi-Civita connection with the Tanaka-Webster connection ∇ of the base manifold] and then requests that X be a critical point for (2.104), though only with respect to smooth 1-parameter variations of X through unit vector fields. As shown in [103] such X (a *subelliptic harmonic vector field*, according to the terminology adopted in [103]) is a solution to

$$\Delta_b X - |\nabla^H X|^2_\theta \, X = 0 \tag{2.108}$$

where Δ_b is the sublaplacian on vector fields i.e., locally

$$\Delta_b X = -\sum_{a=1}^{2n} \{ \nabla_{X_a} \nabla_{X_a} X - \nabla_{\nabla_{X_a} X_a} X \}$$

and $\nabla^H X$ is the bundle morphism $Y \in H(M) \to \nabla_Y X \in T(M)$. Based on the knowledge on the boundary behavior of a solution to the Dirichlet problem (2.106)–(2.107) one may expect that that the boundary values of harmonic vector fields on (Ω, g) (which are C^∞ up to the boundary) be solutions to (2.108). The contents of Theorem 2.36 is that a different phenomenon occurs and the boundary values should rather be $X = 0$. The perhaps surprising result in Theorem 2.36 may be anticipated by looking at the particular case of parallel vector fields X (clearly such X satisfies (2.54) provided it is C^2 in Ω) which are C^1 up to the boundary. If for instance $X \in H(\mathcal{F})$ and $\nabla^g X = 0$ then $\nabla^g_N X = 0$ and (2.70) imply

$$\varphi \nabla_N X = X - \frac{\varphi f}{2}\{(\phi X)(r)T - X(r)N\}$$

hence $X = O(\varphi)$ as $\varphi \to 0$, and then $X = 0$ on $\partial\Omega$. In general

Proposition 2.41 *Let X be a tangent vector field on Ω which is C^1 up to the boundary. If X is parallel (with respect to the Bergman metric) in Ω then $X = 0$ on $\partial\Omega$.*

Proof. X decomposes as $X = Y + \theta(X)T + \frac{1}{2}X(\varphi)N$ where $Y = \pi_H X \in H(\mathcal{F})$. Let us assume that $\nabla^g X = 0$. Then

$$\nabla^g_T Y + T(\theta(X))T + \theta(X)\nabla^g_T T + \frac{1}{2}T(X(\varphi))N + \frac{1}{2}X(\varphi)\nabla^g_T N = 0$$

resulting (by (2.69) and (2.72)–(2.73)) in three equations: the $H(\mathcal{F})$-component is

$$\nabla_T Y - \frac{1}{f}\phi Y - \frac{1}{2}\theta(X)\nabla^H r + \frac{1}{4}X(\varphi)\phi\nabla^H r = 0$$

or $\phi Y = O(\varphi)$ so that $Y = 0$ on $\partial\Omega$. Moreover the T-component is

$$\frac{\varphi f}{2}(\phi Y)(r) + \varphi T(\theta(X)) - \frac{\varphi f}{2}\theta(X)T(r)$$
$$- \frac{1}{4}\{f\varphi N(r) + \frac{4f}{\varphi} - 6fr + 4f\varphi r^2\} = 0$$

yielding $X(\varphi) = O(\varphi)$ so that $X(\varphi) = 0$ on $\partial\Omega$. A similar asymptotic analysis of the N-component furnishes $\theta(X) = O(\varphi)$ as $\varphi \to 0$. ∎

The moral conclusion of Theorems 2.36 and 2.37 is that given a harmonic vector field X on a bounded strictly pseudoconvex domain $\Omega \subset \mathbb{C}^n$ such that X is C^2 up to the boundary, X must vanish at $\partial\Omega$. It is a natural question whether solutions to (2.54) satisfy some sort of maximum modulus principle (and then Ω together with the Bergman metric would carry no nonzero harmonic vector fields smooth up to $\partial\Omega$). The difficulties arise mainly from the fact that the Bergman metric g doesn't extend to the boundary. If in turn g were the Euclidean metric then

Proposition 2.42 *Let $\Omega \subset \mathbb{R}^N$ be a smoothly bounded domain on which Green's lemma holds. Let X be a solution to the Dirichlet problem $\Delta_0 X + |\nabla^0 X|^2 X = 0$ in Ω and $X = 0$ on $\partial\Omega$, such that X is C^2 up to $\partial\Omega$, where the Laplacian $\Delta_0 X$ and the covariant derivative $\nabla^0 X$ are defined in terms of the Euclidean metric g_0 on $\mathbb{R}^N$. Assume that $\|X\|_\infty < 1$. Then for any $x \in \Omega$ one has $(\nabla^0 X)_x = 0$ or $X_x = 0$.*

Here $\|X\|_\infty = \sup_{x\in\overline{\Omega}}|X|_x$. To prove Proposition 2.42 note first that

$$
\begin{aligned}
g_0(\Delta_0 X, X) &= \sum_j \left\{ g_0\left(\nabla^0_{X_j}\nabla^0_{X_j}, X\right) - g_0\left(\nabla^0_{\nabla^0_{X_j}X_j} X, X\right)\right\} \\
&= \sum_j \left\{ X_j\left(g_0\left(\nabla^0_{X_j}X, X\right)\right) - g_0\left(\nabla^0_{X_j}X, \nabla^0_{X_j}X\right) - \frac{1}{2}\left(\nabla^0_{X_j}X_j\right)|X|^2\right\} \\
&= \frac{1}{2}\sum_j \left\{ X_j^2|X|^2 - (\nabla^0_{X_j}X_j)|X|^2\right\} - \sum_j |\nabla^0_{X_j}X|^2 \\
&= \frac{1}{2}\Delta\left(|X|^2\right) - |\nabla^0 X|^2,
\end{aligned}
$$

where Δ is the ordinary Laplacian on functions. Taking the inner product of (2.54) with X (and using the previous identity)

$$\Delta\left(|X|^2\right) + 2|\nabla^0 X|^2(|X|^2 - 1) = 0.$$

Let us multiply by $|X|^2$ and integrate over Ω. Then (by Green's lemma)

$$\int_{\partial\Omega} |X|^2 g_0(\nabla^0|X|^2, \nu) - \int_\Omega |\nabla^0\left(|X|^2\right)|^2 + 2\int_\Omega |\nabla^0 X|^2|X|^2(|X|^2 - 1) = 0$$

where ν is the outward unit normal on $\partial\Omega$, hence $|\nabla^0 X|^2|X|^2 = 0$ in Ω.

2.10. CONFORMAL CHANGE OF METRIC ON THE TORUS

Let (M,g) be a compact orientable manifold and $u \in C^\infty(M)$. Let $\tilde{g} = e^{2u}g$ be another Riemannian metric on M, conformal to g. If $X \in \Gamma^\infty(S(M,g))$ then $\tilde{X} = e^{-u}X \in \Gamma^\infty(S(M,\tilde{g}))$. We wish to compute $\|\tilde{\nabla}\tilde{X}\|_{\tilde{g}}$. Let $\{E_i : 1 \le i \le n\}$ be a local orthonormal frame of $(T(M),g)$. If

$\tilde{E}_i = e^{-u} E_i$ then $\{\tilde{E}_i : 1 \le i \le n\}$ is a local orthonormal frame of $(T(M), \tilde{g})$. The Levi-Civita connections ∇ and $\tilde{\nabla}$ of (M,g) and $(M,\tilde{g})$ are related by

$$\tilde{\nabla}_Y Z = \nabla_Y Z + Y(u)Z + Z(u)Y - g(Y,Z)\,(du)^{\sharp}$$

for any $Y, Z \in \mathfrak{X}(M)$. Therefore

$$\tilde{\nabla}_{\tilde{E}_i} \tilde{X} = e^{-2u}\left\{\nabla_{E_i} X + X(u)E_i - g(X,E_i)\,(du)^{\sharp}\right\}$$

and then

$$e^{2u}\|\tilde{\nabla}_{\tilde{E}_i}\tilde{X}\|_{\tilde{g}}^2 = \|\nabla_{E_i}X\|_g^2 + X(u)^2 + g(X,E_i)^2\|du\|_g^2$$
$$+\,2\left\{X(u)g(\nabla_{E_i}X, E_i) - g(X,E_i)(\nabla_{E_i}X)(u) - X(u)g(X,E_i)E_i(u)\right\}.$$

Taking the sum over $i \in \{1,\dots,n\}$ one obtains

$$\|\tilde{\nabla}\tilde{X}\|_{\tilde{g}}^2 = e^{-2u}\Big\{\|\nabla X\|_g^2 + \|\nabla u\|_g^2 + (n-2)X(u)^2$$
$$+\,2\,(\operatorname{div}(X)X - \nabla_X X)\,u\} \qquad (2.109)$$

for any $X \in \Gamma^{\infty}(S(M,g))$. Also it is immediate that

$$d\,\mathrm{vol}(\tilde{g}) = e^{nu}\, d\,\mathrm{vol}(g).$$

Let $n = 2$ and let us use (2.109) to compute $\int_{T^2}\|\tilde{\nabla}\tilde{X}\|_{\tilde{g}}^2\, d\,\mathrm{vol}(\tilde{g})$. As $d\,\mathrm{vol}(\tilde{g}) = e^{2u}\, d\,\mathrm{vol}(g)$ we obtain

$$\tilde{\mathcal{B}}(\tilde{X}) = \mathcal{B}(X)$$
$$+\int_{T^2}\Big\{\|\nabla u\|_g^2 + 2\,(\operatorname{div}(X)X - \nabla_X X)\,u\Big\}\, d\,\mathrm{vol}(g). \qquad (2.110)$$

As a consequence of (2.110) we obtain the following

Theorem 2.43 (G. Wiegmink, [309]) *Let T^2 be a torus, g an arbitrary Riemannian metric on T^2, and K_g the Gaussian curvature of (T^2, g). Let $u \in C^{\infty}(T^2)$ and $\tilde{g} = e^{2u}g$. Given $X \in \mathcal{E}$ let $\tilde{X} = e^{-u}X \in \tilde{\mathcal{E}} = \Gamma^{\infty}(S(T^2,\tilde{g}))$. Let $\mathcal{B} : \mathcal{E} \to [0,+\infty)$ and $\tilde{\mathcal{B}} : \tilde{\mathcal{E}} \to [0,+\infty)$ be respectively the total bending functionals of (T^2,g) and $(T^2,\tilde{g})$. Then*

$$\tilde{\mathcal{B}}(\tilde{X}) = \mathcal{B}(X) + \int_{T^2}\Big\{\|\nabla u\|_g^2 - 2uK_g\Big\}\, d\,\mathrm{vol}(g) \qquad (2.111)$$

hence the difference $\tilde{B}(\tilde{X}) - \mathcal{B}(X)$ is a constant depending only on the two metrics g and $\tilde{g}$. Moreover $\inf_{X\in\mathcal{E}}\mathcal{B}(X)$ is achieved in $\mathcal{E}$.

Proof. It may be shown (cf. [309], p. 337) that for any $U \in \mathcal{E}$

$$\operatorname{div}((\operatorname{div} U)U - \nabla_U U) = K_g .$$

Then (2.111) follows from (2.110) and Green's lemma. In particular if $\inf_{X\in\mathcal{E}} \mathcal{B}(X)$ is achieved in $\mathcal{E}$ then $\inf_{Y\in\tilde{\mathcal{E}}} \tilde{\mathcal{B}}(Y)$ is achieved in $\tilde{\mathcal{E}}$ for any metric $\tilde{g}$ conformal to g. On the other hand, by a result of J.L. Kazdan & F.W. Warner, [185], there is $u \in C^\infty(T^2)$ such that $\tilde{g} = e^{2u} g$ is flat. Let $\mathbb{H}^2 = \{z \in \mathbb{C} : \operatorname{Im}(z) > 0\}$ be the upper half plane and $\mathrm{SL}(2,\mathbb{Z}) = \left\{ \begin{pmatrix} a & b \\ c & d \end{pmatrix} : a, b, c, d \in \mathbb{Z}, \; ac - bd = 1 \right\}$ the modular group. There is a natural action of the modular group on the upper half plane given by

$$\begin{pmatrix} a & b \\ c & d \end{pmatrix} \cdot z = \frac{az+b}{cz+d}, \quad z \in \mathbb{H}^2, \quad \begin{pmatrix} a & b \\ c & d \end{pmatrix} \in \mathrm{SL}(2,\mathbb{Z}).$$

By the well-known Proposition 2.5.11 in [311], p. 79, the isometry classes of flat tori are parametrized by pairs (a, ϕ) where a is a positive real number and $\phi \in \mathbb{H}^2/\mathrm{SL}(2,\mathbb{Z})$. Precisely there are $a \in (0, +\infty)$ and $z \in \mathbb{H}^2$ such that $(T^2, \tilde{g})$ belongs to the isometry class of the flat torus $\mathbb{R}^2/\{mr + nrz : (m,n) \in \mathbb{Z}^2\}$ where $r = \sqrt{a}$ (and $\phi = \mathrm{SL}(2,\mathbb{Z}) \cdot z$). Therefore on $(T^2, \tilde{g})$ there exist globally defined parallel unit vector fields so that the total bending functional achieves its greatest lower bound $\inf_{Y\in\tilde{\mathcal{E}}} \tilde{\mathcal{B}}(Y) = 0$. ■

2.11. SOBOLEV SPACES OF VECTOR FIELDS

As indicated in Section 2.2 of this monograph, for each C^∞ vector field X on M one may think of ∇X as a smooth section in the vector bundle $T^*(M) \otimes T(M) \to M$. This gives a map $\nabla : \mathfrak{X}(M) \to \Gamma^\infty(T^*(M) \otimes T(M))$. Let ∇^* be the formal adjoint of ∇ i.e.,

$$(\nabla^* \varphi, X) = (\varphi, \nabla X) \tag{2.112}$$

for any $\varphi \in \Gamma_0^\infty(T^*(M) \otimes T(M))$. As customary we denote by $\Gamma_0^\infty(E)$ the space of all C^∞ sections of compact support in the vector bundle $E \to M$. For instance $\Gamma_0^\infty(T^{1,0}(M)) = \mathfrak{X}_0^\infty(M)$ (the space of all C^∞ vector fields with compact support). The L^2 inner products in (2.112) are given by

$$(X,Y) = \int_M g(X,Y)\,d\,\mathrm{vol}(g),$$

$$(\varphi,\psi) = \int_M g^*(\varphi,\psi)\,d\,\mathrm{vol}(g).$$

Of course the integrals are convergent when M is compact or at least one of the vector fields $X, Y \in \mathfrak{X}(M)$ (respectively at least one of the endomorphisms $\varphi, \psi \in \Gamma^\infty(T^*(M) \otimes T(M))$) has compact support.

Let $\{E_i : 1 \leq i \leq n\}$ be a local orthonormal frame of $T(M)$. Let φ be a C^1 section in $T(M)^* \otimes T(M)$. For any $Y \in \mathfrak{X}_0^\infty(M)$

$$\begin{aligned}
(\nabla^*\varphi, Y) &= \int_M g(\varphi, \nabla Y)\,d\,\mathrm{vol}(g) \\
&= \int_M \sum_{i=1}^n g(\varphi E_i, \nabla_{E_i} Y)\,d\,\mathrm{vol}(g) \\
&= \int_M \sum_i \{E_i(g(\varphi E_i, Y)) - g(\nabla_{E_i}\varphi E_i, Y)\}\,d\,\mathrm{vol}(g).
\end{aligned}$$

Let $X_\varphi \in T(M)$ be uniquely determined by

$$g(X_\varphi, Z) = g(\varphi Z, Y), \quad Z \in T(M).$$

Then (by $\nabla g = 0$ and Green's lemma)

$$\begin{aligned}
(\nabla^*\varphi, Y) &= \int_M \sum_i \{g(\nabla_{E_i} X_\varphi, E_i) + g(X_\varphi, \nabla_{E_i} E_i) - g(\nabla_{E_i}\varphi E_i, Y)\}\,d\,\mathrm{vol}(g) \\
&= -\int_M g(Y, \nabla_{E_i}\varphi E_i - \varphi \nabla_{E_i} E_i)\,d\,\mathrm{vol}(g)
\end{aligned}$$

hence

$$\nabla^*\varphi = -\sum_{i=1}^n (\nabla_{E_i}\varphi)E_i. \tag{2.113}$$

Consequently

$$\Delta_g X = \nabla^* \nabla X \tag{2.114}$$

for any C^2 vector field X on M. The equation (2.114) suggests the natural notion of weak solution to the nonlinear system $\Delta_g X = \|\nabla X\|^2 X$ (to be introduced shortly in this section).

As M is a Riemannian manifold, the vector bundle $\mathcal{T}^{r,s}(M) \to M$ of all tangent (r,s)-tensors is a Riemannian bundle in a natural way hence the pointwise norm $\|\varphi\| = g^*(\varphi,\varphi)^{1/2} : M \to [0,+\infty)$ of each measurable (r,s)-tensor field φ on M is well defined. Let $L^1(M)$ be the space of all integrable functions on M, with respect to the Riemannian measure. We set

$$(\varphi,\psi) = \int_M g^*(\varphi,\psi)\, d\,\mathrm{vol}(g)$$

for any measurable sections φ, ψ in $\mathcal{T}^{r,s}(M) \to M$ such that $g^*(\varphi,\psi) \in L^1(M)$. For any smooth section φ in $\mathcal{T}^{r,s}(M) \to M$ we denote by $\nabla\varphi$ its covariant derivative hence we have an operator

$$\nabla : \Gamma^\infty(\mathcal{T}^{r,s}(M)) \to \Gamma^\infty(\mathcal{T}^{r,s+1}(M)).$$

The *formal adjoint* of ∇ is

$$\nabla^* : \Gamma_0^\infty(\mathcal{T}^{r,s+1}(M)) \to \Gamma_0^\infty(\mathcal{T}^{r,s}(M)),$$

$$(\nabla^* h,\varphi) = (h,\nabla\varphi), \quad h \in \Gamma_0^\infty(\mathcal{T}^{r,s+1}(M)), \ \varphi \in \Gamma_0^\infty(\mathcal{T}^{r,s}(M)).$$

Let $p \geq 1$ and let $L^p(\mathcal{T}^{r,s}(M))$ consist of all measurable (r,s)-tensor fields φ on M such that $\int_M \|\varphi\|^p d\,\mathrm{vol}(g) < \infty$. We set as usual

$$\|\varphi\|_{L^p} = \|\varphi\|_{L^p(\mathcal{T}^{r,s}(M))} = \left(\int_M \|\varphi\|^p d\,\mathrm{vol}(g) \right)^{1/p}$$

for any $\varphi \in L^p(\mathcal{T}^{r,s}(M))$. Then $L^p_{\mathrm{loc}}(\mathcal{T}^{r,s}(M))$ consists of all measurable sections in $\mathcal{T}^{r,s}(M) \to M$ such that $\chi_A\varphi \in L^p(\mathcal{T}^{r,s}(M))$ for any relatively compact subset $A \subset\subset M$ (here χ_A is the characteristic function of A).

Definition 2.44 A locally integrable section $\varphi \in L^1_{\mathrm{loc}}(\mathcal{T}^{r,s}(M))$ is said to be *weakly differentiable* if there is $\psi \in L^1_{\mathrm{loc}}(\mathcal{T}^{r,s+1}(M))$ such that

$$(\psi,h) = (\varphi,\nabla^* h), \quad h \in \Gamma_0^\infty(\mathcal{T}^{r,s+1}(M)).$$

The section ψ (which is of course uniquely determined up to a set of measure zero) is denoted by $\nabla\varphi$ and referred to as the (*weak*) *covariant derivative* of φ. ■

In particular a measurable function f on M is weakly differentiable if there is $\omega \in L^1_{\text{loc}}(T^*(M))$ such that $(\omega,\alpha) = (f, \nabla^*\alpha)$ for any $\alpha \in \Gamma_0^\infty(T^*(M))$ and the (weak) covariant derivative of f is the locally integrable 1-form $\nabla f = \omega$. This is of course equivalent to the definition in [224], p. 1089. Indeed we have

Proposition 2.45 *Let M be an oriented Riemannian manifold and f a locally integrable function on M. If f is weakly differentiable then for each $X \in \mathfrak{X}^\infty(M) = \Gamma^\infty(T^{1,0}(M))$ there is $h \in L^1_{\text{loc}}(M)$ such that*

$$(h, \phi) = -\int_M f\,[X(\phi) + \phi\,\text{div}(X)]\,d\,\text{vol}(g), \quad \phi \in C_0^\infty(M). \tag{2.115}$$

In particular $h = (\nabla f)(X)$ almost everywhere in M for each $X \in \mathfrak{X}(M)$.

Following the terminology in [224], if (2.115) holds for some $h \in L^1_{\text{loc}}(M)$ then f *has a weak derivative* with respect to X. Of course h in (2.115) is uniquely determined up to a set of measure zero and one adopts the customary notation $h = X(f)$.

Let us prove Proposition 2.45. If $\omega = \nabla f$ is the (weak) covariant derivative of f and $X \in \mathfrak{X}(M)$ an arbitrary smooth vector field then the candidate for $X(f)$ is $h = (\nabla f)(X)$. For any test function $\phi \in C_0^\infty(M)$

$$(h, \phi) = \int_M h\phi\,d\,\text{vol}(g) = \int_M \omega(X)\phi\,d\,\text{vol}(g).$$

Let (U, x^i) be a local coordinate system on M so that $\omega(X)\phi = \omega_i \phi X^i$ on U. Thus we may consider the (globally defined) differential 1-form $\alpha \in \Gamma_0^\infty(T^*(M))$ given by

$$\alpha|_U = \phi g_{ij} X^j\,dx^i \tag{2.116}$$

on U. Therefore $\omega(X)\phi = \omega_i \alpha_j g^{ij}$ on U so that

$$(h, \phi) = \int_M g^*(\omega, \alpha)\,d\,\text{vol}(g) = (\omega, \alpha) = (f, \nabla^*\alpha).$$

On the other hand for any $u \in C_0^\infty(M)$

$$(\nabla^*\alpha, u) = (\alpha, \nabla u) = \int_M g^*(\alpha, \nabla u)\,d\,\text{vol}(g)$$

and (by (2.116))

$$g^*(\alpha, \nabla u)\big|_U = \alpha_i (\nabla u)_j g^{ij} = \alpha_i \frac{\partial u}{\partial x^j} g^{ij}$$

$$= \phi g_{ik} X^k \frac{\partial u}{\partial x^j} g^{ij} = \phi X^j \frac{\partial u}{\partial x^j} = (\phi X(u))|_U$$

so that (by Green's lemma)

$$(\nabla^* \alpha, u) = \int_M \phi X(u) d\operatorname{vol}(g) = -\int_M u\{X(\phi) + \phi \operatorname{div}(X)\} d\operatorname{vol}(g).$$

Consequently $\nabla^* \alpha = -X(\phi) - \phi \operatorname{div}(X)$ and (2.115) is proved.

Note that (2.115) may be written $(h, \phi) = (f, X^*(\phi))$ for any $\phi \in C_0^\infty(M)$ where X^* is the formal adjoint of X. The converse of Proposition 2.45 is a bit more involved. We show that

Proposition 2.46 *Let M be an oriented Riemannian manifold and $f \in L^1_{\text{loc}}(M)$. If f has a weak derivative with respect to any $X \in \mathfrak{X}^\infty(M)$ then f is weakly differentiable. Moreover, under the same assumption, for every local coordinate system (U, x^i) on M the function $f\big|_U$ has a weak derivative with respect to $\partial/\partial x^i$ for any $i \in \{1, \dots, n\}$ and*

$$(\nabla f)\big|_U = \frac{\partial}{\partial x^i}\left(f\big|_U\right) dx^i \tag{2.117}$$

a.e. in U.

We need

Lemma 2.47 *Let $\psi \in C^\infty(M)$ and assume that $f \in L^1_{\text{loc}}(M)$ has a weak derivative with respect to $X \in \mathfrak{X}^\infty(M)$. Then $\psi f \in L^1_{\text{loc}}(M)$ and ψf has a weak derivative with respect to X. In particular*

$$X(\psi f) = X(\psi) f + \psi X(f). \tag{2.118}$$

Proof. For each $A \subset\subset M$

$$\int_A |\psi f| d\operatorname{vol}(g) \le \sup_A |\psi| \, \|f\|_{L^1(A)} < \infty$$

so that ψf is locally integrable. On the other hand, for any $\phi \in C_0^\infty(M)$

$$-\int_M \psi f\{X(\phi) + \phi \operatorname{div}(X)\} d\operatorname{vol}(g)$$

$$= -\int_M f\{X(\phi\psi) - \phi X(\psi) + \phi\psi \operatorname{div}(X)\} d\operatorname{vol}(g)$$

$$= \int_M X(f)\phi\psi \, d\operatorname{vol}(g) + \int_M fX(\psi)\phi \, d\operatorname{vol}(g)$$

$$= \int_M \{X(f)\psi + fX(\psi)\} d\operatorname{vol}(g)$$

so that ψf is weakly derivable with respect to X and the weak derivative is given by (2.118). ■

Proof of Proposition 2.46. Let $f \in L^1_{\text{loc}}(M)$ admit weak derivatives with respect to each $X \in \mathfrak{X}^\infty(M)$. We claim that for any open set $U \subset M$ and any $Y \in \mathfrak{X}^\infty(U)$, the restriction $f\big|_U$ admits a weak derivative with respect to Y. To prove the claim let $X \in \mathfrak{X}^\infty(M)$ be a smooth extension of Y to the whole of M and let $\phi \in C_0^\infty(M)$ be a test function such that $\operatorname{supp}(\phi) \subset U$. Then (by (2.115))

$$\int_U X(f)\Big|_U \phi \, d\operatorname{vol}(g) = -\int_U f\big|_U \{X(\phi) + \phi \operatorname{div}(X)\}\big|_U \, d\operatorname{vol}(g)$$
$$= -\int_U f\big|_U \{Y(\phi) + \phi \operatorname{div}(Y)\} \, d\operatorname{vol}(g).$$

All test functions in $C_0^\infty(U)$ are restrictions to U of such ϕ hence $f\big|_U$ admits a weak derivative with respect to Y and $Y(f) = X(f)\big|_U$. In particular $(\partial/\partial x^i)\left(f\big|_U\right)$ are well-defined elements of $L^1_{\text{loc}}(U)$ (for each $1 \le i \le n$). Let $\alpha \in \Gamma_0^\infty(T^*(M))$ be locally given by

$$\alpha|_U = \alpha_i \, dx^i .$$

Then (by Green's lemma)

$$(\nabla^*\alpha)\big|_U = -\frac{\partial \alpha^i}{\partial x^i} - \alpha^i \operatorname{div}\left(\partial/\partial x^i\right), \quad \alpha^i = g^{ij}\alpha_j .$$

Let $\{\varphi_U : U \in \mathcal{U}\}$ be a smooth partition of unity subordinated to the locally finite open covering $\mathcal{U}$ of M. We set $K = \operatorname{supp}(\alpha) \subset M$. We also assume that each $U \in \mathcal{U}$ is the domain of a local chart (U, x^i) on M. Then

$$\begin{aligned}(f, \nabla^*\alpha) &= \int_M f\,\nabla^*\alpha\, d\,\mathrm{vol}(g) = \sum_{U\in\mathcal{U}} \int_K \varphi_U f\, \nabla^*\alpha\, d\,\mathrm{vol}(g) \\ &= -\sum_{U\in\mathcal{U}} \int_{U\cap K} \varphi_U f\big|_U \left\{ \frac{\partial \alpha^i}{\partial x^i} + \alpha^i \operatorname{div}\left(\frac{\partial}{\partial x^i}\right)\right\} d\,\mathrm{vol}(g).\end{aligned}$$

By Lemma 2.47 and the first part of the proof of Proposition 2.46 one has $\varphi_U f\big|_U \in L^1_{\mathrm{loc}}(U)$ and $\varphi_U f\big|_U$ is weakly derivable at the direction $\partial/\partial x^i$ hence (by (2.118))

$$\begin{aligned}(f, \nabla^*\alpha) &= \sum_U \int_{U\cap K} \frac{\partial}{\partial x^i}\left(\varphi_U f\big|_U\right) \alpha^i d\,\mathrm{vol}(g) \\ &= \sum_U \int_{U\cap K} \left\{ \frac{\partial \varphi_U}{\partial x^i} f\big|_U + \varphi_U \frac{\partial}{\partial x^i}\left(f\big|_U\right)\right\} \alpha^i d\,\mathrm{vol}(g) \\ &= \sum_U \int_M \alpha\left(\mathrm{grad}(\varphi_U)\right) d\,\mathrm{vol}(g) + \sum_U \int_{U\cap K} \varphi_U \frac{\partial}{\partial x^i}\left(f\big|_U\right) \alpha^i d\,\mathrm{vol}(g)\end{aligned}$$

and the first integral vanishes (because of $\sum_U \varphi_U = 1$). Next we claim that there is $\omega \in L^1_{\mathrm{loc}}(T^*(M))$ such that $\omega|_U = (\partial/\partial x^i)\left(f\big|_U\right) dx^i$. If the claim is true then

$$(f, \nabla^*\alpha) = \sum_U \int_M \varphi_U g^*(\omega, \alpha) d\,\mathrm{vol}(g) = (\omega, \alpha)$$

for any $\alpha \in \Gamma^\infty_0(T^*(M))$ so that f is weakly differentiable and $\nabla f = \omega$. This would prove the first statement in Proposition 2.46. To prove the claim let us check first that the local 1-forms $\{(\partial/\partial x^i)\left(f\big|_U\right) : U \in \mathcal{U}\}$ glue up to a (globally defined) 1-form ω on M. To this end let $U, U' \in \mathcal{U}$ such that $U \cap U' \neq \emptyset$ and let $\phi \in C^\infty_0(U \cap U')$ be an arbitrary test function on $U \cap U'$. Let $J^i_j = \partial x'^i/\partial x^j \in C^\infty(U \cap U')$ be the Jacobian matrix of the local coordinate transformation $x'^i = x'^i(x^1, \dots, x^n)$. Then (again by Lemma 2.47)

$$\begin{aligned}&\int_{U\cap U'} \frac{\partial}{\partial x^i}\left(f\big|_U\right) \phi\, d\,\mathrm{vol}(g) \\ &\quad = -\int_{U\cap U'} f\big|_U \left\{\frac{\partial \phi}{\partial x^i} + \phi \operatorname{div}\left(\frac{\partial}{\partial x^i}\right)\right\} d\,\mathrm{vol}(g)\end{aligned}$$

$$= - \int_{U\cap U'} f\big|_U \left\{ J_i^j \frac{\partial \phi}{\partial x'^j} + \phi \operatorname{div}\left(J_i^j \frac{\partial}{\partial x'^j} \right) \right\} d\operatorname{vol}(g)$$

$$= - \int_{U\cap U'} f\big|_{U'} J_i^j \left\{ \frac{\partial \phi}{\partial x'^j} + \phi \operatorname{div}\left(\frac{\partial}{\partial x'^j} \right) \right\} d\operatorname{vol}(g)$$

$$- \int_{U\cap U'} f\big|_U \, \phi \, \frac{\partial J_i^j}{\partial x'^j} d\operatorname{vol}(g)$$

$$= \int_{U\cap U'} \frac{\partial}{\partial x'^j}\left(f\big|_{U'} J_i^j \right) \phi \, d\operatorname{vol}(g) - \int_{U\cap U'} f\big|_{U'} \, \phi \, \frac{\partial J_i^j}{\partial x'^j} d\operatorname{vol}(g)$$

$$= \int_{U\cap U'} J_i^j \frac{\partial}{\partial x'^j}\left(f\big|_{U'} \right) \phi \, d\operatorname{vol}(g)$$

hence

$$\frac{\partial}{\partial x^i}\left(f\big|_U \right) = J_i^j \frac{\partial}{\partial x'^j}\left(f\big|_{U'} \right)$$

a.e. in $U \cap U'$. Finally ω is locally integrable as

$$\int_A g^*(\omega,\omega)^{1/2} d\operatorname{vol}(g) = \sum_{U\in\mathcal{U}} \int_A \varphi_U g^*(\omega,\omega)^{1/2} d\operatorname{vol}(g)$$

$$= \sum_U \int_{U\cap A} \varphi_U \left| g^{ij} \frac{\partial}{\partial x^i}\left(f\big|_U \right) \frac{\partial}{\partial x^j}\left(f\big|_U \right) \right|^{1/2} d\operatorname{vol}(g)$$

$$\leq \sum_U \sqrt{C_U} \int_{U\cap A} \sum_{i=1}^n \left| \frac{\partial}{\partial x^i}\left(f\big|_U \right) \right| d\operatorname{vol}(g)$$

$$\leq \sum_{U,i} \sqrt{C_U} \left\| \frac{\partial}{\partial x^i}\left(f\big|_U \right) \right\|_{L^1(U\cap A)} < \infty$$

for any $A \subset\subset M$ i.e., $g^*(\omega,\omega) \in L^1_{\text{loc}}(M)$. Here

$$C_U = \max\left\{ \sup_{U\cap A} \left| g^{ij} \right| : 1 \leq i,j \leq n \right\}$$

and we exploited the fact that $(\partial/\partial x^i)(f\big|_U) \in L^1_{\text{loc}}(U)$ for any $1 \leq i \leq n$. It remains that we justify (2.117). This is rather obvious for we may consider a test form $\alpha \in \Gamma_0^\infty(T^*(M))$ such that $K = \mathrm{supp}(\alpha) \subset U$ and then

$$\int_U g^*\left((\nabla f)\big|_U, \alpha\right) d\,\mathrm{vol}(g) = \int_M g^*(\nabla f, \alpha) d\,\mathrm{vol}(g)$$

$$= \int_K f\,\nabla^*\alpha = -\int_K f\big|_U \left\{\frac{\partial \alpha^i}{\partial x^i} + \alpha^i \,\mathrm{div}(\partial/\partial x^i)\right\} d\,\mathrm{vol}(g)$$

$$= \int_K \frac{\partial}{\partial x^i}\left(f\big|_U\right) \alpha^i d\,\mathrm{vol}(g) = \int_U g^*\left(\frac{\partial}{\partial x^i}\left(f\big|_U\right) dx^i, \alpha\right) d\,\mathrm{vol}(g).$$

All test forms in $\Gamma_0^\infty(T^*(U))$ are restrictions to U of such α hence (2.117) follows. ■

Sobolev spaces of tangent tensor fields on a Riemannian manifold may be introduced recursively as follows. We set

$$\mathcal{H}_g^{0,p}(T^{r,s}(M)) = L^p(T^{r,s}(M))$$

and we denote by $\mathcal{H}_g^{1,p}(T^{r,s-1}(M))$ the space of all weakly differentiable $\varphi \in \mathcal{H}_g^{0,p}(T^{r,s-1}(M))$ such that $\nabla\varphi \in L^p(T^{r,s}(M))$. Recursively, for any $k \geq 2$ let $\mathcal{H}_g^{k,p}(T^{r,s-1}(M))$ consist of all $\varphi \in \mathcal{H}_g^{k-1,p}(T^{r,s-1}(M))$ such that $\nabla\varphi \in \mathcal{H}_g^{k-1,p}(T^*(M) \otimes T^{r,s}(M))$. Then $\mathcal{H}_g^{k,p}(T^{r,s}(M))$ is a Banach space with the norm

$$\|\varphi\|_{\mathcal{H}_g^{k,p}(T^{r,s}(M))} = \left(\sum_{j=0}^{k} \|\nabla^j \varphi\|^p_{L^p(T^{0,j}(M)\otimes T^{r,s}(M))}\right)^{1/p}.$$

It is customary to adopt the abbreviated notation $\mathcal{H}_g^k(T^{r,s}(M)) = \mathcal{H}_g^{k,2}(T^{r,s}(M))$ when $p = 2$. We wish to detail this construction in the case of vector fields (i.e., $r = 1$ and $s = 0$).

Definition 2.48 We say a vector field $X \in L^p(T(M))$ is *weakly differentiable* if there is a locally integrable section φ_X in $T(M)^* \otimes T(M)$ such that

$$(\varphi_X, \psi) = (X, \nabla^*\psi), \quad \psi \in \Gamma_0^\infty(T(M)^* \otimes T(M)).$$

Then φ_X is uniquely determined, except for a set of measure zero, and denoted by ∇X. ■

Let $\mathcal{H}_g^{1,p}(T(M))$ consist of all weakly differentiable $X \in L^p(T(M))$ such that $\nabla X \in L^p(T(M)^* \otimes T(M))$. We endow $\mathcal{H}_g^{1,p}(T(M))$ with the norm

$$\|X\|_{1,p} = \|X\|_{\mathcal{H}_g^{1,p}(T(M))}$$
$$= \left(\|X\|^p_{L^p(T(M))} + \|\nabla X\|^p_{L^p(T^*(M)\otimes T(M))} \right)^{1/p}.$$

Theorem 2.49 *$\mathcal{H}_g^{1,p}(T(M))$ is a Banach space for $1 \le p < \infty$ which is reflexive*[2] *for $1 < p < \infty$ and separable*[3] *for $1 \le p < \infty$. Also $\mathcal{H}_g^{1,2}(T(M))$ is a separable Hilbert space.*

Proof. Let $\{X_\nu\}_{\nu \ge 1}$ be a Cauchy sequence in $\mathcal{H}_g^{1,p}(T(M))$. It follows that $\{X_\nu\}_{\nu\ge 1}$ and $\{\nabla X_\nu\}_{\nu \ge 1}$ are Cauchy sequences in $L^p(T(M))$ and $L^p(T(M)^* \otimes T(M))$, respectively. Thus there are a vector field $X \in L^p(T(M))$ and a bundle morphism $\varphi \in L^p(T(M)^* \otimes T(M))$ such that $\|X_\nu - X\|_{L^p} \to 0$ and $\|\nabla X_\nu - \varphi\|_{L^p} \to 0$ for $\nu \to \infty$. On the other hand

$$\int_M g(\nabla X_\nu, \psi)\, d\,\mathrm{vol}(g) = \int_M g(X_\nu, \nabla^* \psi)\, d\,\mathrm{vol}(g) \tag{2.119}$$

for any $\psi \in \Gamma_0^\infty(T(M)^* \otimes T(M))$. Then (2.119) for $\nu \to \infty$ yields

$$\int_M g(\varphi, \psi)\, d\,\mathrm{vol}(g) = \int_M g(X, \nabla^* \psi)\, d\,\mathrm{vol}(g)$$

so that $X \in \mathcal{H}_g^{1,p}(T(M))$. Next we set

$$E = L^p(T(M)) \times L^p(T(M)^* \otimes T(M))$$

and consider the map $\Phi : \mathcal{H}_g^{1,p}(T(M)) \to E$ given by $\Phi(X) = (X, \nabla X)$. At this point the remainder of the proof of Theorem 2.49 is imitative of that of Theorem 1 in [322], p. 149. ∎

Again to draw a parallel among our presentation (of weakly differentiable functions and vector fields and the corresponding Sobolev type spaces) and the approach in [224], p. 1090, we shall prove

[2] Let $\mathfrak{X}$ be a normed vector space. Let $\chi : \mathfrak{X} \to \mathfrak{X}^{**} = (\mathfrak{X}^*)^*$ be the map given by $\langle \chi(v), \Lambda \rangle = \Lambda(v)$ for any $v \in \mathfrak{X}$ and any $\Lambda \in \mathfrak{X}^*$. Then χ is linear, continuous, injective and $\|v\|_{\mathfrak{X}} = \|\chi(v)\|_{\mathfrak{X}^{**}}$ for any $v \in \mathfrak{X}$. The space $\mathfrak{X}$ is *reflexive* if $\chi(\mathfrak{X}) = \mathfrak{X}^{**}$.

[3] A Banach space $\mathfrak{X}$ is *separable* if $\mathfrak{X}$ admits a countable dense subset.

Proposition 2.50 *Let M be a compact orientable Riemannian manifold. Let $H(M)$ consist of all $f \in L^1(M)$ such that f is weakly derivable at each direction $X \in \mathfrak{X}^\infty(M)$ and $f^2, X(f)^2 \in L^1(M)$. Also let $H(T(M))$ consist of all measurable sections $X : M \to T(M)$ such that $g(X,Z) \in H(M)$ for any $Z \in \mathfrak{X}^\infty(M)$. Then*

$$H(M) = \mathcal{H}^{1,2}_g(\mathcal{T}^{0,0}(M)), \quad H(T(M)) = \mathcal{H}^{1,2}_g(\mathcal{T}^{1,0}(M)). \qquad (2.120)$$

Proof. Let $f \in H(M)$. Then (by Proposition 2.45) f is weakly differentiable. Moreover for any $U \in \mathcal{U}$ the restriction $f\big|_U$ has a weak derivative with respect to each $\partial/\partial x^i$. The notations are those in the proof of Proposition 2.46. Additionally we assume that $\mathcal{U}$ is a refinement of an open covering $\mathcal{V}$ of M with local coordinate systems such that for any $U \in \mathcal{U}$ there is $V \in \mathcal{V}$ with $U \subset\subset V$ (and the local coordinates on U are the restrictions to U of the local coordinates on V). As M is compact, $\mathcal{U}$ may be chosen to be finite to start with. Then

$$\|\nabla f\|^2_{L^2(T^*(M))} = \int_M \|\nabla f\|^2 d\,\mathrm{vol}(g) = \sum_{U \in \mathcal{U}} \int_M \varphi_U \|\nabla g\|^2 d\,\mathrm{vol}(g)$$

$$= \sum_U \int_U \varphi_U g^{ij} \frac{\partial}{\partial x^i}\left(f\big|_U\right) \frac{\partial}{\partial x^j}\left(f\big|_U\right) d\,\mathrm{vol}(g)$$

$$\leq \sum_U \int_U \varphi_U \Lambda \sum_i \left[\frac{\partial}{\partial x^i}\left(f\big|_U\right)\right]^2 \leq \sum_{U,i} C_U \left\|\frac{\partial}{\partial x^i}\left(f\big|_U\right)\right\|^2_{L^2(U)} < \infty$$

where $\Lambda(x)$ is the maximum eigenvalue of $g^{ij}(x)$ for any $x \in V$ and $C_U = \sup\{\Lambda(x) : x \in \overline{U}\}$. We have shown that $\nabla f \in L^2(T^*(M))$ hence $f \in \mathcal{H}^{1,2}_g(\mathcal{T}^{0,0}(M))$.

Vice versa if $f \in \mathcal{H}^{1,2}_g(\mathcal{T}^{0,0}(M))$ then $f \in L^2(M)$ and f is weakly differentiable. Then (by Proposition 2.46) f is weakly derivable at any direction $X \in \mathfrak{X}^\infty(M)$ and $X(f) = (\nabla f)(X)$. Therefore

$$\|X(f)\|_{L^2(M)} = \int_M |X(f)|^2 d\,\mathrm{vol}(g) = \int_M |(\nabla f)(X)|^2 d\,\mathrm{vol}(g)$$

$$\leq \int_M \|\nabla X\|^2 \|X\|^2 d\,\mathrm{vol}(g) \leq C \|\nabla X\|^2_{L^2(T^*(M))} < \infty$$

where $C = \sup_M \|X\|^2$. We have shown that $X(f) \in L^2(M)$ and we may conclude that $f \in H(M)$. ■

To prove the second equality in (2.120) we need some preparation.

Lemma 2.51 *The restriction map $X \mapsto X|_U$ is a continuous map $H(T(M)) \to H(T(U))$. Moreover let $X \in H(T(M))$ be locally represented as $X|_U = X^i \partial/\partial x^i$ a.e. in U. Then $X^i \in H(U)$ for any $1 \le i \le n$.*

Proof. By the first statement in Lemma 2.51 if $X \in H(T(M))$ then $X|_U \in H(T(U))$ hence $g(X|_U, Z) \in H(U)$ for any $Z \in \mathfrak{X}^\infty(U)$. In particular for $Z = g^{ij}\partial/\partial x^j$ one has

$$X^i = g\left(X|_U, g^{ij}\frac{\partial}{\partial x^j}\right) \in H(U), \quad 1 \le i \le n.$$

Lemma 2.51 is proved. ■

As a first consequence of Lemma 2.51, one may define the covariant derivative ∇X of each $X \in H(T(M))$ by observing that the locally defined $(1,1)$-tensor fields $(\nabla X)_U$ given by

$$(\nabla X)_U = \left[\frac{\partial}{\partial x^i}(X^j) + \Gamma^j_{ik}X^k\right] dx^i \otimes \frac{\partial}{\partial x^j} \tag{2.121}$$

glue up to a (globally defined) $(1,1)$-tensor field ∇X on M such that $(\nabla X)|_U = (\nabla X)_U$. Note that the expression in the right hand side of (2.121) makes sense precisely because X^i (as an element of $H(U)$) is weakly derivable in the direction $\partial/\partial x^i$. Let us go back to the proof of (2.120). To this end let $\psi \in \Gamma^\infty(\mathcal{T}^{1,1}(M))$. Then (by (2.113))

$$\begin{aligned}(X, \nabla^*\psi) &= \int_M g(X, \nabla^*\psi)\, d\,\mathrm{vol}(g) \\ &= \sum_{U\in\mathcal{U}} \int_M \varphi_U g(X, \nabla^*\psi)\, d\,\mathrm{vol}(g) = -\sum_U \{I_U - J_U\}\end{aligned}$$

where

$$I_U = \int_U \varphi_U g^{ij} g\left(X, \nabla_{\partial/\partial x^i}\left(\psi\frac{\partial}{\partial x^j}\right)\right) d\,\mathrm{vol}(g),$$

$$J_U = \int_U \varphi_U g^{ij} g\left(X, \psi\, \nabla_{\partial/\partial x^i}\frac{\partial}{\partial x^j}\right).$$

We may represent ψ locally as

$$\psi|_U = \psi_i^j \, dx^i \otimes \frac{\partial}{\partial x^j}$$

for some $\psi_i^j \in C^\infty(U)$. For the sake of simplicity we set $\partial_i = \partial/\partial x^i$. We start by computing I_U i.e.,

$$I_U = \int_U \varphi_U g^{ij} g\left(X, \nabla_{\partial_i}\left(\psi_j^k \partial_k\right)\right) d\,\mathrm{vol}(g) = I'_U + I''_U$$

where

$$I'_U = \int_U \varphi_U g^{ij} \frac{\partial \psi_j^k}{\partial x^i} g(X, \partial_k) d\,\mathrm{vol}(g),$$

$$I''_U = \int_U \varphi_U g^{ij} \psi_j^k \Gamma^\ell_{ik} g(X, \partial_\ell).$$

Here Γ^i_{jk} are the local coefficients of the Levi-Civita connection of (M,g). To compute I'_U we use Lemma 2.47 and obtain

$$I'_U = \int_U \left\{ \frac{\partial}{\partial x^i} \left[\varphi_U g^{ij} \psi_j^k g(X, \partial_k)\right] - \psi_j^k \frac{\partial}{\partial x^i} \left[\varphi_U g^{ij} g(X, \partial_k)\right] \right\} d\,\mathrm{vol}(g)$$

(by the very definition of the weak derivative of $\varphi_U g^{ij} \psi_j^k g(X, \partial_k)$ in the direction ∂_i and again by Lemma 2.47)

$$= -\int_U \Big\{ \varphi_U g^{ij} \psi_j^k g(X, \partial_k) \operatorname{div}(\partial_i)$$

$$+ \psi_j^k \left[\frac{\partial \varphi_U}{\partial x^i} g^{ij} g(X, \partial_k) + \varphi_U (\partial_i g^{ij}) g(X, \partial_k) + \varphi_U g^{ij} \partial_i [g(X, \partial_k)] \right] \Big\}$$

(as $\operatorname{div}(\partial_i) = \Gamma^\ell_{i\ell}$ and $dg^{ij} = -g^{i\ell}\omega^j_\ell - \omega^i_\ell g^{\ell j}$ where $\omega^i_j = \Gamma^i_{kj} dx^k$ are the connection 1-forms)

$$= -\int_U \Big\{ \varphi_U g^{ij} \psi_j^k g(X, \partial_k) \Gamma^\ell_{i\ell}$$

$$+ \psi_j^k \Big[\frac{\partial \varphi_U}{\partial x^i} g^{ij} g(X, \partial_k) - \varphi_U \left(g^{i\ell} \Gamma^j_{i\ell} + \Gamma^i_{i\ell} g^{\ell j} \right) g(X, \partial_k)$$

$$+ \varphi_U g^{ij} \partial_i [g(X, \partial_k)] \Big] \Big\} \, d\,\mathrm{vol}(g).$$

Also

$$J_U = \int_U \varphi_U g^{ij} \psi_k^\ell \Gamma_{ij}^k g(X, \partial_\ell)\, d\mathrm{vol}(g).$$

Summing up (and observing the cancelation of the Christoffel symbols)

$$\begin{aligned}
(X, \nabla^* \psi) &= \sum_U \{J_U - I_U\} \\
&= \sum_U \int_U \varphi_U g^{ij} \psi_k^\ell \Gamma_{ij}^k g(X, \partial_\ell) d\mathrm{vol}(g) - \sum_U \{I'_U + I''_U\} \\
&= \sum_U \int_U \varphi_U g^{ij} \left(\psi_k^\ell \Gamma_{ij}^k - \psi_j^k \Gamma_{ik}^\ell\right) g(X, \partial_\ell) d\mathrm{vol}(g) - \sum_U I'_U \\
&= \sum_U \int_U \varphi_U g^{ij} \left(\psi_k^\ell \Gamma_{ij}^k - \psi_j^k \Gamma_{ik}^\ell\right) g(X, \partial_\ell) d\mathrm{vol}(g) \\
&\quad + \int_U \Big\{ \varphi_U g^{ij} \psi_j^k g(X, \partial_k) \Gamma_{i\ell}^\ell \\
&\quad + \psi_j^k \Big[\frac{\partial \varphi_U}{\partial x^i} g^{ij} g(X, \partial_k) - \varphi_U \left(g^{i\ell} \Gamma_{i\ell}^j + \Gamma_{i\ell}^i g^{\ell j}\right) g(X, \partial_k) \\
&\quad + \varphi_U g^{ij} \partial_i [g(X, \partial_k)] \Big] \Big\} d\mathrm{vol}(g) \\
&= -\sum_U \int_U \varphi_U g^{ij} \psi_j^k \Gamma_{ik}^\ell g(X, \partial_\ell) d\mathrm{vol}(g) \\
&\quad + \sum_U \int_M g\left(X, \psi(\mathrm{grad}(\varphi_U))\right) d\mathrm{vol}(g) \\
&\quad + \sum_U \int_U \varphi_U g^{ij} \psi_j^k \partial_i [g(X, \partial_k)] d\mathrm{vol}(g)
\end{aligned}$$

and $\sum_U g\left(X, \psi(\mathrm{grad}(\varphi_U))\right) = 0$ because of $\sum_U \varphi_U = 1$. On the other hand (by Lemma 2.51)

$$\partial_i [g(X, \partial_k)] = \partial_i \left(X^m g_{mk}\right) = \frac{\partial g_{mk}}{\partial x^i} X^m + g_{mk} \partial_i \left(X^m\right).$$

Let us differentiate in $g_{mk}g^{k\ell} = \delta^\ell_m$ to get

$$\begin{aligned}\frac{\partial g_{mk}}{\partial x^i} &= -g_{\ell k} g_{mj} \frac{\partial g^{j\ell}}{\partial x^i} = -g_{\ell k} g_{mj} (dg^{j\ell})\partial_i \\ &= g_{\ell k} g_{mj} \left(g^{js}\omega^\ell_s + \omega^j_s g^{s\ell} \right) \partial_i = g_{\ell k} g_{mj} \left(g^{js}\Gamma^\ell_{is} + \Gamma^j_{is} g^{s\ell} \right) \\ &= g_{\ell k}\Gamma^\ell_{im} + g_{mj}\Gamma^j_{ik}\end{aligned}$$

from which

$$\begin{aligned}\partial_i[g(X,\partial_k)] &= \left(g_{\ell k}\Gamma^\ell_{im} + g_{mj}\Gamma^j_{ik} \right) X^m + g_{mk}\partial_i\left(X^m\right) \\ &= g_{\ell k}\left[\partial_i\left(X^\ell\right) + \Gamma^\ell_{im}X^m\right] + g_{m\ell}\Gamma^\ell_{ik}X^m.\end{aligned}$$

Therefore

$$\begin{aligned}\left(X, \nabla^*\psi\right) = &-\sum_U \int_U \varphi_U g^{ij}\psi^k_j \Gamma^\ell_{ik} g(X,\partial_\ell)\, d\,\mathrm{vol}(g) \\ &+\sum_U \int_U \varphi_U g^{ij}\psi^k_j \partial_i[g(X,\partial_k)]\, d\,\mathrm{vol}(g) \\ = &-\sum_U \int_U \varphi_U g^{ij}\psi^k_j \Gamma^\ell_{ik} g(X,\partial_\ell)\, d\,\mathrm{vol}(g) \\ &+\sum_U \int_U \varphi_U g^{ij}\psi^k_j \left(g_{\ell k}\nabla_i X^\ell + g_{m\ell}\Gamma^\ell_{ik}X^m \right) d\,\mathrm{vol}(g)\end{aligned}$$

where we have set

$$\nabla_i X^\ell = \partial_i\left(X^\ell\right) + \Gamma^\ell_{im}X^m$$

so that (by (2.121))

$$(\nabla X)|_U = \left(\nabla_i X^\ell\right) dx^i \otimes \frac{\partial}{\partial x^\ell}$$

a.e. in U. Once again we observe the cancellation of the Christoffel symbols and obtain

$$\begin{aligned}\left(X, \nabla^*\psi\right) &= \sum_U \int_U \varphi_U g^{ij}\psi^k_j g_{\ell k}\, \nabla_i X^\ell\, d\,\mathrm{vol}(g) \\ &= \sum_U \int_M \varphi_U\, g^*(\nabla X, \psi)\big|_U\, d\,\mathrm{vol}(g) = (\nabla X, \psi)\end{aligned}$$

for any $\psi \in \Gamma^\infty(\mathcal{T}^{1,1}(M))$. Here we are under the assumptions of Proposition 2.50 so M is compact. Also (by (2.121)) the local coefficients of ∇X are locally integrable so that $\nabla X \in L^1_{\text{loc}}(\mathcal{T}^{1,1}(M))$. Since ψ is arbitrary it follows that X is weakly differentiable and its covariant derivative is the $(1,1)$-tensor field ∇X (locally given by (2.121)). To show that $X \in \mathcal{H}^{1,2}_g(T(M))$, it remains to be checked that $\nabla X \in L^2(T(M))$. To this end one observes first that X^i, $\partial_i\left(X^j\right) \in L^2(U)$ yields $\nabla_i X^j \in L^2(U)$ and then performs the estimates

$$\begin{aligned}
\|\nabla X\|^2_{L^2(M)} &= \sum_U \int_U \varphi_U g^{ij} \nabla_i X^k \nabla_j X^\ell g_{k\ell}\, d\,\text{vol}(g) \\
&\leq \sum_U C_U \int_U \varphi_U \left(\sum_{i,j} \nabla_i X^j \right)^2 d\,\text{vol}(g) \\
&\leq 2 \sum_{U,i,j} C_U \left\| \nabla_i X^j \right\|^2_{L^2(U)} < \infty
\end{aligned}$$

(where $C = \max\{\sup_{\overline{U}} \left| g^{ij} g_{k\ell} \right| : 1 \leq i,j,k,\ell \leq n\}$) and we may conclude. The opposite inclusion $\mathcal{H}^{1,2}_g(\mathcal{T}^{1,0}(M)) \subseteq H(T(M))$ is left as an exercise to the reader.

We may adopt the following

Definition 2.52 We say $X \in \mathcal{H}^1_g(T(M))$ is a *weak solution* to the harmonic vector fields system (2.28) if

$$\int_M \{g(\nabla X, \nabla Y) - \|\nabla X\|^2 g(X, Y)\} d\,\text{vol}(g) = 0$$

for any $Y \in \mathfrak{X}^\infty_0(M)$. A *weak harmonic vector field* is a unit vector field $X \in \mathcal{H}^1_g(T(M))$ which is a weak solution to (2.28). ■

The study of weak harmonic vector fields (existence, local properties, etc.) is an open problem. As well as for harmonic maps, one should not expect smoothness for a weak harmonic vector field in general. Indeed if $n \geq 3$ and

$$X = \frac{x^i}{|x|} \frac{\partial}{\partial x^i}, \quad |x| = \left(\sum_{i=1}^n x_i^2 \right)^{1/2},$$

then $X \in \mathcal{H}^1_{g_0}(T(\mathbb{R}^n))$ and X is a weak harmonic vector field on $\mathbb{R}^n$ (endowed with the standard flat metric g_0) yet X is not smooth. The fact that X is a weak solution to $\Delta_{g_0} X - \|\nabla X\|^2 X = 0$ may be easily checked, as follows. First, it should be observed that given an open set $\Omega \subset \mathbb{R}^n$ and a smooth vector field $X \in \Gamma^\infty(S(\Omega))$ written as

$$X = u^i \frac{\partial}{\partial x^i}, \quad \sum_{i=1}^n u_i^2 = 1,$$

then X is a harmonic vector field if and only if $u = (u^1, \ldots, u^n) : \Omega \to S^{n-1}$ is a harmonic map. Let $Y \in \mathfrak{X}_0^\infty(\mathbb{R}^n)$ and $\epsilon > 0$. As the map

$$u : \mathbb{R}^n \setminus \{0\} \to S^{n-1}, \quad u(x) = |x|^{-1} x, \quad x \neq 0,$$

is harmonic it follows that X is harmonic, and in particular weakly harmonic, as a unit vector field on $\mathbb{R}^n \setminus B_\epsilon$. Here B_ϵ is the Euclidean ball $B_\epsilon = \{x \in \mathbb{R}^n : |x| < \epsilon\}$. Hence

$$\int_{\mathbb{R}^n} \{g_0(\nabla X, \nabla Y) - \|\nabla X\|^2 g_0(X, Y)\} d\,\mathrm{vol}(g_0)$$
$$= \int_{B_\epsilon} \{g_0(\nabla X, \nabla Y) - \|\nabla X\|^2 g_0(X, Y)\} d\,\mathrm{vol}(g_0).$$

Moreover

$$\left| \int_{B_\epsilon} \{g_0(\nabla X, \nabla Y) - \|\nabla X\|^2 g_0(X, Y)\} d\,\mathrm{vol}(g_0) \right|$$
$$\leq \int_{B_\epsilon} \{ |g_0(\nabla X, \nabla Y)| + \|\nabla X\|^2 \, |g_0(X, Y)| \} d\,\mathrm{vol}(g_0)$$

(by the Cauchy-Schwartz inequality)

$$\leq \int_{B_\epsilon} \{ \|\nabla X\| \, \|\nabla Y\| + \|\nabla X\|^2 \, \|Y\| \} d\,\mathrm{vol}(g_0)$$
$$\leq C \int_{B_\epsilon} \|\nabla X\| \, (1 + \|\nabla X\|) d\,\mathrm{vol}(g_0)$$

where

$$C = \max\left\{ \sup_{x \in \Gamma} \|Y\|_x, \sup_{x \in \Gamma} \|\nabla Y\|_x \right\} > 0, \quad \Gamma = \mathrm{supp}(Y).$$

On the other hand

$$\|\nabla X\|_x = \frac{\sqrt{n-1}}{|x|}, \quad x \in \mathbb{R}^n \setminus \{0\},$$

hence

$$\int_{B_\epsilon} \|\nabla X\| \, (1 + \|\nabla X\|) d\mathrm{vol}(g_0)$$

$$= \int_0^\epsilon d\rho \int_{|x|=\rho} \|\nabla X\| \, (1 + \|\nabla X\|) dA$$

$$= \sqrt{n-1} \int_0^\epsilon d\rho \int_{|x|=\rho} \frac{1}{\rho}\left(1 + \frac{\sqrt{n-1}}{\rho}\right) dA$$

where dA is the Riemannian volume form on the sphere $|x| = \rho$. Let ω_n be the area of the unit $(n-1)$-dimensional sphere in $\mathbb{R}^n$. Finally

$$\int_{B_\epsilon} \|\nabla X\| \, (1 + \|\nabla X\|) \, d\mathrm{vol}(g_0)$$

$$= \omega_n \left(\frac{\epsilon}{\sqrt{n-1}} + \frac{n-1}{n-2} \right) \epsilon^{n-2} \to 0, \quad \epsilon \to 0.$$

Let (M, g) be a compact orientable Riemannian manifold. The study (e.g., existence of minimizers) of functionals (Dirichlet energy integrals with variable exponent) of the form

$$\int_M \|\nabla X\|^{p(x)} d\mathrm{vol}(g)(x),$$

where $p : M \to [1, +\infty)$ is a given measurable function, is an open problem. See P.A. Hästö, [162], for the scalar case. Let us look at the total bending functional $\mathcal{B} : \mathcal{H}^1_g(T(M)) \to [0, +\infty)$ (where $p(x) = 2$ for any $x \in M$). Clearly

$$\mathcal{B}(tX + (1-t)Y) \leq t\mathcal{B}(X) + (1-t)\mathcal{B}(Y),$$

for any $t \in [0, 1]$ and any $X, Y \in \mathcal{H}^1_g(T(M))$ i.e., $\mathcal{B}$ is convex. Moreover let $\{X_\nu\}_{\nu \geq 1}$ be a sequence in $\mathcal{H}^1_g(T(M))$ such that $X_\nu \to X$ as $\nu \to \infty$ for some $X \in \mathcal{H}^1_g(T(M))$. Then

$$\left| \int_M g(\nabla X_\nu - \nabla X, \nabla X) d\,\mathrm{vol}(g) \right| \le \int_M \|\nabla X_\nu - \nabla X\| \, \|\nabla X\| d\,\mathrm{vol}(g) \le \|\nabla X_\nu - \nabla X\|_{L^2(T(M))} \, \mathcal{B}(X)^{1/2} \to 0,$$

as $\nu \to \infty$. That is

$$\lim_{\nu \to \infty} \int_M g(\nabla X_\nu - \nabla X, \nabla X) d\,\mathrm{vol}(g) = 0.$$

Also

$$\mathcal{B}(X_\nu) = \int_M \|\nabla X_\nu\|^2 \, d\,\mathrm{vol}(g) = \mathcal{B}(X) + \|\nabla X_\nu - \nabla X\|^2_{L^2(T(M))} + 2 \int_M g(\nabla X_\nu - \nabla X, \nabla X) \, d\,\mathrm{vol}(g)$$

so that $\mathcal{B}(X_\nu) \to \mathcal{B}(X)$ as $\nu \to \infty$. In particular $\mathcal{B}$ is lower semi-continuous.[4] Existence of minima for $\mathcal{B} : \mathcal{H}^1_g(T(M)) \to [0, +\infty)$ would then follow from general results[5] in the calculus of variations provided that $\mathcal{B}$ is also coercive.[6]

Let us set

$$\mathcal{H}^1_g(S(M)) = \left\{ X \in \mathcal{H}^1_g(T(M)) : g(X, X) = 1 \quad \text{a.e.} \quad \text{in } M \right\}.$$

As to the existence of minimizers to $\mathcal{B} : \mathcal{H}^1_g(S(M)) \to \mathbb{R}$ we may quote the following

Theorem 2.53 (G. Nunes & J. Ripoll, [224]) *Let (M, g) be a compact orientable n-dimensional Riemannian manifold with $n \geq 3$.*

[4] A functional $I : \mathfrak{X} \to \mathbb{R}$ defined on a Banach space $\mathfrak{X}$ is *lower semi-continuous* if $I(u) \le \liminf_{i \to \infty} I(u_i)$ for any sequence $u_i \in \mathfrak{X}$ converging in B to $u \in \mathfrak{X}$ as $i \to \infty$.

[5] Let X be a reflexive Banach space. If $I : X \to [0, +\infty)$ is a convex, lower semi-continuous and coercive functional, then there is an element in X which minimizes I (cf. e.g., Theorem 2.1 in [188]). The problem of finding conditions (presumably some sort of Poincaré inequality) which ensure coercivity of $\mathcal{B}$ is left open.

[6] A functional $I : \mathfrak{X} \to \mathbb{R}$ defined on a Banach space $\mathfrak{X}$ is *coercive* if $I(u_i) \to +\infty$ whenever $\|u_i\|_{\mathfrak{X}} \to +\infty$.

i. *There is a finite subset* $\{x_1, \ldots, x_m\} \subset M$ *and a unit vector field* $X \in \mathcal{H}^1_g(S(M))$ *defined on* $M \setminus \{x_1, \ldots, x_m\}$. *In particular* $\mathcal{H}^1_g(S(M))$ *is nonempty.*

ii. *There is* $X_0 \in \mathcal{H}^1_g(S(M))$ *such that*

$$\mathcal{B}(X_0) = \inf\left\{\mathcal{B}(X) : X \in \mathcal{H}^1_g(S(M))\right\}.$$

iii. *If* M *is parallelizable then*

$$\inf\left\{\mathcal{B}(X) : X \in \Gamma^\infty(S(M))\right\} = \inf\left\{\mathcal{B}(X) : X \in \mathcal{H}^1_g(S(M))\right\}.$$

We only prove (ii). For the proof of (i) and (iii), the reader may see [224], p. 1096–1097. Let $\{X_\nu\}_{\nu \ge 1}$ be a minimizing sequence[7] for $\mathcal{B}$ in $\mathcal{H}^1_g(S(M))$. As

$$\|X_\nu\|_{L^2(T(M))} = \mathrm{Vol}(M), \quad \nu \ge 1,$$

the sequence $\{X_\nu\}_{\nu \ge 1}$ is bounded in $\mathcal{H}^1_g(T(M))$. On the other hand $\mathcal{H}^1_g(T(M))$ is a Hilbert space, hence it is reflexive. Therefore we may apply a classical result in functional analysis (i.e., the Eberlein-Smulian theorem[8]) to conclude that there is a subsequence $\{X_{\nu_k}\}_{k \ge 1}$ converging weakly[9] to some $X_0 \in \mathcal{H}^1_g(T(M))$. We set $Y_k = X_{\nu_k}$ for simplicity. By Kondrakov's theorem[10] for Riemannian manifolds $\mathcal{H}^1_g(T(M))$ is compactly embedded[11] in $L^2(T(M))$. Hence there is a subsequence $\{Y_{k_j}\}_{j \ge 1}$

[7] Let $\mathfrak{X}$ be a vector space and $\sigma \subset \mathfrak{X}$ a subset. Let $f : \mathfrak{X} \to \mathbb{R}$ be a function such that $d = \inf_{x \in \sigma} f(x) > -\infty$. A sequence $\{x_\nu\}_{\nu \ge 1} \subset \sigma$ is *minimizing* for f if $f(x_\nu) \to d$ as $\nu \to \infty$.

[8] A Banach space $\mathfrak{X}$ is reflexive if and only if any bounded sequence in $\mathfrak{X}$ admits a weakly convergent subsequence (cf. e.g., Theorem A.27 in [97], p. 623).

[9] That is $\lim_{k \to \infty} \left(X_{\nu_k}, X\right)_{\mathcal{H}^1_g(T(M))} = (X_0, X)_{\mathcal{H}^1_g(T(M))}$ for any $X \in \mathcal{H}^1_g(T(M))$.

[10] Cf. e.g., Theorem 2.34 in [18], p. 55. Precisely if M is a compact n-dimensional Riemannian manifold without boundary and $1 \ge 1/p > 1/q - k/n > 0$ then the embedding operator $\mathcal{H}^{k,q}_g(M) \hookrightarrow L^p(M)$ is compact. A slight modification of the proof (in [18], p. 55–56) shows that $\mathcal{H}^{1,2}_g(T(M)) \hookrightarrow L^2(T(M))$ is compact as well.

[11] Let $\mathfrak{X}$ be a normed vector space. A subset $A \subset \mathfrak{X}$ is *compact* if every sequence of points in A has a subsequence converging in $\mathfrak{X}$ to an element of A. Also $A \subset \mathfrak{X}$ is *precompact* if its closure $\overline{A}$ (in the norm topology) is compact. An operator $f : \mathfrak{X} \to \mathfrak{Y}$ of normed vector spaces is *compact* if $f(A)$ is precompact in $\mathfrak{Y}$ for any bounded subset $A \subset \mathfrak{X}$. A normed vector space $\mathfrak{X}$ is *embedded* in the normed vector space $\mathfrak{Y}$ if $\mathfrak{X}$ is a vector subspace of Y and the identity operator $I : \mathfrak{X} \to \mathfrak{Y}$, $I(x) = x$, $x \in \mathfrak{X}$, is continuous (equivalently there is a constant $C > 0$ such that $\|x\|_{\mathfrak{Y}} \le C\|x\|_{\mathfrak{X}}$ for any $x \in \mathfrak{X}$). Then $\mathfrak{X}$ is *compactly embedded* in $\mathfrak{X}$ if the embedding operator $I : \mathfrak{X} \to \mathfrak{Y}$ is compact.

of $\{Y_k\}_{k\geq 1}$ converging to X_0 in the L^2 norm. We set $Z_j = Y_{k_j}$ for simplicity. As $\|Z_j - X_0\|_{L^2(T(M))} \to 0$ for $j \to \infty$ there is a subsequence $\{Z_{j_k}\}_{k\geq 1}$ of $\{Z_j\}_{j\geq 1}$ converging to X_0 almost everywhere in M. Thus $g(X_0, X_0) = 1$ a.e. in M so that $X_0 \in \mathcal{H}^1_g(S(M))$. The lower semicontinuity of the total bending functional $\mathcal{B} : \mathcal{H}^1_g(S(M)) \to \mathbb{R}$ then yields

$$\int_M g(X_0, X_0)\, d\,\mathrm{vol}(g) = \inf_{X \in \mathcal{H}^1_g(S(M))} \mathcal{B}(X).$$

CHAPTER THREE

Harmonicity and Stability

Contents

One of the purposes of this chapter is to discuss Hopf and unit Killing vector fields in the context of the theory of harmonic vector fields on Riemannian manifolds. The two classes of vector fields are related (cf. Theorem 3.5). Also unit Killing vector fields on Einstein manifolds are harmonic (cf. Proposition 3.7). More can be said on unit Killing vector fields on real space forms $M^n(c)$ of (constant) sectional curvature $c \geq 0$. Indeed these are harmonic maps (cf. Theorem 3.9) and are actually parallel when $c = 0$. Starting from the result by S.D. Han & J.W. Yim, [157] (that the only harmonic vector fields on a sphere S^3 are the Hopf vector fields, cf. Theorem 3.10) we report on recent findings by D. Perrone (cf. [244]) on unit Killing vector fields on 3-dimensional Riemannian manifolds (cf. Theorem 3.14). Motivated by a result of Y.L. Xin, [321], in harmonic map theory (cf. Theorem 3.17 in this chapter) one expects instability of Hopf vector fields on spheres. The issue is addressed in Theorem 3.18 and as it turns out, this is indeed the case for Hopf vector fields on S^{2m+1} for $m \geq 2$. As we argue in this chapter, a possible way to circumvent this difficulty (resulting in a harmonious further development of the theory) is to

© 2012 Elsevier Inc. All rights reserved.

replace the Dirichlet functional E by *Brito's functional*

$$\tilde{E}(V) = E(V) + \frac{(n-1)(n-3)}{2} \int_M \|H_V\|^2 d\,\mathrm{vol}(g)$$

where H_V is the mean curvature vector of the distribution $(\mathbb{R}V)^\perp$ (cf. Definition 3.29 below). Then Hopf vector fields on S^{2m+1} are absolute minima of $\tilde{E} : \Gamma^\infty(S(S^{2m+1})) \to [0, +\infty)$ (cf. F. Brito, [71], and Theorem 3.30). Section 3.6 of Chapter 3 is a report on vector fields with singularities, related to the work by E. Boeckx & L. Vanhecke, [51], and F. Brito & P.G. Walczak, [72]. Chapter 3 ends with the presentation of the results of G. Wiegmink, [309], on harmonic vector fields on Riemannian tori.

3.1. HOPF VECTOR FIELDS ON SPHERES

Let $(\mathbb{R}^{2m+2}, g_0)$ be the Euclidean space carrying the standard complex structure $J_0 : \mathbb{R}^{2m+2} \to \mathbb{R}^{2m+2}$ i.e.,

$$J_0(x_1, y_1, \ldots, x_{m+1}, y_{m+1}) = (-y_1, x_1, \ldots, -y_{m+1}, x_{m+1}),$$

for any $(x_1, y_1, \ldots, x_{m+1}, y_{m+1}) \in \mathbb{R}^{2m+2}$. We identify as customary $\mathbb{R}^{2m+2}$ and $\mathbb{C}^{m+1}$ by setting $z_j = x_j + iy_j$ for any $1 \leq j \leq m+1$, where $i = \sqrt{-1}$. Then $J_0 z = iz$ for any $z \in \mathbb{C}^{m+1}$. The same symbol J_0 denotes the natural almost complex structure

$$J_0 : T(\mathbb{C}^{m+1}) \to T(\mathbb{C}^{m+1}),$$

$$J_0\left(\frac{\partial}{\partial x_j}\right) = \frac{\partial}{\partial y_j}, \quad J_0\left(\frac{\partial}{\partial y^j}\right) = -\frac{\partial}{\partial x_j}, \quad 1 \leq j \leq m+1.$$

Then $J_0^2 = -I$ (where I denotes the identical transformation of $T(\mathbb{C}^{m+1})$) and J_0 is compatible to g_0 i.e., $g_0(J_0 X, J_0 Y) = g_0(X, Y)$ for any $X, Y \in \mathfrak{X}(\mathbb{C}^{m+1})$. Also $\nabla^0 J_0 = 0$ where ∇^0 is the Levi-Civita connection of $(\mathbb{C}^{m+1}, g_0)$ i.e., the metric g_0 is Kaehlerian. We define as usual

$$\frac{\partial}{\partial z_j}, \frac{\partial}{\partial \bar{z}_j} \in \Gamma^\infty(T(\mathbb{C}^{m+1}) \otimes \mathbb{C}), \quad 1 \leq j \leq m+1,$$

$$\frac{\partial}{\partial z_j} = \frac{1}{2}\left(\frac{\partial}{\partial x^j} - i\frac{\partial}{\partial y^j}\right), \quad \frac{\partial}{\partial \bar{z}_j} = \frac{1}{2}\left(\frac{\partial}{\partial x_j} + i\frac{\partial}{\partial y_j}\right).$$

Let us extend J_0 to $T(\mathbb{C}^{m+1})$ by $\mathbb{C}$-linearity (and denote the extension by the same symbol J_0). Then

$$J_0\left(\frac{\partial}{\partial z_j}\right) = i\frac{\partial}{\partial z_j}, \quad J_0\left(\frac{\partial}{\partial \overline{z}^j}\right) = -i\frac{\partial}{\partial \overline{z}_j}, \quad 1 \leq j \leq m+1.$$

Let $S^{2m+1} = \{(z_1, \ldots, z_{m+1}) \in \mathbb{C}^{m+1} : \sum_{j=1}^{m+1} |z_j|^2 = 1\}$ be the standard unit sphere in $\mathbb{C}^{m+1}$ and let g be the induced Riemannian metric on S^{2m+1} i.e., $g = \iota^* g_0$ where $\iota : S^{2m+1} \to \mathbb{C}^{m+1}$ is the inclusion. Let ν be the unit normal field on S^{2m+1} given by

$$\nu(p) = \vec{p}, \quad p \in S^{2m+1}.$$

Here for each $p \in \mathbb{R}^{2m+2}$ we set

$$\vec{p} = \sum_{j=1}^{m+1} \left\{ x_j(p) \frac{\partial}{\partial x_j}\bigg|_p + y_j(p) \frac{\partial}{\partial y_j}\bigg|_p \right\} \in T_p(\mathbb{R}^{2m+2}).$$

Let us consider $\xi_0 = J_0 \nu$. Then ξ_0 is tangent to S^{2m+1} and $\|\xi_0\| = 1$. The fibres of the canonical map $\pi : S^{2m+1} \to \mathbb{C}P^m$ are the leaves of a foliation $\mathcal{F}$ of S^{2m+1} by great circles and ξ_0 spans $T(\mathcal{F})$. The tangent bundle to $\mathcal{F}$ is the vertical bundle $\mathrm{Ker}(d\pi)$.

Definition 3.1 The tangent vector field ξ_0 is referred to as the *(standard) Hopf vector* field on S^{2m+1}. ■

For each $X \in \mathfrak{X}(S^{2m+1})$ we denote by $-\varphi X$ the tangential component of $J_0 X$ so that

$$J_0 X = -\varphi X + g_0(J_0 X, \nu)\nu = -\varphi X - g(X, \xi_0)\nu. \tag{3.1}$$

Then φ is an endomorphism of $T(S^{2m+1})$ such that $\varphi(\xi_0) = 0$ and $\varphi X = -J_0 X$ for any $X \in \mathfrak{X}(S^{2m+1})$ which is orthogonal to ξ_0. Moreover

$$g(\varphi X, Y) = -g_0(J_0 X, Y) = g_0(X, J_0 Y) = -g(X, \varphi Y)$$

for any $X, Y \in T(S^{2m+1})$, hence φ is a skew-symmetric $(1,1)$-tensor field on S^{2m+1}.

Proposition 3.2 *The standard Hopf vector field $\xi_0 \in T(S^{2m+1})$ is a unit, vertical and Killing vector field.*

Proof. We shall need the Gauss formula of $S^{2m+1} \hookrightarrow \mathbb{R}^{2m+2}$

$$\nabla^0_X Y = \nabla_X Y + g_0(\nabla^0_X Y, \nu)\nu, \quad X, Y \in \mathfrak{X}(S^{2m+1}),$$

where ∇ is the Levi-Civita connection of (S^{2m+1}, g). An elementary calculation shows that

$$\nabla^0_X \nu = X, \quad X \in \mathfrak{X}(\mathbb{R}^{2m+2}), \tag{3.2}$$

so that the Gauss formula may also be written as

$$\nabla^0_X Y = \nabla_X Y - g(X, Y)\nu, \quad X, Y \in \mathfrak{X}(S^{2m+1}).$$

In particular (for $Y = \xi_0$)

$$\nabla^0_X \xi_0 = \nabla_X \xi_0 - g(X, \xi_0)\nu. \tag{3.3}$$

Let us use the identities (3.1)–(3.2) and $(\nabla^0_X J_0)\nu = 0$ to compute

$$\nabla^0_X \xi_0 = \nabla_X J_0 \nu = J_0 \nabla_X \nu = J_0 X = -\varphi X - g(X, \xi_0)\nu.$$

Together with (3.3) this yields

$$\varphi X = -\nabla_X \xi_0, \quad X \in \mathfrak{X}(S^{2m+1}).$$

Yet, as previously observed, φ is skew-symmetric hence ξ_0 is a Killing vector field on S^{2m+1}. We refer to [253], Chapter 9, for a study on Killing vector fields. ■

Definition 3.3 An *orthogonal complex* structure on $\mathbb{R}^{2m+2}$ is an orthogonal matrix $J \in O(2m+2)$ such that $J^2 = -I_{2m+2}$. ■

Here I_n denotes the $n \times n$ unit matrix. An orthogonal complex structure J on $\mathbb{R}^{2m+2}$ determines a $(1,1)$-tensor field on $\mathbb{R}^{2m+2}$, denoted by the same symbol J, such that $J^2 = -I$ and $g_0(JX, JY) = g_0(X, Y)$ for any $X, Y \in T(\mathbb{R}^{2m+2})$. Indeed if $J = [J^A_B]$ we may set

$$J\frac{\partial}{\partial x^A} = J^B_A \frac{\partial}{\partial x^B},$$

where $(x^1, \ldots, x^{2m+2})$ are the Cartesian coordinates on $\mathbb{R}^{2m+2}$ and the range of indices is

$$1 \leq A, B, \cdots \leq 2m+2.$$

Note that $g_0(JX, JY) = g_0(X, Y)$ may equivalently be written as $J^t \circ J = I$ where J^t is determined by $g_0(J^t X, Y) = g_0(X, JY)$ for any $X, Y \in T(\mathbb{R}^{2m+2})$.

Definition 3.4 A tangent vector field $\xi \in \mathfrak{X}(S^{2m+1})$ is said to be a *Hopf vector field* on S^{2m+1} if $\xi = J\nu$ for some orthogonal complex structure J on $\mathbb{R}^{2m+2}$. ■

Theorem 3.5 (G. Wiegmink, [310]) *The Hopf vector fields on S^{2m+1} are precisely the unit Killing vector fields.*

Proof. Let ξ be a unit Killing vector field on S^{2m+1}. Let us show ξ is also a Hopf vector field. Let

$$\Phi : \mathbb{R} \times S^{2m+1} \to S^{2m+1}, \quad (t,p) \mapsto \Phi(t,p),$$

be the global 1-parameter group of global transformations obtained by integrating ξ. As ξ is a Killing vector field the transformations $\Phi_t : S^{2m+1} \to S^{2m+1}$ given by $\Phi_t(p) = \Phi(t,p)$, for any $p \in S^{2m+1}$, are isometries of (S^{2m+1}, g) for any $t \in \mathbb{R}$. Consequently if $\{e_j : 1 \le j \le 2m+2\}$ is the canonical basis in $\mathbb{R}^{2m+2}$ and $\Phi_t(e_j) = a^i_j(t) e_i$ then $[a^i_j(t)] \in O(2m+2)$ for any $t \in \mathbb{R}$. Therefore we may consider the smooth curve

$$\gamma : \mathbb{R} \to O(2m+2), \quad \gamma(t) = \left[a^i_j(t)\right]_{1 \le i,j \le 2m+2}, \quad t \in \mathbb{R}.$$

As $\Phi_0 = 1_{S^{2m+1}}$ it follows that $\gamma(0) = [\delta^i_j] = I_{2m+2}$ (the unit matrix). Then

$$\dot{\gamma}(0) \in T_{I_{2m+2}}(O(2m+2)) = \mathbf{so}(2m+2).$$

Here $\mathbf{so}(2m+2)$ is the Lie algebra of $O(2m+2)$ (the skew-symmetric $(2m+2) \times (2m+2)$ matrices). If $p \in \mathbb{R}^{2m+2}$ we define

$$J_p : T_p(\mathbb{R}^{2m+2}) \to T_p(\mathbb{R}^{2m+2}), \quad J_p \left.\frac{\partial}{\partial x_j}\right|_p = A^i_j \left.\frac{\partial}{\partial x^i}\right|_p,$$

where the matrix $[A^i_j]$ is given by

$$A^i_j = \frac{da^i_j}{dt}(0), \quad 1 \le i,j \le 2m+2.$$

Then

$$(J\nu)_p = J_p \vec{p} = x^j(p) A^i_j \left.\frac{\partial}{\partial x^i}\right|_p = \xi_p$$

because of

$$\xi_p = \frac{dC_p}{dt}(0), \quad C_p(t) = \Phi_t(p), \quad p \in S^{2m+1}, \ t \in \mathbb{R}.$$

Finally J is an orthogonal transformation because

$$g_{0,p}(J_p\vec{p}, J_p\vec{p}) = g_0(J\nu, J\nu)_p = g_0(\xi,\xi)_p = 1 = g_{0,p}(\vec{p},\vec{p}),$$

for any $p \in S^{2m+1}$ hence $g_{0,p}(J_p\vec{p}, J_p\vec{p}) = g_{0,p}(\vec{p},\vec{p})$ for any $p \in \mathbb{R}^{2m+2}$.

Vice versa, let $\xi \in \mathfrak{X}(S^{2m+1})$ be a Hopf vector field on S^{2m+1} i.e., there is an orthogonal complex structure J on $\mathbb{R}^{2m+2}$ such that $J\nu = \xi$. Then (as J is an isometry of $(\mathbb{R}^{2m+2}, g_0)$)

$$g(\xi,\xi) = g_0(J\nu, J\nu) = g_0(\nu,\nu) = 1$$

hence ξ is a unit vector field. Let us set

$$J\frac{\partial}{\partial x^j} = J_j^i \frac{\partial}{\partial x^i}, \quad [J_j^i] \in O(2m+2) \cap \mathbf{so}(2m+2).$$

Then for any $p \in S^{2m+1}$

$$\xi_p = J_p \vec{p} = x^j(p) J_j^i \left. \frac{\partial}{\partial x^i} \right|_p$$

i.e., $\xi^i = x^j J_j^i$, $1 \leq i \leq 2m+2$. Then for any $X \in \mathfrak{X}(\mathbb{R}^{2m+2})$

$$X(\xi^i) = X^j J_j^i + x^j X(J_j^i) = X^j J_j^i$$

hence

$$\nabla^0_X \xi = X(\xi^i) \frac{\partial}{\partial x^i} = JX.$$

Therefore for each $X \in \mathfrak{X}(S^{2m+1})$

$$(\nabla^0_X J)\nu = \nabla^0_X J\nu - J\nabla^0_X \nu = \nabla^0_X \xi - JX = 0.$$

Now we may show that ξ is a Killing vector field along the lines of the proof of Proposition 3.2. ∎

Remark 3.6

i. Let $\xi = J\nu$ be a Hopf vector field on S^{2m+1}. For any $X \in \mathfrak{X}(S^{2m+1})$ (by the Gauss formula)

$$\begin{aligned}\nabla^0_X \xi &= \nabla_X \xi + g(\nabla^0_X \xi, \nu)\nu = \nabla_X \xi + g(JX, \nu)\nu \\ &= \nabla_X \xi - g(X, J\nu)\nu = \nabla_X \xi - g(X,\xi)\nu\end{aligned}$$

hence

$$\nabla_X \xi = JX, \quad X \in (\mathbb{R}\xi)^\perp. \tag{3.4}$$

ii. If ξ is a Hopf vector field on S^{2m+1} then it may be easily shown that (φ, ξ, η, g) is a Sasakian structure on S^{2m+1} where $\varphi = -\nabla\xi$ and $\eta = g(\xi, \cdot)$.

iii. For a sphere $S^{2m+1}(r)$ of radius r the notions of standard Hopf vector field ξ_0 and arbitrary Hopf vector field ξ may be defined as on a unit sphere. However in that case $\nu_p = (1/r)\vec{p}$ for any $p \in S^{2m+1}(r)$. Moreover Wiegmink's result (cf. Theorem 3.5 above) that the Hopf vector fields are precisely the unit Killing vector fields holds on $S^{2m+1}(r)$ as well (the proof is a straightforward adaptation of the proof of Theorem 3.5). ■

Proposition 3.7 *Let (M,g) be a real n-dimensional Riemannian manifold. If V is a Killing vector field on M then $\Delta_g V = QV$ where Δ_g is the rough Laplacian and Q the Ricci operator. In particular if (M,g) is an Einstein manifold then any unit Killing vector field V on M is harmonic, $\|\nabla V\|^2 = \rho/n$, and if M is compact then $E(V) = \frac{1}{2}(n + \rho/n)$ where ρ is the scalar curvature of (M,g).*

Proof. For each $p \in M$ let $\{E_i : 1 \le i \le n\}$ be a local orthonormal frame on $T(M)$ defined on the open neighborhood $U \subseteq M$ of p such that $(\nabla E_i)_p = 0$ for any $1 \le i \le n$. Next, for each $v \in T_p(M)$ let $X \in \mathfrak{X}(U)$ such that $X_p = v$ and $(\nabla X)_p = 0$. Note that

$$[X, E_i]_p = (\nabla_X E_i)_p - (\nabla_{E_i} X)_p = 0.$$

Let us set $e_i = E_i(p)$ for simplicity. Then

$$g_p((\Delta_g V)_p, v) = g(\Delta_g V, X)_p = -\sum_{i=1}^{n} g(\nabla_{E_i}\nabla_{E_i} V - \nabla_{\nabla_{E_i} E_i} V, X)_p$$

(as $\left(\nabla_{\nabla_{E_i} E_i} V\right)_p$ depends only on the value of $\nabla_{E_i} E_i$ at p, which is zero)

$$= -\sum_i g(\nabla_{E_i}\nabla_{E_i} V, X)_p = -\sum_i \left\{e_i(g(\nabla_{E_i} V, X)) - g(\nabla_{E_i} V, \nabla_{E_i} X)_p\right\}$$
$$= -\sum_i e_i(g(\nabla_{E_i} V, X)).$$

On the other hand

$$g(QV, X)_p = \sum_{i=1}^{n} R(V, E_i, X, E_i)_p = \sum_i g(R(X, E_i)V, E_i)_p$$
$$= \sum_i g(-\nabla_X \nabla_{E_i} V + \nabla_{E_i}\nabla_X V + \nabla_{[X,E_i]} V, E_i)_p$$

(as $[X, E_i]_p = 0$ yields $\left(\nabla_{[X,E_i]} V\right)_p = 0$)

$$= \sum_i g(-\nabla_X \nabla_{E_i} V + \nabla_{E_i} \nabla_X V, E_i)_p$$

$$= -\sum_i \{X(g(\nabla_{E_i} V, E_i)) - g(\nabla_{E_i} V, \nabla_X E_i)$$

$$-E_i(g(\nabla_X V, E_i)) + g(\nabla_X V, \nabla_{E_i} E_i)\}_p$$

(as $(\nabla E_i)_p = 0$)

$$= -\sum_i \{X(g(\nabla_{E_i} V, E_i)) - E_i(g(\nabla_X V, E_i))\}_p$$

$$= -X_p(\mathrm{div}(V)) + \sum_i e_i \left((\mathcal{L}_V g)(X, E_i) - g(\nabla_{E_i} V, X)\right)$$

hence

$$g(QV, X)_p = g(\Delta_g V, X)_p - v(\mathrm{div}(V)) + \sum_i e_i((\mathcal{L}_V g)(X, E_i)). \tag{3.5}$$

As V is a Killing vector field one has $\mathcal{L}_V g = 0$ and $\mathrm{div}(V) = 0$ hence (3.5) yields $\Delta_g V = QV$. ■

Remark 3.8

i. When M is compact one may show (cf. [258], p. 171) that $\mathrm{div}(V) = 0$ and $\Delta_g V = QV$ imply that V is a Killing vector field.

ii. Let h_V be the $(1,1)$-tensor field determined by $(\mathcal{L}_V g)(X, Y) = g(h_V X, Y)$ for any $X, Y \in \mathfrak{X}(M)$. Then

$$\sum_{i=1}^n e_i((\mathcal{L}_V g)(X, E_i)) = \sum_i e_i(g(h_V X, E_i))$$

$$= \sum_i \left\{g(\nabla_{E_i} h_V X, E_i) + g(h_V X, \nabla_{E_i} E_i)\right\}_p$$

(as $(\nabla E_i)_p = 0$)

$$= \sum_i g(\nabla_{E_i} h_V X, E_i)_p = \mathrm{div}(h_V X)_p$$

hence the identity (3.5) becomes

$$g(\Delta_g V, X)_p = g(QV, X)_p + g(\nabla \mathrm{div}(V), X)_p - \mathrm{div}(h_V X)_p. \tag{3.6}$$

In particular if V is an infinitesimal conformal transformation i.e., $\mathcal{L}_V g = 2\sigma g$ for some $\sigma \in C^\infty(M)$ then

$$\operatorname{div}(V) = n\sigma, \quad h_V = 2\sigma I, \quad \operatorname{div}(h_V X) = 2g(\nabla\sigma, X)$$

hence

$$\Delta_g V = QV + (n-2)\nabla\sigma. \tag{3.7}$$

iii. Let V be a unit Killing vector field and X a unit vector field orthogonal to V. Then $\varphi = \nabla V$ is skew-symmetric and $\varphi V = 0$. Consequently for any $p \in M$ and any $v \in T_p(M) \setminus \{0\}$ such that $g_p(V_p, v) = 0$ the sectional curvature of the 2-plane $\alpha \subset T_p(M)$ spanned by $\{V_p, v\}$ is nonnegative. Indeed let $X \in \mathfrak{X}(M)$ such that $X_p = v$. Then

$$\begin{aligned} k(\alpha) &= R(V,X,V,X)_p = g(R(V,X)V,X)_p \\ &= -g(\nabla_V\nabla_X V - \nabla_X\nabla_V V - \nabla_{[V,X]}V, X)_p \end{aligned}$$

(as $(\nabla_V g)(\nabla_X V, X) = 0$)

$$\begin{aligned} &= -V_p(g(\nabla_X V, X)) + g(\nabla_X V, \nabla_V X)_p \\ &\quad + g(\nabla_X \varphi V, X)_p + g(\varphi[V,X], X)_p \end{aligned}$$

(as $\varphi V = 0$ and φ is skew)

$$= g(\varphi X, \nabla_V X)_p - g([V,X], \varphi X)_p = \|\nabla_X V\|_p^2 \geq 0.$$

Therefore, if $k(\alpha) > 0$ for any $v \in \left(\mathbb{R}V_p\right)^\perp$ and any $p \in M$ then $\varphi_p : \left(\mathbb{R}V_p\right)^\perp \to \left(\mathbb{R}V_p\right)^\perp$ is a linear isomorphism, hence M must be odd-dimensional. Consequently if a real space form $M^n(c)$ with $c \neq 0$ admits a unit Killing vector field then n is odd and $c > 0$. ∎

Theorem 3.9 *Let $M = M^n(c)$ be an n-dimensional Riemannian manifold of constant sectional curvature c and V a unit Killing vector field on M. Then $c \geq 0$ and $V : (M,g) \to (S(M), G_s)$ is a harmonic map. Moreover*

a. *If M is compact and orientable then*

$$E(V) = \left(\frac{n}{2} + \frac{n-1}{2}c\right)\operatorname{Vol}(M).$$

b. *If $c = 0$ then V is parallel.*

c. *If $c > 0$ then n is odd.*

Proof. As V is a Killing vector field we may apply Proposition 3.7 to conclude that V is a harmonic vector field (as any space of sectional curvature is in particular an Einstein manifold). Also, as M has constant sectional curvature c, its scalar curvature is $\rho = n(n-1)c$ and $\|\nabla V\|^2 = (n-1)c$. When M is compact and orientable (again by Proposition 3.7)

$$E(V) = \left(\frac{n}{2} + \frac{\rho}{2n}\right)\mathrm{Vol}(M) = \left(\frac{n}{2} + \frac{n-1}{2}c\right)\mathrm{Vol}(M)$$

and if $c = 0$ then $\|\nabla V\| = 0$ thus yielding (b) in Theorem 3.9. The statement (c) follows part (iii) of Remark 3.8. Moreover $\mathrm{div}(V) = 0$ and $\nabla_V V = 0$ (as V is a Killing vector field) yield $\mathrm{trace}_g\,\{R(\nabla.V, V)\cdot\} = 0$ hence (by Corollary 2.24) V is a harmonic map of (M,g) into $(S(M), G_s)$. ■

As Hopf vector fields on the sphere S^{2m+1} are Killing we obtain

Theorem 3.10 (S.D. Han & J.W. Yim, [157]) *The Hopf vector fields ξ on S^{2m+1} are harmonic maps $(S^{2m+1}, g) \to (S(S^{2m+1}), G_s)$ where $g = \iota^* g_0$ is the canonical Riemannian metric on S^{2m+1}. Moreover*

$$E(\xi) = \left(2m + \frac{1}{2}\right)\mathrm{Vol}(S^{2m+1}).$$

The following result (providing the converse of Theorem 3.10 for $m = 1$) is also due to S.D. Han & J.W. Yim, [157]

Theorem 3.11 *Let ξ be a unit vector field tangent to S^3. Then $\xi : (S^3, g) \to (S(S^3), G_s)$ is a harmonic map if and only if ξ is a Hopf vector field on S^3.*

It is unknown whether on a sphere S^{2m+1}, $m > 1$, there is any non-Killing unit vector field which is a harmonic map.

Remark 3.12 By Theorem 3.9 it follows that the hyperbolic space $(\mathbb{R}^n_+, g_{\mathrm{hyp}})$ of constant sectional curvature $-c^2 < 0$

$$\mathbb{R}^n_+ = \{y = (y_1, \dots, y_n) \in \mathbb{R}^n : y_n > 0\},$$

$$g_{\mathrm{hyp}} = \frac{1}{(cy_n)^2} \sum_{i=1}^{n} dy_i \otimes dy_i,$$

admits no unit Killing vector fields. Nevertheless we may show that

$$V = cy_n \frac{\partial}{\partial y_n} \in \Gamma^\infty\left(S\left(\mathbb{R}^n_+\right)\right)$$

is a harmonic vector field (which is not a harmonic map). Indeed

$$E_i = cy_n \frac{\partial}{\partial y_i}, \quad E_n = V = cy_n \frac{\partial}{\partial y_n}, \quad 1 \le i \le n-1,$$

is a (global) orthonormal frame of $(T(\mathbb{R}^n_+), g_{\text{hyp}})$ and

$$[V, E_i] = cE_i, \quad [E_i, E_j] = 0, \quad 1 \leq i, j \leq n-1.$$

Let ∇ be the Levi-Civita connection of $(\mathbb{R}^n_+, g_{\text{hyp}})$. Since

$$2g_{\text{hyp}}(\nabla_X Y, Z) = g_{\text{hyp}}([X, Y], Z) - g_{\text{hyp}}([Y, Z], X) + g_{\text{hyp}}([Z, X], Y)$$

it follows that

$$\nabla_{E_i} E_j = c\delta_{ij} V, \quad \nabla_{E_i} V = -cE_i, \quad \nabla_V E_i = 0, \quad \nabla_V V = 0$$

for any $1 \leq i, j \leq n-1$. In particular $\|\nabla V\|^2 = (n-1)c^2$. Then

$$\begin{aligned}\Delta_{g_{\text{hyp}}} V &= -\sum_{i=1}^{n-1} \{\nabla_{E_i} \nabla_{E_i} V - \nabla_{\nabla_{E_i} E_i} V\} \\ &= -\sum_i \left\{\nabla_{E_i}(-cE_i) - c\nabla_V V\right\} = (n-1)c^2 V = \|\nabla V\|^2 V\end{aligned}$$

and we may conclude that V is a harmonic vector field.

Since $(\mathbb{R}^n_+, g_{\text{hyp}})$ is a space form $M^n(-c^2)$, i.e., a Riemannian manifold of constant sectional curvature $-c^2 < 0$, and V is a harmonic vector field which is also geodesic (i.e., $\nabla_V V = 0$) one may use Proposition 2.20 (or rather the identity (2.31) in its proof) together with Corollary 2.24 to conclude that V is a harmonic map of $\mathbb{R}^n_+$ into $S(\mathbb{R}^n_+)$ if and only if $\text{div}(V) = 0$. Yet

$$\text{div}(V) = \sum_{i=1}^{n} g_{\text{hyp}}(\nabla_{E_i} V, E_i) = \sum_{i=1}^{n-1} g_{\text{hyp}}((-cE_i), E_i) = -(n-1)c \neq 0$$

hence V is not a harmonic map. Note that the geodesic flow determined by V consists of the half-lines in $\mathbb{R}^n_+$ parallel to the axis y_n. The vector fields E_j $(1 \leq j \leq n-1)$ above are also harmonic. Indeed

$$\begin{aligned}\Delta_{g_{\text{hyp}}} E_j &= -\sum_{i=1}^{n} \left\{\nabla_{E_i} \nabla_{E_i} E_j - \nabla_{\nabla_{E_i} E_i} E_j\right\} \\ &= -\sum_{i=1}^{n-1} \left\{\nabla_{E_i} \nabla_{E_i} E_j - \nabla_{\nabla_{E_i} E_i} E_j\right\} = -\sum_{i=1}^{n-1} \{\nabla_{E_i} c\delta_{ij} V - c\nabla_V E_j\} \\ &= -c\nabla_{E_j} V = c^2 E_j = \|\nabla E_j\|^2 E_j.\end{aligned}$$

Moreover the vector fields E_j are not harmonic maps (although one has $\text{div}(E_j) = 0$) because of $\nabla_{E_j} E_j = cV$. It is unknown whether the hyperbolic space $(\mathbb{R}^n_+, g_{\rm hyp})$ admits unit vector fields which are harmonic maps of $\mathbb{R}^n_+$ into $S(\mathbb{R}^n_+)$. ■

3.2. THE ENERGY OF UNIT KILLING FIELDS IN DIMENSION 3

Let M be a compact orientable n-dimensional Riemannian manifold. The greatest lower bound of $E : \Gamma^\infty(S(M)) \to [0, +\infty)$ is $\frac{n}{2}\,\mathrm{Vol}(M)$ and this is achieved solely by the parallel unit tangent vector fields (when these exist). The problem of minimizing E is therefore more interesting when M admits no parallel unit vector fields, as for instance when $M = S^{2m+1}$ or, more generally, when M is a compact Riemannian manifold of nonzero constant sectional curvature.

Our previous Theorem 3.9 shows that unit Killing fields on a space from $M = M^n(c)$, $c > 0$, are in particular harmonic vector fields, hence critical points of $E : \Gamma^\infty(S(M)) \to [0, +\infty)$. Hence the first candidates of minimum points of E are the Killing fields which are critical points of the energy functional. The scope of this section is to examine the issue in dimension 3.

G. Wiegmink, [310], has shown that the unit Killing vector fields on S^3, i.e., the Hopf vector fields on S^3, are stable critical points for $E : \Gamma^\infty(S(S^3)) \to [0, +\infty)$. More recently F. Brito, [71], has shown that these vector fields are the *only* vector fields to give absolute extrema of $E : \Gamma^\infty(S(S^3)) \to [0, +\infty)$. We shall need the following

Lemma 3.13 *Let (M, g) be an n-dimensional Riemannian manifold and X a tangent vector field on M. Then*

$$\mathrm{Ric}(X, X) = -\mathrm{trace}_g\{(\nabla X) \circ (\nabla X)\} + (\mathrm{div}\, X)^2 + \mathrm{div}(\nabla_X X) - \mathrm{div}((\mathrm{div}\, X)X). \qquad (3.8)$$

Proof. For each point $p \in M$ let $\{E_1, \ldots, E_n\}$ be a local orthonormal frame of $T(M)$ defined on the open subset $U \subseteq M$ such that $p \in U$ and $(\nabla E_i)_p = 0$ for any $1 \le i \le n$. Then (the calculation is carried out at the point p yet one omits the point p for the simplicity of the notation)

$$\mathrm{Ric}(X, X) = \sum_{i=1}^{n} R(X, E_i, X, E_i) = \sum_i g(R(X, E_i)X, E_i)$$
$$= -\sum_i g(\nabla_X \nabla_{E_i} X - \nabla_{E_i} \nabla_X X - \nabla_{[X, E_i]} X,\ E_i)$$

$$= -\sum_i \left\{X(g(\nabla_{E_i}X, E_i)) - g(\nabla_{E_i}X, \nabla_X E_i)\right\}$$

$$+ \operatorname{div}(\nabla_X X) - \sum_i g(\nabla_{\nabla_{E_i}X}X, E_i)$$

$$= -X(\operatorname{div}(X)) + \operatorname{div}(\nabla_X X) - \operatorname{trace}_g \{(\nabla X) \circ (\nabla X)\}$$

yielding (3.8) due to the identity $\operatorname{div}(fX) = f \operatorname{div}(X) + X(f)$.

When M is compact Lemma 3.13 implies the identity (cf. also [258], p. 170)

$$\int_M \operatorname{Ric}(X,X) d\operatorname{vol}(g) = \int_M \left\{(\operatorname{div} X)^2 - \operatorname{trace}_g (\nabla X) \circ (\nabla X)\right\} d\operatorname{vol}(g). \tag{3.9}$$

Let us consider a unit vector field U on an n-dimensional Riemannian manifold M with $n \geq 3$. Let H_U and $\sigma_2(U)$ be respectively the mean curvature vector and the second mean curvature of the distribution $\mathcal{D} = (\mathbb{R}U)^\perp \subset T(M)$. These are respectively given by

$$H_U = -\frac{1}{n-1}(\operatorname{div} U)U,$$

$$\sigma_2(U) = \frac{1}{2}\left\{(\operatorname{div} U)^2 - \operatorname{trace}_g (\nabla U) \circ (\nabla U)\right\}.$$

With these notations the identity (3.8) in Lemma 3.13 may be written

$$\operatorname{Ric}(U,U) - 2\sigma_2(U) = \operatorname{div}(\nabla_U U + (n-1)H_U). \tag{3.10}$$

Moreover, let us assume that M is compact and orientable and integrate (3.10) over M. By Green's lemma

$$\int_M \operatorname{Ric}(U,U) d\operatorname{vol}(g) = 2\int_M \sigma_2(U) d\operatorname{vol}(g). \tag{3.11}$$

We shall need the following local expression of the second mean curvature $\sigma_2(U)$. Let $\{E_1, \ldots, E_n\}$ be a local orthonormal frame of $T(M)$ with $E_n = U$, defined on the open set $\Omega \subseteq M$, and let us set

$$s_{ij} = g(\nabla_{E_i} U, E_j), \quad 1 \leq i, j \leq n.$$

A calculation then shows that

$$\sigma_2(U) = \sum_{1 \leq i < j \leq n} \left(s_{ii}s_{jj} - s_{ij}s_{ji}\right)$$

on Ω. In particular if $\dim_{\mathbb{R}} M = 3$ then

$$\sigma_2(U) = s_{11}s_{22} - s_{12}s_{21}$$

on Ω. When $\dim_{\mathbb{R}} M = 2$ we adopt the convention $\sigma_2(U) = 0$. We shall prove the following ■

Theorem 3.14 (D. Perrone, [244]) *Let M be a real 3-dimensional compact oriented Riemannian manifold and U a unit tangent vector field on M. Let us assume that U is an eigenvector of the Ricci operator with the corresponding eigenvalue $\lambda \in C^\infty(M)$ (i.e., $QU = \lambda U$). Then the following statements are equivalent*

a. *λ is constant along the integral curves of U and*

$$E(U) = \int_M \left(\frac{3}{2} + \frac{1}{2}\mathrm{Ric}(U,U) \right) d\,\mathrm{vol}(g).$$

b. *U is a Killing vector field.*

Moreover, if this is the case (i.e., one of the equivalent statements (a)–(b) *holds) then $\lambda = $ const. ≥ 0 and* i) *U is parallel (equivalently M is locally isometric to a Riemmanian product $N \times \mathbb{R}$) and* ii) *(M, g) is homothetic to a Sasakian manifold i.e., if*

$$c = \sqrt{\lambda/2}, \quad \bar{g} = c^2 g, \quad \xi = (1/c)\,U, \quad \eta = \bar{g}(\xi, \cdot), \quad \varphi = -\nabla\xi,$$

then $(\varphi, \xi, \eta, \bar{g})$ is a Sasakian structure.

Proof. Let $\{E_1, E_2, E_3\}$ be a local orthonormal frame of $T(M)$ with $E_3 = U$. We set as above $s_{ij} = g(\nabla_{E_i} U, E_j)$ for any $1 \leq i, j \leq 3$. Note that $s_{i3} = 0$ for any $i \in \{1, 2, 3\}$. Then

$$\begin{aligned}
\|\nabla U\|^2 &= \sum_{i=1}^{3} \|\nabla_{E_i} U\|^2 \geq \sum_{i=1}^{2} \|\nabla_{E_i} U\|^2 = s_{11}^2 + s_{12}^2 + s_{21}^2 + s_{22}^2 \\
&= 2(s_{11}s_{22} - s_{12}s_{21}) + (s_{11} - s_{22})^2 + (s_{12} + s_{21})^2 \\
&= 2\sigma_2(U) + (s_{11} - s_{22})^2 + (s_{12} + s_{21})^2 \geq 2\sigma_2(U).
\end{aligned}$$

Yet (by (3.11)) $2\int_M \sigma_2(U)\, d\,\mathrm{vol}(g) = \int_M \mathrm{Ric}(U,U)\, d\,\mathrm{vol}(g)$ hence

$$\int_M \|\nabla U\|^2 d\,\mathrm{vol}(g) \geq \int_M \mathrm{Ric}(U,U) d\,\mathrm{vol}(g). \tag{3.12}$$

The inequality (3.12) is due to F. Brito, [71]. Next (by (3.12))

$$E(U) = \frac{3}{2}\,\mathrm{Vol}(M) + \frac{1}{2}\int_M \|\nabla U\|^2 d\,\mathrm{vol}(g)$$
$$\geq \int_M \left(\frac{3}{2} + \frac{1}{2}\,\mathrm{Ric}(U,U)\right) d\,\mathrm{vol}(g)$$

with equality if and only if

$$\nabla_U U = 0, \quad s_{11} - s_{22} = 0, \quad s_{12} + s_{21} = 0. \tag{3.13}$$

Indeed equality in the inequality above reads $\int_M \|\nabla U\|^2 d\,\mathrm{vol}(g) = 2\int_M \sigma_2(U) d\,\mathrm{vol}(g)$ hence we may integrate in $\|\nabla U\|^2 \geq 2\sigma_2(U) + (s_{11} - s_{22})^2 + (s_{12} + s_{21})^2$ and conclude that $s_{11} - s_{22} = 0$ and $s_{12} + s_{21} = 0$. Moreover, if these relations are satisfied then we may integrate in $\|\nabla U\|^2 = \|\nabla_U U\|^2 + 2\sigma_2(U)$ to conclude that $\nabla_U U = 0$ as well. Thus equality in $E(U) \geq \int_M \left(\frac{3}{2} + \frac{1}{2}\mathrm{Ric}(U,U)\right) d\,\mathrm{vol}(g)$ does imply the relations (3.13). The converse is left as an exercise to the reader. At this point we may prove (a) $\Longrightarrow$ (b). We have

$$\begin{aligned}(\mathcal{L}_U g)(E_i, E_j) &= U(g(E_i, E_j)) - g([U, E_i], E_j) - g(E_i, [U, E_j]) \\ &= U(g(E_i, E_j)) - g(\nabla_U E_i - \nabla_{E_i} U, E_j) - g(E_i, \nabla_U E_j - \nabla_{E_j} U) \\ &= (\nabla_U g)(E_i, E_j) + s_{ij} + s_{ji}\end{aligned}$$

that is

$$(\nabla_U g)(E_i, E_j) = s_{ij} + s_{ji}.$$

Therefore to show that U is Killing it suffices to check that $s_{11} = 0$. We set for simplicity

$$f_1 = s_{11} = s_{22}, \quad f_2 = s_{12} = -s_{21},$$
$$\alpha = g(\nabla_U E_1, E_2), \quad \beta = g(\nabla_{E_1} E_2, E_1), \quad \gamma = g(\nabla_{E_2} E_2, E_1).$$

Then

$$\nabla_{E_1} U = f_1 E_1 + f_2 E_2, \quad \nabla_{E_2} U = -f_2 E_1 + f_1 E_2, \quad \nabla_U U = 0,$$
$$\nabla_{E_1} E_1 = -f_1 U - \beta E_2, \quad \nabla_{E_2} E_1 = f_2 U - \gamma E_2, \quad \nabla_U E_1 = \alpha E_2,$$
$$\nabla_{E_1} E_2 = -f_2 U + \beta E_1, \quad \nabla_{E_2} E_2 = -f_1 U + \gamma E_1, \quad \nabla_U E_2 = -\alpha E_1,$$
$$[E_1, U] = f_1 E_1 + (f_2 - \alpha) E_2, \quad [E_2, U] = (\alpha - f_2) E_1 + f_1 E_2,$$
$$[E_1, E_2] = -2 f_2 U + \beta E_1 + \gamma E_2.$$

Consequently

$$R(E_1,U)U = (U(f_1)+f_1^2-f_2^2)E_1 + (U(f_2)+2f_1f_2)E_2, \tag{3.14}$$

$$R(E_2,U)U = -(U(f_2)+2f_1f_2)E_1 + (U(f_1)+f_1^2-f_2^2)E_2, \tag{3.15}$$

$$R(E_1,E_2)U = (E_1(f_2)+E_2(f_1))E_1 + (-E_1(f_1)+E_2(f_2))E_2. \tag{3.16}$$

Since $QU = \lambda U$

$$\lambda = \mathrm{Ric}(U,U) = R(E_1,U,E_1,U) + R(E_2,U,E_2,U)$$

and the identities (3.14)–(3.15) imply

$$U(f_1)+f_1^2-f_2^2 = -\frac{\lambda}{2}. \tag{3.17}$$

As

$$R(E_1,E_2,U,E_2) = \mathrm{Ric}(E_1,U) = 0,$$
$$R(E_1,E_2,U,E_1) = -\mathrm{Ric}(E_2,U) = 0,$$

the identity (3.16) implies

$$E_1(f_2)+E_2(f_1) = 0, \quad E_1(f_1)-E_2(f_2) = 0. \tag{3.18}$$

Next (3.17)–(3.18) and the condition $U(\lambda) = 0$ (as λ is constant along the integral curves of U, by our assumption (a)) imply that

$$\sum_{i=1}^{3} E_iE_i(f_1) = -2f_1U(f_1) + \beta E_1(f_2) + \gamma E_2(f_2).$$

On the other hand

$$(\nabla_U U)(f_1) = 0, \quad (\nabla_{E_1}E_1)(f_1) = -f_1U(f_1) - \beta E_2(f_1),$$
$$(\nabla_{E_2}E_2)(f_1) = -f_1U(f_1) + \gamma E_1(f_1).$$

Hence

$$\Delta f_1 = -\sum_{i=1}^{3}\{E_iE_i(f_1) - (\nabla_{E_i}E_i)(f_1)\} = 0$$

i.e., f_1 is a harmonic function (on the compact connected manifold M) so that $f_1 = \text{constant}$. Yet $\mathrm{div}(U) = 2f_1$ hence (by Green's lemma) the constant in discussion is actually zero i.e., $f_1 = 0$ which proves statement (b). Note at this point that, under the assumptions (a), the identities (3.17)–(3.18) also imply

$$\lambda = 2f_2^2 = \text{constant} \geq 0.$$

Let us prove the implication (b) $\Longrightarrow$ (a). Assuming that U is a Killing vector field one easily proves (3.13) and therefore the equality $E(U) = \frac{3}{2}\text{Vol}(M) + \frac{1}{2}\int_M \text{Ric}(U,U)\,d\,\text{vol}(g)$. Also, it is an easy matter that when U is Killing the equality $QU = \lambda U$ yields $\lambda = \text{constant} \geq 0$. Statement (a) is thereby proved.

Let us prove now the second part of Theorem 3.14 (i.e., the statements (i)–(ii) there). Assume that one of the equivalent statements (a)–(b) holds good. Let us set

$$c = \begin{cases} -f_2 = \sqrt{\lambda/2}, & f_2 \leq 0, \\ f_2, & f_2 \geq 0. \end{cases}$$

When $f_2 = 0$ one checks easily that U is parallel hence statement (i) is proved. When $f_2 \neq 0$ we consider

$$\overline{g} = c^2 g, \quad \xi = (1/c)U, \quad \eta = cg(U,\cdot), \quad \varphi = -\nabla\xi.$$

These tensor fields are easily seen to satisfy the relations

$$\eta(\xi) = 1, \quad \varphi^2 = -I + \eta \otimes \xi, \quad d\eta = g(\cdot,\varphi\cdot), \quad g(\varphi\cdot,\varphi\cdot) = g - \eta\otimes\eta,$$

hence (φ,ξ,η,g) is a contact metric structure on M which is Sasakian precisely when (cf. [42]) ξ is a Killing vector field. ■

Theorem 3.14 and its proof admit the following corollaries

Corollary 3.15 *Let $M = M^3(c)$ be a compact orientable 3-dimensional real space form of (constant) sectional curvature $c \geq 0$. Then for any unit vector field U tangent to M*

$$E(U) \geq \left(\frac{3}{2} + c\right)\text{Vol}(M),$$

with equality if and only if U is a Killing vector field.

Corollary 3.16 (F. Brito, [71]) *The unit tangent vector fields of minimum energy on the sphere $S^3(r)$ are precisely the Hopf vector fields on $S^3(r)$.*

Cf. also A. Higuchi & B.S. Kay & C.M. Wood, [164]. The uniqueness part in F. Brito's theorem (cf. [71]) follows from the uniqueness part in the result by H. Gluck & W. Ziller, [133], that unit vector fields of minimum volume on S^3 are precisely the Hopf vector fields (equivalently the unit Killing vector fields). H. Gluck & W. Ziller use (cf. *op. cit.*) methods within calibrated geometry. Another approach (similar to the arguments in the proof of Theorem 3.14, yet applied to the volume functional) shows that on each compact orientable space form $M^3(c)$, $c \geq 0$, the unit vector fields of minimum volume $(c+1)\mathrm{Vol}(M)$ are precisely the unit Killing vector fields (cf. [244]).

By a result of J.C. Gonzàles-Dàvila & L. Vanhecke, [146], given a compact oriented 3-dimensional Riemannian manifold M if V is a unit vector field tangent to M which is both Killing and harmonic and if the sectional curvatures along the 2-planes tangent to $(\mathbb{R}V)^\perp$ are $\geq c^2 > 0$ then $\pm V$ minimize the energy. Also if $M = G$ is a 3-dimensional Lie group a classification of left invariant unit vector fields which are harmonic or determine harmonic maps is given in [146].

3.3. INSTABILITY OF HOPF VECTOR FIELDS

By Corollary 3.16 the Hopf vector fields on S^3 are absolute minima for $E : \Gamma^\infty(S(S^3)) \to [0, +\infty)$ hence stable critical points of the energy functional. This is a rather surprising bias from the theory of harmonic maps where it is known that

Theorem 3.17 (Y.L. Xin, [321]) *Let N be Riemannian manifold. Then any nonconstant harmonic map $f : S^n \to N$ is unstable provided that $n \geq 3$.*

In particular the Hopf vector fields ξ on S^3 are unstable as harmonic maps $\xi : S^3 \to S(S^3)$. What about the stability of Hopf vector fields on a sphere S^{2m+1} (with $m > 1$)? To answer this question we report on results by C.M. Wood, [316].

An inspection of the proof of Xin's theorem (Theorem 3.17 above) shows that instability of $f : S^n \to N$ follows from

$$(\mathrm{Hess}\, E)_f(A_a) = -(n-2) \int_{S^n} \hat{h}(f_* A_a, f_* A_a)\, d\,\mathrm{vol}(g)$$

where h is the Riemannian metric on N (and $\hat{h} = f^{-1}h$ the metric induced by h in $f^{-1}TN \to S^n$) and $A_a \in \mathfrak{X}(S^n)$ are the vector fields given by

$$A_a(p) = \vec{a} - \langle a, p \rangle \vec{p}, \quad p \in S^n, \; a \in \mathbb{R}^{n+1},$$

and $\langle \cdot, \cdot \rangle$ is the Euclidean inner product on $\mathbb{R}^{n+1}$. As to the case of Hopf vector fields ξ on the sphere S^n with $n = 2m + 1$ let W_a be the component of A_a along $(\mathbb{R}\xi)^\perp$ i.e.,

$$W_a = A_a - g(A_a, \xi)\xi = \vec{a} - g_0(\vec{a}, \nu)\nu - g(\vec{a}, \xi)\xi.$$

Then we may state

Theorem 3.18 (C.M. Wood, [316], O. Gil-Medrano & E. Llinares-Fuster, [131]) *Let ξ be a Hopf vector field on S^{2m+1}. Then for any $a \in \mathbb{R}^{2m+2}$ the Hessian of the energy functional $E : \Gamma^\infty(S(S^{2m+1})) \to [0, +\infty)$ at the critical point ξ satisfies*

$$(\mathrm{Hess}\, E)_\xi (W_a, W_a) = -\frac{m}{m+1} \|a\|^2 (2m - 3) \, \mathrm{Vol}(S^{2m+1}).$$

In particular the Hopf vector fields on S^{2m+1}, $m > 1$, are unstable of index at least $2m + 2$.

To prove Theorem 3.18, we need some preparation. For the remainder of this section ξ will denote a Hopf vector field on the sphere S^{2m+1}. Also let J_ξ be the Jacobi operator associated to ξ. Let

$$\{E_1, \ldots, E_n\} = \{E_i, E_{i^*} = JE_i, \xi : 1 \le i \le m\}$$

be a local orthonormal frame on $T(S^{2m+1})$ (with $n = 2m + 1$ and $E_n = \xi$ and $E_{i+m} = E_{i^*} = JE_i$ for any $1 \le i \le m$). Here J is an orthogonal almost complex structure on $\mathbb{C}^{m+1}$.

Lemma 3.19 *For any $W \in (\mathbb{R}\xi)^\perp \subset T(S^n)$*

$$(\mathrm{Hess}\, E)_\xi (W, W)$$
$$= -\int_{S^n} \left\{ (2m - 1) \| W \|^2 - \sum_{j=1}^{2m} (B_{nj})^2 - \sum_{i,j=1}^{2m} (B_{ij})^2 \right\} d\,\mathrm{vol}(g)$$

where $B_{ij} = g(\nabla_{E_i} W, E_j)$ for any $1 \le i,j \le n = 2m + 1$.

Proof. For any $W \in (\mathbb{R}\xi)^\perp$ we may apply Definition 2.30

$$(\mathrm{Hess}\; E)_\xi (W, W) = \int_{S^n} g(J_\xi W, W) d\,\mathrm{vol}(g)$$
$$= \int_{S^n} \left(\|\nabla W\|^2 - \| W \|^2 \|\nabla \xi \|^2 \right) d\,\mathrm{vol}(g). \qquad (3.19)$$

As ξ is a unit Killing vector field we may apply Proposition 3.7 to obtain

$$\|\nabla\xi\|^2 = g(\Delta_g\xi,\xi) = g(Q\xi,\xi) = \mathrm{Ric}(\xi,\xi) = 2m. \tag{3.20}$$

Moreover

$$\nabla_{E_i} W = \sum_{j=1}^{m} (B_{ij}E_j + B_{ij^*}E_{j^*}) + B_{in}\xi,$$

$$\nabla_{E_{i^*}} W = \sum_{j=1}^{m} (B_{i^*j}E_j + B_{i^*j^*}E_{j^*}) + B_{i^*n}\xi,$$

$$\nabla_{\xi} W = \sum_{j=1}^{m} (B_{nj}E_j + B_{nj^*}E_{j^*}).$$

To prove the last equality one makes use of $g(\nabla_\xi W,\xi) = -g(W,\nabla_\xi\xi) = 0$ as ξ is a Killing vector field. Moreover, as W and ξ are orthogonal one has (cf. (3.4)) $\nabla\xi = J$ on $(\mathbb{R}\xi)^\perp$ so that

$$B_{in} = g(\nabla_{E_i} W,\xi) = -g(\nabla_{E_i}\xi, W) = -g(E_{i^*}, W) = -W^{i^*},$$

$$B_{i^*n} = g(\nabla_{E_{i^*}} W,\xi) = -g(\nabla_{E_{i^*}}\xi, W) = g(W,E_i) = W^i.$$

Hence

$$\|W\|^2 = \sum_{i=1}^{m}\left((W^i)^2 + (W^{i^*})^2\right) = \sum_{i=1}^{m}\left(B_{in}^2 + B_{i^*n}^2\right).$$

Consequently

$$\begin{aligned}\|\nabla W\|^2 &= \sum_{i=1}^{m}\{\|\nabla_{E_i} W\|^2 + \|\nabla_{E_{i^*}} W\|^2\} + \|\nabla_\xi W\|^2 \\ &= \sum_{i,j=1}^{m}\left\{B_{ij}^2 + B_{ij^*}^2 + B_{i^*j}^2 + B_{i^*j^*}^2\right\} \\ &\quad + \sum_{i=1}^{m}\{B_{in}^2 + B_{i^*n}^2\} + \sum_{j=1}^{m}\left\{B_{nj}^2 + B_{nj^*}^2\right\}\end{aligned}$$

that is

$$\|\nabla W\|^2 = \sum_{i,j=1}^{2m} B_{ij}^2 + \sum_{j=1}^{2m} B_{nj}^2 + \|W\|^2. \tag{3.21}$$

Now Lemma 3.19 follows from the identities (3.19)–(3.21). ■

Lemma 3.20

For any $a = (a_1, \ldots, a_{n+1}) \in \mathbb{R}^{n+1}$ *with* $n = 2m+1$

$$(\text{Hess } E)_\xi(W_a, W_a) = -(2m-1)\|a\|^2 \text{Vol}(S^n) + (4m-1)\int_{S^n} \left(f_a^2 + \overline{f}_a^2\right) d\,\text{vol}(g) \tag{3.22}$$

where $f_a, \overline{f}_a \in C^\infty(S^n, \mathbb{R})$ *are given by*

$$f_a(p) = g_0(\vec{a}, \nu)_p, \quad \overline{f}_a(p) = g_0(\vec{a}, \xi)_p, \quad p \in S^n.$$

Also $\vec{a} = \sum_{j=1}^{n+1} a_j\, \partial/\partial x_j \in \mathfrak{X}(\mathbb{R}^{n+1})$.

Proof. Using the very definition of the vector fields W_a

$$g_0(W_a, \nu)_p = g(W_a, \xi)_p = 0$$

hence $W_a \in (\mathbb{R}\xi)^\perp \subset T(S^n)$. We recall (cf. the identity (2.26) in H. Urakawa, [292], p. 168) that the vector fields $A_a = \sum_{j=1}^{n+1} a_j\, \partial/\partial x_j - f_a \nu$ satisfy

$$\nabla_{E_\gamma} A_a = -f_a E_\gamma, \quad \gamma = 1, 2, \ldots, n = 2m+1.$$

Then

$$\nabla_{E_\gamma} W_a = -f_a E_\gamma - E_\gamma(\overline{f}_a)\xi + \overline{f}_a \varphi E_\gamma, \quad \gamma = 1, 2, \ldots, n,$$

so that

$$B_{ij} = -f_a \delta_{ij}, \quad B_{ij^*} = \overline{f}_a \delta_{ij}, \quad B_{i^*j} = -\overline{f}_a \delta_{ij}, \quad B_{i^*j^*} = -f_a \delta_{ij},$$
$$B_{ni} = B_{ni^*} = 0, \quad i, j = 1, \ldots, m.$$

Consequently

$$\sum_{\beta,\gamma=1}^{2m} B_{\beta\gamma}^2 + \sum_{\gamma=1}^{2m} B_{n\gamma}^2 = \sum_{\beta,\gamma=1}^{2m} B_{\beta\gamma}^2 = 2\sum_{i,j=1}^{m} (f_a^2 + \overline{f}_a^2)\delta_{ij} = 2m(f_a^2 + \overline{f}_a^2).$$

On the other hand

$$\|W_a\|^2 = \|a\|^2 - \left(f_a^2 + \overline{f}_a^2\right).$$

Therefore (by Lemma 3.19 above) we obtain the identity (3.22) in Lemma 3.20. ■

Lemma 3.21 *For any* $a \in \mathbb{R}^{n+1}$

$$\int_{S^n} \bar{f}_a^2 \, d\,\mathrm{vol}(g) = \int_{S^n} f_a^2 \, d\,\mathrm{vol}(g) = \frac{\|a\|^2}{n+1} \mathrm{Vol}(S^n).$$

Proof. Let J be the complex structure entering the definition of ξ i.e., $\xi = J\nu$. Then

$$\int_{S^n} \bar{f}_a^2 \, d\,\mathrm{vol}(g) = \int_{S^n} \bar{f}_a g_0(\vec{a}, \xi) \, d\,\mathrm{vol}(g)$$

$$= - \int_{S^n} \bar{f}_a g_0(J\vec{a}, \nu) \, d\,\mathrm{vol}(g)$$

where $\vec{a} = \sum_{j=1}^{n+1} x_j(a) \partial / \partial x_j \in \mathfrak{X}(\mathbb{R}^{n+1})$. By Green's lemma

$$\int_{S^n} g_0(-\bar{f}_a J\vec{a}, \nu) \, d\,\mathrm{vol}(g) = \int_{B^{n+1}} \mathrm{div}(-\bar{f}_a J\vec{a}) \; \omega_{n+1}$$

where

$$\begin{aligned} \mathrm{div}\left(-\bar{f}_a J\vec{a}\right) &= -\bar{f}_a \mathrm{div}(J\vec{a}) - (J\vec{a})(\bar{f}_a) = -(J\vec{a})(\bar{f}_a) \\ &= -(J\vec{a}) \left(g_0(-J\vec{a}, \nu)\right) = g_0\left(J\vec{a}, \nabla^0_{J\vec{a}} \nu\right) = \| J\vec{a}\|^2 = \|a\|^2 \end{aligned}$$

and

$$\int_{B^{n+1}} \omega_{n+1} = \mathrm{Vol}(B^{n+1}) = \frac{\mathrm{Vol}(S^n)}{n+1}.$$

The calculation of the integral $\int_{S^n} f_a^2 \, d\,\mathrm{vol}(g)$ is similar and therefore left as an exercise to the reader. ∎

At this point the proof of Theorem 3.18 follows easily from the Lemmas 3.20 and 3.21.

Remark 3.22 In the proof of Lemma 3.20 one made use of the fact that

$$\nabla_X A_a = -f_a X, \quad X \in \mathfrak{X}(S^n),$$

hence the vector fields A_a are conformal yet not Killing

$$(\mathcal{L}_{A_a} g)(X, Y) = g(\nabla_X A_a, Y) + g(X, \nabla_Y A_a) = -2 f_a g(X, Y).$$

Let $\mathcal{K}$ and $\mathcal{C}$ be respectively the linear spaces of all Killing vector fields on S^n and of all conformal vector fields (i.e., infinitesimal conformal transformations) on S^n. Moreover, let us set

$$\mathcal{A} = \{A_a : a \in \mathbb{R}^{n+1}\}.$$

As $\dim_{\mathbb{R}}\mathcal{K} = n(n+1)/2$ and $\dim_{\mathbb{R}}\mathcal{C} = (n+1)(n+2)/2$ it follows that $\mathcal{C} = \mathcal{K} \oplus \mathcal{A}$. ■

Remark 3.23 Let (M,g) be a compact orientable Riemannian manifold. If V is a unit vector field tangent to M then $V_c = (1/\sqrt{c})V$ is a unit vector field tangent to $(M,\overline{g})$ with $\overline{g} = cg$, $c > 0$. Then (by Remark 2.4 in Chapter 2)

$$E_{\overline{g}}(V_c) = c^{\frac{n}{2}-1} E_g(V) + \text{const.}, \quad n = \dim(M).$$

Consequently V is a critical point of E_g if and only if V_c is a critical point of $E_{\overline{g}}$ and the corresponding Hessian forms have the same index. Therefore Theorem 3.18 implies that the *Hopf vector fields on* $S^{2m+1}(r)$, $m > 1$, *are unstable critical points of index at least* $2m+2$. ■

3.4. EXISTENCE OF MINIMA IN DIMENSION > 3

Let (M,g) be a compact orientable Riemannian manifold. As well known the identity map $1_M : M \to M$ is a harmonic map.

Definition 3.24 (T. Nagano, [221]) A Riemannian manifold (M,g) is said to be *stable* if the identity map is stable. Otherwise M is said to be *unstable.* ■

By a result of R.T. Smith, [273], if (M,g) is a compact n-dimensional Enstein manifold then M is stable if and only if $\lambda_1 \geq 2\rho/n$ where λ_1 is the first nonzero eigenvalue of the Laplace-Beltrami operator on functions and ρ is the scalar curvature. As reported on earlier (cf. Section 3.2) in this monograph F. Brito has shown (cf. [71]) that the Hopf vector fields in S^3 (i.e., the unit vector fields tangent to the fibres of the Hopf fibration) are the only minimizers of the energy functional $E : S(S^3) \to [0, +\infty)$ and in particular they are stable. However Hopf vector fields on S^{2m+1}, $m > 1$, were seen (cf. Section 3.3) to be unstable. To the knowledge of the authors of this monograph, in dimension > 3 there are neither examples of unit vector fields realizing the absolute minimum for the energy nor criteria of existence of minima.

The sphere S^{2m+1} is a standard example of Sasakian space form of constant sectional curvature $+1$. Let (η_0, g_0) be the natural Sasakian structure of S^{2m+1}. Let (M,g) be any $(2m+1)$-dimensional compact Riemannian manifold of constant sectional curvature $+1$. Then M is a spherical space form $(S^{2m+1}/\Gamma, g)$ for some finite subgroup $\Gamma \subset O(2m+2)$ (in which the identity is the only element corresponding to the eigenvalue $+1$). Here g is the Riemannian metric induced by g_0 on the quotient S^{2m+1}/Γ. We may identify $U(m+1)$ with the subgroup of $O(2m+2)$ preserving η_0. Then Γ is conjugate in $O(2m+2)$ to a subgroup of $U(m+1)$. Therefore η_0 induces a contact form η on M. See J.A. Wolf, [312]. As g is induced by g_0 it follows that (η, g) is a Sasakian structure on M (cf. S. Tanno, [283]). Thus every odd dimensional compact orientable Riemannian manifold of constant sectional curvature $+1$ is a Sasakian manifold.

Let $M^{2m+1}(c)$ be a Sasakian space form i.e., a real $(2m+1)$-dimensional Sasakian manifold of constant φ-sectional curvature c. Any such manifold has constant scalar curvature $\rho = m[(2m+1)(c+3)+c-1]/2$ and is η-Einstein (cf. [42], p. 113) i.e., $\mathrm{Ric} = \alpha g + \beta \eta \otimes \eta$ with $a = [(m+1)c + 3m-1]/2$ and $b = -[(m+1)(c-1)]/2$. Also $c = 1$ if and only if M^{2m+1} has constant sectional curvature $+1$. On the other hand, by a result of H. Urakawa (cf. [295], p. 572) the first nonzero eigenvalue λ_1 of the Laplace-Beltrami operator on $(S^{2m+1}/\Gamma, g)$ with $\Gamma \neq \{1_M\}$ is $\geq 4m$, that is $(S^{2m+1}/\Gamma, g)$ with $\Gamma \neq \{1_M\}$ is stable. As an outgrowth of the ideas above one has

Theorem 3.25 (D. Perrone & L. Vergori, [254]) *Let (M, η, g) be a $(2m+1)$-dimensional $(m \geq 1)$ compact Sasakian space form of constant φ-sectional curvature c.* i) *If $\mu_1 < \min\{(m+1)c + 3m - 1, 4m\}$ then M is unstable.* ii) *If $c \geq 1$ and ξ is unstable then M is unstable.* iii) *If $c = 1$ then the following statements are equivalent* a) *$\mu_1 = 4m$,* b) *M is stable, and* c) *$\pi_1(M) \neq 0$. Here μ_1 is the first nonzero eigenvalue of the Laplacian on 1-forms.*

By Theorem 3.25, the stability of the Sasakian manifold M implies the stability of the Reeb vector field. It is therefore a natural problem to study the existence of minima of $E : \Gamma^\infty(S(M)) \to [0, +\infty)$ when M is a stable compact Riemannian manifold. H. Urakawa has classified (cf. [294]) the compact simply connected irreducible Riemannian symmetric spaces which are stable (these are in particular Einstein).

Let us discuss the existence of minima for $E : \Gamma^\infty(S(M)) \to [0, +\infty)$ where (M, g) is a stable compact Einstein manifold of dimension n i.e.,

$\mathrm{Ric} = \kappa g$. By applying the second variation formula

$$(\mathrm{Hess}\, E)_{1_M}(X,X) = \int_M g(J_{1_M}X, X)\, d\,\mathrm{vol}(g) = \int_M g(\Delta_g X - QX, X)\, d\,\mathrm{vol}(g) \tag{3.23}$$

for any $X \in \mathfrak{X}(M)$, where Δ_g is the rough Laplacian, Q is the Ricci operator, and $J_{1_M} = \Delta_g - Q$ is the Jacobi operator of the identity map. As M is stable (by (3.23) and Lemma 2.15)

$$(\mathrm{Hess}\, E)_{1_M}(X,X) = \int_M \left(\|\nabla X\|^2 - \mathrm{Ric}(X,X)\right) d\,\mathrm{vol}(g) \geq 0 \tag{3.24}$$

for any $X \in \mathfrak{X}(M)$. Then (by Proposition 2.3)

$$E(U) = \frac{1}{2}\int_M \|\nabla U\|^2 d\,\mathrm{vol}(g) + \frac{n}{2}\,\mathrm{Vol}(M) \geq \frac{\kappa + n}{2}\,\mathrm{Vol}(M) \tag{3.25}$$

for any $U \in \Gamma^\infty(S(M))$. Let us recall that a *Jacobi vector field* is a smooth solution X to $J_{1_M}X = 0$. Jacobi vector fields are studied by K. Yano & T. Nagano, [326] (there Jacobi vector fields are referred to as *geodesic vector fields*). If U_0 is a unit Jacobi vector field then

$$\begin{aligned}\int_M \|\nabla U_0\|^2 d\,\mathrm{vol}(g) &= \int_M g(\Delta_g U_0, U_0)\, d\,\mathrm{vol}(g) \\ &= \int_M \mathrm{Ric}(U_0, U_0)\, d\,\mathrm{vol}(g) = \kappa\,\mathrm{Vol}(M).\end{aligned}$$

Hence

$$E(U_0) = \frac{\kappa + n}{2}\,\mathrm{Vol}(M) \leq E(U)$$

for any $U \in \Gamma^\infty(S(M))$ i.e., U_0 minimizes the energy.

Conversely, let U be a unit vector field such that $E(U) = (\kappa + n)\,\mathrm{Vol}(M)/2$. Then (by (3.25)) $\int_M \|\nabla U\|^2\, d\,\mathrm{vol}(g) = \kappa\,\mathrm{Vol}(M)$ so that (by (3.24)) $(\mathrm{Hess}\, E)_{1_M}(U,U) = 0$. Since M is stable we may expand U as $U = \sum_{i=1}^\infty E_i$ where $\{E_i : i \geq 1\}$ are such that

$$J_{1_M}E_i = a_i E_i, \quad \int_M g(E_i, E_j)\, d\,\mathrm{vol}(g) = 0, \;\; i \neq j, \;\; a_i \geq 0.$$

Thus

$$J1_M U = \sum_{i=p+1}^{\infty} a_i E_i, \quad p = \dim \mathrm{Ker}(J1_M), \;\; a_i > 0, \;\; i \geq p+1,$$

so that

$$0 = (\mathrm{Hess}\, E)_{1_M}(U,U) = \int_M g(J1_M U, U)\, d\,\mathrm{vol}(g)$$

$$= \sum_{i=p+1}^{\infty} a_i \int_M g(E_i, E_i)\, d\,\mathrm{vol}(g).$$

It follows that $E_i = 0$ for any $i \geq p+1$ so that $J1_M U = 0$ i.e., U is a Jacobi vector field. If $U \in \Gamma^\infty(S(M))$ is a unit vector field whose flow is volume preserving (i.e., $\mathrm{div}(U) = 0$) then it may be easily shown that U is a Killing vector field if and only if U is a Jacobi vector field (one may see for instance [258], p. 171). Hence

Proposition 3.26 *Let (M,g) be a stable compact Einstein manifold. Then the unit Jacobi vector fields are the only minimizers of the energy. In particular if $U \in \Gamma^\infty(S(M))$ is volume preserving then U minimizes the energy if and only if U is a Killing vector field.*

Consequently one obtains

Corollary 3.27 *Let (M,η,g) be a stable compact Einstein contact metric manifold. Then the Reeb vector ξ minimizes the energy if and only if M is a Sasakian manifold.*

Proof. The flow of ξ is volume preserving. Moreover (by a result of C. Boyer et al., [66]) a compact Einstein K-contact manifold is Sasaki-Einstein. At this point the result follows from Proposition 3.26.

We end this section with the following ■

Theorem 3.28 (D. Perrone & L. Vergori, [254]) *Let (M,η,g) be a compact Sasakian manifold of constant sectional curvature $+1$ and dimension $\dim(M) = 2m+1 > 3$. Then the Reeb vector ξ minimizes the energy if and only if $\pi_1(M) \neq 0$.*

Proof. Let us assume that ξ minimizes the energy and $\pi_1(M) = 0$. Then M is the sphere S^{2m+1}. On the other hand on S^{2m+1} the Hopf vector fields are precisely the unit Killing vector fields (cf. our Section 3.1). Then ξ is a Hopf vector field and as such unstable, a contradiction. The converse follows from Theorem 3.25 and Corollary 3.27. ■

3.5. BRITO'S FUNCTIONAL

Since the Hopf vector fields on S^{2m+1}, $m > 1$, are unstable critical points of the energy functional $E : \Gamma^\infty(S(S^{2m+1})) \to [0,+\infty)$ F. Brito, [71], was led to the construction of a "correction" of E whose precise description is given by the following

Definition 3.29 Let (M,g) be a compact orientable real n-dimensional Riemannian manifold and V a tangent vector field on M. We set

$$\tilde{E}(V) = E(V) + \frac{(n-1)(n-3)}{2} \int_M \|H_V\|^2 d\,\mathrm{vol}(g)$$

$$= \frac{n}{2}\mathrm{Vol}(M) + \frac{1}{2}\int_M \left\{\|\nabla V\|^2 + (n-1)(n-3)\|H_V\|^2\right\} d\,\mathrm{vol}(g)$$

where $H_V = -\frac{1}{n-1}(\mathrm{div}\, V)\, V$ is the mean curvature vector of the distribution $(\mathbb{R}V)^\perp \subset T(M)$. The functional $\tilde{E} : \mathfrak{X}(M) \to [0,+\infty)$ is referred to as *Brito's energy functional.* ■

We may state

Theorem 3.30 (F. Brito, [71]) *Let M be a compact orientable n-dimensional Riemannian manifold and V a unit vector field on M. Then*

$$\tilde{E}(V) \geq \frac{n}{2}\mathrm{Vol}(M) + \frac{1}{2}\int_M \mathrm{Ric}(V,V)\, d\,\mathrm{vol}(g). \tag{3.26}$$

In particular if $M = S^{2m+1}$ and ξ is a Hopf vector field on S^{2m+1} then

$$\tilde{E}(V) \geq \tilde{E}(\xi) = E(\xi) = \frac{4m+1}{2}\mathrm{Vol}(S^{2m+1}).$$

Proof. Let us consider a local orthnormal frame $\{E_i : 1 \leq i \leq n\}$ of $T(M)$ with $E_n = V$. Let us set

$$s_{ij} = g(\nabla_{E_i} V, E_j), \quad 1 \leq i,j \leq n.$$

Note that $s_{in} = 0$ for any $1 \leq i \leq n$. Then

$$\|\nabla V\|^2 = \sum_{i=1}^{n-1} \|\nabla_{E_i} V\|^2 + \|\nabla_V V\|^2 = \sum_{i,j=1}^{n-1} s_{ij}^2 + \|\nabla_V V\|^2$$

so that

$$\begin{aligned}
&\|\nabla V\|^2 + (n-1)(n-3)\|H_V\|^2 - \|\nabla_V V\|^2 \\
&= \sum_{i,j=1}^{n-1} s_{ij}^2 + \frac{n-3}{n-1}\left(\sum_{i=1}^{n-1} s_{ii}\right)^2 \\
&= \sum_{i=1}^{n-1} s_{ii}^2 + \sum_{1 \leq i \neq j \leq n-1} s_{ij}^2 + \frac{n-3}{n-1}\left(\sum_{i=1}^{n-1} s_{ii}^2 + 2 \sum_{1 \leq i < j \leq n-1} s_{ii}s_{jj}\right) \\
&= \frac{2(n-2)}{n-1}\sum_{i=1}^{n-1} s_{ii}^2 + \sum_{i \neq j} s_{ij}^2 + \frac{2(n-3)}{n-1}\sum_{i<j} s_{ii}s_{jj} \\
&= \frac{2(n-2)}{n-1}\frac{1}{n-2}\left(\sum_{i<j}(s_{ii} - s_{jj})^2 + 2\sum_{i<j} s_{ii}s_{jj}\right) \\
&\quad + \sum_{i<j}(s_{ij} + s_{ji})^2 - 2\sum_{i<j} s_{ij}s_{ji} + \frac{2(n-3)}{n-1}\sum_{i<j} s_{ii}s_{jj} \\
&= \frac{2}{n-1}\sum_{i<j}(s_{ii} - s_{jj})^2 + \frac{4}{n-1}\sum_{i<j} s_{ii}s_{jj} + \sum_{i<j}(s_{ij} + s_{ji})^2 \\
&\quad - 2\sum_{i<j} s_{ij}s_{ji} + \frac{2(n-3)}{n-1}\sum_{i<j} s_{ii}s_{jj}
\end{aligned}$$

or

$$\|\nabla V\|^2 = \frac{2}{n-1}\sum_{i<j}(s_{ii} - s_{jj})^2 + \sum_{i<j}(s_{ij} + s_{ji})^2 + 2\sum_{i<j}(s_{ii}s_{jj} - s_{ij}s_{ji}). \tag{3.27}$$

Since

$$(\operatorname{div} V)^2 = \left(\sum_{i=1}^{n-1} s_{ii}\right)^2 = \sum_{i=1}^{n-1} s_{ii}^2 + 2\sum_{i<j} s_{ii}s_{jj},$$

$$\begin{aligned}\operatorname{trace}_g\{(\nabla V)\circ(\nabla V)\} &= \sum_{j=1}^{n} g(\nabla_{\nabla_{E_j}V} V, E_j)\\ &= \sum_{i,j=1}^{n-1} s_{ji}s_{ij} = \sum_{i=1}^{n-1} s_{ii}^2 + 2\sum_{i<j} s_{ij}s_{ji},\end{aligned}$$

one obtains

$$2\sum_{i<j}(s_{ii}s_{jj} - s_{ij}s_{ji}) = (\operatorname{div} V)^2 - \operatorname{trace}_g\{(\nabla V)\circ(\nabla V)\} = 2\sigma_2(V).$$

Therefore the identity (3.27) becomes

$$\begin{aligned}\|\nabla V\|^2 + (n-1)(n-3)\|H_V\|^2 &= 2\sigma_2(V) + \|\nabla_V V\|^2\\ &+ \frac{2}{n-1}\sum_{i<j}(s_{ii} - s_{jj})^2 + \sum_{i<j}(s_{ij} + s_{ji})^2. \qquad (3.28)\end{aligned}$$

At this point the inequality (3.26) follows from (3.28) and (3.9). Let us assume now that $M = S^{2m+1}$ and that $V = \xi$ is a Hopf vector field. Such ξ is a unit Killing vector field hence

$$\begin{aligned}s_{ij} + s_{ji} &= g(\nabla_{E_i}\xi, E_j) + g(\nabla_{E_j}\xi, E_i) = (\mathcal{L}_\xi g)(E_i, E_j) = 0,\\ s_{ii} &= g(\nabla_{E_i}\xi, E_i) = 2(\mathcal{L}_\xi g)(E_i, E_i) = 0,\\ \operatorname{div}\xi &= \sum_i s_{ii} = 0 \quad (\text{hence } H_\xi = 0),\\ \nabla_\xi \xi &= \sum_i g(\nabla_\xi \xi, E_i)E_i = 0.\end{aligned}$$

Consequently (3.28) and (3.26) imply

$$\begin{aligned}\frac{2n-1}{2}\operatorname{Vol}(S^n) = \tilde{E}(\xi) &= \frac{n}{2}\operatorname{Vol}(S^n) + \frac{1}{2}\int_{S^n} \operatorname{Ric}(\xi,\xi)\, d\operatorname{vol}(g)\\ &= \frac{n}{2}\operatorname{Vol}(S^n) + \frac{1}{2}\int_{S^n} \operatorname{Ric}(V, V)\, d\operatorname{vol}(g) \le \tilde{E}(V)\end{aligned}$$

for any unit tangent vector field V on S^n, $n = 2m+1$. Theorem 3.30 is proved. ∎

3.6. THE BRITO ENERGY OF THE REEB VECTOR

Let S^{2n+1} be the standard odd-dimensional sphere with the canonical metric g and let ξ be a Hopf vector field on S^{2n+1}. If η is the 1-form on S^{2n+1} given by $\eta = g(\xi, \cdot)$ then (η, g) is a K-contact metric structure on S^{2n+1} and ξ is the corresponding Reeb vector field (cf. the terminology in [42]). In the sequel, we extend Brito's result (cf. Theorem 3.30 above) to the case of a compact K-contact manifold and its associated Reeb vector field. To start with we show that

Proposition 3.31 *Let $(M, (\phi, \xi, \eta, g))$ be a compact K-contact real $(2n+1)$-dimensional manifold. Let* Ric *be the Ricci tensor of (M, g).*

i. *One has*

$$\tilde{E}(V) \geq \tilde{E}(\xi) = E(\xi) = \left(2n + \frac{1}{2}\right)\mathrm{Vol}(M),$$

for any unit tangent vector field V on (M, g) provided that $\mathrm{Ric}(V, V) \geq 2n$. Here $\mathrm{Vol}(M)$ is short for $\mathrm{Vol}(M, g) = \int_M d\,\mathrm{vol}(g)$.

ii. *There is a D-homothetic K-contact structure $(\phi_t, \xi_t, \eta_t, g_t)$ on M such that*

$$\tilde{E}(V) \geq \tilde{E}(\xi_t) = E(\xi_t) = t^{n+1} E(\xi),$$

for any unit tangent vector field V on (M, g_t) provided that $\mathrm{Ric}(X, X) \geq k\|X\|^2$ for some $k \in \mathbb{R}$, $k > -2$, and for any $X \in \mathfrak{X}(M)$. Here $(\phi_t, \xi_t, \eta_t, g_t)$ is given by

$$\phi_t = \phi, \quad \xi_t = (1/t)\xi, \quad \eta_t = t\eta, \quad g_t = tg + t(t-1)\eta \otimes \eta, \quad t > 0.$$

For D-homothetic contact metric structures see S. Tanno, [280].

Proof of Proposition 3.31.

i. Let V be a unit vector field tangent to (M, g). By Brito's inequality (3.26), one has

$$\tilde{E}(V) \geq \frac{2n+1}{2}\mathrm{Vol}(M) + \frac{1}{2}\int_M \mathrm{Ric}(V, V)\,d\,\mathrm{vol}(g).$$

As (by hypothesis) $\mathrm{Ric}(V, V) \geq 2n$ one has

$$\tilde{E}(V) \geq \left(2n + \frac{1}{2}\right)\mathrm{Vol}(M).$$

We recall (cf. e.g., [42], p. 62) that for any contact metric structure (ϕ, ξ, η, g) one has $\operatorname{div}\xi = 0$ hence the mean curvature vector of the contact distribution $\operatorname{Ker}(\eta) = (\mathbb{R}\xi)^\perp$ vanishes

$$H_\xi = -\frac{1}{2n}(\operatorname{div}\xi)\xi = 0.$$

In particular for the given K-contact structure

$$\tilde{E}(\xi) = E(\xi) = \left(2n + \frac{1}{2}\right)\operatorname{Vol}(M)$$

hence the result.

ii. Let $t > 0$ and let Ric_t be the Ricci tensor of the Riemannian manifold (M, g_t). A straightforward calculation based on $g_t = tg + t(t-1)\,\eta \otimes \eta$ shows that

$$\operatorname{Ric}_t = \operatorname{Ric} - 2(t-1)g + 2(t-1)(nt+n+1)\,\eta \otimes \eta. \qquad (3.29)$$

Cf. also S.I. Goldberg, [139], p. 653. Let us fix a value of the parameter $t > 0$ such that

$$0 < t \leq \frac{k+2}{2n+2}.$$

We shall show that $\operatorname{Ric}_t(W, W) \geq 2n$ for any unit tangent vector field W on (M, g_t). Indeed any such W may be written as $W = V + f\xi$ where $V \in \operatorname{Ker}(\eta)$ and $f \in C^\infty(M)$ is given by $f = \eta(W)$. Then (by (3.29))

$$\operatorname{Ric}_t(W, W) = \operatorname{Ric}(W, W) - 2(t-1)\|W\|^2 + 2(t-1)(nt+n+1)f^2$$

(as the identity $Q\xi = 2n\xi$ holds on any K-contact metric manifold, cf. [42], and in particular $\operatorname{Ric}(V, \xi) = 0$)

$$= \operatorname{Ric}(V, V) + 2nf^2$$
$$- 2(t-1)\|V\|^2 - 2(t-1)f^2 + 2(t-1)(nt+n+1)f^2$$
$$= \operatorname{Ric}(V, V) - 2(t-1)\|V\|^2 + 2nt^2f^2.$$

On the other hand, as W is a unit vector field with respect to g_t

$$1 = g_t(W, W) = t\|W\|^2 + (t^2 - t)f^2 = t\|V\|^2 + t^2f^2$$

i.e., $t^2f^2 = 1 - t\|V\|^2$. Therefore

$$\operatorname{Ric}_t(W, W) = \operatorname{Ric}(V, V) - 2(nt+t-1)\|V\|^2 + 2n$$
$$\geq [k + 2 - 2t(n+1)]\|V\|^2 + 2n \geq 2n.$$

Finally

$$d\,\mathrm{vol}(g_t) = (2^n n!)^{-1}\eta_t \wedge (d\eta_t)^n = t^{n+1}\,d\,\mathrm{vol}(g),$$

$$E(\xi_t) = \left(2n + \frac{1}{2}\right)\mathrm{Vol}(M, g_t) = \left(2n + \frac{1}{2}\right) t^{n+1}\,\mathrm{Vol}(M) = t^{n+1} E(\xi)$$

hence (ii) follows by applying part (i) of Proposition 3.31.

■

Any $(2n+1)$-dimensional K-contact Einstein manifold has constant scalar curvature $\rho = 2n(2n+1)$. Then Proposition 3.31 implies

Corollary 3.32 (D. Perrone, [243]) *The Reeb vector field of each compact K-contact Einstein manifold is an absolute minimum for Brito's energy functional.*

C. Boyer & K. Galicki, [67], exhibited a large class of compact Sasakian-Einstein manifolds, including Tanno's example of a Sasakian-Einstein structure (η, g) on $S^3 \times S^2$ (cf. [280]) where g is *not* a Riemannian product of constant curvature metrics. In these examples the Reeb vector is an absolute minimum of Brito's energy functional. On the other hand any compact K-contact Einstein manifold is Sasakian (cf. C. Boyer & K. Galicki, [66]).

Theorem 3.33 (D. Perrone, [243]) *Let (M, ϕ, ξ, η, g) be a compact K-contact η-Einstein manifold of real dimension $2n+1 > 3$. Let W be the Webster scalar curvature.*

- **i.** *If $W \geq n(n+1)/2$ then the Reeb vector ξ is an absolute minimum of Brito's energy functional $\tilde{E}$.*
- **ii.** *If $W > 0$ then for any $0 < t \leq 2W/[n(n+1)]$ the Reeb vector ξ_t is an absolute minimum of Brito's energy $\tilde{E}_t$ defined in terms of the D-homothetic metric g_t.*
- **iii.** *If $W = 0$ then in general ξ is not a minimum for both energy functionals $\tilde{E}$ and E.*

We recall (cf. e.g., K. Yano & M. Kon, [325]) that a contact metric manifold (M, η, g) is *η-Einstein* if $\mathrm{Ric} = \alpha g + \beta \eta \otimes \eta$ for some real valued smooth functions $\alpha, \beta \in C^\infty(M)$. In dimension $2n+1 \geq 5$ for any η-Einstein K-contact manifold (M, η, g), the functions α, β are actually constant

$$\alpha = \frac{\rho}{2n} - 1, \quad \beta = -\frac{\rho}{2n} + 2n + 1.$$

Moreover a $(2n+1)$-dimensional η-Einstein K-contact manifold is Einstein if and only if $\beta = 0$ i.e., $\mathrm{Ric} = 2ng$. On the other hand, the Webster scalar curvature of a $(2n+1)$-dimensional K-contact manifold is given by

$$W = \frac{1}{8}\left(\rho - \mathrm{Ric}(\xi,\xi) + 4n\right) = \frac{\rho + 2n}{8}.$$

The first equality may be taken as a definition of the *generalized Tanaka-Webster scalar curvature* as introduced by S. Tanno, [281], p. 362, on an arbitrary contact metric manifold. See also S. Dragomir & G. Tomassini, [110], p. 50. Then the Ricci tensor of a η-Einstein K-contact manifold of real dimension $2n+1$ with $n > 1$ is given by

$$\mathrm{Ric} = \frac{4W - 2n}{n} g + \frac{2n^2 + 2n - 4W}{n} \eta \otimes \eta$$

(where W is constant). Consequently

$$\mathrm{Ric}(X,X) \geq 2n\|X\|^2 \Longleftrightarrow W \geq \frac{n(n+1)}{2}.$$

At this point statement (i) in Theorem 3.33 follows from part (i) in Proposition 3.31.

In order to prove statement (ii) in Theorem 3.33 we consider the D-homothetic K-contact metric structure $(\phi_t, \xi_t, \eta_t, g_t)$ for a fixed value of the parameter $t > 0$ satisfying

$$0 < t \leq \frac{2W}{n(n+1)}$$

(provided that $W > 0$). It may be easily checked (along the lines of the proof of part (ii) in Proposition 3.31) that the Ricci tensor Ric_t of the metric $g_t = tg + (t^2 - t)\eta \otimes \eta$ satisfies $\mathrm{Ric}_t(V, V) \geq 2n$ for any unit vector field V on (M, g_t). Then part (ii) in Proposition 3.31 implies that ξ_t is a minimum point of $\tilde{E}_t$.

Let us prove (iii) in Theorem 3.33. It suffices to build an example where ξ is not a minimum of $\tilde{E}$ (or E). Let $\mathbb{R}^{2n+1}$ carry the standard Sasakian structure (ϕ, ξ, η, g)

$$\phi \frac{\partial}{\partial x^i} = -\frac{\partial}{\partial y^i}, \quad \phi \frac{\partial}{\partial y^i} = \frac{\partial}{\partial x^i} + y^i \frac{\partial}{\partial z}, \quad \phi \frac{\partial}{\partial z} = 0,$$

$$\xi = 2\frac{\partial}{\partial z}, \quad \eta = \frac{1}{2}\left(dz - \sum_{i=1}^{n} y^i dx^i\right),$$

$$g = \eta \otimes \eta + \frac{1}{4}\sum_{i=1}^{n}\left(dx^i \otimes dx^i + dy^i \otimes dy^i\right).$$

The contact bundle $\mathrm{Ker}(\eta)$ is the span of

$$\left\{ \frac{\partial}{\partial x^i} + y^i \frac{\partial}{\partial z}, \ \frac{\partial}{\partial y^i} : 1 \leq i \leq n \right\}$$

and the vector fields

$$E_i = 2\frac{\partial}{\partial y^i}, \quad E_{i+n} = 2\left(\frac{\partial}{\partial x^i} + y^i \frac{\partial}{\partial z} \right), \quad \xi = 2\frac{\partial}{\partial z},$$

form a ϕ-basis (see also [42], p. 48). $\mathbb{R}^{2n+1}$ equipped with this contact metric structure is a Sasakian space form of constant ϕ-sectional curvature $c = -3$ while its Ricci tensor is given by

$$\mathrm{Ric} = -2g + 2(n+1)\,\eta \otimes \eta$$

(cf. [42], p. 113–114). In particular $(\mathbb{R}^{2n+1}, (\phi, \xi, \eta, g))$ is η-Einstein of Webster scalar curvature $W = 0$. One may identify $\mathbb{R}^{2n+1}$ with the Heisenberg group $\mathbb{H}_n$ of all matrices of the form

$$\mathcal{A} = \begin{pmatrix} 1 & Y & z \\ O^t & I_n & X^t \\ 0 & O & 1 \end{pmatrix},$$

$$X = (x_1, \ldots, x_n), \quad Y = (y_1, \ldots, y_n), \quad O = (0, \ldots, 0),$$
$$X,\, Y,\, O \in \mathbb{R}^n, \quad z \in \mathbb{R}.$$

The coordinates (x_i, y_i, z) provide a global chart of $\mathbb{H}_n$. It may be easily checked that the vector fields

$$\{E_i, \ E_{i+n} = \phi E_i, \ \xi : 1 \leq i \leq n\} \tag{3.30}$$

are left invariant. Moreover the ϕ-basis (3.30) (an orthonormal frame relative to the contact metric g) satisfies

$$[E_i, \phi E_i] = \xi, \quad 1 \leq i \leq n, \tag{3.31}$$

while the remaining Lie brackets are zero. Summing up (ϕ, ξ, η, g) is a left invariant Sasakian structure on the unimodular Lie group $\mathbb{H}_n$. Note that for $n = 1$ one recovers the left invariant Sasakian structure on the lowest dimensional Heisenberg group as introduced in [42], p. 49. Using (3.31) one may easily check that the Levi-Civita connection ∇ of $(\mathbb{H}_n, g)$ is expressed by

$$\nabla_{E_i}\xi = -\phi E_i, \quad \nabla_{\phi E_i}\xi = E_i, \quad \nabla_{E_i}\phi E_i = \xi, \tag{3.32}$$

$$\nabla_\xi E_i = -\phi E_i, \quad \nabla_\xi \phi E_i = E_i, \quad \nabla_{\phi E_i} E_i = -\xi, \tag{3.33}$$

for any $1 \leq i \leq n$, while the remaining covariant derivatives are zero. Let $\Gamma \subset \mathbb{H}_n$ be the discrete subgroup of all matrices $\mathcal{A} \in \mathbb{H}_n$ with integer entries. The space of right cosets $M = \Gamma \setminus \mathbb{H}_n$ is a compact differentiable manifold and the natural projection $\pi : \mathbb{H}_n \to M$ is smooth. Each left invariant vector field on $\mathbb{H}_n$ descends to M. Equivalently if $X \in \mathfrak{X}(\mathbb{H}_n)$ is left invariant, then $(d_{ba}\pi)X_{ba} = (d_a\pi)X_a$ for any $a \in \mathbb{H}_n$ and any $b \in \Gamma$. Indeed if L_b is the left translation with b then

$$(\pi \circ L_b)(a) = \pi(ba) = \Gamma\, ba = \Gamma\, a = \pi(a)$$

hence (using the left invariance of X)

$$(d_{ba}\pi)X_{ba} = [d_a(\pi \circ L_b)]X_a = (d_a\pi)X_a .$$

Similarly any left invariant tensor field on $\mathbb{H}_n$, and in particular its standard Sasakian structure, descends to the quotient space M. For the remainder of this section left invariant tensor fields on $\mathbb{H}_n$ and their projections on M are denoted by the same symbols. The identities (3.32)–(3.33) imply

$$\|\nabla E_i\|^2 = \|\nabla_{\phi E_i} E_i\|^2 + \|\nabla_\xi E_i\|^2 = 2, \quad \|\nabla \xi\|^2 = 2n,$$

and

$$\operatorname{div} \xi = 0, \quad \operatorname{div} E_i = 0,$$

that is the mean curvature vectors H_ξ and H_{E_i} (of the distributions $(\mathbb{R}\xi)^\perp$ and $(\mathbb{R}E_i)^\perp$ vanish). Then

$$\begin{aligned}
&\tilde{E}(E_i) - \tilde{E}(\xi) = E(E_i) - E(\xi) \\
&= \frac{1}{2} \int_M \left(\|\nabla E_i\|^2 - \|\nabla \xi\|^2 \right) d\,\mathrm{vol}(g) = -(n-1)\ \mathrm{Vol}(M).
\end{aligned}$$

Consequently, as $n > 1$, ξ can be a minimum of neither $\tilde{E}$ nor E.

Remark 3.34 The proof of part (iii) in Theorem 3.33 yields

$$\begin{aligned}
&(\mathrm{Hess}\ E)_\xi(E_i, E_i) \\
&= \int_M \left(\|\nabla E_i\|^2 - \|E_i\|^2 \|\nabla \xi\|^2 \right) d\,\mathrm{vol}(g) = -2(n-1)\ \mathrm{Vol}(M).
\end{aligned}$$

It follows that the Reeb vector ξ of the Sasakian manifold $M = \Gamma \setminus \mathbb{H}_n$ with $n > 1$ is E-unstable. Nevertheless, as we shall show in Chapter 4, the Reeb vector of $\Gamma \setminus \mathbb{H}_1$ is E-stable. Therefore the Reeb vector ξ enjoys the same stability properties as Hopf vector fields on odd-dimensional spheres. Note, however, the bias in the behavior of ξ on $\Gamma \setminus \mathbb{H}_n$ with respect to Brito's energy functional. ■

3.7. VECTOR FIELDS WITH SINGULARITIES

One of the conclusions of the previous section is that Hopf vector fields on S^n with $n = 2m+1$ aren't minima of the Dirichlet energy functional $E : \Gamma^\infty(S(M)) \to [0, +\infty)$. The remaining problem is of course to compute $\inf_{X \in \Gamma^\infty(S(M))} E(X)$ and investigate whether this is achieved for some vector field perhaps allowed to possess (a finite number of) singular points. For instance, radial vector fields on S^n are examples of unit vector fields with only isolated singularities.

3.7.1. Geodesic Distance

Let (M, g) be a Riemannian manifold and $p \in M$ a given point. If $\exp_p$ is a diffeomorphism of a neighborhood V of the origin in $T_p(M)$ then $U = \exp_p(V)$ is a *normal neighborhood* of p. If $B_\epsilon(0) = \{v \in T_p(M) : g_p(v, v) < \epsilon^2\}$ is such that $\overline{B_\epsilon(0)} \subset V$ then $B(p, \epsilon) = \exp_p[B_\epsilon(0)]$ is the *geodesic ball* of center p and radius $\epsilon > 0$. By the Gauss lemma (cf. e.g., Lemma 3.5 in [98], p. 69) the boundary $S(p, \epsilon)$ of a geodesic ball $B(p, \epsilon)$ is a smooth real hypersurface in M, orthogonal to the geodesics that issue at p. $S(p, \epsilon)$ is a *geodesic sphere* at p. The geodesics in $B(p, \epsilon)$ that start at p are the *radial geodesics*. Given $v \in T_p(M)$ with $\|v\| = 1$ let us set $\gamma_v(t) = \exp_p(tv)$. When t is small γ_v is the unique minimal unit speed geodesic joining p and $\exp_p(tv)$. Let A be the set of all $t > 0$ such that γ_v is the unique minimal geodesic joining p and $\gamma_v(t)$. Also let $t_0 = \sup A$. When t_0 is finite $\gamma_v(t_0)$ is a *cut point* of p. Let Cut(p) be the set of all cut points of p. If $d : M \times M \to [0, +\infty)$ is the distance associated to the Riemannian structure g, the function (the distance from p)

$$r : M \to [0, +\infty), \quad r(q) = d(p, q), \quad q \in M,$$

is only Lipschitz on M yet r is actually smooth on $M \setminus \text{Cut}(p)$. In applications one may differentiate $r : M \setminus \text{Cut}(p) \to (0, +\infty)$, perhaps several times. The resulting calculations will involve the curvature of (M, g) (cf. e.g., R. Schoen, [268]). As well known (cf. e.g., Corollary 2.8 in [98], p. 271) $M \setminus \text{Cut}(p)$ is homeomorphic to an open ball in the Euclidean space. Loosely speaking, the topology of M is contained in its cut locus.

Let $u : \mathbb{R}^n \to T_p(M)$ be a $\mathbb{R}$-linear isomorphism such that

$$g_p(u(e_i), u(e_j)) = \delta_{ij}, \quad 1 \le i,j \le n,$$

where $\{e_1, \dots, e_n\} \subset \mathbb{R}^n$ is the canonical linear basis. If $U = \exp_p(V) \subseteq M$ is a normal neighborhood of p then

$$\chi = (x^1, \dots, x^n) : U \to \mathbb{R}^n, \quad \chi(q) = \left(\exp_p \circ u\right)^{-1}(q), \quad q \in U,$$

is a local chart on U and $(x^1, \dots, x^n)$ is a *normal coordinate system* on M at p. If $\overline{B_\epsilon(0)} = \{v \in T_p(M) : \|v\| \le \epsilon\} \subset V$ and $B(p,\epsilon) = \exp_p[B_\epsilon(0)]$ as above then

$$B(p,\epsilon) = \{q \in U : \sum_{i=1}^{n} x^i(q)^2 < \epsilon^2\}.$$

We end this short reminder of Riemannian geometry by recalling that i) every point $q \in B(p,\epsilon)$ can be joined to p by a geodesic lying in $B(p,\epsilon)$ and such a geodesic is unique, ii) the length of the geodesic in (i) is equal to the distance $d(p,q)$, and iii) $B(p,\epsilon)$ equals the metric ball $\{q \in M : d(p,q) < \epsilon\}$ (cf. e.g., Proposition 3.4 in [189], p. 165). The reader might find it useful to look at N. Shimakura, [270], p. 263–268, and at G. De Rham, [96], p. 132–143.

3.7.2. F. Brito & P.G. Walczak's Theorem

Let $p \in M$ and $U \subseteq M$ be a normal neighborhood of p. The distance from p function $r = (p, \cdot) : U \setminus \{p\} \to \mathbb{R}$ is smooth. Let then $\partial/\partial r \in \mathfrak{X}(U \setminus \{p\})$ be the tangent vector field dual to dr with respect to g i.e.,

$$g\left(\frac{\partial}{\partial r}, X\right) = X(r), \quad X \in \mathfrak{X}(U \setminus \{p\}).$$

$\partial/\partial r$ is a unit vector field tangent to the geodesics issuing at p. Also $\partial/\partial r$ is the outward normal at each point of any small geodesic sphere $S(p,a)$. Moreover the distribution $(\mathbb{R}(\partial/\partial r))^\perp$ is completely integrable and the geodesic spheres $S(p,a)$ are its maximal integral manifolds.

Let us look now at unit radial vector fields on the standard sphere. Let $p \in S^n$ and let T be the unit tangent vector field tangent to the radial geodesics issuing at p. The vector field T is defined on $S^n \setminus \{\pm p\}$. Clearly T determines a geodesic flow and the distribution $(\mathbb{R}T)^\perp$ is completely integrable and its leaves are totally umbilical. As a straightforward consequence it may be shown that the unit radial vector field $T \in \mathfrak{X}(S^n \setminus \{\pm p\})$ is a harmonic vector field (cf. E. Boeckx & L. Vanhecke, [51]).

Theorem 3.35 (F. Brito & P.G. Walczak, [72]) *Let M be a real n-dimensional, $n \geq 3$, compact orientable Riemannian manifold. Let V be a unit vector field on M possessing a finite number of singularities. Then*

$$E(V) \geq \frac{1}{2} \int_M \left(n + \frac{1}{n-2} \mathrm{Ric}(V,V) \right) d\,\mathrm{vol}(g). \tag{3.34}$$

If $n \geq 4$ then equality is achieved in (3.34) if and only if V is geodesic and $(\mathbb{R}V)^{\perp}$ is a completely integrable distribution whose leaves are totally umbilical in M.

To prove Theorem 3.35 we need some preparation. Let $A \subset M$ be a finite subset (eventually empty). Then $\Omega = M \setminus A$ is an open dense subset of M whose closure is compact. For sufficiently small $a > 0$ let $B(p,a)$ and $S(p,a)$ be respectively the geodesic ball and the geodesic sphere of center p and radius a.

Lemma 3.36 *Let $A = \{p_1, \ldots, p_k\} \subset M$ and $\Omega = M \setminus A$. Let $p \in A$ and $f : \Omega \to [0, +\infty)$ be a continuous function. If*

$$\liminf_{a \to 0^+} \int_{S(p,a)} f d\,\mathrm{vol}(g_{S(p,a)}) > 0$$

then

$$\int_M f^2 d\,\mathrm{vol}(g) = \infty.$$

Here $g_{S(p,a)}$ is the first fundamental form of $S(p,a)$ in (M,g).

Proof. Let $C > 0$ such that $\mathrm{Vol}(S(p,a)) \leq C a^{n-1}$ for sufficiently small $a > 0$. Under the assumptions in Lemma 3.36 there is $\epsilon > 0$ such that

$$\int_{S(p,a)} f\, d\,\mathrm{vol}(g_{S(p,a)}) \geq \epsilon$$

for sufficiently small $a > 0$. By Hölder's inequality

$$\epsilon \leq \int_{S(p,a)} f\, d\,\mathrm{vol}(g_{S(p,a)}) \leq \mathrm{Vol}(S(p,a))^{1/2} \left(\int_{S(p,a)} f^2 d\,\mathrm{vol}(g_{S(p,a)}) \right)^{1/2}$$

hence

$$\int_{S(p,a)} f^2 d\,\mathrm{vol}(g_{S(p,a)}) \geq \frac{\epsilon^2}{\mathrm{Vol}(S(p,a))} \geq \frac{\epsilon^2}{Ca^{n-1}}$$

for sufficiently small $a > 0$. Moreover, by Fubini's theorem

$$\int_{B(p,a)} f^2 d\,\mathrm{vol}(g) = \int_0^a \left(\int_{S(p,t)} f^2 d\,\mathrm{vol}(g_{S(p,a)}) \right) dt \geq \frac{\epsilon^2}{C} \int_0^a t^{1-n}\, dt = \infty.$$

■

Lemma 3.37 *Let (M,g) be a compact orientable Riemannian manifold of real dimension $n \geq 3$. Let $V \in \mathfrak{X}(\Omega)$ with $\Omega = M \setminus A$ and $A = \{p_1, \ldots, p_k\} \subset M$. If $E(V) < \infty$ i.e., $\int_M \|\nabla V\|^2 d\,\mathrm{vol}(g) < \infty$ then*

$$\int_M \{\mathrm{Ric}(V,V) - 2\sigma_2(V)\} d\,\mathrm{vol}(g) = 0,$$

where $\sigma_2(V)$ is the second mean curvature of the distribution $(\mathbb{R}V)^\perp$.

Proof. Let $a_i > 0$ be sufficiently small positive numbers and $S_i = S(p_i, a_i)$ and $B_i = B(p_i, a_i)$ respectively the geodesic sphere and ball of center p_i and radius a_i, for each $1 \leq i \leq k$. Let ν_i be the outward unit normal on S_i. The identity (3.8) in Lemma 3.13 applies to V so that

$$\begin{aligned} \mathrm{Ric}(V,V) - 2\sigma_2(V) &= \mathrm{div}(\nabla_V V - (\mathrm{div}\, V)\, V) \\ &= \mathrm{div}(\nabla_V V + (n-1)H_V). \end{aligned}$$

Then (by Green's lemma)

$$\left| \int_{M\setminus \cup_i B_i} \{\mathrm{Ric}(V,V) - 2\sigma_2(V)\} d\,\mathrm{vol}(g) \right|$$

$$= \left| \int_{M\setminus \cup_i B_i} \mathrm{div}\,(\nabla_V V + (n-1)H_V) d\,\mathrm{vol}(g) \right|$$

$$= \left| \sum_i \int_{S_i} g(\nabla_V V + (n-1)H_V, \nu_i) d\,\mathrm{vol}(g_{S_i}) \right|$$

$$\leq \sum_i \int_{S_i} \|\nabla_V V + (n-1)H_V\| d\mathrm{vol}(g_{S_i})$$

$$\left| \int_{M \setminus \cup_i B_i} \{\mathrm{Ric}(V,V) - 2\sigma_2(V)\} d\mathrm{vol}(g) \right|$$

$$\leq C_n \sum_i \int_{S_i} \|\nabla V\| d\mathrm{vol}(g_{S_i}) \tag{3.35}$$

for some constant $C_n > 0$ depending only on n. At this point we may apply Lemma 3.36 to the function $f = \|\nabla V\|$ (by taking into account that $\int_M \|\nabla V\|^2 < \infty$) to conclude that

$$\liminf_{a_i \to 0^+} \int_{S_i} \|\nabla V\| d\mathrm{vol}(g_{S_i}) = 0.$$

Therefore (by (3.35)) the integral $\int_M \{\mathrm{Ric}(V,V) - 2\sigma_2(V)\} d\mathrm{vol}(g)$ is well defined as a principal value and vanishes. ■

Proof of Theorem 3.35. Let $\{E_1, \dots, E_n\}$ be a local orthonormal frame with $E_n = V$ (defined on some open subset $U \subseteq \Omega$). Let us set

$$s_{ij} = g(\nabla_{E_i} V, E_j), \quad 1 \leq i,j \leq n-1.$$

Then

$$\|\nabla V\|^2 = \|\nabla_V V\|^2 + \sum_{i,j=1}^{n-1} s_{ij}^2$$

and

$$\sum_{i,j=1}^{n-1} s_{ij}^2 = \frac{1}{n-2} \left\{ \sum_{i<j} (s_{ii} - s_{jj})^2 + \sum_{i<j} (s_{ij} + s_{ji})^2 \right.$$

$$\left. + (n-3) \sum_{\neq j} s_{ij}^2 + 2 \sum_{i<j} (s_{ii}s_{jj} - s_{ij}s_{ji}) \right\},$$

$$2 \sum_{i<j} \{s_{ii}s_{jj} - s_{ij}s_{ji}\} = 2\sigma_2(V).$$

Thus

$$\|\nabla V\|^2 \geq \frac{2}{n-2}\sigma_2(V). \tag{3.36}$$

When $E(V) = \infty$ then (3.34) is trivially satisfied. If $E(V) < \infty$ then (3.34) follows from (3.36) and Lemma 3.37. Indeed

$$\begin{aligned} 0 &= \int_M \{\mathrm{Ric}(V,V) - 2\sigma_2(V)\} d\,\mathrm{vol}(g) \\ &\geq \int_M \{\mathrm{Ric}(V,V) - (n-2)\|\nabla V\|^2\} d\,\mathrm{vol}(g) \\ &= \int_M \mathrm{Ric}(V,V) d\,\mathrm{vol}(g) - (n-2)\{2E(V) - n\mathrm{Vol}(M)\}. \end{aligned}$$

If $n > 3$ and equality is achieved in (3.34) then $\nabla_V V = 0$, $s_{ij} = 0$ for any $i \neq j$, and $s_{ii} = s_{jj}$ for any $1 \leq i,j \leq n-1$. ∎

Theorem 3.38 (F. Brito & P.G. Walczak, [72]) *Let V be a unit vector field on S^{2m+1}, $m > 1$, possessing a finite number of singularities. Then*

$$E(V) \geq \frac{4m^2 + 2m - 1}{2(2m-1)} \mathrm{Vol}(S^{2m+1}) \tag{3.37}$$

and equality is achieved if and only if V is a unit radial vector field on $S^{2m+1} \setminus \{\pm p\}$, for some $p \in S^{2m+1}$.

Proof. When $M = S^n$ with $n = 2m+1$ one has $\mathrm{Ric}(V,V) = n-1 = 2m$ and the inequality (3.34) may be written in the form (3.37). Applying once again Theorem 3.35 equality is achieved in (3.37) if and only if V is geodesic, the distribution $(\mathbb{R}V)^\perp$ is completely integrable, and the leaves of the foliation $\mathcal{F}$ tangent to $(\mathbb{R}V)^\perp$ are totally umbilical in S^{2m+1}. The leaves of $\mathcal{F}$ are therefore spheres at a constant geodesic distance one from another, and hence lie on $(2m+1)$-dimensional hyperplanes in $\mathbb{R}^{2m+2}$. Consequently the maximal integral curves of V are great circles of S^{2m+1} passing through two fixed antipodal points $\{-p,p\}$, so that V must be a unit radial vector field on $S^{2m+1} \setminus \{\pm p\}$ (if $q \in S^{2m+1} \setminus \{\pm p\}$ then V_q is the unit tangent vector to the unique geodesic of S^{2m+1} connecting p and q). ∎

For a compact Einstein manifold the inequality (3.34) becomes

$$E(V) \geq \frac{1}{2}\left(n + \frac{\rho}{n(n-2)}\right)\mathrm{Vol}(M)$$

where ρ is the scalar curvature of (M,g). On the other hand, by a result of B-Y. Chen, [83], the only irreducible locally symmetric spaces admitting totally umbilical hypersurfaces are the real space forms. Therefore

Corollary 3.39 *Let (M,g) be a real n-dimensional, $n \geq 4$, irreducible locally symmetric space of nonconstant sectional curvature. Let V be a unit tangent vector field on M possessing a finite number of singularities. Then*

$$E(V) > \frac{1}{2}\left(n + \frac{\rho}{n(n-2)}\right)\mathrm{Vol}(M)$$

(*strict inequality*).

As another consequence of Lemma 3.37 we may state (cf. also [56])

Corollary 3.40 *The tori are the only compact oriented surfaces admitting unit vector fields V such that $E(V) < \infty$ and V is globally defined except perhaps on a finite set A.*

Proof. Let us assume that there is $V \in \Gamma^\infty(S(\Omega))$ such that $E(V) < \infty$, where $\Omega = M \setminus A$. As $\dim(M) = 2$ one has $\sigma_2(V) = 0$. Then (by Lemma 3.37)

$$\int_M K_g \, d\,\mathrm{vol}(g) = 0,$$

where K_g denotes the Gaussian curvature. Now the Gauss-Bonnet theorem together with the (well-known) classification of compact oriented surfaces shows that M must be a torus. The converse is easy. ■

By Theorem 3.38 the energy of a radial vector field is a lower bound of the energy functional $E : \Gamma^\infty(S(S^{2m+1})) \to [0,+\infty)$ yet this lower bound is not achieved in $\Gamma^\infty(S(S^{2m+1}))$. V. Borrelli & F. Brito & O. Gil-Medrano, [62], have shown that this lower bound is actually $\inf\{E(X) : X \in \Gamma^\infty(S(S^{2m+1}))\}$ provided $m > 1$ (their method is to build a sequence of globally defined smooth unit vector fields tangent to S^{2m+1} whose energy converges to that of a radial vector field).

Let P be an embedded submanifold of (M,g). For a point $q \in M$ which is sufficiently close to P there is a unique geodesic of (M,g) connecting q to a point $p \in P$ and meeting P orthogonally at p. Let us consider, for some tubular neighborhood U of P, the function $r : U \to [0,+\infty)$ given by $r(q) = \mathrm{dist}(q,P) = d(q,p)$ for any $q \in U$, where $d : M \times M \to [0,+\infty)$ is the distance function associated to the metric g. Then r is smooth on $U \setminus P$.

The vector field $\partial/\partial r$ is again a unit geodesic vector field tangent to the geodesics normal to P. It is also the outward normal to the geodesic tubes $P(r)$ about P. E. Boeckx & L. Vanhecke, [51], studied radial vector fields about points and about totally geodesic submanifolds of rank one symmetric spaces. Their results may be stated as follows

A. Let M be a nonflat two-point homogeneous space. Then any radial vector field defined in a (pointed) normal neighborhood of a point is a harmonic vector field but not a harmonic map.

B. Let M be a Riemannian manifold and P a totally geodesic submanifold. Let us assume that one of the following assumptions holds

 i. M has constant sectional curvature;

 ii. M is a two-point homogeneous space with a complex structure J and P is J-invariant;

 iii. M is a $2m$-dimensional Kähler manifold of constant holomorphic sectional curvature and P is an m-dimensional anti-invariant submanifold.

 Then a radial unit vector field V defined on a tubular neighborhood of P is a harmonic vector field. Moreover V is a harmonic map if and only if M is flat.

C. On any Sasakian manifold of constant φ-sectional curvature each radial vector field defined on a tubular neighborhood of a characteristic line is a harmonic vector field but not a harmonic map.

3.7.3. Harmonic Radial Vector Fields

Besides being an important source of natural examples, harmonicity of radial vector fields leads to a new characterization of harmonic spaces. We recall that a *harmonic space* is a Riemannian manifold all of whose small geodesic spheres have constant mean curvature. By a result of E. Boeckx & L. Vanhecke, [52], a Riemannian manifold is a harmonic space if and only if each radial unit vector field defined on a pointed normal neighborhood is a harmonic vector field. These vector fields are not harmonic maps unless the manifold is flat.

E. Boeckx & J.C. Gonzales-Davila & L. Vanhecke, [56], provided generalizations of Theorems 3.35 and 3.38 by explicitly computing the energy of radial vector fields about points and about specific totally geodesic submanifolds in a compact rank one symmetric space. The main technical ingredients in [56] are the facts that a radial unit vector field is geodesic and the corresponding orthogonal distribution at a point of the given geodesic sphere, or tube, coincides with the tangent space to that geodesic sphere, or tube. As a further application, it may be shown that *a compact rank one symmetric space* (M, g) *with* $\dim(M) \geq 3$ *is a complex projective space if and only if the energy of radial vector fields about points is infinite.*

Let (M, g) be an n-dimensional Riemannian manifold. Let $\xi \in \Gamma^\infty(S(M))$ be a unit vector field on M. We set $A_\xi = -\nabla\xi$ and

$$\nu_\xi(X) = \operatorname{trace}\{Z \mapsto \left(\nabla_Z A_\xi^t\right) X\}, \quad X \in \mathfrak{X}(M).$$

We shall need the following

Lemma 3.41 *ξ is a harmonic vector field if and only if $\nu_\xi(X) = 0$ for any $X \in (\mathbb{R}\xi)^\perp$.*

Proof. Let $\{E_i : 1 \leq i \leq n\}$ be a local orthonormal frame on $T(M)$, defined on the open set $U \subseteq M$. For any $X \in \mathfrak{X}(M)$ (by $\nabla g = 0$)

$$\begin{aligned}
\nu_\xi(X) &= \sum_{i=1}^n g((\nabla_{E_i} A_\xi^t)X, E_i) \\
&= \sum_{i=1}^n \{g(\nabla_{E_i} A_\xi^t X, E_i) - g(A_\xi^t \nabla_{E_i} X, E_i)\} \\
&= \sum_{i=1}^n \{E_i(g(A_\xi^t X, E_i)) - g(A_\xi^t X, \nabla_{E_i} E_i) - g(\nabla_{E_i} X, A_\xi E_i)\} \\
&= \sum_i \{-E_i(g(X, \nabla_{E_i}\xi)) + g(X, \nabla_{\nabla_{E_i} E_i}\xi) + g(\nabla_{E_i} X, \nabla_{E_i}\xi)\} \\
&= \sum_i \{g(X, \nabla_{\nabla_{E_i} E_i} X) - g(X, \nabla_{E_i} \nabla_{E_i}\xi)\}
\end{aligned}$$

on U. We may conclude that

$$\nu_\xi(X) = g(X, \Delta_g \xi), \quad X \in \mathfrak{X}(M). \tag{3.38}$$

By (3.38) ν_ξ vanishes on $(\mathbb{R}\xi)^\perp$ if and only if $\Delta_g\xi$ and ξ are collinear i.e., $\Delta_g \xi = \lambda\xi$ for some $\lambda \in C^\infty(M)$. Taking the inner product with ξ now shows that $\lambda = \|\nabla\xi\|^2$. ∎

From now on, besides from $\|\xi\| = 1$ and $\nabla_\xi \xi = 0$ we assume that the distribution $(\mathbb{R}\xi)^\perp$ is involutive. Thus there is a codimension 1 foliation $\mathcal{F}$ of M such that $T(\mathcal{F}) = (\mathbb{R}\xi)^\perp$. In the sequel we use a few rudimental notions of foliation theory (cf. e.g., [288]). Let $\nu(\mathcal{F}) = T(M)/T(\mathcal{F})$ be the normal bundle of the foliation. We set

$$\alpha(X,Y) = \pi\,\nabla_X Y, \quad X,Y \in T(\mathcal{F}),$$

where $\pi : T(M) \to \nu(\mathcal{F})$ is the natural projection. One may consider the orthogonal complement $T(\mathcal{F})^{\perp}$ of $T(\mathcal{F})$ in $(T(M),g)$ and identify $\nu(\mathcal{F})$ and $T(\mathcal{F})^{\perp}$. Indeed

$$\sigma : \nu(\mathcal{F}) \to T(\mathcal{F})^{\perp}, \quad \sigma(s) = Y_s^{\perp}, \quad s \in \nu(\mathcal{F}),$$
$$Y_s \in T(M), \quad s = \pi(Y_s),$$

gives a well-defined bundle isomorphism $\nu(\mathcal{F}) \approx T(\mathcal{F})^{\perp}$. Then the restriction of α to each leaf $L \in M/\mathcal{F}$ is the second fundamental form of the immersion $L \hookrightarrow M$. For each $Z \in T(\mathcal{F})^{\perp}$ we consider the bundle endomorphism $W(Z) : T(\mathcal{F}) \to T(\mathcal{F})$ determined by

$$g(\sigma\,\alpha(X,Y),Z) = g(W(Z)X,Y), \quad X,Y \in T(\mathcal{F}).$$

The restriction of $W(Z)$ to a leaf L is the Weingarten operator of $L \hookrightarrow M$, corresponding to the normal vector field Z. The second fundamental form α of $\mathcal{F}$ is symmetric hence $W(Z) : T(\mathcal{F}) \to T(\mathcal{F})$ is self-adjoint. Moreover, one sets

$$\kappa(Z) = \operatorname{trace} W(Z), \quad Z \in T(\mathcal{F})^{\perp},$$
$$\kappa(X) = 0, \quad X \in T(\mathcal{F}).$$

The resulting 1-form $\kappa \in \Omega^1(M)$ is the *mean curvature form* of $\mathcal{F}$. Dually one may consider the vector field $\tau \in T(\mathcal{F})^{\perp}$ determined by

$$g(\tau,Z) = \kappa(Z), \quad Z \in T(\mathcal{F})^{\perp}.$$

The restriction of τ to a leaf of $\mathcal{F}$ is the mean curvature vector of that leaf (for simplicity taken here without the customary $1/(n-1)$ factor). For further use we set $H = \|\tau\|$.

Lemma 3.42 (E. Boeckx & L. Vanhecke, [51]) *Let (M,g) be a Riemannian manifold. Let $\xi \in \mathfrak{X}^1(M)$ be a unit geodesic vector field such that the distribution $(\mathbb{R}\xi)^{\perp}$ is involutive. Let H be the mean curvature of the corresponding foliation $\mathcal{F}$. Then*

$$\nu_\xi(X) = X(H) - \operatorname{Ric}(X,\xi) \tag{3.39}$$

for any $X \in T(\mathcal{F})$. Consequently ξ is harmonic if and only if $dH + \operatorname{Ric}(\xi,\cdot)$ vanishes along $(\mathbb{R}\xi)^{\perp}$.

Proof. Note that $T(\mathcal{F})^\perp = \mathbb{R}\xi$. The corresponding Weingarten operator $W(\xi)$ is given by

$$g(W(\xi)X, Y) = g(\sigma\alpha(X, Y), \xi) = g(\pi^\perp \nabla_X Y, \xi)$$

where $\pi^\perp : T(M) \to T(\mathcal{F})^\perp$ is the projection associated to the decomposition $T(M) = T(\mathcal{F}) \oplus T(\mathcal{F})^\perp$. Next (by $\nabla g = 0$)

$$g(W(\xi)X, Y) = g(\nabla_X Y, \xi) = X(g(Y, \xi)) - g(Y, \nabla_X \xi) = g(A_\xi X, Y)$$

for any $X, Y \in T(\mathcal{F})$. On the other hand

$$g(A_\xi X, \xi) = -g(\nabla_X \xi, \xi) = -\frac{1}{2} X\left(\|\xi\|^2\right) = 0,$$

hence A_ξ maps $T(\mathcal{F})$ into itself and the restriction of A_ξ to $T(\mathcal{F})$ is precisely the Weingarten operator $W(\xi)$. Hence $A_\xi : T(\mathcal{F}) \to T(\mathcal{F})$ is self-adjoint. Let then $\{E_i : 1 \le i \le n-1\}$ be a local orthonormal frame of $T(\mathcal{F})$ defined on an open set U and consisting of eigenvectors of A_ξ i.e.,

$$A_\xi E_i = \lambda_i E_i, \quad 1 \le i \le n-1,$$

for some C^∞ functions $\lambda_i : U \to \mathbb{R}$. Note that $A_\xi^t E_i = \lambda_i E_i$ as well. Let us set as usual

$$L_\xi = I + A_\xi^t \circ A_\xi .$$

Then $A_\xi \xi = 0$ yields $L_\xi \xi = \xi$. Also

$$L_\xi E_i = \left(1 + \lambda_i^2\right) E_i, \quad 1 \le i \le n-1.$$

Next we wish to compute $\nu_\xi(E_j)$. As $\{E_1, \dots, E_{n-1}, \xi\}$ is an orthonormal frame of $T(M)$ on U we have

$$\begin{aligned} \nu_\xi(E_j) &= \operatorname{trace}\{Z \mapsto (\nabla_Z A_\xi^t) E_j\} \\ &= \sum_{i=1}^{n-1} g((\nabla_{E_i} A_\xi^t) E_j, E_i) + g((\nabla_\xi A_\xi^t) E_j, \xi). \end{aligned}$$

The last term vanishes

$$\begin{aligned} g((\nabla_\xi A_\xi^t) E_j, \xi) &= g(\nabla_\xi \left(\lambda_j E_j\right), \xi) - g(A_\xi^t \nabla_\xi E_j, \xi) \\ &= \xi(\lambda_j g(E_j, \xi)) - \lambda_j g(E_j, \nabla_\xi \xi) = 0. \end{aligned}$$

Therefore

$$\begin{aligned}\nu_\xi(E_j) &= \sum_{i=1}^{n-1} g\left(\left(\nabla_{E_i} A_\xi^t\right) E_j, E_i\right)\\ &= \sum_i \left\{g\left(\nabla_{E_i} A_\xi^t E_j, E_i\right) - g\left(A_\xi^t \nabla_{E_i} E_j, E_i\right)\right\}\\ &= \sum_i \left\{g(\nabla_{E_i}(\lambda_j E_j), E_i) - \lambda_i g\left(\nabla_{E_i} E_j, E_i\right)\right\}\end{aligned}$$

and one obtains

$$\nu_\xi(E_j) = E_j(\lambda_j) + \sum_{i=1}^{n-1} (\lambda_j - \lambda_i) g(\nabla_{E_i} E_j, E_i) \tag{3.40}$$

for any $1 \leq j \leq n-1$. At this point we need to use the Codazzi equation

$$g(R(X,Y)Z,\xi) = -g((\nabla_X W(\xi))Y, Z) + g((\nabla_Y W(\xi))X, Z), \tag{3.41}$$

for any $X, Y, Z \in T(\mathcal{F})$. By setting $X = Z = E_i$ and $Y = E_j$ in (3.41) one gets

$$\begin{aligned}-g(R(E_i,E_j)E_i, \xi) &= g((\nabla_{E_i} A_\xi)E_j, E_i) - g((\nabla_{E_j} A_\xi)E_i, E_i)\\ &= g(\nabla_{E_i}(\lambda_j E_j) - A_\xi \nabla_{E_i} E_j, E_i) - g(\nabla_{E_j}(\lambda_i E_i) - A_\xi \nabla_{E_j} E_i, E_i)\end{aligned}$$

hence

$$-g(R(E_i,E_j)E_i, \xi) = (\lambda_j - \lambda_i) g(\nabla_{E_i} E_j, E_i) + \delta_{ij} E_i(\lambda_j) - E_j(\lambda_i), \tag{3.42}$$

for any $1 \leq i,j \leq n-1$. Let us express the term $(\lambda_j - \lambda_i) g(\nabla_{E_i} E_j, E_i)$ from (3.42) and substitute into (3.40). We have

$$\begin{aligned}\nu_\xi(E_j) &= E_j(\lambda_j) - \sum_i \{g(R(E_i,E_j)E_i, \xi) + \delta_{ij} E_i(\lambda_j) - E_j(\lambda_i)\}\\ &= E_j\left(\sum_i \lambda_i\right) - \mathrm{Ric}(E_j, \xi).\end{aligned}$$

As locally $H = \sum_{i=1}^{n-1} \lambda_i$ we may conclude that $\nu_\xi(X) = X(H) - \mathrm{Ric}(X,\xi)$ for any $X \in (\mathbb{R}\xi)^\perp$. Lemma 3.42 is proved. We may state ∎

Theorem 3.43 (E. Boeckx & L. Vanhecke, [51]) *Let (M,g) be a harmonic Riemannian manifold. If $p \in M$ and $U \subseteq M$ is a normal neighborhood of*

p let $\xi = \partial/\partial r \in \mathfrak{X}(U \setminus \{p\})$ be a radial vector field. Then ξ is a harmonic vector field. Conversely if (M,g) is an Einstein manifold such that in any normal neighborhood the radial vector field is harmonic then (M,g) is a harmonic manifold.

Proof. Let $\xi \in \mathfrak{X}(U \setminus \{p\})$ be a radial vector field. Then the distribution $(\mathbb{R}\xi)^\perp$ is integrable thus giving rise to a foliation $\mathcal{F}$ of $U \setminus \{p\}$ by geodesic spheres. In a harmonic manifold all geodesic spheres have constant mean curvature (cf. J. Berndt et al., [39], L. Vanhecke, [298]) hence $H \in \mathbb{R}$. On the other hand any harmonic manifold is Einstein so that $X(H) - \text{Ric}(X,\xi) = 0$ for any $X \in T(\mathcal{F})$. Then (by Lemma 3.42) the radial vector field ξ is harmonic. Vice versa, under the assumptions of Theorem 3.43, one has (by (3.39)) $X(H) = 0$ hence $H =$ constant i.e., (M,g) is a harmonic manifold. ∎

We end the section by showing that on any harmonic manifold $\xi = \partial/\partial r$ is a weak harmonic vector field on U, provided that $n \geq 3$. To this end we ought to check that $\xi \in \mathcal{H}^1_g(T(U))$ satisfies

$$\int_U \{g^*(\nabla\xi, \nabla X) - \|\nabla\xi\|^2 g(\xi, X)\}\, d\,\text{vol}(g) = 0 \tag{3.43}$$

for any $X \in \mathfrak{X}^\infty_0(U)$. Let $\epsilon > 0$ such that $\overline{B(p,\epsilon)} \subset U$. The integral in the left hand side of (3.43) may be written as $I_\epsilon(X) + J_\epsilon(X)$ where

$$I_\epsilon(X) = \int_{U \setminus B(p,\epsilon)} \{g^*(\nabla\xi, \nabla X) - \|\nabla\xi\|^2 g(\xi, X)\}\, d\,\text{vol}(g),$$

$$J_\epsilon(X) = \int_{B(p,\epsilon)} \{g^*(\nabla\xi, \nabla X) - \|\nabla\xi\|^2 g(\xi, X)\}\, d\,\text{vol}(g).$$

We wish to show that $I_\epsilon(X) = 0$. Let $\{E_i : 1 \leq i \leq n-1\}$ be a local orthonormal frame of $(\mathbb{R}\xi)^\perp$ consisting of eigenvectors of A_ξ as in the proof of Lemma 3.42. Clearly $I_\epsilon(X)$ is additive with respect to X hence it suffices to show that $I_\epsilon(\varphi E_j) = 0$ and $I_\epsilon(\varphi\xi) = 0$ for any $\varphi \in C^\infty_0(U)$ and any $1 \leq j \leq n-1$. An easy calculation shows that the integrand in $I_\epsilon(X)$ vanishes for $X = \varphi\xi$. Moreover

$$g^*(\nabla\xi, \nabla(\varphi E_j)) - \|\nabla\xi\|^2 g(\xi, \varphi E_j)$$

(as ξ is geodesic)

$$= \sum_{i=1}^{n-1} g(\nabla_{E_i}\xi, \nabla_{E_i}(\varphi E_j)) = -\sum_i \lambda_i g(E_i, E_i(\varphi)E_j + \varphi \nabla_{E_i} E_j)$$
$$= -\lambda_j E_j(\varphi) - \varphi \sum_i \lambda_i g(E_i, \nabla_{E_i} E_j) = -\lambda_j E_j(\varphi) - \varphi S_j$$

where we set $S_j = \sum_{i=1}^{n-1} \lambda_i g(\nabla_{E_i} E_j, E_i)$. Let us sum over $1 \le i \le n-1$ in (3.42) to get

$$\mathrm{Ric}(E_j, \xi) = \lambda_j \sum_i g(\nabla_{E_i} E_j, E_i) - S_j + E_j(\lambda_j) - E_j(H)$$

or

$$S_j = \lambda_j \sum_i g(\nabla_{E_i} E_j, E_i) + E_j(\lambda_j)$$

as $H \in \mathbb{R}$ and $\mathrm{Ric}(E_j, \xi) = (\rho/n) g(E_j, \xi) = 0$ on $U \setminus B(p, \epsilon)$. Hence

$$g^*(\nabla\xi, \nabla(\varphi E_j)) - \|\nabla\xi\|^2 g(\xi, \varphi E_j)$$
$$= -\lambda_j E_j(\varphi) - \lambda_j \varphi \sum_i g(\nabla_{E_i} E_j, E_i) - \varphi E_j(\lambda_j)$$
$$= -\sum_i g(\nabla_{E_i}(\varphi\lambda_j E_j), E_i) = -\mathrm{div}(\varphi\lambda_j E_j) + g(\nabla_\xi(\varphi\lambda_j E_j), \xi)$$

and the last term vanishes because $\nabla_\xi \xi = 0$. Therefore (by Green's lemma)

$$I_\epsilon(\varphi E_j) = -\int_{U \setminus B(p,\epsilon)} \mathrm{div}(\varphi\lambda_j E_j)\, d\,\mathrm{vol}(g) = \int_{S(p,\epsilon)} \varphi\lambda_j g(\xi, E_j)\, dA = 0$$

as ξ is the outward unit normal on $S(p, \epsilon)$. As to the second integral (by the Cauchy-Schwartz inequality)

$$|J_\epsilon(X)| \le \int_{B(p,\epsilon)} \left\{ |g^*(\nabla\xi, \nabla X)| + \|\nabla\xi\|^2 |g(\xi, X)| \right\} d\,\mathrm{vol}(g)$$

$$\leq \int_{B(p,\epsilon)} \{\|\nabla\xi\| \, \|\nabla X\| + \|\nabla\xi\|^2 \|X\|\} d\,\mathrm{vol}(g)$$

$$\leq C \int_{B(p,\epsilon)} \|\nabla\xi\| \, (1 + \|\nabla\xi\|) \, d\,\mathrm{vol}(g)$$

where

$$C = \max\{\sup_{q\in\Gamma} \|X\|_q, \sup_{q\in\Gamma} \|\nabla X\|_q\}, \quad \Gamma = \mathrm{supp}(X).$$

Let $(x^1, \ldots, x^n)$ be a normal coordinate system on U such that $x^i(p) = 0$. For any $0 < \rho \leq \epsilon$ the radial vector field ξ on $S(p, \rho)$ is given by $\xi = (x^i/\rho)\, \partial/\partial x^i$ hence

$$\nabla\xi = \frac{1}{\rho}\left(\delta^i_j + \Gamma^i_{jk} x^k\right) dx^j \otimes \frac{\partial}{\partial x^i}$$

and then

$$\|\nabla\xi\| \leq C_n \frac{1+\rho}{\rho}, \tag{3.44}$$

on $S(p, \rho)$ for some constant $C_n > 0$. Indeed

$$\|\nabla\xi\|^2 = \rho^{-2} g^{ij} g_{k\ell} \left(\delta^k_i + \Gamma^k_{ir} x^r\right)\left(\delta^\ell_j + \Gamma^\ell_{js} x^s\right) \leq \sum_{i,j,k,\ell} \frac{A}{\rho^2}\left(1 + B \sum_r |x^r|\right)^2$$

where

$$A = \max_{i,j,k,\ell} \sup\{\left|g^{ij}(q) g_{k\ell}(q)\right| : q \in \overline{B(p, \epsilon_0)}\},$$

$$B = \max_{i,j,k} \sup\{\left|\Gamma^i_{jk}(q)\right| : q \in \overline{B(p, \epsilon_0)}\},$$

where $\epsilon_0 > 0$ is fixed such that $\overline{B(p, \epsilon_0)} \subset U$. Then for $0 < \delta \leq \epsilon_0$ one obtains the inequality (3.44). Finally for $0 < \epsilon \leq \epsilon_0$

$$\left|J_\epsilon(X)\right| \leq C \int_0^\epsilon d\rho \int_{S(p,\rho)} C_n \frac{1+\rho}{\rho}\left(1 + C_n \frac{1+\rho}{\rho}\right) dA$$

$$\leq C'_n \int_0^\epsilon \rho^{n-3} (1+\rho)(1+2\rho)\, d\rho \to 0, \quad \epsilon \to 0,$$

as $\mathrm{Vol}(S(p,\rho)) \leq a_n \rho^{n-1}$ for some $a_n > 0$.

3.8. NORMAL VECTOR FIELDS ON PRINCIPAL ORBITS

The purpose of this section is to report on a result by G. Nunes & J. Ripoll, [224]. Precisely we shall prove the following

Theorem 3.44 ([224], p. 1091) *Let (M,g) be an n-dimensional $(n \geq 3)$ compact orientable Riemannian manifold and $G \subset \mathrm{Isom}(M,g)$ a compact Lie group acting on M with cohomogeneity one. Let us assume that either G has no singular orbits or each singular orbit of G has dimension $\leq n-3$. Let N be a unit normal vector field orthogonal to the principal orbits of G. Then $N \in \mathcal{H}^{1,2}_g(T(M))$ and N is a critical point of the total bending functional $\mathcal{B} : \mathcal{H}^{1,2}_g(T(M)) \to \mathbb{R}$. Let $M^* \subset M$ be the union of all principal orbits of G and let $H : M^* \to \mathbb{R}$ be the function given by $H(x) =$ the mean curvature of the orbit $G(x)$ with respect to N. Then $H \in L^2(M)$ and*

$$\mathcal{B}(N) = -\int_M \mathrm{Ric}(N,N)\, d\,\mathrm{vol}(g) + \int_M H^2\, d\,\mathrm{vol}(g). \tag{3.45}$$

By a *principal orbit* we mean one of maximal dimension ($n-1$ under the assumptions of Theorem 3.44). A brief review of compact transformation groups and cohomogeneity one actions is given in Appendix D of this monograph. To prove Theorem 3.44 we need

Lemma 3.45 *Let $(u_1(t), \ldots, u_m(t))$ and $(v_1(t), \ldots, v_m(t))$ be continuous 1-parameter families of vectors in $\mathbb{R}^m$ with $m \geq 2$ defined for $0 \leq t \leq \ell$. Let us assume that*

a. *$B_t = \{v_1(t), \ldots, v_m(t)\}$ is a positive[1] linear basis of $\mathbb{R}^m$ for each $0 \leq t < \ell$.*
b. *Let $W \subset \mathbb{R}^m$ be spanned by B_ℓ. Then $\dim_{\mathbb{R}} W \leq m-2$.*
c. *If $v_i(\ell) = 0$ and $v_j(\ell) = 0$ then*

$$\lim_{t \to \ell} \frac{|v_i(t)|}{|v_j(t)|} = 1. \tag{3.46}$$

For each $0 \leq t < \ell$ we consider the $m \times m$ matrices

$$D_t = \left[\langle v_i(t), v_j(t)\rangle\right]_{1 \leq i,j \leq m}$$

[1] If $v_j(t) = v^i_j(t) e_i$ then $\det\left[v^i_j(t)\right] > 0$ for any $0 \leq t < \ell$, where $\{e_1, \ldots, e_m\}$ is the canonical linear basis in $\mathbb{R}^m$.

and $M_t = \left[a_i^j(t)\right]_{1 \le i,j \le m}$ *given by*

$$u_i(t) = a_i^j(t) v_j(t), \quad 1 \le i \le m.$$

Then

$$\sup\left\{(\text{trace } M_t)^2 \sqrt{det\ D_t}\ :\ 0 \le t < \ell\right\} < \infty, \tag{3.47}$$

$$\lim_{t \to \ell+}\left[(\text{trace } M_t)\ \sqrt{det\ D_t}\right] = 0. \tag{3.48}$$

Proof. We look first at the case where B_t is an orthogonal system i.e., $\langle v_i(t), v_j(t)\rangle = 0$ for any $i \neq j$ and any $0 \le t < \ell$. Then

$$D_t = \text{diag}\left(|v_1(t)|^2, \ldots, |v_m(t)|^2\right), \quad \text{trace}(M_t) = \sum_{i=1}^m a_i^i,$$

hence

$$(\text{trace } M_t)^p \sqrt{\det D_t} = \left|\prod_{k=1}^m v_k(t)\right| \left(\sum_{j=1}^m a_j^j\right)^p, \quad p \in \{1, 2\}. \tag{3.49}$$

Let us take the inner product of $u_i(t) = a_i^j(t) v_j(t)$ with $v_k(t)$ to get

$$a_i^k(t) = \frac{\langle u_i(t), v_k(t)\rangle}{|v_k(t)|^2}$$

and let us contract the indices i and k. We obtain

$$\sum_{i=1}^m a_i^i(t) = \sum_{i=1}^m \frac{\langle u_i(t), v_i(t)\rangle}{|v_i(t)|^2}$$

so that (3.49) becomes

$$(\text{trace } M_t)\ \sqrt{\det D_t} = \left(\sum_{i=1}^m \frac{\langle u_i(t), v_i(t)\rangle}{|v_i(t)|^2}\right) \prod_{k=1}^m |v_k(t)|$$

$$= \sum_{i=1}^m \langle u_i(t), \frac{v_i(t)}{|v_i(t)|}\rangle \prod_{k \neq i} |v_k(t)|$$

for any $0 \le t < \ell$. Recall that

$$W = \mathbb{R}B_\ell, \quad \dim_{\mathbb{R}} W \le m - 2.$$

We claim that for any $i \in \{1, \dots, m\}$ there is an index $k_i \in \{1, \dots, m\} \setminus \{i\}$ such that $v_{k_i}(\ell) = 0$. Indeed if there was $i \in \{1, \dots, m\}$ such that $v_k(\ell) = 0$ for any $k \in \{1, \dots, m\} \setminus \{i\}$ then $\{v_k(\ell) : k \in \{1, \dots, m\} \setminus \{i\}\}$ would be an orthogonal system consisting of nonzero vectors hence $\dim_{\mathbb{R}} W \geq m - 1$, a contradiction. Consequently

$$\lim_{t \to \ell^+} \prod_{k \neq i} |v_k(t)| = 0, \quad 1 \leq i \leq m,$$

and (3.48) holds good. On the other hand

$$\begin{aligned}
(\operatorname{trace} M_t)^2 \sqrt{\det D_t} &= \left[\sum_{i,j=1}^{m} \frac{\langle u_i(t), v_i(t) \rangle}{|v_i(t)|^2} \frac{\langle u_j(t), v_j(t) \rangle}{|v_j(t)|^2} \right] \prod_{k=1}^{m} |v_k(t)| \\
&= \sum_{i,j=1}^{m} \left[\left\langle u_i(t), \frac{v_i(t)}{|v_i(t)|} \right\rangle \left\langle u_j(t), \frac{v_j(t)}{|v_j(t)|} \right\rangle \frac{\prod_{k=1}^{m} |v_k(t)|}{|v_i(t)|\,|v_j(t)|} \right] \\
&= \sum_{i=1}^{m} \left[\left\langle u_i(t), \frac{v_i(t)}{|v_i(t)|} \right\rangle^2 \prod_{k \neq i} \frac{|v_k(t)|}{|v_i(t)|} \right] \\
&\quad + \sum_{i \neq j} \left[\left\langle u_i(t), \frac{v_i(t)}{|v_i(t)|} \right\rangle \left\langle u_j(t), \frac{v_j(t)}{|v_j(t)|} \right\rangle \prod_{k \neq i,\, k \neq j} |v_k(t)| \right].
\end{aligned}$$

For each $i \in \{1, \dots, m\}$ we fix an index $k_i \in \{1, \dots, m\} \setminus \{i\}$ such that $v_{k_i}(\ell) = 0$. Then

$$\begin{aligned}
(\operatorname{trace} M_t)^2 \sqrt{\det D_t} &= \sum_{i=1}^{m} \left[\left\langle u_i(t), \frac{v_i(t)}{|v_i(t)|} \right\rangle^2 \frac{|v_{k_i}(t)|}{|v_i(t)|} \prod_{k \neq i,\, k \neq k_i} |v_k(t)| \right] \\
&\quad + \sum_{i \neq j} \left[\left\langle u_i(t), \frac{v_i(t)}{|v_i(t)|} \right\rangle \left\langle u_j(t), \frac{v_j(t)}{|v_j(t)|} \right\rangle \prod_{k \neq i,\, k \neq j} |v_k(t)| \right].
\end{aligned} \tag{3.50}$$

Therefore (by (3.46) and (3.50)) the set

$$\left\{ (\operatorname{trace} M_t)^2 \sqrt{\det D_t} : 0 \leq t < \ell \right\}$$

is bounded. Let us go back to the general case (where B_t is not necessarily orthogonal). Yet D_t is symmetric hence there is an orthogonal matrix

$O_t \in O(m)$ such that

$$O_t D_t O_t^{-1} = \operatorname{diag}(\lambda_1(t), \ldots, \lambda_m(t)), \quad 0 \le t < \ell.$$

Next we set

$$\tilde{v}_i(t) = \left(O_t^{-1}\right)_i^j v_j(t), \quad 0 \le t \le \ell,$$
$$\tilde{u}_i(t) = d_i^j \tilde{v}_j(t), \quad 0 \le t < \ell,$$

so that

$$\langle \tilde{v}_i(t), \tilde{v}_j(t)\rangle = \left(O_t^{-1}\right)_i^r \left(O_t^{-1}\right)_j^s \langle v_r(t), v_s(t)\rangle$$
$$= \sum_{s,r=1}^{m} (O_t)_s^j \langle (D_t)_{sr}\rangle \left(O_t^{-1}\right)_i^r = \left(O_t D_t O_t^{-1}\right)_i^j = \lambda_i \delta_i^j$$

i.e., $\tilde{B}_t = \{\tilde{v}_1(t), \ldots, \tilde{v}_m(t)\}$ is an orthogonal system for any $0 \le t < \ell$. At this point we may apply the first part of the proof of Lemma 3.45 to the systems $\{\tilde{u}_j(t) : 1 \le j \le m\}$ and $\{\tilde{v}_j(t) : 1 \le j \le m\}$ and conclude that

$$\sup\{\left(\operatorname{trace} \tilde{M}_t\right)^2 \sqrt{\det \tilde{D}_t} \, : \, 0 \le t < \ell\} < \infty,$$
$$\lim_{t \to \ell+} \left(\operatorname{trace} \tilde{M}_t\right) \sqrt{\det \tilde{D}_t} = 0,$$

where

$$\tilde{M}_t = \left[\tilde{a}_i^j\right], \quad \tilde{u}_i(t) = \tilde{a}_i^j \tilde{v}_j(t), \quad \tilde{D}_t = \left[\langle \tilde{v}_i, \tilde{v}_j\rangle\right]_{1 \le i,j \le m}.$$

Yet $\tilde{M}_t = M_t$ and $\det\left(\tilde{D}_t\right) = \det(D_t)$ and Lemma 3.45 is proved in full generality. ■

Another ingredient needed in the proof of Theorem 3.44 is the theory of cohomogeneity one manifolds (cf. Appendix D in this monograph). It is known that M/G is diffeomorphic either to a circle or to a closed interval. In the first case $F = \emptyset$ (with the notations of Appendix D) and N is globally smooth. If M/G is a closed interval then G has two singular orbits $\mathcal{O}_1$ and $\mathcal{O}_2$. By the assumption in Theorem 3.44

$$\dim(\mathcal{O}_i) \le n - 3, \quad i \in \{1, 2\}.$$

Let $Z \in C^\infty(T(M))$ and $f = g(N, Z)$. For every $\phi \in C^\infty(M)$ and $X \in \mathfrak{X}^\infty(M)$

$$\operatorname{div}(\phi f X) = \phi X(f) + f X(\phi) + f\phi \operatorname{div}(X) \tag{3.51}$$

on $M \setminus F$ (here $F = \mathcal{O}_1 \cup \mathcal{O}_2$). We set

$$M^s = \{x \in M : \text{dist}(x, F) \geq s\}$$

with respect to the Riemannian distance function d on M. Let us integrate (3.51) over M^s. Then (by Green's lemma)

$$\int_{\partial M^s} f\phi g(X, \eta^s)\, d\,\text{vol}(g^s) = \int_{M^s} \big\{\phi X(f) + fX(\phi) + f\phi\, \text{div}(X)\big\} d\,\text{vol}(g)$$

where g^s is the first fundamental form of M^s in (M, g) and η^s is the exterior unit normal on ∂M^s. Let $s \to 0$. We obtain

$$\int_M X(f)\phi\, d\,\text{vol}(g) = -\int_M f\{X(\phi) + \phi\, \text{div}(X)\} d\,\text{vol}(g)$$

for any $\phi \in C^\infty(M)$ hence f is weakly derivable at any direction $X \in \mathfrak{X}^\infty(M)$ and hence (by Proposition 2.46 in Chapter 2) f is weakly differentiable. We ought to show that $f \in H(M)$ (which in turn implies that $N \in H(T(M))$ by the very definition of the Sobolev type space $H(T(M))$). By the proof of Proposition 2.50 (cf. also Lemma 2.51) the weak differentiability of $f = g(N, Z)$ for any $Z \in \mathfrak{X}^\infty(M)$ implies that N is weakly differentiable. Then (again by Proposition 2.50) it suffices to show that $\nabla N \in L^2(T(M))$. To this end let P be a principal orbit of G and let us set $\ell = \text{dist}(P, \mathcal{O}_1)$. Let M^+ be the connected component of $M \setminus P$ containing $\mathcal{O}_1$. Let $\delta : M \to \mathbb{R}$ be the distance to P i.e.,

$$\delta(x) = \min\{d(x, y) : y \in P\}, \quad x \in M.$$

Next, one observes that for each $t \in \mathbb{R}$ either $P_t = \delta^{-1}(t)$ is empty or it is an orbit of G. Let $t \in [0, \ell]$ and let us set

$$M_t = \{x \in M : \delta(x) \leq t\}.$$

Then

$$\int_{M^+} \|\nabla N\|^2 d\,\text{vol}(g) = \lim_{t \to \ell} \int_{M_t} \|\nabla N\|^2 d\,\text{vol}(g). \tag{3.52}$$

Moreover it should be observed that $\|\text{grad}(\delta)\| = 1$ and $\text{grad}(\delta)$ is orthogonal to the orbits of G so that one may assume that $N = \text{grad}(\delta)$. Then

$$\|\nabla N\| = \|\text{Hess}(\delta)\|, \tag{3.53}$$

$$(\Delta\delta)(x) = H_\delta, \quad x \in M^+ \setminus \mathcal{O}_1, \tag{3.54}$$

where H_δ is the mean curvature with respect to N of the orbit $G(x)$ passing through x. Let $t \in [0, \ell)$. We wish to apply Reilly's formula (E.25) in Appendix E to the function $\delta : M_t \to \mathbb{R}$ i.e.,

$$\int_{M_t} \left\{ (\Delta\delta)^2 - \|\mathrm{Hess}(\delta)\|^2 - \mathrm{Ric}(\mathrm{grad}(\delta), \mathrm{grad}(\delta)) \right\} d\,\mathrm{vol}(g)$$

$$= \int_{\partial M_t} \{ (-\Delta_t z_t + u_t H_t)\, u_t$$

$$- g_t(\mathrm{grad}_t(z_t), \mathrm{grad}_t(u_t)) + g_t(a_t \mathrm{grad}_t z_t, \mathrm{grad}_t z_t) \big\} d\,\mathrm{vol}(g_t)$$

where g_t and a_t are respectively the first fundamental form and the Weingarten operator of ∂M_t in (M_t, g). Also $H_t = \mathrm{trace}(a_t)$ is the (non-normalized) mean curvature of ∂M_t in M_t. Moreover Δ_t and grad_t are respectively the Laplace-Beltrami and the gradient operators of the Riemannian manifold $(\partial M_t, g_t)$. Finally $z_t = \delta|_{\partial M_t}$ while u_t is the normal derivative of δ i.e., $u_t = g(\mathrm{grad}(\delta), N)\big|_{\partial M_t}$. It follows that

$$z_t = \delta|_{\partial M_t} = t = \mathrm{const.}, \quad u_t = g(N,N)\big|_{\partial M_t} = 1, \quad \partial M_t = P_t \cup P,$$

so that (by taking into account (3.54) and the orientation of the boundary)

$$\int_{M_t} \left\{ H_\delta^2 - \|\mathrm{Hess}(\delta)\|^2 - \mathrm{Ric}(N,N) \right\} d\,\mathrm{vol}(g)$$

$$= \int_{P_t} H_t d\,\mathrm{vol}(g_t) - \int_P H_0 d\,\mathrm{vol}(g_0) \tag{3.55}$$

for each $0 \le t < \ell$. Next (by Theorem 3.2.12 in [115], p. 429)

$$\int_{M_t} H_\delta^2 d\,\mathrm{vol}(g) = \int_0^t \left(\int_{P_s} H_s^2 \, d\,\mathrm{vol}(g_s) \right) ds$$

hence (by passing to the limit with $t \to \ell$ in (3.55) and exploiting (3.52))

$$\int_{M^+} \|\nabla N\|^2 d\,\mathrm{vol}(g) = -\int_{M^+} \mathrm{Ric}(N,N) d\,\mathrm{vol}(g) + \int_P H_0 d\,\mathrm{vol}(g_0)$$

$$+ \lim_{t \to \ell} \left\{ -\int_{P_t} H_t d\,\mathrm{vol}(g_t) + \int_0^t \left[\int_{P_s} H_s^2 d\,\mathrm{vol}(g_s) \right] ds \right\}. \tag{3.56}$$

Next, for every $0 \le t \le \ell$ we consider

$$\phi_t : P \to P_t, \quad \phi_t(x) = \exp_x(tN_x), \quad x \in P.$$

For each $0 \le t < \ell$ the map $\phi_t : P \to P_t$ is an orientation preserving diffeomorphism. Then

$$\int_{P_t} H_t^2 \, d\,\mathrm{vol}(g_t) = \int_P (H_t \circ \phi_t)^2 \phi_t^* d\,\mathrm{vol}(g_t)$$

$$= \int_P (H_t \circ \phi_t)^2 \lambda_t \, d\,\mathrm{vol}(g_0), \quad 0 \le t < \ell, \tag{3.57}$$

where λ_t is the Jacobian of ϕ_t. Let $x \in P$ and let $\gamma(t) = \exp_x(tN_x) \in M$ be the geodesic of initial data $\gamma(0) = x$ and $\dot{\gamma}(0) = N_x$. Moreover, let $\tau_t : T_{\gamma(t)}(M) \to T_{\gamma(0)}(M)$ be the parallel displacement operator along γ from $\gamma(t)$ to $\gamma(0)$. Given a tangent vector $w \in T_x(P)$ we consider a curve $\alpha : (-\epsilon, \epsilon) \to P$ such that $\alpha(0) = x$ and $\dot{\alpha}(0) = w$. We use α to build the variation of γ given by

$$f(s,t) = \exp_{\alpha(s)}\left(tN_{\alpha(s)}\right), \quad |s| < \epsilon, \quad 0 \le t \le \ell. \tag{3.58}$$

Let us set $\gamma_t(s) = f(s,t)$ and consider the infinitesimal variation J_w of γ (induced by the variation (3.58)) defined by

$$J_{w,\gamma(t)} = \frac{d\gamma_t}{ds}(0), \quad 0 \le t \le \ell.$$

Then (cf. e.g., Theorem 1.2 in [189], Vol. II, p. 64) $J_w \in \mathcal{J}_\gamma$ where $\mathcal{J}_\gamma$ is the space of all Jacobi fields along γ (so that $\dim_{\mathbb{R}} \mathcal{J}_\gamma = 2n$). For each $X \in \mathcal{J}_\gamma$ we set as customary $X' = \nabla_{\dot{\gamma}} X$ and $X'' = \nabla_{\dot{\gamma}} X'$. Then

$$J_w'' + R(J_w, \dot{\gamma})\dot{\gamma} = 0. \tag{3.59}$$

Let $\{w_1, \dots, w_{n-1}\} \subset T_x(P)$ be an orthonormal basis and let us set

$$v_i = \tau_t\left(J_{w_i}(\gamma(t))\right), \quad u_i = \tau_t\left(J_{w_i}'(\gamma(t))\right), \quad 1 \le i \le n-1.$$

Note that

$$J_{w_i}(\gamma(t)) = (d_x \phi_t) w_i, \quad 1 \le i \le n-1.$$

Since $\phi_t : P \to P_t$ is a diffeomorphism for each $0 \le t < \ell$ and $\phi_\ell : P \to \mathcal{O}_1$ has (by the assumption in Theorem 3.44 on the dimension of the singular orbit $\mathcal{O}_1$) rank at most $n-3$, it follows that $B_t = \{v_i(t) : 1 \le i \le n-1\}$ is a linear basis of $T_x(P)$ for any $0 \le t < \ell$ while the dimension of the space spanned by $B_\ell = \{v_i(\ell) : 1 \le i \le n-1\}$ (over $\mathbb{R}$) is at most $n-3$. Let us check that the system B_t satisfies the assumption (c) in Lemma 3.45. Indeed if $v_i(\ell) = 0$ and $v_j(\ell) = 0$ then $J_{w_i}(\gamma(\ell)) = 0$ and $J_{w_j}(\gamma(\ell)) = 0$ so that

$$\lim_{t\to\ell} \frac{\|v_i(t)\|}{\|v_j(t)\|} = \lim_{t\to\ell} \frac{\|J_{w_i}(\gamma(t))\|}{\|J_{w_j}(\gamma(t))\|} = 1.$$

Also (cf. e.g., [305], p. 342)

$$J'_{w_i}(\gamma(t)) - \left(\nabla_{J_{w_i}} N\right)_{\gamma(t)} \in T_{\gamma(t)}(P_t)^\perp$$

so that

$$g_{\gamma(t)}\left(J'_{w_i}(\gamma(t)), u\right) = g_{\gamma(t)}\left(\left(\nabla_{J_{w_i}} N\right)_{\gamma(t)}, u\right)$$
$$= -g_{t,\gamma(t)}\left((a_t)_{\gamma(t)}\, u, J_{w_i}(\gamma(t))\right)$$

where g_t and a_t are respectively the first fundamental form and the Weingarten operator of P_t in (M, g). In particular for $u = J_{w_j}(\gamma(t))$ (recall that $\{J_{w_j}(\gamma(t)); 1 \le j \le n-1\}$ is a linear basis in $T_{\gamma(t)}(P_t)$)

$$g_t\left(J'_{w_i}, J_{w_j}\right)_{\gamma(t)} = -g_t\left(a_t J_{w_j}, J_{w_i}\right)_{\gamma(t)}$$

or (by $(a_t)_{\gamma(t)} J_{w_j}(\gamma(t)) = (a_t)_j^k J_{w_k} J_{w_k}(\gamma(t))$ and the fact that $\tau_t : T_{\gamma(t)}(P_t) \to T_x(P)$ is a linear isometry)

$$\langle u_i(t), v_j(t)\rangle = -(a_t)_j^k \langle v_i(t), v_k(t)\rangle$$

where $\langle\ ,\ \rangle = g_{0,x}$. We may identify $T_x(M)$ with $\mathbb{R}^n$ by choosing a local coordinate system about x and use Lemma 3.45. With the notations there

$$(a_t)_j^m = \left(D_t^{-1}\right)^{im} (M_t)_i^k (D_t)_{kj}$$

hence

$$H_t = -\text{trace}(M_t).$$

Also

$$\lambda_t = \sqrt{\det(D_t)}$$

follows easily from $\phi_t^* d\,\text{vol}(g_t) = \lambda_t\, d\,\text{vol}(g_0)$. Thus (by Lemma 3.45)

$$K(x) := \sup_{0\le t<\ell} \left[(\text{trace}\, M_t)^2 \sqrt{\det(D_t)}\right] < \infty$$

so that

$$\int_P (H_t \circ \phi_t)^2 \lambda_t \, d\text{vol}(g_0) \le \int_P K \, d\text{vol}(g_0).$$

Yet P_t is an orbit of G (for each $0 \le t < \ell$) hence H_t and λ_t are constant along P_t so that $K =$ constant on P implying that

$$\int_P (H_t \circ \phi_t)^2 \lambda_t \, d\text{vol}(g_0) \le K \,\text{Vol}(P) < \infty.$$

Consequently (by taking into account (3.57)) we may set

$$\int_{M^+} H^2 d\text{vol}(g) := \lim_{t\to\ell} \int_0^t \left[\int_{P_s} H_s^2 d\text{vol}(g_s) \right] ds < \infty.$$

Also (again by Lemma 3.45)

$$\lim_{t\to\ell} (H_t \circ \phi_t)\, \lambda_t = -\lim_{t\to\ell} (\text{trace}\, M_t) \sqrt{\det(D_t)} = 0$$

hence

$$\lim_{t\to\ell} \int_{P_t} H_t d\text{vol}(g_t) = \lim_{t\to\ell} \int_P (H_t \circ \phi_t)\, \lambda_t d\text{vol}(g_0) = 0.$$

Finally (by (3.56))

$$\int_{M^+} \|\nabla N\|^2 d\text{vol}(g) = \int_P H_0 d\text{vol}(g_0) - \int_{M^+} \text{Ric}(N,N) d\text{vol}(g) + \int_{M^+} H^2 d\text{vol}(g) < \infty \tag{3.60}$$

i.e., $\|\nabla N\| \in L^2(M)$ and then $N \in \mathcal{H}_g^1(T(M))$. If M^- is the other connected component of $M \setminus P$ (the one containing $\mathcal{O}_2$) one may conduct similar calculations leading to (the following analog to (3.60))

$$\int_{M^-} \|\nabla N\|^2 d\text{vol}(g) = -\int_P H_0 d\text{vol}(g_0) - \int_{M^-} \text{Ric}(N,N) d\text{vol}(g) + \int_{M^-} H^2 d\text{vol}(g) < \infty. \tag{3.61}$$

The sum of (3.60) and (3.61) furnishes (3.45). Let us check that N is a weak solution to the harmonic vector field system. To this end a straightforward calculation shows that

$$\Delta_g N - \|\nabla N\|^2 N = 0 \quad \text{in} \quad M \setminus F. \tag{3.62}$$

On the other hand F is a set of measure zero hence for any $Z \in \mathfrak{X}^\infty(M)$ (by (3.62) and (2.114) in Chapter 2)

$$\int_M g(\nabla N, \nabla Z) d\,\mathrm{vol}(g) = \int_{M\setminus F} g(\nabla N, \nabla Z) d\,\mathrm{vol}(g)$$

$$= \int_{M\setminus F} g(\nabla^* \nabla N, Z) d\,\mathrm{vol}(g) = \int_{M\setminus F} g(\Delta_g N, Z) d\,\mathrm{vol}(g)$$

$$= \int_{M\setminus F} \|\nabla N\|^2 g(N, Z) d\,\mathrm{vol}(g) = \int_M \|\nabla N\|^2 g(N, Z) d\,\mathrm{vol}(g).$$

The reader should of course keep in mind Definition 2.52 in Chapter 2. ■

3.9. RIEMANNIAN TORI

Let $\Gamma = \{md_1 + nd_2 \in \mathbb{R}^2 : m, n \in \mathbb{Z}\}$ be a lattice in $\mathbb{R}^2$ and $T^2 = \mathbb{R}^2/\Gamma$ the corresponding torus. Let $\pi : \mathbb{R}^2 \to T^2$ be the natural projection (the universal covering of T^2). Throughout Section 3.9 we adopt the notations and conventions in Section 1.5 of Chapter 1 and Sections 2.6 and 2.10 of Chapter 2. Let g be an arbitrary Riemannian metric on T^2 and let $\{S, W\}$ be an orthonormal frame such that $JS = W$ where J is the complex structure on T^2 induced by a fixed orientation of T^2 (chosen such that π is orientation preserving). Let us recall (cf. Section 2.6 in Chapter 2) that

$$\nabla_S S = aW, \quad \nabla_S W = -aS,$$
$$\nabla_W S = bW, \quad \nabla_W W = -bS.$$

If $u \in C^2(T^2)$ then (with the convention $\{E_1, E_2\} = \{S, W\}$)

$$\Delta u = -\sum_{i=1}^{2} \left\{E_i^2 - (\nabla_{E_i} E_i)\right\} u = -S^2 u + (\nabla_S S) u - W^2 u + (\nabla_W W) u$$

hence

$$\Delta u = \left(-S^2 - W^2 - bS + aW\right)u. \tag{3.63}$$

Also the following formulae may be easily checked

$$\hat{\Delta} f = \left(-\hat{S}^2 - \hat{W}^2 - (b \circ \pi)\hat{S} + (a \circ \pi)\hat{W}\right)f, \quad f \in C^2(\mathbb{R}^2). \tag{3.64}$$

Objects with a hat indicate the dependence on the pullback metric $\hat{g} = \pi^* g$. Moreover

$$\hat{\Delta}(u \circ \pi) = (\Delta u) \circ \pi, \quad u \in C^2(T^2), \tag{3.65}$$

$$\widehat{\operatorname{div}}\left(\hat{Y}\right) = (\operatorname{div} Y) \circ \pi, \tag{3.66}$$

where $\hat{Y} \in \mathfrak{X}(\mathbb{R}^2)$ and $Y \in \mathfrak{X}(T^2)$ are π-related i.e., $(d_\xi \pi)\hat{Y}_\xi = Y_{\pi(\xi)}$ for any $\xi \in \mathbb{R}^2$. Moreover given $\varphi, \psi \in C^\infty(T^2)$ and $\alpha \in C^\infty(\mathbb{R}^2)$ such that $\cos\alpha = \varphi \circ \pi$ and $\sin\alpha = \psi \circ \pi$ a straightforward calculation leads to

$$(\varphi \Delta \psi - \psi \Delta \varphi) \circ \pi = \hat{\Delta}\alpha. \tag{3.67}$$

3.9.1. Harmonic Vector Fields on Riemannian Tori

We recall (cf. Section 2.6 in Chapter 2) that the total bending functional $\mathcal{B} : \mathcal{E} \to [0, +\infty)$ of a Riemannian torus (T^2, g) was recast as a functional defined in terms of the angle functions associated to unit vector fields $X \in \mathcal{E}$

$$\mathcal{B} : \mathcal{W} \to [0, +\infty), \quad \mathcal{B}(\alpha) = 2 \int_Q \|\hat{\nabla}\alpha + \hat{Z}\|_{\hat{g}}^2 \pi^* \omega_T$$

for any angle function $\alpha \in \mathcal{W}$ (cf. (2.43) in Chapter 2). Here $Z = aS + bW$ and $\hat{Z} \in \mathfrak{X}(\mathbb{R}^2)$ is π-related to Z. The purpose of this section is to establish the following

Theorem 3.46 (G. Wiegmink, [309]) *Let (T^2, g) be a real 2-dimensional Riemannian torus and $\mathcal{E} = \Gamma^\infty(S(T^2, g))$ the set of all unit tangent vector fields on (T^2, g). Let $\{S, W\}$ be an orthonormal frame of $(T(T^2), g)$ such that $JS = W$. Then* a) *the following statements are equivalent*

i. *$X \in \mathcal{E}$ is a critical point of $\mathcal{B}$ i.e., $\{d\mathcal{B}(X_t)/dt\}_{t=0} = 0$ for any $\{X_t\}_{t \in I} \in \mathcal{E}_I$ such that $X_0 = X$.*

ii. *Any angle function α (with respect to $\{S, W\}$) of X satisfies*

$$\hat{\Delta}\alpha - (Sa + Wb) \circ \pi = 0. \tag{3.68}$$

iii. *The (S,W)-coordinates (φ,ψ) of X satisfy*

$$\varphi\Delta\psi-\psi\Delta\varphi-Sa-Wb=0.$$

iv. *X satisfies the identity*

$$g(\nabla_X\nabla_X X, JX)=g(\nabla_{JX}\nabla_{JX}JX, X).$$

Moreover b) *if $X\in\mathcal{E}$ is a critical point of $\mathcal{B}$ then the full orbit of X under any smooth action of a Lie group on $\mathcal{E}$ leaving $\mathcal{B}$ invariant consists of critical points. The set of critical points of $\mathcal{B}$ intersects each homotopy class $\mathcal{E}^{(S,W)}_{(m,n)}\in\{[X]:X\in\mathcal{E}\}\subset\pi(T^2,S(T^2))$ exactly in one orbit of the $SO(2)$-action on $\mathcal{E}$. Therefore, up to this action, there is but one critical point in each class $\mathcal{E}^{(S,W)}_{(m,n)}$.*

c) *Let $u\in C^\infty(T^2)$ and let $\tilde{g}=e^{2u}g$ be a metric on T^2 in the conformal class of g. Then a unit vector field $X\in\mathcal{E}$ is a critical point of $\mathcal{B}$ if and only if $e^{-u}X$ is a critical point of $\tilde{\mathcal{B}}$.*

d) *Let h be a flat metric on T^2 and ∇^h its Levi-Civita connection. There is an h-orthonormal frame $\{S_0,W_0\}$ which is parallel with respect to ∇^h. Let $\langle\cdot,\cdot\rangle$ and $\|\cdot\|$ be the Euclidean inner product and norm on $\mathbb{E}^2$ and let us set $D=\|d_1\|^2\|d_2\|^2-\langle d_1,d_2\rangle^2$. The (S_0,W_0)-angle functions $\lambda_{m,n}:\mathbb{R}^2\to\mathbb{R}$ of the critical points of $\mathcal{B}$ on (T^2,h) in the homotopy class $\mathcal{E}^{(S_0,W_0)}_{(m,n)}$ are given by*

$$\begin{aligned}\lambda_{m,n}(\xi)=\frac{2\pi}{D}\big[&\langle d_1,\xi\rangle\left(m\|d_2\|^2-n\langle d_1,d_2\rangle\right)\\&+\langle d_2,\xi\rangle\left(n\|d_1\|^2-m\langle d_1,d_2\rangle\right)\big]+s,\quad s\in\mathbb{R},\end{aligned}\tag{3.69}$$

for any $\xi\in\mathbb{R}^2$. Also

$$\mathcal{B}(\lambda_{m,n})=\frac{\pi}{D}\left(m^2\|d_2\|^2+n^2\|d_1\|^2-2mn\langle d_1,d_2\rangle\right).$$

e) *For any critical point $X\in\mathcal{E}$ of $\mathcal{B}$ on (T^2,g) and for any $q\in T^2$ there is a conformal coordinate chart $f:U\to\mathbb{E}^2$ (where $U\subseteq T^2$ is an open neighborhood of q) such that*

$$X=\left(\cos\lambda_{m,n}\right)\frac{\partial}{\partial f_1}+\left(\sin\lambda_{m,n}\right)\frac{\partial}{\partial f_2}$$

in terms of the Gaussian frame field of f with $\lambda_{m,n}$ as in (3.69) for suitable $(m,n)\in\mathbb{Z}^2$. When $(m,n)=(0,0)$ the f-coordinates of X are constant.

The beautiful Theorem 3.46 is perhaps the main result in [309] (and shows that on a Riemannian torus a quite accurate description of harmonic vector fields is available).

To prove Theorem 3.46 we need to establish a few preparatory results.

Lemma 3.47 *For any semiperiodic function $\beta \in Per(m,n)$, there exists a tangent vector field $Y \in \mathfrak{X}(T^2)$ such that*

$$\hat{\Delta}\beta = -(\mathrm{div}\ Y) \circ \pi, \qquad \int_Q (\hat{\Delta}\beta)\pi^* \omega_T = 0,$$

where $Q = \{sd_1 + td_2 : (s,t) \in [0,1]^2\}$.

Proof. Let $\beta \in \mathrm{Per}(m,n)$ be a semiperiodic function and let $V = \hat{\nabla}\beta \in \mathfrak{X}(\mathbb{R}^2)$ be the gradient of β with respect to the metric $\hat{g}$. By Lemma 1.35 the function $\xi \in \mathbb{R}^2 \mapsto (d_\xi \pi) V_\xi \in T_{\pi(\xi)}(T^2)$ is periodic, hence its restriction to the fibre $\pi^{-1}(p)$ is constant so that the tangent vector field $Y \in \mathfrak{X}(T^2)$ given by

$$Y_p = (d_\xi \pi) V_\xi, \quad \xi \in \pi^{-1}(p), \quad p \in T^2,$$

is well defined and π-related to V. Therefore, according to our conventions, the vector field V may be denoted by $\hat{Y}$. Then (by (3.66))

$$\hat{\Delta}\beta = -\widehat{\mathrm{div}}(\hat{Y}) = -(\mathrm{div}\, Y) \circ \pi,$$

$$\int_Q \left(\hat{\Delta}\beta\right) \pi^* \omega_T = -\int_Q \pi^* \left(\mathrm{div}(Y)\ \omega_T\right) = -\int_{T^2} \mathrm{div}(Y)\,\omega_T = 0$$

by Green's lemma. Lemma 3.47 is proved. ∎

Let $X : I \to \mathcal{E}$ be a smooth path in $\mathcal{E}$ i.e., a C^∞ family $\{X(t)\}_{t \in I} \in \mathcal{E}^I$ (see Section 1.5 in Chapter 1). Let $\alpha \in C^\infty(\mathbb{R}^2 \times I)$ be an angle function for X and let us set $\dot{\alpha} = \partial\alpha/\partial t$. Then

$$\frac{d}{dt}\{\mathcal{B}(X(t))\} = 2\frac{d}{dt}\int_Q \|\hat{\nabla}\alpha + \hat{Z}\|_{\hat{g}}^2 \pi^* \omega_T = 4\int_Q \hat{g}(\hat{\nabla}\alpha + \hat{Z}, \hat{\nabla}\dot{\alpha})\pi^*\omega_T.$$

On the other hand

$$\hat{g}(\hat{\nabla}\alpha + \hat{Z}, \hat{\nabla}\dot{\alpha}) = \left(\hat{\nabla}\alpha + \overline{Z}\right)\dot{\alpha} = \widehat{\mathrm{div}}\left(\dot{\alpha}\left(\hat{\nabla}\alpha + \hat{Z}\right)\right) - \dot{\alpha}\,\widehat{\mathrm{div}}\left(\hat{\nabla}\alpha + \hat{Z}\right).$$

Note that $\dot{\alpha}\left(\hat{\nabla}\alpha + \hat{Z}\right)$ is π-related to some $U \in \mathfrak{X}(T^2)$ (because α is semiperiodic and $\hat{Z}$ is π-related to Z). Hence

$$\int_Q \widehat{\mathrm{div}}\left(\dot{\alpha}\left(\hat{\nabla}\alpha + \hat{Z}\right)\right)\pi^*\omega_T = \int_Q \widehat{\mathrm{div}}(\hat{U})\,\pi^*\omega_T = \int_{T^2} \mathrm{div}(U)\,\omega_T = 0.$$

We may conclude (as $\operatorname{div}(Z) = S(a) + W(b)$) that

$$\frac{d}{dt}\{\mathcal{B}(X(t))\} = 4\int_Q \dot{\alpha}\left\{\hat{\Delta}\alpha - (Sa + Wb)\circ\pi\right\}\pi^*\omega_T. \tag{3.70}$$

At this point we may prove Theorem 3.46. To establish the implication (ii) $\Longrightarrow$ (i) let $X \in \mathcal{E}$ be a unit vector field on T^2 and $\alpha \in \mathcal{W}$ an angle function for X such that (3.68) is satisfied. Let $\{X(t)\}_{t\in I}$ be a smooth variation of X through unit vector fields such that $X(0) = X$. This of course amounts to considering a smooth 1-parameter variation $\{\alpha_t\}_{t\in I}$ of α (i.e., $\alpha_0 = \alpha$) through semiperiodic functions. On the other hand we may evaluate (3.70) at $t = 0$ to get (by (3.68))

$$\frac{d}{dt}\{\mathcal{B}(X(t))\}_{t=0} = 4\int_Q \beta\left\{\hat{\Delta}\alpha - (Sa + Wb)\circ\pi\right\}\pi^*\omega_T = 0$$

where $\beta(\xi) = \dot{\alpha}(\xi, 0)$ for any $\xi \in \mathbb{R}^2$ i.e., X is a critical point of $\mathcal{B}: \mathcal{E} \to [0, +\infty)$. Vice versa if $\{d\mathcal{B}(X(t))/dt\}_{t=0} = 0$ for any smooth 1-parameter variation $X(t) \in \mathcal{E}$ of X, in other words for any smooth 1-parameter variation $\alpha_t \in \mathcal{W}$ of α, then we may choose the variation α_t such that $\dot{\alpha}(\xi, 0) = \left(\hat{\Delta}\alpha\right)(\xi) - (Sa + Wb)_{\pi(\xi)}$ so that

$$\begin{aligned} 0 &= \int_Q \beta(\xi)\left(\hat{\Delta}\alpha - (Sa + Wb)\circ\pi\right)(\xi)\, d\operatorname{vol}(\hat{g})(\xi) \\ &= \int_Q \|\hat{\Delta}\alpha - (Sa + Wb)\circ\pi\|(\xi)^2\, d\operatorname{vol}(\hat{g})(\xi) \end{aligned}$$

hence $\left(\hat{\Delta}\alpha\right)(\xi) - (Sa + Wb)_{\pi(\xi)} = 0$ that is (3.68) holds and the implication (i) $\Longrightarrow$ (ii) is proved.

Next the equivalence (ii) $\Longleftrightarrow$ (iii) is a consequence of the identity (3.67) above.

To prove the implication (iii) $\Longrightarrow$ (iv) we choose the orthonormal frame $\{S, W\} = \{X, JX\}$ so that $\varphi = 1$ and $\psi = 0$. Then (by the very assumption in (iii))

$$\begin{aligned} 0 &= -\varphi\Delta\psi + \psi\Delta\varphi + (Sa + Wb) = Sa + Wb \\ &= X(g(\nabla_X X, JX)) + (JX)(g(\nabla_{JX} X, JX)). \end{aligned}$$

Vice versa (iv) implies (iii) because (T^2, J, g) is a Kähler manifold. Finally (i) follows from (iv) (because (iv) is easily seen to imply $\Delta_g X - \|X\|^2 X = 0$). Statement (a) in Theorem 3.46 is proved.

Let us prove statement (b) in Theorem 3.46. Let $G \times \mathcal{E} \to \mathcal{E}$ be a smooth action of a Lie group G on $\mathcal{E}$ such that $\mathcal{B}(a \cdot X) = \mathcal{B}(X)$ for any $a \in G$ and any $X \in \mathcal{E}$. Let $\text{Crit}(\mathcal{B})$ be the set of all critical points of $\mathcal{B} : \mathcal{E} \to [0, +\infty)$. Clearly if $X \in \text{Crit}(\mathcal{B})$ the orbit $G \cdot X = \{a \cdot X : a \in G\}$ is contained in $\text{Crit}(\mathcal{B})$. Let

$$a = \begin{pmatrix} \cos r & -\sin r \\ \sin r & \cos r \end{pmatrix} \in \text{SO}(2), \quad r \in \mathbb{R}. \tag{3.71}$$

and let us set $a \cdot X = (\cos r)\, X + (\sin r) JX$ (as in Theorem 2.26 in Chapter 2 of this book). Let $(m, n) \in \mathbb{Z}^2$ and let $\mathcal{E}^{(S,W)}_{(m,n)} \in \pi(T^2, S(T^2))$ be the homotopy class consisting of all $X \in \mathcal{E}$ with $\text{htp}^{(S,W)}(X) = (m, n)$. We shall show that if $\text{Crit}(\mathcal{B}) \cap \mathcal{E}^{(S,W)}_{(m,n)} \neq \emptyset$ then $\text{Crit}(\mathcal{B}) \cap \mathcal{E}^{(S,W)}_{(m,n)}$ is a SO(2)-orbit. Precisely if $X \in \text{Crit}(\mathcal{B}) \cap \mathcal{E}^{(S,W)}_{(m,n)}$ then we shall show that

$$\text{Crit}(\mathcal{B}) \cap \mathcal{E}^{(S,W)}_{(m,n)} = \text{SO}(2) \cdot X. \tag{3.72}$$

As $X \in \text{Crit}(\mathcal{B})$ and $\mathcal{B}$ is SO(2)-invariant it follows that

$$\text{SO}(2) \cdot X \subseteq \text{Crit}(\mathcal{B}). \tag{3.73}$$

Let $\alpha \in \text{Per}(m, n)$ be an angle function for X and $a \in \text{SO}(2)$ (determined by a number $r \in \mathbb{R}$ as in (3.71)). Using the fact that $X \circ \pi = (\cos\alpha)\, S \circ \pi + (\sin\alpha)\, W \circ \pi$ one has

$$\begin{aligned} (a \cdot X) \circ \pi &= (\cos r)\, X \circ \pi + (\sin r)\, (JX) \circ \pi \\ &= \cos(\alpha + r)\, S \circ \pi + \sin(\alpha + r) W \circ \pi \end{aligned}$$

that is $\alpha + r$ is an angle function for $a \cdot X$. The fact that α is semiperiodic obviously implies that $\alpha + r$ is semiperiodic as well, and precisely $\alpha + r \in \text{Per}(m, n)$ hence $\text{htp}^{(S,W)}(a \cdot X) = (m, n)$. Therefore $a \cdot X \in \mathcal{E}^{(S,W)}_{(m,n)}$ and we may conclude that

$$\text{SO}(2) \cdot X \subseteq \mathcal{E}^{(S,W)}_{(m,n)}. \tag{3.74}$$

By (3.73)–(3.74) one has

$$\text{SO}(2) \cdot X \subseteq \text{Crit}(\mathcal{B}) \cap \mathcal{E}^{(S,W)}_{(m,n)}. \tag{3.75}$$

To check the opposite inclusion let $Y \in \text{Crit}(\mathcal{B}) \cap \mathcal{E}^{(S,W)}_{(m,n)}$. Then all angle functions $\beta \in C^\infty(\mathbb{R}^2)$ of Y belong to $\text{Per}(m,n)$ and satisfy (as a consequence of statement (a), cf. description (ii) of the critical point Y) $\hat{\Delta}\beta = f$ where $f = (Sa + Wb) \circ \pi \in C^\infty(\mathbb{R}^2)$. On the other hand (by (3.68)) $\hat{\Delta}\alpha = f$ hence

$$\hat{\Delta}(\beta - \alpha) = 0 \tag{3.76}$$

in $\mathbb{R}^2$ and in particular in Q. Let us recall that on any Riemannian manifold (M,g) with the Laplace-Beltrami operator Δ one has $\Delta(uv) = u\Delta v + v\Delta u - 2g(\nabla u, \nabla v)$ for any $u, v \in C^2(M)$. In particular for $v = u$ one has $\Delta(u^2) = 2\left(u\Delta u - \|\nabla u\|^2\right)$. If moreover u is a harmonic function then $\Delta(u^2) = -2\|\nabla u\|^2$. For $M = \mathbb{R}^2$ and $u = \beta - \alpha$ this gives (by (3.76))

$$\int_Q \hat{\Delta}\left[(\beta - \alpha)^2\right]\pi^*\omega_T = -2\int_Q \|\hat{\nabla}(\beta - \alpha)\|^2_g \,\pi^*\omega_T.$$

Note that $\beta - \alpha \in \text{Per}(m,n)$ so that $(\beta - \alpha)^2 \in \text{Per}(0,0)$ as well. Then (by Lemma 3.47)

$$\int_Q \hat{\Delta}\left[(\beta - \alpha)^2\right]\pi^*\omega_T = 0.$$

Consequently $\int_Q \|\hat{\nabla}(\beta - \alpha)\|^2_g \,\pi^*\omega_T = 0$ so that $\beta - \alpha = r$ for some $r \in \mathbb{R}$ everywhere in Q. Thus

$$Y \circ \pi = (\cos\beta) S \circ \pi + (\sin\beta) W \circ \pi = (a \cdot X) \circ \pi$$

in Q, that is $Y = a \cdot X$ in $\pi(Q) = T^2$ so that $Y \in SO(2) \cdot X$. We may conclude that

$$\text{Crit}(\mathcal{B}) \cap \mathcal{E}^{(S,W)}_{(m,n)} \subseteq SO(2) \cdot X.$$

Together with the inclusion (3.75), this implies the desired equality (3.72). At this point it remains to be seen that for any $(m,n) \in \mathbb{Z}^2$

$$\text{Crit}(\mathcal{B}) \cap \mathcal{E}^{(S,W)}_{(m,n)} \neq \emptyset.$$

The proof is based on the following well-known result in the theory of partial differential equations

Lemma 3.48 *Let M be a compact orientable Riemannian manifold and $f \in C^0(M)$. Then the equation $\Delta u = f$ has a weak solution $u \in W^{1,2}(M)$ if and only if $\int_M f \, d\text{vol}(g) = 0$.*

Let $(m,n) \in \mathbb{Z}^2$ and let $\beta \in \text{Per}(m,n)$. Then (by Lemma 3.47) there is $Y \in \mathfrak{X}(T^2)$ such that $\hat{\Delta}\beta = -(\text{div } Y) \circ \pi$. Let $f = \text{div}(Y + Z)$ where $Z = aS + bW$. Clearly $\int_{T^2} f \,\omega_T = 0$ hence (by Lemma 3.48) there is a weak solution u_0 to $\Delta u = f$. Also $u \in C^\infty(T^2)$ because $f \in C^\infty(T^2)$ and the Laplacian Δ is hypoelliptic. Let us set by definition

$$\alpha = u_0 \circ \pi + \beta \in C^\infty(\mathbb{R}^2).$$

As $u_0 \circ \pi$ is periodic and β semiperiodic it follows that α is semiperiodic, too. Precisely $\alpha \in \text{Per}(m,n)$. Moreover

$$\begin{aligned}\hat{\Delta}\alpha &= \hat{\Delta}(u_0 \circ \pi) + \hat{\Delta}\beta = (\Delta u_0) \circ \pi - (\text{div } Y) \circ \pi \\ &= \text{div}(Z) \circ \pi = (Sa + Wb) \circ \pi\end{aligned}$$

i.e., α is a solution to (3.68). Let us define $X \in \mathfrak{X}(T^2)$ by setting

$$X_p = \cos\alpha(\xi) S_p + \sin\alpha(\xi) W_p, \quad \xi \in \pi^{-1}(p), \quad p \in T^2.$$

The definition of X_p doesn't depend upon the choice of $\xi \in \pi^{-1}(p)$ because α is semiperiodic. Also $\alpha \in \text{Per}(m,n)$ is an angle function for X hence $\text{htp}^{(S,W)}(X) = (m,n)$. Finally, due to (3.70) and (3.68), it follows that X is a critical point of $\mathcal{B}$ i.e., $X \in \text{Crit}(\mathcal{B}) \cap \mathcal{E}^{(S,W)}_{(m,n)}$.

Statement (c) in Theorem 3.46 follows easily from the identity (2.110) in Chapter 2.

Let us discuss statement (d) in Theorem 3.46. If h is a flat metric on T^2 and ∇^h is the Levi-Civita connection of (T^2, h) then we may build a globally defined orthonormal frame $\{S_0, W_0\}$ of $T(T^2)$ (i.e., $h(S_0, S_0) = h(W_0, W_0) = 1$ and $h(S_0, W_0) = 0$) such that $\{S_0, W_0\}$ are ∇^h-parallel (i.e., $\nabla^h S_0 = \nabla^h W_0 = 0$). The construction of $\{S_0, W_0\}$ is standard.[2] With respect to the chosen frame $\{S, W\} = \{S_0, W_0\}$ one has $a = b = 0$ and $\hat{\Delta} = \partial^2/\partial^2 x + \partial^2/\partial^2 y$ where (x, y) are the natural coordinates on $\mathbb{R}^2$. Moreover, it is straightforward that the functions (3.69) are precisely the (m,n)-semiperiodic solutions to (3.68) in the flat case at hand.

Let us justify statement (e) in Theorem 3.46. By the J.L. Kazdan & F.W. Warner result (cf. [185]) quoted in Section 2.10 of Chapter 2, given a Riemannian metric g on T^2 there is a flat Riemannian metric h in the conformal class of g. Let $\{S_0, W_0\}$ be/an h-orthonormal frame of $T(T^2)$ such

[2] One merely starts with a h_p-orthonormal basis $\{s_0, w_0\} \subset T_p(T^2)$ at a point $p \in T^2$ and parallel translates $\{s_0, w_0\}$ (with respect to ∇^h) along the geodesics of (T^2, h) issuing at p.

that $\nabla^h S_0 = \nabla^h W_0 = 0$, as before. Let $\Theta_{S_0} = h(S_0, \cdot)$ and $\Theta_{W_0} = h(W_0, \cdot)$. Then

$$2(d\Theta_{S_0})(X,Y) = X(\Theta_{S_0}(Y)) - Y(\Theta_{S_0}(X)) - \Theta_{S_0}([X,Y])$$
$$= (\nabla^h_X \Theta_{S_0})Y - (\nabla^h_Y \Theta_{S_0})X = 0.$$

Similarly $d\Theta_{W_0} = 0$. By the Poincaré lemma for any $q \in T^2$ there is an open neighborhood $U \subseteq T^2$ of q and there exist smooth functions $f_1, f_2 : U \to \mathbb{R}$ such that $\Theta_{S_0} = df_1$ and $\Theta_{W_0} = df_2$ on U. Let $f : U \to \mathbb{R}^2$ be given by $f = (f_1, f_2)$. Let $\mathbb{E}^2$ denote $\mathbb{R}^2$ endowed with the canonical flat metric $dx \otimes dx + dy \otimes dy$. Since

$$S_0(f_1) = 1, \;\; S_0(f_2) = 0, \;\; W_0(f_1) = 0, \;\; W_0(f_2) = 1,$$

it follows that

$$f^*(dx \otimes dx + dy \otimes dy) = \Theta_{S_0} \otimes \Theta_{S_0} + \Theta_{W_0} \otimes \Theta_{W_0} = h$$

hence f is an injective isometric immersion of (U,h) into $\mathbb{E}^2$ hence a conformal local chart of (T^2, g).

Corollary 3.49 *Let (T^2, g) be a Riemannian torus and X a nowhere zero Killing vector field on (T^2, g). Then $X/\|X\|$ is a critical point of $\mathcal{B}$. If additionally $\|X\|$ is constant then X is parallel and g is flat.*

Proof. Let us set $f = \|X\| \in C^\infty(T^2)$. Also let us consider the unit vector fields $S, W \in \mathcal{E}$ given by

$$S = (1/f)\,X, \quad W = (1/f)\,JX.$$

Then (as X is Killing)

$$0 = (\mathcal{L}_X g)(S,W) = g(\nabla_S X, W) + g(\nabla_W X, S)$$
$$= S(f) g(S,W) + f\, g(\nabla_S S, W) + (1/f)\, g(\nabla_W X, X)$$
$$= f\,a + \frac{1}{2f} W(\|X\|^2) = f\,a + W(f)$$

so that $a = -W(\log f)$. Similar calculations show that $b = 0$ and $Sa + Wb = 0$. Then by Theorem 3.46 it follows that S is a critical point of $\mathcal{B}$.

3.9.2. Stability

Let $X_0 \in \mathcal{E}$ be a unit vector field on the torus (T^2, g) and let $X = \{X(t)\}_{|t|<\epsilon} \in \mathcal{E}_{(-\epsilon,\epsilon)}$ be a 1-parameter variation of X_0 (such that $X(0) = X_0$) of class C^∞.

Definition 3.50 A 1-parameter variation $Y = \{Y(t)\}_{|t|<\epsilon} \in \mathcal{E}_{(-\epsilon,\epsilon)}$ of class C^∞ is said to be in *first order contact* with X at $t = t_0$ if $X(t_0) = Y(t_0)$ and $(dX/dt)(t_0) = (dY/dt)(t_0)$. ■

The scope of this section is to prove the following result (which slightly reformulates and completes Proposition 2.34 in Chapter 2 of this book)

Theorem 3.51 (G. Wiegmink, [309]) *Let us consider a Riemannian torus (T^2, g) and let $X_0 \in \mathrm{Crit}(\mathcal{B}) \subset \mathcal{E}$ be a critical point of the total bending functional. Then*

$$\frac{d^2}{dt^2}\{\mathcal{B}(X(t))\}_{t=0} \geq 0 \tag{3.77}$$

for any smooth 1-parameter variation $X = \{X(t)\}_{|t|<\epsilon} \in \mathcal{E}_{(-\epsilon,\epsilon)})$ of X_0 (such that $X(0) = X_0$). Moreover one has equality in (3.77) if and only if X is in first order contact at $t = 0$ with a variation $Y = \{Y(t)\}_{|t|<\epsilon}$ of X_0 such that $Y(t) \in SO(2) \cdot X_0$ for any $|t| < \epsilon$.

Here $\mathrm{SO}(2) \cdot X_0$ denotes the SO(2)-orbit of $X_0 \in \mathcal{E}$ with respect to the action of SO(2) on $\mathcal{E}$ given by (2.44) in Chapter 2.

Proof of Theorem 3.51. Let $X \in \mathcal{E}_{(-\epsilon,\epsilon)}$ with $X(0) = X_0$ and let $\alpha \in C^\infty(\mathbb{R}^2 \times (-\epsilon,\epsilon))$ be an angle function for X. We recall (cf. (3.70)) that

$$\frac{d}{dt}\{\mathcal{B}(X(t))\} = 4\int_Q \dot{\alpha}\left[\hat{\Delta}\alpha - (Sa + Wb)\circ\pi\right]\pi^*\omega_T.$$

Let us differentiate with respect to t

$$\frac{d^2}{dt^2}\{\mathcal{B}(X(t))\} = 4\int_Q \left\{\dot{\beta}\left[\hat{\Delta}\alpha - (Sa + Wb)\circ\pi\right] + \beta\hat{\Delta}\beta\right\}\pi^*\omega_T$$

where $\beta = \dot{\alpha}$. Let us set $\alpha_0(\xi) = \alpha(\xi, 0)$ and $\beta_0(\xi) = \beta(\xi, 0)$ for any $\xi \in \mathbb{R}^2$. Then for $t = 0$ (as $X_0 \in \mathrm{Crit}(\mathcal{B})$ implies that $\hat{\Delta}\alpha_0 = (Sa + Wb)\circ\pi$, cf. Theorem 3.46)

$$\frac{d^2}{dt^2}\{\mathcal{B}(X(t))\}_{t=0} = 4\int_Q \beta_0\left(\hat{\Delta}\beta_0\right)\pi^*\omega_T = 4\int_{T^2} v(\Delta v)\omega_T \tag{3.78}$$

where $v \in C^\infty(T^2)$ is given by $v(p) = \beta_0(\xi)$ for any $\xi \in \pi^{-1}(p)$ and any $p \in T^2$. Then (by Green's lemma)

$$\frac{d^2}{dt^2}\{\mathcal{B}(X(t))\}_{t=0} = 4\int_{T^2}\left\{-\operatorname{div}(v\nabla v) + \|\nabla v\|^2\right\}\omega_T = 4\int_{T^2}\|\nabla v\|^2 \geq 0$$

and (3.77) is proved. Let us assume that equality takes place in (3.77). Then $v = r$ on T^2 for some $r \in \mathbb{R}$ i.e., $\beta_0 = r$ on $\mathbb{R}^2$. We set by definition

$$\gamma(\xi, t) = \alpha_0(\xi) + t\beta_0(\xi), \quad \xi \in \mathbb{R}^2, \ |t| < \epsilon,$$

so that $\gamma \in \mathrm{Per}_{(-\epsilon,\epsilon)}(m,n)$ where $(m,n) = \mathrm{htp}^{(S,W)}(X_0)$. Moreover let $Y = \{Y(t)\}_{|t|<\epsilon}$ be given by

$$Y(t)_p = (\cos\gamma(\xi,t))S_p + (\sin\gamma(\xi,t))W_p, \quad \xi \in \pi^{-1}(p), \ p \in T^2.$$

Then $Y(0) = X_0 = X(0)$ and $(dY/dt)(0) = (dX/dt)(0)$. Also

$$Y(t) = (\cos(tr))\, X_0 + (\sin(tr))\, JX_0 \in \mathrm{SO}(2)\cdot X_0$$

for any $|t| < \epsilon$. Vice versa if $X(t)$ stays in $\mathrm{SO}(2)\cdot X_0$ then $\mathcal{B}(X(t))$ is constant so that (3.77) holds good. ■

Corollary 3.52 *Let (T^2, g) be a Riemannian torus. For any $(m,n) \in \mathbb{Z}^2$ the restriction $\mathcal{B} : \mathcal{E}^{(S,W)}_{(m,n)} \to [0, +\infty)$ of $\mathcal{B}$ to the homotopy class $\mathcal{E}^{(S,W)}_{(m,n)}$ attains an absolute minimum. Moreover* $\inf\left\{\mathcal{B}(X) : X \in \mathcal{E}^{(S,W)}_{(m,n)}\right\}$ *is achieved precisely at the critical points in* $\mathrm{Crit}(\mathcal{B}) \cap \mathcal{E}^{(S,W)}_{(m,n)}$.

Proof. Let $(m,n) \in \mathbb{Z}^2$ and let $\mathcal{B}_{(m,n)}$ be the restriction of $\mathcal{B}$ to $\mathcal{E}^{(S,W)}_{(m,n)} \in \{[X] : X \in \mathcal{E}\} \subset \pi(T^2, S(T^2))$. Let Y be a critical point of $\mathcal{B}$ in $\mathcal{E}^{(S,W)}_{(m,n)}$ and let $X \in \mathcal{E}^{(S,W)}_{(m,n)}$ be arbitrary. Let $\lambda, \beta \in \mathrm{Per}(m,n)$ be respectively angle functions for Y and X. Let $\alpha \in \mathrm{Per}_{\mathbb{R}}(m,n)$ be defined by

$$\alpha(\xi,t) = (1-t)\lambda(\xi) + t\beta(\xi), \quad \xi \in \mathbb{R}^2, \ t \in \mathbb{R}.$$

Then $\hat{\Delta}\alpha_t = (1-t)\hat{\Delta}\lambda + t\hat{\Delta}\beta$ and $\dot{\alpha}(\xi,t) = \beta(\xi) - \lambda(\xi) = \dot{\alpha}(\xi,0)$ for any $\xi \in \mathbb{R}^2$ and any $t \in \mathbb{R}$. Moreover, $\hat{\Delta}\lambda = (Sa + Wb)\circ\pi$ because

$Y \in \text{Crit}(\mathcal{B})$ (cf. Theorem 3.46). Then (by (3.70))

$$\frac{d}{dt}\{\mathcal{B}(\alpha_t)\} = 4\int_Q \dot{\alpha}\left[\hat{\Delta}\alpha - (Sa + Wb)\circ\pi\right]\pi^*\omega_T$$
$$= 4\int_Q (\beta-\lambda)t\hat{\Delta}(\beta-\lambda)\,\pi^*\omega_T = 4t\int_Q \dot{\alpha}_0\hat{\Delta}\dot{\alpha}_0\pi^*\omega_T$$

where $\dot{\alpha}_0(\xi) = \dot{\alpha}(\xi,0)$ for any $\xi \in \mathbb{R}^2$. Yet $\alpha_0 = \lambda$ is a critical point of $\mathcal{B}$ hence (by (3.78) and (3.77))

$$\frac{d}{dt}\{\mathcal{B}(\alpha_t)\} = t\frac{d^2}{ds^2}\{\mathcal{B}(\alpha_s)\}_{s=0} \geq 0.$$

Thus $\mathcal{B}(Y) = \mathcal{B}(\alpha_0) \leq \mathcal{B}(\alpha_1) = \mathcal{B}(X)$ so that $\inf\mathcal{B}_{(m,n)} = \mathcal{B}(Y)$. Corollary 3.52 is proved. ∎

3.9.3. Examples and Open Problems

Let $L \in \mathbb{R}_+$ and let $c = (c_1, c_2) : \mathbb{R} \to \mathbb{R}_+ \times \mathbb{R} \subset \mathbb{R}^2$ be a C^∞ immersion such that both c_i are L-periodic functions. Next we consider $f : \mathbb{R}^2 \to \mathbb{R}^3$ given by

$$f(x,y) = ((\cos x)\,c_1(y),\ (\sin x)\,c_1(y),\ c_2(y)), \quad (x,y) \in \mathbb{R}^2.$$

Clearly f is a C^∞ immersion. We set $d_1 = (2\pi, 0)$ and $d_2 = (0, L)$ and $\Gamma = \{md_1 + nd_2 : (m,n) \in \mathbb{Z}^2\}$. Let $T^2 = \mathbb{R}^2/\Gamma$ and let $\pi : \mathbb{R}^2 \to T^2$ be the natural projection. Clearly f induces an immersion $\Psi : T^2 \to \mathbb{R}^3$ given by $\Psi(p) = f(\xi)$ for any $\xi \in \pi^{-1}(p)$ and any $p \in T^2$. We endow T^2 with the Riemannian metric $g = \Psi^* g_0$ where $g_0 = dx^2 + dy^2 + dz^2$ is the standard flat metric on $\mathbb{R}^3$. That is to say we endow T^2 with a Riemannian metric g arising from an immersion of T^2 in $\mathbb{E}^3 = (\mathbb{R}^3, g_0)$ as a torus of revolution with an arbitrary profile curve. The maps $\xi \in \mathbb{R}^2 \mapsto (d_\xi\pi)(\partial/\partial x)_\xi$ and $\xi \in \mathbb{R}^2 \mapsto (d_\xi\pi)(\partial/\partial y)_\xi$ are periodic hence $\{\partial/\partial x,\ \partial/\partial y\}$ are projectable. Let $X, Y \in \mathfrak{X}(T^2)$ be given by

$$X_p = (d_\xi\pi)\left.\frac{\partial}{\partial x}\right|_\xi, \quad Y_p = (d_\xi\pi)\left.\frac{\partial}{\partial y}\right|_\xi, \quad \xi \in \pi^{-1}(p), \quad p \in T^2.$$

Next we set

$$S = \frac{1}{c_1}X, \quad W = \frac{1}{\|\dot{c}\|}Y,$$

so that $\{S, W\}$ is a global g-orthonormal frame of $T(T^2)$. It should be observed that the standard Euclidean torus may be obtained for

$$c(y) = (R + r\cos y, r\sin y), \quad y \in \mathbb{R}. \tag{3.79}$$

Indeed if this is the case $\Psi(T^2)$ is given by the equations

$$X = (R + r\cos y)\cos x, \quad Y = (R + r\cos y)\sin x, \quad Z = r\sin y,$$

where (X, Y, Z) are the Cartesian coordinates in $\mathbb{R}^3$. Eliminating the parameter $y \in \mathbb{R}$ one obtains the fourth order equation

$$(X^2 + Y^2 + Z^2 - R^2 - r^2)^2 = 4R^2(r^2 - Z^2). \tag{3.80}$$

Directly on the familiar equation (3.80), one may read off the two families of circles lying on $\Psi(T^2)$ the first of which is obtained by intersecting $\Psi(T^2)$ with a plane passing through the axis $X = 0$, $Y = 0$. Such a plane intersects the torus in two circles of radius r which are symmetric with respect to the Z-axis. Also the planes $Z = a$ $(a \in \mathbb{R})$ intersect the torus in two real circles if $|a| \leq r$, centered at the origin (and coinciding when $Z = \pm r$). If $Z = 0$ one gets the maximal and minimal circles $X^2 + Y^2 = (R \pm r)^2$. It may be easily seen that $(d\Psi)S$ and $(d\Psi)W$ are the unit tangents to the two families of circles above, so that $\Psi(T^2)$ possesses two unit tangent vector fields which are *regular* (in the sense of R. Palais, [233]). It is a classical fact that the torus is the only closed surface possessing regular vector fields (cf. e.g., G. Vrânceanu, [304]).

Let us go back to the case of an arbitrary profile curve where g is given by

$$g = c_1(y)^2\, dx^2 + \|\dot{c}(y)\|^2\, dy^2. \tag{3.81}$$

A straightforward calculation based on (3.81) shows that

$$\nabla_W S = 0, \quad \nabla_W W = 0,$$

$$\nabla_S S = -\frac{c_1'(y)}{c_1(y)\|\dot{c}(y)\|} W, \quad \nabla_S W = \frac{c_1'(y)}{c_1(y)\|\dot{c}(y)\|} S.$$

Consequently

$$a = g(\nabla_S S, W) = -\frac{c_1'(y)}{c_1(y)\|\dot{c}(y)\|}, \quad b = g(\nabla_W S, W) = 0, \quad Sa + Wb = 0.$$

On the other hand for any angle function α of S one has $\alpha(\xi) \in \{2m\pi : m \in \mathbb{Z}\}$ hence (by the continuity of α) it must be $\alpha =$ constant. Hence

$$\hat{\Delta}\alpha = 0 = (Sa + Wb) \circ \pi$$

and we may use Theorem 3.46 to conclude that S is a critical point of $\mathcal{B}$ and actually (by Corollary 3.52) the absolute minimum of $\mathcal{B}$ restricted to $\mathcal{E}_{(0,0)}^{(S,W)}$. The same is of course true for W. More can be said for the particular metric $g = \Psi^* g_0$ on T^2.

Proposition 3.53 (G. Wiegmink, [309]) *Let us consider the torus* $T^2 = \mathbb{R}^2/\{(2\pi m, Ln) : (m,n) \in \mathbb{Z}^2\}$ *endowed with the metric* $g = c_1(y)^2\, dx^2 + \|\dot{c}(y)\|^2\, dy^2$. *Then* $S, W \in \mathrm{Crit}(\mathcal{B}) \cap \mathcal{E}_{(0,0)}^{(S,W)}$ *and*

$$\mathcal{B}(S) = \mathcal{B}(W) = \int_{T^2} a^2 d\,\mathrm{vol}(g) = 2\pi \int_0^L \frac{c_1'(y)^2}{c_1(y)\|\dot{c}(y)\|}\, dy = 2\pi A_0.$$

The constant A_0 *may be determined*[3] *when* (T^2, g) *is the standard Euclidean torus with* $c(y)$ *given by* (3.79) *for* $R > r > 0$, *that is* $A_0 = 2\pi R/r$.

It is an open problem to extend G. Wiegmink's results to Finslerian tori, cf. e.g., P. Dazord, [94].

We end this section with the following remark and open problem. Let $\phi : T^2 \to N$ be a smooth map of a Riemannian torus (T^2, g) into a *warped product* $N = L \times_w \mathbb{R}$ for some C^∞ function $w : N \to (0, +\infty)$ that is $N = L \times \mathbb{R}$ where L is a $(\nu - 1)$-dimensional Riemannian manifold carrying the metric g_L and N is endowed with the Riemannian metric

$$g_N = \pi_1^* g_L + w^2\, dt \otimes dt \tag{3.82}$$

where $\pi_1 : N \to L$ is the natural projection. Let $f = \pi_1 \circ \phi : T^2 \to L$ and $\theta = \pi_2 \circ \phi \in C^\infty(T^2)$ where $\pi_2 : N \to \mathbb{R}$ is the natural projection. We wish to compute the Hilbert-Schmitd norm of $d\phi$. To this end we use a g-orthonormal frame $\{S, W\}$ as before. Indeed for any $p \in T^2$

$$\begin{aligned}\|d\phi\|_p^2 &= \mathrm{trace}_g\left(\phi^* g_N\right)_p = \left(\phi^* g_N\right)(S,S)_p + \left(\phi^* g_N\right)(W,W)_p \\ &= \|df\|_p^2 + w(\phi(p))^2 \left\{\left[(\phi^* dt)_p S_p\right]^2 + \left[(\phi^* dt)_p W_p\right]^2\right\}\end{aligned}$$

and $\phi^* dt = d\theta$ so that

$$\|d\phi\|^2 = \|df\|^2 + (w \circ \phi)^2 \|\nabla\theta\|^2. \tag{3.83}$$

[3] We take the opportunity to correct a misprint in G. Wiegmink, [309], p. 342. Indeed one has $\int r[(\sin^2 y)/(R + r\cos y)]\, dy = \tau y - \sin y - 2\sqrt{\tau^2 - 1}\arctan\sqrt{[(\tau-1)/(\tau+1)](1-\cos y)/(1+\cos y)}$ + const., where $\tau = R/r$, and $L = 2\pi$.

Let us assume that $\phi : T^2 \to N$ is a harmonic map of (T^2, g) into (N, g_N) where g_N is given by (3.82). Let us consider the following smooth 1-parameter variation of ϕ

$$\phi_s : T^2 \to N, \quad \phi_s(p) = (f(p), \theta_s(p)),$$
$$\theta_s : T^2 \to \mathbb{R}, \quad \theta_s(p) = \theta(p) + s\varphi(p), \quad p \in T^2, \quad |s| < \epsilon,$$

where $\varphi \in C^\infty(T^2)$ is a given smooth function. As ϕ is harmonic (by replacing ϕ with ϕ_s in (3.83))

$$0 = \frac{d}{ds}\{E(\phi_s)\}_{s=0} = \frac{d}{ds}\left\{\int_{T^2} \|d\phi_s\|^2 d\,\mathrm{vol}(g)\right\}_{s=0}$$
$$= \frac{d}{ds}\left\{\int_{T^2} \left(\|df\|^2 + (w \circ \phi_s)^2 \|\nabla\theta_s\|^2\right) d\,\mathrm{vol}(g)\right\}_{s=0}$$
$$= \int_{T^2} \left\{(2w \circ \phi)^2 g(\nabla\varphi, \nabla\theta) + \varphi\left(\frac{\partial w^2}{\partial t} \circ \phi\right) \|\nabla\theta\|^2\right\} d\,\mathrm{vol}(g).$$

Next we integrate by parts

$$\int_{T^2} (w^2 \circ \phi) g(\nabla\varphi, \nabla\theta) d\,\mathrm{vol}(g) = \int_{T^2} (w^2 \circ \phi)(\nabla\theta)(\varphi) d\,\mathrm{vol}(g)$$
$$= \int_{T^2} \left\{\mathrm{div}\left(\varphi(w^2 \circ \phi)\,\nabla\theta\right) - \varphi\,\mathrm{div}((w^2 \circ \phi)\,\nabla\theta)\right\} d\,\mathrm{vol}(g)$$
$$= -\int_{T^2} \varphi\,\mathrm{div}((w^2 \circ \phi)\,\nabla\theta) d\,\mathrm{vol}(g)$$

so that

$$\mathrm{div}\left((w^2 \circ \phi)\,\nabla\theta\right) = \frac{1}{2}\left(\frac{\partial w^2}{\partial t} \circ \phi\right) \|\nabla\theta\|^2. \tag{3.84}$$

The fact that the component $\theta = \pi_2 \circ \phi$ of a harmonic map $\phi : T^2 \to L \times_w \mathbb{R}$ satisfies (3.84) was first observed by B. Solomon, [271] (and the result holds for harmonic maps from an *arbitrary* Riemannian manifold (M, g) into $L \times_w \mathbb{R}$, cf. Lemma 1 in [271], p. 153). The equation (3.84) is elliptic hence any smooth solution θ to (3.84) satisfies the strong maximum principle (cf. e.g., E. Calabi, [76]). Consequently for any harmonic map $\phi : T^2 \to L \times_w \mathbb{R}$ the image $\phi(T^2)$ lies in $L \times \{t_\phi\}$ for some $t_\phi \in \mathbb{R}$. To formulate the open problem alluded to above we need to recall a few notions of topology. Let

$$\Sigma = S^\nu \cap \{(x_1, \ldots, x_{\nu+1}) \in \mathbb{R}^{\nu+1} : x_1^2 + x_2^2 = 0\}$$

be a $(\nu - 2)$-dimensional sphere embedded in S^ν as a totally geodesic submanifold. A continuous map $\phi : T^2 \to S^\nu$ is said to *meet* Σ if $\phi(T^2) \cap \Sigma \neq \emptyset$. Also $\phi : T^2 \to S^\nu$ is said to *link* Σ if it does not meet Σ and the map $\phi : T^2 \to S^n \setminus \Sigma$ is not homotopically nontrivial. By a beautiful result of B. Solomon (cf. Theorem 1 in *op. cit.*, p. 155)

Theorem 3.54 *A nonconstant harmonic map $\phi : T^2 \to S^\nu$ either links or meets Σ.*

The key ingredient in the proof of Theorem 3.54 is to observe that $S^\nu \setminus \Sigma$ is isometric to a warped product $S_+^{\nu-1} \times_w S^1$ where $S_+^{\nu-1} = \{(y_1, \ldots, y_\nu) \in \mathbb{R}^\nu : y_1^2 + \cdots + y_\nu^2 = 1,\ y_\nu > 0\}$. Then whenever ϕ neither meets nor links Σ it must be constant. Indeed if $\phi : T^2 \to S_+^{\nu-1} \times_w S^1$ is a null-homotopic harmonic map then ϕ lifts to a harmonic map $\tilde{\phi} : T^2 \to S_+^{\nu-1} \times_w \mathbb{R}$ (one also exploits the fact that the former target is the universal covering space of $S_+^{\nu-1} \times_w S^1$). Then one applies the argument above (based on the strong maximum principle) to conclude that $\tilde{\phi}$ is actually a harmonic map $T^2 \to S_+^{\nu-1}$ and as such (by a result in J. Eells & L. Lemaire, [112]) a constant.

We come to our final comment. Note that $S_+^{\nu-1} \times S^1$ is homotopy equivalent to S^1. Consequently a continuous map $\phi : T^2 \to S_+^{\nu-1} \times S^1$ is null-homotopic if and only if $p \circ \phi : T^2 \to S^1$ is null-homotopic, where $p : S_+^{\nu-1} \times S^1 \to S^1$ is the natural projection. Yet the homotopy classes of continuous maps $T^2 \to S^1$ form the Bruschlinsky group $\pi^1(T^2)$ of the torus, computed in Chapter 1 as identified with the set of homotopy classes of unit tangent vector fields $\{[X] : X \in \mathcal{E}\}$ (where $\mathcal{E} = \Gamma^\infty(S(T^2))$). We expect a nontrivial link among the geometry of smooth maps $T^2 \to S^\nu \setminus \Sigma$ and the geometry of unit tangent vector fields on T^2.

CHAPTER FOUR

Harmonicity and Contact Metric Structures

Contents

Our first examples of harmonic vector fields were Hopf vector fields on a sphere S^{2m+1}, that is Reeb vector fields underlying Sasakian structures on S^{2m+1}. It is therefore a natural problem to study harmonicity of Reeb vector fields on arbitrary contact Riemannian manifolds. As it turns out, looking at the harmonicity of the Reeb vector field of a contact metric manifold M is relevant for understanding the geometry of M itself. After a brief preparation of contact Riemannian geometry (cf. D.E. Blair, [42]) one introduces one of the main object of study in this chapter, that is the notion of an *H-contact manifold* (a contact metric manifold whose Reeb vector is harmonic, cf. Definition 4.2 below). H-contact manifolds may be described as in Theorem 4.5 [due to D. Perrone, [242], and generalizing a result by J.C. Gonzàlez-Dàvila & L. Vanhecke (cf. [145]) holding in dimension 3]. Conclusive results may be obtained on 3-dimensional contact

© 2012 Elsevier Inc. All rights reserved.

metric manifolds with $\|\tau\| =$ constant, where $\tau = \mathcal{L}_\xi g$. Any such manifold M is H-contact if and only if M is either Sasakian or M is locally isometric to a unimodular Lie group G (equipped with a non-Sasakian left invariant contact metric structure) whose universal covering belongs to the list $\{SU(2), \widetilde{E(2)}, \widetilde{SL(2,\mathbb{R})}, E(1,1)\}$ (cf. Theorem 4.24 below). It is noteworthy that one distinguishes among the various candidates in the list by looking at $p = 4\sqrt{2}W/\|\tau\|$ where W is the Webster scalar curvature, so that the discussion in Theorem 4.24 is related to the work by S.S. Chern & R.S. Hamilton, [86], and S. Tanno, [281].

4.1. H-CONTACT MANIFOLDS

4.1.1. Contact Metric Manifolds

We start by reviewing a few facts of contact Riemannian geometry. Our main reference is the monograph [42], yet we also rely on the exposition in [286]. Let $\mathbb{R}^{2n+1}$ with the Cartesian coordinates $(t, x^1, \ldots, x^n, x^{n+1}, \ldots, x^{2n+1})$. A *contact transformation* is a local diffeomorphism of $\mathbb{R}^{2n+1}$ preserving the 1-form

$$\eta = dt - \sum_{i=1}^{n} x^{i+n}\, dx^i \tag{4.1}$$

up to multiplication by a nonzero real valued function. The family $\mathcal{G}_{2n+1,\eta}$ of such local diffeomorphisms contains the identities and the inverses, and admits a partial composition law ($\mathcal{G}_{2n+1,\eta}$ is the *contact pseudogroup*). Taking germs of local diffeomorphisms, one obtains a topological groupoid $\Gamma_{2n+1,\eta}$ to which there corresponds (cf. e.g., the method of construction presented in [64]) a classifying space $B\Gamma_{2n+1,\eta}$. The space $B\Gamma_{2n+1,\eta}$ classifies codimension $2n+1$ foliations which admit locally a contact structure transverse to the leaves. The case of interest in this chapter is that of the foliation by points and the underlying manifold admits a global contact form. For a precise treatment we need a few definitions.

A real 1-form η on a real $(2n+1)$-dimensional manifold M is a *contact form* if $\eta \wedge (d\eta)^n$ is a volume form on M. By the Darboux theorem (cf. [42]) for any point of M there is a local coordinate system (U, t, x^A) with respect to which η is given by (4.1). A *contact structure* on M is a maximal atlas $\mathcal{A}$ whose transition functions are elements of $\mathcal{G}_{2n+1,\eta}$. A contact structure $\mathcal{A}$ determines a contact form provided the transition functions of $\mathcal{A}$ lie in $\mathcal{G}^+_{2n+1,\eta}$, the pseudogroup of orientation preserving transformations in $\mathcal{G}_{2n+1,\eta}$ (if n is odd then $\mathcal{G}^+_{2n+1,\eta} = \mathcal{G}_{2n+1,\eta}$). Therefore on oriented manifolds the two notions are equivalent, and a pair (M,η) or $(M,\mathcal{A})$ is referred to merely as a *contact manifold*. If orientability is not assumed and n is even, then M may admit a contact structure $\mathcal{A}$ without a globally defined form η (an example is $M = \mathbb{R}^{n+1} \times \mathbb{R}P^n$, cf. R. Stong, [276], for a comparative discussion of the two definitions).

Two contact forms η_0 and η_1 are *equivalent* if there is a transformation $f : M \to M$ and a nowhere zero function $\lambda \in C^\infty(M)$ such that $f^*\eta_1 = \lambda\eta_0$. Therefore, on an oriented manifold, an equivalence class of contact structures is defined by a section of $\mathbb{P}(T^*(M))$ (the projective bundle associated to the cotangent bundle).

The condition $\eta \wedge (d\eta)^n \neq 0$ implies that η has maximal rank $2n$ so that there is a smooth distribution $H(M)$ of rank $2n$ such that $(d\eta)^n$ is never zero on $H(M)$. Next there is an associated 1-dimensional smooth foliation on M defined by a vector field $\xi \in \mathfrak{X}(M)$ satisfying $\eta(\xi) = 1$ and $\xi \rfloor d\eta = 0$, the *characteristic direction* or *Reeb vector* field of (M,η). Let us give a few examples of contact manifolds.

a. Let N be a real n-dimensional manifold and $M = T^*(N) \times \mathbb{R}$. Let $\pi : T^*(N) \to N$ be the projection, $\alpha \in T^*(N)$, and $Z \in T_\alpha(T^*(N))$. We consider the 1-form $\omega \in \Omega^1(T^*(N))$ given by

$$\omega_\alpha(Z) = \alpha\left((d_\alpha\pi)Z\right).$$

If t is the coordinate function on $\mathbb{R}$ then $\eta = \Pi^*\omega + dt$ is a contact form on M, where $\Pi : M \to T^*(N)$ is the natural projection.

b. A projective algebraic variety X^{2n} has natural symplectic structure whose defining 2-form Ω belongs to an integral cohomology class. Let $S^1 \to M \xrightarrow{\pi} X^{2n}$ be the principal circle bundle over X^{2n} determined by this class. There is a contact form η on M such that $\pi^*\Omega = d\eta$. The leaves of the associated 1-dimensional foliation are the fibres of π. See also [284].

c. Let $a_j \in \mathbb{Z}$, $a_j \geq 1$, and let $V = V(a_0, \ldots, a_n)$ be given by

$$V = S^{2n+1} \cap \{z = (z_0, \ldots, z_n) \in \mathbb{C}^{n+1} : z_0^{a_0} + \cdots + z_n^{a_n} = 0\}.$$

Then the real 1-form $\eta \in \Omega^1(\mathbb{C}^{n+1})$

$$\eta = \frac{i}{2}\sum_{j=0}^{n} \frac{1}{a_j}\left(z_j\, d\overline{z}_j - \overline{z}_j\, dz_j\right)$$

induces a contact form on V, cf. C.B. Thomas, [287].

A necessary condition for the existence of a contact structure on M is that $\mathrm{GL}(2n+1,\mathbb{R})$ (the structure group of $T(M)$) be reducible to $\mathrm{SO}(2n+1)$. By a result in [286], if $M = S^{2n+1}$ then there are contact forms corresponding to each reduction of the structure group of the tangent bundle.

Equivalence classes of contact forms are stable, so that they may be studied by topological methods (cf. again [286]). Precisely if M is a compact orientable $(2n+1)$-dimensional manifold and $[\eta]$ is an equivalence class of contact forms on M then there is a neighborhood $\mathcal{U}$ of $[\eta]$ in $\mathbb{P}(T^*(M))$ such that the 1-forms in each $[\eta'] \in \mathcal{U}$ satisfy the contact condition and there is a transformation $h: M \to M$ such that $h^*\eta$ is equivalent to η'.

An *associated metric* on (M,η) is a Riemannian metric g on M such that $\eta(X) = g(\xi, X)$ for any $X \in T(M)$ and there is a $(1,1)$-tensor field ϕ on M satisfying

$$\phi^2 = -I + \eta \otimes \xi,$$
$$(d\eta)(X,Y) = g(X, \phi Y), \quad X, Y \in T(M).$$

The pair (η, g) (as well as the synthetic object (ϕ, ξ, η, g)) is referred to as a *contact metric structure* on M (and $(M, (\phi, \xi, \eta, g))$ is a *contact metric manifold*).

If (M,η) is a contact manifold and g an associated metric we denote by ∇ and R the Levi-Civita connection and the curvature tensor field of (M,g). Also let Ric, Q and ρ be the Ricci curvature, the Ricci operator ($g(QX,Y) = \mathrm{Ric}(X,Y)$ for any $X, Y \in T(M)$) and the scalar curvature of (M,g). The tensor field $h = \frac{1}{2}\mathcal{L}_\xi \phi$ (where $\mathcal{L}$ is the Lie derivative) is symmetric i.e., $g(hX,Y) = g(X,hY)$ for any $X, Y \in T(M)$ and

$$\nabla \xi = -\phi - \phi h, \quad \nabla_\xi \phi = 0, \quad h\phi = -\phi h, \quad h\xi = 0. \tag{4.2}$$

Let us set $\tau = \mathcal{L}_\xi g$. Then

$$\|\tau\|^2 = 4\,\mathrm{trace}_g(h^2) = 8n - 4\mathrm{Ric}(\xi,\xi),$$
$$\tau(X,Y) = 2g(h\phi X, Y), \quad X, Y \in T(M).$$

Since h anti-commutes with ϕ, if E is an eigenvector of h corresponding to the eigenvalue λ, i.e., $h(E) = \lambda E$ where λ is a smooth function, then ϕE is also an eigenvector of h corresponding to the eigenvalue $-\lambda$. Consequently for any point $p \in M$ there is an open neighborhood $U \subseteq M$ and an orthonormal frame of the form

$$\{E_0 = \xi, E_i, E_{i+n} = \phi E_i : 1 \le i \le n\}$$

(referred to as a local *ϕ-basis*) defined on U and consisting of eigenvectors of h. A contact metric manifold M is a *η-Einstein manifold* if the Ricci operator Q is of the form

$$Q = aI + b\eta \otimes \xi$$

for some $a, b \in C^\infty(M)$. A contact metric manifold is a *K-contact manifold* if ξ is a Killing vector field (equivalently $h = 0$). A contact metric structure (ϕ, ξ, η, g) is *normal*, or *Sasakian*, if

$$(\nabla_X \phi)Y = g(X, Y)\xi - \eta(Y)X, \quad X, Y \in T(M).$$

Any Sasakian manifold is K-contact and the converse is also true when $n = 1$ i.e., when M is 3-dimensional. On any K-contact manifold, ξ is an eigenvector of the Ricci operator. Precisely

$$Q\xi = 2n\xi.$$

We shall need the following notion.

Definition 4.1 (D.E. Blair et al., [46]) A contact metric manifold $(M, (\phi, \xi, \eta, g))$ is called a *(k, μ)-space* if

$$R(X, Y)\xi = k\{\eta(X)Y - \eta(Y)X\} + \mu\{\eta(X)hY - \eta(Y)hX\} \tag{4.3}$$

for some $k, \mu \in \mathbb{R}$ and any $X, Y \in T(M)$. ■

If M is a (k, μ)-space then

$$Q\xi = 2nk\xi, \quad k \leq 1, \quad h^2 = (k - 1)\phi^2,$$

and if $k = 1$ or $h = 0$ then M is Sasakian. Contact metric manifolds satisfying the requirement (4.3) for some nonconstant smooth functions $k, \mu \in C^\infty(M)$ are referred to as *generalized (k, μ)-spaces*. However, generalized (k, μ)-spaces exist only in dimension three, cf. T. Koufogiorgos & C. Tsichlias, [197].

4.1.2. *H*-Contact Manifolds

We adopt the following

Definition 4.2 A contact metric manifold whose underlying Reeb vector is a harmonic vector field is called an *H-contact manifold*. ■

Sasakian manifolds are (cf. G. Wiegmink, [309]) examples of H-contact manifolds. K-contact manifolds are (cf. J.C. Gonzàles-Dàvila & L. Vanhecke, [145]) examples of H-contact manifolds, as well. It is also shown in [145] that the Reeb vector of a Sasakian manifold is actually a harmonic map. When M is a K-contact manifold P. Rukimbira, [262], showed that the Reeb vector $\xi : M \to S(M)$ is a ϕ-holomorphic map (i.e., $\phi_{1,\xi(x)} \circ (d_x\xi) = (d_x\xi) \circ \phi_x$ for any $x \in M$, where ϕ_1 is the $(1,1)$-tensor field underlying the standard contact metric structure of $S(M)$ as a real hypersurface in the almost Kähler manifold $T(M)$) hence (by a result of S. Ianuş & A.M. Pastore, [175]) ξ is a harmonic map. It should be observed that, in spite of the adopted terminology (cf. again [175]) the notion of a ϕ-holomorphic map (among two given almost contact metric manifolds) involves.

Implicitly the metric structures (and isn't the proper CR analog to the notion of a holomorphic map in complex analysis, which is the notion of CR map, cf. e.g., Definition 1.3 in [110], p. 4). A natural question in this context, to which we shall come back later on (see Chapter 8 of this monograph) is under which conditions is the Reeb vector ξ (underlying a given almost contact metric structure) a CR map? By a result of D. Perrone et al., [107], if ξ is a CR map, then the natural almost CR structure on M (associated to the given almost contact metric structure, cf. S. Ianuş, [173]) is strictly pseudoconvex and *a posteriori* ξ is a pseudohermitian map. As it turns out, if ξ is also geodesic then it is a harmonic map.

In the sequel, we address the problem of characterizing a H-contact manifold. We shall need the following

Theorem 4.3 (D. Perrone, [242]) *Let $(M,(\phi,\xi,\eta,g))$ be a real $(2n+1)$-dimensional contact metric manifold. Then*

$$\Delta_g\xi = 4n\xi - Q\xi = \|\nabla\xi\|^2\xi - \pi_H Q\xi$$

where $\|\nabla\xi\|^2 = 2n + \text{trace}\,(h^2)$ and $\pi_H : T(M) \to H(M) = \text{Ker}(\eta)$ is the natural projection associated to the direct sum decomposition $T(M) = H(M) \oplus \mathbb{R}\xi$.

To prove Theorem 4.3 we need

Lemma 4.4 *Let M be a real $(2n+1)$-dimensional contact metric manifold and $\{E_0 = \xi,\ E_j,\ E_{n+j} = \phi E_j : 1 \le j \le n\}$ a local orthonormal ϕ-basis. Then*

$$\sum_{i=0}^{2n} g((\nabla_X\phi)E_i,\ E_i) = 0, \tag{4.4}$$

$$\sum_{i=0}^{2n} g((\nabla_X h\phi)E_i, E_i) = 0. \tag{4.5}$$

for any $X \in T(M)$.

Proof. As $X(g(\phi E_i, E_i)) = 0$ it follows that

$$g(\nabla_X \phi E_i, E_i) - g(E_i, \phi \nabla_X E_i) = 0$$

hence (4.4) holds good. Moreover, as

$$g((\nabla_X h\phi)\xi, \xi) = 0$$

the identity (4.5) is equivalent to

$$\sum_{i=1}^{n} g((\nabla_X h\phi)E_i, E_i) + \sum_{i=1}^{n} g((\nabla_X h\phi)\phi E_i, \phi E_i) = 0.$$

If the chosen ϕ-basis consists of eigenvectors of h corresponding to the eigenvalues $\{\lambda_0 = 0,\ \lambda_j,\ -\lambda_j : 1 \leq j \leq n\}$ then

$$g((\nabla_X h\phi)E_i, E_i) = 2\lambda_i g(\nabla_X E_i, \phi E_i),$$
$$g((\nabla_X h\phi)\phi E_i, \phi E_i) = -2\lambda_i g(\nabla_X E_i, \phi E_i),$$

hence one obtains (4.5). Lemma 4.4 is proved. ∎

Proof of Theorem 4.3. Using (2.114) and (4.2) we may write

$$\Delta_g \xi = -\nabla^* \phi - \nabla^*(\phi h) = -\nabla^* \phi + \nabla^*(h\phi).$$

We need the formula (cf. the identity $(3.4)_a$ by S. Olszak, [227])

$$\sum_{i=0}^{2n} (\nabla_{E_i} \phi) E_i = 2n\xi. \tag{4.6}$$

Thus

$$\Delta_g \xi = 2n\xi + \nabla^*(h\phi). \tag{4.7}$$

Next we compute the Ricci curvature $\text{Ric}(X, \xi)$. By (4.2)

$$\begin{aligned} R(X, Y)\xi &= -\nabla_X \nabla_Y \xi + \nabla_Y \nabla_X \xi + \nabla_{[X,Y]} \xi \\ &= (\nabla_X \phi)Y + (\nabla_X \phi h)Y - (\nabla_Y \phi)X - (\nabla_Y \phi h)X \end{aligned}$$

so that

$$\begin{aligned}\mathrm{Ric}(X,\xi) &= \sum_{i=0}^{2n} R(E_i,X,E_i,\xi) = \sum_{i=0}^{2n} g(R(X,E_i)\xi,E_i)\\ &= \sum_{i=0}^{2n}\{g((\nabla_X\phi)E_i,E_i)+g((\nabla_X\phi h)E_i,E_i)\\ &\quad -g((\nabla_{E_i}\phi)X,E_i)-g((\nabla_{E_i}\phi h)X,E_i)\}\\ &= \sum_{i=0}^{2n}\{(\nabla_X\phi)E_i,E_i)-g((\nabla h\phi)E_i,E_i)\\ &\quad +g((\nabla_{E_i}\phi)E_i,X)+g((\nabla_{E_i}h\phi)X,E_i)\}\end{aligned}$$

hence (by (4.4)–(4.5) and (4.6), as $h\phi$ is symmetric)

$$g(X,Q\xi) = 2ng(\xi,X) - g(\nabla^* h\phi, X).$$

Thus

$$\begin{aligned}\nabla^* h\phi &= -Q\xi + 2n\xi = -\pi_H Q\xi - g(Q\xi,\xi)\xi + 2n\xi\\ &= -\pi_H Q\xi + \frac{1}{4}\|\tau\|^2\xi.\end{aligned}$$

Finally (by (4.7))

$$\begin{aligned}\Delta_g\xi &= 4n\xi - Q\xi = \left(2n + \frac{\|\tau\|^2}{4}\right)\xi - \pi_H Q\xi\\ &= \|\nabla\xi\|^2\xi - \pi_H Q\xi.\end{aligned}$$

Theorem 4.3 yields ■

Theorem 4.5 (D. Perrone, [242]) *A contact metric manifold is an H-contact manifold if and only if the Reeb vector field is an eigenvector of the Ricci operator.*

For $n=1$ Theorem 4.5 was proved by J.C. Gonzàles-Dàvila & L. Vanhecke, [145]. Theorem 4.5 implies that Einstein contact metric manifolds, and more generally η-Einstein manifolds, are H-contact manifolds.

If M is a (k,μ)-space then $Q\xi = 2nk\xi$. Moreover, a calculation (in the presence of a local orthonormal frame consisting of eigenvectors of h) based on (4.3) shows that $\mathrm{trace}_g R(\nabla.\xi,\xi)\cdot = 0$. Then (by Theorem 4.3 above)

$$\|\nabla\xi\|^2 = g(\Delta_g\xi,\xi) = 2n(2n-k).$$

Then a (k,μ)-space is an H-contact manifold. Precisely, ξ is a harmonic map and its energy is

$$E(\xi) = \frac{2n+1}{2}\operatorname{Vol}(M) + \frac{1}{2}\int_M \|\nabla\xi\|^2 \, d\operatorname{vol}(g)$$

$$= \frac{1}{2}(4n+1+2n(1-k)) \operatorname{Vol}(M).$$

If $k=1$, then M is Sasakian and $E(\xi) = \frac{1}{2}(4n+1)\operatorname{Vol}(M)$.

Remark 4.6 As previously recalled, a D-homothetic deformation of a contact metric structure (ϕ,ξ,η,g) is given by

$$\phi_t = \phi, \;\; \xi_t = (1/t)\xi, \quad \eta_t = t\eta, \quad g_t = tg + t(t-1)\eta\otimes\eta, \quad t>0.$$

By a result of S. Tanno, [280], $(\phi_t,\xi_t,\eta_t,g_t)$ is a contact metric structure for any $t>0$. Moreover, the property ξ is an eigenvector of the Ricci operator is invariant (cf. S.I. Goldberg & D. Perrone, [141]) under D-homothetic deformations. Then Theorem 4.3 implies that *the class of H-contact manifolds is invariant under D-homothetic deformations.* ■

Let (M,g) be a Riemannian manifold and $\omega\in\Omega^1(M)$ a differential 1-form on M. We recall (cf. e.g., [258]) the Weitzenböck formula

$$\Delta_1\omega = \Delta_g\omega + \operatorname{Ric}(X_\omega,\cdot)$$

where Δ_1 is the Hodge-de Rham Laplacian on 1-forms and $X_\omega = \omega^\sharp \in \mathfrak{X}(M)$ is the tangent vector field determined by $\omega = g(X_\omega,\cdot)$. If M is a contact metric manifold, then Theorem 4.3 together with the Weitzenböck formula lead to

$$\Delta_1\eta = \Delta_g\eta + \operatorname{Ric}(\xi,\cdot) = g(\Delta_g\xi + Q\xi,\cdot) = 4ng(\xi,\cdot) = 4n\eta$$

i.e., the contact form is an eigenform of the Hodge-de Rham Laplacian on $\Omega^1(M)$.

Definition 4.7 Let (M,g) be a Riemannian manifold and let $1_M : M \to M$ be the identity map. A vector field $X\in\mathfrak{X}(M)$ is said to be a *Jacobi vector* field if it is a solution to $J_{1_M}X = 0$ where J_{1_M} is the Jacobi operator associated to $1_M : M\to M$. ■

As $J_{1_M}X = \Delta_g X - QX$, Theorem 4.3 yields $J_{1_M}\xi = 4n\xi - 2Q\xi$. This leads to the following description of K-contact manifolds.

Corollary 4.8 *A contact metric manifold is a K-contact manifold if and only if ξ is a Jacobi vector field.*

By a result of C. Boyer & K. Galicki, [66], any compact Einstein K-contact manifold is Sasakian. Therefore (by Corollary 4.8) a compact Einstein contact metric manifold is Sasakian if and only if ξ is a Jacobi vector field.

Definition 4.9 An almost contact metric manifold $(M,(\phi,\xi,\eta,g))$ is said to be an *almost cosymplectic* manifold if both the 2-form

$$\Phi(X,Y) = g(\phi X, Y), \quad X, Y \in T(M),$$

and the 1-form η are closed i.e., $d\Phi = 0$ and $d\eta = 0$. An almost cosymplectic manifold is *cosymplectic* if the underlying almost contact metric structure is normal i.e., $[\phi,\phi] = 0$. ■

For the normality condition and its geometric interpretation see S.I. Goldberg & K. Yano, [143], and D.E. Blair, [42], p. 77. Cf. also E. Barletta et al., [23], p. 74. Normality is known to imply $\nabla\xi = 0$ (as a consequence of $\phi(\xi) = 0$ and $\nabla\phi = 0$). A cosymplectic manifold is locally the product of a Kähler manifold and an interval in $\mathbb{R}$. There are however examples of cosymplectic manifolds which aren't globally the product of a Kähler manifold and a real 1-dimensional manifold (cf. again [42], p. 77). Products of almost Kähler manifolds and 1-dimensional manifolds are examples of almost cosymplectic manifolds with ξ parallel. The converse is not true, in general: Z. Olszak, [228], has shown that almost cosymplectic structures with a nonparallel vector ξ exist on certain odd-dimensional Lie groups. In dimension three, an almost cosymplectic manifold is cosymplectic if and only if ξ is parallel (cf. again [228]).

Remark 4.10 S.I. Goldberg & K. Yano, [143], showed that the 1-form η of an almost cosymplectic manifold is harmonic i.e., $\Delta_1\eta = 0$. Hence (by the Wietzenböck formula) $\Delta_g\xi = -Q\xi$. It is therefore implicit in S.I. Goldberg & K. Yano's work (cf. *op. cit.*) the following result similar to Theorem 4.5: *the characteristic vector field of an almost cosymplectic manifold is a harmonic vector field if and only if it is an eigenvector of the Ricci operator.*

For an almost cosymplectic manifold the formula $\Delta_g\xi = -Q\xi$ implies $J_{1M}\xi = -2Q\xi$. Moreover one has $g(Q\xi,\xi) + \|\nabla\xi\|^2 = 0$ (cf. Z. Olszak, [228]) so that

Proposition 4.11

a. *Let M be an almost cosymplectic manifold. Then the characteristic vector field ξ is a Jacobi field if and only if it is parallel or equivalently $\xi : (M,g) \to (T(M), G_s)$ is a harmonic map.*

b. *Let M be an almost cosymplectic metric 3-dimensional manifold. Then ξ is a Jacobi field if and only if M is a cosymplectic manifold.*

What relationship is there among the class of H-contact manifolds and the other classical classes of contact metric manifolds? If we set

$$\mathcal{M}_S : \text{the set of all Sasakian manifolds,}$$

$$\mathcal{M}_K : \text{the set of all } K\text{-contact manifolds,}$$

$$\mathcal{M}_{k,\mu} : \text{the set of all } (k,\mu)\text{-spaces,}$$

$$\mathcal{M}_{\phi s} : \text{the set of all strongly } \phi\text{-symmetric spaces,}$$

$$\mathcal{M}_H : \text{the set of all } H\text{-contact manifolds,}$$

we already observed that $\mathcal{M}_K \subseteq \mathcal{M}_H$ and $\mathcal{M}_{k,\mu} \subseteq \mathcal{M}_H$. The study of the classes $\mathcal{M}_S$, $\mathcal{M}_K$ and $\mathcal{M}_{k,\mu}$ was a rather popular differential geometric subject for the last forty years (cf. e.g., D.E. Blair, [42]). Note that $\mathcal{M}_K \cap \mathcal{M}_{k,\mu} = \mathcal{M}_S$.

Let $S(M)$ be the total space of the tangent sphere bundle with the standard contact metric structure (cf. [42]). By a result of D.E. Blair & T. Koufogiorgos & B.J. Papantoniou, [46], $S(M)$ is a (k,μ)-space if and only if the base manifold M is a space of constant sectional curvature $c \in \mathbb{R}$. Moreover, by a result of S. Tashiro (cf. [42], p. 144) $S(M)$ is a K-contact manifold if and only if M is a space of constant sectional curvature $c = 1$. Therefore the total space $S(M)$ of the tangent sphere bundle over a real space form M of sectional curvature $c \neq 1$ is an example of a (k,μ)-space which is not K-contact i.e., $S(M) \in \mathcal{M}_{k,\mu} \setminus \mathcal{M}_K$. Also, there are examples of K-contact manifolds which are not (k,μ)-spaces i.e., $\mathcal{M}_K \setminus \mathcal{M}_{k,\mu} \neq \emptyset$. Indeed it is known (cf. D.E. Blair & D. Perrone, [47]) that there are examples of K-contact manifolds which are not Sasakian (i.e., $\mathcal{M}_K \setminus \mathcal{M}_S \neq \emptyset$) and these examples are not (k,μ)-spaces (as emphasized above, if a manifold is both K-contact and a (k,μ)-space it must be Sasakian). On the other hand if M is a two-point homogeneous space, then the Reeb vector field of $S(M)$ is a harmonic vector field (see Theorem 4.76 in this chapter). Therefore the total space $S(M)$ of the tangent sphere bundle over a two-point homogeneous space M of nonconstant sectional curvature is an example of an element of $\mathcal{M}_H$ which belongs to neither $\mathcal{M}_K$ nor $\mathcal{M}_{k,\mu}$.

The class $\mathcal{M}_{\phi s}$ of all strongly locally ϕ-symmetric spaces was introduced by E. Boeckx & L. Vanhecke, [53]. The formula (5.3) in [53] with $k = 0$ shows that the characteristic vector field ξ of a space in the class $\mathcal{M}_{\phi s}$ is an eigenvector of the Ricci operator. Therefore a strongly locally ϕ-symmetric space is an H-contact manifold i.e., $\mathcal{M}_{k,\mu} \subseteq \mathcal{M}_H$. The converse is false in general, i.e., the previous inclusion is strict. Indeed $S(M)$ is a strongly locally ϕ-symmetric space if and only if M is a space of constant sectional curvature (cf. [53]). Hence the total space $S(M)$ of the tangent sphere bundle over a two-point homogeneous space of nonconstant sectional curvature is an example of an element in the class $\mathcal{M}_H$ which is not strongly locally ϕ-symmetric. The Reeb vector field of a K-contact manifold defines a harmonic map (cf. [262]). Moreover for any (k,μ)-space a calculation (in the presence of a local orthonormal ϕ-basis consisting of eigenvectors of h) together with (4.3) leads to

$$\text{trace}_g \{R(\xi, \nabla.\xi)\cdot\} = 0.$$

Summing up the facts discussed so far we obtain

Proposition 4.12

i. $\mathcal{M}_H \supset \mathcal{M}_K \supset \mathcal{M}_S$ *and* $\mathcal{M}_H \supset \mathcal{M}_{k,\mu} \supset \mathcal{M}_S$ *and* $\mathcal{M}_H \supset \mathcal{M}_{\phi s}$.

ii. $\mathcal{M}_K \not\supseteq \mathcal{M}_{k,\mu}$ *and* $\mathcal{M}_{k,\mu} \not\supseteq \mathcal{M}_K$.

Moreover the Reeb vector field of any space in $\mathcal{M}_{k,\mu} \cup \mathcal{M}_K \supset \mathcal{M}_S$ *is a harmonic map.*

The condition on the Ricci tensor characterizing the class of H-contact manifolds (cf. Theorem 4.5 above) appears naturally in an array of practical problems. For instance it occurs (cf. S.I. Goldberg & D. Perrone & G. Toth, [142]) in connection with the vanishing of the first Betti number of a contact metric manifold whose underlying metric is critical in the sense of the following

Definition 4.13 Let (M, η) be a real $(2n+1)$-dimensional compact contact manifold. Let us consider the functional

$$F(g) = \frac{1}{2} \int_M \|\tau\|^2 \, d\text{vol}(g)$$

defined on the set $\mathcal{A}(\eta)$ of all associated Riemannian metrics on (M, η). F is called the *Chern-Hamilton functional.* A critical point $g \in \mathcal{A}(\eta)$ of the functional $F : \mathcal{A}(\eta) \to [0, +\infty)$ is referred to as a *critical metric* on (M, η). ∎

The functional $F : \mathcal{A}(\eta) \to [0, +\infty)$ was studied by S.S. Chern & R.S. Hamilton, [86], for $n = 3$ and by D.E. Blair, [42], p. 167, under the additional condition that η is regular (in the sense of R.S. Palais, [233]). Cf. also

S. Tanno, [281], for a study of F in the higher dimension case. A critical metric $g \in \mathcal{A}(\eta)$ is a smooth solution to

$$\nabla_\xi \tau = 2\tau\phi \tag{4.8}$$

Equations (4.8) (i.e., the Euler-Lagrange equations of the variational principle $\delta F(g) = 0$) were derived by S. Tanno, [281]. Together with [23], we adopt the following

Definition 4.14 Equations (4.8) is referred to as *Tanno's equation.* ∎

The critical points of an ample class of functionals (i.e., $F_f(g) = \int_M f(\|\tau\|^2)\, d\,\mathrm{vol}(g)$ with $f: \mathbb{R} \to \mathbb{R}$ real analytic) are actually described by Tanno's equation (cf. Theorem 1 in E. Barletta et al., [23], p. 69).

The Chern-Hamilton functional F is related to the Dirichlet functional E. It should be noted that a contact metric structure (ϕ, ξ, η, g) is determined by the tensors g and ξ. Indeed once g and ξ are given the contact form is given by $\eta = g(\xi, \cdot)$ and the $(1,1)$-tensor ϕ is determined by $d\eta = g(\cdot, \phi\cdot)$ (by the nondegeneracy of g). Let $L(V, g) \in \mathbb{R}$ be the Dirichlet energy of a given unit vector field V tangent to M

$$L(V,g) = \frac{1}{2}\int_M \|dV\|^2\, d\,\mathrm{vol}(g), \quad (V,g) \in \Gamma^\infty(S(M)) \times \mathcal{A}(\eta).$$

The restriction of L to $\Gamma^\infty(S(M)) \times \{g\}$ is precisely E. Moreover, a metric $g \in \mathcal{A}(\eta)$ is a critical point of $L(\xi, \cdot): \mathcal{A}(\eta) \to [0, +\infty)$ if and only if g is a critical point of the Chern-Hamilton functional $F: \mathcal{A}(\eta) \to [0, +\infty)$ because of

$$\|\nabla\xi\|^2 = 2n + \frac{1}{4}\|\tau\|^2.$$

Definition 4.15 A contact metric structure (ξ, g) on M is said to be *critical* if ξ is a critical point of $L(\cdot, g): \Gamma^\infty(S(M)) \to [0, +\infty)$ while g is a critical point of $L(\xi, \cdot): \mathcal{A}(\eta) \to [0, +\infty)$. ∎

By slightly reformulating a result of S.I. Goldberg & D. Perrone & G. Toth, [142], we obtain

Theorem 4.16 *Let $(M, (\phi, \xi, \eta, g))$ be a real $(2n+1)$-dimensional compact contact metric manifold whose underlying contact metric structure (ξ, g) is critical. If* $\mathrm{Ric} + cg$ *is positive definite for some $0 < c < 2 - \|\tau\|/\sqrt{2n}$ then the first Betti number of M is zero (i.e., $b_1(M) = 0$).*

In the K-contact case (i.e., when $\tau = 0$), the result is due to S. Tanno, [280]: if M is compact and $\mathrm{Ric} + cg$ is positive definite then $b_1(M) = 0$. A K-contact structure is of course critical, yet the converse is false in general. Indeed the standard contact metric structure on $S(M^n(-1))$ (the total space of the tangent sphere bundle over a real space form $M^n(-1)$ of sectional curvature -1) is not K-contact yet ξ is a critical point of E and g is a critical point of F (cf. D.E. Blair, [42]). Therefore the class of compact contact metric manifolds whose underlying contact metric structure is critical contains strictly the class of compact K-contact manifolds.

We close this section by recalling that (by a result of K. Bang & D.E. Blair, [43]) a real $(2n+1)$-dimensional conformally flat H-contact manifold is a space of constant sectional curvature c and $n = 1 \Longrightarrow c \in \{0,1\}$ while $n > 1 \Longrightarrow c = 1$. For $n = 1$ the result was proved independently by G. Calvaruso & D. Perrone & L. Vanhecke, [78].

4.2. THREE-DIMENSIONAL H-CONTACT MANIFOLDS

The scope of this section is to give a classification of 3-dimensional contact metric manifolds whose Reeb vector field is a harmonic map.

Let $(M, (\phi, \xi, \eta, g))$ be a 3-dimensional contact metric manifold. For each point $m \in M$ there is a local orthonormal frame of the form $\{\xi, E_1, E_2 = \phi E_1\}$ defined on some open neighborhood U of m in M and consisting of eigenvectors of h. Let $U_1 = \{x \in M : h(x) \neq 0\}$. Then U_1 is an open subset of M. Also let U_2 consist of all $x \in M$ such that $h = 0$ in a neighborhood of x. Then U_2 is open and $U_1 \cup U_2$ is an open dense subset of M. If $hE_1 = \lambda E_1$ for some $\lambda \in C^\infty(U)$ then $hE_2 = -\lambda E_2$ (and $\lambda(x) \neq 0$ for any $x \in U \cap U_1$). We shall need the following

Lemma 4.17 (G. Calvaruso & D. Perrone & L. Vanhecke, [78]) *The following identities*

$$\begin{aligned} &\nabla_\xi E_1 = -aE_2, \quad \nabla_\xi E_2 = aE_1, \\ &\nabla_{E_1} \xi = -(\lambda+1)E_2, \quad \nabla_{E_2}\xi = -(\lambda-1)E_1, \\ &\nabla_{E_1} E_1 = \frac{1}{2\lambda}\left(E_2(\lambda) + A\right)E_2, \quad \nabla_{E_2} E_2 = \frac{1}{2\lambda}\left(E_1(\lambda) + B\right)E_1, \end{aligned} \tag{4.9}$$

$$\nabla_{E_1} E_2 = -\frac{1}{2\lambda} (E_2(\lambda) + A) E_1 + (\lambda + 1)\xi,$$
$$\nabla_{E_2} E_1 = -\frac{1}{2\lambda} (E_1(\lambda) + B) E_2 + (\lambda - 1)\xi,$$

hold on $U \cap U_1$ *for some* C^∞ *function* a. *Here* $A = \mathrm{Ric}(\xi, E_1)$ *and* $B = \mathrm{Ric}(\xi, E_2)$. *Moreover*

$$[E_1, E_2] = -\frac{1}{2\lambda} (E_2(\lambda) + A) E_1 + \frac{1}{2\lambda} (E_1(\lambda) + B) E_2 + 2\xi, \tag{4.10}$$
$$\nabla_\xi h = 2ah\phi + \frac{\xi(\lambda)}{\lambda} h. \tag{4.11}$$

The proof of Lemma 4.17 is straightforward. We shall also need the expression of the Ricci operator Q with respect to the local frame $\{\xi,\ E_1,\ E_2 = \phi E_1\}$ (cf. D. Perrone, [238])

$$Q\xi = 2(1 - \lambda^2)\xi + AE_1 + BE_2,$$
$$QE_1 = A\xi + \left(\frac{\rho}{2} - 1 + \lambda^2 + 2a\lambda\right) E_1 + \xi(\lambda)E_2,$$
$$QE_2 = B\xi + \xi(\lambda)E_1 + \left(\frac{\rho}{2} - 1 + \lambda^2 - 2a\lambda\right) E_2. \tag{4.12}$$

By Lemma 4.17

$$\mathrm{trace}_g \{R(\nabla.\xi, \xi)\cdot\} = -2\lambda\xi(\lambda)\xi + (\lambda - 1)BE_1 + (\lambda + 1)AE_2$$

hence

Proposition 4.18 *Let* $(M, (\phi, \xi, \eta, g))$ *be a 3-dimensional contact metric manifold. Then* ξ *is a harmonic map if and only if* M *is an* H*-contact manifold and* $\xi(\|\tau\|^2) = 0$.

Our next purpose is to establish the following

Theorem 4.19 (D. Perrone, [241]) *Let us consider a 3-dimensional contact metric manifold* $(M, (\phi, \xi, \eta, g))$. *The characteristic vector field* ξ *is a harmonic map if and only if the generalized* (k, μ)*-space condition holds on the open dense subset* $U_1 \cup U_2$.

We start by proving

Lemma 4.20 *Let* $(M, (\phi, \xi, \eta, g))$ *be a generalized* (k, μ)*-space. Then*

$$\nabla_\xi h = \mu h\phi, \quad \mu = 2a, \quad \xi(\lambda) = 0, \quad k = 1 - \lambda^2,$$

on $U \cap (U_1 \cup U_2)$ *where* λ *and* a *are the functions appearing in Lemma* 4.17.

Proof. We adopt the notations in Lemma 4.17. If $U \cap U_2 \neq \emptyset$ then the restriction of the contact structure to $U \cap U_2$ is Sasakian and $h = 0$, $\lambda = a = \mu = 0$ and $k = 1$ on $U \cap U_2$. Moreover (by (4.3)) the operator $\ell = \mathrm{Ric}(\xi, \cdot)\xi$ satisfies

$$\ell = -k\phi^2 + \mu h = k(I - \eta \otimes \xi) + \mu h \tag{4.13}$$

on $U \cap U_1$. Then (by (4.13))

$$\ell\phi = k\phi + \mu h\phi, \quad \phi\ell = k\phi - \mu h\phi,$$

from which

$$\ell\phi + \phi\ell = 2k\phi, \tag{4.14}$$

$$\ell\phi - \phi\ell = 2\mu h\phi. \tag{4.15}$$

We recall (cf. the identity (2.6) in [237]) that on any contact metric manifold

$$2\nabla_\xi h = \ell\phi - \phi\ell.$$

Hence (by (4.15)) $\nabla_\xi h = \mu h\phi$ or equivalently

$$\phi\nabla_\xi h = \mu h. \tag{4.16}$$

Next (by (4.11) in Lemma 4.17 and (4.16)) $\mu = 2a$ and $\xi(\lambda) = 0$. Then (4.13) and (4.16) yield

$$\ell = -k\phi^2 + \phi\nabla_\xi h, \tag{4.17}$$

or equivalently (cf. Remark 2.4 in [237]) $h^2 = (k-1)\phi^2$ so that $k = 1 - \lambda^2$. ■

Lemma 4.21 *If M is a generalized (k, μ)-space then*

$$\mathrm{Ric}(\xi, \cdot)|_{\mathrm{Ker}(\eta)} = 0$$

so that ξ is a harmonic vector field.

Proof. By (4.3) one has $R(X, Y)\xi = 0$ for any $X, Y \in \mathrm{Ker}(\eta)$ hence $\mathrm{Ric}(X, \xi) = 0$ for any $X \in \mathrm{Ker}(\eta)$. ■

Proof of Theorem 4.19. Once again we adopt the notations in Lemma 4.17. As emphasized above, if $U \cap U_2 \neq \emptyset$ then the restriction of the contact metric structure to $U \cap U_2$ is Sasakian and then the statement in Theorem 4.19 is trivially satisfied. Next, let us assume that $U \cap U_1 \neq \emptyset$ and consider the local ϕ-basis $\{\xi, E_1, E_2 = \phi E_1\}$ (as appearing in Lemma 4.17). If M is a generalized (k,μ)-space, Proposition 4.18 and Lemmas 4.20 and 4.21 imply that ξ is a harmonic map.

Vice versa, let us assume that ξ is a harmonic map. We shall show that the generalized (k,μ)-space condition (4.3) is satisfied on $U\cap U_1$. We start by recalling (cf. [238]) that the Ricci operator Q of an arbitrary 3-dimensional contact metric manifold is locally given by

$$Q=\alpha I+\beta\eta\otimes\xi+\phi\nabla_\xi h-\sigma(\phi^2)\otimes\xi+\sigma(E_1)\eta\otimes E_1+\sigma(E_2)\eta\otimes E_2$$

hence (by (4.11))

$$\begin{aligned}Q=\alpha I+\beta\eta\otimes\xi+2ah+\frac{\xi(\lambda)}{\lambda}\phi h\\ -\sigma(\phi^2)\otimes\xi+\sigma(E_1)\eta\otimes E_1+\sigma(E_2)\eta\otimes E_2\end{aligned}\tag{4.18}$$

on $U\cap U_1$, where

$$\sigma=\mathrm{Ric}(\xi,\cdot)|_{\mathrm{Ker}(\eta)},\quad \alpha=\frac{\rho}{2}-1+\lambda^2,\quad \beta=-\frac{\rho}{2}+3-3\lambda^2.$$

By (4.11) and (4.18) it follows that $\sigma=0$ and $\xi(\lambda)=0$ on $U\cap U_1$ if and only if

$$Q=\left(\frac{\rho}{2}-1+\lambda^2\right)I+\left(-\frac{\rho}{2}+3-3\lambda^2\right)\eta\otimes\xi+2ah.\tag{4.19}$$

In the 3-dimensional case $R(X,Y)\xi$ is determined by Q. Precisely

$$\begin{aligned}R(X,Y)\xi&=\eta(X)QY-\eta(Y)QX\\ &\quad-g(QY,\xi)X+g(QX,\xi)Y-\frac{\rho}{2}\{\eta(X)Y-\eta(Y)X\}\end{aligned}$$

hence (by (4.19))

$$R(X,Y)\xi=(1-\lambda^2)\{\eta(X)Y-\eta(Y)X\}+2a\{\eta(X)hY-\eta(Y)hX\}.$$

Therefore the generalized (k,μ)-space condition (4.3) is satisfied on $U\cap U_1$ with $k=1-\lambda^2$ and $\mu=2a$. ■

Example 4.22 Let us consider

$$M=\{x=(x_1,x_2,x_3)\in\mathbb{R}^3 : x_3\neq 0\}$$

and the vector fields

$$E_1=\frac{\partial}{\partial x_1},\quad E_2=\frac{1}{x_3^2}\frac{\partial}{\partial x_2},\quad E_3=2x_2x_3^2\frac{\partial}{\partial x_1}+\frac{2x_1}{x_3^6}\frac{\partial}{\partial x_2}+\frac{1}{x_3^6}\frac{\partial}{\partial x_3}.$$

We define (ϕ,ξ,η,g) by setting

$$\phi(E_1)=0, \quad \phi(E_2)=E_3, \quad \phi(E_3)=-E_2,$$
$$\xi=E_1, \quad \eta(X)=g(\xi,X), \quad g(E_i,E_j)=\delta_{ij}, \quad 1\leq i,j\leq 3.$$

By a result of T. Koufogiorgos & C. Tsichlias, [197], $(M,(\phi,\xi,\eta,g))$ is a generalized (k,μ)-space so that ξ is a harmonic map. On the other hand the requirements $\mathrm{Ric}(\xi,\cdot)|_{\mathrm{Ker}(\eta)}=0$ and $\xi(\lambda)=0$ are invariant under D-homothetic deformations hence for any $t>0$ the contact vector ξ_t underlying the contact metric structure

$$\phi_t=\phi, \quad \xi_t=(1/t)\xi, \quad \eta_t=t\eta, \quad g_t=tg+t(t-1)\,\eta\otimes\eta,$$

is a harmonic map.

T. Koufogiorgos & C. Tsichlias classify (cf. [198]) the 3-dimensional generalized (k,μ)-spaces satisfying $\|\nabla k\|=$ constant $\neq 0$. As it turns out, the class is parameterized by two arbitrary smooth functions of one real variable. T. Koufogiorgos & C. Tsichlias' classification (cf. *op. cit.*) yields a classification of 3-dimensional H-contact manifolds with $\|\nabla\|\tau\|\|=$ constant $\neq 0$. The manifolds in this class are noncompact and parallelizable.

Definition 4.23 (S.S. Chern & R.S. Hamilton, [86], S. Tanno, [281]) Let $(M,(\phi,\xi,\eta,g))$ be a 3-dimensional contact metric manifold. The *Webster scalar curvature* is defined by

$$W=\frac{1}{8}\left(\rho-\mathrm{Ric}(\xi,\xi)+4\right)$$

where ρ and Ric are the scalar curvature and Ricci tensor of (M,g). ■

Moreover if $\tau=\mathcal{L}_\xi g$ and $\|\tau\|\neq 0$ we set

$$p=\frac{4\sqrt{2}\,W}{\|\tau\|}.$$

Our next task is to classify 3-dimensional H-contact metric manifolds with $\|\tau\|=$ constant. It will be useful to recall (cf. D. Perrone, [241]) that the following properties are equivalent

1. M is an H-contact manifold with $\|\tau\|=$ constant.
2. ξ is minimal (i.e., a critical point of the volume functional) and harmonic.
3. ξ is a strongly normal unit vector field.

The notion of a strongly normal unit vector field was introduced by J.C. Gonzàles-Dàvila & L. Vanhecke, [144]. A crucial technical ingredient in [144] is the fact that any strongly normal unit vector field is minimal.

Theorem 4.24 (D. Perrone, [241]) *Let $(M,(\phi,\xi,\eta,g))$ be a 3-dimensional contact metric manifold with $\|\tau\| =$ constant. Then M is an H-contact manifold if and only if either M is a Sasakian manifold or M is locally isometric to a unimodular Lie group G equipped with a non-Sasakian left invariant contact metric structure. Precisely*

- **i.** *If $p > 1$ then $\tilde{G}$ is the 3-sphere group* $\mathrm{SU}(2)$.
- **ii.** *If $p = 1$ then $\tilde{G}$ is the group $\widetilde{E(2)}$ i.e., the universal covering of the group of rigid motions of the Euclidean 2-space.*
- **iii.** *If $-1 \neq p < 1$ then $\tilde{G}$ is the group* $\widetilde{\mathrm{SL}(2,\mathbb{R})}$.
- **iv.** *If $p = -1$ then $\tilde{G}$ is the group $E(1,1)$ of rigid motions of the Minkowski 2-space.*

Here $\tilde{G}$ is the universal covering of G. In all cases ξ is a harmonic map.

Proof. Let $(M,(\phi,\xi,\eta,g))$ be a 3-dimensional contact metric manifold. We adopt the notations in Lemma 4.17. If $U\cap U_2 \neq \emptyset$ then the restriction to $U\cap U_2$ of the contact metric structure is Sasakian hence ξ is harmonic and $\|\tau\| = 0$. Next let us assume that $U\cap U_1 \neq \emptyset$ and let $\{\xi, E_1, E_2 = \phi E_1\}$ be a local ϕ-basis as in Lemma 4.17. Then (on $U\cap U_1$) ξ is harmonic if and only if $A = B = 0$. Let us assume that ξ is harmonic. Then (by (4.12))

$$(\nabla_\xi Q)\xi = 0,$$

$$(\nabla_{E_1} Q)E_1 = \left\{E_1\left(\frac{\rho}{2}\right) + 2\lambda E_1(a)\right\} E_1,$$

$$(\nabla_{E_2} Q)E_2 = \left\{E_2\left(\frac{\rho}{2}\right) + 2\lambda E_2(a)\right\} E_2.$$

Next the identity

$$\frac{1}{2}X(\rho) = \sum_i g((\nabla_{E_i} Q)E_i,\, X)$$

yields

$$E_1\left(\frac{\rho}{2}\right) = E_1\left(\frac{\rho}{2}\right) + 2\lambda E_1(a)$$

hence $E_1(a) = 0$. Similarly $E_2(a) = 0$. Hence (by $[E_1,E_2] = 2\xi$) a is constant on $U\cap U_1$. Then Lemma 4.17 gives

$$[\xi, E_1] = c_2 E_2, \quad [\xi, E_2] = c_1 E_1, \quad [E_1, E_2] = 2\xi,$$

where $c_1 = \lambda + a - 1$ and $c_2 = \lambda - a + 1$, $c_1, c_2 \in \mathbb{R}$. This implies that M is locally isometric to a unimodular Lie group carrying a left invariant contact metric structure, cf. F. Tricerri & L. Vanhecke, [289], and Theorem 3.1 in D. Perrone, [239]. The quoted Theorem 3.1 in [239] classifies unimodular Lie groups equipped with a left invariant contact metric structure in terms of the Webster scalar curvature W and the pseudohermitian invariant p. In the non-Sasakian case this implies the classification stated in Theorem 4.24. The last statement in Theorem 4.24 then follows from Proposition 4.18. ■

Corollary 4.25 *Let* G *be one of the Lie groups* $\mathrm{SU}(2)$, $\mathbb{H}_1$ *(the lowest dimensional Heisenberg group),* $\widetilde{\mathrm{SL}(2,\mathbb{R})}$, $\widetilde{E(2)}$ *or* $E(1,1)$. *A compact 3-dimensional manifold* M *admits an H-contact metric structure* (ϕ, ξ, η, g) *with* $\|\mathcal{L}_\xi g\| =$ *constant if and only if* M *is diffeomorphic to the left quotient of* G *by a discrete subgroup.*

H. Geiges, [122], has shown that a compact 3-dimensional manifold admits a normal contact form (i.e., a Sasakian structure) if and only if it is diffeomorphic to a left quotient G/Γ of a Lie group G by a discrete subgroup $\Gamma \subset G$ where $G \in \left\{\mathrm{SU}(2), \mathbb{H}_1, \widetilde{\mathrm{SL}(2,\mathbb{R})}\right\}$. H. Geiges' result (cf. *op. cit.*) together with Theorem 4.24 above imply Corollary 4.25.

Remark 4.26 Let G be one of the following simply connected unimodular Lie groups

$$\left\{\mathrm{SU}(2), \mathbb{H}_1, \widetilde{\mathrm{SL}(2,\mathbb{R})}, \widetilde{E(2)}, E(1,1)\right\}.$$

Then G admits a discrete subgroup Γ such that the quotient space G/Γ is a manifold and the projection $\pi : G \to G/\Gamma$ is smooth (cf. e.g., [59], p. 97–100). Any left invariant tensor field on G, and in particular a left invariant contact metric structure on G, descends to G/Γ. The contact metric structure induced on G/Γ by a given left invariant contact metric structure on G has the same curvature properties. Also, properties of the (left invariant) Reeb vector on G such as minimality or harmonicity remain invariant under projection on the quotient space. A 3-dimensional Lie group G admits a discrete subgroup Γ such that G/Γ is compact if and only if G is unimodular (cf. J. Milnor, [210]). ■

As previously emphasized, the class of all compact contact metric manifolds with critical contact metric structure contains strictly the class of all K-contact metric manifolds. The purpose of Theorem 4.27 below is to show that, in dimension 3, the two classes may be identified from a topological viewpoint.

Theorem 4.27 (D. Perrone, [242]) *Let us consider a 3-dimensional contact metric manifold $(M,(\xi,g))$. Then the contact metric structure (ξ,g) is critical if and only if either M is a K-contact manifold or M is locally isometric to $\mathrm{SL}(2,\mathbb{R})$ equipped with a non K-contact left invariant contact metric structure.*

Proof. Let us recall that the contact metric structure is critical if ξ is harmonic and τ satisfies Tanno's equation (4.8). We adopt the notations in Lemma 4.17. Then on $U \cap U_1$

$$\nabla_\xi h = 2ah\phi \Longleftrightarrow \xi(\lambda) = 0. \tag{4.20}$$

Let us assume that M is a 3-dimensional H-contact manifold satisfying Tanno's equation yet not K-contact. Then (4.8) and (4.20) imply $\xi(\lambda) = 0$. Hence ξ is a harmonic map (see Proposition 4.18). Then the generalized (k,μ)-space condition is satisfied on the dense open set $U_1 \cup U_2$ (see Theorem 4.19). Therefore on $U_1 \cup U_2$

$$\nabla_\xi \tau = \mu\tau\phi \tag{4.21}$$

and (cf. T. Koufogiorgos & C. Tsichlias, [198])

$$h\nabla\mu = \nabla k. \tag{4.22}$$

On a generalized (k,μ)-space the requirements (4.8) and (4.20) imply that $\mu = 2$. Moreover (by (4.22)) $k =$ constant. Thus M is a (k,μ)-space with $\mu = 2$. A 3-dimensional (k,μ)-space is a 3-dimensional locally homogeneous contact metric manifold (cf. the study of 3-dimensional manifolds admitting a homogeneous contact metric structure performed by D. Perrone, [239]). Precisely (M,g) is locally isometric to a unimodular Lie group G equipped with a left invariant contact metric structure. Due to left invariance it suffices to describe these structures at the level of the Lie algebra $\mathfrak{g}$ of G. On a unimodular Lie group G as above there is an orthonormal frame $\{E_1 = \xi,\ E_2,\ E_3 = \phi E_2\}$ such that

$$[E_1,E_2] = \lambda_3 E_3, \quad [E_2,E_3] = 2E_1, \quad [E_3,E_1] = \lambda_2 E_2. \tag{4.23}$$

Using (4.23) and the first Cartan structure equation one may compute the Levi-Civita connection

$$\left(\nabla_{E_i} E_j\right)_{1\le i,j\le 3}$$
$$= \begin{pmatrix} 0 & \dfrac{\lambda_2+\lambda_3-2}{2}E_3 & \dfrac{-\lambda_2-\lambda_3+2}{2}E_2 \\ \dfrac{\lambda_2-\lambda_3-2}{2}E_3 & 0 & \dfrac{-\lambda_2+\lambda_3+2}{2}E_1 \\ \dfrac{\lambda_2-\lambda_3+2}{2}E_2 & \dfrac{-\lambda_2+\lambda_3-2}{2}E_1 & 0 \end{pmatrix}. \tag{4.24}$$

Using (4.23) one obtains

$$\tau = (\lambda_2-\lambda_3)(\theta^2\otimes\theta^3+\theta^3\otimes\theta^2) \tag{4.25}$$

where $\lambda_2,\lambda_3\in\mathbb{R}$ and $\{\theta^1=\eta,\ \theta^2,\ \theta^3\}$ is the coframe dual to $\{\xi,\ E_2,\ E_3\}$ i.e., $\theta^i(E_j)=\delta^i_j$. Since $\left(\nabla_{E_i}\theta^j\right)^\sharp=\nabla_{E_i}E_j$ one finds (by (4.24)–(4.25))

$$\nabla_\xi \tau = (2-\lambda_2-\lambda_3)\tau\phi. \tag{4.26}$$

Therefore

$$\nabla_\xi\tau = 2\tau\phi \Longleftrightarrow \tau=0 \ \text{ or } \ \lambda_2+\lambda_3=0. \tag{4.27}$$

Then (by a result of D. Perrone, [239], p. 249–250) we may conclude that M is locally isometric to $\widetilde{\mathrm{SL}(2,\mathbb{R})}$ equipped with a left invariant contact metric structure which is K-contact (i.e., $\tau=0$) if and only if $\lambda_2=\lambda_3<0$. ∎

At this point we may prove the converse. Let us consider the Lie group $\widetilde{\mathrm{SL}(2,\mathbb{R})}$ with a left invariant metric g. Let us fix an orthonormal basis $\{E_1,E_2,E_3\}$ in the Lie algebra $\mathfrak{g}$ such that

$$[E_1,E_2]=\lambda_3E_3,\ \ [E_2,E_3]=\lambda_1E_1,\ \ [E_3,E_1]=\lambda_2E_2,$$

where $\lambda_1,\lambda_2,\lambda_3\in\mathbb{R}$ are constants such that $\lambda_1>0$, $\lambda_2>0$ and $\lambda_3<0$ (such a choice is always possible, cf. J. Milnor, [210]). Let η be the 1-form dual to E_1 i.e., $\eta^\sharp=E_1$ with respect to g. We may assume without loss of generality that $\lambda_1=2$ and define ϕ by $\phi E_1=0$, $\phi E_2=E_3$ and $\phi E_3=-E_2$ to conclude that $(\phi,\xi=E_1,\eta,g)$ is a left invariant contact metric structure satisfying (4.24)–(4.27). A calculation based on (4.24) shows that

$$\mathrm{Ric}(\xi,E_1)=\mathrm{Ric}(\xi,E_2)=0.$$

Hence, by assuming that $\lambda_2=-\lambda_3>0$ we may conclude that (ξ,g) is critical.

Theorem 4.27 and the above mentioned result by H. Geiges (cf. [122]) imply

Theorem 4.28 *A compact 3-dimensional manifold M admits a critical contact metric structure if and only if M is diffeomorphic to a left quotient of the Lie group G by a discrete subgroup, where* $G \in \{\mathrm{SU}(2),\ \mathbb{H}_1,\ \widetilde{\mathrm{SL}(2,\mathbb{R})}\}$.

We end this section by mentioning that the study of 3-dimensional H-contact manifolds is related to the study of strongly (and weakly) locally φ-symmetric spaces (cf. G. Calvaruso & D. Perrone & L. Vanhecke, [78], D. Perrone, [243]).

4.2.1. A Characterization of H-Contact Three-Manifolds and New Examples

In this section we aim to characterize 3-dimensional H-contact manifolds and give examples of H-contact three-manifolds whose Reeb vector fields is not a harmonic map.

Definition 4.29 Let $(M,(\varphi,\xi,\eta,g))$ be a contact metric manifold. Then $(M,(\varphi,\xi,\eta,g))$ is called a (κ,μ,ν)*-contact metric manifold* if there are $\kappa,\mu,\nu \in C^\infty(M)$ such that

$$R(X,Y)\xi = \kappa\{\eta(X)Y - \eta(Y)X\} + \mu\{\eta(X)hY - \eta(Y)hX\} + \nu\{\eta(X)\varphi hY - \eta(Y)\varphi hX\}$$

for any $X, Y \in \mathfrak{X}(M)$. ∎

By the arguments in the proof of Theorem 4.19 one may show that

Theorem 4.30 (T. Koufogiorgos & M. Markellos & V.J. Papantoniou, [199]) *Let $(M,(\varphi,\xi,\eta,g))$ be a 3-dimensional contact metric manifold. If M is a (κ,μ,ν)-contact metric manifold then it is an H-contact manifold. Vice versa if M is an H-contact manifold then (φ,ξ,η,g) is a (κ,μ,ν)-contact metric manifold structure on an everywhere dense open subset of M.*

By another result in [199] every (κ,μ,ν)-contact metric manifold of dimension > 3 is either Sasakian or a (κ,μ)-contact metric manifold i.e., the functions κ,μ are constant and $\nu = 0$.

Example 4.31 Let $M = \{(x,y,z) \in \mathbb{R}^3 : x > 0,\ y > 0,\ z > 0\}$ and let us consider the vector fields

$$E_1 = \frac{\partial}{\partial x}, \quad E_2 = \frac{\partial}{\partial y},$$
$$E_3 = -\frac{4}{z}\, e^{G(y,z)}\, G_y(y,z)\, \frac{\partial}{\partial x} + \beta(x,y,z)\frac{\partial}{\partial y} + e^{G(y,z)/2}\frac{\partial}{\partial z},$$

where $G(y,z) < 0$ for any $(y,z) \in \mathbb{R}^2$ and

$$2G_{yy} + G_y^2 = -ze^{-G} \tag{4.28}$$

while $\beta(x,y,z)$ satisfies

$$\beta_x = \frac{4}{x^2 z}\, e^G, \quad \beta_y = \frac{1}{2z}\, e^{G/2} - \frac{1}{2}\, G_z\, e^{G/2} - \frac{4}{xz}\, G_y\, e^G.$$

In particular a solution to (4.28) is given by

$$y = 4\sqrt{\frac{\pi}{z}}\,\mathrm{erf}\sqrt{-\frac{G}{2}}, \quad \mathrm{erf}(x) = \frac{2}{\sqrt{\pi}}\int_0^x e^{-t^2}\, dt.$$

Example 4.32 Let $M = \{(x,y,z) \in \mathbb{R}^3 : z > 0\}$ and

$$E_1 = \frac{\partial}{\partial x}, \quad E_2 = \frac{\partial}{\partial y},$$
$$E_3 = 2(y+z)\frac{\partial}{\partial x} + \left\{\frac{1}{c}e^{cx}z - \left(\frac{c}{2}y + cz + \frac{1}{2z}\right)y\right\}\frac{\partial}{\partial y} + \frac{\partial}{\partial z},$$

where $c \in \mathbb{R} \setminus \{0\}$.

In both Examples 4.31 and 4.32, we consider the synthetic object (φ, ξ, η, g) given by

$$\varphi(E_1) = 0, \quad \varphi(E_2) = E_3, \quad \varphi(E_3) = -E_2,$$
$$\xi = E_1, \quad \eta(X) = g(\xi, X), \quad g(E_i, E_j) = \delta_{ij}.$$

By a result in [199] it follows that $(M, (\varphi, \xi, \eta, g))$ is a (κ, μ, ν)-contact metric manifold with

$$\kappa = 1 - \frac{4}{x^4 z^2}\, e^{2G}, \quad \mu = 2\left(1 + \frac{2}{x^2 z}\, e^G\right), \quad \nu = -\frac{2}{x}$$

in Example 4.31 and

$$\kappa = 1 - \frac{z^2}{4}e^{2cx}, \quad \mu = 2 + ze^{cx}, \quad \nu = c,$$

in Example 4.32. As $\nu \neq 0$ (by Theorem 4.19 and Theorem 4.30) the Reeb vector ξ in Examples 4.31 and 4.32 is a harmonic vector field yet fails to be a harmonic map.

Let us recall the notion that an H-contact metric manifold is invariant under D-homothetic deformations (cf. Remark 4.6 above). This leads (by a D-homothetic deformation of the structures in Examples 4.31 and 4.32) to new examples of H-contact metric manifolds whose underlying Reeb vector is not a harmonic map.

4.2.2. Taut Contact Circles and H-Contact Structures

We shall need the following

Definition 4.33 (H. Geiges & J. Gonzalo, [123]) Let M be a 3-dimensional manifold. A pair of contact forms (θ_1, θ_2) on M is called *a contact circle* if for any $a = a_1 + ia_2 \in S^1 \subset \mathbb{C}$ the linear combination $a_1\theta_1 + a_2\theta_2$ is a contact form, too. ∎

If (θ_1, θ_2) is a contact circle then *any* nontrivial linear combination $\lambda_1\theta_1 + \lambda_2\theta_2$ with $(\lambda_1, \lambda_2) \in \mathbb{R}^2 \setminus \{(0,0)\}$ is also a contact form on M. By a result of H. Geiges & J. Gonzalo, [124], any compact orientable 3-dimensional manifold admits a contact circle.

Definition 4.34 A *taut contact circle* is a pair (θ_1, θ_2) of contact forms such that

$$\theta_1 \wedge d\theta_1 = \theta_2 \wedge d\theta_2, \quad \theta_1 \wedge d\theta_2 = -\theta_2 \wedge d\theta_1 .$$

A taut contact circle (θ_1, θ_2) is called a *Cartan structure* if $\theta_1 \wedge d\theta_2 = 0$. ∎

Theorem 4.35 (H. Geiges & J. Gonzalo, [123]) *Let M be a compact 3-dimensional manifold. Then M admits a taut contact circle if and only if M is diffeomorphic to a left quotient of the Lie group G by a discrete subgroup $\Gamma \subset G$ where $G \in \{\mathrm{SU}(2), \widetilde{\mathrm{SL}(2,\mathbb{R})}, \widetilde{E(2)}\}$.*

On the other hand $\mathbb{H}_1$ admits a Sasakian structure. Therefore, *in general* 3-dimensional Sasakian manifolds fail to admit taut contact circles.

Let $(M, (\phi, \xi, \eta, g))$ be a non-Sasakian contact metric 3-dimensional manifold (i.e., $\tau \neq 0$) and let $\{\xi, E_1, E_2 = \phi E_1\}$ be a local orthonormal frame of $T(M)$ consisting of eigenvectors of h with $hE_1 = \lambda E_1$ and $\lambda > 0$. Let $\{\omega_1, \omega_2\}$ be 1-forms dual to E_1 and E_2 respectively (i.e., $\omega_i^\sharp = E_i$ with respect to g, $i \in \{1,2\}$). The three eigenvalues $\{0, \lambda, -\lambda\}$ of h are everywhere distinct (as $h \neq 0$ everywhere). Hence the corresponding line fields are global. The 1-forms $\eta_1 = \omega_2 + \omega_1 = g(E_2 + E_1, \cdot)$ and $\eta_2 = g(\phi(E_2 + E_1), \cdot) = \omega_2 - \omega_1$ are orthogonal to the contact form η. At this point we may establish the following

Theorem 4.36 (D. Perrone, [246]) *Let us consider a 3-dimensional non-Sasakian H-contact manifold $(M, (\eta, g))$. Then the following statements are equivalent*

1. *The pair (η, η_1) is a taut contact circle.*

2. *The pair (η, η_2) is a taut contact circle.*

3. *The Riemannian manifold (M, g) is locally isometric to one of the following Lie groups $SU(2)$, $\widetilde{SL}(2,\mathbb{R})$, $\widetilde{E}(2)$ equipped with a left invariant non-Sasakian H-contact metric structure satisfying $\nabla_\xi h = 0$ (equivalently $\nabla_\xi \tau = 0$).*

Moreover (η, η_i) is a taut contact circle if and only if (η, η_i) is a Cartan structure, $i \in \{1, 2\}$.

Proof. As M is H-contact, ξ is an eigenvector of the Ricci operator i.e., $A_1 = \mathrm{Ric}(\xi, E_1) = 0$ and $A_2 = \mathrm{Ric}(\xi, E_2) = 0$. Then a calculation based on Lemma 4.17 leads to

$$(\eta \wedge d\eta)(\xi, E_1, E_2) = (d\eta)(E_1, E_2) = g(E_1, \phi E_2) = -1,$$

$$(\eta \wedge d\omega_1)(\xi, E_1, E_2) = (d\omega_1)(E_1, E_2) = \frac{E_2(\lambda)}{4\lambda},$$

$$(\eta \wedge d\omega_2)(\xi, E_1, E_2) = (d\omega_2)(E_1, E_2) = -\frac{E_1(\lambda)}{4\lambda},$$

$$(\omega_1 \wedge d\eta)(\xi, E_1, E_2) = (d\eta)(E_2, \xi) = \frac{1}{2} g(\xi, \nabla_\xi E_2) = 0,$$

$$(\omega_2 \wedge d\eta)(\xi, E_1, E_2) = (d\eta)(\xi, E_1) = \frac{1}{2} g(\xi, \nabla_\xi E_1) = 0.$$

Thus

$$\begin{aligned}(\eta \wedge d\eta_1)(\xi, E_1, E_2) &= \frac{E_2(\lambda) - E_1(\lambda)}{4\lambda} \\ &= (\eta \wedge d\eta_2)(\xi, E_1, E_2) + \frac{E_2(\lambda)}{2\lambda}\end{aligned} \tag{4.29}$$

and

$$(\eta_1 \wedge d\eta)(\xi, E_1, E_2) = (\eta_2 \wedge d\eta)(\xi, E_1, E_2) = 0. \tag{4.30}$$

Moreover

$$(\omega_1 \wedge d\omega_1)(\xi, E_1, E_2) = \frac{\lambda - 1 + a}{2},$$

$$(\omega_2 \wedge d\omega_2)(\xi, E_1, E_2) = \frac{a - \lambda - 1}{2},$$
$$(\omega_1 \wedge d\omega_2)(\xi, E_1, E_2) = (\omega_2 \wedge d\omega_1)(\xi, E_1, E_2) = 0,$$

from which

$$\begin{aligned}(\eta_1 \wedge d\eta_1)(\xi, E_1, E_2) &= (\eta_2 \wedge d\eta_2)(\xi, E_1, E_2) \\ &= (\eta \wedge d\eta)(\xi, E_1, E_2) + a. \end{aligned} \tag{4.31}$$

Let us prove the implication (1) $\Longrightarrow$ (3) in Theorem 4.36. The identities (4.29)–(4.31) imply that the pair (η, η_1) is a taut contact circle (or equivalently a Cartan structure) if and only if

$$E_2(\lambda) = E_1(\lambda), \quad a = 0. \tag{4.32}$$

Using Lemma 4.17 and the previous requirement (4.32) one has

$$\begin{aligned} R(\xi, E_1)E_1 &= -\nabla_\xi \nabla_{E_1} E_1 + \nabla_{E_1} \nabla_\xi E_1 + \nabla_{[\xi, E_1]} E_1 \\ &= -\nabla_\xi \left(\frac{E_2(\lambda)}{2\lambda} E_2 \right) + (\lambda + 1) \nabla_{E_2} E_1 \\ &= \left\{ -\xi \left(\frac{E_2(\lambda)}{2\lambda} \right) - \frac{\lambda + 1}{2\lambda} E_1(\lambda) \right\} E_2 + (\lambda^2 - 1)\xi, \\ R(\xi, E_2)E_2 &= -\nabla_\xi \nabla_{E_2} E_2 + \nabla_{E_2} \nabla_\xi E_2 + \nabla_{[\xi, E_2]} E_2 \\ &- \nabla_\xi \left(\frac{E_1(\lambda)}{2\lambda} E_1 \right) + (\lambda - 1) \nabla_{E_1} E_2 \\ &= \left\{ -\xi \left(\frac{E_1(\lambda)}{2\lambda} \right) - \frac{\lambda - 1}{2\lambda} E_2(\lambda) \right\} E_1 + (\lambda^2 - 1)\xi. \end{aligned}$$

Then

$$g(R(\xi, E_1)E_2, E_1) = \mathrm{Ric}(\xi, E_2) = 0,$$
$$g(R(\xi, E_2)E_1, E_2) = \mathrm{Ric}(\xi, E_1) = 0,$$

imply

$$\xi \left(\frac{E_2(\lambda)}{2\lambda} \right) + \frac{\lambda + 1}{2\lambda} E_1(\lambda) = 0, \tag{4.33}$$

$$\xi \left(\frac{E_1(\lambda)}{2\lambda} \right) + \frac{\lambda - 1}{2\lambda} E_2(\lambda) = 0. \tag{4.34}$$

The identities (4.32)–(4.34) imply

$$E_1(\lambda) = E_2(\lambda) = 0.$$

Moreover $2\xi(\lambda) = [E_1, E_2](\lambda) = 0$ hence λ is locally constant, hence constant (as λ is continuous and M connected). Then

$$[\xi, E_1] = (\lambda + 1)E_2, \quad [E_2, \xi] = -(\lambda - 1)E_1, \quad [E_1, E_2] = 2\xi, \tag{4.35}$$

where $\lambda \in \mathbb{R} \setminus \{0\}$. Then (4.35) implies that M must be locally isometric to a Lie group carrying a left invariant contact metric structure (cf. D. Perrone, [239]). Moreover (by Lemma 4.17) $\nabla_\xi h = 0$. Thus (by Theorem 3.1 in [239]) M must be locally isometric to one of the Lie groups

$$SU(2), \quad \widetilde{SL(2, \mathbb{R})}, \quad \widetilde{E(2)},$$

equipped with a left invariant non-Sasakian H-contact metric structure satisfying $\nabla_\xi h = 0$. ∎

Let us prove the implication (2) $\Longrightarrow$ (3). The identities (4.29)–(4.31) imply that (η, η_2) is a taut contact circle (or equivalently a Cartan structure) if and only if

$$E_2(\lambda) = -E_1(\lambda), \quad a = 0. \tag{4.36}$$

Then, as well as in the proof of the implication (1) $\Longrightarrow$ (3), the requirement (4.36) yields (4.35) hence the result.

Let us prove the implications (3) $\Longrightarrow$ (1) and (3) $\Longrightarrow$ (2). Under the assumption (3) the universal covering of M is one of the unimodular Lie groups $SU(2)$, $\widetilde{SL(2, \mathbb{R})}$ or $\widetilde{E(2)}$. Let G be one of these Lie groups. Then G admits a left invariant non-Sasakian contact metric structure (η, g) whose underlying contact vector ξ is an eigenvector of the Ricci operator and satisfies $\nabla_\xi h = 0$. Also there exist local orthonormal frames of the form $\{\xi, E_1, E_2 = \phi E_1\}$, consisting of eiegenvectors of h and satisfying

$$[E_1, E_2] = 2\xi, \quad [E_2, \xi] = \lambda_1 E_1, \quad [\xi, E_1] = \lambda_2 E_2,$$

with $\lambda_1 + \lambda_2 = 0$ (equivalently $\nabla_\xi h = 0$) and $\lambda_1 \neq \lambda_2$ (equivalently $h \neq 0$). Cf. D. Perrone, [239], for the explicit construction of (η, g). A straightforward calculation shows that

$$(\eta_1 \wedge d\eta_1)(\xi, E_1, E_2) = (\eta_2 \wedge d\eta_2)(\xi, E_1, E_2) = (\eta \wedge d\eta)(\xi, E_1, E_2) = -1,$$

$$(\eta_1 \wedge d\eta)(\xi, E_1, E_2) = (\eta \wedge d\eta_1)(\xi, E_1, E_2) = 0,$$

$$(\eta_2 \wedge d\eta)(\xi, E_1, E_2) = (\eta \wedge d\eta_2)(\xi, E_1, E_2),$$

so that (η, η_i), $i \in \{1,2\}$, are Cartan structures and in particular taut contact circles. Theorem 4.36 is proved.

Let M be a 3-dimensional compact manifold. If M admits a non-Sasakian H-contact structure satisfying $\nabla_\xi h = 0$ then Theorem 4.36 implies that M admits a taut circle. Conversely, let us assume that M admits a taut contact circle. Then (by Theorem 4.35) M is diffeomorphic to a left quotient of the Lie group G by a discrete subgroup $\Gamma \subset G$ where $G \in \{SU(2), \widetilde{SL(2,\mathbb{R})}, \widetilde{E(2)}\}$. Left invariant tensor fields on G descend to the quotient G/Γ hence any left invariant non-Sasakian H-contact structure with $\nabla_\xi h = 0$ on G induces a contact metric structure on G/Γ which enjoys the same properties (cf. D. Perrone, [239]). Therefore

Corollary 4.37 *A 3-dimensional compact manifold M admits a taut contact circle if and only if it admits a non-Sasakian H-contact structure satisfying $\nabla_\xi h = 0$.*

The requirement $\nabla_\xi h = 0$ (equivalently $\nabla_\xi \tau = 0$) was examined for the first time by S.S. Chern & R.S. Hamilton, [86]. Several differential geometric problems (in Riemanninan geometry) in a close relationship to the requirement $\nabla_\xi h = 0$ are described in the survey paper [245].

4.3. STABILITY OF THE REEB VECTOR FIELD

Few results concerning the stability of the Reeb vector field are known so far. E. Boeckx & J.C. Gonzales-Davila & L. Vanhecke, [56], studied the stability of the Reeb vector underlying the standard contact metric structure on $S(M)$ when M is a compact quotient of a two-point homogeneous space. V. Borrelli, [61], studied the stability of the Reeb vector field of an arbitrary Sasakian manifold. Precisely

Theorem 4.38 (V. Borrelli, [61]) *Let $(M, (\phi, \xi, \eta, g))$ be a compact Sasakian manifold and*

$$\xi_k = \sqrt{k}\,\xi, \quad g_k = k^{-1} g, \quad k \in (0, +\infty).$$

Then

i. *Assume that ξ is stable with respect to the energy functional E. Then there is $k_s \in (0, +\infty)$ such that ξ_k is stable with respect to the volume functional if and only if $0 < k \le k_s$.*

ii. *If ξ is unstable with respect to the energy functional E then there are k_s, $k_s' \in [0, +\infty)$ such that $k_s \le k_s'$ and ξ_k is stable with respect to the volume functional if and only if $k_s \le k \le k_s'$.*

The remainder of this section is devoted to a discussion of the results in [243] on the stability of the Reeb vector field of a 3-dimensional compact H-manifold M with respect to the energy functional $E : \Gamma^\infty(S(M)) \to [0, +\infty)$. We distinguish between the following cases: 1) M is Sasakian, 2) M is a generalized (k,μ)-space, and 3) M is non-Sasakian and ξ is strongly normal.

4.3.1. Stability of ξ for Sasakian 3-Manifolds and Generalized (k,μ)-Spaces

As previously shown, Hopf vector fields on S^3, i.e., unit Killing vector fields on S^3, are the only absolute minimizers of the energy and hence they are stable. On the other hand, a unit Killing vector field on S^3 may be looked at as the Reeb vector of a Sasakian structure (η, g) on S^3 with Webster scalar curvature $W = 1$. The main result of S.S. Chern & R.S. Hamilton, [86], states that any 3-dimensional compact contact manifold (M, η) admits a compatible metric g whose Webster scalar curvature W is either positive or a nonnegative constant. The Ricci tensor field of a 3-dimensional contact metric manifold is given by

$$\mathrm{Ric} = 2(2W - 1)g + 4(1 - W)\,\eta \otimes \eta, \tag{4.37}$$

where (due to $\mathrm{Ric}(\xi,\xi) = 2 - \|\tau\|^2/4$) W is given by (cf. also Definition 4.23)

$$W = \frac{1}{8}\left(\rho + 2 + \frac{\|\tau\|^2}{4}\right). \tag{4.38}$$

Theorem 4.39 (D. Perrone, [243]–[244]) *Let $(M, (\eta, g))$ be a 3-dimensional compact Sasakian manifold.*

i. *If $W \geq 1$ then*

$$E(V) \geq E(\xi) = \frac{5}{2}\mathrm{Vol}(M,g), \quad V \in \Gamma^\infty(S(M)),$$

and equality is achieved if and only if V is a Killing vector field and an eigenvector of the Ricci operator corresponding to the eigenvalue 2. Also, the equality is achieved for some $V \in \Gamma^\infty(S(M))$ which is not collinear to ξ if and only if (M,g) has constant sectional curvature $+1$, so that the universal covering space of M is $\tilde{M} = S^3$.

ii. *If $W > 0$ then there is a D-homothetic deformation of (η, g) into a Sasakian structure (η_t, g_t) such that*

$$E(V) \geq E(\xi_t) = t^2 E(\xi), \quad V \in \Gamma^\infty(S(M, g_t)),$$

and equality is achieved if and only if V is a Killing vector field and an eigenvector of the Ricci operator corresponding to the eigenvalue 2 with respect to g_t. Also the equality is achieved for some $V \in \Gamma^\infty(S(M,g_t))$, which is not collinear to ξ if and only if (M,g_t) has constant sectional curvature $+1$.

iii. *If $W = 0$ then the universal covering space $\tilde{M}$ of M is the Heisenberg group $\mathbb{H}_1$ and ξ is E-stable.*

Proof of (i). The Reeb vector ξ is a unit Killing vector field and an eigenvector of the Ricci operator Q such that $Q\xi = 2\xi$. Moreover the eigenvalues of Q are $\lambda_0 = 2 = \mathrm{Ric}(\xi,\xi)$ and $\lambda_1 = \lambda_2 = 2(2W-1)$. Therefore if $W \geq 1$ then $\mathrm{Ric}(V,V) \geq 2$ for any $V \in \Gamma^\infty(S(M))$. We must show that

$$E(V) \geq E(\xi) = \frac{5}{2}\mathrm{Vol}(M), \quad V \in \Gamma^\infty(S(M)),$$

and $E(V) = E(\xi)$ if and only if V is Killing and $QV = 2V$. As ξ is Killing and $Q\xi = 2\xi$ Theorem 3.14 implies that $E(\xi) = 5\,\mathrm{Vol}(M)/2$. Using $\mathrm{Ric}(V,V) \geq 2 = \mathrm{Ric}(\xi,\xi)$ and the inequality (3.12) we get

$$E(V) \geq \frac{1}{2}\int_M (3 + \mathrm{Ric}(V,V))\, d\mathrm{vol}(g) \geq E(\xi).$$

If V is a unit Killing vector field and $QV = 2V$ then (again by Theorem 3.14) it should be that $E(V) = E(\xi)$.

Let V now be a unit vector field such that $E(V) = E(\xi)$ i.e.,

$$\int_M (\mathrm{Ric}(V,V) - 2)\, d\mathrm{vol}(g) = 0, \quad \mathrm{Ric}(V,V) \geq 2.$$

Then $\mathrm{Ric}(V,V) = 2 = \mathrm{Ric}(\xi,\xi)$ hence (by (4.37))

$$(W-1)[\eta(V)^2 - 1] = 0. \tag{4.39}$$

Then $E(V) = 5\,\mathrm{Vol}(M)/2 = (1/2)\int_M (3 + \mathrm{Ric}(V,V))\, d\mathrm{vol}(g)$ implies (3.13). Then, as well as in the proof of Theorem 3.14, V is Killing if and only if $f_1 := (1/2)\,\mathrm{div}(V) = 0$. Let $\mathcal{A}_1 = \{x \in M : W(x) - 1 \neq 0\}$ (an open set) and $\mathcal{A}_2$ be the set of all $x \in M$ such that $W - 1 = 0$ in a neighborhood of x. Then $\mathcal{A}_1 \cup \mathcal{A}_2$ is an open dense subset of M. By (4.39) it follows that $\eta(V)^2 = 1$ on $\mathcal{A}_1$ so that $V = \pm\xi$ (and then V is Killing and $QV = 2V$). In particular $f_1 = 0$ on $\mathcal{A}_1$. By (4.37) it follows that g is Einstein on $\mathcal{A}_2$ and then $QV = 2V$. Therefore, by arguments similar to those in the proof of Theorem 3.14, it follows that f_1 is a harmonic function on $\mathcal{A}_2$. Hence f_1 is harmonic on $\mathcal{A}_1 \cup \mathcal{A}_2$ and then on M. Yet M is compact and connected hence f_1 must be a constant so that $f_1 = 0$. Thus V is a Killing vector field satisfying $QV = 2V$.

Let now V be a unit vector field not collinear to ξ such that $E(V) = E(\xi)$. Then $QV = 2V$. At each point $x \in M$ there is a local orthonormal φ-basis $\{E_0 = \xi,\ E_1,\ E_2 = \varphi E_1\}$ consisting of eigenvectors of the Ricci operator i.e., $QE_i = \lambda_i E_i$ for $i \in \{0,1,2\}$. Let us set $V_x = v_0 e_0 + v_1 e_1 + v_2 e_2$ where $e_i = E_i(x)$. By $W \geq 1$ and (4.37)

$$\begin{aligned} 2 &= \mathrm{Ric}(V_x, V_x) = \lambda_0 + v_1^2(\lambda_1 - \lambda_0) + v_2^2(\lambda_2 - \lambda_0) \\ &= 2 + 4v_1^2(W-1) + 4v_2^2(W-1) \geq 2. \end{aligned}$$

Then (as V_x and ξ_x are not collinear), $\lambda_2 = \lambda_1 = \lambda_0 = 0$ so that (M,g) has constant sectional curvature 1.

Conversely, if M has constant sectional curvature 1, then (by Theorem 3.14) one has $E(V) = E(\xi)$ for any unit Killing vector field V on M.

Proof of (ii). Let us assume that $W > 0$. Then $W_0 = \inf\{W(p) : p \in M\} > 0$ and we may consider the D-homothetic deformation

$$\phi_t = \phi, \ \ \xi_t = (1/t)\xi, \ \ \eta_t = t\eta, \ \ g_t = tg + (t^2 - t)\eta \otimes \eta,$$

with $t \in (0, W_0]$. Then (η_t, g_t) is a K-contact structure of scalar curvature $\rho_t = (\rho + 2)/t - 2$ and of Webster scalar curvature (by (4.38))

$$W_t = \frac{1}{8}(\rho_t + 2) = (1/t)W.$$

Moreover a calculation shows that $d\,\mathrm{vol}(g_t) = t^2\, d\,\mathrm{vol}(g)$ so that $E(\xi_t) = t^2 E(\xi)$. As $W \geq W_0 \geq t$ one has $W_t \geq 1$ and (ii) follows from (i).

Proof of (iii). Let us assume that $W = 0$. Then M is a Sasakian manifold of constant scalar curvature. By a result of Y. Watanabe, [306], M must be a (3-dimensional) locally ϕ-symmetric Sasakian manifold hence M is a (3-dimensional) locally homogeneous Sasakian manifold of Webster scalar curvature $W = 0$. Then (by Theorem 3.1 in [239]) it follows that M is locally isometric to the Heisenberg group $\mathbb{H}_1$ equipped with a left invariant Sasakian structure. Precisely M is the space of right cosets $\mathbb{H}_1/\Gamma$ for some discrete subgroup $\Gamma \subset \mathbb{H}_1$ and the Sasakian structure of $\mathbb{H}_1/\Gamma$ is induced by the left invariant Sasakian structure of $\mathbb{H}_1$. Then there is a globally defined orthonormal frame of the form $\{E_0 = \xi,\ E_1,\ E_2 = \phi E_1\}$ such that (by the identity (3.2) in [239] with $\lambda_2 = \lambda_3 = 0$)

$$\begin{cases} \nabla_\xi \xi = 0, & \nabla_\xi E_1 = -E_2, \quad \nabla_\xi E_2 = E_1, \\ \nabla_{E_1} \xi = -E_2, & \nabla_{E_1} E_1 = 0, \quad \nabla_{E_1} E_2 = \xi, \\ \nabla_{E_2} \xi = E_1, & \nabla_{E_2} E_1 = -\xi, \quad \nabla_{E_2} E_2 = 0. \end{cases} \tag{4.40}$$

Let $X \in \mathrm{Ker}(\eta)$ so that $X = f_1 E_1 + f_2 E_2$ for some $f_i \in C^\infty(M)$, $i \in \{1,2\}$. By (4.40)

$$\begin{cases} \nabla_\xi X = \left(\xi(f_1) + f_2\right) E_1 + \left(\xi(f_2) - f_1\right) E_2, \\ \nabla_{E_1} X = f_2 \xi + E_1(f_1) E_1 + E_1(f_2) E_2, \\ \nabla_{E_2} X = -f_1 \xi + E_2(f_1) E_1 + E_2(f_2) E_2. \end{cases}$$

Then

$$\begin{aligned} \|\nabla X\|^2 &= 2(f_1^2 + f_2^2) + \xi(f_1)^2 + \xi(f_2)^2 \\ &\quad + E_1(f_1)^2 + E_1(f_2)^2 + E_2(f_1)^2 + E_2(f_2)^2 + 2\left(f_2 \xi(f_1) - f_1 \xi(f_2)\right). \end{aligned} \tag{4.41}$$

By (4.40) one has $\mathrm{div}(E_i) = 0$, $i \in \{0,1,2\}$, hence

$$\mathrm{div}(\varphi_1 \varphi_2 E_i) = \varphi_1 \varphi_2 \mathrm{div}(E_i) + E_i(\varphi_1 \varphi_2) = \varphi_1 E_i(\varphi_2) + \varphi_2 E_i(\varphi_1)$$

for any $\varphi_1, \varphi_2 \in C^\infty(M)$. Thus

$$\int_M \varphi_1 E_i(\varphi_2)\, d\,\mathrm{vol}(g) = -\int_M \varphi_2 E_i(\varphi_1)\, d\,\mathrm{vol}(g). \tag{4.42}$$

By (4.42) and (4.40)

$$\begin{aligned} &\int_M \left(E_2(f_2) E_1(f_1) - E_1(f_2) E_2(f_1)\right) d\,\mathrm{vol}(g) \\ &\quad = \int_M \left(-f_2 E_2 E_1(f_1) + f_2 E_1 E_2(f_1)\right) d\,\mathrm{vol}(g) \\ &\quad = \int_M f_2 [E_1, E_2](f_1)\, d\,\mathrm{vol}(g) = \int_M \left(f_2 \xi(f_1) - f_1 \xi(f_2)\right) d\,\mathrm{vol}(g). \end{aligned}$$

Then (by (4.41))

$$\begin{aligned}(\mathrm{Hess}\,E)_\xi(X,X) &= \int_M \left(\|\nabla X\|^2 - \|\nabla\xi\|^2\|X\|^2\right)\,d\mathrm{vol}(g)\\ &= \int_M \left(\|\nabla X\|^2 - 2\|X\|^2\right)\,d\mathrm{vol}(g)\\ &= \int_M \left\{\left(E_1(f_2)-E_2(f_1)\right)^2 + \left(E_1(f_1)+E_2(f_2)\right)^2\right\}\,d\mathrm{vol}(g)\\ &\quad + \int_M \left\{\xi(f_1)^2+\xi(f_2)^2\right\}\,d\mathrm{vol}(g) \geq 0,\end{aligned}$$

that is ξ is stable. Theorem 4.39 is proved.

Theorem 4.40 *Let $(M,(\eta,g))$ be a compact generalized (k,μ)-space. If*

$$W \geq max\left\{\frac{3-k}{2}+\frac{\mu}{4}\sqrt{1-k},\ \frac{3-k}{2}-\frac{\mu}{4}\sqrt{1-k}\right\} \tag{4.43}$$

then the Reeb vector field ξ is stable. If additionally M is non-Sasakian (k,μ)-space then the requirement (4.43) *may be written $\mu \leq 2(k-2)/[1-\sqrt{1-k}] < 0$.*

For a proof of Theorem 4.40 one may see D. Perrone, [243]. As a consequence of Theorem 4.40 we may deal with the following (an example of a stable harmonic vector field which is not Killing)

Example 4.41 Let $\mathfrak{g}$ be a 3-dimensional Lie algebra. Let us consider a linear basis $\{E_1,E_2,E_3\}\subset\mathfrak{g}$ such that

$$[E_2,E_3]=\lambda_1 E_1,\quad [E_3,E_1]=\lambda_2 E_2,\quad [E_1,E_2]=\lambda_3 E_3, \tag{4.44}$$

for some $\lambda_i\in(0,+\infty)$, $i\in\{1,2,3\}$. Then the Lie group G associated to $\mathfrak{g}$ is the 3-sphere group SU(2) or the rotation group SO(3) $\approx$ SU(2)/$\{\pm I\}$ (cf. J. Milnor, [210], p. 307). Let g be the Riemannian metric on G defined by requiring that $\{E_1,E_2,E_3\}$ be orthonormal. Let $\{\eta^1,\eta^2,\eta^3\}$ be the dual frame i.e., $(\theta^i)^\sharp = E_i$, $i\in\{1,2,3\}$. As $\lambda_i\neq 0$ each η^i is a contact form and E_i is the corresponding Reeb field. We may assume $\lambda_1=2$ and set by definition

$$\phi(E_1)=0,\ \phi(E_2)=E_3,\ \phi(E_3)=-E_2\,.$$

Then $d\eta^1 = g(\cdot, \phi\cdot)$ so that $(\phi,\ \xi = E_1,\ \eta = \eta^1,\ g)$ is a contact metric structure on G. Moreover $h = \frac{1}{2}\mathcal{L}_\xi \phi$ and the curvature tensor field R satisfy (cf. D.E. Blair & T. Koufogiorgos & B.J. Papantoniou, [46])

$$hE_2 = \lambda E_2, \quad hE_3 = -\lambda E_3$$

$$R(X,Y)Z = k\{\eta(X)Y - \eta(Y)X\} + \mu\{\eta(X)hY - \eta(Y)hX\},$$

$$\lambda = \frac{\lambda_3 - \lambda_2}{2}, \quad k = 1 - \frac{(\lambda_3 - \lambda_2)^2}{4} = 1 - \lambda^2, \quad \mu = 2 - \lambda_2 - \lambda_3.$$

Given $0 < \lambda < 1$ let us consider a constant μ such that $\mu \le 2(1+\lambda)^2/(\lambda - 1) < 0$ and set

$$\lambda_1 = 2, \quad \lambda_2 = 1 - \lambda - \frac{\mu}{2} > 0, \quad \lambda_3 = 1 + \lambda - \frac{\mu}{2} > 0. \tag{4.45}$$

The Lie algebra structure (4.44) with λ_i, $i \in \{1,2,3\}$, given by (4.45) gives rise to a contact metric structure on G which organizes G as a non-Sasakian (k,μ)-space satisfying the requirement (4.43) in Theorem 4.40 above. Thus *ξ is a stable harmonic non-Killing vector field.*

4.3.2. Stability of Strongly Normal Reeb Vector Fields

Definition 4.42 (J.C. Gonzàles-Dàvila & L. Vanhecke, [144]) A unit vector field V on a Riemannian manifold (M,g) is said to be *strongly normal* if

$$g(\nabla_X(\nabla V)Y, Z) = 0$$

for any $X, Y, Z \in \mathfrak{X}(M)$ orthogonal to V. ■

Contact metric 3-dimensional manifolds whose Reeb vector field is strongly normal are H-contact ([241], Theorem 1.2). In the sequel, we study the stability of the Reeb vector field ξ under the additional assumptions that ξ is strongly normal and the manifold is non-Sasakian (i.e., $\tau \neq 0$). The discussion is in terms of $p = 4\sqrt{2}W/\|\tau\|$ and relies on the classification in Theorem 4.24.

Theorem 4.43 *Let $(M,(\eta,g))$ be a compact non-Sasakian 3-dimensional manifold whose Reeb vector field ξ is strongly normal.*

i. *If $-1 < p < 1 + 4\sqrt{2}/\|\tau\|$ then ξ is E-unstable.*

ii. *If $p \ge 1 + 4\sqrt{2}/\|\tau\|$ and $t > (p-1)\|\tau\|/[4\sqrt{2}]$ then under D-homothetic deformation (η_t, g_t) of (η, g) the Reeb vector field $\xi_t = (1/t)\xi$ is E-unstable.*

The case $p \le -1$ is open as yet. To prove Theorem 4.43, let $(M,(\eta,g))$ be a non-Sasakian contact metric 3-dimensional manifold such that the underlying Reeb vector ξ is strongly normal. By Theorem 4.24 the manifold M must be the space of right cosets G/Γ where G is a unimodular Lie group and $\Gamma \subset G$ a discrete subgroup and the contact metric structure of G/Γ is induced by a left invariant contact metric structure on G. Also G is $\widetilde{E(2)}$ (the universal covering of the group of rigid motions of the Euclidean 2-space) when $p = 1$, the 3-sphere group SU(2) when $p > 1$, the group $\widetilde{\mathrm{SL}(2,\mathbb{R})}$ when $-1 \neq p < 1$, or the group $E(1,1)$ (the group of rigid motions of the Minkowski 2-space) when $p = -1$. There is a globally defined orthonormal frame of the form $\{E_0 = \xi,\ E_1,\ E_2 = \phi E_1\}$ consisting of eigenvectors of h i.e., $hE_1 = \lambda_1 E_1$ and $hE_2 = -\lambda E_2$ with $\lambda \in (0, +\infty)$ such that (cf. (3.2) in D. Perrone, [239], for $2 - \lambda_2 - \lambda_3 = 2a$)

$$\left(\nabla_{E_i} E_j\right)_{0 \le i,j \le 2} = \begin{pmatrix} 0 & -aE_2 & aE_1 \\ -(\lambda+1)E_2 & 0 & (\lambda+1)\xi \\ (1-\lambda)E_1 & (\lambda-1)\xi & 0 \end{pmatrix}, \tag{4.46}$$

for some $a \in \mathbb{R}$. Then (by (4.46))

$$(\mathrm{Ric}(E_i, E_i))_{0 \le i \le 2} = \left(2(1-\lambda^2),\ 2a(\lambda - 1),\ -2a(\lambda+1)\right),$$

so that

$$\rho = \mathrm{trace}_g\,\mathrm{Ric} = 4(1-a) - 2(1+\lambda^2).$$

Moreover

$$\|\tau\| = 2\sqrt{2}\lambda, \quad W = \frac{1-a}{2}, \quad p = \frac{2W}{\lambda}.$$

By (4.2) one finds $\|\nabla\xi\|^2 = 2 + \|\tau\|^2/4$ hence

$$\begin{aligned}(\mathrm{Hess}\, E)_\xi(X, X) &= \int_M \left(\|\nabla X\|^2 - \|X\|^2\|\nabla\xi\|^2\right) d\,\mathrm{vol}(g) \\ &= \int_M \left(\|\nabla X\|^2 - (2 + 2\lambda^2)\|X\|^2\right) d\,\mathrm{vol}(g), \quad X \in \mathrm{Ker}(\eta).\end{aligned} \tag{4.47}$$

Let $(x_1, x_2) \in \mathbb{R}^2$ and $X = x_1 E_1 + x_2 E_2 \in \mathrm{Ker}(\eta)$. Then (by (4.46))

$$\begin{cases} \nabla_\xi X = -(1-2W)x_1 E_2 + (1-2W)x_2 E_1, \\ \nabla_{E_1} X = (1+\lambda)x_2\xi, \\ \nabla_{E_2} X = (\lambda - 1)x_1\xi, \end{cases} \tag{4.48}$$

hence

$$\|\nabla X\|^2 = \left((1-2W)^2 + (\lambda-1)^2\right)x_1^2 + \left((1-2W)^2 + (\lambda+1)^2\right)x_2^2.$$

Then

$$\|\nabla X\|^2 - 2(1+\lambda^2)\|X\|^2 = Ax_1^2 + Bx_2^2$$

where

$$A = \frac{\|\tau\|^2}{8}(p+1)\left(p-1-\frac{4\sqrt{2}}{\|\tau\|}\right), \quad B = \frac{\|\tau\|^2}{8}(p-1)\left(p+1-\frac{4\sqrt{2}}{\|\tau\|}\right).$$

If $B \geq 0$ and $A < 0$ we may choose $(x_1, x_2) \in \mathbb{R}^2$ such that $x_1^2 > -(B/A)x_2^2$ so that $Ax_1^2 + Bx_2^2 < 0$. Then (by (4.47)) ξ is unstable when $A < 0$. Let us observe that $-1 < p < 1 + 4\sqrt{2}/\|\tau\|$ implies $A < 0$ so that statement (i) in Theorem 4.43 is proved. If $p \geq 1 + 4\sqrt{2}/\|\tau\|$ then we may consider $t > (p-1)\|\tau\|/[4\sqrt{2}] > 0$ and

$$\phi_t = \phi, \quad \xi_t = (1/t)\xi, \quad \eta_t = t\eta, \quad g_t = tg + t(t-1)\eta\otimes\eta,$$

so that (η_t, g_t) is a contact metric structure whose underlying Reeb vector field ξ_t is strongly normal. Indeed ξ_t is harmonic (by Remark 4.6) and $\|\tau_t\| = (1/t)\|\tau\| =$ constant hence ξ_t is strongly normal. A calculation shows that (cf. [141])

$$\rho_t = \frac{\rho+2}{t} - 2 + (t-1)\frac{\|\tau\|^2}{4t^2}$$

hence $W_t = (1/t)W$. Therefore

$$p_t = p < 1 + \frac{4\sqrt{2}}{\|\tau\|}t = 1 + \frac{4\sqrt{2}}{\|\tau_t\|}$$

hence (by part (i) in Theorem 4.43) ξ_t is unstable.

Corollary 4.44 *Let $(M, (\eta, g))$ be a compact contact 3-dimensional manifold whose Reeb vector field ξ is strongly normal. If $W = 0$ then either* i) *the universal covering space $\tilde{M}$ is the Heisenberg group $\mathbb{H}_1$ and ξ is a stable harmonic Killing vector field, or* ii) *the universal covering space $\tilde{M}$ is the group $\widetilde{\mathrm{SL}(2,\mathbb{R})}$ and ξ is an unstable harmonic non-Killing vector field.*

Corollary 4.44 follows from part (iii) in Theorem 4.39 (when $\tau = 0$) and from part (i) in Theorem 4.43 (when $\tau \neq 0$).

G. Wiegmink has shown that on each 2-dimensional Riemannian torus $(\mathbb{T}^2, g)$ the total bending $\mathcal{B}$ functional achieves its minimum (cf. Theorem 4 in [309] or Theorem 2.43 in Chapter 2 of this monograph) and all the critical points of $\mathcal{B}$ are stable (cf. Theorem 6 in [309] or Theorem 3.51 in Chapter 3 of this monograph).

The situation is considerably different in dimension three. The 3-dimensional torus $\mathbb{T}^3$ admits a natural flat contact metric structure (cf. [42], p. 68 and p. 100). As $\mathbb{T}^3$ is Ricci flat its Reeb vector field ξ is harmonic. Also $0 = \mathrm{Ric}(\xi,\xi) = 2 - 2\lambda^2$ hence $\lambda = 1$. Thus ξ is strongly normal and $2W = 1$ so that $p = 1$. Therefore (by part (i) in Theorem 4.43) ξ is unstable. Moreover (by (4.46)) the vector field E_1 (spanning the eigenbundle of h corresponding to the eigenvalue 1) is parallel. Hence $E(E_1) = (1/2)\,\mathrm{Vol}(\mathbb{T}^3)$ and for any $V \in \Gamma^\infty(S(\mathbb{T}^3))$

$$E(V) = \frac{1}{2}\mathrm{Vol}(\mathbb{T}^3) + \frac{1}{2}\int_{\mathbb{T}^3} \|\nabla V\|^2 \,\mathrm{vol}(g) \geq \frac{1}{2}\mathrm{Vol}(\mathbb{T}^3) = E(E_1).$$

We may conclude that

Corollary 4.45 *The Reeb vector field underlying the natural flat contact metric structure on $\mathbb{T}^3$ is an unstable critical point and the energy functional achieves its minimum*

$$\mathit{inf}\{E(V) : V \in \Gamma^\infty(S(\mathbb{T}^3))\} = \frac{1}{2}\mathrm{Vol}(\mathbb{T}^3).$$

Remark 4.46

a. J.C. Gonzàles-Dàvila & L. Vanhecke studied (cf. [146]) the stability of harmonic unit vector fields and the existence of absolute minima for the energy functional on 3-dimensional manifolds and in particular on compact quotients of unimodular Lie groups.

b. Let (M,g) be a Riemannian manifold and V a unit vector field on M. We recall that $V : (M,g) \to (S(M), G_s)$ is a harmonic map, i.e., a critical point of $E : C^\infty(M, S(M)) \to [0,+\infty)$, if and only if V is a harmonic vector field and $\mathrm{trace}_g\{R(\nabla . V, V)\cdot\} = 0$ (cf. Theorem 2.19). Clearly instability with respect to $E : \Gamma^\infty(S(M)) \to [0,+\infty)$ implies instability with respect to $E : C^\infty(M, S(M)) \to [0,+\infty)$. For the Reeb vector field ξ of a contact metric 3-dimensional manifold as in Theorem 4.43 one checks easily (by (4.46)) that $\mathrm{trace}_g\{R(\nabla . \xi, \xi)\cdot\} = 0$. Then all instability results proved in this section may be reformulated for ξ as a harmonic map of (M,g) into $(S(M), G_s)$. ■

4.4. HARMONIC ALMOST CONTACT STRUCTURES

The purpose of this section is to report briefly on recent results by E. Vergara-Diaz & C.M. Wood, [299], on the harmonicity of an almost contact structure.

Let (M,g) be a Riemannian manifold and $P \to M$ a principal G-bundle. Let $H \subset G$ be a Lie subgroup such that the homogeneous space G/H is reductive. Let us assume that G/H is equipped with a G-invariant Riemannian metric and that $P \to M$ is endowed with a connection Γ. Let

$$N = P \times_G (G/H) = (P \times (G/H))/G \approx P/H$$

be the associated bundle with standard fibre G/H and let Γ^* be the associated connection so that (cf. e.g., R. Crittenden, [92])

$$T_w(N) = \Gamma^*_w \oplus \mathrm{Ker}(d_w \rho), \quad w \in N,$$

where $\rho : N \to M$ is the natural projection. Let $g^\uparrow$ be the horizontal lift of g i.e.,

$$g^\uparrow_w(A,B) = g_{\rho(w)}((d_w\rho)A, (d_w\rho)B), \quad A, B \in \Gamma^*_w, \ w \in N.$$

Let h be the Riemannian metric on N defined as the product of $g^\uparrow$ and the fibre metric on G/H.

Definition 4.47 (C.M. Wood, [318]) A section $\sigma : M \to N$ is said to be *harmonic* if σ is a critical point of the energy functional $E : \Gamma^\infty(N) \to [0, +\infty)$

$$E(\sigma) = \frac{1}{2} \int_\Omega \|d\sigma\|^2 d\,\mathrm{vol}(g), \quad \sigma \in \Gamma^\infty(N),$$

for any relatively compact domain $\Omega \subset M$, i.e., $\{dE(\sigma_t)/dt\}_{t=0} = 0$ for any smooth 1-parameter variation $\{\sigma_t\}_{|t|<\epsilon}$ of σ (i.e., $\sigma_0 = \sigma$) through smooth sections $\sigma_t \in \Gamma^\infty(N)$, $|t| < \epsilon$, supported in Ω. ■

The Euler-Lagrange equations of the variational principle $\delta E(\sigma) = 0$ were derived by C.M. Wood, [318].

The problem of studying harmonicity of smooth sections in an associated bundle with an *arbitrary* standard fibre F (equipped with a Riemannian metric) is open. A covariant derivative of such sections is well defined (cf. R. Crittenden, [92]) relative to a given connection Γ in $P \to M$. See also S. Dragomir, [100].

In particular, let $G = \mathrm{SO}(2n+1)$ and let $P \to M$ be the principal $\mathrm{SO}(2n+1)$-bundle of all positively oriented orthonormal frames tangent to a real $(2n+1)$-dimensional Riemannian manifold (M,g). An almost contact metric structure (ϕ,ξ,η,g) on M may be thought of as a C^∞ section σ in the associated bundle $N = P/\mathrm{U}(n) \to M$ with standard fibre $F = \mathrm{SO}(2n+1)/U(n)$. The harmonicity of such σ was investigated by E. Vergara-Diaz & C.M. Wood, [299]. We may state

Theorem 4.48 *Let (ϕ,ξ,η,g) be an almost contact metric structure on M and $H(M) = (\mathbb{R}\xi)^\perp$. Let $J : H(M) \to H(M)$ be the restriction of ϕ to $H(M)$. Let $\sigma \in \Gamma^\infty(P/\mathrm{U}(n))$ be the section associated to (ϕ,ξ,η,g). Then σ is harmonic if and only if $[D^*DJ,J] = 0$ (that is, J is a harmonic almost complex structure in H, in the sense of [320]) and*

$$\Delta_g \xi = \|\xi\|^2 \xi - \frac{1}{2} J\, T(\phi)$$

where $T(\phi) = \mathrm{trace}_g\{DJ \otimes \nabla\xi\}$ and D is the connection in $H(M) \to M$ defined as the orthogonal projection of ∇ on $H(M)$. In particular, if both ξ and J are harmonic then σ is harmonic if and only if $T(\phi) = 0$.

Therefore the study of the harmonicity of almost contact metric structures σ is related to the study of H-contact structures. When $(H(M),J,D)$ is a *Kaehler bundle* (i.e., the Hermitian structure induced by g on $H(M)$ is parallel) it may be shown that σ is a harmonic section (respectively a harmonic map) if and only if ξ is a harmonic vector field (respectively a harmonic map). Two instances where $(H(M),J,D)$ is a Kaehler bundle are indicated in the following

Theorem 4.49 (E. Vergara-Diaz & C.M. Wood, [299]) *Let us assume that* a) *M is a 3-dimensional almost contact metric manifold, or* b) *M is an oriented real hypersurface in a Kaehler manifold, equipped with the induced almost contact metric structure. Then $(H(M),J,D)$ is a Kaehler bundle.*

Another recent result is

Theorem 4.50 (E. Vergara-Diaz & C.M. Wood, [300]) *An H-contact structure on M is harmonic if and only if the ρ^*-Ricci curvature of M is symmetric (equivalently ϕ-invariant), where $\rho^*(X,Y) = g(R(X,Ei)\phi Y, \phi E_i)$ for any $X, Y \in T(M)$, and R is the Riemann curvature tensor of (M,g).*

The (κ, μ)-manifolds are H-contact [242]. To provide an example, one may show that all contact metric structures satisfying the (κ, μ)-nullity condition (4.3) are harmonic (that is the ρ^*-Ricci curvature is symmetric), cf. [300]. Such structures include unit tangent bundles over spaces of constant curvature.

A new family of examples of harmonic unit vector fields is provided by the Reeb vector fields of the warped products $\mathbb{R} \times_f \tilde{M}$ and $\tilde{M} \times_f \mathbb{R}$ carrying the almost contact metric structure induced by an almost Hermitian structure on $\tilde{M}$ (cf. [300]). A Kenmotsu manifold is locally isometric to a warped product $\mathbb{R} \times_f \tilde{M}$ where $f(t) = ce^t$ $(c \in \mathbb{R})$ and $\tilde{M}$ is Kählerian (cf. [42], p. 80).

4.5. REEB VECTOR FIELDS ON REAL HYPERSURFACES

Let M be an orientable real hypersurface in a complex $(n+1)$-dimensional Kaehler manifold (M_0, J_0, g_0), where J_0 is the complex structure and g_0 the Kaehlerian metric. As customary $M_0 = M^{n+1}(c)$ denotes a complex $(n+1)$-dimensional complex space form of (constant) holomorphic sectional curvature c. Let ν be a unit normal vector field on M. Then $g_0(J_0\nu, \nu) = 0$ hence $\xi = J_0\nu$ is a unit vector field[1] tangent to M. Let us set

$$J_0 X = \phi X - \eta(X)\nu, \quad X \in T(M),$$

so that ϕ and η are respectively a $(1,1)$-tensor field and a differential 1-form on M. Let g be the first fundamental form of M in M_0 i.e., $g = \iota^* g_0$ where $\iota : M \to M_0$ is the inclusion. Then (ϕ, ξ, η, g) is an almost contact metric structure on M. Let $A = A_\nu$ be the shape operator associated to the normal section ν.

Definition 4.51 Let M be an orientable real hypersurface of a Kaehler manifold with the naturally induced almost contact metric structure (ϕ, ξ, η, g). We call $(M, (\phi, \xi, \eta, g))$ a *Hopf hypersurface* if the Reeb vector field is an eigenvector of the shape operator A i.e., $A\xi = \alpha\xi$ for some $\alpha \in C^\infty(M)$. ■

We recall the Gauss and Weingarten formulae

$$\nabla^0_X Y = \nabla_X Y + B(X, Y)\nu, \quad \nabla^0_X \nu = -AX,$$

[1] According to the conventions adopted in [290]–[291] the Reeb vector underlying the naturally induced almost contact metric structure on M is $-\xi$.

for any $X, Y \in \mathfrak{X}(M)$, where ∇^0 and ∇ are the Levi-Civita connections of (M_0, g_0) and (M, g) while $B(X, Y) = g(AX, Y)$ is the second fundamental form of the given immersion. Then

$$(\nabla_X \phi) Y = g(AX, Y)\xi - \eta(Y)AX, \quad \nabla_X \xi = -\phi AX.$$

As $\nabla_\xi \xi = -\phi A\xi$ it follows that $\nabla_\xi \xi = 0$ if and only if $A\xi = \alpha\xi$ for some $\alpha \in C^\infty(M)$. We recall the Gauss-Codazzi-Ricci equations

$$R(X, Y)Z = \tan\{R_0(X, Y)Z\} + g(AX, Z) - g(AY, Z)AX,$$
$$R_0(X, Y)\nu = (\nabla_X A)Y - (\nabla_Y A)X,$$

where R_0 and R are the curvature tensor fields of ∇^0 and ∇. As usual $\tan_x : T_x(M_0) \to T_x(M)$ is the projection associated to the direct sum decomposition $T_x(M_0) = [(d_x\iota)T_x(M)] \oplus \mathbb{R}\nu_x$ for any $x \in M$. Let $\{E_i : 1 \le i \le 2n+1\}$ be a local orthonormal frame of $T(M)$. By the Ricci equation

$$g(R_0(X, E_i)\nu, E_i) = g((\nabla_X A)E_i, E_i) - g((\nabla_{E_i} A)X, E_i)$$

or (by summing over $1 \le i \le 2n+1$)

$$\mathrm{Ric}_0(X, \nu) = \sum_{i=1}^{2n+1} \{g((\nabla_X A)E_i, E_i) - g((\nabla_{E_i} A)X, E_i)\} \tag{4.49}$$
$$= X(H) - (\operatorname{div} A)\, X,$$

where Ric_0 is the Ricci tensor field of (M_0, g_0) and $H = \mathrm{trace}(A) \in C^\infty(M)$ is the mean curvature of M in M_0. Also the 1-form $\operatorname{div} A \in \Omega^1(M)$ is given by

$$(\operatorname{div} A)X = g(\mathrm{trace}_g \nabla A, X), \quad X \in \mathfrak{X}(M).$$

Let ∇u be the gradient of $u \in C^1(M)$ with respect to g i.e., locally

$$\nabla u = \sum_{i=1}^{2n+1} E_i(u)E_i.$$

Let $\nabla^* : \Gamma^\infty(T^*(M) \otimes T(M)) \to \mathfrak{X}(M)$ be the formal adjoint of $\nabla : \mathfrak{X}(M) \to \Gamma^\infty(T^*(M) \otimes T(M))$. The shape operator A is a C^∞ section in the vector bundle $T^*(M) \otimes T(M) \to M$. Then (locally)

$$\nabla^* A = -\sum_{i=1}^{2n+1} (\nabla_{E_i} A)E_i = \mathrm{trace}_g \nabla A$$

so that $(\operatorname{div} A)^\sharp = -\nabla^* A$ i.e., the 1-form $\operatorname{div} A$ and the tangent vector field $-\nabla^* A$ are dual with respect to g. Throughout $\sharp : \Omega^1(M) \to \mathfrak{X}(M)$ denotes "raising of indices" with respect to g i.e., $g(\omega^\sharp, X) = \omega(X)$ for any $\omega \in \Omega^1(M)$ and any $X \in \mathfrak{X}(M)$. By the Gauss-Codazzi equations we get

$$\begin{aligned}\operatorname{Ric}(X,Y) &= \operatorname{Ric}_0(X,Y) \\ &\quad - g_0(R_0(\nu,X)Z,\nu) - g(AX,AZ) + Hg(AX,Z) \qquad (4.50)\end{aligned}$$

and in particular

$$Q\xi = Q_0\xi - g(Q_0\xi,\nu)\nu + R_0(\nu,\xi)\nu - A^2\xi + HA\xi, \qquad (4.51)$$

where Q_0 is the Ricci operator of (M_0, g_0).

4.5.1. The Rough Laplacian and Criteria of Harmonicity

Let Δ_g be the rough Laplacian on $\mathfrak{X}(M)$. We wish to compute $\Delta_g\xi$. Let $\{E_i : 1 \le i \le 2n+1\}$ be a local orthonormal frame of $T(M)$ consisting of eigenvectors of the shape operator. Then

$$\Delta_g\xi = -\sum_{i=1}^{2n+1} \{\nabla_{E_i}\nabla_{E_i}\xi - \nabla_{\nabla_{E_i}E_i}\xi\}.$$

Moreover we consider a local orthonormal frame of the form

$$\{\xi, V_\alpha, \phi V_\alpha : 1 \le \alpha \le n\}$$

with $V_\alpha \in H(M) = \operatorname{Ker}(\eta)$ for any $1 \le \alpha \le n$. Without loss of generality, we may assume that both local frames are defined on the same open set. Then

$$\Delta_g\xi = \sum_{\alpha=1}^{n} \left\{g(\Delta_g\xi, V_\alpha)V_\alpha + g(\Delta_g\xi, \phi V_\alpha)\phi V_\alpha\right\} + g(\Delta_g\xi,\xi)\xi. \qquad (4.52)$$

Note that (as a consequence of $\|\xi\| = 1$)

$$g(\nabla_{\nabla_{E_i}E_i}\xi, \xi) = 0.$$

Hence

$$\begin{aligned}g(\Delta_g\xi,\xi) &= -\sum_i g(\nabla_{E_i}\nabla_{E_i}\xi,\xi) \\ &= -\sum_i \left\{E_i\left(\frac{1}{2}E_i(\|\xi\|^2)\right) - \|\nabla_{E_i}\xi\|^2\right\} = \|\nabla\xi\|^2.\end{aligned}$$

Let us compute the remaining terms in (4.52). As $\nabla_X \xi = -\phi A X$

$$g(\nabla_{E_i}\nabla_{E_i}\xi, V) = g(\nabla_{E_i}(-\phi A)E_i, V)$$
$$= -g((\nabla_{E_i}\phi A)E_i + \phi A \nabla_{E_i} E_i, V)$$
$$= -g((\nabla_{E_i}\phi A)E_i - \nabla_{\nabla_{E_i} E_i}\xi, V)$$

hence

$$g(\nabla_{E_i}\nabla_{E_i}\xi - \nabla_{\nabla_{E_i} E_i}\xi, V) = -g((\nabla_{E_i}\phi A)E_i, V) \tag{4.53}$$

for any tangent vector field V on M. Using (4.53) for $V \in \{V_\alpha, \phi V_\alpha\}$ the identity (4.52) may be written

$$\Delta_g \xi = \|\nabla \xi\|^2 \xi + \sum_{i,\alpha} \left\{ g((\nabla_{E_i}\phi A)E_i, V_\alpha) V_\alpha + g((\nabla_{E_i}\phi A)E_i, \phi V_\alpha)\phi V_\alpha \right\}. \tag{4.54}$$

As the adjoint (with respect to g) of the operator $\nabla_X \phi A$ is precisely $-\nabla_X A\phi$ one has

$$g((\nabla_{E_i}\phi A)E_i, V) = -g(E_i, (\nabla_{E_i} A\phi) V)$$
$$= -g(E_i, \nabla_{E_i} A\phi V - A\phi \nabla_{E_i} V)$$

(by adding and subtracting $A\nabla_{E_i}\phi V$)

$$= -g(E_i, \nabla_{E_i} A\phi V - A\nabla_{E_i}\phi V + A\nabla_{E_i}\phi V - A\phi\nabla_{E_i} V)$$

so that

$$g((\nabla_{E_i}\phi A)E_i, V) = -g(E_i, (\nabla_{E_i} A)\phi V + A(\nabla_{E_i}\phi) V). \tag{4.55}$$

At this point, one may apply (4.55) for $V \in \{V_\alpha, \phi V_\alpha\}$ so that (4.54) may be written as

$$\Delta_g \xi = \|\nabla \xi\|^2 \xi - (f_1 + f_2) \tag{4.56}$$

where

$$f_1 = \sum_{i,\alpha} \left\{ g(E_i, (\nabla_{E_i} A)\phi V_\alpha) V_\alpha - g(E_i, (\nabla_{E_i} A) V_\alpha)\phi V_\alpha \right\},$$

$$f_2 = \sum_{i,\alpha} \left\{ g(E_i, A(\nabla_{E_i}\phi) V_\alpha) V_\alpha + g(E_i, A(\nabla_{E_i}\phi)\phi V_\alpha)\phi V_\alpha \right\}.$$

We shall need the following

Lemma 4.52 (D. Perrone, [247]) *Let M be an orientable real hypersurface in a Kaehler manifold M_0. Then*

$$\begin{aligned}\Delta_g\xi &= -\phi\nabla^* A+\phi^2A^2\xi+\|\nabla\xi\|^2\xi\\ &= -\phi\nabla H+\pi_H\, Q_0\xi+\phi^2A^2\xi+\|\nabla\xi\|^2\xi \end{aligned}\tag{4.57}$$

where $\pi_H : T(M)\to \mathrm{Ker}(\eta)$ is the natural projection associated to the decomposition $T(M)=\mathrm{Ker}(\eta)\oplus\mathbb{R}\xi$.

Proof. Using (4.49) one may compute f_1 and find

$$f_1 = -\phi\nabla H+\pi_H JQ_0\nu. \tag{4.58}$$

Moreover (again by (4.49)) $g_0(Q_0\nu, X) = X(\mathrm{trace}\,A)-(\mathrm{div}\,A)\,X$ hence

$$\begin{aligned}\pi_H JQ_0\nu &= \sum_{\alpha=1}^{n}\{g_0(JQ_0\nu,\, V_\alpha)V_\alpha+g_0(JQ_0\nu,\,\phi V_\alpha)\phi V_\alpha\}\\ &= \sum_\alpha\{-g_0(Q_0\nu,\phi V_\alpha)V_\alpha+g_0(Q_0\nu,\,V_\alpha)\phi V_\alpha\}\\ &= \sum_\alpha\{[V_\alpha(\mathrm{trace}\,A)-(\mathrm{div}\,A)(V_\alpha)]\phi V_\alpha\\ &\quad -[(\phi V_\alpha)(\mathrm{trace}\,A)-(\mathrm{div}\,A)(\phi V_\alpha)]V_\alpha\}\\ &= \phi\nabla H-\sum_\alpha \phi\{(\mathrm{div}\,A)(V_\alpha)V_\alpha+(\mathrm{div}\,A)(\phi V_\alpha)\phi V_\alpha\\ &\quad +(\mathrm{div}\,A)(\xi)\xi\} = \phi\nabla H-\phi\,\mathrm{trace}\,\nabla A.\end{aligned}$$

Thus (4.58) becomes

$$f_1 = -\phi\,\mathrm{trace}\,\nabla A = \phi\,\nabla^* A. \tag{4.59}$$

Let us compute f_2. One has

$$\begin{aligned} f_2 &= \sum_{i=1}^{2n+1}\sum_{\alpha=1}^{n}\{g(A(\nabla_{E_i}\phi)V_\alpha, E_i)V_\alpha+g(A(\nabla_{E_i}\phi)\phi V_\alpha, E_i)\phi V_\alpha\}\\ &= \sum_i\sum_\alpha\{g(AE_i,\,V_\alpha)g(AE_i,\xi)V_\alpha+g(AE_i,\phi V_\alpha)g(AE_i,\xi)\phi V_\alpha\}\\ &= \sum_\alpha\{g(AV_\alpha, A\xi)V_\alpha+g(A\phi V_\alpha, A\xi)\phi V_\alpha\}\\ &= \sum_\alpha\{g(A^2\xi,\,V_\alpha)V_\alpha+g(A^2\xi,\phi V_\alpha)\phi V_\alpha\}\end{aligned}$$

or

$$f_2 = -\phi^2 A^2 \xi. \tag{4.60}$$

Finally the identities (4.56)–(4.60) yield (4.57) in Lemma 4.52. ■

We may state

Theorem 4.53 (K. Tsukada & L. Vanhecke, [290], D. Perrone, [247]) *Let M be an orientable real hypersurface of a Kaehler manifold. Then*

i. *the Reeb vector field ξ is harmonic if and only if $\phi\,\nabla^* A - \phi^2 A^2 \xi = 0$.*

ii. *Let us assume that* $\operatorname{div} A = 0$. *Then ξ is harmonic if and only if ξ is an eigenvector of A^2.*

iii. *Let us assume that M is a Hopf surface. Then ξ is harmonic if and only if $\nabla^* A$ is proportional to ξ.*

Theorem 4.54 (D. Perrone, [247]) *Let M be an orientable Hopf hypersurface of a Kaehler manifold M_0. Let us assume that the unit normal vector ν on M is an eigenvector of Q_0. Then*

i. *ξ is harmonic if and only if $\nabla H = \xi(H)\xi$.*

ii. *Let us assume that the principal curvature α corresponding to the principal direction ξ is constant along the integral curves of ξ. Then ξ is harmonic if and only if the mean curvature H is constant.*

Proof. Under the assumptions in Theorem 4.54 (i.e., M is a Hopf surface and ξ is an eigenvector of the Ricci operator Q_0) the identity (4.57) in Lemma 4.52 may be written as

$$\Delta_g \xi = \phi\,\nabla H + \|\nabla \xi\|^2 \xi. \tag{4.61}$$

Statement (i) in Theorem 4.54 follows from (4.61). To prove (ii) let α be constant along the integral curves of ξ. Let $\{E_i : 1 \le i \le 2n+1\}$ be a local orthonormal frame of $T(M)$ consisting of eigenvectors of the shape operator A i.e., $AE_i = \lambda_i E_i$. Also we assume that $E_1 = \xi$. As ν is an eigenvector of Q_0

$$\sum_{i=2}^{2n+1} g_0(R_0(\xi, E_i)\nu, E_i) = \operatorname{Ric}_0(\xi, \nu) = 0.$$

By $\nabla_X \xi = -\phi A X$ and the Codazzi equation

$$0 = \sum_{i=2}^{2n+1} g((\nabla_\xi A)E_i - (\nabla_{E_i} A)\xi, E_i)$$

$$= \sum_{i=2}^{2n+1} g(\xi(\lambda_i)E_i + \lambda_i \nabla_\xi E_i - E_i(\alpha)\xi + \alpha \nabla_{E_i}\xi, E_i)$$

$$+ \sum_{i=2}^{2n+1} (-\nabla_\xi E_i + \nabla_{E_i}\xi, AE_i)$$

$$= \sum_{i=2}^{2n+1} \{\xi(\lambda_i)\alpha g(-\phi AE_i, E_i) + g(-\phi AE_i, AE_i) + \lambda_i g(\nabla_\xi E_i, E_i)\}$$

$$= \xi\left(\sum_{i=2}^{2n+1} \lambda_i\right) = \xi(H - \alpha) = \xi(H).$$

Then (by (4.61)) $\Delta_g \xi$ is collinear to ξ if and only if H is constant. ■

As a consequence of Theorem 4.54

Corollary 4.55 (K. Tsukada & L. Vanhecke, [290], D. Perrone, [247]) *Let M be an orientable Hopf hypersurface of a complex space form $M^{n+1}(c)$ with $c \neq 0$. Then ξ is a harmonic map of (M,g) into $(S(M), G_s)$ if and only if the mean curvature H of M in $M^{n+1}(c)$ is constant.*

Proof. Hopf hypersurfaces of a complex space form $M^{n+1}(c)$ of sectional curvature $c \neq 0$ enjoy the property that $A\xi = \alpha\xi$ for some $\alpha \in \mathbb{R}$ (cf. e.g., K. Tsukada & L. Vanhecke, [291]). Moreover

$$\tan\{R_0(X,Y)Z\} = \frac{c}{4}\{g(X,Z)Y - g(Y,Z)X - g(\phi Y, Z)\phi X$$
$$+ g(\phi X, Z)\phi Y - 2g(X, \phi Y)\phi Z\}. \qquad (4.62)$$

Let $\{E_1 = \xi, E_i : 2 \leq i \leq 2n+1\}$ be a local orthonormal frame of $T(M)$ consisting of eigenvectors of A. Then (by the Gauss equation)

$$\operatorname{trace}_g\{R(\nabla.\xi, \xi)\cdot\} = \sum_{i=2}^{2n+1} R(\nabla_{E_i}\xi, \xi)E_i = \sum_{i=2}^{2n+1} R(\phi AE_i, \xi)E_i$$

$$= \sum_{i=2}^{2n+1} \{g(A\phi AE_i, E_i)A\xi - g(AE_i, \xi)A\phi AE_i\} = 0.$$

Then (by Theorem 4.54), ξ is a harmonic map if and only if H is constant. ■

Let $\tau = \mathcal{L}_\xi g$ and let T be the symmetric $(1,1)$-tensor field on M defined by $g(TX, Y) = \tau(X, Y)$ for any $X, Y \in \mathfrak{X}(M)$.

Proposition 4.56 *Let M an orientable real hypersurface of a Kaehler manifold. Then*

$$T = [A, \phi], \quad \operatorname{div}\xi = \frac{1}{2} \operatorname{trace} T = 0.$$

Moreover $T\xi = 0$ if and only if M is a Hopf hypersurface. Also if $TX = 0$ for any $X \in \operatorname{Ker}(\eta)$ then $T = 0$.

Proof.

$$\begin{aligned}\tau(X, Y) = (\mathcal{L}_\xi g)(X, Y) &= g(\nabla_X \xi, Y) + g(X, \nabla_Y \xi)\\ &= g((A\phi - \phi A)X, Y).\end{aligned}$$

In particular $\operatorname{trace}[A, \phi] = 0$ so that (by $\phi\xi = 0$) $A\xi = \alpha\xi$ if and only if $T\xi = 0$. Moreover, with respect to a local orthonormal frame $\{E_i : 1 \leq i \leq 2n+1\}$ consisting of eigenvectors of A

$$\operatorname{div}\xi = \sum_{i=1}^{2n+1} g(\nabla_{E_i}\xi, E_i) = \frac{1}{2} \sum_{i=1}^{2n+1} (\mathcal{L}_\xi g)(E_i, E_i) = \frac{1}{2} \operatorname{trace} T = 0.$$

The last statement in Proposition 4.56 follows from

$$g(T\xi, \xi) = g(A\xi, \phi\xi) = 0$$

and

$$g(T\xi, X) = g(\xi, TX), \quad X \in \operatorname{Ker}(\eta).$$

■

Lemma 4.57 (D. Perrone, [247]) *Let M be an orientable real hypersurface of a Kaehler manifold M_0. Then*

$$\Delta_g \xi = Q\xi + \nabla^* T = Q\xi - \operatorname{trace}_g \nabla T. \tag{4.63}$$

Proof. Let $\{E_i : 1 \leq i \leq 2n+1\}$ be a local orthonormal frame of $T(M)$ as above. The identity $\nabla\xi = -\phi A$ implies

$$g(\nabla_X \xi, Y) + g(\nabla_Y \xi, X) = g((A\phi - \phi A)X, Y) = g(TX, Y). \tag{4.64}$$

Next (4.64) together with

$$\begin{aligned}\operatorname{Ric}(\xi, X) &= \sum_{i=1}^{2n+1} g(R(X, E_i)\xi, E_i)\\ &= \sum_{i=1}^{2n+1} \{\nabla_{[X,E_i]}\xi - \nabla_X \nabla_{E_i}\xi - \nabla_{E_i}\nabla_X \xi\}\end{aligned}$$

imply that

$$\begin{aligned}
\operatorname{Ric}(\xi, X) = & \sum_{i=1}^{2n+1} \{g(T[X, E_i], E_i) - g(\nabla_{E_i}\xi, [X, E_i]) \\
& - X(g(\nabla_{E_i}\xi, E_i)) + g(\nabla_{E_i}\xi, \nabla_X E_i)\} \\
& + \sum_{i=1}^{2n+1} \{E_i(g(\nabla_X \xi, E_i)) - g(\nabla_X \xi, \nabla_{E_i} E_i)\} \\
= & \sum_{i=1}^{2n+1} \{g(\nabla_{E_i}\xi, \nabla_{E_i}\xi) + g(T[X, E_i], E_i) - X(\operatorname{div}\xi)\} \\
& + \sum_{i=1}^{2n+1} \{E_i(-g(\nabla_{E_i}\xi, \nabla_{E_i} X) + g(TX, E_i)) \\
& + g(\nabla_{\nabla_{E_i} E_i}\xi, X) - g(TX, \nabla_{E_i} E_i)\} \\
= & \sum_{i=1}^{2n+1} \{g(\nabla_{\nabla_{E_i} E_i}\xi, X) - g(\nabla_{E_i}\nabla_{E_i}\xi, X) \\
& + g(T[X, E_i], E_i) - g(TX, \nabla_{E_i} E_i) + E_i(g(TX, E_i))\}
\end{aligned}$$

or (by the very definition of Δ_g)

$$\operatorname{Ric}(\xi, X) = g(\Delta_g \xi, X) + \sum_{i=1}^{2n+1} \{g(T[X, E_i], E_i) + g(\nabla_{E_i} TX, E_i)\}. \tag{4.65}$$

As T is symmetric one may choose from the very beginning the local frame $\{E_i : 1 \leq i \leq 2n+1\}$ to consist of eigenvectors of T. Thus

$$\begin{aligned}
& \sum_{i=1}^{2n+1} \{g(T[X, E_i], E_i) + g(\nabla_{E_i} TX, E_i)\} \\
& \quad = \sum_{i=1}^{2n+1} \{g(\nabla_X E_i, TE_i) + g((\nabla_{E_i} T)X, E_i)\} \\
& \quad = (\operatorname{div} T)\, X = g(\operatorname{trace}_g \nabla T, X).
\end{aligned}$$

The last identity together with (4.65) implies (4.63) in Lemma 4.57. ■

Remark 4.58 Let us assume that the induced almost contact metric structure (ϕ, ξ, η, g) on M is actually a contact metric structure. Then (by the contact condition) $T = 2h\phi$ hence (cf. [242])

$$\text{trace}_g \nabla T = -2\nabla^* h\phi = 2Q\xi - 4n\xi$$

and (4.63) becomes

$$\Delta_g \xi = -Q\xi + 4n\xi. \tag{4.66}$$

It should be observed that this is precisely the contents of Theorem 4.3 (hence Lemma 4.57 may be thought of as a generalization of the result in Theorem 4.3). ■

We may state the following

Theorem 4.59 *Let M be an orientable real hypersurface of a Kaehler manifold M_0. Then*

i. *ξ is harmonic if and only if $Q\xi = \lambda\xi + \nabla^* T$, and if ξ is an eigenvector of Q then*

ii. *ξ is harmonic if and only if $\nabla^* T$ is proportional to ξ.*

Definition 4.60 (M. Kon, [193]) The real hypersurface M is said to be *pseudo-Einstein* if $\text{Ric} = ag + b\eta \otimes \eta$ for some $a, b \in \mathbb{R}$. ■

According to the terminology of S. Tanno, [284], a contact Riemannian manifold M is an *η-Einstein space* if the Ricci curvature is given by $\text{Ric} = ag + b\eta \otimes \eta$ for some $a, b \in C^\infty(M)$. When M is an m-dimensional, $m > 3$, η-Einstein space and a K-contact Riemannian manifold then a are b are constants.

M. Kon classified (cf. [193]) complete pseudo-Einstein real hypersurfaces in $\mathbb{C}^{n+1}$ and in the complex projective space $\mathbb{C}P^{n+1}$ ($n \geq 2$). For each hypersurface in these classes the Reeb vector ξ is an eigenvector of the Ricci operator Q so that the following harmonicity criterium holds

Corollary 4.61 *Let M be an orientable pseudo-Einstein real hypersurface in a Kaehler manifold M_0. Then ξ is harmonic if and only if $\nabla^* T$ is proportinal to ξ.*

Real hypersurfaces satisfying $T = 0$ (i.e., $[A, \phi] = 0$) have been classified by M. Okumura, [226], when $M_0 = \mathbb{C}P^{n+1}$, by S. Montiel & A. Romero, [214], when $M_0 = \mathbb{C}H^{n+1}$ (the complex hyperbolic space), and by J. Berndt & Y.J. Suh, [38], when $M_0 = G_2(\mathbb{C}^{n+2})$ (a complex 2-plane Grassmannian manifold). In an arbitrary Kaehlerian ambient space M_0, any *quasi-umbilical* real hypersurface M (i.e., one whose shape operator is of the form $A = aI + b\eta \otimes \xi$ for some $a, b \in C^\infty(M)$) obeys to $T = 0$. By a result of A. Banyaga, [20], the characteristic foliation of a compact simply connected real hypersurface satisfying $T = 0$ has a compact leaf. By Proposition 4.56 the condition $T = 0$ may be replaced by $T = 0$ along $\mathrm{Ker}(\eta)$. On such hypersurfaces one may derive a harmonicity criterium similar to that obtained in the context of abstract (i.e., not embedded) contact metric structures.

Corollary 4.62 *Let M be an orientable real hypersurface of a Kaehler manifold M_0 satisfying $[A, \phi] = 0$ on $\mathrm{Ker}(\eta)$. Then ξ is harmonic if and only if ξ is an eigenvector of the Ricci operator Q.*

Remark 4.63 a) By a result of S. Tachibana, [278], i) any harmonic 1-form ω on a compact Sasakian manifold is orthogonal to the contact form i.e., $i(\eta)\omega = 0$, ii) for any harmonic 1-form ω the 1-form $\tilde{\omega}$ given by $\tilde{\omega}_j = \varphi^i_j \omega_i$ is harmonic as well, and as a corollary iii) the first Betti number of a compact Sasakian manifold is zero or even. No analog for harmonic vector fields is known so far (see also (ii) in Remark 2.25 of Chapter 2).

b) By a result of S. Tanno, [284], if M is a real m-dimensional compact K-contact η-Einstein space with $a > 0$ (equivalently the scalar curvature ρ is $> m-1$) then $b_1(M) = 0$. It is an open problem to study harmonic vector fields on a K-contact η-Einstein space. See also S. Tanno, [280], D.E. Blair & S.I. Goldberg, [45], and S.I. Goldberg, [140].

4.5.2. Ruled Hypersurfaces

We adopt the following

Definition 4.64 (M. Kimura, [186], M. Kimura & S. Maeda, [187]) Let M be an orientable real hypersurface of a complex space form $M^{n+1}(c)$ of holomorphic sectional curvature $c \neq 0$. If the distribution $H(M) = \mathrm{Ker}(\eta) = (\mathbb{R}\xi)^\perp$ is completely integrable and its leaves are totally geodesic submanifolds of $M^{n+1}(c)$ then M is said to be a *ruled* real hypersurface. ∎

By a result of M. Kimura & S. Maeda, [187], the shape operator of a ruled real hypersurface is given by

$$A\xi = a\xi + bE, \quad AE = b\xi, \quad AX = 0, \tag{4.67}$$

for some $a, b \in C^\infty(M)$, $b \neq 0$, and some unit vector field $E \in \mathrm{Ker}(\eta)$, and any $X \in \mathrm{Ker}(\eta)$ orthogonal to E. Then the mean curvature of M in $M^{n+1}(c)$ is $H = a$ and the Reeb vector is not a principal direction (i.e., M is not a Hopf hypersurface). Moreover (by $\nabla\xi = -\phi A$ and by (4.67))

$$\nabla_\xi \xi = -b\phi E, \quad \nabla_E \xi = 0, \quad \nabla_X \xi = 0,$$

for any $X \in \mathrm{Ker}(\eta)$ orthogonal to E, hence $\|\nabla\xi\|^2 = b^2$. As $\phi^2 A^2 \xi = -abE$ and $M^{n+1}(c)$ is Kaehler-Einstein equation (4.57) in Lemma 4.52 may be written

$$\Delta_g \xi = \phi \nabla a - abE + b^2 \xi.$$

Furthermore $\phi \nabla a = abE$ if and only if $(\phi E)(a) = -ab$ and $X(a) = 0$ for any $X \in \mathrm{Ker}(\eta)$ orthogonal to ϕE. We may conclude that

Theorem 4.65 (K. Tsukada & L. Vanhecke, [291]) *Let M be a ruled hypersurface in $M^{n+1}(c)$. Then*

i. *ξ is harmonic if and only if $(\phi E)(a) = -ab$ and $X(a) = 0$ for any vector field $X \in \mathrm{Ker}(\eta)$ orthogonal to ϕE.*

ii. *The Reeb vector field of a minimal ruled hypersurface is a harmonic vector field, but it is not a harmonic map.*

4.5.3. Real Hypersurfaces of Contact Type

Let M be a real hypersurface of a complex $(n+1)$-dimensional Kaehler manifold (M_0, J_0, g_0). Orientability, or equivalently the existence of a smooth globally defined unit normal field on M, is assumed throughout without further mention. Let (ϕ, ξ, η, g) be the induced almost contact metric structure. Let $\Omega_0(X, Y) = g_0(X, J_0 Y)$ be the Kaehler 2-form on M_0 $(X, Y \in \mathfrak{X}(M_0))$. Let $\iota : M \to M_0$ be the canonical inclusion.

Definition 4.66 (M. Okumura, [225]) M is said to be a hypersurface of *contact type*[2] if $(d\eta)(X, Y) = r\, g(X, \phi Y)$ for some C^∞ function $r : M \to \mathbb{R} \setminus \{0\}$ and any $X, Y \in \mathfrak{X}(M)$. ■

If M is a hypersurface of contact type as in Definition 4.66 then $d\eta = r\iota^* \Omega_0$. Indeed (as $g = \iota^* g_0$)

$$\begin{aligned} g(X, \phi Y) &= g_0(X, \phi Y) = g_0(X, JY - \eta(Y)\nu) \\ &= g_0(X, JY) = \Omega_0(X, Y) \end{aligned}$$

for any $X, Y \in T(M)$. As $\Phi \in \Omega^2(M)$ (given by $\Phi(X, Y) = g(X, \phi Y)$) has rank $2n$ it follows that η is a contact form i.e., $\eta \wedge (d\eta)^n$ is a volume form on M. When $r = 1$, the induced almost contact metric structure on M is a contact metric structure.

Lemma 4.67 *Any hypersurface of contact type is a Hopf hypersurface.*

Proof. As M is a hypersurface of contact type, for any $Y \in \mathrm{Ker}(\eta)$

$$(d\eta)(\xi, Y) = rg(\xi, \phi Y) = 0$$

[2] As a matter of terminology in use, it should be noted that a real hypersurface $M \to M_0$ of a symplectic manifold (M_0, ω_0), where ω_0 is the symplectic form, is referred to by A. Banyaga (cf. [20]) as a hypersurface of contact type whenever M admits a contact from η such that $d\eta = \iota^* \omega_0$.

(as $\mathrm{Ker}(\eta)$ is ϕ-invariant). On the other hand

$$2(d\eta)(\xi, Y) = -\eta([\xi, Y])$$
$$= g(\nabla_Y \xi - \nabla_\xi Y, \xi) = g(\phi A\xi, Y) = -g(A\xi, \phi Y),$$

so that $A\xi$ is proportional to ξ. ■

Lemma 4.68 *Let (M, η) be a $(2n+1)$-dimensional contact manifold. If $f \in C^\infty(M)$ satisfies $df = \xi(f)\eta$ then f is a constant.*

Proof. By the assumption $df = \xi(f)\eta$ the function f is constant if and only if $\xi(f) = 0$. One has

$$0 = d^2 f = (d\xi(f)) \wedge \eta + \xi(f)\, d\eta. \tag{4.68}$$

The reminder of the proof is by contradiction. Let us assume that $\xi(f)_{x_0} \neq 0$ for some $x_0 \in M$. By continuity there is an open neighborhood U of x_0 in M such that $\xi(f)_x \neq 0$ for any $x \in U$. Then (by (4.68))

$$d\eta = \frac{1}{\xi(f)}\, \eta \wedge d\xi(f)$$

on U, hence $\eta \wedge (d\eta)^n = 0$ on U, a contradiction (as η is a contact form). ■

Lemma 4.69 *Let M be a hypersurface of contact type of a Kaehler manifold M_0. If $\dim(M) = 2n+1 > 3$, i.e., $n \geq 2$, then r is a constant. If $\dim(M) = 3$, i.e., $n = 1$, then r is constant along the integral curves of ξ.*

Proof. As $\Phi = \iota^*\Omega_0$ is a closed 2-form on M we may differentiate the identity $d\eta = r\Phi$ and obtain

$$0 = d^2\eta = dr \wedge \Phi + r\, d\Phi = dr \wedge \Phi$$

that is $dr \wedge \Phi = 0$. Let $X, Y \in \mathrm{Ker}(\eta)$ such that X is orthogonal to both $\{Y, \phi Y\}$. Then

$$(dr \wedge \Phi)(X, Y, \phi Y) = X(r) g(Y, \phi^2 Y) = -X(r)\|Y\|^2$$

and

$$(dr \wedge \Phi)(\xi, Y, \phi Y) = \xi(r) g(Y, \phi^2 Y) = -\xi(r)\|Y\|^2.$$

We may conclude that r is constant when $n \geq 2$ and merely that $\xi(r) = 0$ when $n = 1$. ■

Remark 4.70 (M. Okumura, [225]) If M is a hypersurface of contact type in a complex space form $M^{n+1}(c)$ then r is a constant in all dimensions $n \geq 1$. ■

Theorem 4.71 *Let M be a hypersurface of contact type in a Kaehler manifold M_0. Let us assume that ν is an eigenvector of the Ricci operator Q_0. If $\dim(M) = 2n+1 > 3$ i.e., $n \geq 2$ then the following statements are equivalent*

i. *ξ is harmonic.*
ii. *The mean curvature H is constant.*
iii. *ξ is an eigenvector of the Ricci operator Q.*

The statements (i) *and* (ii) *are equivalent when $n = 1$ as well.*

Proof. Let M be a hypersurface of contact type in the Kaehler manifold M_0. Let us assume that ν is an eigenvector of Q_0 and that $n \geq 2$. By Lemma 4.67 it follows that M is a Hopf hypersurface. By Theorem 4.54 we know that ξ is harmonic if and only if $\nabla H = \xi(H)\xi$. Then the implication (ii) $\Longrightarrow$ (i) is immediate. Moreover, if ξ is harmonic then $dH = \xi(H)\eta$ and then (by Lemma 4.68) H is constant. Let us prove the equivalence (i) $\Longleftrightarrow$ (iii) (under the assumption that $n \geq 2$). By Lemma 4.69 the contact type condition is satisfied with $r =$ constant. For $r \neq 1$ the Riemannian metric g is not associated to η yet this is easy to adjust. Indeed

$$\hat{\phi} = \phi, \ \hat{\xi} = (1/r)\xi, \ \hat{\eta} = r\eta, \ \hat{g} = r^2 g,$$

is a contact metric structure. Next

$$\Delta_{\hat{g}}\hat{\xi} = r^{-3}\,\Delta_g\xi.$$

Moreover the Ricci operators of $(M,\hat{g})$ and (M,g) are related by

$$\hat{Q} = r^{-2}\,Q.$$

By Theorem 4.3

$$\Delta_{\hat{g}}\hat{\xi} = \|\nabla\hat{\xi}\|_{\hat{g}}^2\,\hat{\xi} + \phi^2\hat{Q}\hat{\xi}.$$

Thus

$$\begin{aligned}\Delta_g\xi &= r^3\Delta_{\hat{g}}\hat{\xi} = r^3\left\{\|\nabla\hat{\xi}\|_{\hat{g}}^2\,\hat{\xi} + \phi^2\hat{Q}\hat{\xi}\right\}\\ &= r^3\left\{r^{-2}\|\nabla\xi\|_g^2(1/r)\xi + r^{-2}\phi^2 Q(1/r)\xi\right\} = \|\nabla\xi\|_g^2\xi + \phi^2 Q\xi\end{aligned}$$

and one may conclude that ξ is harmonic if and only if ξ is an eigenvector of Q. Theorem 4.71 is proved. ■

Hypersurfaces of contact type in a complex space form $M^{n+1}(c)$ have been classified by M. Okumura, [225], when $c = 0$, by M. Kon, [193], when

$c > 0$, and by M.H. Vernon, [302], when $c < 0$. The Hopf hypersurfaces in a complex 2-plane Grasmann manifold (studied by J. Berndt & Y.J. Suh, [38]) are of contact type and their Reeb vector ξ is a harmonic map.

Corollary 4.72 *The Reeb vector of a hypersurface of contact type in a complex space form is a harmonic map.*

Proof. The identity (4.51) implies that statement (iii) in Theorem 4.71 is equivalent to the statement: ξ is an eigenvector of $R_0(\cdot,\nu)\nu$. At this point Corollary 4.72 follows from Theorem 4.71 and the Gauss equation (of M in $M^{n+1}(c)$). ■

Corollary 4.73 *Let M be a real $(2n+1)$-dimensional hypersurface of contact type in a Kaehler-Einstein manifold M_0 and let (ϕ,ξ,η,g) be the induced almost contact structure. Then*

i. *if $n \geq 1$ then $(\hat{\phi},\hat{\xi},\hat{\eta},\hat{g})$ is an H-contact structure if and only if M has constant mean curvature in M_0. Moreover*

ii. *if $n > 1$ then $(\hat{\phi},\hat{\xi},\hat{\eta},\hat{g})$ is an H-contact structure if and only if ξ is an eigenvector of the Ricci operator of (M,g).*

Remark 4.74 Let $SO(2n+1) \to P \to M$ be the principal bundle of all positively oriented orthonormal frames tangent to a $(2n+1)$-dimensional Riemannian manifold (M,g). We recall that an almost contact metric structure on M may be thought of as a section σ in the associated bundle $P/U(n) \to M$ (with standard fibre $SO(2n+1)/U(n)$). Let us assume that M is an oriented real hypersurface in a *nearly Kaehler* manifold (M_0,J_0,g_0) (i.e., the almost Hermitian structure (J_0,g_0) satisfies $(\nabla^0_X J_0)Y + (\nabla^0_Y J)X = 0$ for any $X,Y \in \mathfrak{X}(M_0)$). Let σ be the induced almost contact metric structure on M. By a result of E. Vergara-Diaz & C.M. Wood, [299], if M is a contact metric hypersurface then σ is a harmonic section if and only if M is an H-contact manifold. Also (cf. *op. cit.*), if M is an oriented real hypersurface of a Kähler manifold then σ is a harmonic section (respectively a harmonic map) if and only if ξ is harmonic as a unit vector field (respectively ξ is a harmonic map). ■

4.6. HARMONICITY AND STABILITY OF THE GEODESIC FLOW

The present section is devoted to the study of the geometry of Riemannian manifolds (M,g) whose (total space of the) tangent sphere bundle

$S(M)$ is an H-contact manifold with respect to its standard contact metric structure.

4.6.1. The Ricci Curvature

Let (M,g) be a real n-dimensional Riemannian manifold and let (ϕ,ξ,η,G_{cs}) be the standard contact metric structure on $S(M)$. We wish to compute the Ricci tensor of $(S(M),G_{cs})$. To this end one chooses first a local orthonormal frame $\{\sigma_1,\ldots,\sigma_n\}$ of $(\pi^{-1}TM,\hat{g})$, defined on the open set $\pi^{-1}(U)$, such that $\sigma_n = L^{-1}\mathcal{L}$ where $L=\hat{g}(\mathcal{L},\mathcal{L})\in C^\infty(T(M))$. Note that $L=1$ everywhere on $S(M)$. If $\nu=-\gamma\,\mathcal{L}$ (the unit normal vector on $S(M)$) then $\xi' = -J\nu = \beta\,\mathcal{L}$ so that

$$\{2\gamma\,\sigma_1,\ldots,2\gamma\,\sigma_{n-1},2\beta\,\sigma_1,\ldots,2\beta\,\sigma_{n-1},\xi=2\xi'\} \tag{4.69}$$

is a local orthonormal frame of $T(S(M)) = [\gamma\,\mathrm{Ker}(\omega)]\oplus[\mathcal{H}|_{S(M)}]$. Here $\omega(\sigma)=\hat{g}(\sigma,\mathcal{L})$ for any $\sigma\in\pi^{-1}TM$.

Let $X\in\mathfrak{X}(M)$ and let us consider $X^T\in\Gamma^\infty(\mathcal{V})$ (the *tangential lift* of X) locally given by

$$X^T_u = X^V_u - \omega(\hat{X})_u\,y^i(u)\,\left.\frac{\partial}{\partial y^i}\right|_u,$$

for any $u\in\pi^{-1}(U)$. Here (U,x^i) is a local coordinate system on M and $(\pi^{-1}(U),x^i,y^i)$ are the induced local coordinates on $T(M)$. Clearly $X^T_u\in T_u(S(M))$ for any $u\in S(M)$. We also adopt the notation $\overline{X}=\hat{X}-\omega(\hat{X})\,\mathcal{L}\in\Gamma^\infty(\pi^{-1}TM)$. Then $\overline{X}\in\mathrm{Ker}(\omega)$ and $\gamma\,\overline{X}=X^T$. Using the local frame (4.69) one may compute the Ricci tensor field $\mathrm{Ric}_{G_{cs}}$ of $(S(M),G_{cs})$ and obtain (cf. e.g., [53])

$$\begin{aligned}\mathrm{Ric}_{G_{cs}}(X^T,Y^T)_u &= (n-2)\{g(X,Y)_x - g_x(X_x,u)g_x(Y_x,u)\}\\ &\quad + \frac{1}{4}\sum_{i=1}^{n} g_x(R_x(u,X_x)e_i,\,R_x(u,Y_x)e_i),\end{aligned} \tag{4.70}$$

$$\mathrm{Ric}_{G_{cs}}(X^T,Y^H) = \frac{1}{2}\{(\nabla\mathrm{Ric})_x(u,X_x,Y_x) - (\nabla\mathrm{Ric})_x(X_x,u,Y_x)\}, \tag{4.71}$$

$$\mathrm{Ric}_{G_{cs}}(X^H,Y^H) = \mathrm{Ric}(X,Y)_x - \frac{1}{2}\sum_{i=1}^{n} g_x(R_x(u,e_i)X_x,\,R_x(u,e_i)Y_x), \tag{4.72}$$

for any $X, Y \in \mathfrak{X}(M)$ and any $u \in S(M)_x$, $x \in M$, where $e_i = \sigma_i(u) \in (\pi^{-1}TM)_u = T_x(M)$ for any $1 \leq i \leq n$. Also, R and Ric are the curvature and the Ricci tensor of (M,g). By Theorem 4.5 it follows that $(S(M), (\phi, \xi, \eta, G_{cs}))$ is an H-contact manifold if and only if the characteristic vector ξ is an eigenvector of $\mathrm{Ric}_{G_{cs}}$ i.e.,

$$\mathrm{Ric}_{G_{cs}}(\xi, X^T)_u = 0, \quad \mathrm{Ric}_{G_{cs}}(\xi, Y^H)_u = 0,$$

for any $X, Y \in \mathfrak{X}(M)$ and any $u \in S(M)_x$ such that Y_x and u are orthogonal. Then (by (4.71)–(4.72))

$$(\nabla \mathrm{Ric})_x(u, u, X_x) = (\nabla \mathrm{Ric})_x(X_x, u, u), \tag{4.73}$$

$$2\mathrm{Ric}(X_x, u) = \sum_{i=1}^{n} g_x(R_x(u, e_i)X_x, R_x(u, e_i)u), \tag{4.74}$$

for any $X \in \mathfrak{X}(M)$ and any $u \in S(M)_x$ such that X_x and u are orthogonal. As u is an arbitrary unit vector tangent to M at x the identity (4.73) is equivalent to

$$(\nabla_X \mathrm{Ric})(Y, Z) = (\nabla_Y \mathrm{Ric})(X, Z), \quad X, Y, Z \in \mathfrak{X}(M), \tag{4.75}$$

i.e., Ric is a *Codazzi tensor* field on M. We may conclude that

Proposition 4.75 *The total space $S(M)$ of the tangent sphere bundle over a Riemannian manifold (M,g) is an H-contact manifold if and only if* a) *the Ricci tensor* Ric *of (M,g) is a Codazzi tensor and* b) *the identity* (4.74) *holds for any $u \in S(M)_x$ and any $x \in M$.*

Let $x \in M$ and let $u \in S(M)_x$ be a fixed unit vector. The *Jacobi operator*

$$R_u = R_x(u, \cdot)u$$

is symmetric, it is diagonalizable. Thus we may complete u to a g_x-orthonormal basis $\{e_1, \ldots, e_n\} \subset T_x(M)$ such that $e_n = u$ and $R_u(e_i) = \lambda_i e_i$ for some $\lambda_i \in \mathbb{R}$ and any $1 \leq i \leq n$. Then

$$\mathrm{Ric}_x(e_j, u) = -\sum_{i=1}^{n-1} g_x(R_x(u, e_i)e_i, e_j) = -\sum_{i=1}^{n-1} g_x(R_{e_i}u, e_j),$$

$$\sum_{i=1}^{n-1} g_x(R_x(u, e_j)e_j, R_x(u, e_i)u)$$

$$= \sum_{i=1}^{n-1} \lambda_i g_x(R_x(u, e_i)e_j, e_i) = -\sum_{i=1}^{n-1} \lambda_i g(R_{e_i}u, e_j),$$

and (4.74) becomes

$$\sum_{i=1}^{n-1}(2-\lambda_i)g_x(R_{e_i}u,\, e_j)=0, \quad 1\le j\le n-1. \tag{4.76}$$

Moreover, ξ is a harmonic map of $S(M)$ into $S(S(M))$ if additionally

$$\sum_{i=1}^{n-1}\left\{R^{G_{cs}}\left(\beta\mathcal{L},\nabla^{G_{cs}}_{\beta\sigma_i}\xi\right)\beta\sigma_i+R^{G_{cs}}\left(\beta\mathcal{L},\nabla^{G_{cs}}_{\sigma_i^T}\xi\right)\sigma_i{}^T\right\}=0. \tag{4.77}$$

Here $\nabla^{G_{cs}}$ is the Levi-Civita connection of $(S(M),G_{cs})$ and $R^{G_{cs}}$ its curvature tensor field. Using

$$\left(\nabla^{G_{cs}}_{X^H}\xi\right)_u=(R_x(X_x,u)u)^V_u,$$

$$\left(\nabla^{G_{cs}}_{X^T}\xi\right)_u=2\left(X_x-g_x(X_x,u)u\right)^H_u+(R_x(X_x,u)u)^H_u,$$

and the explicit expression of the curvature (cf. e.g., [42], p. 140) it may be shown that (4.77) is equivalent to

$$\sum_{i=1}^{n-1}(1-\lambda_i)g_x((\nabla R)_x(u,u,e_i,e_i),u)=0, \tag{4.78}$$

$$\sum_{i=1}^{n-1}2(1-\lambda_i)g_x((\nabla R)(u,u,e_i,e_i),e_j)$$
$$+\sum_{i=1}^{n-1}(2-\lambda_i)g_x(\nabla R)_x(e_i,e_i,u,u),e_j)=0, \tag{4.79}$$

$$\sum_{i=1}^{n-1}((4-3\lambda_i)+\lambda_j(1-\lambda_i))g_x(R_{e_i}u,e_j)=0, \tag{4.80}$$

for any $1\le j\le n-1$.

Let us recall that a Riemannian manifold (M,g) is a *two-point homogeneous space* if for any (x,y), $(x',y')\in M\times M$ such that $d(x,y)=d(x',y')$ there is an isometry $f\in \mathrm{Isom}(M,g)$ such that $f(x)=x'$ and $f(y)=y'$. Here $d: M\times M\to[0,+\infty)$ is the distance function associated to the Riemannian metric g. Two-point homogeneous spaces were classified: the complete list consists of the Euclidean spaces, the compact rank one symmetric spaces and their noncompact duals. The compact rank one symmetric spaces are the sphere S^n, the projective spaces $\mathbb{R}P^n$, $\mathbb{C}P^n$ and $\mathbb{H}P^n$, and the Cayley plane CayP^2. By a result in [55], a Riemannian manifold is locally isometric to a two-point homogeneous space if and only if the tangent sphere bundle $(S(M), (\phi, \xi, \eta, G_{cs}))$ obeys to

$$\nabla_{\xi}^{G_{cs}} h = 2ah\phi + 2b\phi S$$

for some $a, b \in C^{\infty}(S(M))$ which are constant on the fibres of π. Here $h = (1/2)\, \mathcal{L}_{\xi}\phi$ and S is the identity on horizontal vectors and minus the identity on vertical vectors.

Theorem 4.76 (E. Boeckx & L. Vanhecke, [54]) *If (M,g) is a two-point homogeneous space then $(S(M), (\phi, \xi, \eta, G_{cs}))$ is an H-contact manifold and ξ is a harmonic map of $S(M)$ into $S(S(M))$.*

Proof. One should check that the requirements (4.75)–(4.80) are satisfied. The proof relies on two basic ingredients. First any two-point homogeneous space is locally symmetric hence (4.75) and (4.77)–(4.78) hold good. Second the Jacobi operator is subject to the following property: if u, v are two orthogonal unit tangent vectors such that $R_u v = \lambda v$ then $R_v u = \lambda u$ as well (cf. e.g., [88]). Therefore the tangent vector $R_{e_i} u$ appearing in (4.76) and (4.79) is proportional to u so that (4.76) and (4.79) are satisfied. ■

Clearly the statement in Theorem 4.76 holds for the geodesic flow ξ' as well. Also, Theorem 4.76 remains true if M is but locally isometric to a two-point homogeneous space.

Let (M, J, g) be a complex m-dimensional Kähler manifold and let us consider the tangent vector fields $\xi_1, \xi_2 \in \mathfrak{X}(S(M))$ locally given by

$$\xi_1 = \sum_{i=1}^{2m} y^i \left(J \frac{\partial}{\partial x^i} \right)^T, \quad \xi_2 = \sum_{i=1}^{2m} y^i \left(J \frac{\partial}{\partial x^i} \right)^H.$$

Then

Theorem 4.77 (E. Boeckx & L. Vanhecke, [54]) *If (M, J, g) is a complex space form then the unit vector fields ξ_1 and ξ_2 are harmonic. Also each $\xi_i : S(M) \to S(S(M))$ is a harmonic map.*

On the pointed tangent bundle $(T(M) \setminus M,\ G_s)$ one may consider the outward unit normal to $S_r(M)$ and the normalized geodesic flow

$$\nu_1 = L^{-1}\, \gamma\, \mathcal{L}, \quad \nu_2 = L^{-1}\, \beta\, \mathcal{L}, \quad L = \hat{g}(\mathcal{L}, \mathcal{L})^{1/2}.$$

Then

Theorem 4.78 (E. Boeckx & L. Vanhecke, [54]) *Let M be a Riemannian manifold. Then* a) ν_1 *is a harmonic vector field and a harmonic map. Also* b) *if M is a two-point homogeneous space then ν_2 is a harmonic vector field and a harmonic map.*

4.6.2. H-Contact Tangent Sphere Bundles

The purpose of this section is to report on the state-of-art results related to the following problem (raised by E. Boeckx & L. Vanhecke, [54]): Are two-point homogenous spaces the only Riemannian manifolds M whose total space $S(M)$ of the tangent sphere bundle is an H-contact manifold? The problem is open in general. Nevertheless partial positive solutions are already available. For instance if M is locally reducible then (by a result of G. Calvaruso & D. Perrone, [77]) $S(M)$ is an H-contact manifold if and only if M is flat. Also when $\dim(M) \in \{2,3\}$ then (by a result in [54]) $S(M)$ is an H-contact manifold if and only if M has constant sectional curvature.

We recall that a Riemannian manifold (M,g) is *(locally) conformally flat* if for any $x \in M$ there is an open neighborhood $U \subseteq M$ of x and an everywhere positive smooth function $\varphi : U \to \mathbb{R}$ such that φg is a flat metric on U. The study of conformally flat Riemannian manifolds is a classical issue in differential geometry. It is well known that the curvature tensor field of a real n-dimensional conformally flat Riemannian manifold (M,g) is locally given by

$$\begin{aligned} R_{ijk\ell} = & \frac{1}{n-2}\left\{g_{ik}R_{j\ell} + g_{j\ell}R_{ik} - g_{i\ell}R_{jk} - g_{jk}R_{i\ell}\right\} \\ & - \frac{\rho}{(n-1)(n-2)}\left\{g_{ik}g_{j\ell} - g_{i\ell}g_{jk}\right\} \end{aligned} \tag{4.81}$$

provided that $n \geq 3$. The identity (4.81) is the explicit form of $W = 0$ where W is the Weyl conformal curvature tensor field. By a classical result when $n \geq 4$ the converse holds (i.e., $W = 0$ implies that M is conformally flat) while $W = 0$ always holds on a 3-dimensional Riemannian manifold. We may state

Theorem 4.79 (G. Calvaruso & D. Perrone, [77]) *Let (M,g) be a Riemannian manifold such that $W = 0$. Then $(S(M), (\phi, \xi, \eta, G_{cs}))$ is an H-contact manifold if and only if (M,g) has constant sectional curvature.*

Proof. When (M,g) has constant sectional curvature its Ricci tensor is a Codazzi tensor and a straightforward calculation shows that requirement (b) in Proposition 4.75 is satisfied (as both sides in (4.74) vanish identically).

Conversely let M be a Riemannian manifold such that $W = 0$ and (4.74) holds. Let $x \in M$ and let $\{e'_1, \dots, e'_n\}$ be an orthonormal basis in $T_x(M)$ consisting of eigenvectors of the Ricci operator Q_x i.e., $\text{Ric}_x(e'_k, e'_\ell) = \rho_k \delta_{k\ell}$ for any $1 \le k, \ell \le n$ where ρ_k are the corresponding eigenvalues. Let $i \ne j$ be two arbitrary indices. For any $\theta \in \mathbb{R}$ the system $\{e_k : 1 \le k \le n\}$ given by

$$\begin{aligned} \{e_1, \dots, e_n\} &= \{e'_k : k \in \{1, \dots, n\} \setminus \{i,j\}\} \cup \{u, v\}, \\ u &= (\cos\theta)\, e'_i + (\sin\theta)\, e'_j, \quad v = (-\sin\theta)\, e'_i + (\cos\theta)\, e'_j, \end{aligned}$$

is another orthonormal basis in $T_x(M)$. As u and v are orthogonal and $\{e_k : 1 \le i \le n\} \subset T_x(M)$ is an orthonormal basis ($e_n = u$) we may use (4.74) to compute $\text{Ric}_x(u,u)$ and we obtain

$$2\,\text{Ric}_x(u, v) = \sum_{k=1}^{n} g_x(R_x(u, e_k)v,\, R_x(u, e_k)u). \tag{4.82}$$

Let us use (4.81) to compute the right hand side of (4.82). As $\{e'_1, \dots, e'_n\}$ consists of eigenvectors of the Ricci operators one may easily see that (4.81) yields

$$\begin{aligned} R_x(e'_r, e'_s)e'_s &= K_{rs}\, e'_r, \quad 1 \le r, s \le n, \quad r \ne s, \\ K_{rs} &= \frac{\rho_r + \rho_s}{n-2} - \frac{\rho}{(n-1)(n-2)}. \end{aligned} \tag{4.83}$$

Next we compute $g_x(R_x(u, e_r)v,\, R_x(u, e_r)u)$ for any $1 \le r \le n$. As

$$e_r = \begin{cases} e_k, & \text{for some } k \in \{1, \dots, n\} \setminus \{i,j\} \text{ if } r \le n-2, \\ v, & \text{if } r = n-1, \\ u, & \text{if } r = n, \end{cases}$$

it follows that

$$\begin{aligned} & g_x(R_x(u, e_r)v,\, R_x(u, e_r)u) \\ & \quad = (K_{jk} - K_{ik})(\sin\theta)(\cos\theta)(K_{ik}\cos^2\theta + K_{jk}\sin^2\theta), \end{aligned} \tag{4.84}$$

if $r < n-1$ and $e_r = e'_k$, and

$$g_x(R_x(u, e_r)v,\, R_x(u, e_r)u) = 0, \quad r \in \{n-1, n\}. \tag{4.85}$$

Let us substitute from (4.84)–(4.85) into (4.82). We obtain

$$2\,\mathrm{Ric}_x(u,v) = \sum_{\substack{k \in \{1,\ldots,n\} \\ k \neq i, \quad k \neq j}} (K_{jk} - K_{ik})(\sin\theta)(\cos\theta)(K_{ik}\cos^2\theta + K_{jk}\sin^2\theta). \tag{4.86}$$

On the other hand

$$\mathrm{Ric}_x(u,v) = (\rho_j - \rho_i)\ \sin\theta\ \cos\theta$$

hence (by (4.86))

$$2(\rho_j - \rho_i) = \sum_{\substack{k \in \{1,\ldots,n\} \\ k \neq i, \quad k \neq j}} (K_{jk} - K_{ik})(K_{ik}\cos^2\theta + K_{jk}\sin^2\theta) \tag{4.87}$$

for all $\theta \in \mathbb{R}$ such that $\sin\theta\ \cos\theta \neq 0$. Note that (4.83) implies that

$$K_{ji} - K_{ik} = \frac{\rho_j - \rho_i}{n-2}, \quad k \in \{1,\ldots,n\} \setminus \{i,j\},$$

so that (4.87) becomes

$$(\rho_j - \rho_i)\left\{2 + \frac{1}{n-2}\sum_{k \notin \{i,j\}} (K_{ik}\cos^2\theta + K_{jk}\sin^2\theta)\right\} = 0. \tag{4.88}$$

As θ is an arbitrary number such that $\sin\theta\ \cos\theta \neq 0$ the identity (4.88) implies that $\rho_i = \rho_j$. Hence the eigenvalues of the Ricci operator Q_x coincide so that M is an Einstein manifold. Since M is by assumption conformally flat we may conclude that M has constant sectional curvature. ∎

Other results related to the E. Boeckx & L. Vanhecke problem were obtained in [77]. For instance a) let M be a 4-dimensional non Ricci flat Kähler manifold. Then $S(M)$ is an H-contact manifold if and only if M has constant holomorphic sectional curvature. Also b) if M is a compact Kähler manifold of nonnegative sectional curvature and $S(M)$ is H-contact then M is a Kähler-Einstein locally symmetric space. Finally c) if M is a Sasakian manifold then $S(M)$ is H-contact if and only if M has constant sectional curvature $+1$.

Recently S.C. Chun et al. [90], obtained the following result. Let (M,g) be an n-dimensional Einstein manifold ($n \geq 3$). Then $S(M)$ is H-contact if and only if M is 2-Stein. Also, they classify simply connected irreducible symmetric spaces (M,g) for which $S(M)$ is H-contact.

4.6.3. The Stability of the Geodesic Flow

Let (M,g) be a compact orientable Riemannian manifold and $S(M) = S(M,g)$ the total space of the tangent sphere bundle. If $V \in \Gamma^\infty(S(M))$ is a harmonic vector field the Hessian form at V is given by

$$(\text{Hess } E)_V(X,X) = \int_M \left(\|\nabla X\|^2 - \|X\|^2\|\nabla V\|^2\right) d\,\text{vol}(g), \quad X \in V^\perp,$$

and V is stable when $(\text{Hess } E)_V$ is positive definite (otherwise V is unstable, see Chapter 2 of this monograph).

Let $c > 0$. If $V \in \Gamma^\infty(S(M,g))$ is a harmonic vector field then the vector field $V_c \in \Gamma^\infty(S(M,g_c))$ given by $V_c = (1/\sqrt{c})\, V$ is harmonic too, where $g_c = cg$. Also $E_{g_c}(V_c) = c^{n/2-1}\left\{[n(c-1)/2]\,\text{Vol}(M,g) + E_g(V)\right\}$ hence

$$(\text{Hess } E_{g_c})_{V_c}(X,X) = c^{n/2}(\text{Hess } E_g)_V(X,X), \quad X \in V^\perp.$$

Consequently, V_c is E_{g_c}-stable if and only if V is E_g-stable.

There are few stability results on the stability of the Reeb vector field of an H-contact manifold of dimension > 3. Let $S(M)$ be equipped with the standard contact metric structure

$$(\phi, \xi, \eta, G_{cs}) = (\phi', 2\xi', \frac{1}{2}\eta', \frac{1}{4}\tilde{G}_s).$$

In the previous section we have shown that $(S(M), (\phi, \xi, \eta, G_{cs}))$ is an H-contact manifold whenever the base manifold (M,g) is an n-dimensional Riemannian manifold locally isometric to a two-point homogeneous space. Let us study the stability of ξ when (M,g) is a compact quotient of a two-point homogeneous space (compact quotients always exist, cf. [60]). The results we report on are due to E. Boeckx & J.C. Gonzales-Davila & L. Vanhecke, [57]. By the very definition of Hess E

$$\begin{aligned}(\text{Hess } E_{G_{cs}})_\xi(A,A) &= 2^{-(2n-1)}(\text{Hess } E_{\tilde{G}_s})_{\xi'}(A,A)\\ &= 2^{-(2n-1)} \int_{S(M)} \left(\|\nabla^{\tilde{G}_s} A\|^2 - \|A\|^2\|\nabla^{\tilde{G}_s}\xi'\|^2\right) d\,\text{vol}(\tilde{G}_s)\end{aligned}$$

for any $A \in \mathfrak{X}(S(M))$ such that $\tilde{G}_s(A, \xi') = 0$. The norm in the identity above is of course associated to the metric $\tilde{G}_s$. Recall that $\xi' = \beta \mathcal{L}$ is a unit vector field with respect to $\tilde{G}_s$. For each $X \in \mathfrak{X}(M)$ the tangential lift X^T is orthogonal to ξ'. As customary in all local calculations we use a local orthonormal frame of the form

$$\{\gamma\sigma_1^T, \ldots, \gamma\sigma_{n-1}^T, \beta\sigma_1, \ldots, \beta\sigma_{n-1}, \xi'\}$$

where $\{\sigma_1, \ldots, \sigma_n\}$ is a local orthonormal frame in $\pi^{-1}TM$ with $\sigma_n = L^{-1}\mathcal{L}$. Using the expressions of $\nabla^{\tilde{G}_s}_{X^T}\xi'$ and $\nabla^{\tilde{G}_s}_{X^H}\xi'$ (reported in the previous section) one obtains

$$\|\nabla^{\tilde{G}_s}\xi'\|_u^2 = (n-1) - \mathrm{Ric}_x(u,u) + \frac{1}{2}\|R_u\|^2.$$

As M is locally isometric to a two-point homogeneous space, one has $\mathrm{Ric} = (\rho/n)g$ and

$$\|R_u\|^2 = \frac{1}{n(n+2)}\left(\|\mathrm{Ric}\|^2 + \frac{3}{2}\|R\|^2\right)_x, \qquad u \in S(M)_x.$$

Hence

$$\|\nabla^{\tilde{G}_s}\xi'\|^2 = (n-1) - \frac{\rho}{n} + \frac{\rho^2}{2n^2(n+2)} + \frac{3}{4n(n+2)}\|R\|^2. \tag{4.89}$$

Moroever

$$\|X^T\|^2 = \|\hat{X}\|^2 - \hat{g}(\hat{X}, \mathcal{L})^2.$$

To compute $\|\nabla^{\tilde{G}_s}X^T\|$ we recall that

$$\nabla^{\tilde{G}_s}_{Y^T}X^T = -\hat{g}(\hat{X}, \mathcal{L})Y^T, \quad \nabla^{\tilde{G}_s}_{Y^H}X^T = -(\nabla_Y X)^T + \frac{1}{2}\beta\hat{R}(X^H, \beta\mathcal{L})\hat{Y}.$$

Let $u \in S(M)_x$ and $e_i = \sigma_i(u)$. Thus

$$\begin{aligned}\|\nabla^{\tilde{G}_s}X^T\|_u^2 &= (n-1)g_x(X_x,u)^2 + \|\nabla X\|_x^2 \\ &\quad - \sum_{i=1}^{n}\left\{g_x\left(u, \left(\hat{\nabla}_{\beta\sigma_i}\hat{X}\right)_u\right)^2 + \frac{1}{4}\|R_x(X_x,u)e_i\|^2\right\}.\end{aligned}$$

Then (by integrating along the fibre)

$$\int_{S(M)} \|X^T\|^2 d\,\mathrm{vol}(\tilde{G}_s) = \frac{n-1}{n} c_{n-1} \int_M \|X\|^2 d\,\mathrm{vol}(g), \tag{4.90}$$

$$\int_{S(M)} \|\nabla^{\tilde{G}_s} X^T\|^2 d\,\mathrm{vol}(\tilde{G}_s) = \frac{c_{n-1}}{n} \left\{ (n-1) \int_M \|\nabla X\|^2 d\,\mathrm{vol}(g) \right.$$

$$\left. + \left(\frac{\|R\|^2}{4n} + n - 1 \right) \int_M \|X\|^2 d\,\mathrm{vol}(g) \right\}, \tag{4.91}$$

where $c_{n-1} = \mathrm{Vol}(S^{n-1})$. Using (4.89)–(4.91) one obtains

$$(\mathrm{Hess}\, E_{\tilde{G}_s})_{\xi'}(X^T, X^T)$$

$$= \frac{(n-1)c_{n-1}}{n} \left(\int_M \|\nabla X\|^2 d\,\mathrm{vol}(g) + \delta \int_M \|X\|^2 d\,\mathrm{vol}(g) \right) \tag{4.92}$$

where

$$\delta = \frac{5-2n}{4n(n-1)(n-2)} \|R\|^2 - \frac{\rho^2}{2n^2(n+2)} + \frac{\rho}{n} - (n-2).$$

Let us assume that the first Betti number is nonzero ($b_1(M) \neq 0$) and consider a vector field X_0 dual to some nonzero harmonic form on M (i.e., $X_0 = \omega^\sharp$ for some $\omega \in \Omega^1(M)$ such that $\Delta\omega = 0$ and $\omega \neq 0$). Then (by the Weitzenbök formula)

$$\Delta_1 X_0 = \Delta_g X_0 + \mathrm{Ric}(X_0, \cdot)$$

where Δ_1 is the Hodge-de Rham Laplacian acting (by duality) on vector fields. Then

$$\int_M \|\nabla X_0\|^2 d\,\mathrm{vol}(g) = -\int_M \mathrm{Ric}(X_0, X_0)\, d\,\mathrm{vol}(g).$$

If M has constant sectional curvature $\lambda \leq 0$ then

$$\rho = n(n-1)\lambda, \quad \|R\|^2 = 2n(n-1)\lambda^2,$$

$$\int_M \|\nabla X_0\|^2 d\,\mathrm{vol}(g) = -(n-1)\lambda \int_M \|X_0\|^2 d\,\mathrm{vol}(g).$$

By (4.92) one finds

$$\left(\text{Hess}\, E_{\tilde{G}_s}\right)_{\xi'}(X_0^T, X_0^T)$$
$$= -\frac{(n-1)(n-2)(\lambda^2+2)c_{n-1}}{2n}\int_M \|X_0\|^2 d\,\text{vol}(g) < 0$$

provided that $n \geq 3$. If M is a compact Kähler manifold of constant negative holomorphic sectional curvature $\mu < 0$ of dimension $\dim_{\mathbb{R}}(M) = n = 2m$ then

$$\rho = m(m+1)\mu, \quad \|R\|^2 = 2m(m+1)\mu^2,$$
$$\int_M \|\nabla X_0\|^2 d\,\text{vol}(g) = -\frac{(m+1)\mu}{2}\int_M \|X_0\|^2 d\,\text{vol}(g).$$

Then (again by (4.92))

$$\left(\text{Hess}\, E_{\tilde{G}_s}\right)_{\xi'}(X_0^T, X_0^T) = -\frac{(m-1)c_{2m-1}}{2m}\left[\frac{(2m+11)\mu^2}{16} + 2(2m-1)\right]\int_M \|X_0\|^2 d\,\text{vol}(g) < 0$$

provided that $m \geq 2$. If M is a compact quaternionic Kähler manifold of constant negative Q-sectional curvature $\nu < 0$ and dimension $\dim_{\mathbb{R}}(M) = n = 4m$ then

$$\rho = 4m(m+2)\nu, \quad \|R\|^2 = 4m(5m+1)\nu^2,$$
$$\int_M \|\nabla X_0\|^2 d\,\text{vol}(g) = -(m+2)\nu\int_M \|X_0\|^2 d\,\text{vol}(g).$$

Then (again by (4.92))

$$\left(\text{Hess}\, E_{\tilde{G}_s}\right)_{\xi'}(X_0^T, X_0^T) = -\frac{c_{4m-1}}{4m}\left[\frac{4m^2+33m-13}{8}\nu^2 + 2(2m-1)(4m-1)\right]\int_M \|X_0\|^2 d\,\text{vol}(g)$$

provided that $m \geq 1$. Finally if M is a compact quotient of the Cayley plane $\mathrm{Cay}\, H^2(\zeta)$ of minimal sectional curvature $\zeta < 0$ then

$$\rho = 144\zeta, \quad \|R\|^2 = 576\zeta^2,$$

$$\int_M \|\nabla X_0\|^2 d\,\mathrm{vol}(g) = -9\zeta \int_M \|X_0\|^2 d\,\mathrm{vol}(g)$$

and then (by (4.92))

$$\left(\mathrm{Hess}\, E_{\tilde{G}_s}\right)_{\xi'} (X_0^T, X_0^T) = -\frac{21c_{15}}{64}\left(9\zeta^2 + 40\right) \int_M \|X_0\|^2 d\,\mathrm{vol}(g) < 0.$$

Summing up the information obtained so far we may state the following

Theorem 4.80 (E. Boeckx & J.C. Gonzales-Davila & L. Vanhecke, [57]) *Let M be a real n-dimensional ($n \geq 3$) compact quotient of a two-point homogeneous space of nonpositive curvature with $b_1(M) \neq 0$. Then the Reeb vector field ξ of $S(M)$ is an unstable critical point of the energy functional and the index of ξ is at least $b_1(M)$.*

In the positive curvature case one may prove (cf. [57]) a similar yet weaker result. Indeed on any real n-dimensional ($n \geq 3$) compact quotient of a two-point homogeneous space of positive curvature the existence of nonzero Killing vector fields implies the instability of ξ as a critical point of the energy functional, in certain ranges of the dimension and of the curvature. The geodesic flow ξ' on $S(S^2)$ is however stable. In this case $(S(S^2), \frac{1}{4}\tilde{G}_s)$ is locally isometric to S^3 and ξ is the Hopf vector field. Except for few particular results the question of stability of ξ remains open, particularly in the case of positive curvature. An intriguing case (according to [57]) is that of S^n with $n \geq 3$ (whose natural contact metric structure on $S(S^n)$ is actually Sasakian). By a result in [252] the geodesic flow on $S(S^n)$ (with $n > 6$) is an unstable harmonic vector field.

CHAPTER FIVE

Harmonicity with Respect to g-Natural Metrics

Contents

Let (M,g) be an n-dimensional Riemannian manifold. The Sasaki metric G_s is a Riemannian metric on $T(M)$ naturally associated to g yet exhibiting a certain "rigidity" (according to the philosophy in M.T.K. Abbassi & M. Sarih, [6]) in the sense that most natural requirements imposed to G_s imply drastic limitations on the choice of g. For instance (by a result of Y. Tashiro, [285]) the natural contact metric structure on $S(M)$ is K-contact if and only if (M,g) has constant sectional curvature 1. Also, as demonstrated earlier in this monograph (cf. Theorem 2.10 in Chapter 2), O. Nouhaud, [223], and T. Ishihara, [176], showed that parallel vector fields are, among all smooth vector fields, the only harmonic maps. Subsequently O. Gil-Medrano, [126], showed that the critical points of the energy functional restricted to $\mathfrak{X}(M)$ are again the parallel vector fields (cf. Theorem 2.17 in Chapter 2). It is our point of view that the quoted (negative) results arise from the choice of the metric on the target space (the Sasaki metric G_s) and that the emerging difficulties may be circumnavigated by a new choice of metric on $T(M)$, possibly among the

© 2012 Elsevier Inc. All rights reserved.

so called *g-natural metrics* (cf. [6]). Indeed it is known that G_s belongs to a wide family of Riemannian metrics on $T(M)$, the family of g-natural metrics, parametrized by $C^\infty(\mathbb{R}_0^+, \mathbb{R}^6)$, where $\mathbb{R}_0^+ = [0, +\infty)$.

According to the above philosophy one endows $T(M)$ with an arbitrary Riemannian g-natural metric G and looks at smooth vector fields V on M thought as maps of (M,g) into $(T(M), G)$, cf. [2]. Several results discussed earlier in this book carry over to this context (e.g., O. Nouhaud and T. Ishihara's Theorem 2.10 turns out to remain true under a certain 2-parameter and 1-function deformation of the Sasaki metric) and new phenomena arise. For instance, examples of non-parallel vector fields which are harmonic maps with respect to a certain 2-parameter family of g-natural metrics may be indicated (such as the Reeb vector fields of Sasakian manifolds, Hopf vector fields on unit odd-dimensional spheres and conformal gradient vector fields on S^2). M. Benyounes & E. Loubeau & C.M. Wood, [32]-[33], studied harmonic sections in a Riemannian vector bundle with respect to the generalized Cheeger-Gromoll metrics $h_{p,q}$. When $E = T(M)$ the metrics $h_{p,q}$ are g-natural (see below).

Given a unit vector field V, one looks at the harmonicity of the map from (M,g) into the tangent sphere bundle $(S(M), \tilde{G})$ where $\tilde{G}$ is induced by a g-natural metric on $T(M)$ (cf. [3]). By decomposing the tension field into vertical and horizontal components one derives two equations describing the harmonicity of $V : (M,g) \to (S(M), \tilde{G})$. As it turns out, one of these equations does not depend upon the choice of $\tilde{G}$. In particular the class of harmonic vector fields (cf. Theorem 2.23 in Chapter 2) is invariant under a 4-parameter deformation of the Sasaki metric. The remaining equation is a natural generalization of the situation encountered in Theorem 2.19. For unit Killing vector fields, both harmonic map equations are independent of the choice of $\tilde{G}$. An example of a (non-Killing) harmonic unit vector field which is not a harmonic map with respect to the Sasaki metric yet is a harmonic map with respect to a certain 2-parameter family of g-natural metrics (none of which is of the Kaluza-Klein type) is provided by the unit normal to the horoball foliation of the hyperbolic space.

The harmonic map equations for unit vector fields are also written for the particular case of Reeb fields in contact metric geometry. In particular Hopf vector fields on the unit sphere S^{2m+1} are harmonic maps for any Riemannian g-natural metric $\tilde{G}$ on $S(S^{2m+1})$. In dimension three we give an analogue to Theorem 3.10 (Han & Yim's theorem) for a Riemannian manifold of constant sectional curvature $\kappa \neq 0$. We also extend this result to Riemannian $(2n+1)$-manifolds M of constant sectional curvature $\kappa > 0$ with $\pi_1(M) \neq 0$.

When the tangent sphere bundle $S(M)$ over a two-point homogeneous space (M,g) as well as the tangent sphere bundle $S(S(M))$ are equipped with the Sasaki metrics, a result by E. Boeckx & L. Vanhecke, [54], shows that the geodesic flow is harmonic and provides a harmonic map $S(M) \to S(S(M))$ (cf. also Theorem 4.76 in this monograph). If one endows $S(M)$ and $S_\rho(S(M))$ with arbitrary Riemannian g-natural metrics then the geodesic flow (which has constant length equal to some $\rho > 0$ not necessarily 1) is always a harmonic vector field and it is a harmonic map under appropriate conditions on the coefficients determining the fixed g-natural metric, cf. [4], allowing one to exhibit large families of harmonic maps defined on a compact Riemannian manifold and having a target space with a highly nontrivial geometry.

5.1. g-NATURAL METRICS

According to [6] a g-natural metric G is given by

$$G(X^H, Y^H)_u = (\alpha_1 + \alpha_3)(r^2) g(X,Y)_x + (\beta_1 + \beta_3)(r^2) g_x(X_x, u) g_x(Y_x, u), \tag{5.1}$$

$$G(X^H, Y^V)_u = \alpha_2(r^2) g(X,Y)_x + \beta_2(r^2) g_x(X_x, u) g_x(Y_x, u), \tag{5.2}$$

$$G(X^V, Y^H)_u = \alpha_2(r^2) g(X,Y)_x + \beta_2(r^2) g_x(X_x, u) g_x(Y_x, u), \tag{5.3}$$

$$G(X^V, Y^V)_u = \alpha_1(r^2) g(X,Y)_x + \beta_1(r^2) g_x(X_x, u) g_x(Y_x, u), \tag{5.4}$$

for any $X, Y \in \mathfrak{X}(M)$ and any $u \in T(M)$ with $x = \pi(u) \in M$. Here α_i, $\beta_i : \mathbb{R}_0^+ \to \mathbb{R}$, $i \in \{1,2,3\}$, is an arbitrary choice of C^∞ functions and $r^2 = g_x(u,u)$. For $\alpha_2 = \beta_2 = 0$ one obtains precisely the g-natural metrics on $T(M)$ with respect to which the horizontal and vertical distributions are orthogonal. See also [191].

We adopt the following notations for further use.

$$\phi_i(t) = \alpha_i(t) + t\beta_i(t),$$

$$\alpha(t) = \alpha_1(t)(\alpha_1(t) + \alpha_3(t)) - \alpha_2(t)^2,$$

$$\phi(t) = \phi_1(t)(\phi_1(t) + \phi_3(t)) - \phi_2(t)^2,$$

for any $t \in \mathbb{R}_0^+$. Then

Lemma 5.1 *A g-natural metric G on the tangent bundle of a Riemannian manifold (M,g) is a Riemannian metric if and only if*

$$\alpha_1(t) > 0, \quad \phi_1(t) > 0, \quad \alpha(t) > 0, \quad \phi(t) > 0, \tag{5.5}$$

for any $t \in \mathbb{R}_0^+$.

The Sasaki metric G_s corresponds to the choice

$$\alpha_1(t) = 1, \quad \alpha_2(t) = \alpha_3(t) = \beta_1(t) = \beta_2(t) = \beta_3(t) = 0, \tag{5.6}$$

while the *Cheeger-Gromoll metric* G_{cg} is obtained for

$$\alpha_2(t) = \beta_2(t) = 0, \tag{5.7}$$

$$\alpha_1(t) = \beta_1(t) = -\beta_3(t) = \frac{1}{1+t}, \quad \alpha_3(t) = \frac{t}{1+t}. \tag{5.8}$$

The 2-parameter family of metrics investigated by V. Oproiu, [231], is a family of Riemannian g-natural metrics on $T(M)$. A Riemannian g-natural metric G on TM is a Kaluza-Klein metric (as commonly defined on principal bundles [315]) if $\alpha_2 = \beta_2 = \beta_1 + \beta_3 = 0$. All examples (i.e., G_s, G_{cg}, V. Oproiu's metrics and the Kaluza-Klein metrics) satisfy $\alpha_2 = \beta_2 = 0$ (hence the horizontal and vertical distributions are orthogonal). See also [34].

Proposition 5.2 (Cf. [6]) *Let (M,g) be a Riemannian manifold. Let G be a Riemannian g-natural metric on $T(M)$. Then the Levi-Civita connection $\bar{\nabla}$ of $(T(M),G)$ is given by*

$$(\bar{\nabla}_{X^H} Y^H)_u = (\nabla_X Y)^H_u + h\{A(u;X_x,Y_x)\} + v\{B(u;X_x,Y_x)\}, \tag{5.9}$$

$$(\bar{\nabla}_{X^H} Y^V)_u = (\nabla_X Y)^V_u + h\{C(u;X_x,Y_x)\} + v\{D(u;X_x,Y_x)\}, \tag{5.10}$$

$$(\bar{\nabla}_{X^V} Y^H)_u = h\{C(u;Y_x,X_x)\} + v\{D(u;Y_x,X_x)\}, \tag{5.11}$$

$$(\bar{\nabla}_{X^V} Y^V)_u = h\{E(u;X_x,Y_x)\} + v\{F(u;X_x,Y_x)\}, \tag{5.12}$$

for any $X,Y \in \mathfrak{X}(M)$ and any $u \in TM$ with $x = \pi(u) \in M$. Here $h\{z\}$ and $v\{z\}$ are the horizontal and vertical lifts of $z \in T_x(M)$. Also A, B, C, D, E, and F are given by

$$\begin{aligned} A(u;X_x,Y_x) = &-A_1\,[R_x(X_x,u)Y_x + R_x(Y_x,u)X_x] \\ &+ A_2\,[g_x(Y_x,u)X_x + g_x(X_x,u)Y_x] - A_3 g_x(R_x(X_x,u)Y_x,u)u \\ &+ A_4 g_x(X_x,Y_x)u + A_5 g_x(X_x,u)g_x(Y_x,u)u, \end{aligned} \tag{5.13}$$

with

$$A_1 = -\frac{\alpha_1\alpha_2}{2\alpha}, \quad A_2 = \frac{\alpha_2}{2\alpha}(\beta_1+\beta_3),$$
$$A_3 = \frac{1}{\alpha\phi}\alpha_2\{\alpha_1[\phi_1(\beta_1+\beta_3)-\phi_2\beta_2]+\alpha_2(\beta_1\alpha_2-\beta_2\alpha_1)\},$$
$$A_4 = \frac{\phi_2}{\phi}(\alpha_1+\alpha_3)',$$
$$A_5 = \frac{\phi_2}{\phi}(\beta_1+\beta_3)' + \frac{\alpha_2}{\alpha\phi}(\beta_1+\beta_3)[\phi_2\beta_2-\phi_1(\beta_1+\beta_3)] + \frac{1}{\alpha\phi}(\alpha_1+\alpha_3)(\alpha_1\beta_2-\alpha_2\beta_1),$$

$$\begin{aligned} B(u;X_x,Y_x) = &-B_1R_x(X_x,u)Y_x - B_2R_x(X_x,Y_x)u \\ &+ B_3[g_x(Y_x,u)X_x + g_x(X_x,u)Y_x] \\ &- B_4g_x(R_x(X_x,u)Y_x,u)u + B_5g_x(X_x,Y_x)u \\ &+ B_6g_x(X_x,u)g_x(Y_x,u)u, \end{aligned} \tag{5.14}$$

with

$$B_1 = \frac{\alpha_2^2}{\alpha}, \quad B_2 = -\frac{\alpha_1}{2\alpha}(\alpha_1+\alpha_3), \quad B_3 = -\frac{1}{2\alpha}(\alpha_1+\alpha_3)(\beta_1+\beta_3),$$
$$B_4 = \frac{\alpha_2}{\alpha\phi}\{\alpha_2[\phi_2\beta_2-\phi_1(\beta_1+\beta_3)]+(\alpha_1+\alpha_3)(\beta_2\alpha_1-\beta_1\alpha_2)\},$$
$$B_5 = -\frac{1}{\phi}(\phi_1+\phi_3)(\alpha_1+\alpha_3)',$$
$$\begin{aligned} B_6 = &-\frac{1}{\phi}(\phi_1+\phi_3)(\beta_1+\beta_3)' \\ &+\frac{1}{\alpha\phi}(\beta_1+\beta_3)(\alpha_1+\alpha_3)[(\phi_1+\phi_3)\beta_1-\phi_2\beta_2] \\ &+\frac{\alpha_2}{\alpha\phi}(\beta_1+\beta_3)[\alpha_2(\beta_1+\beta_3)-(\alpha_1+\alpha_3)\beta_2], \end{aligned}$$

$$\begin{aligned} C(u;X_x,Y_x) = &-C_1R(Y_x,u)X_x + C_2g_x(X_x,u)Y_x + C_3g_x(Y_x,u)X_x \\ &- C_4g_x(R_x(X_x,u)Y_x,u)u + C_5g_x(X_x,Y_x)u \\ &+ C_6g_x(X_x,u)g_x(Y_x,u)u, \end{aligned} \tag{5.15}$$

with

$$C_1 = -\frac{\alpha_1^2}{2\alpha}, \quad C_2 = -\frac{\alpha_1}{2\alpha}(\beta_1+\beta_3),$$
$$C_3 = \frac{\alpha_1}{\alpha}(\alpha_1+\alpha_3)' - \frac{\alpha_2}{\alpha}\left(\alpha_2' - \frac{\beta_2}{2}\right),$$
$$C_4 = \frac{\alpha_1}{2\alpha\phi}\{\alpha_2(\alpha_2\beta_1 - \alpha_1\beta_2) + \alpha_1[\phi_1(\beta_1+\beta_3) - \phi_2\beta_2]\},$$
$$C_5 = \frac{\phi_1}{2\phi}(\beta_1+\beta_3) + \frac{\phi_2}{2\phi}(2\alpha_2' - \beta_2),$$
$$C_6 = \frac{\phi_1}{\phi}(\beta_1+\beta_3)' + \frac{1}{\alpha\phi}\left[(\alpha_1+\alpha_3)' + \frac{\beta_1+\beta_3}{2}\right]$$
$$\times \{\alpha_2(\alpha_1\beta_2 - \alpha_2\beta_1) + \alpha_1[\phi_2\beta_2 - (\beta_1+\beta_3)\phi_1]\} + \frac{1}{\alpha\phi}\left[\alpha_2' - \frac{\beta_2}{2}\right]$$
$$\times \{\alpha_2[\beta_1(\phi_1+\phi_3) - \beta_2\phi_2] - \alpha_1[\beta_2(\alpha_1+\alpha_3) - \alpha_2(\beta_1+\beta_3)]\},$$

$$\begin{aligned} D(u; X_x, Y_x) = &-D_1 R_x(Y_x, u)X_x + D_2 g_x(X_x, u)Y_x + D_3 g_x(Y_x, u)X_x \\ &- D_4 g_x(R_x(X_x, u)Y_x, u)u + D_5 g_x(X_x, Y_x)u \\ &+ D_6 g_x(X_x, u) g_x(Y_x, u)u, \end{aligned} \tag{5.16}$$

with

$$D_1 = \frac{\alpha_1\alpha_2}{2\alpha}, \quad D_2 = \frac{\alpha_2}{2\alpha}(\beta_1+\beta_3),$$
$$D_3 = -\frac{\alpha_2}{\alpha}(\alpha_1+\alpha_3)' + \frac{1}{\alpha}(\alpha_1+\alpha_3)\left[\alpha_2' - \frac{\beta_2}{2}\right],$$
$$D_4 = \frac{\alpha_1}{2\alpha\phi}\{(\alpha_1+\alpha_3)(\alpha_1\beta_2 - \alpha_2\beta_1) + \alpha_2[\phi_2\beta_2 - \phi_1(\beta_1+\beta_3)]\},$$
$$D_5 = -\frac{\phi_2}{2\phi}(\beta_1+\beta_3) + \frac{1}{2\phi}(\phi_1+\phi_3)(2\alpha_2' - \beta_2),$$
$$D_6 = -\frac{\phi_2}{\phi}(\beta_1+\beta_3)' + \frac{1}{\alpha\phi}\left[(\alpha_1+\alpha_3)' + \frac{\beta_1+\beta_3}{2}\right]$$
$$\times \{(\alpha_1+\alpha_3)(\alpha_2\beta_1 - \alpha_1\beta_2) + \alpha_2[\phi_1(\beta_1+\beta_3) - \phi_2\beta_2]\}$$
$$+ \frac{1}{\alpha\phi}\left[\alpha_2' - \frac{\beta_2}{2}\right]\{(\alpha_1+\alpha_3)[\beta_2\phi_2 - \beta_1(\phi_1+\phi_3)]$$
$$+ \alpha_2[\beta_2(\alpha_1+\alpha_3) - \alpha_2(\beta_1+\beta_3)]\},$$

$$E(u;X_x,Y_x) = E_1[g_x(Y_x,u)X_x + g_x(X_x,u)Y_x] + E_2 g_x(X_x,Y_x)u + E_3 g_x(X_x,u)g_x(Y_x,u)u, \qquad (5.17)$$

with

$$E_1 = \frac{\alpha_1}{\alpha}\left[\alpha_2' + \frac{\beta_2}{2}\right] - \frac{\alpha_2}{\alpha}\alpha_1', \quad E_2 = \frac{1}{\phi}\left[\phi_1\beta_2 - \phi_2(\beta_1 - \alpha_1')\right],$$

$$E_3 = \frac{1}{\phi}\left(2\phi_1\beta_2' - \phi_2\beta_1'\right) + \frac{2}{\alpha\phi}\alpha_1'\{\alpha_1[\alpha_2(\beta_1+\beta_3) - \beta_2(\alpha_1+\alpha_3)]$$
$$+\alpha_2[\beta_1(\phi_1+\phi_3) - \beta_2\phi_2]\} + \frac{1}{\alpha\phi}(2\alpha_2' + \beta_2)\{\alpha_1[\phi_2\beta_2 - \phi_1(\beta_1+\beta_3)]$$
$$+\alpha_2(\alpha_1\beta_2 - \alpha_2\beta_1)\},$$

$$F(u;X_x,Y_x) = F_1[g_x(Y_x,u)X_x + g_x(X_x,u)Y_x] + F_2 g_x(X_x,Y_x)u + F_3 g_x(X_x,u)g_x(Y_x,u)u, \qquad (5.18)$$

with

$$F_1 = -\frac{\alpha_2}{\alpha}\left[\alpha_2' + \frac{\beta_2}{2}\right] + \frac{1}{\alpha}(\alpha_1+\alpha_3)\alpha_1',$$

$$F_2 = \frac{1}{\phi}\left[(\phi_1+\phi_3)(\beta_1 - \alpha_1') - \phi_2\beta_2\right],$$

$$F_3 = \frac{1}{\phi}\left[(\phi_1+\phi_3)\beta_1' - 2\phi_2\beta_2'\right] + \frac{2}{\alpha\phi}\alpha_1'\{\alpha_2[\beta_2(\alpha_1+\alpha_3) - \alpha_2(\beta_1+\beta_3)]$$
$$+ (\alpha_1+\alpha_3)[\beta_2\phi_2 - \beta_1(\phi_1+\phi_3)]\}$$
$$+ \frac{1}{\alpha\phi}(2\alpha_2' + \beta_2)\{\alpha_2[\phi_1(\beta_1+\beta_3) - \phi_2\beta_2]$$
$$+ (\alpha_1+\alpha_3)(\alpha_2\beta_1 - \alpha_1\beta_2)\}.$$

Here, ∇ and R are the Levi-Civita connection and curvature tensor field of (M,g).

5.1.1. Generalized Cheeger-Gromoll Metrics

Let (M,g) be a Riemannian manifold and $\pi : E \to M$ be a Riemannian vector bundle endowed with a connection ∇ and a holonomy-invariant fibre metric $\langle , \rangle$. Let us denote by K the Dombrowski map associated to ∇. M. Benyounes et al., [32], introduced the *generalized Cheeger-Gromoll metrics* $h_{p,q}$ on E and studied harmonic sections in E. Given $p,q \in \mathbb{R}$ $z \in E$ the generalized Cheeger-Gromoll metrics $h_{p,q}$ are given by

$$h_{p,q}(A,B) = g((d_z\pi)A,(d_z\pi)B) + \frac{1}{(1+\|u\|^2)^p}\left[\langle K_zA,K_zB\rangle + q\langle K_zA,z\rangle\langle K_zB,z\rangle\right]$$

for any $z \in E$ and any $A,B \in T_z(E)$. If $q \geq 0$ then $h_{p,q}$ is a Riemannian metric. If $q < 0$ then $h_{p,q}$ has varying signature and is a Riemannian metric only in a tubular neighborhood of the zero section. When $E = T(M)$ the generalized Cheeger-Gromoll metrics $h_{p,q}$ are given by

$$h_{p,q}(X^H,Y^H) = g_x(X,Y),$$
$$h_{p,q}(X^H,Y^V) = h_{p,q}(X^V,Y^H) = 0,$$
$$h_{p,q}(X^V,Y^V) = \frac{1}{(1+\|u\|^2)^p}\left[g_x(X,Y) + qg_x(X,u)g_x(Y,u)\right],$$

for any $u \in T(M)$ with $x = \pi(u)$ and any $X,Y \in T_x(M)$. Let us denote by $G_{\alpha_1,\beta_1,k}$ the g-natural Riemannian metric determined by the smooth functions α_i, β_i such that

$$\alpha_1 > 0,\ \alpha_1 + \alpha_3 = k,\ \beta_1 + \beta_3 = 0,\ \beta_1 \geq 0,\ \alpha_2 = \beta_2 = 0, \tag{5.19}$$

where $k > 0$ is a constant. Note that $G_{\alpha_1,\beta_1,k}$ are Kaluza-Klein metrics. The functions (5.19) satisfy the inequalities (5.5) that is

$$\alpha_1(t) > 0, \quad \phi_1(t) > 0, \quad \alpha(t) = k\alpha_1(t) > 0, \quad \phi(t) = k\phi_1(t) > 0,$$

for any $t \in \mathbb{R}_0^+$. In the Riemannian case (i.e., when $q \geq 0$) the generalized Cheeger-Gromoll metrics $h_{p,q}$ are of type $G_{\alpha_1,\beta_1,k}$ where

$$\alpha_1 = \frac{1}{(1+t)^p}, \quad \beta_1 = \frac{q}{(1+t)^p} \quad \text{and} \quad k = 1.$$

It should be observed that $G_s = h_{0,0}$ and $G_{cg} = h_{1,1}$.

5.1.2. g-Natural Riemannian Metrics on $S_\rho(M)$

Given a Riemannian manifold (M,g) we set as usual $S_\rho(M)_x = \{u \in T_x(M) : g_x(u,u) = \rho^2\}$ for any $x \in M$.

Definition 5.3 A *g-natural metric* on $S_\rho(M)$ is the restriction to $S_\rho(M)$ of a g-natural metric on $T(M)$. ■

As shown in [7], every Riemannian g-natural metric $\tilde{G}$ on $S(M) = S_1(M)$ is necessarily induced by a Riemannian g-natural metric G on $T(M)$ of the form

$$\begin{cases} G_u(X^H, Y^H) = (a+c)\, g_x(X,Y) + \beta\, g_x(X,u) g_x(Y,u), \\ G_u(X^H, Y^V) = b\, g_x(X,Y), \\ G_u(X^V, Y^V) = a g_x(X,Y), \end{cases} \tag{5.20}$$

with $a, b, c \in \mathbb{R}$ and $\beta \in C^\infty(\mathbb{R}_0^+, \mathbb{R})$. A comparison to the general expressions (5.1)–(5.4) shows that

$$\alpha_1 = a, \quad \alpha_2 = b, \quad \alpha_3 = c, \quad \beta_1 = \beta_2 = 0, \quad \beta_3 = \beta. \tag{5.21}$$

Such $\tilde{G}$ depends solely on $d = \beta(1)$ (rather than on the full β). The identities (5.5) and (5.21) show that $\tilde{G}$ is a Riemannian metric if and only if

$$a > 0, \quad \alpha = a(a+c) - b^2 > 0, \quad \phi = a(a+c+d) - b^2 > 0. \tag{5.22}$$

The Sasaki metric $\tilde{G}_s$ and the Cheeger-Gromoll metric $\tilde{G}_{cg}$ are Riemannian g-natural metrics on $S(M)$ and satisfy $b = 0$. One may easily check that the vector field $N^G \in \mathfrak{X}(T(M))$ given by

$$N_u^G = \frac{1}{\sqrt{(a+c+d)\phi}} [-bu^H + (a+c+d)u^V], \quad u \in T(M),$$

is a unit normal vector field on $S(M)$.

Let X^{tG} be the *tangential lift* of $X \in T_x(M)$ i.e., the tangential projection of the vertical lift of X to $u \in S(M)$

$$\begin{aligned} X^{tG} &= X^V - G_u(X^V, N_u^G)\, N_u^G \\ &= X^V - \sqrt{\frac{\phi}{a+c+d}}\, g_x(X,u)\, N_u^G, \quad u \in S(M), \quad x = \pi(u). \end{aligned}$$

If $X \in T_x(M)$ is orthogonal to u then $X^{tG} = X^V$. The tangent space $T_u(S(M))$ is spanned by vectors of the form X^H and Y^{tG} with $X, Y \in T_x(M)$. Consequently for all $x \in M$,

$$\begin{cases} \tilde{G}_u(X^H, Y^H) = (a+c)\, g_x(X,Y) + d\, g_x(X,u) g_x(Y,u), \\ \tilde{G}_u(X^H, Y^{tG}) = b\, g_x(X,Y), \\ \tilde{G}_u(X^{tG}, Y^{tG}) = a g_x(X,Y) - \dfrac{\phi}{a+c+d} g_x(X,u) g_x(Y,u), \end{cases} \tag{5.23}$$

for any $u \in S(M)$ with $x = \pi(u)$ and any $X, Y \in T_x(M)$. Also $a, b, c, d \in \mathbb{R}$ must satisfy the inequalities (5.22). Should we replace $S(M)$ by $S_\rho(M)$ in the above considerations the term $a+c+d$ must be replaced by $\rho^2(a+c+\rho^2 d)$. Note that (by (5.23)) horizontal and vertical vectors are orthogonal with respect to $\tilde{G}$ if and only if $b = 0$.

Remark 5.4 As a contact metric manifold $S(M)$ has been traditionally equipped with the Riemannian metric $\bar{g}$ homothetic to $\tilde{G}_s$ with the homothety factor $1/4$, inducing the standard contact metric structure (η, G_{cs}) on $S(M)$. By a recent result of M.T.K. Abbassi et al., [8], there is a family of contact metric structures $(\tilde{G}, \tilde{\eta}, \tilde{\varphi}, \tilde{\xi})$ over $S(M)$ (referred to as *natural contact metric structures* (or *g-natural contact metric structures*)) depending on three parameters $a, b, c \in \mathbb{R}$ where $\tilde{G}$ is given by (5.23) and

$$\tilde{\xi}_{(x,u)} = r u^H, \quad 1/r^2 = 4[a(a+c) - b^2] = a+c+d,$$

$$\tilde{\eta}(X^H) = \frac{1}{r} g(X, u), \quad \tilde{\eta}(X^{tG}) = brg(X, u),$$

$$\tilde{\varphi}(X^H) = \frac{1}{2r\alpha}\left[-bX^H + (a+c)X^{tG} + \frac{bd}{a+c+d} g(X, u)u^H\right],$$

$$\tilde{\varphi}(X^{tG}) = \frac{1}{2r\alpha}\left[-aX^H + bX^{tG} + \frac{\phi}{a+c+d} g(X, u)u^H\right],$$

for every $X \in T(M)$. For $a = 1/4$, $b = c = d = 0$ and $r = 2$ we get the standard contact metric structure on $S(M)$ induced from the Sasaki metric. ■

5.2. NATURALLY HARMONIC VECTOR FIELDS

5.2.1. The Energy of $V : (M, g) \to (T(M), G)$

Let (M, g) be a compact n-dimensional Riemannian manifold and $(T(M), G)$ its tangent bundle equipped with an arbitrary Riemannian g-natural metric G. Each vector field $V \in \mathfrak{X}(M)$ may be looked at as a smooth map of Riemannian manifolds $V : (M, g) \to (T(M), G)$. As

$(dV)X = X^H + (\nabla_X V)^V$ for any $X \in \mathfrak{X}(M)$ one has (by (5.4))

$$\begin{aligned}(V^*G)(X,Y) &= G(X^H + (\nabla_X V)^V, Y^H + (\nabla_Y V)^V)\\ &= (\alpha_1+\alpha_3)(r^2)g(X,Y) + (\beta_1+\beta_3)(r^2)g(X,V)g(Y,V)\\ &\quad + \alpha_2(r^2)\,[g(X,\nabla_Y V) + g(Y,\nabla_X V)]\\ &\quad + \beta_2(r^2)\,[g(X,V)g(\nabla_Y V,V) + g(Y,V)g(\nabla_X V,V)]\\ &\quad + \alpha_1(r^2)g(\nabla_X V,\nabla_Y V) + \beta_1(r^2)g(\nabla_X V,V)g(\nabla_Y V,V),\end{aligned} \tag{5.24}$$

for any $X, Y \in \mathfrak{X}(M)$, where $r = \|V\|$. It should be observed (cf. (5.24)) that in general V^*G depends on the length of V.

As usual the energy $E(V)$ of V is the energy of the map $V : (M,g) \to (T(M),G)$. That is $E(V) = \int_M e(V)\,d\mathrm{vol}(g)$ where the density $e(V)$ is given by

$$e_x(V) = \frac{1}{2}\|d_x V\|^2 = \frac{1}{2}\mathrm{trace}_g(V^*G)_x. \tag{5.25}$$

By (5.24) one obtains (cf. [2])

$$\begin{aligned}E(V) = \frac{1}{2}\int_M \Big\{ & n(\alpha_1+\alpha_3)(r^2) + (\beta_1+\beta_3)(r^2)r^2 + 2\alpha_2(r^2)\mathrm{div}(V)\\ & + 2\beta_2(r^2)\,V(r^2) + \alpha_1(r^2)\|\nabla V\|^2 + \frac{1}{4}\beta_1(r^2)\|\nabla r^2\|^2 \Big\}\, d\mathrm{vol}(g)\end{aligned} \tag{5.26}$$

where $r = \|V\|$. When $T(M)$ is equipped with a g-natural Riemannian metric of type $G_{\alpha_1,\beta_1,k}$ then (by (5.26))

$$\begin{aligned}E_{\alpha_1,\beta_1,k}(V) &= \frac{1}{2}\int_M \Big\{ nk + \alpha_1(r^2)\|\nabla V\|^2 + \frac{1}{4}\beta_1(r^2)\|\nabla r^2\|^2 \Big\}\, d\mathrm{vol}(g)\\ &\geq \frac{nk}{2}\,\mathrm{Vol}(M),\end{aligned} \tag{5.27}$$

and the equality holds if and only if V is parallel. Then for each of the metrics $G_{\alpha_1,\beta_1,k}$ (including G_s and $h_{p,q}$) parallel vector fields are precisely the absolute minima for the energy functional (this extends Corollary 2.8 in Chapter 2).

The vertical energy of a section σ in a Riemannian vector bundle E equipped with a generalized Cheeger-Gromoll metric $h_{p,q}$ is computed in [32]. One has

$$E^v_{p,q}(\sigma) = \frac{1}{2}\int_M \|\sigma_*^V\|^2 \, d\mathrm{vol}(g)$$
$$= \frac{1}{2}\int_M \frac{1}{(1+r^2)^p}\left(\|\nabla\sigma\|^2 + \frac{q}{4}\|\nabla r^2\|^2\right) d\mathrm{vol}(g) \tag{5.28}$$

where $r = \|\sigma\|$. The full energy and the vertical energy differ by an additive constant (cf. Remark 2.6 in Chapter 2).

Let us equip $T(M)$ with an arbitrary Riemannian g-natural metric G and compute $E(V)$ for each $V \in \mathfrak{X}^\rho(M)$. Here

$$\mathfrak{X}^\rho(M) = \left\{V \in \mathfrak{X}(M) : \|V\|^2 = \rho\right\}$$

with $\rho =$ constant > 0. One obtains

$$E(V) = \frac{1}{2}[(n-1)(\alpha_1+\alpha_3)+\phi_1+\phi_3](\rho)\,\mathrm{Vol}(M,g)$$
$$+\frac{1}{2}\alpha_1(\rho)\cdot\int_M \|\nabla V\|^2 \, d\mathrm{vol}(g). \tag{5.29}$$

Then (by $\alpha_1 > 0$ and (5.29))

$$E(V) \geq \frac{1}{2}[(n-1)(\alpha_1+\alpha_3)+\phi_1+\phi_3](\rho)\,\mathrm{Vol}(M,g) > 0,$$

where (by (5.5)) $[(n-1)(\alpha_1+\alpha_3)+\phi_1+\phi_3] > 0$. We have proved the following

Proposition 5.5 *Let (M,g) be a compact Riemannian manifold. Let G be a Riemannian g-natural metric on $T(M)$. A vector field $V \in \mathfrak{X}^\rho(M)$ is an absolute minimum for the energy $E : \mathfrak{X}^\rho(M) \to [0,+\infty)$ if and only if V is parallel.*

5.2.2. The Tension Field of $V : (M,g) \to (T(M),G)$

Let (M,g) be a Riemannian manifold, $V \in \mathfrak{X}(M)$ a tangent vector field, and G an arbitrary Riemannian g-natural metric on $T(M)$. The tension field $\tau(V)$ of $V : (M,g) \to (T(M),G)$ has been computed in [2] (the calculation is rather involved). Using the identities (5.9)–(5.12) in Proposition 5.2 one may show that

Theorem 5.6 (Cf. [2]) *Let (M,g) be a Riemannian manifold, $V \in \mathfrak{X}(M)$, and G an arbitrary Riemannian g-natural metric on $T(M)$. Then*

$$\tau(V)_{V_x} = \tau_h(V)^H_{V_x} + \tau_v(V)^V_{V_x}, \quad x \in M, \tag{5.30}$$

where

$$\begin{aligned}\tau_h(V) &= -2A_1 QV - 2C_1 \mathrm{tr}[R(\nabla . V, V)\cdot] + C_3 \nabla r^2 + E_1 \nabla_{\nabla r^2} V \\ &\quad + 2C_2 \nabla_V V + \big[2A_2 - A_3 g(QV, V) + nA_4 + A_5 r^2 \\ &\quad - 2C_4 g(\mathrm{trace}_g[R(\nabla . V, V)\cdot], V) + 2C_5 \mathrm{div}(V) \\ &\quad + C_6 V(r^2) + E_2 \|\nabla V\|^2 + (1/4)E_3 \|\nabla r^2\|^2\big] V \end{aligned} \tag{5.31}$$

and

$$\begin{aligned}\tau_v(V) &= -\Delta_g V - B_1 QV - 2D_1 \mathrm{trace}_g[R(\nabla . V, V)\cdot] + D_3 \nabla r^2 \\ &\quad + F_1 \nabla_{\nabla r^2} V + 2D_2 \nabla_V V + [2B_3 - B_4 g(QV, V) + nB_5 + B_6 r^2 \\ &\quad - 2D_4 g(\mathrm{trace}_g[R(\nabla . V, V)\cdot], V) + 2D_5 \mathrm{div}(V) \\ &\quad + D_6 V(r^2) + F_2 \|\nabla V\|^2 + (1/4)F_3 \|\nabla r^2\|^2\big] V. \end{aligned} \tag{5.32}$$

When $G = G_s$ is the Sasaki metric we obtain Proposition 2.12 in Chapter 2. Except for the case where $\nabla V = 0$ (cf. Remark 5.17 below) and few other special cases the equations $\tau_h(V) = 0$ and $\tau_v(V) = 0$ are difficult to work with in full generality, even for vector fields of constant length. In the special case of a g-natural Riemannian metric G with $\alpha_2(\rho) = \beta_2(\rho) = 0$ we get

Theorem 5.7 (Cf. [2]) *Let (M,g) be a Riemannian manifold and G a g-natural Riemannian metric on $T(M)$ satisfying $\alpha_2(\rho) = \beta_2(\rho) = 0$, $\rho > 0$. Then a vector field $V \in \mathfrak{X}^\rho(M)$ is a harmonic map of (M,g) into $(T(M), G)$ if and only if*

$$\begin{aligned}&(\phi_1 + \phi_3)(\rho)\big[\alpha_1(\rho)\,\mathrm{trace}_g\{R(\nabla . V, V)\cdot\} - (\beta_1 + \beta_3)(\rho)\nabla_V V\big] \\ &\quad + (\beta_1 + \beta_3)(\rho)\,[(\alpha_1 + \alpha_3)(\rho)\mathrm{div}(V) \\ &\quad - \alpha_1(\rho) g(\mathrm{trace}_g\{R(\nabla . V, V)\cdot\}, V)\big] V = 0 \end{aligned} \tag{5.33}$$

and

$$\Delta_g V + \left[\frac{(\beta_1+\beta_3)}{\alpha_1}(\rho) + n\frac{(\alpha_1+\alpha_3)'}{\phi_1}(\rho) \right.$$
$$\left. + \rho\frac{\alpha_1(\beta_1+\beta_3)' - \beta_1(\beta_1+\beta_3)}{\alpha_1\phi_1}(\rho) + \frac{\alpha_1' - \beta_1}{\phi_1}(\rho)\|\nabla V\|^2 \right] V = 0. \tag{5.34}$$

In particular (by (5.34)), $\Delta_g V$ and V are collinear. Therefore V is an eigenvector of the rough Laplacian Δ_g and (since $\sqrt{\rho} = \|V\| =$ constant) one has $\Delta_g V = \frac{1}{\rho}\|\nabla V\|^2 V$ and then (by (5.34))

$$\left(\frac{1}{\rho}\alpha_1 + \alpha_1'\right)(\rho)\|\nabla V\|^2 + [n(\alpha_1+\alpha_3) + t(\beta_1+\beta_3)]'(\rho) = 0.$$

As the metrics $G_{\alpha_1,\beta_1,k}$ satisfy (5.19), Theorem 5.7 admits the following

Corollary 5.8 *Let V be a smooth vector field on M with $\|V\|^2 = \rho =$ constant.*

i. *If $\alpha_1(\rho) + \rho\alpha_1'(\rho) \neq 0$ then $V : (M,g) \to (T(M), G_{\alpha_1,\beta_1,k})$ is a harmonic map if and only if V is parallel.*

ii. *If $\alpha_1(\rho) + \rho\alpha_1'(\rho) = 0$ then $V : (M,g) \to (T(M), G_{\alpha_1,\beta_1,k})$ is a harmonic map if and only if $\text{trace}_g\{R(\nabla . V, V)\cdot\} = 0$ and $\Delta_g V$ is collinear to V. If this is the case $V : (M,g) \to (S_{\sqrt{\rho}}(M), \tilde{G}_s)$ is a harmonic map, too. Here $\tilde{G}_s$ is the Sasaki metric induced on $S_{\sqrt{\rho}}(M)$.*

5.2.3. Naturally Harmonic Vector Fields

By a result of O. Gil-Medrano, [126] (cf. Theorem 2.17 in Chapter 2) a vector field $V : (M,g) \to (T(M), G_s)$ is a critical point of the energy functional $E : \mathfrak{X}(M) \to [0,+\infty)$ if and only if V is parallel. When G_s is replaced by a generalized Cheeger-Gromol metric $h_{p,q}$ the same matter was investigated by M. Benyounes & E. Loubeau & C.M. Wood, [32]–[33]. We look at the case where $T(M)$ is equipped with an arbitrary g-natural Riemannian metric G. Let M be compact. A vector field $V \in \mathfrak{X}(M)$ is a critical point for the energy functional $E : \mathfrak{X}(M) \to [0,+\infty)$ if and only if (cf. [2])

$$\alpha_2\tau_h(V) + \beta_2 g(\tau_h(V), V)V + \alpha_1\tau_v(V) + \beta_1 g(\tau_v(V), V)V = 0. \tag{5.35}$$

By (5.35), the projection of the tension field $\tau(V)$ on the vertical distribution vanishes (for any Riemannian g-natural metric G).

Definition 5.9 A vector field V is called *G-harmonic* if it satisfies the critical point condition (5.35). ∎

Clearly if $V:(M,g)\to(T(M),G)$ is a harmonic map then V is G-harmonic. In general the converse does not hold. Indeed let us look at the case where $\alpha_2=\beta_2=0$. Under this assumption (5.35) is equivalent to $\tau_v(V)=0$. Several consequences of (5.35) are examined in [2]. In particular when $\alpha_2=\beta_2=0$ and $\|V\|$ is constant $\tau_v(V)=0$ is equivalent to (5.34). Then

Theorem 5.10 *Let V be a smooth vector field on M with $\|V\|^2=\rho\in(0,+\infty)$.*

i. *If $\alpha_1(\rho)+\rho\alpha_1'(\rho)\neq 0$ then V is $G_{\alpha_1,\beta_1,k}$-harmonic if and only if $\nabla V=0$.*

ii. *If $\alpha_1(\rho)+\rho\alpha_1'(\rho)=0$ then V is $G_{\alpha_1,\beta_1,k}$-harmonic if and only if $\Delta_g V=(1/\rho)\|\nabla V\|^2 V$. If this is the case V is a harmonic section in $S_{\sqrt{\rho}}(M)$ i.e., it is harmonic with respect to the Sasaki metric for variations through vector fields of length $\sqrt{\rho}$.*

If $G_{\alpha_1,\beta_1,k}=h_{p,q}$ the condition $\alpha_1(\rho)+\rho\alpha_1'(\rho)=0$ becomes $p=1+1/\rho$ (cf. [32]). This condition is satisfied for the metric $G_{\alpha_1,\beta_1,k}$ with $\alpha_1=k_1e^{-(1/\rho)t}$, $k_1\in(0,+\infty)$.

Let us consider the vertical energy for a section σ in a Riemannian vector bundle E (over a compact Riemannian manifold M) equipped with a generalized Cheeger-Gromoll metric $h_{p,q}$. It is given by (5.28). The Euler-Lagrange equations associated to the variational principle $\delta E^v_{p,q}(\sigma)=0$ read (variations are of course through sections in E, cf. M. Benyounes et al., [32])

$$(1+r^2)\Delta_g\sigma+p\nabla_{\nabla r^2}\sigma=\left[p\|\nabla\sigma\|^2-\frac{pq}{4}\|\nabla r^2\|^2-\frac{q}{2}(1+r^2)\Delta r^2\right]\sigma, \tag{5.36}$$

where $\|\sigma\|=r$.

Definition 5.11 A smooth section σ satisfying[1] (5.36) is called $h_{p,q}$-harmonic. ∎

We end the section with the following remarks. a) If $\|\sigma\|^2=\rho=$ constant >0 then σ is $h_{p,q}$-harmonic if and only if $\nabla\sigma=0$ except when $p=1+1/\rho$. In this case, σ is $h_{p,q}$-harmonic if and only if σ is a harmonic

[1] With M not necessarily compact.

section in the sphere bundle $S_{\sqrt{\rho}}(E)$ (i.e., σ is G_s-harmonic with respect to variations through sections in $S_{\sqrt{\rho}}(E)$). Cf. [32].

b) If M is compact and $\|\sigma\|$ is not constant then for each $p \in \mathbb{R}$ there is at most one $q \in \mathbb{R}$ such that σ is $h_{p,q}$-harmonic. Cf. [32].

5.2.4. Vertically Harmonic Vector Fields

Let (M,g) be a compact orientable Riemannian manifold and $(T(M),\tilde{g})$ the total space of its tangent bundle equipped with an arbitrary Riemannian metric $\tilde{g}$. The *vertical energy* of a vector field $V:(M,g)\to(T(M),\tilde{g})$ is given by

$$E^v_{\tilde{g}}(V) = \int_M \| V^v_* \|^2 d\,\mathrm{vol}(g)$$

where V^v_* is the vertical component of $V_*: T(M) \to T(T(M)) = \mathcal{H} \oplus \mathcal{V}$ (a slightly different notation was used in Section 5.1). If $\tilde{g} = G$ is a Riemannian g-natural metric on $T(M)$ then (by (5.4))

$$\begin{aligned} \| V^v_* \|^2 &= \frac{1}{2}\sum_{i=1}^{n} G_V((\nabla_{e_i} V)^v, (\nabla_{e_i} V)^v) \\ &= \frac{1}{2}\left[\alpha_1(r^2)\|\nabla V\|^2 + \frac{1}{4}\beta_1(r^2)\|\nabla r^2\|^2\right] \end{aligned}$$

where $r^2 = \| V \|^2$. Hence

$$E^v_G(V) = \frac{1}{2}\int_M \left[\alpha_1(r^2)\|\nabla V\|^2 + \frac{1}{4}\beta_1(r^2)\|\nabla r^2\|^2\right] d\,\mathrm{vol}(g) \geq 0. \tag{5.37}$$

Therefore (by (5.37) and the fact that parallel vector fields have constant length), any parallel vector field V is an absolute minimum of E^v_G i.e., V is vertically harmonic. If $\nabla V = 0$ the equation (5.35) becomes

$$[n(\alpha_1+\alpha_3)+t(\beta_1+\beta_3)]'(\rho) = 0, \qquad (\rho = \| V \|^2)$$

and one obtains

Theorem 5.12 ([2]) *Let (M,g) be a compact orientable Riemannian manifold and V a parallel vector field on M. Then V is vertically harmonic but not G-harmonic if and only if $\rho = \| V\|^2 \in (0,+\infty)$ satisfies $[n(\alpha_1+\alpha_3)+t(\beta_1+\beta_3)]'(\rho) \neq 0$.*

The feature demonstrated in Theorem 5.12 is not enjoyed by the Sasaki metric, by the Cheeger-Gromoll type metrics in [32], or by the class of

Kaluza-Klein metrics in [34] (all of the above are Riemannian g-natural metrics with $\alpha_1+\alpha_3=$ constant and $\beta_1+\beta_3=0$). One may however exhibit wide classes of Riemannian g-natural metrics G to which Theorem 5.12 applies. For instance let G be determined by the functions α_i and β_i satisfying (5.5) and

$$\alpha_1+\alpha_3=e^{at}, \quad \beta_1+\beta_3=b, \quad a,b\in\mathbb{R}, \ a\neq 0, \quad ab\geq 0.$$

These are the first examples of Riemannian metrics on $T(M)$ for which vertically harmonic vector fields need not be harmonic. Vertical harmonicity with respect to Riemannian g-natural metrics appears to be worth further investigation.

5.2.5. Naturally Harmonic Unit Vector Fields

We start by giving a natural generalization of a result due to C.M. Wood and G. Wiegmink. Here a unit vector field V is thought as a map $V:(M,g)\to(S(M),\tilde{G})$, where $\tilde{G}=\iota^*G$ and $\iota: S(M)\to T(M)$ is the inclusion. Of course the energy densities of $V:(M,g)\to(S(M),\tilde{G})$ and $V:(M,g)\to(T(M),G)$ are the same. The energy of $V:(M,g)\to(T(M),G)$ is given by (5.26). Let now $\tilde{G}$ be an arbitrary Riemannian g-natural metric on $S(M)$. Then $\tilde{G}$ is given by (5.23) with $a,b,c,d\in\mathbb{R}$ satisfying (5.22). Then $\tilde{G}=\iota^*G$ for some Riemannian g-natural metric G on $T(M)$ of the form (5.20). Then (5.26) reads

$$E(V)=\frac{1}{2}[n(a+c)+d]\operatorname{Vol}(M,g)+\frac{a}{2}\int_M \|\nabla V\|^2\, d\operatorname{vol}(g). \tag{5.38}$$

By $a>0$ and (5.38)

$$\begin{aligned} E(V) &\geq \frac{1}{2}[n(a+c)+d]\operatorname{Vol}(M,g) \\ &= \frac{1}{2}[(n-1)(a+c)+a+c+d]\operatorname{Vol}(M,g)>0, \end{aligned} \tag{5.39}$$

for any $V\in\Gamma^\infty(S(M))$. The equality is achieved in (5.39) if and only if V is parallel. We may state the generalization announced above

Theorem 5.13 *Let (M,g) be a compact orientable Riemannian manifold. Let $\tilde{G}$ be a Riemannian g-natural metric on $S(M)$. A unit vector field V is an absolute minimum for the energy $E:\Gamma^\infty(S(M))\to[0,+\infty)$ if and only if V is parallel.*

Definition 5.14 Given a compact orientable Riemannian manifold (M,g) a vector field $V\in\Gamma^\infty(S(M))$ is said to be *harmonic* if and only if the map $V:(M,g)\to(S(M),\tilde{G})$ is a critical point of the energy functional $E:\Gamma^\infty(S(M))\to[0,+\infty)$ associated to g and $\tilde{G}$ (variations are of course through unit vector fields). ■

Let $V(t)$ be a smooth 1-parameter variation of V in $\Gamma^\infty(S(M))$. Then (by (5.38))

$$E(t)=E(V(t))=\frac{n(a+c)+d}{2}\mathrm{Vol}(M,g)+\frac{a}{2}\int_M \|\nabla V(t)\|^2\, d\mathrm{vol}(g). \tag{5.40}$$

Differentiating (5.40) we obtain (as $V(0)=V$)

$$E'(0)=a\int_M g(\nabla V,\nabla V'(0))dv_g=a\int_M g(\Delta_g V,V'(0))\,d\mathrm{vol}(g). \tag{5.41}$$

To derive the last equality we made use of the identity

$$\Delta g(X,Y)=g(\Delta_g X,Y)+g(X,\Delta_g Y)-2g(\nabla X,\nabla Y)$$

for any $X,Y\in\mathfrak{X}(M)$. Note that $V'(0)$ is orthogonal to V. Moreover for any vertical vector field W^V there exists a variation $\{V(t)\}$ of V by unit vector fields such that $W^V=V'(0)$.

By $a>0$ and (5.41) it follows that V is harmonic if and only if the component of $\Delta_g V$ orthogonal to V vanishes, that is $\Delta_g V$ and V are collinear. Thus

Theorem 5.15 (Cf. [3]) *Let (M,g) be a compact orientable Riemannian manifold. Let $\tilde{G}$ be a g-natural Riemannian metric on $S(M)$. A unit vector field V is harmonic if and only if $\Delta_g V$ and V are collinear.*

It is noteworthy (as a consequence of Theorem 5.15) that the harmonicity of V doesn't depend upon the choice of $\tilde{G}$. Another formulation of this fact is that the result by C.M. Wood, [316], and G. Wiegmink, [309] (cf. Theorem 2.23 in this book), is invariant under a 4-parameter deformation of the Sasaki metric.

5.3. VECTOR FIELDS WHICH ARE NATURALLY HARMONIC MAPS

In the theory of harmonic maps, a fundamental question concerns the existence of harmonic maps between two given Riemannian manifolds

(M,g) and (M',g'). If (M,g) is compact and the target space (M',g') is a Riemannian manifold of non-positive sectional curvature, there is a harmonic map $f:(M,g)\to(M',g')$ in each homotopy class of maps from M to M' (cf. J. Eells & J.H. Sampson, [113]). No general existence result is known when (M',g') does not satisfy this condition. It is therefore an interesting task to find examples of harmonic maps into Riemannian manifolds whose sectional curvature isn't necessarily non-positive. In the absence of general existence theorems one will adopt direct *ad hoc* construction methods.

5.3.1. A Generalization of the Ishihara-Nouhaud Theorem

The purpose of this section is to investigate whether the "rigidity" property exhibited in the Ishihara-Nouhaud theorem (i.e., Theorem 2.10 in Chapter 2) is peculiar to the Sasaki metric. Our finding is that other Riemannian g-natural metrics possessing the same property do exist.

Theorem 5.16 *Let (M,g) be a compact Riemannian manifold and G a Riemannian g-natural metric on $T(M)$ satisfying*

$$\begin{cases}\alpha_2=\beta_2=0, \quad \alpha_1,\alpha_3\in\mathbb{R}, \ \alpha_1>0, \quad \alpha_3>-\alpha_1,\\ \beta_1=-\beta_3\geq 0, \quad \beta_1'\leq 0.\end{cases} \tag{5.42}$$

Let $V\in\mathfrak{X}(M)$. The following statements are equivalent

i. *V is parallel.*
ii. *$V:(M,g)\to(T(M),G)$ is a harmonic map.*
iii. *V is an absolute minimum of the energy $E:\mathfrak{X}(M)\to[0,+\infty)$ and $E(V)=\frac{1}{2}n(\alpha_1+\alpha_3)\operatorname{Vol}(M)$.*

Proof. Note first that (5.42) implies (5.5) so that the g-natural metrics described by (5.42) are actually Riemannian. Let us assume now that (5.42) holds. If $V\in\mathfrak{X}(M)$ is parallel let $\rho=\|V\|^2\in(0,+\infty)$. As is well known, the existence of a nonzero parallel vector field V on M yields the local reducibility of M. That is (M,g) is locally isometric to a product manifold $\mathbb{R}\times M'$ equipped with the product metric and V may be identified (locally) with a vector field tangent to the flat term $\mathbb{R}$ of the product. Rewriting (5.31) and (5.32) for a parallel vector field V, one concludes that $\tau(V)=0$ (i.e., V is a harmonic map of (M,g) into $(T(M),G)$) if and only if

$$-2A_1(\rho)QV+[2A_2-A_3g(QV,V)+nA_4+\rho A_5](\rho)V=0 \tag{5.43}$$

and

$$-\Delta_g V - B_1(\rho)QV + [2B_3 - B_4 g(QV,V) + nB_5 + \rho B_6](\rho)V = 0, \tag{5.44}$$

where $\sqrt{\rho} = \|V\|$. Since V is tangent to the term $\mathbb{R}$ of $\mathbb{R}\times M'$ the curvature terms vanish. Moreover $\Delta_g V = 0$. Therefore (5.43) and (5.44) are equivalent to

$$[2A_2 + nA_4 + \rho A_5](\rho) = [2B_3 + nB_5 + \rho B_6](\rho) = 0. \tag{5.45}$$

Next (by (5.42)) $\alpha_1 + \alpha_3$ is a constant and $\beta_1 + \beta_3 = 0$. Then (by exploiting the explicit form of the functions A_i and B_i furnished by Proposition 5.2) it follows that (5.45) holds. This proves the implication (i) $\Longrightarrow$ (ii). As to the converse, if V is a harmonic map of (M,g) into $(T(M),G)$ then (by Theorem 5.6) $\tau_h(V) = 0$ and $\tau_v(V) = 0$. Next (by Proposition 5.2 and (5.42)) one may easily check that $\tau_v(V) = 0$ reads

$$-\Delta_g V + \left[\frac{\beta_1(r^2)}{\phi_1(r^2)}\|\nabla V\|^2 + \frac{\beta_1'(r^2)}{4\phi_1(r^2)}\|\nabla r^2\|^2\right]V = 0, \tag{5.46}$$

where $r = \|V\|$. Let us take the scalar product of (5.46) by V and integrate over M. Since

$$\int_M g(\Delta_g V, V)\,d\mathrm{vol}(g) = \int_M \|\nabla V\|^2\,d\mathrm{vol}(g)$$

we get (by the very definition of ϕ_1)

$$\int_M \frac{\alpha_1(r^2)}{\phi_1(r^2)}\|\nabla V\|^2\,d\mathrm{vol}(g) - \int_M r^2\frac{\beta_1'(r^2)}{4\phi_1(r^2)}\|\nabla r^2\|^2\,d\mathrm{vol}(g) = 0. \tag{5.47}$$

By (5.42) it follows that $\alpha_1 > 0$, $\phi_1 > 0$ and $\beta_1' \le 0$. Therefore (5.47) implies that V is parallel.

Let us check the equivalence (i) $\Longleftrightarrow$ (iii). If G satisfies (5.42) then (by (5.26)) for any $V \in \mathfrak{X}(M)$ the energy of V is

$$\begin{aligned} E(V) &= \frac{1}{2}\int_M \left\{n(\alpha_1+\alpha_3) + \alpha_1\|\nabla V\|^2 + \frac{1}{4}\beta_1(r^2)\|\nabla r^2\|^2\right\}d\mathrm{vol}(g) \\ &\ge \frac{n}{2}(\alpha_1+\alpha_3)\,\mathrm{Vol}(M), \end{aligned} \tag{5.48}$$

where $r = \|V\|$, and the equality holds if and only if V is parallel.

It should be observed that (5.42) determines a family of Riemannian g-natural metrics depending on two parameters $\alpha_1, \alpha_3 \in \mathbb{R}$ and on a smooth function $\beta_1 : \mathbb{R}_0^+ \to \mathbb{R}$ (subject to some restrictions). One may easily exhibit examples of Riemannian g-natural metrics satisfying (5.42). The Sasaki metric G_s is obtained for $\alpha_1 = 1$ and $\alpha_3 = \beta_1 = 0$.

Remark 5.17 Using the explicit expressions of $\tau_h(V)$ and $\tau_v(V)$ in Theorem 5.6, one may conclude (cf. [2]) that a parallel vector field V is a harmonic map of (M,g) into $(T(M),G)$ if and only if $\rho = \| V \|^2$ is a critical point of the function $A(t) = n(\alpha_1 + \alpha_3) + t(\beta_1 + \beta_3)$ that is $A'(\rho) = 0$. To end with, we note that a) for the g-natural metrics $G_{\alpha_1,\beta_1,k}$ the function $A(t)$ is constant and equals nk, and b) for all Riemannian g-natural metrics G on TM such that $A'(t) \neq 0$ for all t, parallel vector fields are not harmonic maps of (M,g) into $(T(M),G)$. ■

5.3.2. Non-parallel Vector Fields Which Are Harmonic Maps

Besides from generalizing the T. Ishihara & O. Nouhaud theorem as above, we may show that there are examples of non-parallel vector fields which are harmonic maps of (M,g) into $(T(M),G)$ for some Riemannian g-natural metric G on the tangent bundle. Our examples are among the Reeb vector fields, the Hopf vector fields, the conformal and Killing vector fields. Using (5.30) one obtains (cf. [2])

Theorem 5.18 *Let $(M,(\xi,\eta,g))$ be a contact metric manifold and G an arbitrary Riemannian g-natural metric on $T(M)$. If $\xi : (M,g) \to (T(M),G)$ is a harmonic map then $(M,(\eta,g))$ is an H-contact manifold.*

Theorem 5.19 *Let $(M,(\xi,\eta,g))$ be a $(2m+1)$-dimensional contact metric manifold and G a Riemannian g-natural metric on $T(M)$ satisfying $\alpha_2(1) = \beta_2(1) = 0$. Then $\xi : (M,g) \to (T(M),G)$ is a harmonic map if and only if* i) *M is an H-contact manifold,* ii) $\text{trace}_g\{R(\nabla.\xi,\xi)\cdot\} = 0$ *and* iii)

$$\| \nabla\xi \|^2 (\alpha_1 + \alpha_1')(1) + [(2m+1)(\alpha_1 + \alpha_3) + t(\beta_1 + \beta_3)]'(1) = 0. \quad (5.49)$$

On a Sasakian manifold both $\text{trace}_g\{R(\nabla.\xi,\xi)\cdot\} = 0$ and the H-contact condition are satisfied. Moreover (5.49) becomes

$$2m(\alpha_1 + \alpha_1')(1) + [(2m+1)(\alpha_1 + \alpha_3) + t(\beta_1 + \beta_3)]'(1) = 0. \quad (5.50)$$

In particular

Theorem 5.20 *Let $(M,(\xi,\eta,g))$ be a Sasakian manifold and G a Riemannian g-natural metric on $T(M)$ satisfying $\alpha_2(1)=\beta_2(1)=0$. Then the following statements are equivalent*

i. *$\xi:(M,g)\to(TM,G)$ is a harmonic map.*
ii. *ξ is G-harmonic.*
iii. *The relation (5.50) is satisfied.*

Since a Hopf vector field ξ on the unit sphere S^{2m+1} (equipped with the canonical metric) may always be looked at as the Reeb vector field underlying a suitable Sasakian structure, we conclude that the result in Theorem 5.20 holds for the Hopf vector fields on the unit sphere S^{2m+1} as well. On the other hand, one may easily exhibit examples of Riemannian g-natural metrics satisfying (5.50). For instance (5.50) holds for all Riemannian g-natural metrics of type $G_{\alpha_1,\beta_1,k}$ where $\alpha_1=k_1e^{-t}$ with $k_1\in(0,+\infty)$.

Next for any vector $a\in\mathbb{R}^{n+1}$, $a\neq 0$, consider the *conformal gradient vector field* A_a defined as $A_a=\nabla\lambda_a$ where $\lambda_a(x)=\langle x,a\rangle$ i.e., $A_a(x)=\vec{a}-\langle x,a\rangle\vec{x}$ (cf. Section 3.3) for any $x\in S^n$. Harmonicity of conformal gradient vector fields was first investigated in [32]–[33] by equipping the tangent bundle with a metric of Cheeger-Gromoll type $h_{p,q}$. Yet the examples exhibited there are not Riemannian (but metrics of varying signature). Moreover, the use of these metrics doesn't lead to examples of harmonic maps defined on S^2. Using Riemannian g-natural metrics we have

Theorem 5.21 ([5]) *If G is a Riemannian g-natural metric on $T(S^2)$ determined by*

$$\begin{cases}\alpha_1>0,\alpha_1+\alpha_3>0,\alpha_1+2(\alpha_1+\alpha_3)'=0,\\ \alpha_2=\beta_1=\beta_2=\beta_3=0,\end{cases}\tag{5.51}$$

then $A_a:(S^2,g)\to(T(S^2),G)$ is a harmonic map, where g is the canonical metric.

For example, we can take explicitly

$$\alpha_1(t)=\mu e^{-\frac{1}{2(\mu_1+1)}t},\quad \alpha_3(t)=\mu_1\alpha_1(t),\quad \alpha_2=\beta_1=\beta_2=\beta_3=0,$$

for any real constants $\mu>0$ and $\mu_1\geq 0$. Note that if G is a Riemannian g-natural metric on TS^2 determined by (5.51), then G is a Kaluza-Klein metric and A_a^*G is conformal to g. Next, we recall the following decreasing property for harmonic immersions of a surface, proved by Sampson ([265],

Theorem 7, p. 217): if $f : (M^2, g) \to (\tilde{M}, \tilde{g})$ is a harmonic immersion and $f^*\tilde{g}$ is conformal to g, then the sectional curvatures of $(M^2, f^*\tilde{g})$ and $(\tilde{M}, \tilde{g})$ satisfy

$$K_{f^*\tilde{g}}(T_x M^2) \leq K_{\tilde{g}}(f_* T_x M^2),$$

for any $x \in M^2$. This result ensures that (TS^2, G) admits some positive sectional curvatures for any Riemannian g-natural metric G on TS^2 satisfying (5.51). In fact, the Gauss-Bonnet Theorem then gives

$$\frac{1}{2\pi} \int_{S^2} K_{V_a^* \tilde{g}}(T_x S^2) = \chi(S^2) = 2 > 0,$$

where $\chi(S^2)$ denotes the Euler number of S^2.

5.3.3. Unit Vector Fields Which Are Harmonic Maps

Our purpose in this section is to give natural generalizations of results by S.D. Han & J.W. Yim, [157]. Let (M, g) be a Riemannian manifold and $V \in \Gamma^\infty(S(M))$ a unit vector field on M. By (5.31) and (5.32) and Proposition 5.2 where $T(M)$ is equipped with a g-natural Riemannian metric G satisfying (5.20) the tension field $\tau(V)$ of $V : (M, g) \to (T(M), G)$ is given by

$$\tau(V) = \tau_h(V)^H + \tau_v(V)^V, \tag{5.52}$$

where

$$\begin{aligned} \tau_h(V) &= \frac{ab}{\alpha} QV + \frac{a^2}{\alpha} \mathrm{trace}_g\{R(\nabla_. V, V)\cdot\} - \frac{ad}{\alpha} \nabla_V V \\ &+ \left[\frac{bd}{\alpha} - \frac{a^2 bd}{\alpha\phi} g(QV, V) + \frac{\alpha b \beta'(1) - abd^2}{\alpha\phi} \right. \\ &\left. - \frac{a^3 d}{\alpha\phi} g(\mathrm{trace}_g\{R(\nabla_. V, V)\cdot\}, V) + \frac{ad}{\phi} \mathrm{div}(V) \right] V, \end{aligned} \tag{5.53}$$

$$\begin{aligned} \tau_v(V) &= -\Delta_g V - \frac{b^2}{\alpha} QV - \frac{ab}{\alpha} \mathrm{trace}_g\{R(\nabla_. V, V)\cdot\} + \frac{bd}{\alpha} \nabla_V V \\ &+ \left[-\frac{(a+c)d}{\alpha} + \frac{ab^2 d}{\alpha\phi} g(QV, V) + \frac{b^2 d^2 - \alpha \ell \beta'(1)}{\alpha\phi} \right. \\ &\left. + \frac{a^2 bd}{\alpha\phi} g(\mathrm{trace}_g\{R(\nabla_. V, V)\cdot\}, V) - \frac{bd}{\phi} \mathrm{div}(V) \right] V, \end{aligned} \tag{5.54}$$

where $\ell = a + c + d$. Let now $\tau_1(V)$ be the tension field of $V : (M,g) \to (S(M), \tilde{G})$. As $\tilde{G} = \iota^* G$ (where $\iota : S(M) \to T(M)$ is the inclusion) $\tau_1(V)$ is the tangential component of $\tau(V)$ with respect to the decomposition $T(T(M)) = T(S(M)) \oplus \mathbb{R}N^G$. We may then state the following (cf. [3])

Theorem 5.22 *Let (M,g) be a Riemannian manifold and $V \in \Gamma^\infty(S(M))$ a unit vector field. Let $\tilde{G}$ be a g-natural metric on $S(M)$. The tension field $\tau_1(V)$ of $V : (M,g) \to (S(M), \tilde{G})$ is given by*

$$\tau_1(V) = \tau_{1h}(V)^H + \tau_{1v}(V)^V \tag{5.55}$$

where

$$\begin{aligned}
\tau_{1h}(V) = {} & \frac{ab}{\alpha} QV + \frac{a^2}{\alpha} \operatorname{trace}_g\{R(\nabla_. V, V)\cdot\} - \frac{ad}{\alpha} \nabla_V V \\
& + \left[-\frac{b(ad+b^2)}{\alpha \ell} g(QV, V) - \frac{b}{\ell} g(\Delta_g V, V) + \frac{d}{\ell} \operatorname{div}(V) \right. \\
& \left. - \frac{a(ad+b^2)}{\alpha \ell} g(\operatorname{trace}_g\{R(\nabla_. V, V)\cdot\}, V) \right] V,
\end{aligned} \tag{5.56}$$

$$\begin{aligned}
\tau_{1v}(V) = {} & -\Delta_g V - \frac{b^2}{\alpha} QV - \frac{ab}{\alpha} \operatorname{trace}_g\{R(\nabla_. V, V)\cdot\} + \frac{bd}{\alpha} \nabla_V V \\
& + \left[\frac{b^2}{\alpha} g(QV, V) + g(\Delta_g V, V) + \frac{ab}{\alpha} g(\operatorname{trace}_g\{R(\nabla_. V, V)\cdot\}, V) \right] V.
\end{aligned} \tag{5.57}$$

In particular $V : (M,g) \to (S(M), \tilde{G})$ is a harmonic map if and only if $\tau_{1h}(V) = \tau_{1v}(V) = 0$. Then (by (5.56)–(5.57))

Theorem 5.23 (Cf. [3]) *Let (M,g) be a Riemannian manifold, V a unit vector field, and $\tilde{G}$ a Riemannian g-natural metric on $S(M)$. Then $V : (M,g) \to (S(M), \tilde{G})$ is a harmonic map if and only if V is a harmonic vector field and*

$$bQV + a \operatorname{trace}_g\{R(\nabla_. V, V)\cdot\} = \left\{ b \|\nabla V\|^2 - d \operatorname{div}(V) \right\} V + d \nabla_V V. \tag{5.58}$$

For the Sasaki metric $\tilde{G}_s$ one obtains (by Theorem 5.23) Theorem 2.19 (the S.D. Han & J.W. Yim theorem) as an immediate corollary. Also, Theorem 5.23 allows one to generalize the quoted S.D. Han & J.W. Yim theorem to a two-parameter family of Riemannian g-natural metrics on $S(M)$ (including $\tilde{G}_s$). Indeed

Corollary 5.24 (Cf. [3]) *Let (M,g) be a Riemannian manifold and $V \in \Gamma^\infty(S(M))$. Let $\tilde{G}$ be a Riemannian g-natural metric on $S(M)$ with $b = d = 0$. Then $V : (M,g) \to (S(M), \tilde{G})$ is a harmonic map if and only if $\Delta_g V$ and V are collinear and* $\mathrm{trace}_g\{R(\nabla.V, V)\cdot\} = 0$.

Harmonicity with respect to g-natural metrics leads to new descriptions of unit Killing vector fields. Indeed (cf. [3])

Theorem 5.25 *A unit vector field V is Killing if and only if the harmonic map equations for $V : (M,g) \to (S(M), \tilde{G})$ do not depend upon the choice of the g-natural Riemannian metric on $S(M)$.*

Let us give an example of a unit vector field $V \in \mathfrak{X}^1(M)$ which is not a Killing vector field and emphasize that the harmonic maps equations for $V : (M,g) \to (S(M), \tilde{G})$ depend explicitely on the choice of Riemannian g-natural metric. Other examples will be given in the sequel within contact metric geometry.

Example 5.26 Let us consider the hyperbolic space of constant negative sectional curvature $-k^2$ i.e., $(\mathbb{R}^n_+, g_{\mathrm{hyp}})$ with $\mathbb{R}^n_+ = \{y \in \mathbb{R}^n : y_n > 0\}$ and $g_{\mathrm{hyp}} = k^{-2} y_n^{-2} (dy_1 \otimes dy_1 + \cdots + dy_n \otimes dy_n)$. Clearly $\mathbb{R}^n_+$ admit no unit Killing vector fields. The vector fields $E_i = k y_n \partial/\partial y_i$, $1 \le i \le n$, form a g_{hyp}-orthonormal frame of $T(\mathbb{R}^n_+)$. Let us set $V = E_n$ (the unit vector field normal to the horoball foliation of $\mathbb{R}^n_+$). A calculation shows that covariant derivatives of E_i are given by (see also Remark 3.12 in this book)

$$\nabla_{E_i} E_j = k\delta_{ij} V, \quad \nabla_{E_i} V = -kE_i, \quad \nabla_V E_i = 0, \quad \nabla_V V = 0, \tag{5.59}$$

for any $1 \le i,j < n$. In particular (5.59) implies that $\|\nabla V\|^2 = (n-1)k^2$ and $\Delta_g V = -\mathrm{trace}\nabla^2 V = \|\nabla V\|^2 V$. By Theorem 5.15 it follows that V is a harmonic (unit) vector field. Moreover, as $\mathbb{R}^n_+$ has constant sectional curvature $-k^2$, one obtains

$$\mathrm{trace}_g\{R(\nabla.V, V)\cdot\} = k^2 \nabla_V V - (\mathrm{div}(V))V = -k^2 \mathrm{div}(V)V$$

and (by (5.59)) $\mathrm{div}(V) = (1-n)k \neq 0$. Hence (by Theorem 5.23) $V : (\mathbb{R}^n_+, g_{\mathrm{hyp}}) \to (S(\mathbb{R}^n_+), \tilde{G})$ is a harmonic map if and only if

$$ak^2 - 2bk - d = 0. \tag{5.60}$$

It must be observed that Riemannian g-natural metrics $\tilde{G}$ for which $V : (\mathbb{R}^n_+, g_{\mathrm{hyp}}) \to (S(\mathbb{R}^n_+), \tilde{G})$ is a harmonic map are (by (5.60)) precisely those satisfying $d = ak^2 - 2bk$ (a three-parameter family of Riemannian g-natural metrics on $S(\mathbb{R}^n_+)$). In particular $V = E_n : (\mathbb{R}^n_+, g_{\text{hyp}}) \to (S(\mathbb{R}^n_+), \tilde{G}_s)$ is not a harmonic map (because for $\tilde{G} = \tilde{G}_s$ one has $b = d = 0$ and $ak^2 \neq 0$).

Let $H^3 = (\mathbb{R}^3_+, g_{\text{hyp}})$ be the hyperbolic 3-space of constant sectional curvature -1. We use the notations in Example 5.26 above. E_3 is a unit vector field whose integral curves are a family of vertical geodesics, and $\{E_1, E_2\}$ is a frame for the orthogonal horosphere foliation. A calculation based on (5.59) shows that E_i is harmonic of constant bending

$$\|\nabla E_1\|^2 = \|\nabla E_2\|^2 = 1, \quad \|\nabla E_3\|^2 = 2. \tag{5.61}$$

The harmonicity of E_i ($i \in \{1,2,3\}$) was proved in [129]. Any horospherical unit vector field of the form

$$X = (\cos t)\, E_1 + (\sin t)\, E_2, \quad t \in \mathbb{R},$$

is also harmonic and $\|\nabla X\|^2 = 1$. C.M. Wood, [319], observed that any such X is parallel when restricted to a horosphere (with respect to the induced (Euclidean) geometry) although clearly not parallel in H^3. Together with $\pm E_3$ these are the only invariant harmonic unit vector fields on H^3 (cf. [129]). By a result in [319]

Theorem 5.27 *Let X be a harmonic horospherical unit vector field. Then either X is invariant or*

$$X = -\frac{y_2}{\sqrt{y_1^2 + y_2^2}} E_1 + \frac{y_1}{\sqrt{y_1^2 + y_2^2}} E_2$$

up to translation in H^3.

Next we wish to discuss the harmonicity of the Reeb and Hopf vector fields. Let $V = \xi$ be the Reeb vector field underlying a contact metric manifold $(M, (\xi, \eta, g))$. By Theorem 5.23, Theorem 4.5, and (4.2) we obtain

Theorem 5.28 (Cf. M.T.K. Abbassi et al., [3]) *Let $(M, (\xi, \eta, g))$ be a $(2m+1)$-dimensional contact metric manifold and $\tilde{G}$ a Riemannian g-natural metric on $S(M)$. Then $\xi : (M, g) \to (S(M), \tilde{G})$ is a harmonic map if and only if $Q\xi$ and ξ are collinear and*

$$a\,\text{trace}_g\{R(\nabla_{\cdot}\xi, \xi)\cdot\} = 2b\,(\|\nabla\xi\|^2 - 2m)\xi.$$

P. Rukimbira, [262], showed that the Reeb vector field of a K-contact manifold (M, g, ξ) is a harmonic map $\xi : (M, g) \to (S(M), \tilde{G}_s)$. As a consequence of Theorem 5.28 one obtains

Theorem 5.29 (Cf. [251]) *The Reeb vector field of a K-contact manifold* (M,g,ξ) *is a harmonic map* $\xi : (M,g) \to (S(M),\tilde{G})$ *for any Riemannian g-natural metric* $\tilde{G}$ *on* $S(M)$.

By a result of Y.L. Xin, [321] (cf. also Theorem 3.17 in this monograph), any stable harmonic map from the sphere to a Riemannian manifold is a constant map. Theorem 5.29 yields

Corollary 5.30 *Let* ξ *be a Hopf vector field on the unit sphere* (S^{2n+1},g). *Then* $\xi : (S^{2n+1},g) \to (S(S^{2n+1}),\tilde{G})$ *is an unstable harmonic map for any Riemannian g-natural metric* $\tilde{G}$ *on* $S(S^{2n+1})$.

Remark 5.31 The result in Theorem 3.10 is invariant under a 4-parameter deformation of the Sasaki metric on $S(S^{2m+1})$. ■

We may state

Theorem 5.32 (cf. [250]) *Let* (M,g) *be a 3-dimensional Riemannian manifold of constant sectional curvature* $\kappa \neq 0$ *and* $S(M)$ *its unit tangent sphere bundle equipped with a Riemannian g-natural metric* $\tilde{G}$ *with* $d \neq -\kappa a$ *and* $b=0$. *Let* $\xi \in \mathfrak{X}^1(M)$. *Then* $\xi : (M,g) \to (S(M),\tilde{G})$ *is a harmonic map if and only if* ξ *is a Killing vector field and* $\kappa > 0$.

Proof. Let us assume that ξ is a unit Killing vector field. Then (cf. [258], p. 169) $\nabla_V V = 0$, $\operatorname{div}(V) = 0$, $\Delta_g V = QV$ and (as V is a unit vector field) $g(\Delta_g V, V) = \|\nabla V\|^2$. As M has constant sectional curvature $\xi : (M,g) \to (S(M),\tilde{G}_s)$ is a harmonic map. Hence (by Theorem 5.23) $\xi : (M,g) \to (S(M),\tilde{G})$ is a harmonic map for any $\tilde{G}$. Conversely, let us assume that $\xi : (M,g) \to (S(M),\tilde{G})$ is a harmonic map, where $\tilde{G}$ is a Riemannian g-natural metric with $d \neq -\kappa a$ and $b=0$. By (5.58)

$$(i)\ \Delta_g \xi = \|\xi\|^2 \xi, \quad (ii)\ \operatorname{trace}_g R(\nabla.\xi,\xi)\cdot = -\frac{d}{a}\big[(\operatorname{div}\xi)\xi - \nabla_\xi \xi\big]. \quad (5.62)$$

As M has constant sectional curvature k

$$\operatorname{trace}_g R(\nabla.\xi\,,\xi)\cdot = k\big[\operatorname{div}(\xi)\xi - \nabla_\xi \xi\big].$$

Next $k \neq -d/a$ and condition (ii) in (5.62) imply

$$\operatorname{div}(\xi) = 0 \quad \text{and} \quad \nabla_\xi \xi = 0. \quad (5.63)$$

Let us set $\tau = \mathcal{L}_\xi g$ where $\mathcal{L}_\xi$ is the Lie derivative. That is

$$\tau(X,Y) = g(\nabla_X \xi, Y) + g(X, \nabla_Y \xi).$$

τ is a symmetric tensor of type $(0,2)$. Let h be the corresponding symmetric (1,1) tensor

$$\tau(X,Y) = g(hX,Y).$$

Then (by (5.63))

$$g(h\xi, Y) = \tau(\xi, Y) = g(\nabla_\xi \xi, Y) + g(\xi, \nabla_Y \xi) = \frac{1}{2} Y(\|\xi\|^2) = 0$$

hence $h\xi = 0$. Let us consider a local orthonormal frame $\{e_1, e_2, e_3 = \xi\}$ consisting of eigenvectors of h i.e., $h\xi = 0$, $he_1 = \lambda_1 e_1$ and $he_2 = \lambda_2 e_2$. Since

$$\operatorname{div}(\xi) = g(\nabla_\xi \xi, \xi) + g(\nabla_{e_1} \xi, e_1) + g(\nabla_{e_2} \xi, e_2) = \frac{1}{2}\operatorname{trace}(h) = \frac{1}{2}(\lambda_1 + \lambda_2),$$

we get (by (5.63)) $\lambda_2 = -\lambda_1$. Therefore ξ is Killing if and only if $\lambda_1 = 0$. Let us set

$$f_1 = \frac{1}{2}\lambda_1, \quad f_2 = g(\nabla_{e_1}\xi, e_2), \quad f_3 = g(\nabla_{e_1} e_2, e_1),$$
$$f_4 = g(\nabla_{e_2} e_2, e_1), \quad f_5 = g(\nabla_\xi e_1, e_2).$$

Then

$$\begin{aligned}
\nabla_{e_2} e_1 &= f_2 \xi - f_4 e_2, & \nabla_{e_1}\xi &= f_1 e_1 + f_2 e_2, & \nabla_\xi \xi &= 0,\\
\nabla_{e_2}\xi &= -f_2 e_1 - f_1 e_2, & \nabla_{e_1} e_1 &= -f_1 \xi - f_3 e_2, & \nabla_\xi e_2 &= -f_5 e_1,\\
\nabla_{e_2} e_2 &= f_1 \xi + f_4 e_1, & \nabla_{e_1} e_2 &= -f_2 \xi + f_3 e_1, & \nabla_\xi e_1 &= f_5 e_2,
\end{aligned}$$

Consequently

$$\begin{aligned}
R(e_1,\xi)\xi &= -\nabla_{e_1}\nabla_\xi \xi + \nabla_\xi \nabla_{e_1}\xi + \nabla_{[e_1,\xi]}\xi \\
&= \left\{f_1^2 - f_2^2 + \xi(f_1)\right\} e_1 + \left\{2 f_1 f_5 + \xi(f_2)\right\} e_2 \qquad (5.64)
\end{aligned}$$

$$\begin{aligned}
R(e_2,\xi)\xi &= -\nabla_{e_2}\nabla_\xi \xi + \nabla_\xi \nabla_{e_2}\xi + \nabla_{[e_2,\xi]}\xi \\
&= \left\{2 f_1 f_5 - \xi(f_2)\right\} e_1 + \left\{f_1^2 - f_2^2 - \xi(f_1)\right\} e_2 \qquad (5.65)
\end{aligned}$$

$$\begin{aligned}
R(e_1,e_2)\xi &= -\nabla_{e_1}\nabla_{e_2}\xi + \nabla_{e_2}\nabla_{e_1}\xi + \nabla_{[e_1,e_2]}\xi \\
&= \left(e_1(f_2) + e_2(f_1) + 2 f_1 f_3\right) e_1 + \left(e_1(f_1) + e_2(f_2) - 2 f_1 f_4\right) e_2. \qquad (5.66)
\end{aligned}$$

$$\begin{aligned}
R(e_1,e_2)e_1 &= -\nabla_{e_1}\nabla_{e_2} e_1 + \nabla_{e_2}\nabla_{e_1} e_1 + \nabla_{[e_1,e_2]} e_1 \\
&= -\left(e_1(f_2) + e_2(f_1) + 2 f_1 f_3\right)\xi \\
&\quad + \left(f_1^2 - f_2^2 + e_1(f_4) - e_2(f_3) - 2 f_2 f_5 - f_3^2 - f_4^2\right) e_2. \qquad (5.67)
\end{aligned}$$

Besides

$$R(e_1,\xi,e_1,\xi) = R(e_2,\xi,e_2,\xi) = R(e_1,e_2,e_1,e_2) = \kappa \tag{5.68}$$

and

$$R(e_1,\xi)e_2 = R(e_1,e_2)\xi = R(e_2,\xi)e_1 = 0. \tag{5.69}$$

By (5.64)–(5.69) one obtains

$$f_2^2 - f_1^2 - \xi(f_1) = \kappa \quad \text{and} \quad 2f_1f_5 + \xi(f_2) = 0, \tag{5.70}$$

$$f_2^2 - f_1^2 + \xi(f_1) = \kappa \quad \text{and} \quad 2f_1f_5 - \xi(f_2) = 0, \tag{5.71}$$

$$e_1(f_2) + e_2(f_1) + 2f_1f_3 = 0 \quad \text{and} \quad e_1(f_1) + e_2(f_2) - 2f_1f_4 = 0, \tag{5.72}$$

$$f_1^2 - f_2^2 + e_1(f_4) - e_2(f_3) - 2f_2f_5 - f_3^2 - f_4^2 = \kappa. \tag{5.73}$$

By (5.70) and (5.71) one gets

$$f_2^2 - f_1^2 = \kappa, \quad f_1f_5 = 0 \quad \text{and} \quad \xi(f_1) = \xi(f_2) = 0. \tag{5.74}$$

Using (5.74) the identity (5.73) becomes

$$e_1(f_4) - e_2(f_3) - 2f_2f_5 - f_3^2 - f_4^2 = 2k. \tag{5.75}$$

Moreover

$$-\Delta_g\xi = -\|\nabla\xi\|^2\xi + \big(e_1(f_1) - e_2(f_2) - 2f_1f_4\big)e_1$$
$$+ \big(e_1(f_2) - e_2(f_1) - 2f_1f_3\big)e_2,$$

where $\|\nabla\xi\|^2 = 2\big(f_1^2 + f_2^2\big)$. Due to (i) in (5.62) and (5.76) one gets

$$e_1(f_1) - e_2(f_2) = 2f_1f_4 \quad \text{and} \quad e_1(f_2) - e_2(f_1) = 2f_1f_3. \tag{5.76}$$

Combining (5.72) and (5.76) one has

$$e_1(f_2) = e_2(f_2) = 0.$$

Moreover, (5.74) implies $\xi(f_2) = 0$ and $f_1^2 = f_2^2 - \kappa$. Hence f_1 and f_2 are constant. If $f_1 \neq 0$ then (5.70) and (5.72) imply $f_3 = f_4 = f_5 = 0$ from which (by (5.73)) $\kappa = 0$. Yet $\kappa \neq 0$ and we may conclude that $f_1 = 0$ so that ξ is Killing and (by (5.71)) $\kappa = f_2^2 > 0$. ■

As an immediate consequence of Theorem 5.32, Theorem 3.10 (Han-Yim's theorem) is invariant under a three-parameter deformation of the Sasaki metric on $S(M)$.

If $H^n(-\kappa)$ ($\kappa > 0$) is the hyperbolic space the problem whether unit vector fields (of course non-Killing) which are harmonic maps $H^n(-\kappa) \to (S(H^n(-\kappa)), \tilde{G}_s)$ exist. Theorem 5.32 yields the following non-existence result in dimension three.

Corollary 5.33 *Let $H^3(-\kappa)$ be the hyperbolic 3-space. There is no unit vector field X on $H^3(-\kappa)$ such that*

$$X : H^3(-\kappa) \to (S(H^3(-\kappa)), \tilde{G}_s)$$

is a harmonic map. The result is invariant under a 3-parameter deformation of the Sasaki metric on $S(H^3(-\kappa))$.

Remark 5.34 The flat three-space and the sphere S^3 equipped with a metric of nonconstant sectional curvature do possess unit vector fields which are not Killing yet are harmonic maps (cf. [250]). ■

Theorem 5.35 (Cf. [250]) *Let (M,g) be a real space form of constant sectional curvature $\kappa > 0$, $\dim(M) = 2m+1$, such that M is not homeomorphic to the sphere S^{2m+1} and let $(S(M), \tilde{G})$ be its unit tangent sphere bundle equipped with a Riemannian g-natural metric $\tilde{G}$ with $b \neq 0$ and $d \neq -ka$. Let $\xi \in \mathfrak{X}^1(M)$. Then*

i. *$\xi : (M,g) \to (S(M), \tilde{G})$ is a harmonic map if and only if ξ is Killing.*

ii. *If ξ is a solenoidal (i.e., divergence free) unit vector field then $\xi : (M,g) \to (S(M), \tilde{G})$ is a harmonic map if and only if ξ has minimum energy $E_{\tilde{G}} : \mathfrak{X}^1(M) \to \mathbb{R}$.*

5.4. GEODESIC FLOW WITH RESPECT TO g-NATURAL METRICS

Let (M,g) be a Riemannian manifold. As usual, let $S_\rho(M)$ be the tangent sphere bundle of radius ρ and $S(M) = S_1(M)$. In the previous sections we studied harmonic (unit) vector fields with respect to an arbitrary Riemannian g-natural metric $\tilde{G}$ on $S(M)$. Similar results hold for vector fields of constant length ρ (i.e., Theorem 5.15 and Theorem 5.23 hold for vector fields of constant length ρ). If this is the case equation (5.58) becomes

$$b\,QV + a\,\mathrm{trace}_g\{R(\nabla.V, V)\cdot\} = \frac{1}{\rho^2}\left\{b\|\nabla V\|^2 - d\rho^2 \mathrm{div}(V)\right\}V + d\nabla_V V. \tag{5.77}$$

Next we look at the geodesic flow vector field $\tilde{\xi}$. Let (M,g) be a two-point homogeneous space (i.e., M is either flat or rank-one symmetric). When both $S(M)$ and $S(S(M))$ are endowed with the Sasaki metrics E. Boeckx & L. Vanhecke, [54], showed that $\tilde{\xi} : (S(M), G_s) \to (S(SM)), (G_s)_s)$ is a harmonic vector field and a harmonic map (cf Section 4.3 in this monograph). Let us replace now the Sasaki metric on $S(M)$ with an arbitrary Riemannian g-natural metric $\tilde{G}$. Then $\tilde{\xi}$ has constant length ρ (not necessarily 1). Let us look at map $\tilde{\xi} : S(M) \to S_\rho(S(M))$. We then equip $S_\rho(S(M))$ with a Riemannian g-natural metric $\tilde{\tilde{G}}$ derived from $\tilde{G}$ and look at the harmonicity of the map $\tilde{\xi} : (S(M), \tilde{G}) \to (S_\rho(S(M)), \tilde{\tilde{G}})$. Two natural questions arise: i) When is $\tilde{\xi} : (S(M), \tilde{G}) \to (S_\rho(S(M)), \tilde{\tilde{G}})$ a harmonic vector field? and ii) When is $\tilde{\xi} : (S(M), \tilde{G}) \to (S_\rho(S(M)), \tilde{\tilde{G}})$ a harmonic map?

Definition 5.36 A vector field of constant length $V \in \mathfrak{X}^r(M)$ is said to be *harmonic* if it is a critical point for the energy functional $E : \mathfrak{X}^r(M) \to [0, +\infty)$ (variations are through vector fields of constant length $\sqrt{r}$). ■

If $\tilde{G}$ is an arbitrary Riemannian g-natural metric on $S_1(M)$ then (by (5.23)) $\|\tilde{\xi}\|^2_{\tilde{G}} = a + c + d$. Note that $a + c + d > 0$ as $a > 0$ and $\phi = a(a + c + d) - b^2 > 0$. Hence $\tilde{\xi}$ has constant length $\rho = \sqrt{a + c + d}$ (not necessarily equal to 1) so that it is a map $\tilde{\xi} : S(M) \to S_\rho(S(M))$. Let us look at the harmonicity of $\tilde{\xi}$, both as a map $S(M) \to S_\rho(S(M))$ and as a vector field (i.e., a critical point of $E : \mathfrak{X}^r(S_1(M)) \to [0, +\infty)$ where $r = \rho^2$). Let us endow $S_\rho(S(M))$ with an arbitrary Riemannian g-natural metric $\tilde{\tilde{G}}$ derived from $\tilde{G}$. Then (by (5.23)) $\tilde{\tilde{G}}$ depends on four constants $\bar{a}, \bar{b}, \bar{c}, \bar{d}$ satisfying

$$\bar{a} > 0, \quad \bar{a}(\bar{a} + \bar{c}) - \bar{b}^2 > 0, \quad \bar{a}(\bar{a} + \bar{c} + \rho^2 \bar{d}) - \bar{b}^2 > 0.$$

By Theorem 5.15

Theorem 5.37 (Cf. [4]) *Let (M,g) be a two-point homogeneous space. Let $\tilde{G}$ be a Riemannian g-natural metric on $S(M)$ and $\tilde{\tilde{G}}$ a $\tilde{G}$-natural Riemannian metric on $S_\rho(S(M))$. Then the geodesic flow vector field $\tilde{\xi}$ on $S(M)$ is a harmonic vector field with respect to $\tilde{G}$ and $\tilde{\tilde{G}}$ i.e., $\Delta_{\tilde{G}}\tilde{\xi}$ and $\tilde{\xi}$ are collinear.*

Theorem 5.37 is the source of a large number of examples of harmonic geodesic flows (as both $\tilde{G}$ and $\tilde{\tilde{G}}$ depend on four parameters).

By Theorem 5.23 the geodesic flow $\tilde{\xi}$ is a harmonic map if and only if (5.77) holds i.e.,

$$\bar{b}\,\tilde{Q}\tilde{\xi} + \bar{a}\,\mathrm{trace}_{\tilde{G}}\{\tilde{R}(\tilde{\nabla}.\tilde{\xi},\tilde{\xi})\cdot\} = \frac{1}{\rho^2}\{\bar{b}\,\|\tilde{\nabla}\tilde{\xi}\|^2_{\tilde{G}} - \bar{d}\,\rho^2 \mathrm{div}_{\tilde{G}}\tilde{\xi}\}\tilde{\xi} + \bar{d}\tilde{\nabla}_{\tilde{\xi}}\tilde{\xi}, \tag{5.78}$$

where $\tilde{Q}$ is the Ricci operator of $(S(M),\tilde{G})$. As $\mathrm{div}_{\tilde{G}}(\tilde{\xi}) = 0$ and $\tilde{\nabla}_{\tilde{\xi}}\tilde{\xi} = 0$ (cf. [4]) the identity (5.78) becomes

$$\bar{a}\,\rho^2\,\mathrm{trace}_{\tilde{G}}\{\tilde{R}(\tilde{\nabla}.\tilde{\xi},\tilde{\xi})\cdot\} = -\bar{b}\left\{\rho^2\tilde{Q}\tilde{\xi} + \|\tilde{\nabla}\tilde{\xi}\|^2_{\tilde{G}}\tilde{\xi}\right\}. \tag{5.79}$$

As (M,g) is two-point homogeneous it is (as well known) globally Ossermann so that the eigenfunctions λ_i of the Jacobi operator $R_u = R(u,\cdot)u$ are constant (cf. [88]). Let us compute $\mathrm{trace}_{\tilde{G}}\{\tilde{R}(\tilde{\nabla}.\tilde{\xi},\tilde{\xi})\cdot\}$, $\tilde{Q}\tilde{\xi}$ and $\|\tilde{\nabla}\tilde{\xi}\|^2_{\tilde{G}}$. Then (5.79) becomes

$$n\,\alpha\,a\bar{b}\sum_{i=1}^{n-1}\lambda_i^2 = \left[\bar{a}b^3 d + 2\bar{b}\,\alpha(\alpha - b^2)\right]S - n(n-1)\bar{b}\,\alpha(a+c)^2, \tag{5.80}$$

where S is the scalar curvature of (M,g). Then

Theorem 5.38 (Cf. [4]) *Let (M,g) be a two-point homogeneous n-space, $\tilde{G}$ a Riemannian g-natural metric on $S(M)$, and $\tilde{\tilde{G}}$ a $\tilde{G}$-natural Riemannian metric on $S_\rho(S(M))$. Then the geodesic flow vector field $\tilde{\xi}$ on $S(M)$ is a harmonic map of $(S(M),\tilde{G})$ into $(S_\rho(S(M)),\tilde{\tilde{G}})$ if and only if* (5.80) *is satisfied.*

Theorem 5.38 may be used to build several examples of harmonic maps (defined by geodesic vector fields). If (M,g) is a two-point homogeneous space and $\tilde{G}$ is a g-natural Riemannian metric on $S(M)$ then there is a 3-parameter family $\mathcal{F}$ of $\tilde{G}$-natural Riemannian metrics on $S_\rho(S(M))$. $\mathcal{F}$ consists of all $\tilde{G}$-natural Riemannian metrics $\tilde{\tilde{G}}$ depending on the parameters $\bar{a},\bar{c},\bar{d} \in \mathbb{R}$, with $\bar{b}$ uniquely determined by (5.80).

Corollary 5.39 *If (M,g) is a space of constant sectional curvature $k > 0$ then there is a 2-parameter family of g-natural Riemannian metrics $\tilde{G}$ on $S(M)$ such that the geodesic flow is a harmonic map $\tilde{\xi} : (S(M),\tilde{G}) \to (S_\rho(S(M)),\tilde{\tilde{G}})$ for any $\tilde{G}$-natural Riemannian metric $\tilde{\tilde{G}}$.*

Proof. It suffices to consider the Riemannian g-natural metrics $\tilde{G}$ determined by $a > 0, b = 0, c = a(k-1)$ and $d > -ak$ and to apply Theorem 5.38.

The following result is an immediate corollary of Theorem 5.38

Corollary 5.40 *Let (M,g) be a flat Riemannian manifold, $\tilde{G}$ a Riemannian g-natural metric on $S(M)$ and $\tilde{\tilde{G}}$ a $\tilde{G}$-natural Riemannian metric on $S_\rho(S(M))$. The geodesic flow is a harmonic map $\tilde{\xi} : (S(M), \tilde{G}) \to (S_\rho(S(M)), \tilde{\tilde{G}})$ if and only if horizontal and tangential distributions are $\tilde{\tilde{G}}$-orthogonal, that is $\bar{b} = 0$.*

The harmonicity of $\tilde{\xi} : (S(M), \tilde{G}) \to (S_\rho(S(M)), \tilde{\tilde{G}})$ is also related to the Killing property with respect to $\tilde{G}$. Theorem 5.38 and its corollaries show that an appropriate choice of the g-natural metrics $\tilde{G}$ and $\tilde{\tilde{G}}$ leads to examples of geodesic flow vector fields which are harmonic maps of $S(M)$ into $S_\rho(S(M))$ (where M is a two-point homogeneous space). However the requirement that the geodesic flow vector field be a harmonic map with respect to *any* Riemannian g-natural metric on the target manifold turns out to be rather restrictive. Indeed (by Theorem 6 in [3]) this only happens when the vector field itself is Killing.

Theorem 5.41 (Cf. [4]) *Let (M,g) be a two-point homogeneous space and $\tilde{G}$ a g-natural Riemannian metric on $S(M)$. The following statements are equivalent*

i. $\tilde{\xi} : (S(M), \tilde{G}) \to (S_\rho(S(M)), \tilde{\tilde{G}})$ *is a harmonic map for any $\tilde{G}$-natural Riemannian metric on $S_\rho(S(M))$.*

ii. *The geodesic flow $\tilde{\xi}$ on $S(M)$ is a Killing vector field.*

iii. (M,g) *has constant sectional curvature $k = \frac{a+c}{a} > 0$ and horizontal and tangential distributions are $\tilde{G}$-orthogonal.*

CHAPTER SIX

The Energy of Sections

Contents

Let (E, π, M) be a (locally trivial) real vector bundle of rank k where $\dim(M) = n$ and $\dim(E) = n + k$ and $\pi : E \to M$ is the projection. Let $\{\sigma_1, \dots, \sigma_k\}$ be a local frame in $E \to M$, defined on the open set $U \subseteq M$. Without loss of generality we may assume that U is also a local coordinate neighborhood in M with coordinates $\chi = (\tilde{x}^1, \dots, \tilde{x}^n) : U \to \mathbb{R}^n$. Let $x^i : \pi^{-1}(U) \to \mathbb{R}$ be defined by $x^i = \tilde{x}^i \circ \pi$ for any $1 \leq i \leq n$. Let $v^\alpha : \pi^{-1}(U) \to \mathbb{R}$ be given by

$$u = v^\alpha(u)\sigma_\alpha(\pi(u)), \quad u \in \pi^{-1}(U), \quad 1 \leq \alpha \leq k.$$

Then $(\pi^{-1}(U), x^i, v^\alpha)$ is a local coordinate system on E. For the elementary considerations below one may also see W.A. Poor, [258].

Definition 6.1

i) A smooth curve $\gamma : (-\epsilon, \epsilon) \to E$ is called *vertical* if $\gamma(t) \in E_x$ for some $x \in M$ and any $|t| < \epsilon$. Let $z \in E$.

ii) A tangent vector $X \in T_z(E)$ is said to be *vertical* if X is tangent to some vertical curve in E i.e., there is a vertical curve $\gamma : (-\epsilon, \epsilon) \to E$ such that $\gamma(0) = z$ and $\dot{\gamma}(0) = X$. A tangent vector field $X \in \mathfrak{X}(E)$ is said to be *vertical* if the tangent vector $X_z \in T_z(E)$ is vertical for any $z \in E$.

© 2012 Elsevier Inc. All rights reserved.

iii) Let $x \in M$ and $z \in \pi^{-1}(x) = E_x$. The *vertical lift* of $u \in E_x$ in z is the tangent vector $u_z^V = \dot{\gamma}(0) \in T_z(E)$ where $\gamma = \gamma_{z,u} : (-\epsilon, \epsilon) \to E$ is the vertical curve given by $\gamma(t) = z + tu$ for any $|t| < \epsilon$.

iv) Given a cross-section $\sigma \in \Gamma^\infty(E)$, its *vertical lift* is the tangent vector field $\sigma^V \in \mathfrak{X}(E)$ given by

$$\sigma^V(z) = (\sigma(\pi(z)))_z^V \in T_z(E), \quad z \in E,$$

where $(\sigma(\pi(z)))_z^V$ is the vertical lift of $\sigma(\pi(z))$ in z. ■

Let $\sigma \in \Gamma^\infty(E)$. We wish to express its vertical lift $\sigma^V \in \mathfrak{X}(E)$ in local coordinates. Let $(U, \tilde{x}^i)$ be a local coordinate system on M and $\{\sigma_1, \ldots, \sigma_k\}$ a local frame in E defined on U. Let $z \in \pi^{-1}(U) \subseteq E$ and $x = \pi(z) \in U$ so that

$$\sigma(x) = v^\alpha(\sigma(x))\sigma_\alpha(x).$$

Let $\gamma = \gamma_{z,\sigma(x)} : (-\epsilon, \epsilon) \to E_x \subset \pi^{-1}(U)$ be the vertical curve given by $\gamma(t) = z + t\sigma(x)$ for any $|t| < \epsilon$. Let

$$\gamma^i(t) = x^i(\gamma(t)), \quad \gamma^{\alpha+n}(t) = v^\alpha(\gamma(t)), \quad |t| < \epsilon,$$

be the local components of γ with respect to the local coordinate system $(\pi^{-1}(U), x^i, v^\alpha)$. By the very definition of γ

$$\gamma^i(t) = \tilde{x}^i(x) = \text{const.}, \quad \gamma^{\alpha+n}(t) = t\,v^\alpha(\sigma(x)), \quad |t| < \epsilon,$$

hence

$$\dot{\gamma}(0) = \frac{d\gamma^i}{dt}(0)\frac{\partial}{\partial x^i}\bigg|_{\gamma(0)} + \frac{d\gamma^{\alpha+n}}{dt}(0)\frac{\partial}{\partial v^\alpha}\bigg|_{\gamma(0)} = v^\alpha(\sigma(x))\frac{\partial}{\partial v^\alpha}\bigg|_z \tag{6.1}$$

so that

$$\sigma^V(z) = (\sigma(x))_z^V = \dot{\gamma}(0) = v^\alpha(\sigma(x))\frac{\partial}{\partial v^\alpha}\bigg|_z$$

or

$$\sigma^V = \left(v^\alpha \circ \sigma \circ \pi\right)\frac{\partial}{\partial v^\alpha} \tag{6.2}$$

everywhere in $\pi^{-1}(U)$. As $v^\alpha \circ \sigma_\beta = \delta^\alpha_\beta$ the identity (6.2) shows that

$$\sigma_\alpha^V = \frac{\partial}{\partial v^\alpha}, \quad 1 \le \alpha \le k. \tag{6.3}$$

The *vertical space* at $z \in E$ is the subspace $\mathcal{V}(E)_z \subset T_z(E)$ consisting of all vertical vectors in $T_z(E)$. We set

$$\mathcal{V}(E) = \bigcup_{z \in E} \mathcal{V}(E)_z$$

(disjoint union) and observe that $\mathcal{V}(E)$ is (the total space of) a vector bundle over E in a natural manner (the *vertical subbundle* of the tangent bundle $T(E) \to E$). It should be observed that

$$\mathcal{V}(E)_z = \text{Ker}(d_z \pi), \quad z \in E.$$

Indeed let $z \in E$ and $x = \pi(z) \in M$ and $(U, \tilde{x}^i)$ be a local coordinate system on M such that $x \in U$, together with a local frame $\{\sigma_\alpha : 1 \le \alpha \le k\}$ in E defined on U. Let us set $\pi^i = \tilde{x}^i \circ \pi$, $1 \le i \le n$. Then $\pi^i = x^i$ hence

$$(d_z \pi) \frac{\partial}{\partial x^i}\bigg|_z = \frac{\partial \pi^j}{\partial x^i}(z) \frac{\partial}{\partial \tilde{x}^j}\bigg|_{\pi(z)} = \frac{\partial}{\partial \tilde{x}^i}\bigg|_x,$$

$$(d_z \pi) \frac{\partial}{\partial v^\alpha}\bigg|_z = \frac{\partial \pi^j}{\partial v^\alpha}(z) \frac{\partial}{\partial \tilde{x}^j}\bigg|_{\pi(z)} = 0,$$

so that

$$\text{Ker}(d_z \pi) = \sum_{\alpha=1}^{k} \mathbb{R} \frac{\partial}{\partial v^\alpha}\bigg|_z .$$

On the other hand (by (6.1)) any vertical vector tangent to E at z is in the span of $\{(\partial/\partial v^\alpha)_z : 1 \le \alpha \le k\}$ hence $\mathcal{V}(E)_z \subseteq \text{Ker}(d_z \pi)$. Vice versa $(\partial/\partial v^\alpha)_z$ is a vertical vector (because $(\partial/\partial v^\alpha)_z = \dot{\gamma}(0)$ where $\gamma(t) = z + t\sigma_\alpha(x)$, $|t| < \epsilon$, which is precisely the contents of the identities (6.3) evaluated at z) hence $\text{Ker}(d_z \pi) \subseteq \mathcal{V}(E)_z$.

6.1. THE HORIZONTAL BUNDLE

Let $D : \Gamma^\infty(E) \to \Gamma^\infty(T^*(M) \otimes E)$ be a connection in the vector bundle $\pi : E \to M$. Let $(\pi^{-1}(U), x^i, v^\alpha)$ be a local coordinate system on E as above (naturally induced by a local coordinate system $(U, \tilde{x}^i)$ on M and a local frame $\{\sigma_\alpha : 1 \le \alpha \le k\}$ in E on U). We set for simplicity

$$\partial_i = \frac{\partial}{\partial \tilde{x}^i}, \quad 1 \le i \le n.$$

Let $\Gamma^{\beta}_{i\alpha} \in C^{\infty}(U)$ be defined by

$$D_{\partial_i}\sigma_\alpha = \Gamma^{\beta}_{i\alpha}\sigma_\beta.$$

Let $X \in \mathfrak{X}(M)$ and $\sigma \in \Gamma^{\infty}(E)$ be respectively a tangent vector field on M and a cross-section in E locally represented by

$$X = X^i\partial_i, \quad \sigma = f^\alpha\sigma_\alpha,$$

for some $X^i, f^\alpha \in C^{\infty}(U)$, so that

$$D_X\sigma = X^i\left(\partial_i f^\beta + f^\alpha\Gamma^{\beta}_{i\alpha}\right)\sigma_\beta.$$

Let $\gamma : (-\epsilon,\epsilon) \to M$ be a smooth curve in M and $\sigma \in \Gamma^{\infty}(E)$ a smooth section in E. With respect to a local frame $\{\sigma_\alpha : 1 \le \alpha \le k\}$ in E on U one has $\sigma = f^\alpha\sigma_\alpha$ for some $f^\alpha \in C^{\infty}(U)$. As γ is continuous there is $0 < \delta \le \epsilon$ such that $\gamma(t) \in U$ for any $|t| < \delta$. One may assume as customary, by eventually shrinking U, that U is the domain of a local chart $\chi = (\tilde{x}^1,\dots,\tilde{x}^n) : U \to \mathbb{R}^n$ on M and set $\gamma^i = \tilde{x}^i \circ \gamma$. Then

$$\left(D_{d\gamma/dt}\sigma\right)_{\gamma(t)} = \left(\frac{dV^\beta}{dt}(t) + V^\alpha(t)\frac{d\gamma^i}{dt}(t)\,\Gamma^{\beta}_{i\alpha}(\gamma(t))\right)\sigma_\beta(\gamma(t)) \qquad (6.4)$$

where $V^\alpha(t) = f^\alpha(\gamma(t))$ for any $|t| < \delta$. The right hand side of (6.4) actually defines $\left(D_{d\gamma/dt}\sigma\right)_{\gamma(t)}$ for any smooth section $\sigma(t)$ defined along the curve γ.

Definition 6.2 Let $D : \Gamma^{\infty}(E) \to \Gamma^{\infty}(T^*(M) \otimes E)$ be a connection in E. i) A smooth cross-section $\sigma(t)$ defined along the smooth curve $\gamma : (-\epsilon,\epsilon) \to M$ in E is a map $\sigma : t \mapsto \sigma(t)$, locally $\sigma(t) = \sum_\alpha f^\alpha(t)\sigma_\alpha(t)$ with $f^\alpha(t)$ smooth functions. Such smooth section is said to be *parallel* (with respect to D) along γ if $\left(D_{d\gamma/dt}\sigma\right)_{\gamma(t)} = 0$ for any $|t| < \epsilon$. ii) Let $C : (-\epsilon,\epsilon) \to E$, $C(t) = (\gamma(t),\sigma(t))$, be a smooth curve in E that is locally $\gamma(t) = (x^1(C(t)),\dots,x^n(C(t)))$ and $\sigma(t) = (v^1(C(t)),\dots,v^k(C(t)))$. We say C is a *horizontal* curve if $\sigma(t)$ is parallel along γ. iii) A tangent vector $v \in T_z(E)$ is said to be *horizontal* (with respect to D) if there is a horizontal curve $C : (-\epsilon,\epsilon) \to E$ such that $C(0) = z$ and $(dC/dt)_{t=0} = v$. iv) A tangent vector field $X \in \mathfrak{X}(E)$ is said to be *horizontal* (with respect to D) if X_z is a horizontal tangent vector for any $z \in E$. ■

Let $z \in E$ and $x = \pi(z) \in M$. Let $\gamma : (-\epsilon,\epsilon) \to M$ be a smooth curve such that $\gamma(0) = x$. There is a unique horizontal curve $\gamma^\uparrow : (-\epsilon,\epsilon) \to E$ such that a) $\gamma^\uparrow(0) = z$ and b) $\pi \circ \gamma^\uparrow = \gamma$ (cf. e.g., [258]).

Definition 6.3 The curve $\gamma^{\uparrow}$ is referred to as the *horizontal lift* of γ issuing at z. ■

Let $z \in E$ and $x = \pi(z) \in M$. Also let $v \in T_x(M)$ and let us consider a smooth curve $\gamma : (-\epsilon, \epsilon) \to M$ with the initial data (x, v) i.e.,

$$\gamma(0) = x, \quad \frac{d\gamma}{dt}(0) = v.$$

Let $\gamma^{\uparrow} : (-\epsilon, \epsilon) \to E$ be the unique horizontal lift of γ issuing at z.

Definition 6.4 The horizontal tangent vector

$$\frac{d\gamma^{\uparrow}}{dt}(0) \in T_z(E)$$

is called the *horizontal lift* of v in z and is denoted by v_z^H. ■

Let us derive the local expression of v_z^H. Let $(U, \tilde{x}^i)$ be a local coordinate system on M such that $x = \pi(z) \in U$ and $\{\sigma_\alpha : 1 \le \alpha \le k\}$ a local frame in E on U. Let $\gamma : (-\epsilon, \epsilon) \to U$ be a smooth curve such that $\gamma(0) = x$ and $\dot{\gamma}(0) = v$. Let $\gamma^{\uparrow} : (-\epsilon, \epsilon) \to \pi^{-1}(U)$ be the unique horizontal lift of γ such that $\gamma^{\uparrow}(0) = z$. As $\gamma^{\uparrow} : (-\epsilon, \epsilon) \to E$ is a horizontal curve, there is $\sigma(t)$ such that $\gamma^{\uparrow} = \sigma \circ \gamma$ and $D_{d\gamma/dt}\sigma = 0$ along γ hence (by (6.4))

$$\frac{dV^\alpha}{dt}(t) = -V^\beta(t)\frac{da^i}{dt}(t)\Gamma^\alpha_{i\beta}(\gamma(t)), \quad 1 \le \alpha \le k, \tag{6.5}$$

where $V^\alpha(t) = f^\alpha(\gamma(t))$ for any $|t| < \epsilon$ and $\sigma = f^\alpha \sigma_\alpha$ on U. Let us set $\gamma^i(t) = \tilde{x}^i(\gamma(t))$. Then

$$\begin{aligned}
\frac{d\gamma^{\uparrow}}{dt}(t) &= \frac{d\gamma^i}{dt}(t)\left.\frac{\partial}{\partial x^i}\right|_{\gamma^{\uparrow}(t)} + \frac{dV^\alpha}{dt}(t)\left.\frac{\partial}{\partial v^\alpha}\right|_{\gamma^{\uparrow}(t)} \\
&= \frac{d\gamma^i}{dt}(t)\left\{\left.\frac{\partial}{\partial x^i}\right|_{\gamma^{\uparrow}(t)} - V^\beta(t)\Gamma^\alpha_{i\beta}(\gamma(t))\left.\frac{\partial}{\partial v^\alpha}\right|_{\gamma^{\uparrow}(t)}\right\}.
\end{aligned}$$

If $v = \lambda^i(\partial/\partial\tilde{x}^i)_x$ for some $\lambda^i \in \mathbb{R}$ then

$$\frac{d\gamma^i}{dt}(0) = \lambda^i, \quad 1 \le i \le n.$$

Also $z = \gamma^{\uparrow}(0) = \sigma(\gamma(0)) = \sigma(x)$ so that

$$V^\alpha(0) = f^\alpha(\gamma(0)) = f^\alpha(x) = v^\alpha(\sigma(x)) = v^\alpha(z).$$

We may conclude that

$$v_z^H = \frac{d\gamma^\uparrow}{dt}(0) = \frac{d\gamma^i}{dt}(0) \left\{ \frac{\partial}{\partial x^i}\bigg|_{\gamma^\uparrow(0)} - V^\beta(0)\Gamma^\alpha_{i\beta}(\gamma(0)) \frac{\partial}{\partial v^\alpha}\bigg|_{\gamma^\uparrow(0)} \right\}$$

$$= \lambda^i \left\{ \frac{\partial}{\partial x^i}\bigg|_z - v^\beta(z)\Gamma^\alpha_{i\beta}(x) \frac{\partial}{\partial v^\alpha}\bigg|_z \right\}.$$

In particular, the definition of v_z^H doesn't depend upon the choice of smooth curve $\gamma : (-\epsilon, \epsilon) \to M$ of initial data (x, v). Let us set

$$\delta_i \equiv \frac{\delta}{\delta x^i} = \frac{\partial}{\partial x^i} - v^\beta \left(\Gamma^\alpha_{i\beta} \circ \pi \right) \frac{\partial}{\partial v^\alpha} \in \mathfrak{X}(\pi^{-1}(U)), \quad 1 \le i \le n. \tag{6.6}$$

Summing up we have shown that given $v = \lambda^i(\partial/\partial \tilde{x}^i)_x \in T_x(M)$ then its horizontal lift in $z \in E_x$ is given by $v_z^H = \lambda^i \delta_i(z)$.

Definition 6.5 If $X \in \mathfrak{X}(M)$ is a tangent vector field on M then its *horizontal lift* (with respect to D) is the tangent vector field $X^H \in \mathfrak{X}(E)$ given by

$$X^H(z) = \left(X_{\pi(z)}\right)_z^H, \quad z \in E,$$

where $(X_z)_z^H$ is the horizontal lift of the tangent vector $v = X_{\pi(z)} \in T_{\pi(z)}(M)$. ■

The horizontal lift of a tangent vector field X on M is locally given by $X^H = (X^i \circ \pi)\delta_i$ where $X = X^i \partial/\partial \tilde{x}^i$ for some $X^i \in C^\infty(U)$.

Let R^D be the curvature tensor field of D

$$R^D(X, Y)\sigma = -D_X D_Y \sigma + D_Y D_X \sigma + D_{[X,Y]}\sigma,$$

for any $X, Y \in \mathfrak{X}(M)$ and any $\sigma \in \Gamma^\infty(E)$. It may be easily checked that

Proposition 6.6 *Let $D : \Gamma^\infty(E) \to \Gamma^\infty(T^*(M) \otimes E)$ be a connection in the real vector bundle $E \to M$ of rank k. Then*

$$[s^V, r^V] = 0, \quad [X^H, s^V] = (D_X s)^V, \tag{6.7}$$

$$[X^H, Y^H] = [X, Y]^H + \omega(X, Y) \tag{6.8}$$

for any $X, Y \in \mathfrak{X}(M)$ and any $s, r \in \Gamma^\infty(E)$. Here $\omega(X, Y)$ is the globally defined smooth section in $\mathcal{V}(E)$ locally given by

$$\omega(X, Y)|_{\pi^{-1}(U)} = v^\alpha \left(R^D(X, Y)\sigma_\alpha\right)^V$$

for any local frame $\{\sigma_\alpha : 1 \le \alpha \le k\}$ on $U \subseteq M$.

We check (6.8) and leave the proof of (6.7) as an exercise to the reader. First for any $z \in \pi^{-1}(U)$

$$[\delta_i, \delta_j]_z = v^\beta(z)\left(\frac{\partial \Gamma^\alpha_{i\beta}}{\partial \tilde{x}^j} - \frac{\partial \Gamma^\alpha_{j\beta}}{\partial \tilde{x}^i} + \Gamma^\lambda_{i\beta}\Gamma^\alpha_{j\lambda} - \Gamma^\lambda_{j\beta}\Gamma^\alpha_{i\lambda}\right)_{\pi(z)} \frac{\partial}{\partial v^\alpha}\bigg|_z$$

hence

$$[\delta_i, \delta_j] = v^\alpha\left(R^D\left(\frac{\partial}{\partial \tilde{x}^i}, \frac{\partial}{\partial \tilde{x}^j}\right)\sigma_\alpha\right)^V.$$

Finally if $X = X^i\partial/\partial\tilde{x}^i$ and $Y = Y^j\partial/\partial\tilde{x}^j$ then

$$\begin{aligned}[X^H, Y^H] &= ([X, Y]^i \circ \pi)\delta_i + (X^i \circ \pi)(Y^j \circ \pi)[\delta_i, \delta_j] \\ &= [X, Y]^H + v^\alpha(R^D(X, Y)\sigma_\alpha)^V.\end{aligned}$$

Remark 6.7 As a consequence of (6.7) and the classical Frobenius theorem the vertical distribution $\mathcal{V}(E)$ is completely integrable. ■

Let $z \in E$ and $(U, \tilde{x}^i)$ be a local coordinate system on M such that $x = \pi(z) \in U$. Let $\{\sigma_\alpha : 1 \le \alpha \le k\}$ be a local frame in E on U. Let $\mathcal{H}(E)_z$ be the linear subspace of $T_z(E)$ spanned by $\{\delta_i(z) : 1 \le i \le n\}$. It may be easily checked that the definition of $\mathcal{H}(E)_z$ doesn't depend upon the choice of local coordinates and local frame at $x = \pi(z)$. Note that $\mathcal{H}(E)_z$ is real n-dimensional. We set

$$\mathcal{H}(E) = \bigcup_{z\in E} \mathcal{H}(E)_z$$

(disjoint union). An inspection of (6.6) reveals that $\mathcal{H}(E)$ is a smooth distribution on E.

Definition 6.8 The linear space $\mathcal{H}(E)_z$ is called the *horizontal space* at z (associated to the connection D). Also $\mathcal{H}(E) \to E$ is the *horizontal distribution* (associated to D). ■

Proposition 6.9 *Let D be a connection in E. The horizontal distribution $\mathcal{H}(E)$ associated to D is complementary to the vertical distribution i.e.,*

$$T_z(E) = \mathcal{H}(E)_z \oplus \mathcal{V}(E)_z, \quad z \in E. \tag{6.9}$$

The horizontal distribution $\mathcal{H}(E)$ is involutive if and only if D is flat (i.e., $R^D = 0$).

Proof. Let X be a tangent vector field on E and $z_0 \in E$ an arbitrary point. Let us choose as customary a local coordinate system $(U, \tilde{x}^i)$ and a local frame $\{\sigma_\alpha : 1 \leq \alpha \leq k\}$ on U such that $x_0 = \pi(z_0) \in U$. Let $(\pi^{-1}(U), x^i, v^\alpha)$ be the induced local coordinates on E. Then

$$X = a^i \frac{\partial}{\partial x^i} + b^\alpha \frac{\partial}{\partial v^\alpha}$$

on $\pi^{-1}(U)$, for some $a^i, b^\alpha \in C^\infty(\pi^{-1}(U))$. Let us set

$$\partial_i = \frac{\partial}{\partial x^i}, \quad \dot{\partial}_\alpha = \frac{\partial}{\partial v^\alpha},$$

for the sake of simplicity. Then $\partial_i = \delta_i + v^\beta \Gamma^\alpha_{i\beta} \dot{\partial}_\alpha$ so that

$$X = a^i \delta_i + \left(b^\alpha + v^\beta \Gamma^\alpha_{i\beta} a^i \right) \dot{\partial}_\alpha$$

on $\pi^{-1}(U)$ and in particular in z_0 so that (6.9) holds at z_0. The last statement in Proposition 6.9 follows easily from the identity (6.8) in Proposition 6.6. ■

Definition 6.10 Let $D : \Gamma^\infty(E) \to \Gamma^\infty(T^*(M) \otimes E)$ be a connection in the real vector bundle $E \to M$ of rank k. For each $z \in E$ we consider the linear map $K_z : T_z(E) \to E_{\pi(z)}$ defined by

$$K_z(X) = 0, \quad X \in \mathcal{H}(E)_z,$$
$$K_z\left(\dot{\partial}_\alpha\big|_z\right) = \sigma_\alpha(\pi(z)), \quad 1 \leq \alpha \leq k,$$

for any local frame $\{\sigma_\alpha : 1 \leq \alpha \leq k\}$ defined on an open neighborhood $U \subseteq M$ of $x = \pi(z)$. The resulting vector-valued 1-form $K \in \Gamma^\infty(T^*(E) \otimes \pi^{-1}E)$ is called the *Dombrowski map.* ■

It may be easily checked that the definition of K_z doesn't depend upon the choice of local frame $\{\sigma_\alpha : 1 \leq \alpha \leq k\}$ at $x = \pi(z)$. The following sequence of vector bundles and vector bundle morphisms

$$0 \to \pi^{-1}E \xrightarrow{\gamma} T(E) \xrightarrow{L} \pi^{-1}T(M) \to 0 \tag{6.10}$$

is exact. The pullback bundles appearing in (6.10) are described by the commutative diagram

$$\begin{array}{ccccc} \pi^{-1}E & \to & E & \leftarrow & \pi^{-1}T(M) \\ \downarrow & & \downarrow & & \downarrow \\ E & \to & M & \leftarrow & T(M) \end{array}$$

Also the vector bundle morphism $\gamma : \pi^{-1}E \to T(E)$ is locally given by

$$\gamma_z(\sigma_\alpha(\pi(z))) = \left.\frac{\partial}{\partial v^\alpha}\right|_z, \quad z \in \pi^{-1}(U), \quad 1 \le \alpha \le k,$$

for any local frame $\{\sigma_\alpha : 1 \le \alpha \le k\}$ of E on U. Finally the vector bundle morphism $L : T(E) \to \pi^{-1}T(M)$ is given by

$$L_z X = (d_z\pi)X, \quad X \in T_z(E), \quad z \in E.$$

Similarly, in the presence of a connection D in E, the sequence of vector bundles and vector bundle morphisms

$$0 \to \mathcal{H}(E) \hookrightarrow T(E) \xrightarrow{K} \pi^{-1}E \to 0 \tag{6.11}$$

is exact. We close this section by exhibiting a geometric interpretation of the Dombrowski map. Let $C : (-\epsilon, \epsilon) \to E$ be a smooth curve in E and let us set $\gamma(t) = \pi(C(t)) \in M$ for any $|t| < \epsilon$. Let us set as customary $\gamma^i(t) = \tilde{x}^i(\gamma(t))$ with respect to a local coordinate system $(U, \tilde{x}^i)$ on M. Also

$$C^i(t) = x^i(C(t)), \quad C^{\alpha+n}(t) = v^\alpha(C(t)), \quad |t| < \epsilon,$$

with respect to the local coordinate system $(\pi^{-1}(U), x^i, v^\alpha)$ (associated to $(U, \tilde{x}^i)$ and to the local frame $\{\sigma_\alpha : 1 \le \alpha \le k\}$ of E on U). By the very definition of the local coordinates $x^i : \pi^{-1}(U) \to \mathbb{R}$ one has $C^i(t) = \tilde{x}^i(\pi(C(t))) = \tilde{x}^i(\gamma(t)) = \gamma^i(t)$. Also we set for simplicity $V^\alpha(t) = C^{\alpha+n}(t)$ for any $1 \le \alpha \le k$. Then

$$\begin{aligned}\frac{dC}{dt}(t) &= \frac{d\gamma^i}{dt}(t) \left.\frac{\partial}{\partial x^i}\right|_{C(t)} + \frac{dV^\alpha}{dt}(t) \left.\frac{\partial}{\partial v^\alpha}\right|_{C(t)} \\ &= \frac{d\gamma^i}{dt}(t)\, \delta_i\Big|_{C(t)} + \left(\frac{dV^\alpha}{dt}(t) + V^\beta(t)\Gamma^\alpha_{i\beta}(\gamma(t))\frac{d\gamma^i}{dt}(t)\right) \left.\frac{\partial}{\partial v^\alpha}\right|_{C(t)}.\end{aligned}$$

Let $\sigma(t)$ be a smooth section in E defined along γ. If $\sigma(t) = V^\alpha(t)\sigma_\alpha(t)$ for any $|t| < \epsilon$ then

$$\frac{dC}{dt}(t) = \frac{d\gamma^i}{dt}(t)\, \delta_i\Big|_{C(t)} + \left(D_{d\gamma/dt}\sigma\right)^\alpha \left.\frac{\partial}{\partial v^\alpha}\right|_{C(t)}$$

and applying $K_{C(t)}$ to both sides gives

$$K_{C(t)} \frac{dC}{dt}(t) = \left(D_{d\gamma/dt}\sigma\right)_{\gamma(t)}, \quad |t| < \epsilon, \tag{6.12}$$

which is the geometric interpretation we were looking for. Moreover, let $\sigma \in \Gamma^\infty(E)$ be an arbitrary smooth section and let us consider the differential of the map $\sigma : M \to E$ at a point $x \in M$

$$d_x\sigma : T_x(M) \to T_{\sigma(x)}(E).$$

Let $v \in T_x(M)$ represented as $v = \lambda^i (\partial/\partial \tilde{x}^i)_x$ for some $\lambda^i \in \mathbb{R}$, with respect to a local coordinate system $(U, \tilde{x}^i)$ at x. We set

$$\sigma^i = x^i \circ \sigma, \quad \sigma^{\alpha+n} = v^\alpha \circ \sigma,$$

i.e., consider the local components of σ with respect to the local coordinate system $(\pi^{-1}(U), x^i, v^\alpha)$. Then

$$(d_x\sigma)v = \lambda^i (d_x\sigma) \left.\frac{\partial}{\partial \tilde{x}^i}\right|_x = \lambda^i \left(\frac{\partial \sigma^j}{\partial \tilde{x}^i}(x) \left.\frac{\partial}{\partial x^j}\right|_{\sigma(x)} + \frac{\partial \sigma^{\alpha+n}}{\partial \tilde{x}^i}(x) \left.\frac{\partial}{\partial v^\alpha}\right|_{\sigma(x)} \right)$$

$$= \lambda^i \left\{ \frac{\partial \sigma^j}{\partial \tilde{x}^i}(x) \left.\frac{\delta}{\delta x^j}\right|_{\sigma(x)} \right.$$

$$\left. + \left(\frac{\partial \sigma^{\alpha+n}}{\partial \tilde{x}^i}(x) + \sigma^{\beta+n}(x)\Gamma^\alpha_{j\beta}(x) \frac{\partial \sigma^j}{\partial \tilde{x}^i}(x) \right) \left.\frac{\partial}{\partial v^\alpha}\right|_{\sigma(x)} \right\}.$$

On the other hand, $\sigma^i = x^i \circ \sigma = \tilde{x}^i \circ \pi \circ \sigma = \tilde{x}^i \circ 1_M = \tilde{x}^i$ hence $\partial\sigma^i/\partial\tilde{x}^j = \delta^i_j$. Consequently

$$(d_x\sigma)v = \lambda^i \left\{ \left.\frac{\delta}{\delta x^i}\right|_{\sigma(x)} + \left(\frac{\partial \sigma^{\alpha+n}}{\partial \tilde{x}^i}(x) + \sigma^{\beta+n}(x)\Gamma^\alpha_{i\beta}(x) \right) \left.\frac{\partial}{\partial v^\alpha}\right|_{\sigma(x)} \right\}$$

$$= \lambda^i \left\{ \left.\frac{\delta}{\delta x^i}\right|_{\sigma(x)} + \left(D_{\partial/\partial\tilde{x}^i}\sigma\right)^\alpha (x) \left.\frac{\partial}{\partial v^\alpha}\right|_{\sigma(x)} \right\}.$$

Let X be a tangent vector field on M extending v i.e., $X_x = v$. Then

$$(d_x\sigma)X_x = X^H_{\sigma(x)} + (D_X\sigma)^V_{\sigma(x)}. \tag{6.13}$$

6.2. THE SASAKI METRIC

Let $\pi : E \to M$ be a real vector bundle of rank k endowed with a Riemannian bundle metric h. Let D be a connection in E such that $Dh = 0$ i.e.,

$$X(h(s,r)) = h(D_X s, r) + h(s, D_X r),$$

for any $X \in \mathfrak{X}(M)$ and any $s, r \in \Gamma^\infty(E)$. Such D is commonly referred to as a *metric connection* in (E, h). Metric connections are not unique (cf. e.g., [258], p. 120). As a consequence of $Dh = 0$ one has $R^D \in \Omega^2(\mathrm{Ad}(E))$ i.e.,

$$h(R^D(X,Y)s, r) = -h(s, R^D(X,Y)r).$$

Let $\mathcal{C}(E,h)$ be the affine subspace of all metric connections in (E,h). Let g be a Riemannian metric on the base manifold M and ∇ its Levi-Civita connection. Given the data (h, D, g) with $D \in \mathcal{C}(E,h)$ one may build a Riemannian metric G_s on E in a natural way (i.e., G_s is the ordinary Sasaki metric when $E = T(M)$ and $h = g$, $D = \nabla$). Indeed we may set

$$G_s(A,B)_z = g_{\pi(z)}((d_z\pi)A_z, (d_z\pi)B_z) + h_{\pi(z)}(K_zA_z, K_zB_z), \qquad (6.14)$$

for any $A, B \in \mathfrak{X}(E)$ and any $z \in E$. Here $K : T(E) \to \pi^{-1}E$ is the Dombrowski map associated to D. It may be easily checked that

Proposition 6.11 *For any $D \in \mathcal{C}(E,h)$ and any Riemannian metric g on M the $(0,2)$-tensor field G_s given by (6.14) is a Riemannian metric on E.*

Definition 6.12 The Riemannian metric G_s on E associated to the data (h, D, g) with $D \in \mathcal{C}(E,h)$ is called the *Sasaki metric* on E. ■

Let $z \in E$ and $A, B \in T_z(E)$. Let then $C_i : (-\epsilon, \epsilon) \to E$ be two smooth curves ($i \in \{1,2\}$) such that $C_i(0) = z$ and $\dot{C}_1(0) = A$, $\dot{C}_2(0) = B$. We set $\gamma_i(t) = \pi(C_i(t))$ for any $|t| < \epsilon$. Let σ_i be the smooth section in E defined along γ_i such that $\sigma_i(\gamma_i(t)) = C_i(t)$ for any $|t| < \epsilon$. Then

$$(d_{C_i(t)}\pi)\frac{dC_i}{dt}(t) = \frac{d\gamma_i}{dt}(t), \quad K_{C_i(t)}\frac{dC_i}{dt}(t) = \left(D_{d\gamma_i/dt}\sigma_i\right)_{\gamma_i(t)}, \quad |t| < \epsilon,$$

hence

$$G_{s\,z}(A,B) = g_x(\dot{\gamma}_1(0), \dot{\gamma}_2(0)) + h_x\left(\left(D_{d\gamma_1/dt}\sigma_1\right)_x, \left(D_{d\gamma_2/dt}\sigma_2\right)_x\right)$$

where $x = \pi(z) \in M$. It may be easily checked that

Proposition 6.13 *Let $D \in \mathcal{C}(E,h)$ be a metric connection in a Riemannian vector bundle over a Riemannian manifold (M, g). Let G_s be the Sasaki metric associated to the data (h, D, g). Then*

$$G_s(X^H, Y^H) = g(X,Y) \circ \pi, \quad G_s(X^H, s^V) = 0, \quad G_s(s^V, r^V) = h(s,r) \circ \pi,$$

for any $X, Y \in \mathfrak{X}(M)$ and any $s, r \in \Omega^0(E)$. In particular the distributions $\mathcal{H}(E)$ and $\mathcal{V}(E)$ are mutually orthogonal with respect to the Sasaki metric.

Remark 6.14 If $\{E_i : 1 \le i \le n\}$ and $\{\sigma_\alpha : 1 \le \alpha \le k\}$ are local orthonormal frames of $(T(M), g)$ and (E, h) respectively, say both defined on the open subset $U \subseteq M$, then $\{E_i^H, \sigma_\alpha^V : 1 \le i \le n,\ 1 \le \alpha \le k\}$ is a local orthonormal frame of $(T(E), G_s)$ defined on the open subset $\pi^{-1}(U) \subseteq E$. ■

The Levi-Civita connection of (E, G_s) may be computed (cf. J.J. Konderak, [194], and D.E. Blair, [42], p. 149) with the result that formulae similar to those in Proposition 1.14 (where $E = T(M)$ and $h = g$, $D = \nabla$) hold good. To state the result, we need to introduce a few differential geometric objects in the pullback bundle $\pi^{-1}E \to E$.

Definition 6.15 The *Liouville vector* is the smooth cross-section $\mathcal{L} \in \Omega^0(\pi^{-1}E)$ defined by

$$\mathcal{L}(z) = z \in E_{\pi(z)} = \left(\pi^{-1}E\right)_z, \quad z \in E.$$

The *natural lift* of a section $\sigma \in \Omega^0(E)$ is the section $\hat{\sigma} \in \Omega^0(\pi^{-1}E)$ defined by $\hat{\sigma} = \sigma \circ \pi$. ■

Locally if $\{\sigma_\alpha : 1 \le \alpha \le k\}$ is a local frame in E on U then

$$\mathcal{L}(z) = z = v^\alpha(z)\sigma_\alpha(z), \quad z \in \pi^{-1}(U),$$

hence $\mathcal{L} = v^\alpha\ \sigma_\alpha$ on $\pi^{-1}(U)$. The Riemannian bundle metric h in $E \to M$ induces a Riemannian bundle metric $\hat{h}$ in $\pi^{-1}E \to E$ determined by

$$\hat{h}(\hat{s}, \hat{r}) = h(s, r) \circ \pi, \quad s, r \in \Omega^0(E). \tag{6.15}$$

Clearly not all sections in $\pi^{-1}E \to E$ are natural lifts of sections in $E \to M$ (the map $\sigma \in \Omega^0(M) \mapsto \hat{\sigma} \in \Omega^0(\pi^{-1}E)$ is injective yet not surjective). However, given a local frame $\{\sigma_\alpha : 1 \le \alpha \le k\}$ in E on U it follows that $\{\hat{\sigma}_\alpha : 1 \le \alpha \le k\}$ is a local frame of $\pi^{-1}E$ on $\pi^{-1}(U)$ hence the formula (6.15) does determine $\hat{h}$ uniquely.

Similarly a connection D in E induces a connection $\hat{D}$ in $\pi^{-1}E$ determined by

$$\hat{D}_{X^H}\hat{r} = \widehat{D_X r}, \quad \hat{D}_{s^V}\hat{r} = 0, \tag{6.16}$$

for any $X \in \mathfrak{X}(M)$ and any $r, s \in \Omega^0(E)$. The formulae (6.16) determine a unique connection $\hat{D}$ in $\pi^{-1}E$ due to the observation above (a mild reformulation of which is that $\{\hat{\sigma}_\alpha : 1 \le \alpha \le k\}$ generates the module $\Gamma^\infty(\pi^{-1}(U), \pi^{-1}E)$ over the ring $C^\infty(\pi^{-1}(U))$) together with the fact that $T(E) = \mathcal{H}(E) \oplus \mathcal{V}(E)$. Also if $D \in \mathcal{C}(E, h)$ then $\hat{D} \in \mathcal{C}(\pi^{-1}E, \hat{h})$ i.e., $\hat{D}$ is a metric connection in $(\pi^{-1}E, \hat{h})$.

Note that $\gamma : \pi^{-1}E \to \mathcal{V}(E) \subset T(E)$ is a bundle isomorphism (referred to as the *vertical lift*, as well as in the case $E = T(M)$). Moreover γ (previously defined by using a suitable local frame) is also determined by

$$\gamma \hat{s} = s^V, \quad s \in \Omega^0(E).$$

Next let $\beta : \pi^{-1}T(M) \to \mathcal{H}(E)$ be the bundle isomorphism (referred to as the *horizontal lift*, as well as in the case $E = T(M)$) determined by

$$\beta \hat{X} = X^H, \quad X \in \mathfrak{X}(M).$$

Here $\hat{X} = \ell(X)$ is the smooth section in $\pi^{-1}T(M) \to E$ given by

$$\hat{X}(z) = X_{\pi(z)}, \quad z \in E.$$

When $E = T(M)$ this coincides with the natural lift of X as introduced in Definition 1.2 in Chapter 1. As a counterpart of (6.11), the following sequence of vector bundles and vector bundle morphisms

$$0 \to \pi^{-1}T(M) \xrightarrow{\beta} T(E) \xrightarrow{K} \pi^{-1}E \to 0 \tag{6.17}$$

is exact.

Proposition 6.16 *The Levi-Civita connection ∇^{G_s} of (E, G_s) is given by*

$$\nabla^{G_s}_{s^V} r^V = 0, \tag{6.18}$$

$$\nabla^{G_s}_{X^H} Y^H = (\nabla_X Y)^H + \frac{1}{2}\gamma R^{\hat{D}}(X^H, Y^H)\mathcal{L}, \tag{6.19}$$

$$\nabla^{G_s}_{X^H} r^V = (D_X r)^V - \frac{1}{2}\left(\hat{h}(R^{\hat{D}}(X^H, \beta\ell\cdot)\mathcal{L}, \hat{r})^\sharp\right)^H, \tag{6.20}$$

$$\nabla^{G_s}_{s^V} Y^H = \frac{1}{2}\left(\hat{h}(\hat{s}, R^{\hat{D}}(\beta\ell\cdot, Y^H)\mathcal{L})^\sharp\right)^H, \tag{6.21}$$

for any $X, Y \in \mathfrak{X}(M)$ and any $s, r \in \Omega^0(E)$, where $R^{\hat{D}}$ is the curvature tensor field of the induced connection $\hat{D}$ in $\pi^{-1}E \to E$ and $\sharp : \Omega^1(M) \to \mathfrak{X}(M)$ the canonical musical isomorphism associated to g.

We emphasize that $\beta\ell X = X^H$ for any $X \in \mathfrak{X}(M)$. Also the 1-form $\hat{h}(R^{\hat{D}}(X^H, \beta\ell\cdot)\mathcal{L}, \hat{r}) \in \Omega^1(M)$ is defined by

$$Y \in \mathfrak{X}(M) \mapsto \hat{h}(R^{\hat{D}}(X^H, Y^H)\mathcal{L}, \hat{r}) \in C^\infty(M).$$

Hence $\hat{h}(R^{\hat{D}}(X^H, \beta\ell\cdot)\mathcal{L}, \hat{r})^\sharp \in \mathfrak{X}(M)$ is given by

$$g(\hat{h}(R^{\hat{D}}(X^H, \beta\ell\cdot)\mathcal{L}, \hat{r})^\sharp, Y) = \hat{h}(R^{\hat{D}}(X^H, Y^H)\mathcal{L}, \hat{r})$$
$$= \sum_{i=1}^{n} \hat{h}(R^{\hat{D}}(X^H, E_i^H)\mathcal{L}, \hat{r})E_i$$

for any local orthonormal frame $\{E_i : 1 \leq i \leq n\}$ in $(T(M), g)$. Proposition 6.16 is due to J.J. Konderak, [194] (the proof is of course similar to that of Proposition 1.14 in Chapter 1).

Remark 6.17 The projection $\pi : E \to M$ is a Riemannian submersion of (E, G_s) onto (M, g). Indeed $(\operatorname{Ker}(d_z\pi))^\perp = \mathcal{V}(E)_z^\perp = \mathcal{H}(E)_z$ and $d_z\pi : \mathcal{H}(E) \to T_{\pi(z)}(M)$ is a linear isometry, for any $z \in E$. ∎

It may be shown (cf. Theorem 3.11 by J.J. Konderak, [194]) that π is a totally geodesic map (i.e., $\beta_\pi = 0$, where β_π is the second fundamental form of π) if and only if D is flat (i.e., $R^D = 0$).

6.3. THE SPHERE BUNDLE $U(E)$

With the conventions in the previous section we set

$$U(E) = \{z \in E : h_{\pi(z)}(z, z) = 1\}.$$

This gives a bundle $S^{k-1} \to U(E) \xrightarrow{\pi} M$ and a real hypersurface $U(E)$ in E. Let us consider the tangent vector field ν on E, defined along $U(E)$, given by

$$\nu(z) = (\gamma\ \mathcal{L})_z \in T_z(E), \quad z \in U(E).$$

Then ν is a unit vector field i.e.,

$$G_s(\nu, \nu)_z = \hat{h}(\mathcal{L}, \mathcal{L})_z = h_{\pi(z)}(z, z) = 1,$$

for any $z \in U(E)$. Let $z \in U(E)$ and $v \in T_z(U(E))$. Let us consider a smooth curve $C : (-\epsilon, \epsilon) \to U(E)$ such that $C(0) = z$ and $\dot{C}(0) = v$. Let us set $\gamma(t) = \pi(C(t))$ for any $|t| < \epsilon$. Let σ be the smooth section in E defined along γ such that $\sigma(\gamma(t)) = C(t)$ for any $|t| < \epsilon$. We may differentiate in $h_{\gamma(t)}(C(t), C(t)) = 1$ so that (by $Dh = 0$)

$$0 = \frac{d}{dt}\left\{h_{\gamma(t)}(C(t), C(t))\right\} = 2h(D_{d\gamma/dt}\sigma, \sigma)_{\gamma(t)}$$
$$= h_{\gamma(t)}(K_{C(t)}\dot{C}(t), K_{C(t)}\sigma^V_{C(t)}) = G_{s\,C(t)}(\dot{C}(t), \sigma^V_{C(t)})$$

hence (at $t=0$) one obtains $G_{sz}(\nu, \nu_z)=0$. Indeed if locally $\sigma(t)=f^\alpha(t)\sigma_\alpha(t)$, then

$$\nu(C(t)) = (\gamma\,\mathcal{L})_{C(t)} = v^\alpha(C(t))\,\frac{\partial}{\partial v^\alpha}\bigg|_{C(t)} = f^\alpha(t)\,\frac{\partial}{\partial v^\alpha}\bigg|_{C(t)} = \sigma^V(C(t))$$

so that $\sigma^V(z)=\nu_z$. Summing up we have shown that

Proposition 6.18 *Let $D\in\mathcal{C}(E,h)$ be a metric connection in a Riemannian vector bundle (E,h) over a Riemannian manifold and G_s the Sasaki metric associated to (h,D,g). Then $\nu=\gamma\,\mathcal{L}$ is a unit normal vector field on $U(E)$ (as a real hypersurface in (E,G_s)).*

Definition 6.19 Given $s\in\Omega^0(E)$ we consider $s^T\in\mathfrak{X}(U(E))$ given by $s^T(z)=\left(\tan s^V\right)_z$ where $\tan_z: T_z(E)\to T_z(U(E))$ as the natural projection associated to the direct sum decomposition $T_z(E)=T_z(U(E))\oplus\mathbb{R}\nu_z$ for any $z\in U(E)$. The tangential vector field s^T on $U(E)$ is called the *tangential lift* of s. ■

As the normal bundle of the given immersion $U(E)\hookrightarrow E$ is spanned by ν i.e.,

$$T_z(U(E))^\perp = \mathbb{R}\nu_z, \quad z\in U(E),$$

it follows that the tangential lift of s is given by

$$s^T(z) = s^V_z - G_s(s^V,\nu)_z\nu_z = s^V_z - \hat{h}(\hat{s},\mathcal{L})_z\nu_z$$

for any $z\in U(E)$. We may state the following

Proposition 6.20 *The horizontal distribution $\mathcal{H}(E)$ consists of tangential vectors i.e., $\mathcal{H}(E)_z\subset T_z(U(E))$ for any $z\in U(E)$. Moreover*

$$T_z(U(E)) = \mathcal{H}(E)_z\oplus\gamma_z\,\mathrm{Ker}(\omega)_z, \quad z\in U(E), \tag{6.22}$$

where ω is given by $\omega(\sigma)=\hat{h}(\sigma,\mathcal{L})$ for any $\sigma\in\Omega^0(\pi^{-1}E)$.

Proof. Let $X\in\mathfrak{X}(M)$. Then

$$G_s(X^H,\nu) = G_s(X^H,\gamma\,\mathcal{L}) = 0$$

as $\mathcal{H}(E)$ and $\mathcal{V}(E)$ are mutually orthogonal (with respect to the Sasaki metric G_s). To prove the second statement in Proposition 6.20 let $z\in U(E)$ and $A\in T_z(U(E))\subset\mathcal{H}(E)_z\oplus\mathcal{V}(E)_z$. Then $A=\beta_zY+\gamma_z\sigma$ for some $Y\in(\pi^{-1}T(M))_z=T_{\pi(z)}(M)$ and $\sigma\in(\pi^{-1}E)_z=E_{\pi(z)}$. Next

$$A = \beta_zY + \sigma^T + G_{sz}(\gamma_z\sigma,\nu_z)\nu_z$$

where we have set by definition $\sigma^T = \gamma_z\sigma - G_{sz}(\gamma_z\sigma, \nu_z)\nu_z \in T_z(U(E))$. Yet A is tangent to $U(E)$ hence its component along ν_z must vanish

$$0 = G_{sz}(\gamma_z\sigma, \nu_z) = \hat{h}_z(\sigma, \mathcal{L}_z) = \omega_z(\sigma).$$

Summing up, we have shown that $A = \beta_z Y + \gamma_z\sigma$ for some $\sigma \in \mathrm{Ker}(\omega_z)$ so that

$$T_z(U(E)) \subseteq \mathcal{H}(E)_z + \gamma_z \,\mathrm{Ker}(\omega_z), \quad z \in U(E).$$

On the other hand, $\mathcal{H}(E)_z \cap \gamma_z \,\mathrm{Ker}(\omega_z) \subseteq \mathcal{H}(E)_z \cap \mathcal{V}(E)_z = (0)$ so that the sum $\mathcal{H}(E)_z + \gamma_z \,\mathrm{Ker}(\omega_z)$ is direct. Proposition 6.20 is proved. ■

Remark 6.21 Let $L = \hat{h}(\mathcal{L}, \mathcal{L})^{1/2} \in C^\infty(E)$. If $\{E_i : 1 \le i \le n\}$ is a local frame of $(T(M),g)$ defined on the open subset $U \subseteq M$ and $\{L^{-1}\mathcal{L}, \sigma_2, \ldots, \sigma_k\}$ is a local orthonormal frame of $(\pi^{-1}E, \hat{h})$ defined on $\pi^{-1}(U) \subseteq E$ then $\{E_i^H : 1 \le i \le n\} \cup \{\gamma\,\sigma_2, \ldots, \gamma\,\sigma_k\}$ is a local orthonormal frame in $(T(U(E)), G_s)$ defined on $\pi^{-1}(U) \cap U(E)$. ■

Proposition 6.22 *Let A be the shape operator of the given immersion $U(E) \hookrightarrow E$. Then*

$$A\beta Y = 0, \quad A\gamma\sigma = -\gamma\sigma, \tag{6.23}$$

for any $Y \in \Omega^0(\pi^{-1}T(M))$ and any $\sigma \in \Gamma^\infty(\mathrm{Ker}(\omega)) \subset \Omega^0(\pi^{-1}E)$. In particular

$$A(X^H) = 0, \quad A(s^T) = -s^T, \tag{6.24}$$

for any $X \in \mathfrak{X}(M)$ and any $s \in \Omega^0(E)$. Also, the mean curvature vector H of $U(E) \hookrightarrow E$ is given by

$$H = \frac{1}{n+k-1}\,\mathrm{trace}\,(A)\,\nu = -\frac{k-1}{n+k-1}\,\nu. \tag{6.25}$$

Proof. Let A be the shape operator of the immersion $U(E) \hookrightarrow E$ i.e., the Weingarten operator $A = A_\nu$ associated to the normal section ν. We recall (the Weingarten formula)

$$AV = -\nabla^{G_s}_V \nu, \quad V \in \mathfrak{X}(U(E)). \tag{6.26}$$

Let $\{E_i^H, \gamma\,\sigma_\alpha : 1 \le i \le n, \ 2 \le \alpha \le k\}$ be a local orthonormal frame in $T(U(E))$ defined on $\pi^{-1}(U) \cap U(E)$, as in Remark 6.21. Let $\{s_\alpha : 1 \le \alpha \le k\}$ be a local frame of E on U and let us set $\sigma_\alpha = \sum_{\beta=1}^k \lambda_\alpha^\beta \hat{s}_\beta$ for some $\lambda_\alpha^\beta \in C^\infty(\pi^{-1}(U))$ with $2 \leq \alpha \leq k$ and $1 \leq \beta \leq k$. Then $\nu = \nu^\beta s_\beta^V$ and $\gamma\,\sigma_\alpha = \lambda_\alpha^\beta s_\beta^V$. Also

$$V = \sum_{i=1}^{n} u^i E_i^H + \sum_{\alpha=2}^{k} f^\alpha \gamma\,\sigma_\alpha$$

for some $u^i, f^\alpha \in C^\infty(\pi^{-1}(U) \cap U(E))$. We may conduct the following calculation

$$\nabla_V^{G_s} \nu = u^i \nabla_{E_i^H} \nu + \sum_{\alpha=2}^{k} f^\alpha \nabla_{\gamma\,\sigma_\alpha}^{G_s} \nu,$$

$$\nabla_{\gamma\,\sigma_\alpha}^{G_s} \nu = \lambda_\alpha^\beta \nabla_{s_\beta^V}^{G_s} \left(\nu^\rho\, s_\rho^V\right) = \lambda_\alpha^\beta s_\beta^V(\nu^\rho)\, s_\rho^V, \quad 2 \leq \alpha \leq k,$$

as (by (6.18) in Proposition 6.16) $\nabla_{s_\beta^V}^{G_s} s_\rho^V = 0$. Moreover

$$\nabla_{\gamma\,\sigma_\alpha}^{G_s} \nu = \lambda_\alpha^\beta \frac{\partial \nu^\rho}{\partial \nu^\beta}\, s_\rho^V = \lambda_\alpha^\beta\, s_\beta^V = \gamma\, \left(\lambda_\alpha^\beta \hat{s}_\beta\right) = \gamma\,\sigma_\alpha,$$

$$\nabla_{X^H}^{G_s} \nu = \nu^\rho \nabla_{X^H}^{G_s} s_\rho^V + X^H(\nu^\rho)\, s_\rho^V,$$

for any $X \in \mathfrak{X}(M)$. If $X = \lambda^i\, \partial/\partial\tilde{x}^i$ then $X^H = \lambda^i \circ \pi\, \delta_i$ so that

$$X^H(\nu^\rho) = \lambda^i \frac{\delta \nu^\rho}{\delta x^i} = -\lambda^i \nu^\beta \Gamma_{i\beta}^\alpha \frac{\partial \nu^\rho}{\partial \nu^\alpha} = -\lambda^i \nu^\beta\, \Gamma_{i\beta}^\rho$$

and then

$$\nabla_{X^H}^{G_s} \nu = -\lambda^i \nu^\beta\, \Gamma_{i\beta}^\rho\, s_\rho^V + \nu^\rho \left\{ \left(D_X s_\rho\right)^V - \frac{1}{2} \left(\hat{h}(R^{\hat{D}}(X^H, \beta\,\ell\,\cdot)\mathcal{L}, \hat{s}_\rho)^\sharp\right)^H \right\}$$

$$= -\frac{1}{2} \left(\hat{h}(R^{\hat{D}}(X^H, \beta\,\ell\,\cdot)\mathcal{L}, \mathcal{L})^\sharp\right)^H = 0$$

as $R^D \in \Omega^2(\mathrm{Ad}(E))$. Summing up

$$\nabla_{\gamma\,\sigma_\alpha}^{G_s} \nu = \gamma\,\sigma_\alpha, \quad 2 \leq \alpha \leq k, \quad \nabla_{X^H}^{G_s} \nu = 0,$$

hence

$$\nabla_V^{G_s} \nu = \sum_{\alpha=2}^{k} f^\alpha\, \gamma\,\sigma_\alpha = V_{\mathcal{V}(E)}. \tag{6.27}$$

Here $V_{\mathcal{V}(E)}$ denotes the $\mathrm{Ker}(\omega)$-component of V with respect to the decomposition $T(U(E)) = \mathcal{H}(E) \oplus \mathrm{Ker}(\omega)$. At this point (6.23) follows easily from (6.27) (and the Weingarten formula (6.26)). It remains that we compute the mean curvature

$$\operatorname{trace}(A) = \sum_{i=1}^{n} G_s(AE_i^H, E_i^H) + \sum_{\alpha=2}^{k} G_s(A\gamma\,\sigma_\alpha, \gamma\,\sigma_\alpha)$$

$$= -\sum_{\alpha=2}^{k} G_s(\gamma\,\sigma_\alpha, \gamma\,\sigma_\alpha) = -\sum_{\alpha=2}^{k} \hat{h}(\sigma_\alpha, \sigma_\alpha) = -(k-1)$$

thus yielding (6.25). ■

6.4. THE ENERGY OF CROSS SECTIONS

Let $\pi : E \to M$ be a Riemannian vector bundle, with the Riemannian bundle metric h, over a compact orientable Riemannian manifold (M,g). Let $D \in \mathcal{C}(E,h)$ be a fixed metric connection in E. Let G_s be the Sasaki metric on E (associated to the data (h,D,g)). Let $\sigma \in \Omega^0(E)$ be a smooth section thought of as a smooth map $\sigma : M \to E$ of the Riemannian manifold (M,g) into the Riemannian manifold (E, G_s). As such we may compute the ordinary Dirichlet energy of σ

$$E(\sigma) = \frac{1}{2}\int_M \operatorname{trace}_g\left(\sigma^* G_s\right)\, d\operatorname{vol}(g).$$

Let $\{E_i : 1 \leq i \leq n\}$ be a local orthonormal frame in $(T(M),g)$ defined on the open subset $U \subseteq M$. Then for any $x \in U$

$$\operatorname{trace}_g\left(\sigma^* G_s\right)_x = \sum_{i=1}^{n} \left(\sigma^* G_s\right)(E_i, E_i)_x = \sum_{i=1}^{n} G_{s\sigma(x)}((d_x\sigma)E_{i,x}, (d_x\sigma)E_{i,x})$$

yet (by (6.13))

$$(d_x\sigma)E_{i,x} = \left(E_i^H\right)_{\sigma(x)} + \left(D_{E_i}\sigma\right)^V_{\sigma(x)}$$

hence (as $\mathcal{H}(E)$ and $\mathcal{V}(E)$ are mutually orthogonal)

$$\operatorname{trace}_g\left(\sigma^* G_s\right)_x = \sum_{i=1}^{n} \left\{ G_s\left(E_i^H, E_i^H\right)_{\sigma(x)} + G_s((D_{E_i}\sigma)^V, (D_{E_i}\sigma)^V)_{\sigma(x)} \right\}$$

$$= \sum_{i=1}^{n} \left\{ g(E_i, E_i)_x + h(D_{E_i}\sigma, D_{E_i}\sigma)_x \right\} = n + \|D\sigma\|_x^2$$

hence

$$E(\sigma) = \frac{n}{2}\operatorname{Vol}(M) + \frac{1}{2}\int_M \|D\sigma\|^2 \, d\operatorname{vol}(g).$$

Consequently $E(\sigma) \geq (n/2)\operatorname{Vol}(M)$ and equality is achieved if and only if $D\sigma = 0$.

Proposition 6.23 (J.J. Konderak, [194]) *The section $\sigma \in \Omega^0(E)$ is parallel with respect to D if and only if the map $\sigma : (M,g) \to (E, G_s)$ is totally geodesic.*

Proof. Let γ be a geodesic in (M,g) and let us set $C = \sigma \circ \gamma$. By the identity (6.13)

$$\dot{C}(t) = (d_{\gamma(t)}\sigma)\dot{\gamma}(t) = \left(\frac{d\gamma}{dt}\right)^H_{C(t)} + \left(D_{d\gamma/dt}\sigma\right)^V_{C(t)}$$

hence (by (6.18)–(6.21))

$$\begin{aligned}
\left(\nabla^{G_s}_{dC/dt}\frac{dC}{dt}\right)_{C(t)} &= \nabla^{G_s}_{(d\gamma/dt)^H}\left(\frac{d\gamma}{dt}\right)^H + \nabla_{(d\gamma/dt)^H}(D_{d\gamma/dt}\sigma)^V \\
&\quad + \nabla^{G_s}_{(D_{d\gamma/dt}\sigma)^V}\left(\frac{d\gamma}{dt}\right)^H + \nabla^{G_s}_{(D_{d\gamma/dt}\sigma)^V}(D_{d\gamma/dt}\sigma)^V \\
&= \left(\nabla_{d\gamma/dt}\frac{d\gamma}{dt}\right)^H + \frac{1}{2}\gamma\, R^{\hat{D}}\left(\left(\frac{d\gamma}{dt}\right)^H, \left(\frac{d\gamma}{dt}\right)^H\right)\mathcal{L} \\
&\quad + \left(D_{d\gamma/dt}D_{d\gamma/dt}\sigma\right)^V \\
&\quad - \frac{1}{2}\left(\hat{h}\left(R^{\hat{D}}\left(\left(\frac{d\gamma}{dt}\right)^H, \beta\,\ell\,\cdot\right)\mathcal{L}, \ell\, D_{d\gamma/dt}\sigma\right)^\sharp\right)^H \\
&\quad + \frac{1}{2}\left(\hat{h}\left(\ell\, D_{d\gamma/dt}\sigma, R^{\hat{D}}\left(\beta\,\ell\,\cdot, \left(\frac{d\gamma}{dt}\right)^H\right)\mathcal{L}\right)^\sharp\right)^H.
\end{aligned}$$

that is

$$\begin{aligned}
\left(\nabla^{G_s}_{dC/dt}\frac{dC}{dt}\right)_{C(t)} &= \left(D^2_{d\gamma/dt}\sigma\right)^V \\
&\quad - \left(\hat{h}\left(R^{\hat{D}}\left(\left(\frac{d\gamma}{dt}\right)^H, \beta\,\ell\,\cdot\right)\mathcal{L}, \ell\, D_{d\gamma/dt}\sigma\right)^\sharp\right)^H.
\end{aligned} \tag{6.28}$$

If $D\sigma = 0$ then (by (6.28)) $\left(D_{\dot{C}}\dot{C}\right)_{C(t)} = 0$ i.e., C is a geodesic in (E, G_s). Therefore any parallel section σ maps geodesics of (M,g) into geodesics of (E, G_s) i.e., σ is a totally geodesic map. ∎

Theorem 6.24 (J.J. Konderak, [194]) *Let $\sigma \in \Omega^0(E)$ be a smooth section in a Riemannian vector bundle (E,h) over a compact orientable Riemannian manifold (M,g). Let $D \in \mathcal{C}(E,h)$ and let G_s be the Sasaki metric on E associated to (h,D,g). Then $\sigma : (M,g) \to (E, G_s)$ is a harmonic map if and only if $D\sigma = 0$.*

The proof of Theorem 6.24 is similar to that of Theorem 2.10 and hence omitted.

Let $\sigma \in \Omega^0(E)$ be a smooth section thought of as a map of Riemannian manifolds $\sigma : (M,g) \to (E,h)$ and let $\tau(\sigma) \in \Omega^0(\sigma^{-1}TE)$ be the tension tensor of σ. By a result of J.J. Konderak (cf. *op. cit.*)

$$\tau(\sigma) = \left\{ \left(\sum_{i=1}^{n} h(R^D(\cdot, E_i)\sigma, D_{E_i}\sigma)^\sharp \right)^H - (\Delta\sigma)^V \right\} \circ \sigma \qquad (6.29)$$

where the Laplacian $\Delta : \Omega^0(E) \to \Omega^0(E)$ is locally given by

$$\Delta\sigma = -\sum_{i=1}^{n} \left(D_{E_i} D_{E_i}\sigma - D_{\nabla_{E_i} E_i}\sigma \right)$$

for any local orthonormal frame $\{E_i : 1 \le i \le n\}$ of $(T(M),g)$ on $U \subseteq M$. By the very definition of $\sharp : \Omega^1(M) \to \mathfrak{X}(M)$

$$\tau(\sigma) = \sum_{i,j=1}^{n} h\left(R^D(E_j, E_i)\sigma, D_{E_i}\sigma\right) \left(E_j^H\right) \circ \sigma - (\Delta\sigma)^V \circ \sigma$$

and we may conclude that

Corollary 6.25 (J.J. Konderak, [194]) $\tau(\sigma) = 0$ *if and only if* i) $\Delta\sigma = 0$ *and* ii) $\operatorname{trace}_g\{h(R^D(X,\cdot)\sigma, D.\sigma)\} = 0$ *for any* $X \in \mathfrak{X}(M)$.

6.5. UNIT SECTIONS

Let $\sigma \in \Omega^0(E)$ be a unit section i.e., $h_x(\sigma(x), \sigma(x)) = 1$ for any $x \in M$. Thus $\sigma \in \Gamma^\infty(U(E))$ and σ may be looked at as a smooth map of Riemannian manifolds $\sigma : (M,g) \to (U(E), G_s)$. The Sasaki metric on E

and the induced metric on $U(E)$ are denoted by the same symbol G_s. The Dirichlet energy of σ is given by

$$E(\sigma) = \frac{n}{2}\operatorname{Vol}(M) + \frac{1}{2}\int_M \|D\sigma\|^2\, d\operatorname{vol}(g).$$

As σ is a unit section and $Dh = 0$ one may derive the constraint

$$h(D_X\sigma, \sigma) = 0, \quad X \in \mathfrak{X}(M). \tag{6.30}$$

Let $\{E_i : 1 \le i \le n\}$ be a local orthonormal frame of $(T(M), g)$ on $U \subseteq M$. Then

$$\begin{aligned} h(\Delta\sigma, \sigma) &= -\sum_i h(D_{E_i}D_{E_i}\sigma - D_{\nabla_{E_i}E_i}\sigma, \sigma) \\ &= -\sum_i \{E_i(h(D_{E_i}\sigma, \sigma)) - h(D_{E_i}\sigma, D_{E_i}\sigma) - \frac{1}{2}\left(\nabla_{E_i}E_i\right)\left(\|\sigma\|^2\right)\} \end{aligned}$$

on U so that

$$h(\Delta\sigma, \sigma) = \|D\sigma\|^2 \tag{6.31}$$

for any $\sigma \in \Gamma^\infty(U(E))$. Let us denote by $\tau(\sigma) \in \Omega^0(\sigma^{-1}TE)$ and by $\tau_1(\sigma) \in \Omega^0(\sigma^{-1}TU(E))$ respectively the tension fields of the maps of Riemannian manifolds $\sigma : (M, g) \to (E, G_s)$ and $\sigma : (M, g) \to (U(E), G_s)$. Then

$$\tau_1(\sigma) = \tan \tau(\sigma)$$

where $\tan_z : T_z(E) \to T_z(U(E))$ is the projection associated to the decomposition $T_z(E) = T_z(U(E)) \oplus \mathbb{R}\nu_z$ for any $z \in U(E)$. Let $\tau(\sigma)_{\mathcal{H}(E)}$ and $\tau(\sigma)_{\mathcal{V}(E)}$ be the components of $\tau(\sigma)$ in $\mathcal{H}(E)$ and $\mathcal{V}(E)$ respectively. Then for any $x \in M$

$$\tau(\sigma)_x = \left\{\tau(\sigma)_{\mathcal{H}(E)}\right\}_{\sigma(x)} + \left\{\tau(\sigma)_{\mathcal{V}(E)}\right\}_{\sigma(x)}$$

hence (by (6.29))

$$\begin{aligned} \tau_1(\sigma)_x &= \left\{\tau(\sigma)_{\mathcal{H}(E)}\right\}_{\sigma(x)} + \left\{\tau(\sigma)_{\mathcal{V}(E)}\right\}_{\sigma(x)} - G_s(\tau(\sigma)_{\mathcal{V}(E)}, \nu)_{\sigma(x)}\, \nu_{\sigma(x)} \\ &= \left\{\tau(\sigma)_{\mathcal{H}(E)}\right\}_{\sigma(x)} + \left\{-(\Delta\sigma)^V\right\}_{\sigma(x)} - G_s(-(\Delta\sigma)^V, \nu)_{\sigma(x)}\, \gamma_{\sigma(x)}\mathcal{L}_{\sigma(x)} \\ &= \left\{\tau(\sigma)_{\mathcal{H}(E)}\right\}_{\sigma(x)} - (\Delta\sigma)^V_{\sigma(x)} + \hat{h}(\widehat{\Delta\sigma}, \mathcal{L})_{\sigma(x)}\, \gamma_{\sigma(x)}\sigma(x) \\ &= \left\{\tau(\sigma)_{\mathcal{H}(E)}\right\}_{\sigma(x)} + \gamma_{\sigma(x)}\left\{-(\widehat{\Delta\sigma})_{\sigma(x)} + h(\Delta\sigma, \sigma)_x\sigma(x)\right\} \end{aligned}$$

so that (by (6.31))

$$\tau_1(\sigma)_x = \left\{\tau(\sigma)_{\mathcal{H}(E)}\right\}_{\sigma(x)} + \gamma_{\sigma(x)}\left\{-(\Delta\sigma)_x + \|D\sigma\|_x^2\,\sigma(x)\right\}$$

and we may conclude that

Theorem 6.26 *$\sigma : (M,g) \to (U(E), G_s)$ is a harmonic map if and only if* i) $\Delta\sigma = \|D\sigma\|^2\sigma$ *and* ii) $\text{trace}_g\{h(R^D(X,\cdot)\sigma, D.\sigma)\} = 0$ *for any* $X \in \mathfrak{X}(M)$.

Definition 6.27 A smooth section $\sigma \in \Omega^0(E)$ is said to be a *harmonic section* if σ is a critical point of the energy functional $E : \Gamma^\infty(U(E)) \to [0, +\infty)$. ■

Theorem 6.28 (C.M. Wood, [317]) *Let $D \in \mathcal{C}(E,h)$ be a metric connection in a Riemannian vector bundle over a compact oriented Riemannian manifold (M,g). Let $\sigma \in \Gamma^\infty(U(E))$ be a unit section. Then σ is a harmonic section if and only if $\Delta\sigma = \|D\sigma\|^2\sigma$.*

The proof of Theorem 6.28 is similar to that of Theorem 2.23 (due to C.M. Wood, [317]) and hence omitted. We close the section by stating the second variation formula for the energy functional $E : \Gamma^\infty(U(E)) \to \mathbb{R}$. Let $\sigma \in \Gamma^\infty(U(E))$ and let $\{\sigma(t)\}_{|t|<\delta}$ be a smooth 1-parameter variation of σ with $\sigma(0) = \sigma$ and $\alpha = \sigma'(0) \in \langle\sigma\rangle^\perp$. Then, as well as in the case $E = T(M)$

$$E'(0) = \left.\frac{dE(t)}{dt}\right|_{t=0} = \frac{1}{2}\int_M \left(\frac{d}{dt}\|\nabla\sigma(t)\|^2\right)_{t=0} d\,\text{vol}(g)$$

$$= \int_M h(\nabla\sigma, \nabla\alpha)\,d\,\text{vol}(g) = \int_M h(\Delta\sigma, \alpha)\,d\,\text{vol}(g).$$

Moreover (cf. C.M. Wood, [317], p. 74)

$$(\text{Hess}\,E)_\sigma(\alpha) = \left.\frac{d^2E(t)}{dt^2}\right|_{t=0} = \int_M \frac{d}{dt}\left\{h(\nabla\sigma(t), \nabla\sigma'(t))\right\}_{t=0} d\,\text{vol}(g)$$

$$= \int_M \left(\|\nabla\alpha\|^2 - \|\alpha\|^2\|\nabla\sigma\|^2\right) d\,\text{vol}(g).$$

6.6. HARMONIC SECTIONS IN NORMAL BUNDLES

Let $M \to M_0$ be an isometric immersion of a Riemannian manifold (M,g) into the Riemannian manifold (M_0,g_0). Let us assume that $\dim(M) = n$ and $\dim(M_0) = n+k$. Let ∇ and ∇^0 be the Levi-Civita connections of (M,g) and (M_0,g_0) respectively. Let $\pi : T(M)^\perp \to M$ be the normal bundle of the given immersion $M \to M_0$. For each $\nu \in \Omega^0(T(M)^\perp)$ let A_ν be the corresponding Weingarten operator. One has the Weingarten formula

$$\nabla^0_X \nu = -A_\nu X + \nabla^\perp_X \nu, \quad X \in \mathfrak{X}(M),$$

where $\nabla^\perp$ is the normal connection, a connection in $T(M)^\perp \to M$. Let $g^\perp$ be the restriction of g_0 to $T(M)^\perp \otimes T(M)^\perp$. Then $(T(M)^\perp, g^\perp)$ is a Riemannian vector bundle and $\nabla^\perp \in \mathcal{C}(T(M)^\perp, g^\perp)$ i.e., $\nabla^\perp g^\perp = 0$. Therefore the general theory developed in the previous sections applies. In particular we may consider the Sasaki metric G_s on $T(M)^\perp$ (the total space of the normal bundle) associated to the data $(g^\perp, \nabla^\perp, g)$ and study the geometry of smooth maps $\nu : (M,g) \to (T(M)^\perp, g^\perp)$.

Definition 6.29 (K. Hasegawa, [159]) Let $(M_0, (\phi, \xi, \eta, g_0))$ be a Sasakian manifold of real dimension $2n+1$. Let (M,g) be a real n-dimensional Riemannian manifold isometrically immersed in (M_0,g_0). We say M is a *Legendrian* submanifold of $(M_0, (\phi, \xi, \eta, g_0))$ if $T_x(M) \subseteq H(M)_x = \mathrm{Ker}(\eta_x)$ for any $x \in M$. ■

See also D.E. Blair, [42], p. 55 and p. 128. On each Legendrian submanifold ξ is a section in the normal bundle. It should also be observed that $\phi_x T_x(M) \subseteq T_x(M)^\perp$ for any $x \in M$. Consequently, given a local orthonormal frame $\{E_i : 1 \le i \le n\}$ of $T(M)$ on $U \subseteq M$ the system $\{\phi E_1, \ldots, \phi E_n, \xi\}$ is a local orthonormal frame in $T(M)^\perp$ defined on the same open subset U. As (ϕ, ξ, η, g_0) is a Sasakian structure it is in particular K-contact i.e., $\nabla^0_X \xi = -\phi X$ for any $X \in \mathfrak{X}(M_0)$. Consequently (by the Weingarten formula of M in M_0) $A_\xi = 0$ and

$$\nabla^\perp_X \xi = -\phi X, \quad X \in \mathfrak{X}(M). \tag{6.32}$$

We need to recall (cf. Lemma 8.2 in [42], p. 128) that

$$A_{\phi X} Y = A_{\phi Y} X, \quad X, Y \in \mathfrak{X}(M). \tag{6.33}$$

Let α be the second fundamental form of the given immersion $M \hookrightarrow M_0$. Then (by (6.33))

$$\begin{aligned}\alpha(X,Y) &= \sum_{i=1}^{n} g_0(\alpha(X,Y),\phi E_i)\phi E_i + g_0(\alpha(X,Y),\xi)\xi \\ &= \sum_i g(A_{\phi E_i}X,Y)\phi E_i + g(A_\xi X,Y)\xi \\ &= \sum_i g(A_{\phi X}E_i,Y)\phi E_i\end{aligned}$$

hence

$$\alpha(X,Y) = \sum_{i=1}^{n} g(E_i, A_{\phi X}Y)\phi E_i. \tag{6.34}$$

On the other hand for any $X, Y \in \mathfrak{X}(M)$ (by the Gauss and Weingarten formulae)

$$\begin{aligned}(\nabla^0_X \phi)Y &= \nabla^0_X \phi Y - \phi \nabla^0_X Y \\ &= -A_{\phi Y}X + \nabla^\perp_X \phi Y - \phi\,(\nabla_X Y + \alpha(X,Y))\end{aligned}$$

(by (6.34))

$$\begin{aligned}&= -A_{\phi Y}X + \nabla^\perp_X \phi Y - \phi \nabla_X Y + \sum_{i=1}^{n} g(A_{\phi X}Y, E_i)E_i \\ &= -A_{\phi Y}X + \nabla^\perp_X \phi Y - \phi \nabla_X Y + A_{\phi X}Y\end{aligned}$$

(by (6.33))

$$= \nabla^\perp_X \phi Y - \phi \nabla_X Y.$$

Moreover (as g_0 is a Sasakian metric)

$$\left(\nabla^0_X \phi\right) Y = g(X,Y)\xi - \eta(Y)X.$$

Therefore

$$\nabla^\perp_X \phi Y = \phi \nabla_X Y + g(X,Y)\xi. \tag{6.35}$$

Let $\Delta : \Omega^0(T(M)^\perp) \to \Omega^0(T(M)^\perp)$ be the rough Laplacian associated to the data $(\nabla^\perp, g)$. Summing up the information obtained so far

$$\Delta\xi = -\sum_{i=1}^{n}\left(\nabla^\perp_{E_i}\nabla^\perp_{E_i}\xi - \nabla^\perp_{\nabla_{E_i}E_i}\xi\right)$$
$$= -\sum_i\left(-\nabla^\perp_{E_i}\phi E_i + \phi\nabla_{E_i}E_i\right)$$
$$= -\sum_i\left(-\phi\nabla_{E_i}E_i - g(E_i, E_i)\xi + \phi\nabla_{E_i}E_i\right) = n\xi$$

i.e., $\Delta\xi = n\xi$. Therefore (by Theorem 6.28), ξ is a harmonic section in $T(M)^\perp$ (cf. also K. Hasegawa, [159], p. 61). Moreover, the curvature tensor field $R^\perp$ of $\nabla^\perp$ satisfies

$$R^\perp(X, Y)\xi = -\nabla^\perp_X\nabla^\perp_Y\xi + \nabla^\perp_Y\nabla^\perp_X\xi + \nabla^\perp_{[X,Y]}\xi$$

(by (6.32))

$$= \nabla^\perp_X\phi Y - \nabla^\perp_Y\phi X - \phi[X, Y]$$

(by (6.35))

$$= \phi\nabla_X Y + g(X, Y)\xi - \phi\nabla_Y X - g(X, Y)\xi - \phi[X, Y] = 0.$$

In particular

$$\text{trace}_g g_0(R^\perp(X,\cdot)\xi, \nabla^\perp_\cdot\xi) = 0, \qquad X \in \mathfrak{X}(M).$$

Then (by Theorem 6.26), we may conclude that

Theorem 6.30 *Let M be a real n-dimensional Legendrian submanifold of a real $(2n+1)$-dimensional Sasakian manifold M_0. The Reeb vector ξ of M_0 is a harmonic map of (M, g) into $(U(T(M)^\perp), G_s)$. Moreover*

$$E(\xi) = \frac{n}{2}\text{Vol}(M) + \frac{1}{2}\int_M \|\nabla^\perp\xi\|^2\, d\,\text{vol}(g) = n\,\text{Vol}(M).$$

Remark 6.31 As a map of (M_0, g_0) into $(S(M_0), G_{s0})$ the energy of the Reeb vector ξ is $E(\xi) = [(4n+1)/2]\text{Vol}(M_0)$ (because of $\dim(M_0) = 2n+1$ and $\|\nabla^0\xi\| = 2n$). Here $S(M_0) = U(T(M_0))$. ∎

Example 6.32 Let $S^{2n+1}(1)$ carry the standard Sasakian structure. Then $S^n(1)$ is a Legendrian (totally geodesic) submanifold of $S^{2n+1}(1)$ (cf. [42], p. 59). Also the flat torus T^2 is a Legendrian (minimal) submanifold of $S^5(1)$ (cf. [42], p. 59).

6.7. THE ENERGY OF ORIENTED DISTRIBUTIONS

Let M be a compact oriented Riemannian manifold and let us denote by $\pi : \mathcal{T}^{r,s}(M) \to M$ the vector bundle of all tangent tensors of type (r,s) on M. The Riemannian metric g induces a Riemannian bundle metric in $\mathcal{T}^{r,s}(M) \to M$ denoted by the same symbol g. Also the Levi-Civita connection ∇ induces a connection (compatible to the induced bundle metric) in $\mathcal{T}^{r,s}(M) \to M$ (denoted again by ∇). Therefore we may endow the total space $\mathcal{T}^{r,s}(M)$ with the Sasaki metric $G = G_{r,s}$ (so that $G_{1,0}$ is the ordinary Sasaki metric on $T(M)$). In this context one may consider the energy of a (r,s)-tensor field (thought of as a map of Riemannian manifolds $(M,g) \to (\mathcal{T}^{r,s}(M), G_{r,s})$). This section is devoted to applying these notions to the study of the energy of an oriented distribution on a compact oriented Riemannian manifold, following P.B. Chacon & A.M. Naveira & J.M. Weston, [81], and O. Gil-Medrano & J.C. Gonzales-Davila & L. Vanhecke, [128]. Let $\Lambda^k T(M) \subset \mathcal{T}^{k,0}(M)$ be the vector bundle of all skew-symmetric tangent $(k,0)$-tensors on M. Let $G(k, T_p(M))$ be the Grassmann manifold of all oriented k-planes in $T_p(M)$. As $\dim(M) = n$ it follows that $\dim G(k, T_p(M)) = k(n-k)$. We set as usual $G(k,M) = \bigcup_{p\in M} G(k, T_p(M))$ (disjoint union) so that $G(k,M)$ is the Grassmann manifold of all oriented k-planes. If $V_p \subset T_p(M)$ is an oriented k-plane then V_p may be identified with the decomposable k-vector $\sigma_p = e_1 \wedge \cdots \wedge e_k \in \Lambda^k T_p(M)$ where $\{e_1, \ldots, e_k\}$ is a positively oriented orthonormal basis in V_p. *A posteriori* V_p is referred to as the subspace associated to the k-vector σ_p. Under this identification $G(k, T_p(M))$ may be thought of as a submanifold of $\Lambda^k T_p(M)$. Precisely $G(k, T_p(M))$ may be identified with the submanifold $\{\sigma_p \in \Sigma^k(T_p(M)) : \|\sigma_p\| = 1\}$ where $\Sigma^k(T_p(M))$ consists of the decomposable k-vectors at p. Accordingly, the tangent space $T_{V_p}(G(k, T_p(M)))$ may be identified to the subspace $T_{\sigma_p}(G(k, T_p(M))) \subset \Lambda^k T_p(M)$ spanned by

$$\left\{\sigma_\alpha^i = e_1 \wedge \cdots \wedge e_{i-1} \wedge e_\alpha \wedge e_{i+1} \wedge \cdots \wedge e_k : 1 \le i \le k,\ k+1 \le \alpha \le n\right\}$$

where $\{e_{k+1}, \ldots, e_n\}$ completes $\{e_1, \ldots, e_k\}$ such that $\{e_1, \ldots, e_n\}$ is a positively oriented orthonormal basis in $T_p(M)$. Indeed if $\sigma(t) = e_1(t) \wedge \cdots \wedge e_k(t) \in G(k, T_p(M)) \subset \Lambda^k T_p(M)$ is a smooth curve with $\sigma(0) = \sigma_p$ then

$$\sigma'(0) = \sum_{i=1}^{k} e_1 \wedge \cdots \wedge e_{i-1} \wedge e_i'(0) \wedge e_{i+1} \wedge \cdots \wedge e_k.$$

Since $e_i(t)$ is a unit vector $e_i'(0)$ and e_i are orthogonal hence $\sigma'(0)$ lies in the span of $\{\sigma_\alpha^i : 1 \le i \le k,\ k+1 \le \alpha \le n\}$.

Under the identifications above, the Grassmann bundle $G(k,M) \to M$ is the *decomposable unit subbundle* of $\Lambda^k T(M) \to M$ (whose total space $\Lambda^k T(M)$ is thought of as endowed with the Sasaki metric $G = G_{k,0}$ induced by g). Clearly when $k = 1$ one has $\Lambda^1 T(M) = T(M)$ and $G(1,M)$ may be identified to $S(M)$.

A smooth distribution ν of rank k on M, that is a smooth section in the Grassmann bundle $G(k,M)$, may be looked at as a smooth section σ in $\Lambda^k T(M)$. If $\{E_1,\dots,E_n\}$ is a positively oriented local orthonormal frame in $T(M)$ such that $\{E_1,\dots,E_k\}$ spans ν then $\sigma = E_1 \wedge \cdots \wedge E_k$ while $\{E_{k+1},\dots,E_n\}$ span the orthogonal complement $\mathcal{H} = \nu^\perp \subset T(M)$, a section in $G(n-k,M)$ represented by the section $\sigma^\perp = E_{k+1} \wedge \cdots \wedge E_n$ in $\Lambda^{n-k} T(M)$. Such a local frame $\{E_i : 1 \le i \le n\}$ is said to be *adapted* to the given distribution ν.

The energy of the distribution ν is then defined as the energy of the corresponding section $\sigma : M \to G(k,M)$ where $G(k,M)$ is endowed with the Riemannian metric induced by the Sasaki metric G on $\Lambda^k T(M)$. Hence ν is looked at as a map of Riemannian manifolds $(M,g) \to (G(k,M),G) \hookrightarrow (\Lambda^k T(M), G)$ and

$$E(\nu) = E(\sigma) = \frac{n}{2}\mathrm{Vol}(M) + \frac{1}{2}\int_M \|\nabla\sigma\|^2 \, d\,\mathrm{vol}(g).$$

Proposition 6.33 (O. Gil-Medrano & J.C. Gonzales-Davila & L. Vanhecke, [128]) *Let σ be a smooth oriented distribution on a compact orientable Riemannian manifold and let $\sigma^\perp$ be the orthogonal complement of σ. Then $\|\nabla\sigma\| = \|\nabla\sigma^\perp\|$ hence $E(\sigma) = E(\sigma^\perp)$.*

Proof. Let $\{E_i : 1 \le i \le n\}$ be a local orthonormal frame adapted to the distribution σ. Then

$$\sigma = E_1 \wedge \cdots \wedge E_k, \quad \sigma^\perp = E_{k+1} \wedge \cdots \wedge E_n,$$

hence

$$\nabla_X \sigma = \sum_{j=1}^{k} E_1 \wedge \cdots \wedge E_{j-1} \wedge \nabla_X E_j \wedge E_{j+1} \wedge \cdots \wedge E_k$$

so that

$$\nabla_X \sigma = \sum_{i=1}^{k} \sum_{\alpha=k+1}^{n} g(\nabla_X E_i, E_\alpha)\sigma_\alpha^i. \tag{6.36}$$

Next the identities

$$\|\nabla\sigma\|^2 = \sum_{a=1}^{n} \|\nabla_{E_a}\sigma\|^2, \quad \nabla_{E_a}\sigma = \sum_{i=1}^{k}\sum_{\alpha=k+1}^{n} g(\nabla_{E_a}E_i, E_\alpha)\sigma_\alpha^i,$$

yield

$$\|\nabla\sigma\|^2 = \sum_{a=1}^{n}\sum_{i=1}^{k}\sum_{\alpha,\beta=k+1}^{n} g(\nabla_{E_a}E_i, E_\alpha)g(\nabla_{E_a}E_j, E_\beta)g\left(\sigma_\alpha^i, \sigma_\beta^j\right).$$

Moreover by

$$g\left(\sigma_\alpha^i, \sigma_\beta^j\right) = \delta_j^i\delta_\beta^\alpha$$

one gets

$$\|\nabla\sigma\|^2 = \sum_{a=1}^{n}\sum_{i=1}^{k}\sum_{\alpha=k+1}^{n} g(\nabla_{E_a}E_i, E_\alpha)^2.$$

Let us interchange α and j, respectively i and β, to get

$$\|\nabla\sigma\|^2 = \sum_{a=1}^{n}\sum_{j=k+1}^{n}\sum_{\beta=1}^{k} g(\nabla_{E_a}E_\beta, E_j)^2 = \|\nabla\sigma^\perp\|^2.$$

■

Lemma 6.34 *Let σ be an oriented smooth distribution of rank k on (M,g). Then*

$$g\left(R(X,Y)\sigma, \sigma_\alpha^i\right) = g(R(X,Y)E_i, E_\alpha), \tag{6.37}$$

$$g(R(X,Y)\sigma, \nabla_Y\sigma) = \sum_{i=1}^{k}\sum_{\alpha=k+1}^{n} g(\nabla_Y E_i, E_\alpha)g(R(X,Y)E_i, E_\alpha), \tag{6.38}$$

for any $X, Y \in \mathfrak{X}(M)$.

Proof. Let $\{E_1, \ldots, E_n\}$ be an adapted frame as above so that $\sigma = E_1 \wedge \cdots \wedge E_k$. As $R(X,Y)$ is a derivation (cf. e.g., [258], p. 85)

$$R(X,Y)\sigma = \sum_{j=1}^{k} E_1 \wedge \cdots \wedge R(X,Y)E_j \wedge \cdots \wedge E_k$$

hence (6.37) follows (since $\sigma_\alpha^i = E_1 \wedge \cdots \wedge E_{i-1} \wedge E_\alpha \wedge E_{i+1} \wedge \cdots \wedge E_k$ for any $1 \le i \le k$ and $k+1 \le \alpha \le n$). Finally, (6.38) follows from (6.36)–(6.37). Lemma 6.34 is proved.

The identities (6.37)–(6.38) in Lemma 6.34 were first obtained by O. Gil-Medrano & J.C. Gonzales-Davila & L. Vanhecke, [128], by a different approach. ■

Theorem 6.35 (O. Gil-Medrano & J.C. Gonzales-Davila & L. Vanhecke, [128], B-Y. Choi & J.W. Yim, [89]) *Let σ be an oriented smooth distribution of rank k on a compact orientable Riemannian menifold (M,g). The map $\sigma : (M,g) \to (G(k,M), G)$ is harmonic if and only if* i) $\operatorname{trace}_g g(R(X,\cdot)\sigma, \nabla_{\cdot}\sigma) = 0$ *for any $X \in \mathfrak{X}(M)$ and* ii) $g(\Delta\sigma, \sigma_\alpha^i) = 0$ *for any $1 \le i \le k$ and any $k+1 \le \alpha \le n$.*

It should be noted that statement (i) in Theorem 6.35 is equivalent to

$$\sum_{i=1}^{k}\sum_{b=1}^{n} R(E_i, (\nabla_{E_b}E_i)^{\perp})E_b = 0. \tag{6.39}$$

To prove this statement one should first observe that (i) is equivalent to

$$\sum_{b=1}^{n} g(R(E_a, E_b)\sigma, \nabla_{E_b}\sigma) = 0, \qquad 1 \le a \le n.$$

Let us use the identity (6.38) in Lemma 6.34. We have

$$\begin{aligned}\sum_{b=1}^{n} g(R(E_a, E_b)\sigma, \nabla_{E_b}\sigma) &= \sum_{b=1}^{n}\sum_{i=1}^{k}\sum_{\alpha=k+1}^{n} g(R(E_a, E_b)E_i, E_\alpha)\, g(\nabla_{E_b}E_i, E_\alpha)\\ &= \sum_{\substack{1\le i\le k\\ 1\le b\le n}} g(R(E_a, E_b)E_i, (\nabla_{E_b}E_i)^{\perp})\\ &= \sum_{i,b} g(R((\nabla_{E_b}E_i)^{\perp}, E_i)E_b, E_a)\\ &= g\left(\sum_{i,b} R((\nabla_{E_b}E_i)^{\perp}, E_i)E_b, E_a\right)\end{aligned}$$

and the last expression vanishes if and only if (6.39) holds good.

Proof of Theorem 6.35. Let us look at σ as a section in $\Lambda^k T(M)$ and let $\tau(\sigma)$ be its tension field as a map of (M,g) into $(\Lambda^k T(M), G)$. By (6.29) $\tau(\sigma) = 0$ if and only if (a) $\operatorname{trace}_g R(X,\cdot)\sigma, \nabla_{\cdot}\sigma) = 0$ and (b) $\Delta\sigma = 0$ (the two conditions are equivalent to the requirement that the horizontal and vertical components of $\tau(\sigma)$ should vanish). Let $\tau_1(\sigma)$ be the tension tensor of σ thought of as a map of (M,g) into $(G(k,M),G)$. Then $\tau_1(\sigma)=0$ if and only if requirement (a) is fulfilled (that is (i) in Theorem 6.35 holds) and the orthogonal projection of $(\Delta\sigma)_p$ on $T_{\sigma_p}(G(k,T_p(M))$ vanishes. The second requirement is equivalent to (ii) in Theorem 6.35 as $\sigma^i_\alpha(p) = (E_1\wedge\cdots\wedge E_{i-1}\wedge E_\alpha\wedge E_{i+1}\wedge\cdots\wedge E_k)_p$ for $1\leq i\leq k$ and $k+1\leq\alpha\leq n$ is an orthonormal basis of $T_{\sigma_p}(G(k,T_p(M)))$. ∎

Proposition 6.36 (O. Gil-Medrano & J.C. Gonzales-Davila & L. Vanhecke, [128], B-Y. Choi & J.W. Yim, [89]) *Let σ be an oriented smooth distribution of rank k on a compact orientable Riemannian manifold (M,g). Then $\sigma:(M,g)\to(G(k,M),G)$ is a harmonic map if and only if $\sigma^\perp:(M,g)\to(G(n-k,M),G)$ is a harmonic map. Moreover $E(\sigma)=E(\sigma^\perp)$.*

Proof. Let us assume that σ is a harmonic map. By Theorem 6.35 we obtain

$$0=\sum_{b=1}^{n}\sum_{i=1}^{k}\sum_{\alpha=k+1}^{n} g(R(E_a,E_b)E_i,E_\alpha)g(\nabla_{E_b}E_i,E_\alpha)$$
$$=\sum_{b,i,\alpha} g(R(E_a,E_b)E_\alpha,E_i)g(\nabla_{E_b}E_\alpha,E_i)$$

hence $\sigma^\perp$ satisfies condition (i) in Theorem 6.35. It may be also checked (cf. [128], p. 26) that

$$g\left(\Delta\sigma,\sigma^i_\alpha\right) = -g\left(\Delta\sigma^\perp,\left(\sigma^\perp\right)^\alpha_i\right)$$

and (i) holds for σ if and only if it holds for $\sigma^\perp$. ∎

It should be observed that condition (ii) in Theorem 6.35 characterizes the distributions which are critical points of the energy functional restricted to the space $\Gamma^\infty(G(k,M))$. Such critical points are termed *harmonic distributions*. This remark (and definition) together with Theorem 6.35 and Proposition 6.36 imply

Corollary 6.37 *Let (M,g) be a compact orientable Riemannian manifold and $U\in\Gamma^\infty(S(M))$. Then* i) *$E(U)=E(U^\perp)$,* ii) *U is a harmonic vector field if and only if $U^\perp$ is a harmonic distribution, and* iii) *$U:(M,g)\to(S(M),G)$ is a harmonic map if and only if $U^\perp:(M,g)\to(G(n-1,M),G)$ is a harmonic map. Here $S(M)=G(1,M)$.*

6.8. EXAMPLES OF HARMONIC DISTRIBUTIONS

Let $(M,(\phi,\xi,\eta,g))$ be a contact metric manifold of real dimension $2n+1$. Then $\xi^\perp$ is a smooth distribution of rank $2n$ on M. By applying the previously developed theory it follows that $\xi^\perp$ is a harmonic distribution if and only if M is an H-contact manifold i.e., $Q\xi = \lambda\xi$. Moreover $\xi^\perp : (M,g) \to (G(2n,M),G)$ is a harmonic map if and only if $\xi : (M,g) \to (S(M),G)$ is a harmonic map. In particular if M is a Sasakian manifold (or more generally M is a (k,μ)-space, or M is a K-contact manifold), then $\xi^\perp$ is a harmonic map of (M,g) into $(G(2n,M),G)$. If M is one of the 3-dimensional contact manifolds classified in Theorem 4.24 then $\xi^\perp$ (a 2-dimensional distribution) is a harmonic map.

By a result of O. Gil-Medrano & J.C. Gonzales-Davila & L. Vanhecke, [128], the vertical distribution associated to the Hopf fibration $S^3 \to S^{4m+3} \longrightarrow \mathbb{H}P^{m+1}$ (commonly referred to as the *Hopf distribution*) is a harmonic map $(S^{4m+3},g_0) \to (G(3,S^{4m+3}),G)$. P.B. Chacon & A.M. Naveira & J.M. Weston also derive (cf. [81]) the second variation formula and show that the Hopf distribution is an unstable critical point of the energy functional. The Hopf distribution (of rank 3) on S^{4m+3} is but a particular case of the Reeb distribution (of rank 3) such as encountered on a 3-Sasakian manifold. Precisely

Definition 6.38 Let M be a real $(4m+3)$-dimensional manifold $(m \geq 0)$. The synthetic object $\{(\phi_i,\xi_i,\eta_i) : i \in \{1,2,3\}\}$ is called an *almost contact 3-structure* if each (ϕ_i,ξ_i,η_i) is an almost contact structure on M and

$$\phi_i\phi_j - \xi_i \otimes \eta_j = \phi_k = -\phi_j\phi_i + \xi_j \otimes \eta_i,$$
$$\xi_k = \phi_i\xi_j = -\phi_j\xi_i, \quad \eta_k = \eta_i \circ \phi_j = -\eta_j \circ \phi_i$$

for any cyclic permutation (i,j,k) of $(1,2,3)$. ■

Given an almost contact 3-structure as in Definition 6.38 there is a Riemannian metric g compatible to each of the three almost contact structures i.e., (ϕ_i,ξ_i,η_i,g) is an almost contact metric structure.

Definition 6.39 If each (ϕ_i,ξ_i,η_i,g) is Sasakian then $\{(\phi_i,\xi_i,\eta_i,g) : i \in \{1,2,3\}\}$ is referred to as a *3-Sasakian structure* on M (and M is called a *3-Sasakian manifold*). ■

Examples of 3-Sasakian manifolds (besides from S^{4m+3}) abound (cf. e.g., C. Boyer & K. Galicki, [68]).

Definition 6.40 The smooth distribution of rank 3 on a 3-Sasakian manifold M defined by $\xi = \xi_1 \wedge \xi_2 \wedge \xi_3$ is called the *Reeb distribution* on M. ■

The Reeb distribution ξ may be looked at as a smooth map $\xi : (M, g) \to (G(3, M), G)$. Let us consider a local orthonormal frame

$$\{E_a : 1 \leq a \leq 4m+3\} = \{\xi_i, E_\alpha : 1 \leq i \leq 3,\ 4 \leq \alpha \leq 4m+3\}$$

defined on $U \subseteq M$. Let us define $\{\xi^i_\alpha : 1 \leq i \leq 3,\ 4 \leq \alpha \leq 4m+3\}$ by setting

$$\xi^1_\alpha = E_\alpha \wedge \xi_2 \wedge \xi_3,\ \ \xi^2_\alpha = \xi_1 \wedge E_\alpha \wedge \xi_3,\ \ \xi^3_\alpha = \xi_1 \wedge \xi_2 \wedge E_\alpha.$$

Then (by (6.36))

$$\nabla_{E_a} \xi = \sum_{i=1}^{3} \sum_{\alpha=4}^{4m+3} g(\nabla_{E_a} \xi_i, E_\alpha) \xi^i_\alpha, \quad 1 \leq a \leq 4m+3,$$

hence $\left(\text{by} g\left(\xi^i_\alpha, \xi^j_\beta\right) = \delta^{ij}\delta_{\alpha\beta}\right)$

$$\|\nabla_{E_a} \xi\|^2 = \sum_{i=1}^{3} \sum_{\alpha=4}^{4m+3} g(\nabla_{E_a} \xi_i, E_\alpha)^2, \quad 1 \leq a \leq 4m+3.$$

Next we may use $\nabla_{\xi_j} \xi_i = -\phi_i \xi_j$ and $\phi_i \xi_j = \pm \xi_k$ for distinct i, j, k together with $\nabla_{\xi_i} \xi_i = 0$ to obtain

$$\|\nabla_{\xi_j} \xi\|^2 = \sum_{i=1}^{3} \sum_{\alpha=4}^{4m+3} g(\nabla_{\xi_j} \xi_i, E_\alpha)^2 = \sum_{i,\alpha} g(\phi_i \xi_j, E_\alpha)^2 = 0.$$

Moreover (by $\nabla_{E_\beta} \xi_i = -\phi_i E_\beta$)

$$\begin{aligned} \|\nabla_{E_\beta} \xi\|^2 &= \sum_{i=1}^{3} \sum_{\alpha=4}^{4m+3} g(\nabla_{E_\beta} \xi_i, E_\alpha)^2 \\ &= \sum_{i,\alpha} g(-\phi_i E_\beta, E_\alpha)^2 = \sum_{i} g(-\phi_i E_\beta, \phi_i E_\beta)^2 = 3. \end{aligned}$$

Summing up

$$\|\nabla \xi\|^2 = \sum_{a=1}^{4m+3} \|\nabla_{E_a} \xi\|^2 = \sum_{\beta=4}^{4m+3} \|\nabla_{E_\beta} \xi\|^2 = 12m. \tag{6.40}$$

The identity (6.40) allows one to compute the energy of ξ. Let us look at the requirements which imply the harmonicity of ξ. Since $(\phi_i, \xi_i, \eta_i, g)$ is a Sasakian structure one has (cf. e.g., [42])

$$R(X,Y)\xi_i = g(X,\xi_i)Y - g(Y,\xi_i)X = \eta_i(X)Y - \eta_i(Y)X$$

and $\nabla \xi_i = -\phi_i$. Consequently (by (6.38) in Lemma 6.34)

$$\begin{aligned}\sum_{a=1}^{4m+3} R(X,E_a)\xi, \nabla_{E_a}\xi) &= \sum_{a=1}^{4m+3}\sum_{i=1}^{3}\sum_{\alpha=4}^{4m+3} g(\nabla_{E_a}\xi_i, E_\alpha) g(R(X,E_a)\xi_i, E_\alpha)\\ &= \sum_{a,i,\alpha} g(-\phi_i E_a, E_\alpha)\eta_i(X) g(E_a, E_\alpha)\\ &\quad - \sum_{a,i,\alpha} g(-\phi_i E_a, E_\alpha)\eta_i(E_a) g(X, E_a)\\ &= \sum_{a,i,\alpha}\{g(-\phi_i E_a, E_\alpha)\delta^a_\alpha \eta_i(X)\\ &\qquad + g(\phi_i E_a, E_\alpha)\eta_i(E_a) g(X,E_a)\}.\end{aligned}$$

Since $g(\phi_i E_a, E_\alpha)\delta^a_\alpha = g(\phi_i E_\alpha, E_\alpha) = 0$ and $\eta_i(E_a) = g(\xi_i, E_a) = 0$, for any $1 \le a \le 4m+3$ such that $E_a \ne \xi_i$, and $g(\phi_i E_a, E_\alpha) = g(\phi_i \xi_i, E_\alpha) = 0$ whenever $E_a = \xi_i$ it follows that

$$\operatorname{trace}_g g(R(X,\cdot)\xi, \nabla_\cdot \xi) = 0, \quad X \in \mathfrak{X}(M). \tag{6.41}$$

Let $S_{\xi_p} \subset \Lambda^3 T_p(M)$ be the subspace spanned by $\{\xi^i_\alpha : 1 \le i \le 3,\ 4 \le \alpha \le 4m+3\}$. We wish to compute the projection of $(\Delta\xi)_p$ on S_{ξ_p}. By (6.36)

$$\begin{aligned}\nabla_X\nabla_X\xi &= \nabla_X\left(\sum_{i=1}^{3}\sum_{\alpha=4}^{4m+3} g(\nabla_X\xi_i, E_\alpha)\xi^i_\alpha\right)\\ &= \sum_{i,\alpha}\{g(\nabla_X\nabla_X\xi_i, E_\alpha)\xi^i_\alpha + g(\nabla_X\xi_i, \nabla_X E_\alpha)\xi^i_\alpha + g(\nabla_X\xi_i, E_\alpha)\nabla_X\xi^i_\alpha\}\end{aligned}$$

hence

$$\begin{aligned}g\left(\nabla_X\nabla_X\xi, \xi^j_\beta\right) &= g(\nabla_X\nabla_X\xi_j, E_\beta) + g(\nabla_X\xi_j, \nabla_X E_\beta)\\ &\quad + \sum_{i,\alpha} g(\nabla_X\xi_i, E_\alpha) g\left(\nabla_X\xi^i_\alpha, \xi^j_\beta\right).\end{aligned}$$

Moreover the projection of $\nabla_X \xi^i_\alpha$ on S_{ξ_p} is

$$\sum_\beta g(\nabla_X E_\alpha, E_\beta)\xi^i_\beta - \sum_j g(\nabla_X \xi_j, \xi_i)\xi^j_\alpha$$

hence the preceding formula becomes

$$\begin{aligned} g\left(\nabla_X \nabla_X \xi, \xi^j_\beta\right) &= g(\nabla_X \nabla_X \xi_j, E_\beta) + g(\nabla_X \xi_j, \nabla_X E_\beta) \\ &\quad + \sum_\alpha g(\nabla_X \xi_j, E_\alpha) g(\nabla_X E_\alpha, E_\beta) \\ &\quad - \sum_i g(\nabla_X \xi_i, E_\beta) g(\nabla_X \xi_j, \xi_i). \end{aligned} \tag{6.42}$$

Using

$$\nabla_X E_\beta = \sum_i g(\nabla_X E_\beta, \xi_i)\xi_i + \sum_{\alpha=4}^{4m+3} g(\nabla_X E_\beta, E_\alpha) E_\alpha$$

the identity (6.42) may be written as

$$g\left(\nabla_X \nabla_X \xi, \xi^j_\beta\right) = g(\nabla_X \nabla_X \xi_j, E_\beta) + 2\sum_i g(\nabla_X E_\beta, \xi_i) g(\nabla_X \xi_j, \xi_i). \tag{6.43}$$

Moreover (again by (6.36))

$$g(\nabla_{\nabla_X X} \xi, \xi^j_\beta) = g\left(\nabla_{\nabla_X X} \xi_j, E_\beta\right). \tag{6.44}$$

Then (by (6.43)–(6.44))

$$\begin{aligned} g\left(\Delta \xi, \xi^j_\beta\right) &= -\sum_{a=1}^{4m+3} g\left(\nabla_{E_a} \nabla_{E_a} \xi - \nabla_{\nabla_{E_a} E_a} \xi, \xi^j_\beta\right) \\ &= -\sum_a g\left(\nabla_{E_a} \nabla_{E_a} \xi_j - \nabla_{\nabla_{E_a} E_a} \xi_j, E_\beta\right) \\ &\quad - 2\sum_{i,a} g(\nabla_{E_a} E_\beta, \xi_i) g(\nabla_{E_a} \xi_j, \xi_i) \\ &= g(\Delta \xi_j, E_\beta) - 2\sum_{i,a} g(\nabla_{E_a} \xi_i, E_\beta) g(\nabla_{E_a} \xi_i, \xi_j). \end{aligned}$$

On the other hand ξ_i is a Killing vector field, i.e., $\nabla\xi_i$ is skew-symmetric, hence

$$g\left(\Delta\xi, \xi^j_\beta\right) = g(\Delta\xi_j, E_\beta) - 2\sum_{i,a} g(\nabla_{E_\beta}\xi_i, E_a) g(\nabla_{\xi_j}\xi_i, E_a)$$
$$= g(\Delta\xi_j, E_\beta) - 2\sum_i g(\nabla_{E_\beta}\xi_i, \nabla_{\xi_j}\xi_i)$$
$$= g(\Delta\xi_j, E_\beta) - 2\sum_i g(\phi_i E_\beta, \phi_i\xi_j) = g(\Delta\xi_j, E_\beta).$$

Yet $(\phi_j, \xi_j, \eta_j, g)$ is a Sasakian structure so that $\Delta\xi_j = (4m+2)\xi_j$ (for any $j \in \{1,2,3\}$) and then

$$g\left(\Delta\xi, \xi^j_\beta\right) = 0, \quad 1 \le j \le 3, \ 4 \le \beta \le 4m+3. \tag{6.45}$$

At this point the identities (6.40)–(6.41) and (6.45) together with Theorem 6.35 imply

Theorem 6.41 (O. Gil-Medrano & J.C. Gonzales-Davila & L. Vanhecke, [128]) *The Reeb distribution ξ of a real $(4m+3)$-dimensional 3-Sasakian manifold M is a harmonic map of (M,g) into $(G(3,M), G)$. Moreover, its energy is*

$$E(\xi) = \frac{4m+3}{2}\mathrm{Vol}(M) + \frac{1}{2}\int_M \|\nabla\xi\|^2 \, d\,\mathrm{vol}(g) = \frac{16m+3}{2}\mathrm{Vol}(M).$$

Remark 6.42 By Proposition 6.36 it follows that the statement in Theorem 6.41 holds good for the $4m$-dimensional distribution $\xi^\perp$, as well. ■

Other examples of harmonic distributions may be obtained by considering Hopf distributions on real hypersurfaces of a quaternionic Kähler manifold of dimension $n = 4m$ with $m > 1$.

Definition 6.43 A *quaternionic Kähler* structure on a Riemannian manifold (M_0, g_0) is a vector subbundle $\mathcal{J}_0 \subset \mathrm{End}(T(M_0))$ of rank 3 such that the following requirements are fulfilled.

a) for any point $p \in M_0$ there is an open neighborhood $U_0 \subseteq M_0$ of p and there are smooth sections $J_1, J_2, J_3 \in \Gamma^\infty(U_0, \mathcal{J}_0)$ such that for any $i \in \{1,2,3\}$ one has 1) (J_i, g_0) is an almost Hermitian structure on U_0

i.e., $J_i^2 = -I$ and $g_0(J_iX, Y) = -g_0(X, J_iY)$ for any $X, Y \in \mathfrak{X}(U_0)$, and thinking for a moment of the almost complex structures J_i as indexed in $\mathbb{Z}_3$ one has 2) $J_iJ_{i+1} = J_{i+2} = -J_{i+1}J_i$ for any $i \in \mathbb{Z}_3$.

b) $\nabla^0_X J \in \Gamma^\infty(\mathcal{J}_0)$ for any $X \in \mathfrak{X}(M_0)$ and any $J \in \Gamma^\infty(\mathcal{J}_0)$, where ∇^0 is the Levi-Civita connection of (M_0, g_0). A triple $\{J_1, J_2, J_3\}$ satisfying these conditions is commonly referred to as a *canonical basis* of $\Gamma^\infty(\mathcal{J}_0)$. A Riemannian manifold (M_0, g_0) carrying a quaternionic Kähler structure $\mathcal{J}_0$ is said to be a *quaternionic Kähler* manifold. ■

Any quaternionic Kähler manifold M_0 is oriented and has real dimension $4m$ for some $m \geq 1$. When $m > 1$ any such M_0 is an Einstein manifold.

Definition 6.44 Let $(M_0, g_0, \mathcal{J}_0)$ be a quaternionic Kähler manifold. If for any $p \in M_0$ and any unit tangent vector $u \in T_p(M)$ and any $J \in \mathcal{J}_{0,p}$ the sectional curvature of the 2-plane spanned by $\{u, Ju\}$ is a constant $c \in \mathbb{R}$ then M_0 is said to be a *quaternionic space form.* ■

Let M_0 be a quaternionic Kähler manifold and M an orientable real hypersurface in M_0. Let ν be a unit normal field on M. Let $\tan_p : T_p(M_0) \to T_p(M)$ be the natural projection associated to the decomposition $T_p(M_0) = T_p(M) \oplus \mathbb{R}\nu_p$ for any $p \in M$. Let us set

$$\mathcal{J}_p = \{\tan_p \circ (J\big|_{T_p(M)}) : J \in \mathcal{J}_{0,p}\}, \quad p \in M.$$

If $\{J_1, J_2, J_3\}$ is a local frame of $\mathcal{J}_0$ defined on the open set $U_0 \subseteq M_0$ then by setting

$$\phi_{i,p} = \tan_p \circ J_{i,p}\big|_{T_p(M)}, \quad p \in U = U_0 \cap M, \quad i \in \{1, 2, 3\},$$

one builds a local fame $\{\phi_1, \phi_2, \phi_3\}$ in $\mathcal{J}$ defined on U. Clearly ϕ_i is a skew-symmetric $(1,1)$-tensor field on U such that

$$J_iX = \phi_iX + \eta_i(X)\nu, \quad X \in \mathfrak{X}(M),$$

where η_i is a 1-form on U such that $\eta_i(X) = g_0(X, \xi_i)$ for any $X \in \mathfrak{X}(U)$, where $\xi_i = -J_i\nu$. Let g be the induced metric on M. Then

$$\eta_i(\xi_i) = 1, \ \ \phi_i^2 = -I + \xi_i \otimes \eta_i,$$
$$g(\phi_iX, \phi_iY) = g(X, Y) - \eta_i(X)\eta_i(Y),$$
$$\phi_i\xi_i = 0, \ \ \phi_i\xi_j = \xi_k, \ \ \phi_i\xi_k = -\xi_j,$$
$$\phi_k = -\phi_j\phi_i + \xi_j \otimes \eta_i,$$

for any cyclic permutation (i,j,k) of $(1,2,3)$. Hence M carries an almost contact 3-structure. By the very definition of $\mathcal{J}_0$ there exist three 1-forms α_i, $i \in \{1,2,3\}$, defined on U_0 such that

$$\nabla^0_X J_i = \alpha_k(X)J_j - \alpha_j(X)J_k, \quad X \in \mathfrak{X}(U_0),$$

for any cyclic permutation (i,j,k) of $(1,2,3)$. Thus

$$\nabla_X \xi_i = \alpha_k(X)\xi_j - \alpha_j(X)\xi_k + \phi_i SX$$

where S is the Weingarten operator i.e., $\nabla^0_X \nu = -SX$.

Definition 6.45 The 3-dimensional distribution $\mathcal{D}$ on M given by

$$\mathcal{D}_p = \{J\nu_p : J \in \mathcal{J}_{0,p}\}, \quad p \in M,$$

is referred to as the *(quaternionic) Hopf distribution*. If $\mathcal{D}$ is invariant under S (i.e., $S_p(\mathcal{D}_p) \subseteq \mathcal{D}_p$ for any $p \in M$) then M is said to be a *Hopf hypersurface* of M_0. ■

Let ξ be the section in $G(3,M)$ corresponding to the Hopf distribution $\mathcal{D}$ (as in Definition 6.45) so that locally $\xi = \xi_1 \wedge \xi_2 \wedge \xi_3$. It should be observed that the (quaternionic) Hopf distribution looks locally like the Reeb distribution associated to a contact metric 3-structure (and enjoys similar properties, [128]).

Let $G_2(\mathbb{C}^{m+2})$ be the complex Grassmann manifold consisting of all complex 2-dimensional linear subspaces of $\mathbb{C}^{m+2}$. Let $G_2(\mathbb{C}^{m+2})$ carry the canonical Riemannian metric g. Then $G_2(\mathbb{C}^{m+2})$ and its noncompact dual $G_2(\mathbb{C}^{m+2})^*$ are Hermitian symmetric spaces and Einstein manifolds. They carry both a Kählerian structure and a quaternionic Kählerian structure. The real hypersurfaces M_0 in $G_2(\mathbb{C}^{m+2})$ or $G_2(\mathbb{C}^{m+2})^*$, which are Hopf hypersurfaces with respect to both structures, were classified by J. Berndt & Y.J. Suh, [37] (any such Hopf hypersurface turns out to be a tube about some totally geodesic submanifold all of whose principal curvatures are constant). It may be shown that the Hopf distribution of M_0 is a harmonic map of (M_0, g) in $(G(3,M_0), G)$. We may also state

Theorem 6.46 (O. Gil-Medrano & J.C. Gonzales-Davila & L. Vanhecke, [128]) *The Hopf distribution of a Hopf hypersurface in the quaternionic projective space $\mathbb{H}P^m(c)$, $m \geq 2$, is a harmonic map.*

The Hopf hypersurfaces considered in Theorem 6.46 were classified by J. Berndt, [36]. L. Ornea & L. Vanhecke considered (cf. [232]) locally conformal Kähler manifolds (M, J, g) with a parallel Lee form (cf. e.g., [106])

and showed that the canonical distribution spanned by the Lee and anti-Lee vector fields is a harmonic map of (M,g) into $(G(2,M),G)$. In particular L. Ornea & L. Vanhecke consider (cf. *op. cit.*) locally conformal hyperkähler manifolds and show that, whenever the Lee form is parallel, the associated 3- and 4-dimensional distributions are harmonic maps of (M,g) into the appropriate Grassmannians. Also the Lee and anti-Lee vector fields on a Inoue surface with the Tricerri metric (cf. [106], p. 23–25) are harmonic (cf. [232]).

6.9. THE CHACON-NAVEIRA ENERGY

6.9.1. Energy of Smooth Distributions

Let (M,g) be a real n-dimensional oriented Riemannian manifold and ν a smooth distribution on M of rank k. Let $\mathcal{H} = \nu^{\perp}$ be the orthgonal complement of ν in $(T(M),g)$. Let us consider a local orthonormal frame $\{E_a : 1 \leq a \leq n\}$ in $T(M)$ adapted to ν i.e., $\{E_1,\dots,E_k\}$ is a local frame of ν while $\{E_{k+1},\dots,E_n\}$ is a local frame of $\mathcal{H}$. Let ∇ be the Levi-Civita connection of (M,g). We adopt the following convention as to the range of indices

$$1 \leq i,j \leq k, \quad k+1 \leq \alpha,\beta \leq n, \quad 1 \leq a,b \leq n.$$

The functions

$$h^{\alpha}_{ij} = -g(\nabla_{E_i} E_\alpha,\, E_j) = g(E_\alpha,\, \nabla_{E_i} E_j)$$

are the local manifestation of the second fundamental form of ν along the vector fields E_α. Similarly the functions

$$h^{i}_{\alpha\beta} = -g(\nabla_{E_\alpha} E_i,\, E_\beta) = g(E_i,\, \nabla_{E_\alpha} E_\beta)$$

describe locally the second fundamental form of $\mathcal{H}$ along the vector fields E_i. Note that ν is integrable (equivalently $[E_i, E_j] \in \nu$, $1 \leq i, j \leq k$) if and only if $h^{\alpha}_{ij} = h^{\alpha}_{ji}$. Also ν is totally geodesic if and only if $h^{\alpha}_{ij} = 0$.

Definition 6.47 A vector field Z is said to be $\mathcal{H}$-*conformal* if

$$(\mathcal{L}_Z g)(X,Y) = f\, g(X,Y), \qquad X,Y \in \mathcal{H},$$

for some $f \in C^\infty(M)$. ■

Of course a Killing vector field is $\mathcal{H}$-conformal with $f = 0$.

The distribution ν may be looked at as a smooth section ξ of the Grassmann bundle $G(k,M)$ whose total space carries the Sasaki metric G induced by g

$$\xi : (M,g) \to (G(k,M),G), \quad \xi = E_1 \wedge \cdots \wedge E_k.$$

Let M be compact. The energy of ν is

$$E(\nu) = E(\xi) = \frac{n}{2}\operatorname{Vol}(M) + \frac{1}{2}\int_M \|\nabla\xi\|^2 \, d\operatorname{vol}(g). \tag{6.46}$$

P.B. Chacon & A.M. Naveira & J.M. Weston proved (cf. [81]) that the Hopf distribution $S^3 \to S^{4m+3} \to \mathbb{H}P^m$ is an unstable critical point of the energy functional for any $m \geq 1$. Subsequently P.B. Chacon & A.M. Naveira considered (cf. [82]) the following *total bending* of ν

$$\mathcal{D}(\nu) = \int_M \left\{\|\nabla\xi\|^2 + (n-k)(n-k-2)\|\vec{H}_{\mathcal{H}}\|^2 + k^2\|\vec{H}_\nu\|^2\right\} d\operatorname{vol}(g)$$

where

$$\vec{H}_{\mathcal{H}} = \frac{1}{n-k}\sum_{i=1}^{k}\sum_{\alpha=k+1}^{n} h^i_{\alpha\alpha}\, E_i, \quad \vec{H}_\nu = \frac{1}{k}\sum_{\alpha=k+1}^{n}\sum_{i=1}^{k} h^\alpha_{ii}\, E_\alpha.$$

$\vec{H}_{\mathcal{H}}$ and $\vec{H}_\nu$ are respectively the *mean curvature vectors* of $\mathcal{H}$ and ν. Note that

$$\|\nabla\xi\|^2 = \sum_{a=1}^{n}\|\nabla_{E_a}\xi\|^2 = \sum_{i,j,\alpha}\left(h^\alpha_{ij}\right)^2 + \sum_{i,\alpha,\beta}\left(h^i_{\alpha\beta}\right)^2,$$

$$\|\vec{H}_{\mathcal{H}}\|^2 = \frac{1}{(n-k)^2}\sum_i\left(\sum_\alpha h^i_{\alpha\alpha}\right)^2, \quad \|\vec{H}_\nu\|^2 = \frac{1}{k^2}\sum_\alpha\left(\sum_i h^\alpha_{ii}\right)^2.$$

Definition 6.48 We call

$$E^*(\nu) = E^*(\xi) = \frac{n}{2}\operatorname{Vol}(M) + \frac{1}{2}\mathcal{D}(\nu)$$

the *Chacon-Naveira energy* of ν. ■

Note that E^* does not generalize the Brito functional $\tilde{E}$. Indeed if ν is 1-dimensional ($k = 1$) and we identify ν with $\xi \in \Gamma^\infty(S(M))$ then

$$\vec{H}_\nu = -\nabla_\xi\xi, \quad \vec{H}_{\mathcal{H}} = \vec{H}_{\xi^\perp} = -\frac{1}{n-1}(\operatorname{div}\xi)\xi,$$

so that

$$
\begin{aligned}
E^*(\xi) &= \frac{n}{2}\,\mathrm{Vol}(M) \\
&\quad + \frac{1}{2}\int_M \left(\|\nabla\xi\|^2 + (n-3)(n-1)\|\vec{H}_{\xi^{b}{}_{ot}}\|^2 + \|\nabla_\xi \xi\|^2\right) d\,\mathrm{vol}(g) \\
&= \tilde{E}(\xi) + \frac{1}{2}\int_M \|\nabla_\xi \xi\|^2\, d\,\mathrm{vol}(g).
\end{aligned}
$$

Let us go back to the case of an arbitrary smooth distribution of rank $k \geq 1$. The Chacon-Naveira energy attains its minimum $(n/2)\,\mathrm{Vol}(M)$ when both distributions $\mathcal{H}$ and ν are totally geodesic. The main result in [82] may be stated as

Theorem 6.49 (P.B. Chacon & A.M. Naveira, [82]) *If ν is integrable then*

$$
\mathcal{D}(\nu) \geq \int_M \sum_{i,\alpha} K(E_i, E_\alpha)\, d\,\mathrm{vol}(g) \tag{6.47}
$$

where $K(E_i, E_\alpha)$ is the sectional curvature of the 2-plane spanned by $E_i \in \nu$ and $E_\alpha \in \mathcal{H}$. Moreover equality is achieved in (6.47) if and only if ν is totally geodesic and the vector fields $E_1, \ldots, E_k$ are $\mathcal{H}$-conformal.

6.9.2. The Chacon-Naveira Energy of Distributions on 3-Sasakian Manifolds and Normal Complex Contact Manifolds

Let M be a compact real $(4m+3)$-dimensional manifold M endowed with a 3-Sasakian structure $\{(\phi_i, \xi_i, \eta_i, g) : i \in \{1,2,3\}\}$. For the Reeb distribution $\xi = \xi_1 \wedge \xi_2 \wedge \xi_3$ our previous finding was that

$$
E(\xi) = \frac{16m+3}{2}\,\mathrm{Vol}(M)
$$

that is $\int_M \|\nabla\xi\|^2\, d\,\mathrm{vol}(g) = 12m\,\mathrm{Vol}(M)$. Moreover the vector fields ξ_i are Killing and

$$
h_{ij}^{\alpha} = -g(\nabla_{\xi_i} E_\alpha, \xi_j) = g(\nabla_{\xi_i}\xi_j, E_\alpha) = -g(\phi_j \xi_i, E_\alpha) = g(\xi_h, E_\alpha) = 0,
$$

for any cyclic permutation (i,j,h) of $(1,2,3)$. Consequently

$$\mathcal{D}(\xi) = \int_M \|\nabla \xi\|^2 \, d\,\mathrm{vol}(g) = 12m\,\mathrm{Vol}(M), \tag{6.48}$$

$$\mathcal{D}(\xi) = \int_M \sum_{i,\alpha} K_{i\alpha} \, d\,\mathrm{vol}(g). \tag{6.49}$$

Let us consider the unit sphere $S^{4m+3} \subset \mathbb{C}^{2m+2}$. Let ν be the normal unit field on S^{4m+3} (that is $\nu_p = \vec{p} \in T_p(\mathbb{C}^{2m+2})$ for any $p \in S^{4m+3}$). Let $\{J_1, J_2, J_3\}$ be the ordinary complex structures of $\mathbb{R}^{4m+4}$. The vector fields

$$\xi_1 = J_1 \nu, \quad \xi_2 = J_2 \nu, \quad \xi_3 = J_3 \nu,$$

determine the natural 3-Sasakian structure of S^{4m+3}. The vector fields ξ_i are tangent to the vertical bundle of the Hopf fibration $S^3 \to S^{4m+3} \to \mathbb{H}P^m$. By $\sum_{i,\alpha} K_{i\alpha} = 12m$ and by (6.48)–(6.49) one derives the following result (due to P.B. Chacon & A.M. Naveira, [82])

Theorem 6.50 *Among all smooth integrable distributions of rank* 3 *on* S^{4m+3}, *the Reeb distribution* $\xi_1 \wedge \xi_2 \wedge \xi_3$ *of* $S^3 \to S^{4m+3} \to \mathbb{H}P^m$ *minimizes the Chacon-Naveira energy.*

Let ν be an integrable distribution of rank 3 on M. Let us assume that ν is locally expressed by $V = E_1 \wedge E_2 \wedge E_3$ where $\{E_1, \ldots, E_{4m+3}\}$ is a positive orthonormal local frame of $T(M)$.

Definition 6.51 The scalar $K(\nu) = K(E_1, E_2) + K(E_2, E_3) + K(E_3, E_1)$ is called the *curvature* of the distribution ν. ■

In dimension 3, the scalar $2K(\nu)$ is precisely the scalar curvature of M as a Riemannian manifold. Let us show that $K(\nu)$ is well defined i.e., the definition does not depend upon the choice of local positive orthonormal frame. Let $\{\theta^i : 1 \le i \le 4m+3\}$ be the dual coframe i.e., $\theta^i(E_j) = \delta^i_j$ and let us set

$$\Omega_{ab}(X, Y) = g(R(X, Y)E_a, E_b), \quad X, Y \in \mathfrak{X}(M).$$

The sign convention in the definition of the curvature tensor field is of course that adopted in the Chapter 2 of this monograph. Let us consider the $(4m+3)$-form

$$\Omega = \sum_{\sigma \in \sigma_3} \epsilon(\sigma) \sum_{\tau \in \sigma_{4m}} \epsilon(\tau) \Omega_{\sigma(1)\tau(4)} \wedge \theta^{\sigma(2)} \wedge \theta^{\sigma(3)} \wedge \theta^{\tau(5)} \wedge \cdots \wedge \theta^{\tau(4m+3)}$$

where σ_3 is the group of permutations of $\{1,2,3\}$ and σ_{4m} is the group of permutations of the set $\{4,5,\ldots,4m+3\}$. One may easily check that Ω is well defined i.e., its definition doesn't depend upon the chosen positive orthonormal local frame. Also (by (13) in [82], p. 102)

$$\Omega(E_1,\ldots,E_{4m+3}) = 2(4m-1)!\sum_{i,\alpha} K(E_i,E_\alpha). \tag{6.50}$$

Moreover

$$\begin{aligned}\sum_{i,\alpha} K(E_i,E_\alpha) &= \sum_{i,\alpha} R(E_i,E_\alpha,E_i,E_\alpha) \\ &= \sum_{i,a} R(E_i,E_a,E_i,E_a) - \sum_{i,j} R(E_i,E_j,E_i,E_j) \\ &= \sum_i \operatorname{Ric}(E_i,E_i) - 2\sum_{i<j} K(E_i,E_j) = \sum_i \operatorname{Ric}(E_i,E_i) - 2K(\nu).\end{aligned}$$

By a result of T. Kashiwada, [183], any 3-Sasakian manifold is Einstein of scalar curvature $\rho = (4m+2)(4m+3)$. Then

$$\sum_{i,\alpha} K(E_i,E_\alpha) = 3(4m+2) - 2K(\nu). \tag{6.51}$$

By (6.50)–(6.51) it follows that $K(\nu)$ is well defined. If $K(\nu) \leq 3$ then (by (6.48), (6.51)) and Theorem 6.49)

$$\mathcal{D}(\nu) \geq \int_M \sum_{i,\alpha} K(E_i,E_\alpha)\, d\operatorname{vol}(g) \geq 12m\operatorname{Vol}(M) = \mathcal{D}(\xi) \tag{6.52}$$

and equality is achieved in (6.52) if and only if $K(\nu) = 3$, ν is totaly geodesic and E_1, E_2, E_3 are $\mathcal{H}$-conformal. Let us also recall that T. Kashiwada has proved (cf. [184]) the following remarkable result: every contact metric 3-structure is 3-Sasakian (cf. also [248] for a direct proof in dimension three). Therefore we obtain the following result (extending Theorem 6.50)

Theorem 6.52 (D. Perrone, [249]) *Let M be a compact 3-contact metric manifold. Then among all integrable distributions ν on M of rank 3 and curvature $K(\nu) \leq 3$ the Reeb distribution ξ minimizes the energy $\mathcal{D}(\nu)$. Moreover $\mathcal{D} = \mathcal{D}(\xi)$ if and only if ν is totally geodesic, E_1, E_2, E_3 are $\mathcal{H}$-conformal, and $K(\nu) = 3$.*

Let us give an application of Theorem 6.52. Each compact Riemannian manifold of constant sectional curvature +1 and dimension $4m+3$ is a spherical space form $(S^{4m+3}/\Gamma, g)$ with $\Gamma \subset O(4m+4)$ a finite subgroup (where the identity is the only element corresponding to the eigenvalue $+1$) and g is the metric induced on S^{4m+3}/Γ by the canonical metric g_0 of S^{4m+3}. If $(\varphi_i, \xi_i, \eta_i, g_0)$ is the standard 3-Sasakian structure on S^{4m+3}, the spherical space forms S^{4m+3}/Γ which admit a 3-Sasakian structure correspond to the subgroups Γ leaving invariant each of the three Sasakian structures $(\varphi_i, \xi_i, \eta_i)$. Of course on any such manifold $K(\nu) = 3$ for any 3-dimensional distribution ν. Then (by Theorem 6.52)

Theorem 6.53 (D. Perrone, [249]) *Let M be a spherical space form admitting a 3-Sasakian structure. Then among all integrable 3-dimensional distributions on M, the Reeb distribution minimizes the Chacon-Naveira energy $\mathcal{D}$.*

As a 3-Sasakian manifold is Einstein, it follows that any 3-dimensional 3-Sasakian manifold has constant sectional curvature $+1$. Hence on any compact 3-dimensional 3-contact metric manifold the Reeb distribution minimizes $\mathcal{D}$. Let us look at a few examples.

Example 6.54 For a 3-dimensional spherical space form S. Sasaki, [267], has classified the subgroups Γ leaving invariant each of the three Sasakian structures. Precisely each Γ is a finite subgroup of Clifford translations on S^3 and Γ is equivalent to one of the following groups a) $\Gamma = \{I\}$, b) $\Gamma = \{\pm I\}$, c) Γ is the cyclic group of order $q > 2$ generated by $\begin{pmatrix} A & 0 \\ 0 & A \end{pmatrix}$ where

$$A = \begin{pmatrix} \cos\dfrac{2\pi}{q} & -\sin\dfrac{2\pi}{q} \\[2ex] \sin\dfrac{2\pi}{q} & \cos\dfrac{2\pi}{q} \end{pmatrix},$$

and d) Γ is a group of Clifford translations corresponding to a binary dihedral group or the binary polyhedral groups of the regular tetrahedron T^*, octahedron O^*, or icosahedron I^*.

In dimension $4m+3 > 3$ examples of spherical space forms which admit a 3-Sasakian structure are given by $M = S^{4m+3}/\Gamma_r$ where $\Gamma_r = \Gamma \times \cdots \times \Gamma$ ($r = m+1$ terms) with Γ one of the groups in the list (a)–(d) above. In particular the sphere S^{4m+3}, the real projective space $\mathbb{R}P^{4m+3}$, and the lens spaces $L^{4m+3} = S^{4m+3}/\Gamma_r$ with Γ of type (c), admit a 3-Sasakian structure. Therefore the Reeb distribution in these examples minimizes the Chacon-Naveira energy.

Next we look at the characteristic distribution of a complex contact metric manifold. We start by recalling a few notions and basic results on complex contact metric manifolds. For further details the reader may consult D.E. Blair, [42], and B. Kormaz, [196].

Definition 6.55 Let M be a complex manifold of complex dimension $2m+1$. We call M a *complex contact manifold* if it carries an open cover $\mathcal{U}$ of M by local coordinate neighborhoods and a family of holomorphic forms $\{\theta_U \in \Omega^{1,0}(U) : U \in \mathcal{U}\}$ such that i) for any $U \in \mathcal{U}$ one has $\theta_U \wedge (d\theta_U)^m \neq 0$ everywhere in U, and ii) for any $U, U' \in \mathcal{U}$ such that $U \cap U' \neq \emptyset$ there is a nowhere zero holomorphic function $f = f_{UU'} : U \cap U' \to \mathbb{C}^*$ such that $\theta_{U'} = f\theta_U$. ■

Let $(M, \{\theta_U : U \in \mathcal{U}\})$ be a complex contact manifold. The family $\{\theta_U : U \in \mathcal{U}\}$ is referred to as its *complex contact structure*. It determines a smooth distribution $\mathcal{H}_0$ in the following manner. Let us consider the canonical bundle isomorphism $T(M) \to T^{1,0}(M)$ defined by $X \mapsto X - iJX$ for any $X \in T(M)$, where J denotes the complex structure of M and $T^{1,0}(M) = \{X - iJX : X \in T(M)\}$ is the holomorphic tangent bundle. Let $x \in M$ and let $U \in \mathcal{U}$ such that $x \in U$. Then we set by definition

$$\mathcal{H}_0(x) = \{v \in T_x(M) : \theta_{U,x}(v - iJ_x v) = 0\}.$$

By (ii) in Definition 6.55 the definition of $\mathcal{H}_0(x)$ doesn't depend upon the choice of $U \in \mathcal{U}$ such that $x \in U$. Moreover $\mathcal{H}_0(y)$ has constant dimension for any $y \in U$, hence the assignment $x \in M \mapsto \mathcal{H}_0(x) \subset T_x(M)$ is a well defined smooth distribution on M. Also, as a consequence of (i) in Definition 6.55 the distribution $\mathcal{H}_0$ is not involutive. The local 1-forms defining a complex contact structure $\{\theta_U : U \in \mathcal{U}\}$ on M glue up to a (globally defined) 1-form on M if and only if $c_1(M) = 0$, where $c_1(M)$ denotes the first Chern class of M.

Let $(M, \{\theta_U : U \in \mathcal{U}\})$ be a complex contact manifold. Then $\theta = u - iv$ where $u, v \in \Omega^1(U)$ are real 1-forms such that $v = u \circ J$. By (i) in Definition 6.55 the 2-form du is nondegenerate along $\mathcal{H}_0$ hence there is a unique real tangent vector field $B \in \mathfrak{X}(U)$ such that

$$(du)(B, X) = 0, \quad X \in \mathcal{H}_0,$$

$$u(B) = 1, \quad v(B) = 0.$$

Let $A = -JB$ and let us consider the distribution $\mathcal{V}_0$ locally defined by the bi-vector $B \wedge A$. Then $T(M) = \mathcal{H}_0 \oplus \mathcal{V}_0$.

Definition 6.56 The distributions $\mathcal{H}_0$ and $\mathcal{V}_0$ are respectively referred to as the *horizontal distribution* and the *vertical distribution* (or *characteristic distribution*) of the complex contact manifold $(M, \{\theta_U : U \in \mathcal{U}\})$. ■

From now on we assume that $(M, \{\theta_U : U \in \mathcal{U}\})$ is a complex contact manifold whose characteristic distribution $\mathcal{V}_0$ is integrable (this is indeed the case in all known examples).

Definition 6.57 A Hermitian metric g on M is said to be an *associated metric* if 1) for any $U \in \mathcal{U}$ there is a $(1,1)$-tensor field G such that

$$G^2 = -I + u \otimes B + v \otimes A, \quad G \circ J = -J \circ G, \quad GU = 0,$$

$$g(GX, Y) = -g(X, GY), \quad u(X) = g(B, X),$$

$$(du)(X, Y) = g(X, GY) + (\sigma \wedge u)(X, Y),$$

$$(dv)(X, Y) = g(X, HY) - (\sigma \wedge u)(X, Y),$$

for any $X, Y \in T(M)$, where $H = G \circ J$ and $\sigma(X) = g(\nabla_X B, A)$. Also it must be that 2) for any $U, U' \in \mathcal{U}$ such that $U \cap U' \neq \emptyset$ one has

$$u' = au - bv, \quad v' = bu + av,$$

$$G' = aG - bH, \quad H' = bG + aH,$$

for some functions $a, b \in C^\infty(U \cap U')$ with $a^2 + b^2 = 1$. Here $u', v' \in \Omega^1(U)$ are defined by $\theta_{U'} = u' - iv'$. ■

Note that $H^2 = G^2$ hence $H^2 = -I + u \otimes B + v \otimes A$ as well.

Definition 6.58 A synthetic object $(M, J, \{\theta_U : U \in \mathcal{U}\}, g)$ consisting of a complex manifold (M, J), a complex contact structure $\{\theta_U : U \in \mathcal{U}\}$, and an associated Hermitian metric g, is called a *complex contact metric* manifold. ■

By a result of B. Foreman, [119], any complex contact manifold admits associated Hermitian metrics.

Definition 6.59 Let $X \in \mathcal{H}_0(x)$ be a unit vector. Let $p \subset T_x(M)$ be the 2-plane spanned by X and $Y = \lambda GX + \mu HX$ with $\lambda, \mu \in \mathbb{R}$, $\lambda^2 + \mu^2 = 1$. We call p a *GH-plane* while the sectional curvature of p is referred to as the *GH-sectional curvature* of p. ■

Definition 6.60 (D.E. Blair, [42], B. Kormaz, [196]) A complex contact metric manifold M is said to be *normal* if

$$S(X, Y) = 0, \quad T(X, Y) = 0, \quad X, Y \in \mathcal{H}_0,$$

$$S(B, X) = 0, \quad T(A, X) = 0, \quad X \in T(M),$$

where S and T are the $(1,2)$-tensor fields given by

$$
\begin{aligned}
S(X,Y) = {} & [G,G](X,Y) + 2g(X,GY)B \\
& - 2g(X,HY)A + 2v(Y)HX - 2v(X)HY \\
& + \sigma(GY)HX - \sigma(GX)HY + \sigma(X)GHY - \sigma(Y)GHX, \\
T(X,Y) = {} & [H,H](X,Y) + 2g(X,GY)B \\
& + 2g(X,HY)A + 2u(Y)GX - 2u(X)GY \\
& + \sigma(HX)GY - \sigma(HY)GX + \sigma(X)GHY - \sigma(Y)GHX,
\end{aligned}
$$

for any $X, Y \in T(M)$. ■

As a consequence of normality, sectional curvatures of 2-planes spanned by a vector in $\mathcal{V}_0$ and a vector in $\mathcal{H}_0$ equal $+1$. Therefore, given a local orthonormal frame $\{E_\alpha : 1 \le \alpha \le 4m\}$ of $\mathcal{H}_0$ one has

$$\sum_{\alpha=1}^{4m} \{K(B \wedge E_\alpha) + K(A \wedge E_\alpha)\} = 8m. \tag{6.53}$$

Again as a consequence of normality

$$\nabla_X B = -GX + \sigma(X)A, \quad \nabla_X A = -HX - \sigma(X)B.$$

Therefore

$$(\mathcal{L}_B g)(X,Y) = g(\nabla_X B, Y) + g(X, \nabla_Y B) = 0$$

for any $X, Y \in \mathcal{H}_0$. A similar result holds for A. That is both B and A are $\mathcal{H}_0$-Killing vector fields. Moreover

$$g(\nabla_B B, X) = g(\nabla_A A, X) = g(\nabla_B A, X) = g(\nabla_A B, X) = 0$$

for any $X \in \mathcal{H}_0$ i.e., $\mathcal{V}_0$ is totally geodesic. Consequently (as well as in the 3-Sasakian case) Theorem 6.49 and (6.53) yield

$$\mathcal{D}(\mathcal{V}_0) = \int_M \sum_\alpha \{K(B \wedge E_\alpha) + K(A \wedge E_\alpha)\}\, d\,\mathrm{vol}(g) = 8m\,\mathrm{Vol}(M). \tag{6.54}$$

The result in (6.54) is essentially Theorem 2 in [48]. This doesn't imply in general that $\mathcal{V}_0$ minimizes the Chacon-Naveira energy (although this does hold in special instances). Arguments similar to those in the proof of Theorem 6.52 lead to

Theorem 6.61 (D. Perrone, [249]) *Let M be a compact Einstein normal complex contact metric manifold. Then among all integrable 2-dimensional distributions $\mathcal{V}$ on M of curvature $K(\mathcal{V}) \le K(\mathcal{V}_0)$ the characteristic distribution $\mathcal{V}_0$ minimizes the Chacon-Naveira energy $\mathcal{D}(\mathcal{V})$. Moreover $\mathcal{D}(\mathcal{V}) = \mathcal{D}(\mathcal{V}_0)$ if and only if $\mathcal{V}$ is totally geodesic, V_1, V_2 are $\mathcal{H}$-conformal, and $K(\mathcal{V}) = K(\mathcal{V}_0)$.*

Appealing examples of Einstein normal complex contact metric manifolds do exist. A complex contact metric manifold M is normal in the sense of S. Ishihara & M. Konishi (cf. [178]) or briefly M is *IK-normal*, if $S=0$ and $T=0$. Obviously a IK-normal complex contact metric structure is normal in the sense of Definition 6.60. S. Ishihara & M. Konishi, [178], showed that any IK-normal complex contact metric manifold is Kähler-Einstein with $c_1(M) > 0$. Also (by a result of B. Foreman (cf. Theorem 6.1 and Proposition 6.3 in [119]) M is isometric to a twistor space of a quaternionic-Kähler manifold of positive scalar curvature and

$$R(X,Y)B = -u(Y)X + u(X)Y - v(Y)JX + v(X)JY - 2g(JX,Y)A \quad (6.55)$$

where R is the curvature tensor field of (M,g). Then

$$K(\mathcal{V}_0) = g(R(B,A)B,A) = 4, \quad \mathrm{Ric} = (4m+1)g. \quad (6.56)$$

Conversely (again by using results in [119] and [177]), every twistor space $\mathcal{Z}$ of a quaternionic-Kähler manifold of positive scalar curvature admits a IK-normal complex contact metric structure satisfying (6.55).

Let $\mathcal{Z}$ be a compact complex $(2m+1)$-dimensional manifold carrying a complex contact structure. By a result of C. LeBrun, [201], if $\mathcal{Z}$ admits a Kähler-Einstein metric of positive scalar curvature then $\mathcal{Z}$ is the twistor space of a quaternionic-Kähler manifold of positive scalar curvature. Summing up, a complex $(2m+1)$-dimensional compact Kähler-Einstein manifold $\mathcal{Z}$, of positive scalar curvature, carrying a complex contact structure, admits an Einstein normal complex contact metric structure of scalar curvature $\rho = 2(2m+1)(4m+1)$.

Another approach to building twistor spaces admitting an Einstein normal complex contact metric structure may be given as follows. Let $\overline{M}$ be a 3-Sasakian manifold such that one of the Reeb vectors, say ξ_1, is regular. Then the orbit space $M = \overline{M}/\xi_1$ admits a IK-normal complex contact metric structure which is Kähler-Einstein of positive scalar curvature (cf. S. Ishihara & M. Konishi, [177]). Therefore M is isometric to a twistor space of a quaternionic-Kähler manifold of positive scalar curvature. Consequently, the class of twistor spaces of a quaternionic-Kähler manifold of positive scalar curvature consists of Einstein normal complex contact metric manifolds satisfying (6.56). Then (by Theorem 6.61)

Theorem 6.62 (D. Perrone, [249]) *Let $\mathcal{Z}$ be a compact twistor space of a quternionic-Kähler manifold of positive scalar curvature equipped with a IK-normal complex contact metric structure. Then among all 2-dimensional integrable distributions $\mathcal{V}$ on $\mathcal{Z}$ of curvature $K(\mathcal{V}) \leq 4$ the characteristic distribution minimizes the Chacon-Naveira energy.*

The complex projective space $\mathbb{C}P^{2m+1}$ endowed with the Fubini-Study metric g (of constant holomorphic curvature $+4$) enters the above class of examples. Indeed $\mathbb{C}P^{2m+1}$ is the twistor space of the quaternionic-Kähler manifold $\mathbb{Q}P^{2m+1}$. By a result of S. Ishihara & M. Konishi (cf. [177]) $\mathbb{C}P^{2m+1}$ admits a normal complex contact metric structure $(J, \{\theta_U : U \in \mathcal{U}\}, g)$ which appears to be closely related to the standard Sasakian 3-structure on S^{4m+3} (it is induced by the structure on S^{4m+3} via $S^1 \to S^{4m+3} \to \mathbb{C}P^{2m+1}$). Let $\mathcal{V}$ be a 2-dimensional integrable distribution on $\mathbb{C}P^{2m+1}$ locally defined by the bi-vector $V_1 \wedge V_2$. As $\mathbb{C}P^{2m+1}$ has constant holomorphic curvature $+4$ the curvature $K(\mathcal{V})$ satisfies (cf. e.g., [189], p. 167) $K(\mathcal{V}) = 1 + 3\cos\zeta(\mathcal{V}) \leq 4$, where $\cos\zeta(\mathcal{V}) = \left|g(V_1, JV_2)\right|$, and $K(\mathcal{V}) = 4$ if and only if $V_2 = \pm JV_1$. Then (by Theorem 6.62)

Corollary 6.63 *Among all 2-dimensional integrable distributions $\mathcal{V}$ on $\mathbb{C}P^{2m+1}$ the characteristic distribution $\mathcal{V}_0$ of the normal complex contact metric structure induced via the Hopf fibration $S^1 \to S^{4m+3} \to \mathbb{C}P^{2m+1}$ minimizes the Chacon-Naveira energy. Moreover if $\mathcal{V}$ is locally defined by the bi-vector $V_1 \wedge V_2$ then $\mathcal{D}(\mathcal{V}) = \mathcal{D}(\mathcal{V}_0)$ if and only if $\mathcal{V}$ is totally geodesic, $V_2 = \pm JV_1$, and V_1, V_2 are $\mathcal{H}$-conformal.*

CHAPTER SEVEN

Harmonic Vector Fields in CR Geometry

Contents

Let M be a compact orientable n-dimensional Riemannian manifold and let $S(M) \to M$ denote its tangent sphere bundle. As largely exploited in the previous chapters of this monograph $S(M)$ carries a natural Riemannian metric G_s (induced by the Sasaki metric on $T(M)$) naturally associated to the Riemannian metric g of M. A unit vector field X on M is then a smooth map $X : M \to S(M)$. As such, X is a harmonic vector field if it is a critical point of the energy functional

$$E(X) = \frac{1}{2} \int_M \|dX\|^2 d\,\mathrm{vol}(g),$$

(where $\|dX\|^2 = \mathrm{trace}_g(X^* G_s)$) i.e., for any smooth 1-parameter variation $\{X_t\}_{|t|<\epsilon}$ of $X = X_0$ by unit vector fields $X_t : M \to S(M)$ one has

$$\frac{d}{dt}\{E(X_t)\}_{t=0} = 0,$$

cf. Chapter 2 of this monograph or C.M. Wood, [316]. Harmonic vector fields aren't harmonic maps in general (a unit vector field $X : M \to S(M)$ is a harmonic map if and only if X is a harmonic vector field and the additional condition $\mathrm{trace}_g\{R(\nabla.X, X)\cdot\} = 0$ is satisfied, cf. Corollary 2.24

© 2012 Elsevier Inc. All rights reserved.

in Chapter 2 of this monograph or O. Gil-Medrano, [126]). Harmonic vector fields were first studied by G. Wiegmink, [309], though as critical points of the total bending functional

$$\mathcal{B}(X) = \int_M \|\nabla X\|^2 d\,\text{vol}(g),$$

a measure of the failure of a unit vector field X to be parallel with respect to the Levi-Civita connection ∇ of g. However a simple calculation (taking into account the very construction of G_s) showed that $E(X)$ may be also written as

$$E(X) = \frac{1}{2}\mathcal{B}(X) + \frac{n}{2}\text{Vol}(M),$$

(see Proposition 2.3 in Chapter 2 of this monograph) hence the notions in [309] and [316] were seen to be equivalent. Both authors established the first and second variation formulae for $E(X)$ (the variations are through unit vector fields) and gave applications. For instance, by a result in [309], p. 332, the Reeb vector field ξ of a Sasakian manifold is harmonic. Generalizations of this result to arbitrary contact Riemannian manifolds (mainly due to the second author of this monograph) were presented in Chapters 3 and 4 of this book. Any Hopf vector field (i.e., a vector field on the sphere S^{2n+1} tangent to the fibres of the Hopf fibration $S^1 \to S^{2n+1} \to \mathbb{C}P^n$) was shown to be harmonic and unstable provided that $n > 1$, cf. Theorem 3.18 in Chapter 3 or [316], p. 320. The stability part in the previous result is not surprising in light of the result by Y.L. Xin, [321] (that harmonic maps from a sphere to a compact Riemannian manifold are unstable, cf. Theorem 3.17 in Chapter 3). The additional condition $n > 1$ is easy to understand since, as proved by G. Wiegmink, [309], the Hopf vector fields on the sphere S^3 are stable critical points for the energy. The stability problem has not been studied in full generality (i.e., when the source manifold is an arbitrary compact Riemannian manifold) yet in light of a classical result by P.F. Leung, [207] (any nonconstant harmonic map of a compact Riemannian manifold into a sphere is unstable) and taking into account the formal similarity between the shape operator of a sphere and that of $S(M)$ (as a real hypersurface of $(T(M), G_s)$) one may expect instability. For example if (M,g), $n \geq 3$, is a compact quotient of a symmetric space of rank 1 and of non-positive curvature with non-vanishing first Betti number, then the geodesic flow ξ on $S(M)$ is unstable for the energy functional (cf. [56]).

A generalization of harmonic maps to the context of Hörmander systems of vector fields was produced by J. Jost & C-J. Xu, [180], and Z.R. Zhou, [327]. Their notion of a *subelliptic harmonic map* was seen to arise as the local manifestation of the *pseudoharmonic maps* of E. Barletta et al., [25], from a strictly pseudoconvex CR manifold M to a Riemannian manifold. Pseudoharmonic maps themselves admit a nice geometric interpretation in terms of the Fefferman metric F_θ associated to a choice of contact form θ on M (cf. [25], p. 729).

In the present chapter, we report on the development of two analogous theories of *pseudoharmonic vector fields* (or *subelliptic harmonic vector fields*) within CR geometry, following the work by D. Perrone et al., [107], and by Y. Kamishima et al., [103]. Precisely, if X is a unit tangent vector field on a strictly pseudoconvex CR manifold M, lying in the Levi distribution $H(M)$ of M, then its horizontal lift $X^\uparrow$ with respect to the natural connection 1-form σ in the canonical circle bundle $S^1 \to C(M) \to M$ is a map of $C(M)$ into

$$S_1(C(M), F_\theta) = \{V \in T(C(M)) : F_\theta(V, V) = 1\}.$$

F_θ is a Lorentz metric on $C(M)$ hence $S_1(C(M), F_\theta)$ (the Lorentz analog to the tangent sphere bundle over a Riemannian manifold) is a hyperquadric bundle. At this point it is a natural question whether the tangent bundle of a Lorentz manifold admits a natural metric (similar to the Sasaki metric in Riemannian geometry) and almost complex structure. It turns out that this is indeed the case (cf. O. Gil-Medrano & A. Hurtado, [130], and D. Perrone et al., [107]) and one may request that $X^\uparrow$ be a critical point of the functional

$$\mathbb{E}(\mathcal{X}) = \frac{1}{2} \int_{C(M)} \operatorname{trace}_{F_\theta} \left(\mathcal{X}^* G_s\right) d\operatorname{vol}(F_\theta),$$

$$\mathcal{X} \in \Gamma^\infty(S_1(C(M), F_\theta)).$$

It will be shown (following [107]) that given a smooth 1-parameter variation of X through unit vector fields $X_t \in H(M)$

$$\frac{d}{dt}\left\{\mathbb{E}\left(X_t^\uparrow\right)\right\}_{t=0} = \int_{C(M)} F_\theta\left(\tilde{V}, \Box X^\uparrow\right) d\operatorname{vol}(F_\theta), \qquad (7.1)$$

where $\tilde{V} = \left(\partial X_t^\uparrow / \partial t\right)_{t=0}$ and $\Box$ is the wave operator (the Laplacian with respect to the Fefferman metric F_θ). This is a consequence of the more general fact that given a unit vector field X on a compact Lorentz manifold (N,g), one has (similar to the Riemannian case) $\{dE(X_t)/dt\}_{t=0} = \int_N g(V, \Delta_g X) d\,\mathrm{vol}(g)$, where $V = (\partial X_t/\partial t)_{t=0}$ and Δ_g is the rough Laplacian (on vector fields) associated to the Lorentz metric g. When g is a Riemannian metric, $\{dE(X_t)/dt\}_{t=0} = 0$ yields

$$\Delta_g X - \|\nabla X\|^2 X = 0 \tag{7.2}$$

cf. Theorem 2.22 in Chapter 2 of this monograph. This is no longer true when g is a Lorentz metric. However, one may integrate along the fibres in the right hand member of (7.1) to obtain (up to a multiplicative constant)

$$\int_M g_\theta(V, -\Delta_b X + 2\tau JX + 4\nabla_T JX + 6\phi JX)\, \theta \wedge (d\theta)^n$$

which (by taking into account the constraint $g_\theta(V,X) = 0$) leads to equation (7.32) whose principal part is Δ_b, the *sublaplacian* (on vector fields). Equation (7.32) is the CR, or pseudohermitian, analogue to (7.2). One may show that Δ_b is a subelliptic operator (cf. Proposition 7.29) and that (7.32) has a variational interpretation (cf. Theorem 7.32). A theory for the existence and regularity of weak solutions to (7.2), or its CR analogue (7.32), is missing in the present day literature.

The material in Sections 7.1 to 7.3 is organized as follows. In Section 7.1 we describe the canonical metric G_s (a semi-Riemannian metric of signature $(2n - 2\nu, 2\nu)$) and the compatible almost complex structure J on the total space $T(M)$ of the tangent bundle over a semi-Riemannian n-dimensional manifold (M,g) of index $0 \le \nu < n$ (both arise naturally from the semi-Riemannian metric g, as well as in the Riemannian case, cf. [266] and [42]). In Section 7.2 we introduce the bundles of hyperquadrics $Q_\epsilon \to S_\epsilon(M,g) \to M$ (the semi-Riemannian analog to the tangent sphere bundle over a Riemannian manifold) and show that $S_\epsilon(M,g)$ is a semi-Riemannian hypersurface of $(T(M), G_s)$, of index $\epsilon \in \{\pm 1\}$, and compute its Weingarten operator. Also we endow $S_\epsilon(M,g)$ with the natural almost CR structure (induced by the almost complex structure J) and obtain the integrability conditions (7.18). As an application of the material in Section 7.2, we give geometric conditions under which the Reeb vector $\xi : M \to S(M)$ of an almost contact Riemannian manifold is a CR map (cf. Theorem 7.18 and Corollaries 7.19 and 7.20).

In Section 7.3 we consider harmonic vector fields of the form $X^\uparrow$, $X \in H(M)$, from $C(M)$ (the total space of the canonical circle bundle over a strictly pseudoconvex CR manifold M) and project the relevant harmonic vector fields equations on M, cf. (7.32). We show that (7.32) is a subelliptic system of PDEs and that (7.32) are the Euler-Lagrange equations of a variational principle on M, cf. Theorem 7.32.

Building on the work by J. Jost & C-J. Xu, [180], and E. Barletta et al., [25], one may study smooth *pseudoharmonic maps* from a compact strictly pseudoconvex CR manifold and their generalizations. Boundary values of Bergman-harmonic maps $\phi : \Omega \to S$ from a smoothly bounded strictly pseudoconvex domain $\Omega \subset \mathbb{C}^n$ were shown to be pseudoharmonic maps (cf. Y. Kamishima et al., [103]) provided their normal derivatives vanish. It is also known that $\overline{\partial}_b$-pluriharmonic maps are pseudoharmonic maps (cf. again [103]). A smooth vector field $X : M \to T(M)$ is a pseudoharmonic map if and only if X is parallel (with respect to the Tanaka-Webster connection) along the maximally complex, or Levi, distribution. The main purpose of Sections 7.4 to 7.7 is to report on the results by Y. Kamishima et al. (cf. *op. cit.*) on the emerging theory of *subelliptic harmonic vector fields* i.e., unit vector fields $X \in \mathcal{U}(M,\theta)$ which are critical points of the energy functional $E(X) = \frac{1}{2}\int_M \text{trace}_{G_\theta}(\pi_H X^* S_\theta)\,\theta \wedge (d\theta)^n$ relative to variations through unit vector fields. Any such critical point X was shown (cf. again [103]) to satisfy the nonlinear subelliptic system $\Delta_b X - \|\nabla^H X\|^2 X = 0$. Also $\inf_{X \in \mathcal{U}(M,\theta)} E(X) = n\,\text{Vol}(M,\theta)$ yet E is unbounded from above. We also present the derivation of the first and second variation formulae for $E : \mathcal{U}(M,\theta) \to [0,+\infty)$. Subelliptic harmonic vector fields generalize (within CR geometry, or more generally within the theory of Hörmander systems of vector fields) pseudoharmonic (or subelliptic harmonic) maps very much the way harmonic vector fields generalize harmonic maps (within Riemannian or Lorentzian geometry).

7.1. THE CANONICAL METRIC

We start by adapting a few notions of geometry of the tangent bundle (over a Riemannian manifold) to the semi-Riemannian case. As long as tensor calculus is involved, only minor modifications are necessary. Let (M,g) be an n-dimensional semi-Riemannian manifold of index $0 \leq \nu < n$. Let $\pi : T(M) \to M$ be the tangent bundle and $\mathcal{V} = \text{Ker}(d\pi)$ the vertical subbundle. The Levi-Civita connection of (M,g) induces a connection ∇ in the pullback bundle $\pi^{-1}TM \to T(M)$. Indeed let $\left(U, \tilde{x}^i\right)$ be a local coordinate system on M and $\left(\pi^{-1}(U), x^i, y^i\right)$ the induced local coordinates on

$T(M)$. If Γ^i_{jk} are the Christoffel symbols of (M,g) we set

$$\nabla_{\partial_i} X_j = \left(\Gamma^k_{ij} \circ \pi\right) X_k\,, \quad \nabla_{\dot{\partial}_i} X_j = 0.$$

Here X_i is the (locally defined) section in the diagonal bundle given by $X_j(u) = (u, (\partial/\partial\tilde{x}^j)_{\pi(u)})$, for any $u \in \pi^{-1}(U)$ (the *natural lift* of $\partial/\partial\tilde{x}^i$). Note that $\{X_j : 1 \leq j \leq n\}$ is a local frame in $\pi^{-1}TM \to T(M)$. Also $\{\partial_i, \dot{\partial}_i\}$ is short for $\{\partial/\partial x^i, \partial/\partial y^i\}$.

We recall (cf. Chapter 1 of this book) that a nonlinear connection on M is a C^∞ distribution $\mathcal{H} : u \in T(M) \mapsto \mathcal{H}_u \subset T_u(T(M))$ such that

$$T_u(T(M)) = \mathcal{H}_u \oplus \mathcal{V}_u\,, \quad u \in T(M). \tag{7.3}$$

There is a natural nonlinear connection on M, associated to the semi-Riemannian metric, as follows. Also the Liouville vector is the section $\mathcal{L}$ in $\pi^{-1}TM$ given by $\mathcal{L}(u) = (u,u)$, for any $u \in T(M)$.

Lemma 7.1 *Consider the bundle map* $K : T(T(M)) \to \pi^{-1}TM$ *given by* $K\tilde{X} = \nabla_{\tilde{X}}\mathcal{L}$ *for any* $\tilde{X} \in T(T(M))$. *Then* $\mathcal{H} = \mathrm{Ker}(K)$ *is a nonlinear connection on* M. *Locally* $\mathcal{H}$ *is given by the Pfaffian equation*

$$dy^i + N^i_j\, dx^j = 0 \tag{7.4}$$

where $N^i_j = \Gamma^i_{jk} y^k$. *Also* $K(\gamma X) = X$, *for any* $X \in \pi^{-1}TM$.

Lemma 7.1 is an obvious semi-Riemannian analog to Proposition 1.8 in Chapter 1. Here $\gamma : \pi^{-1}TM \to \mathcal{V}$ is the vertical lift i.e., locally $\gamma X_i = \dot{\partial}_i$.

Proof of Lemma 7.1. Let us set

$$\delta_i = \frac{\delta}{\delta x^i} = \partial_i - N_i^{\ j}\, \dot{\partial}_j\,.$$

The Liouville vector is locally given by $\mathcal{L} = y^i X_i$. Then $\nabla_{\delta_i}\mathcal{L} = 0$ hence $\{\delta_i : 1 \leq i \leq n\}$ is a local frame of $\mathcal{H}$. As $\mathcal{V}$ is locally generated by $\{\dot{\partial}_i : 1 \leq i \leq n\}$ it follows that $\mathcal{H}_u \cap \mathcal{V}_u = (0)$, for any $u \in T(M)$. Then (7.3) follows by inspecting dimensions. Finally

$$\nabla_{\gamma X}\mathcal{L} = X^i \nabla_{\dot{\partial}_i}\mathcal{L} = X^i X_i$$

for any $X \in \Gamma^\infty(\pi^{-1}TM)$. ■

The bundle morphism K in Lemma 7.1 is the Dombrowski map. As well as in the Rieamannian case K really depends but on the nonlinear connection $\mathcal{H}$. By the last statement in Lemma 7.1, the Dombrowski map may be recast as

$$K = \gamma^{-1} \circ Q,$$

where $Q: T(T(M)) \to \mathcal{V}$ is the projection associated to the direct sum decomposition (7.3). The horizontal lift is the bundle morphism $\beta: \pi^{-1}TM \to \mathcal{H}$ (locally) given by $\beta X_i = \delta_i$. If $L: T(T(M)) \to \pi^{-1}TM$ is the bundle morphism

$$L_u \tilde{X} = \left(u, (d_u\pi)\tilde{X}\right), \quad \tilde{X} \in T_u(T(M)), \quad u \in T(M),$$

then β may also be described as the inverse of $L: \mathcal{H} \to \pi^{-1}TM$.

Let G_s on $T(M)$ be the $(0,2)$-tensor field given by

$$G_s\left(\tilde{X}, \tilde{Y}\right) = \hat{g}\left(L\tilde{X}, L\tilde{Y}\right) + \hat{g}\left(K\tilde{X}, K\tilde{Y}\right),$$

for any $\tilde{X}, \tilde{Y} \in T(T(M))$. Here $\hat{g}$ is the bundle metric induced in $\pi^{-1}TM \to T(M)$ by the given semi-Riemannian metric i.e., locally $\hat{g}(X_i, X_j) = g_{ij} \circ \pi$, where $g_{ij} = g(\partial/\partial\tilde{x}^i, \partial/\partial\tilde{x}^j)$. Then

Proposition 7.2 *G_s is a semi-Riemannian metric on $T(M)$ of signature $(2n - 2\nu, 2\nu)$. Both $\mathcal{H}_u$ and $\mathcal{V}_u$ have index ν in $(T_u(T(M)), G_{su})$, $u \in T(M)$.*

We refer to G_s as the *canonical metric* on the tangent bundle over the semi-Riemannian manifold (M,g) (this is of course the Sasaki metric in the Riemannian case). Based on our knowledge of the positive definite case, it is likely that G_s is rather rigid e.g., for instance it doesn't possess the hereditary properties (cf. [6]) of the g-natural metrics introduced by O. Kowalski & M. Sekizawa, [200].

Let us consider the almost complex structure J on $T(M)$ given by

$$J(\beta X) = \gamma X, \ J(\gamma X) = -\beta X,$$

for any $X \in \pi^{-1}TM$. J is rarely integrable, in fact only when (M,g) is flat. A remark is in order. The notion of nonlinear connection is due to W. Barthel, [29], and is of current use in Finslerian geometry (cf. e.g., Definition 8.2 in [209], p. 55). We recall (cf. [209], p. 62) that a *Finslerian connection* is a pair $(\nabla, \mathcal{H})$ consisting of a connection ∇ in $\pi^{-1}TM$ and a nonlinear connection $\mathcal{H}$ on M. Note that given a semi-Riemannian bundle metric $\hat{g}$ in $\pi^{-1}TM$ and just any nonlinear connection $\mathcal{H}$, one may produce a metric G_s and an almost complex structure J on $T(M)$ as above and in general the results in Section 8.1 should carry over to Finslerian geometry (by replacing the Levi-Civita connection with one of the canonical Finslerian connections e.g., the

Cartan, Rund, or Berwald connections, cf. [209], p. 108 and p. 115). For the interplay between CR and (complex) Finsler geometry see also [101] and [104].

Proposition 7.3 *Let* $\flat : T(M) \to T^*(M)$ *be the natural diffeomorphism associated to the semi-Riemannian metric* g *and* λ *the Liouville* 1*-form on* $T^*(M)$. *Then* i) $(\flat^*\lambda)^\sharp = \beta\mathcal{L}$ *and* ii) $2\flat^* d\lambda$ *is the Kähler* 2*-form of* $(T(M), J, G_s)$. *In particular* $(T(M), J, G_s)$ *is an indefinite almost Kähler manifold.*

Here $\sharp$ is the inverse of $\flat$. For a study of indefinite Kähler manifolds see M. Barros & A. Romero, [28]. Let (p_i, q^i) be the natural local coordinates on $T^*(M)$ and $\lambda = p_i\, dq^i$ the Liouville 1-form (cf. e.g., C. Godbillon, [137], p. 124). Then

$$p_i \circ \flat = y^j g_{ij}(x), \quad q^i \circ \flat = x^i,$$

hence

$$(d\flat)\delta_j = \frac{\partial}{\partial q^j} + \left(y^k \frac{\partial g_{ki}}{\partial x^j} - N_j^k g_{ki} \right) \frac{\partial}{\partial p_i}, \quad (d\flat)\dot{\partial}_j = g_{ji} \frac{\partial}{\partial p_i},$$

and $G_s((\flat^*\lambda)^\sharp, \tilde{X}) = \lambda(d\flat)\tilde{X}$ yields $(\flat^*\lambda)^\sharp = y^i\delta_i$ which is the first statement in Proposition 7.3. Moreover if $\Omega = dp_i \wedge dq^i$ then

$$\flat^* dp_i = y^k \frac{\partial g_{ki}}{\partial x^j}\, dx^j + g_{ij}\, dy^j, \quad \flat^* dq^i = dx^i,$$

so that

$$\flat^*\Omega = y^k \frac{\partial g_{kj}}{\partial x^i}\, dx^i \wedge dx^j - g_{ij}\, dx^i \wedge dy^j$$

and a straightforward calculation shows that $2\Omega(\tilde{X}, \tilde{Y}) = G_s(\tilde{X}, J\tilde{Y})$ for any $\tilde{X}, \tilde{Y} \in T(T(M))$. Yet Ω is a symplectic 2-form on $T^*(M)$ hence Proposition 7.3 is proved.

Of course, a fundamental question is to describe the holomorphic functions on $(T(M), J)$. Let $\mathbb{R}^n_\nu = (\mathbb{R}^n, b_{n,\nu})$ be the *semi-Euclidean space* i.e., $b_{n,\nu}(x, y) = \sum_{i=1}^n \epsilon_i x^i y^i$, $x, y \in \mathbb{R}^n$, where $\epsilon_i = -1$, $1 \leq i \leq \nu$, and $\epsilon_j = 1$, $\nu + 1 \leq j \leq n$. Let us set

$$T^{1,0}(T(M), J) = \{\tilde{X} - iJ\tilde{X} : \tilde{X} \in T(T(M))\}$$

(where $i = \sqrt{-1}$).

Definition 7.4 A C^1 function $f : T(M) \to \mathbb{C}$ is *holomorphic* if $\overline{Z}(f) = 0$, for any $Z \in T^{1,0}(T(M), J)$. ■

Overbars denote complex conjugation in $T(T(M)) \otimes \mathbb{C}$.

Proposition 7.5 *Let M be an n-dimensional semi-Riemannian manifold of index ν $(0 \leq \nu < n)$. The holomorphic functions $f : T(M) \to \mathbb{C}$ are the C^1 solutions to*

$$\frac{\delta f}{\delta x^j} + i \frac{\partial f}{\partial y^j} = 0, \quad 1 \leq j \leq n. \tag{7.5}$$

Any C^2 holomorphic function $f : T(M) \to \mathbb{C}$ is harmonic $(\Delta f = 0)$. Therefore when (M, g) is Riemannian $(\nu = 0)$ f is C^∞. Moreover, J is integrable if and only if M is locally isometric to $\mathbb{R}^n_\nu$.

The equations (7.5) are the *Cauchy-Riemann* equations on $(T(M), J)$. If we set $\partial/\partial \overline{z}^j = \frac{1}{2}(\partial/\partial x^j + i\partial/\partial y^j)$, then (7.5) may also be written

$$\frac{\partial f}{\partial \overline{z}^j} = \frac{1}{2}\Gamma^k_{j\ell}(x) y^\ell \frac{\partial f}{\partial y^k}. \tag{7.6}$$

The only holomorphic functions on $T(M)$ which are vertical lifts of functions on M are the (complex) constants.

Proof of Proposition 7.5. Let $\tilde{X} = \lambda^j \delta_j + \mu^j \dot{\partial}_j$. Then

$$\tilde{X} - iJ\tilde{X} = \left(\lambda^j + i\mu^j\right)\left(\delta_j - i\dot{\partial}_j\right)$$

hence $\{\delta_j - i\dot{\partial}_j : 1 \leq j \leq n\}$ is a (local) frame of $T^{1,0}(T(M), J)$ on $\pi^{-1}(U)$. The harmonicity of C^2 holomorphic functions $f : T(M) \to \mathbb{C}$ follows from

$$-\Delta f = \delta_i(g^{ij}\delta_j f) + \dot{\partial}_i(g^{ij}\dot{\partial}_j f) + g^{ij}\Gamma^k_{ik}\delta_j f, \tag{7.7}$$

where Δ is the Laplace-Beltrami operator of $(T(M), G_s)$. Indeed, by the Cauchy-Riemann equations (7.5) and by the commutation formula

$$[\delta_i, \dot{\partial}_j] = \Gamma^k_{ij} \dot{\partial}_k$$

one has

$$\Delta f = \sqrt{-1}\left\{\frac{\partial g^{ij}}{\partial x^i} + g^{ik}\Gamma^j_{ik} + g^{ij}\Gamma^k_{ik}\right\}\dot{\partial}_j f = 0$$

(due to $\nabla \hat{g} = 0$). When $\nu = 0$, the canonical metric G_s is Riemannian hence its Laplace-Beltrami operator is elliptic, and then hypoelliptic. Then

C^2 solutions (and actually distribution solutions) to $\Delta f = 0$ are C^∞. The last statement follows from

$$[J, J](\delta_i, \delta_j) = -[J, J](\dot{\partial}_i, \dot{\partial}_j) = R^k_{ij}\,\dot{\partial}_k,$$
$$[J, J](\dot{\partial}_i, \delta_j) = R^k_{ij}\,\delta_k,$$

where $[J, J]$ is the Nijenhuis torsion of J and $R^i_{jk} = R^i_{jk\ell}\,y^\ell$. Also $R^i_{jk\ell}$ is the curvature tensor field of Γ^i_{jk}. Note that $[\delta_i, \delta_j] = -R^k_{ij}\,\dot{\partial}_i$, that is $R^i_{jk} = 0$ is the integrability condition for the Pfaffian system (7.4). We are left with the proof of (7.7). The canonical metric G_s is given by (with respect to the local frame $\{\partial_i,\ \dot{\partial}_i\}$)

$$G_s : \begin{pmatrix} g_{ij} + N^k_i N^\ell_j g_{k\ell} & N^k_i g_{kj} \\ g_{ik} N^k_j & g_{ij} \end{pmatrix}.$$

Consequently, the cometric $G_s{}^*$ induced by G_s on 1-forms is given by

$$G_s{}^* : \begin{pmatrix} g^{ij} & -g^{ik} N_k^{\ j} \\ -N^i_k g^{kj} & g^{ij} + N^i_k N_\ell^{\ j} g^{k\ell} \end{pmatrix}.$$

Then for any C^2 function f on $T(M)$

$$-\Delta f = \frac{\partial}{\partial x^A}\left(G^{AB}\frac{\partial f}{\partial x^B}\right) + Pf$$
$$= \delta_i(g^{ij}\delta_j f) + \dot{\partial}_i(g^{ij}\dot{\partial}_j f) - g^{jk}\Gamma^i_{ki}\delta_j f + Pf,$$

where P is the first order differential operator given by

$$Pf = \frac{\partial}{\partial x^A}\left(\log\sqrt{\underline{G}}\right) G^{AB}\frac{\partial f}{\partial x^B}.$$

Also (x^A) is short for (x^i, y^i) and $G_{AB} = G_s(\partial/\partial x^A, \partial/\partial x^B)$, $\underline{G} = |\det (G_{AB})|$. Moreover

$$Pf = G^{AB}\tilde{\Gamma}^C_{AC}\frac{\partial f}{\partial x^B} = g^{ij}\tilde{\Gamma}^A_{iA}\delta_j f - \tilde{\Gamma}^A_{i+n,A}\left\{N^i_k g^{kj}\delta_j f - g^{ij}\dot{\partial}_j f\right\}$$

where $\tilde{\Gamma}^A_{BC}$ are the local coefficients of D with respect to $\{\partial/\partial x^A : 1 \le A \le 2n\}$. By (7.11)–(7.14)

$$\tilde{\Gamma}^\ell_{ij} = \Gamma^\ell_{ij} + \frac{1}{2} y^s \left(N^k_j R^\ell_{ski} + N^k_i R^\ell_{skj} \right),$$

$$\tilde{\Gamma}^\ell_{i,j+n} = \frac{1}{2} y^k R^\ell_{kji}, \quad \tilde{\Gamma}^\ell_{i+n,j} = \frac{1}{2} y^k R^\ell_{kij},$$

$$\tilde{\Gamma}^{\ell+n}_{i,j+n} = \Gamma^\ell_{ij} - \frac{1}{2} y^s R^k_{sji} N^\ell_k, \quad \tilde{\Gamma}^{\ell+n}_{i+n,j+n} = 0.$$

Note that $R^i_{jki} = 0$. Thus

$$\tilde{\Gamma}^A_{iA} = 2\Gamma^j_{ij}, \quad \tilde{\Gamma}^A_{i+n,A} = 0,$$

so that

$$Pf = 2 g^{ik} \Gamma^\ell_{i\ell} \delta_k f$$

which leads to (7.7). ∎

In the real Finslerian setting, the integrability of the almost complex structure J on $T(M)$ has been discussed by H. Akbar-Zadeh & E. Bonan, [13].

7.2. BUNDLES OF HYPERQUADRICS IN $(T(M), J, G_S)$

Let $q(x) = b_{n,\nu}(x,x) = -\sum_{i=1}^{\nu}(x^i)^2 + \sum_{j=\nu+1}^{n}(x^j)^2$, for $x \in \mathbb{R}^n$. Given $\epsilon \in \{\pm 1\}$ we consider the *(central) hyperquadrics* $Q_\epsilon = q^{-1}(\epsilon)$. Moreover let $S_\epsilon(M,g)_x$ consist of all $u \in T_x(M)$ such that $g_x(u,u) = \epsilon$, for $x \in M$. Let $\{E_i : 1 \le i \le n\}$ be a local orthonormal (i.e., $g(E_i, E_j) = \epsilon_i \delta_{ij}$) frame of $T(M)$ defined on some open neighborhood of $x \in M$. If $u \in S_\epsilon(M,g)_x$ and $u = u^i E_{i,x}$ then $q(u^1, \dots, u^n) = \epsilon$ i.e., the map $u \mapsto (u^1, \dots, u^n)$ is a linear isometry $S_\epsilon(M,g)_x \approx Q_\epsilon$. Therefore $S_\epsilon(M,g) \to M$ is a bundle of hyperquadrics over M. In particular one has

$$S^{n-1}_\nu \to S_1(M,g) \longrightarrow M$$

(a bundle of pseudospheres) and

$$\mathbb{H}^{n-1}_{\nu-1} \to S_{-1}(M,g) \longrightarrow M$$

(a bundle of pseudohyperbolic spaces). Here $S^{n-1}_\nu = \{x \in \mathbb{R}^n_\nu : \sum_{i=1}^n \epsilon_i (x^i)^2 = 1\}$ and $\mathbb{H}^{n-1}_{\nu-1} = \{x \in \mathbb{R}^n_\nu : \sum_{i=1}^n \epsilon_i (x^i)^2 = -1\}$. Finally let us set

$$S_0(M,g) = \{u \in T_x(M) \setminus \{0_x\} : g_x(u,u) = 0\}, \quad x \in M,$$

to obtain $\Lambda \to S_0(M,g) \longrightarrow M$ (a bundle of nullcones) where $\Lambda = q^{-1}(0) \setminus \{0\} \subset \mathbb{R}^n_\nu$. Let (U, x^i) be a local coordinate system on M. The open submanifold $S_\epsilon(M,g) \cap \pi^{-1}(U)$ is given by the equation $g_{ij}(x) y^i y^j = \epsilon$. A vector field $\tilde{X} = a^i \partial_i + b^i \dot{\partial}_i$ is tangent to $S_\epsilon(M,g)$ if and only if

$$y^i \left(a^j y^k \frac{\partial g_{ik}}{\partial x^j} + 2 b^j g_{ij} \right) = 0. \tag{7.8}$$

If $X = \delta_i$ then $a^j = \delta_i^j$, $b^j = -N_i^j$ so that

$$y^j \left(a^\ell y^k \partial_\ell g_{jk} + 2 b^k g_{jk} \right) = y^j y^k \left(\partial_i g_{jk} - 2\Gamma_{ikj} \right) = 0$$

hence (by (7.8)) $\delta_i \in T(S_\epsilon(M,g))$. Moreover (again by (7.8)), $X^i \dot{\partial}_i \in T(S_\epsilon(M,g))$ if and only if $X^i y^j g_{ij} = 0$. Summing up, it is immediate that

Lemma 7.6 *Each horizontal vector on $T(M)$ is tangent to $S_\epsilon(M,g)$ i.e., $\mathcal{H}_u \subset T_u(S_\epsilon(M,g))$ for any $u \in S_\epsilon(M,g)$. Moreover, a vertical vector γX, $X \in \pi^{-1}TM$, is tangent to $S_\epsilon(M,g)$ if and only if X is orthogonal to the Liouville vector. Consequently*

$$T(S_\epsilon(M,g)) = \mathcal{H} \oplus \gamma(\mathbb{R}\mathcal{L})^\perp \tag{7.9}$$

where $(\mathbb{R}\mathcal{L})^\perp = \{X \in \pi^{-1}TM : \hat{g}(X, \mathcal{L}) = 0\}$.

As $\mathbb{R}\mathcal{L}_u$ (with $u \in S_\epsilon(M,g)$) is nondegenerate, so is the corresponding perp space hence (7.9) follows by looking at dimensions.

Recall that on a semi-Riemannian manifold the tangent vectors $v \in T_x(M)$ fall into three types, as $g_x(v,v) < 0$ (v is *timelike* or -1-*like*) or $g_x(v,v) > 0$ (v is *spacelike* or $+1$-*like*) or $g_x(v,v) = 0$ and $v \neq 0_x$ (v is *lightlike* or *null*). Consider the C^∞ function

$$E : T(M) \to \mathbb{R}, \quad E = \hat{g}(\mathcal{L}, \mathcal{L}).$$

Then $E(u) = \epsilon$ for any $u \in S_\epsilon(M,g)$ i.e., $\mathcal{L}_u$ is ϵ-like.

Let us recall that the *sign* ϵ of a semi-Riemannian hypersurface in $(T(M), G_s)$ is $+1$ (respectively -1) if $G_s(z,z) > 0$ (respectively if $G_s(z,z) > 0$) for every normal vector $z \neq 0$. We may state

Theorem 7.7 *Let (M,g) be an n-dimensional semi-Riemannian manifold of index ν $(0 \le \nu < n)$. Let $\epsilon \in \{\pm 1\}$. Then $\gamma\mathcal{L}$ is a unit normal vector field on*

$S_\epsilon(M,g)$ i.e., $S_\epsilon(M,g)$ is a semi-Riemannian hypersurface of $(T(M), G_s)$ of sign ϵ and index $2\nu - 1$ if $\epsilon = -1$, respectively of index 2ν if $\epsilon = 1$. Finally, the Weingarten operator A corresponding to $\gamma\mathcal{L}$ is given by

$$A\beta X = 0, \quad A\gamma X = -\gamma X, \tag{7.10}$$

for any $X \in \pi^{-1}TM$. $S_0(M,g)$ is a real hypersurface of $T(M)$, invariant under the natural action of $\mathbb{R}^* = \mathbb{R} \setminus \{0\}$ (the multiplicative reals) on $T(M)$. The vector field $\gamma\mathcal{L}$ is both tangent and normal to $S_0(M,g)$, hence $S_0(M,g)$ is not a semi-Riemannian hypersurface.

Proof. By Lemma 7.6 each normal section on $S_\epsilon(M,g)$ is vertical. Next, if $Y \in (\mathbb{R}\mathcal{L})^\perp$ then γX is orthogonal to $S_\epsilon(M,g)$ if and only if $\hat{g}(X,Y) = 0$, i.e., γX is proportional to $\gamma\mathcal{L}$. The proof of the next statement in Theorem 7.7 requires the Levi-Civita connection D of $(T(M), G_s)$. It is given by

$$D_{\delta_i}\delta_j = \Gamma^k_{ij}\delta_k - \frac{1}{2}y^k R^\ell_{ijk}\dot{\partial}_\ell, \tag{7.11}$$

$$D_{\delta_i}\dot{\partial}_j = \frac{1}{2}y^k R^\ell_{kji}\delta_\ell + \Gamma^k_{ij}\dot{\partial}_k, \tag{7.12}$$

$$D_{\dot{\partial}_i}\delta_j = \frac{1}{2}y^k R^\ell_{kij}\delta_\ell, \tag{7.13}$$

$$D_{\dot{\partial}_i}\dot{\partial}_j = 0. \tag{7.14}$$

The proof is similar to that in the Riemannian case ($\nu = 0$) and thus omitted. Cf. e.g., S. Sasaki, [266]. Next (by the Weingarten formula)

$$A\delta_i = -D_{\delta_i}\left(y^j\dot{\partial}_j\right) = N_i^{\ j}\dot{\partial}_j - y^j\left(\Gamma^k_{ij}\dot{\partial}_k + \frac{1}{2}y^k R^\ell_{kji}\delta_\ell\right) = 0.$$

Also

$$A\gamma X = -D_{\gamma X}\left(y^j\dot{\partial}_j\right) = -X^j\dot{\partial}_j$$

for any $X \in (\mathbb{R}\mathcal{L})^\perp$. The arguments above apply *verbatim* to show that

$$\mathcal{H} \oplus \gamma(\mathbb{R}\mathcal{L})^\perp \subset T(S_0(M,g)).$$

Yet $E(u) = 0$ for any $u \in S_0(M,g)$ i.e., $\mathcal{L}$ is lightlike. In particular, $\mathcal{L} \in (\mathbb{R}\mathcal{L})^\perp$ hence $\gamma\mathcal{L}$ is tangent to $S_0(M,g)$. Finally (by (7.8))

$$G_s(\tilde{X}, \gamma\mathcal{L}) = \left(b^j + N_i^{\ j}a^i\right)g_{jk}y^k = 0,$$

for any $\tilde{X} = a^i \partial_i + b^i \dot{\partial}_i \in T(S_0(M,g))$ i.e., $\gamma\mathcal{L}$ is normal to $S_0(M,g)$. ■

Let us set

$$T = -J\gamma\mathcal{L} = \beta\mathcal{L}.$$

Then (by Lemma 7.6), T is tangent to $S_\epsilon(M,g)$. Let us consider the differential 1-form θ on $S_\epsilon(M,g)$ given by

$$\theta(\tilde{X}) = G_s(\tilde{X}, T), \quad \tilde{X} \in T(S_\epsilon(M,g)). \tag{7.15}$$

Then

$$\theta(\gamma X) = 0, \ \theta(\beta Y) = \hat{g}(Y, \mathcal{L}), \tag{7.16}$$

for any $X \in (\mathbb{R}\mathcal{L})^\perp$ and $Y \in \pi^{-1}TM$. We set $H(S_\epsilon(M,g)) = \mathrm{Ker}(\theta)$ (the *Levi distribution* of $S_\epsilon(M,g)$) and

$$T_{1,0}(S_\epsilon(M,g)) = \left\{\tilde{X} - iJ\tilde{X} : \tilde{X} \in H(S_\epsilon(M,g))\right\}.$$

At this point, a number of fundamental questions may be asked. For instance i) when is $T_{1,0}(S_\epsilon(M,g))$ a CR structure on $S_\epsilon(M,g)$ (and when is it strictly pseudoconvex, or at least nondegenerate)? ii) What are the CR functions on $S_\epsilon(M,g)$ and when do they extend holomorphically (at least locally)? iii) Are any of the (twisted) Kohn-Rossi cohomology groups of $S_\epsilon(M,g)$ computable? We address (i) (cf. Theorem 7.15 below) and leave (ii)–(iii) as open problems.

We proceed by recalling a few notions of CR and pseudohermitian geometry, needed in the sequel.

Definition 7.8 Let M be a real $(2n+1)$-dimensional C^∞ manifold. An *almost CR structure* on M is a complex subbundle $T_{1,0}(M)$, of complex rank n, of the complexified tangent bundle $T(M) \otimes \mathbb{C}$ such that

$$T_{1,0}(M) \cap T_{0,1}(M) = (0),$$

where $T_{0,1}(M) = \overline{T_{1,0}(M)}$. The integer n is the *CR dimension* of $(M, T_{1,0}(M))$. An almost CR structure is called *CR structure* if it is *integrable*, that is

$$Z, W \in \Gamma^\infty(T_{1,0}(M)) \Longrightarrow [Z, W] \in \Gamma^\infty(T_{1,0}(M)).$$

The *Levi distribution* is $H(M) = \mathrm{Re}\{T_{1,0}(M) \oplus T_{0,1}(M)\}$. ■

The Levi distribution carries the complex structure $J : H(M) \to H(M)$ given by

$$J(Z + \overline{Z}) = i(Z - \overline{Z}), \quad Z \in T_{1,0}(M).$$

Let us set $H(M)_x^{\perp} = \{\omega \in T_x^*(M) : \mathrm{Ker}(\omega) \supseteq H(M)_x\}$ for any $x \in M$. If M is orientable then the real line bundle $\mathbb{R} \to H(M)^{\perp} \to M$ is trivial, hence it admits globally defined nowhere zero C^{∞} sections θ.

Definition 7.9 A smooth globally defined nowhere zero section in $H(M)^{\perp} \to M$ is called a *pseudohermitian structure* on M. ■

Definition 7.10 Let $(M, T_{1,0}(M))$ be a CR manifold. Then M is said to be *nondegenerate* if the *Levi form*

$$L_{\theta}(Z, \overline{W}) = -i(d\theta)(Z, \overline{W}), \quad Z, W \in T_{1,0}(M),$$

is nondegenerate for some pseudohermitian structure θ. ■

It should be observed that if L_{θ} is nondegenerate for some θ then L_{θ} is nondegenerate for all θ. Indeed if $\hat{\theta} = \pm e^{\lambda}\theta$ then $L_{\hat{\theta}} = \pm e^{\lambda} L_{\theta}$. If M is nondegenerate then each pseudohermitian structure θ is a contact form i.e., $\theta \wedge (d\theta)^n$ is a volume form on M, where n is the CR dimension. Given a contact form θ there is a unique tangent vector field $T \in \mathcal{X}(M)$ such that $\theta(T) = 1$ and $T \rfloor d\theta = 0$.

Definition 7.11 T is the *characteristic direction* of $d\theta$. ■

It is customary to extend J (the complex structure along $H(M)$) to an endomorphism of the tangent bundle by requesting that $J(T) = 0$. Then $J^2 = -I + \theta \otimes T$.

Definition 7.12 The *Webster metric* g_{θ} is given by

$$g_{\theta}(X, Y) = (d\theta)(X, JY), \;\; g_{\theta}(X, T) = 0, \;\; g_{\theta}(T, T) = 1,$$

for any $X, Y \in H(M)$. ■

Note that g_{θ} is a semi-Riemannian metric on M of signature $(2r + 1, 2s)$ where (r, s) is the signature of L_{θ}.

Definition 7.13 M is *strictly pseudoconvex* if L_{θ} is positive definite for some θ. ■

If M is strictly pseudoconvex and θ is a contact form such that L_{θ} is positive definite, then g_{θ} is a Riemannian metric.

Definition 7.14 Let $(M, T_{1,0}(M))$ be an almost CR manifold. A *CR function* is a C^1 function $f : M \to \mathbb{C}$ satisfying $\overline{\partial}_b f = 0$ (the *tangential Cauchy-Riemann equations*). Here $\overline{\partial}_b : C^1(M) \to \Gamma^0(T_{0,1}(M)^*)$ is the first order differential operator given by $(\overline{\partial}_b f)\overline{Z} = \overline{Z}(f)$ for any $Z \in T_{1,0}(M)$. ■

Theorem 7.15 *Let (M,g) be an n-dimensional semi-Riemannian manifold of index ν $(0 \leq \nu < n)$. Let $\epsilon \in \{\pm 1\}$. The Levi distribution of $S_\epsilon(M,g)$ decomposes as*

$$H(S_\epsilon(M,g)) = \beta(\mathbb{R}\mathcal{L})^\perp \oplus \gamma(\mathbb{R}\mathcal{L})^\perp. \tag{7.17}$$

In particular $H(S_\epsilon(M,g))$ is J-invariant hence $T_{1,0}(S_\epsilon(M,g))$ is the eigenspace of $J^{\mathbb{C}}$ (the $\mathbb{C}$-linear extension of J to $H(S_\epsilon(M,g)) \otimes \mathbb{C}$) corresponding to the eigenvalue i. Therefore $T_{1,0}(S_\epsilon(M,g))$ is an almost CR structure of CR dimension $n-1$ on $S_\epsilon(M,g)$. It is integrable if and only if

$$Z^i W^j R^k_{ij} = 0, \tag{7.18}$$

for any $Z, W \in (\mathbb{R}\mathcal{L})^\perp \otimes \mathbb{C}$, and then $S_\epsilon(M,g)$ is a strictly pseudoconvex CR manifold. In particular, if M is a semi-Riemannian manifold of constant sectional curvature then $T_{1,0}(S_\epsilon(M,g))$ is integrable.

See also S. Tanno, [282]. One may conjecture that for any semi-Riemannian manifold M with $T_{1,0}(S_\epsilon(M,g))$ integrable one has $M \approx S^n_\nu$ (a local isometry).

Proof of Theorem 7.15. By (7.9) a tangent vector field on $S_\epsilon(M,g)$ is of the form $\tilde{X} = \gamma X + \beta Y$ for some $X \in (\mathbb{R}\mathcal{L})^\perp$ and some $Y \in \pi^{-1}TM$. Then $\tilde{X} \in H(S_\epsilon(M,g))$ if and only if

$$0 = \theta(\tilde{X}) = \hat{g}(Y, \mathcal{L})$$

i.e., $Y \in (\mathbb{R}\mathcal{L})^\perp$ and (7.17) is proved. Note that an element of $T_{1,0}(S_\epsilon(M,g))$ may be written as

$$\tilde{X} - iJ\tilde{Y} = (\beta - i\gamma)(X + iY).$$

Let $Z, W \in (\mathbb{R}\mathcal{L})^\perp \otimes \mathbb{C}$. Then (as D is torsion free)

$$\begin{aligned}
&[(\beta - i\gamma)Z, (\beta - i\gamma)W] \\
&\quad = D_{\beta Z}\beta W - D_{\beta W}\beta Z - D_{\gamma Z}\gamma W + D_{\gamma W}\gamma Z \\
&\qquad - i\{D_{\beta Z}\gamma W - D_{\gamma W}\beta Z + D_{\gamma Z}\beta W - D_{\beta W}\gamma Z\}
\end{aligned}$$

hence (by (7.11)–(7.14))

$$[(\beta - i\gamma)Z, (\beta - i\gamma)W]$$
$$= \left(Z^i \frac{\delta W^j}{\delta x^i} - W^i \frac{\delta Z^i}{\delta x^i}\right)\delta_j - \left(Z^i \frac{\partial W^j}{\partial y^i} - W^i \frac{\partial Z^j}{\partial y^i} + Z^i W^j R^k_{ij}\right)\dot{\partial}_j$$
$$- \sqrt{-1}\left\{\left(Z^i \frac{\partial W^j}{\partial y^i} - W^i \frac{\partial Z^j}{\partial y^i}\right)\delta_j + \left(Z^i \frac{\delta W^j}{\delta x^i} - W^i \frac{\delta Z^j}{\delta x^i}\right)\dot{\partial}_j\right\}.$$

This is of the form $(\beta - i\gamma)V = V^j(\delta_j - i\dot{\partial}_j)$ for some $V \in (\mathbb{R}\mathcal{L})^{\perp} \otimes \mathbb{C}$ if and only if (7.18) holds. The Levi form is given by

$$L_\theta((\beta - i\gamma)Z, (\beta + i\gamma)\overline{W}) = \frac{i}{2}\,\theta([(\beta - i\gamma)Z, (\beta + i\gamma)\overline{W}]),$$

for any $Z, W \in (\mathbb{R}\mathcal{L})^{\perp} \otimes \mathbb{C}$. We wish to compute the $\mathcal{H}$-component of $[(\beta - i\gamma)Z, (\beta + i\gamma)\overline{W}]$. Note that (7.11)–(7.14) may be written as

$$D_{\beta X}\beta Y = \beta\nabla_{\beta X}Y - \frac{1}{2}\gamma\, R(\beta X, \beta Y)\mathcal{L},$$
$$D_{\beta X}\gamma Y = \gamma\nabla_{\beta X}Y + \frac{1}{2}\beta\, R(\beta\mathcal{L}, \beta Y)X,$$
$$D_{\gamma X}\beta Y = \beta\nabla_{\gamma X}Y + \frac{1}{2}\beta\, R(\beta\mathcal{L}, \beta X)Y,$$
$$D_{\gamma X}\gamma Y = \gamma\nabla_{\gamma X}Y,$$

for any $X, Y \in \Gamma^\infty(\pi^{-1}TM)$. Consequently

$$[(\beta - i\gamma)Z, (\beta + i\gamma)\overline{W}] = D_{(\beta - i\gamma)Z}(\beta + i\gamma)\overline{W} - D_{(\beta + i\gamma)\overline{W}}(\beta - i\gamma)Z$$
$$\equiv \beta(\nabla_{\beta Z}\overline{W} - \nabla_{\beta\overline{W}}Z) - i\beta(\nabla_{\gamma Z}\overline{W} + \nabla_{\gamma\overline{W}}Z)$$

modulo vertical terms, hence

$$L_\theta\left((\beta - i\gamma)Z, (\beta + i\gamma)\overline{W}\right) = -\hat{g}\left(Z, \overline{W}\right)$$

i.e., L_θ is nondegenerate (negative definite provided g is Riemannian). Finally, if the curvature tensor field of (M,g) is given by $R^i_{jk\ell} = c(g_{\ell j}\delta^i_k - g_{\ell k}\delta^i_j)$, for some $c \in \mathbb{R}$, then

$$Z^i W^j y^k R^\ell_{ijk} = c\{W^\ell \hat{g}(Z, \mathcal{L}) - Z^\ell \hat{g}(W, \mathcal{L})\} = 0. \qquad \blacksquare$$

As an application of Theorem 7.15, we look at the Reeb vector field of a contact Riemannian manifold. We first recall the notion of a CR map.

Definition 7.16 A smooth map $f : M \to N$ of two almost CR manifolds is a *contact map* if

$$(d_x f)H(M)_x \subseteq H(N)_{f(x)}, \quad x \in M.$$

A contact map is a *CR map* if

$$(d_x f) \circ J_x = J_{N,x}(d_x f), \quad x \in M.$$

Here $J_N : H(N) \to H(N)$ is the complex structure in the Levi distribution of N. ■

Equivalently, a smooth map $f : M \to N$ is a CR map if

$$(d_x f)T_{1,0}(M)_x \subseteq T_{1,0}(N)_{f(x)}, \quad x \in M.$$

If θ and θ_N are pseudohermitian structures on M and N and $f : M \to N$ is a CR map then

$$f^*\theta_N = \lambda\,\theta \tag{7.19}$$

for some $\lambda \in C^\infty(M)$.

Definition 7.17 Let $f : M \to N$ be a CR map and $\lambda \in C^\infty(M)$ given by (7.19). Then f is said to be a *pseudohermitian* map of (M,θ) into (N,θ_N) if λ is a constant. ■

Any almost contact Riemannian manifold $(M,(\phi,\xi,\eta,g))$ carries a natural almost CR structure given by

$$T_{1,0}(M) = \{X - i\phi X : X \in H(M)\},$$

where $H(M) = \mathrm{Ker}(\eta)$ and (by a result of S. Ianuş, [173]) if (ϕ,ξ,η) is normal then $T_{1,0}(M)$ is integrable. We may close a circle of ideas by observing that given a strictly pseudoconvex CR manifold M and a contact form θ on M such that L_θ is positive definite, $(J, -T, -\theta, g_\theta)$ is a contact metric structure on M. It should be emphasized that normality is but a sufficient condition for the integrability of $T_{1,0}(M)$. A characterization of integrability has been produced by S. Tanno, [281], in terms of the so called *Tanno tensor*

$$Q(X,Y) = (D_Y\phi)X + [(D_Y\eta)\phi X]\xi + \eta(X)\phi(D_Y\xi)$$

($T_{1,0}(M)$ is integrable if and only if $Q = 0$). Here D is the Levi-Civita connection of (M,g). A generalization of the basic results in pseudohermitian geometry to the case of contact metric manifolds whose almost CR structure isn't integrable has been started by E. Barletta et al., [23], and D.E. Blair et al., [44].

Let $S(M)$ be short for $S_1(M,g)$ and $\nu = 0$. We need to recall the description of the natural contact metric structure $(\overline{\phi}, \overline{\xi}, \overline{\eta}, \overline{g})$ (induced on $S(M)$ by the almost Kähler structure (J, G_s) of $T(M)$) in some detail. First let $\xi' = \beta\mathcal{L}$ be the geodesic flow on $S(M)$ (remember that horizontal vector fields are tangent to $S(M)$ as a consequence of Lemma 7.6). Let $g' = j^* G_s$ be the induced metric where $j : S(M) \hookrightarrow T(M)$ is the inclusion. Let $\eta' = \theta$ be the contact form on $S(M)$ given by (7.15) or (7.16). Next let us set $\phi' = J - \eta' \otimes \gamma\mathcal{L}$. Since

$$g'(X, \phi' Y) = 2(d\eta')(X, Y), \quad X, Y \in T(S(M)),$$

we need to rescale (ϕ', ξ', η', g') in order to make it into an actual contact metric structure on $S(M)$. Namely we set

$$\overline{\phi} = \phi', \quad \overline{\xi} = 2\xi', \quad \overline{\eta} = \frac{1}{2}\eta', \quad \overline{g} = \frac{1}{4}g'.$$

Summing up, the standard contact metric structure of $S(M)$ is

$$\overline{\phi}\beta X = \gamma X - \hat{g}(X, \mathcal{L})\gamma\mathcal{L}, \quad \overline{\phi}\gamma Z = -\beta Z,$$

$$\overline{\xi} = 2\beta\mathcal{L}, \quad \overline{\eta}(\beta X) = \frac{1}{2}\hat{g}(X, \mathcal{L}), \quad \overline{\eta}(\gamma Z) = 0,$$

$$\overline{g}(\beta X, \beta Y) = \frac{1}{4}\hat{g}(X, Y), \; \overline{g}(\beta X, \gamma Z) = 0, \; \overline{g}(\gamma Z, \gamma W) = \frac{1}{4}\hat{g}(Z, W),$$

for any $X, Y \in \pi^{-1}TM$ and any $Z, W \in (\mathbb{R}\mathcal{L})^{\perp}$. Now we may state

Theorem 7.18 *Let $(M, (\phi, \xi, \eta, g))$ be an almost contact Riemannian manifold and $(S(M), (\overline{\phi}, \overline{\xi}, \overline{\eta}, \overline{g}))$ the tangent sphere bundle, equipped with the standard contact metric structure. Then*

a. *the Reeb vector field $\xi : M \to S(M)$ is a CR map if and only if $D\xi = -J$ where $J = \phi|_{\mathrm{Ker}(\eta)}$. If this is the case then $d\eta = \Phi$ on $\mathrm{Ker}(\eta) \otimes \mathrm{Ker}(\eta)$; in particular the Levi form*

$$G_\eta(X, Y) = -(d\eta)(X, JY), \quad X, Y \in \mathrm{Ker}(\eta),$$

is positive definite (i.e., $T_{1,0}(M)$ is a strictly pseudoconvex almost CR structure) and ξ is a pseudohermitian map.

b. *If ξ is a geodesic vector field (i.e., $D_\xi \xi = 0$), then the following statements are equivalent:* i) *ξ is a CR map,* ii) *(ϕ, ξ, η, g) is a $\mathcal{K}$-contact structure. Finally if* i) *or* ii) *holds, then* ξ *is a minimal (non isometric) immersion and a harmonic vector field.*

Proof of Theorem 7.18. Stated loosely, under the bundle isomorphism $L \oplus K$ of $T(T(M))$ onto $\pi^{-1}TM \oplus \pi^{-1}TM$ the vector $(d\xi)V$ corresponds to $(V, D_V\xi)$. The precise statement is that

$$(L \oplus K)_{\xi(x)}(d_x\xi)V_x = (\xi(x), V_x) \oplus (\xi(x), (D_V\xi)_x), \quad x \in M.$$

Indeed this follows from the decomposition

$$(d_x\xi)V_x = V^i(x)\delta_{i,\xi(x)} + (D_V\xi)^i(x)\dot{\partial}_{i,\xi(x)}$$

for any tangent vector field V on M locally written as $V = V^i\, \partial/\partial\tilde{x}^i$ (and the very definitions of L and K). Let $V \in H(M) = \mathrm{Ker}(\eta)$. Then

$$G_s((D_V\xi)^i\dot{\partial}_i, \gamma\mathcal{L}) = g(D_V\xi, \xi) \circ \pi = 0$$

hence $(D_V\xi)^i\dot{\partial}_i$ is tangent to $S(M)$. Consequently, we may use the formulae (7.16) so that

$$\overline{\eta}_{\xi(x)}(d_x\xi)V_x = \frac{1}{2}\eta(V)_x = 0$$

i.e., ξ is a contact map to start with. Moreover

$$(L \oplus K)_{\xi(x)}(d_x\xi)J_xV_x = (\xi(x), J_xV_x) \oplus (\xi(x), (D_{JV}\xi)_x),$$
$$(L \oplus K)_{\xi(x)}\overline{J}_{\xi(x)}(d_x\xi)V_x = (\xi(x), -(D_V\xi)_x) \oplus (\xi(x), V_x),$$

where $\overline{J} = \overline{\phi}\big|_{\mathrm{Ker}(\overline{\eta})}$. Hence ξ is a CR map if and only if $D\xi = -J$ on $H(M)$. If this is the case, for any $X, Y \in H(M)$

$$2(d\eta)(X,Y) = -\eta([X,Y]) = -g(\xi, D_XY) + g(\xi, D_YX)$$
$$= g(D_X\xi, Y) - g(D_Y\xi, X) = -g(JX, Y) + g(JY, X) = 2\Phi(X,Y)$$

i.e., $d\eta = \Phi$ on $H(M) \otimes H(M)$. At this point we may show that ξ is a pseudohermitian map. As ξ is assumed to be a CR map, one has

$$\xi^*\overline{\eta} = \lambda\eta$$

for some $\lambda \in C^\infty(M)$. Moreover (by (7.16))

$$\lambda(x) = \frac{1}{2}\theta_{\xi(x)}((d_x\xi)\xi_x) = \frac{1}{2}g_{ij}(x)\xi^i(x)\xi^j(x) = \frac{1}{2}$$

so that $\xi^*\overline{\eta} = \frac{1}{2}\eta$. ∎

Let us prove statement (i) $\Longrightarrow$ (ii). As shown above, when ξ is a CR map the condition $d\eta = \Phi$ already holds on $H(M) \otimes H(M)$. On the other hand $\xi \rfloor \Phi = 0$ and for any $Y \in H(M)$

$$2(\xi \rfloor d\eta)(Y) = -\eta([\xi, Y]) = -g(D_\xi Y, \xi) = g(Y, D_\xi \xi) = 0$$

as ξ is assumed to be geodesic. Thus $d\eta = \Phi$ everywhere i.e., (ϕ, ξ, η, g) satisfies the contact condition. Then we may apply Lemma 6.2 in [42], p. 67, to get

$$D_X \xi = -\phi X - \phi h X, \tag{7.20}$$

where $h = \frac{1}{2}\mathcal{L}_\xi \phi$. By statement (a) in Theorem 7.18 and $D_\xi \xi = 0$ one has $D\xi = -\phi$, hence (7.20) implies $h = 0$. Finally one may apply Theorem 6.2 in [42], p. 65, to conclude that ξ is a Killing vector field. The proof of the implication (ii) $\Longrightarrow$ (i) is similar. Next a result by J.C. Gonzáles-Dávila & L. Vanheke, [144], allows one to conclude that ξ is a minimal immersion. Moreover ξ is not isometric as

$$\xi^* \overline{g} = \frac{1}{2} g - \frac{1}{4} \eta \otimes \eta.$$

Finally $\mathrm{Ric}(\xi) = 2n\xi$ (cf. Corollary 7.1 in [42], p. 92) and a result in [242] imply that ξ is a harmonic vector field.

We emphasize that the condition $D_\xi \xi = 0$ is quite natural. It is surely satisfied when the given almost contact metric structure satisfies the contact condition (cf. Theorem 4.5 in [42], p. 37). As to the case of almost contact metric structures not satisfying the contact condition, one may indicate at least the example of Hopf hypersurfaces. We recall that an orientable real hypersurface M of a Kähler manifold is a Hopf hypersurface if the Reeb vector field underlying the naturally induced almost contact metric structure on M is an eigenvector of the Weingarten map (associated to a unit normal field on M). We obtain

Corollary 7.19 *Let M be a Hopf hypersurface of a Kähler manifold $\tilde{M}$. Let (ϕ, η, ξ, g) be the almost contact metric structure induced on M by the Hermitian structure of $\tilde{M}$. Then $\xi : M \to S(M)$ is a CR map if and only if g is a Sasakian metric.*

Proof. If M is a Hopf hypersurface then $D_\xi \xi = 0$ (cf. [247]). Moreover by (3.3) in [247] we have

$$(D_X \phi)(Y) = g(AX, Y)\xi - \eta(Y) AX = g(\phi D_X \xi, Y)\xi - \eta(Y)\phi(D_X \xi).$$

Since $D\xi = -\phi$, the above formula implies that $(D_X\phi)Y = g(X, Y)\xi - \eta(Y)X$ hence (by Theorem 6.3 in [42], p. 69) g is Sasakian. ■

Corollary 7.20 *Let N be a Riemannian manifold. Let us endow $S(N)$ and $U(S(N))$ with the standard contact metric structures. Then the Reeb vector $\overline{\xi} : S(N) \to U(S(N))$ is a CR map if and only if N has constant sectional curvature* $+1$.

Indeed (by Theorem 7.18) $\overline{\xi} : S(N) \to U(S(N))$ is a CR map if and only if $\overline{\xi}$ is Killing, i.e., $S(N)$ is $\mathcal{K}$-contact. By a result of Y. Tashiro, [285], this is equivalent to N having (constant) sectional curvature $+1$. In particular if ξ_0 is the standard contact vector of the unit sphere S^{2n+1} (a Hopf vector field on S^{2n+1}) and ξ_1 the standard contact vector of $U(S^{2n+1})$ then $\xi_1 \circ \xi_0 : S^{2n+1} \to U(U(S^{2n+1}))$ is a CR map (as a composition of CR maps) between a CR manifold and a (non integrable) almost CR manifold.

A remark is in order. By Theorem 7.18 if ξ is geodesic and a CR map, then ξ is a harmonic vector field yet the converse doesn't hold, in general. For instance, if N is a Riemannian symmetric space of rank 1 then the contact vector ξ on $M = S(N)$ is both geodesic and a harmonic vector field (cf. [54]) yet it does not define a CR map unless N has constant sectional curvature $+1$.

Definition 7.21 ([42], p. 77 and p. 80) M is *cosymplectic* if ϕ is parallel ($D\phi = 0$) and M is a *Kenmotsu manifold* if

$$(D_X\phi)Y = g(\phi X, Y)\xi - \eta(Y)\phi X,$$

for any $X, Y \in T(M)$. ■

Proposition 7.22 *Let $(M, (\phi, \xi, \eta, g))$ be an almost contact Riemannian manifold.* i) *If M is a cosymplectic manifold or a Kenmotsu manifold, then $\xi : M \to S(M)$ is not a CR map.* ii) *If (η, g) is a contact metric structure and the natural almost CR structure on $S(M)$ is integrable, then ξ is a CR map.*

Proof. Under the hypothesis of Proposition 7.22 one has either $D_X\xi = 0$ for any $X \in T(M)$ (when M is cosymplectic) or $D_X\xi = -X$ for any $X \in H(M)$ (when M is Kenmotsu). In any case $D_\xi\xi = 0$ and then (by statement (b) in Theorem 7.18) $\xi : M \to S(M)$ is not a CR map. To establish (ii), when $T_{1,0}(S(M))$ is integrable then (by a result of S. Tanno, [282]) (M, g) has constant sectional curvature c. Yet g is a contact metric hence (by a result of Z. Olszak, [227]) c must be $+1$ and hence M is a Sasakian manifold. Finally (again by statement (b) in Theorem 7.18), ξ is a CR map. ■

Remark 7.23 In [251] one studies the Reeb vector ξ as a CR map from M to $S(M)$ when the tangent sphere bundle is equipped with a natural contact metric structure (cf. also Remark 5.4).

7.3. HARMONIC VECTOR FIELDS FROM *C(M)*

Let $(M, T_{1,0}(M))$ be a CR manifold, of CR dimension n. Assume from now on that M is orientable and nondegenerate. For any orientable nondegenerate CR manifold, on which a contact form θ has been fixed, there is a unique linear connection ∇ (the *Tanaka-Webster connection* of (M,θ)) such that i) $H(M)$ is ∇-parallel, ii) $\nabla J = 0$, $\nabla g_\theta = 0$, and iii) the torsion T_∇ of ∇ is *pure*, i.e.,

$$T_\nabla(Z, W) = 0, \quad T_\nabla(Z, \overline{W}) = 2iL_\theta(Z, \overline{W})T,$$

for any $Z, W \in T_{1,0}(M)$, and

$$\tau \circ J + J \circ \tau = 0.$$

Here $\tau(X) = T_\nabla(T, X)$, $X \in T(M)$. When M is strictly pseudoconvex, the contact metric structure (J, T, θ, g_θ) is normal if and only if $\tau = 0$ (i.e., M is K-contact).

Definition 7.24 A complex valued differential p-form η on M is a $(p,0)$-*form* if $T_{0,1}(M) \rfloor \eta = 0$. Let $\Lambda^{p,0}(M) \to M$ be the bundle of all $(p,0)$-forms on M. The *canonical bundle* is $K(M) = \Lambda^{n+1,0}(M)$. Let $s : M \to K(M)$, $s(x) = 0_x \in K(M)_x$, $x \in M$, be the zero section. The *canonical circle bundle* is $C(M) = (K(M) \setminus s(M))/\mathbb{R}_+$ where $\mathbb{R}_+$ are the multiplicative positive reals. ■

Assume from now on that M is strictly pseudoconvex and let us choose a contact form θ such that L_θ is positive definite. Let us consider the (globally defined) 1-form σ on $C(M)$ given by

$$\sigma = \frac{1}{n+2}\left\{d\gamma + \pi^*\left(i\omega^\alpha_\alpha - \frac{i}{2}g^{\alpha\overline{\beta}}dg_{\alpha\overline{\beta}} - \frac{\rho}{4(n+1)}\theta\right)\right\}.$$

Here $\pi : C(M) \to M$ is the projection, $\gamma : \pi^{-1}(U) \to \mathbb{R}$ is a local fibre coordinate, ω^α_β are the connection 1-forms of the Tanaka-Webster connection ∇ with respect to a local frame $\{T_\alpha : 1 \leq \alpha \leq n\}$ of $T_{1,0}(M)$ on $U \subseteq M$ (i.e., $\nabla T_\alpha = \omega^\beta_\alpha \otimes T_\beta$), and $g_{\alpha\overline{\beta}} = L_\theta(T_\alpha, T_{\overline{\beta}})$ are the local components of the Levi form ($T_{\overline{\alpha}} = \overline{T_\alpha}$).

Definition 7.25 $\rho = g^{\alpha\overline{\beta}} R_{\alpha\overline{\beta}}$ is the *pseudohermitian scalar curvature*. Here $R_{\alpha\overline{\beta}} = \text{trace}\{Z \mapsto R^\nabla(Z, T_\alpha)T_{\overline{\beta}}\}$ is the *pseudohermitian Ricci curvature* while R^∇ is the curvature tensor field of ∇. ■

It is noteworthy that σ is a connection 1-form in the principal circle bundle $S^1 \to C(M) \to M$. By a result in [203]

$$F_\theta = \pi^* \tilde{G}_\theta + 2(\pi^*\theta) \odot \sigma$$

is a Lorentzian metric on $C(M)$.

Definition 7.26 F_θ is called the *Fefferman metric* of (M, θ)). ■

Here the $(0,2)$-tensor field $\tilde{G}_\theta$ on M is given by

$$\tilde{G}_\theta(X, Y) = (d\theta)(X, JY), \quad X, Y \in H(M),$$
$$\tilde{G}_\theta(V, T) = 0, \quad V \in T(M).$$

Let $X : M \to S(M)$ be a unit tangent vector field. The tangent sphere bundle $S(M) \to M$ is defined with respect to the Webster metric g_θ. Let $X^\uparrow : C(M) \to T(C(M))$ be the horizontal lift of X with respect to σ, i.e., $X^\uparrow \in \text{Ker}(\sigma)$ and $(d_z\pi)X_z^\uparrow = X_{\pi(z)}$, for any $z \in C(M)$. If additionally $X \in H(M)$, then $X^\uparrow$ is $U_1(C(M), F_\theta)$-valued. Indeed (by the very definition of F_θ)

$$F_\theta\left(X^\uparrow, X^\uparrow\right) = g_\theta(X, X) = 1.$$

Let us assume from now on that M is compact (hence $C(M)$ is compact, as well). The following notion is central to the present section.

Definition 7.27 Given a unit vector field $X \in H(M)$, we say X is *pseudo-harmonic* if its horizontal lift $X^\uparrow : C(M) \to S_1(C(M), F_\theta)$ is harmonic, i.e., $X^\uparrow$ is a critical point of the energy functional

$$\mathbb{E}(\mathcal{X}) = \frac{1}{2} \int_{C(M)} \text{trace}_{F_\theta}\left(\mathcal{X}^* G_s\right) d\,\text{vol}(F_\theta),$$

where $\mathcal{X}$ is a tangent vector field on $C(M)$ [variations are through $S_1(C(M), F_\theta)$-valued vector fields on $C(M)$]. ■

Here G_s is the [semi-Riemannian metric induced on $S_1(C(M), F_\theta)$ by the] canonical metric of $T(C(M))$. In general, if (N, g) is an m-dimensional compact Lorentzian manifold and $X : N \to S_\epsilon(N,g)$ a vector field, then we need to establish the first variation formula for the functional

$$E(X) = \frac{1}{2}\int_N \operatorname{trace}_g(X^* G_s)\, d\operatorname{vol}(g)$$

(G_s is the canonical metric on $T(N)$). Let $\{E_i : 1 \leq i \leq m\}$ be a local g-orthonormal frame at a point $x \in N$. Since

$$(d_x X)E_{i,x} = \lambda_i^j(x)\left(\delta_j + \left[\left(\nabla_j X^k\right)\circ\pi\right]\dot{\partial}_k\right)_{X(x)}$$

(where $E_i = \lambda_i^j(\partial/\partial\tilde{x}^j)$) it follows that

$$\begin{aligned}
\operatorname{trace}_g(X^* G_s)_x &= \sum_{i=1}^m \epsilon_i (X^* G_s)(E_i, E_i)_x \\
&= \sum_{i=1}^m \epsilon_i \{\hat{g}_{X(x)}(L_{X(x)}(d_x X)E_{i,x},\ L_{X(x)}(d_x X)E_{i,x}) \\
&\quad + \hat{g}_{X(x)}(K_{X(x)}(d_x X)E_{i,x},\ K_{X(x)}(d_x X)E_{i,x})\} \\
&= m + g(\nabla X, \nabla X)
\end{aligned}$$

where

$$g(\nabla X, \nabla X) = g^{ij}\left(\nabla_i X^k\right)\left(\nabla_j X^\ell\right) g_{k\ell}.$$

Therefore the energy functional becomes

$$E(X) = \frac{m}{2}\operatorname{Vol}(N) + \frac{1}{2}\int_N g(\nabla X, \nabla X)\, d\operatorname{vol}(g).$$

Let $\mathcal{X} : N \times (-\delta, \delta) \to T(N)$ ($\delta > 0$) be a smooth 1-parameter variation of X ($\mathcal{X}(x,0) = X_x$, $x \in N$) through $S_\epsilon(N,g)$-valued vector fields. Let us set $X_t = \mathcal{X} \circ \alpha_t$, where $\alpha_t : N \to N \times (-\delta,\delta)$, $\alpha_t(x) = (x,t)$, $x \in$, $|t| < \delta$. Let D be the Levi-Civita connection of $h = g + dt \otimes dt$ (a semi-Riemannian metric on $N \times (-\delta, \delta)$) and R^D the curvature tensor field of D. Then

$$\frac{d}{dt} E(X_t) = \frac{1}{2} \frac{d}{dt} \int_N \sum_{i=1}^{m} \epsilon_i g(\nabla_{E_i} X_t, \nabla_{E_i} X_t) \, d\,\mathrm{vol}(g)$$

$$= \int_N \sum_{i=1}^{m} \epsilon_i h(D_{\partial/\partial t} D_{E_i} \mathcal{X}, D_{E_i} \mathcal{X}) \, d\,\mathrm{vol}(g)$$

$$= \int_N \sum_i \epsilon_i h(D_{E_i} D_{\partial/\partial t} \mathcal{X} + R^D(\partial/\partial t, E_i)\mathcal{X}, \nabla_{E_i} \mathcal{X}) \, d\,\mathrm{vol}(g)$$

(as $R^D(\partial/\partial t, E_i) = 0$)

$$= \int_N \sum_i \epsilon_i h(D_{E_i} D_{\partial/\partial t} \mathcal{X}, \nabla_{E_i} \mathcal{X}) \, d\,\mathrm{vol}(g) = (as\ Dh = 0)$$

$$\int_N \sum_i \epsilon_i \{E_i(h(D_{\partial/\partial t} \mathcal{X}, D_{E_i} \mathcal{X})) - h(D_{\partial/\partial t} \mathcal{X}, D_{E_i} D_{E_i} \mathcal{X})\} d\,\mathrm{vol}(g).$$

For each $|t| < \delta$, let us define a vector field Y_t on N by setting

$$g(Y_t, Y)_x = h(D_{\partial/\partial t} \mathcal{X}, D_Y \mathcal{X})_{(x,t)}, \quad x \in N,$$

for any tangent vector field Y on N. Then (by $\nabla g = 0$)

$$\frac{d}{dt} E(X_t) = \int_N \sum_i \epsilon_i \{E_i(g(Y_t, E_i)) - h(D_{\partial/\partial t} \mathcal{X}, D_{E_i} D_{E_i} \mathcal{X})\} \, d\,\mathrm{vol}(g)$$

$$= \int_N \sum_i \epsilon_i \{g(\nabla_{E_i} Y_t, E_i) + g(Y_t, \nabla_{E_i} E_i)$$

$$- h(D_{\partial/\partial t} \mathcal{X}, D_{E_i} D_{E_i} \mathcal{X})\} \, d\,\mathrm{vol}(g) = \int_N \mathrm{div}(Y_t) d\,\mathrm{vol}(g)$$

$$+ \int_N \sum_i \epsilon_i \{h(D_{\partial/\partial t} \mathcal{X}, D_{\nabla_{E_i} E_i} \mathcal{X}) - h(D_{\partial/\partial t} \mathcal{X}, D_{E_i} D_{E_i} \mathcal{X})\} \, d\,\mathrm{vol}(g).$$

By differentiating $E(X_t)$ at $t = 0$, applying Green's lemma and the identity

$$\Delta_g X = -\sum_{i=1}^{m} \epsilon_i \{\nabla_{E_i} \nabla_{E_i} X - \nabla_{\nabla_{E_i} E_i} X\}$$

we obtain

Proposition 7.28 (The *first variation* formula) *Let (N, g) be a compact semi-Riemannian manifold. Let $\epsilon \in \{\pm 1\}$. Let $\{X_t\}_{|t|<\delta}$ be a smooth 1-parameter variation of $X = X_0$ through unit vector fields $X_t : N \to S_\epsilon(N, g)$. Then*

$$\frac{d}{dt}\{E(X_t)\}_{t=0} = \int_N g(V, \Delta_g X)\, d\,\mathrm{vol}(g) \tag{7.21}$$

where $V = (\partial X_t/\partial t)_{t=0}$, with the constraint $g(X, V) = 0$ (obtained by differentiating $g(X_t, X_t) = \epsilon$ at $t = 0$).

Unlike the Riemannian case (when $\nu = 0$, $\epsilon = 1$)

$$\langle X, Y\rangle = \int_N g(X, Y)\, d\,\mathrm{vol}(g)$$

is not an inner product on $\mathcal{X}(N)$, hence one may not conclude (from $\{dE(X_t)/dt\}_{t=0} = 0$) that $\Delta_g X$ and X must be proportional. However, when g is a Lorentzian metric ($\nu = 1$) and $X = X_0$ is a unit timelike vector field, V is spacelike (since $g(V, X) = 0$). Then from (7.21) we get $\{dE(X_t)/dt\}_{t=0} = 0$ if and only if $\Delta_g X$ is proportional to X.

Now let X be a pseudoharmonic vector field on a compact strictly pseudoconvex CR manifold M. Let $\{X_t\}_{|t|<\delta}$ be a smooth 1-parameter variation of X through unit vector fields $X_t \in H(M)$. Then (by (7.21) applied to $N = C(M)$ and $g = F_\theta$)

$$\frac{d}{dt}\{\mathbb{E}(X_t^\uparrow)\}_{t=0} = \int_{C(M)} F_\theta(\tilde{V}, \Box X^\uparrow)\, d\,\mathrm{vol}(F_\theta).$$

Here $\tilde{V} = (\partial X_t^\uparrow/\partial t)_{t=0}$ and

$$-\Box\mathcal{X} = \sum_{a=1}^{2n}\left\{\nabla^{C(M)}_{E_a^\uparrow}\nabla^{C(M)}_{E_a^\uparrow}\mathcal{X} - \nabla^{C(M)}_{\nabla^{C(M)}_{E_a^\uparrow}E_a^\uparrow}\mathcal{X}\right\}$$
$$+\nabla^{C(M)}_{T^\uparrow+S}\nabla^{C(M)}_{T^\uparrow+S}\mathcal{X} - \nabla^{C(M)}_{\nabla^{C(M)}_{T^\uparrow+S}(T^\uparrow+S)}\mathcal{X}$$
$$-\nabla^{C(M)}_{T^\uparrow-S}\nabla^{C(M)}_{T^\uparrow-S}\mathcal{X} + \nabla^{C(M)}_{\nabla^{C(M)}_{T^\uparrow-S}(T^\uparrow-S)}\mathcal{X}$$

for any tangent vector field $\mathcal{X}$ on $C(M)$, where $\nabla^{C(M)}$ is the Levi-Civita connection of $(C(M), F_\theta)$, and S is the (suitably normalized) tangent to the S^1-action (locally $S = ((n+2)/2)\,\partial/\partial\gamma$), and $\{E_a : 1 \le a \le 2n\}$ is a local

g_θ-orthonormal frame of $H(M)$. Note that $T^\uparrow - S$ is a global timelike vector field on $C(M)$ (hence $C(M)$ is a space-time). Clearly $\{E_a^\uparrow : 1 \leq a \leq 2n\} \cup \{T^\uparrow \pm S\}$ is a local orthonormal (with respect to F_θ) frame of $T(C(M))$. We wish to compute $\square X^\uparrow$. To this end, we recall (cf. E. Barletta et al., [26])

$$\nabla^{C(M)}_{X^\uparrow} Y^\uparrow = (\nabla_X Y)^\uparrow - (d\theta)(X,Y)T^\uparrow - \{A(X,Y) + g_\theta(\phi X, Y)\}S, \tag{7.22}$$

$$\nabla^{C(M)}_{X^\uparrow} T^\uparrow = (\tau X + \phi X)^\uparrow, \tag{7.23}$$

$$\nabla^{C(M)}_{T^\uparrow} X^\uparrow = (\nabla_T X + \phi X)^\uparrow - 2G_\theta(X, W)S, \tag{7.24}$$

$$\nabla^{C(M)}_{X^\uparrow} S = \nabla^{C(M)}_{S} X^\uparrow = (JX)^\uparrow, \tag{7.25}$$

$$\nabla^{C(M)}_{T^\uparrow} T^\uparrow = W^\uparrow, \quad \nabla^{C(M)}_{S} S = 0, \tag{7.26}$$

$$\nabla^{C(M)}_{S} T^\uparrow = \nabla^{C(M)}_{T^\uparrow} S = 0, \tag{7.27}$$

for any $X, Y \in H(M)$, where ∇ is the Tanaka-Webster connection of (M, θ) and $\phi : H(M) \to H(M)$ is the skew-symmetric endomorphism given by

$$g_\theta(\phi X, Y) = (d\sigma)\left(X^\uparrow, Y^\uparrow\right), \quad X, Y \in H(M),$$

while $W \in H(M)$ is given by

$$g_\theta(W, Y) = 2(d\sigma)\left(T^\uparrow, Y^\uparrow\right), \quad Y \in H(M).$$

Also $A(X,Y) = g_\theta(X, \tau Y)$. Recall (cf. e.g., [307]) that A is symmetric and trace-less. Consequently, for any $X \in H(M)$

$$-\square X^\uparrow = \sum_{a=1}^{2n} \left\{ \nabla^{C(M)}_{E_a^\uparrow} \nabla^{C(M)}_{E_a^\uparrow} X^\uparrow - \nabla^{C(M)}_{\nabla^{C(M)}_{E_a^\uparrow} E_a^\uparrow} X^\uparrow \right\}$$
$$+ 2 \left\{ \nabla^{C(M)}_{S} \nabla^{C(M)}_{T^\uparrow} X^\uparrow + \nabla^{C(M)}_{T^\uparrow} \nabla^{C(M)}_{S} X^\uparrow \right.$$
$$\left. - \nabla^{C(M)}_{\nabla^{C(M)}_{T^\uparrow} S} X^\uparrow - \nabla^{C(M)}_{\nabla^{C(M)}_{S} T^\uparrow} X^\uparrow \right\}$$

may be written

$$-\Box X^{\uparrow} = (-\Delta_b X + 2\tau JX + 4\nabla_T JX + 6\phi JX + 2\,\mathrm{div}(JX)T)^{\uparrow}$$
$$+ \{2\,\mathrm{div}(\phi X) - 2\,\mathrm{div}(\tau X) - 2g_\theta(JX, W)$$
$$+ \mathrm{trace}((\nabla\tau)X) - \mathrm{trace}((\nabla\phi)X)\}S, \tag{7.28}$$

where

$$\Delta_b X = -\sum_{a=1}^{2n}\left\{\nabla_{E_a}\nabla_{E_a}X - \nabla_{\nabla_{E_a}E_a}X\right\}. \tag{7.29}$$

Indeed (by (7.22)–(7.27))

$$\sum_{a=1}^{2n}\nabla^{C(M)}_{E_a^{\uparrow}}\nabla^{C(M)}_{E_a^{\uparrow}}X^{\uparrow} = \sum_a\{(\nabla_{E_a}\nabla_{E_a}X)^{\uparrow}$$
$$+ g_\theta(E_a, J\nabla_{E_a}X)T^{\uparrow} - g_\theta(E_a, \tau\nabla_{E_a}X - \phi\nabla_{E_a}X)S$$
$$+ E_a(g_\theta(E_a, JX))T^{\uparrow} + g_\theta(E_a, JX)(\tau E_a + \phi E_a)^{\uparrow}$$
$$- E_a(g_\theta(E_a, \tau X - \phi X))S - g_\theta(E_a, \tau X - \phi X)\,(JE_a)^{\uparrow}\big\}.$$

As θ is a contact form, we may consider the divergence operator with respect to the volume form $\omega = \theta \wedge (d\theta)^n$ i.e., $\mathcal{L}_X\omega = \mathrm{div}(X)\,\omega$. As ω is parallel with respect to the Tanaka-Webster connection ∇, the divergence may also be computed as

$$\mathrm{div}(X) = \sum_{a=1}^{2n} g_\theta(\nabla_{E_a}X, E_a).$$

Then (by $\nabla J = 0$)

$$\sum_{a=1}^{2n}\nabla^{C(M)}_{E_a^{\uparrow}}\nabla^{C(M)}_{E_a^{\uparrow}}X^{\uparrow} = \sum_a\{(\nabla_{E_a}\nabla_{E_a}X)^{\uparrow} + E_a(g_\theta(E_a, JX))T^{\uparrow}$$
$$- [E_a(g_\theta(E_a, \tau X - \phi X)) + g_\theta(E_a, \tau\nabla_{E_a}X - \phi\nabla_{E_a}X)]S\}$$
$$+ \mathrm{div}(JX)T^{\uparrow} + (\tau JX + \phi JX)^{\uparrow} - (J\tau X - J\phi X)^{\uparrow}. \tag{7.30}$$

Recall that $\tau \circ J = -J \circ \tau$ (with the corresponding simplification of (7.30)). A similar calculation leads to

$$\sum_{a=1}^{2n} \nabla^{C(M)}_{\nabla^{C(M)}_{E_a^\uparrow} E_a^\uparrow} X^\uparrow = -\operatorname{trace}(\phi)(JX)^\uparrow + \sum_a \{(\nabla_{\nabla_{E_a} E_a} X)^\uparrow$$
$$+ g_\theta(\nabla_{E_a} E_a, JX) T^\uparrow - g_\theta(\nabla_{E_a} E_a, \tau X - \phi X) S\}. \tag{7.31}$$

Taking the exterior differential of σ

$$d\sigma = \frac{1}{n+2} \pi^* \left\{ i\, d\omega_\alpha^\alpha - \frac{i}{2}\, dg^{\alpha\overline{\beta}} \wedge dg_{\alpha\overline{\beta}} - \frac{d(\rho\theta)}{4(n+1)} \right\}$$

we obtain (by the very definition of ϕ) $\phi T_\alpha = \phi_\alpha{}^\beta T_\beta$ where

$$\phi^{\overline{\alpha}\beta} = \frac{i}{2(n+2)} \left\{ R^{\overline{\alpha}\beta} - \frac{\rho}{2(n+1)} g^{\overline{\alpha}\beta} \right\}.$$

Cf. also [26]. Consequently $\operatorname{trace}(\phi) = 0$ and $[\phi, J] = 0$. Using (7.30)–(7.31) together with

$$\nabla^{C(M)}_S \nabla^{C(M)}_{T^\uparrow} X^\uparrow = (J\nabla_T X + J\phi X)^\uparrow,$$
$$\nabla^{C(M)}_{T^\uparrow} \nabla^{C(M)}_S X^\uparrow = (\nabla_T JX + \phi JX)^\uparrow - 2g_\theta(JX, W)S,$$

we obtain (7.28). ■

If $\mathcal{X} : M \times (-\delta, \delta) \to S(M)$ is given by $\mathcal{X}(x,t) = X_t(x)$, $x \in M$, $|t| < \delta$, then $\mathcal{X} = f^i \partial/\partial \tilde{x}^i$, for some $f^i : U \times (-\delta, \delta) \to \mathbb{R}$, with respect to a local coordinate system $(U, \tilde{x}^i)$ on M. Then

$$(d_{(x,t)}\mathcal{X})(\partial/\partial t)_{(x,t)} = \frac{\partial f^i}{\partial t}(x,t)\ \dot{\partial}_i\big|_{\mathcal{X}(x,t)}$$

hence $V = (\partial X_t/\partial t)_{t=0}$ is given (under the identification $\operatorname{Ker}(d\Pi)_{X_x} \approx T_x(M)$, where $\Pi : T(M) \to M$ is the projection) by

$$V_x = \frac{\partial f^i}{\partial t}(x,0)\ \frac{\partial}{\partial \tilde{x}^i}\bigg|_x .$$

Similarly, if $(u^a) = (\tilde{x}^i, \gamma)$ are the induced coordinates on $C(M)$ and $(\partial/\partial \tilde{x}^i)^\uparrow = \lambda_i^a\, \partial/\partial u^a$ then (under the identification $\operatorname{Ker}(d\Pi_C)_{X_z^\uparrow} \approx T_z(C(M))$, where $\Pi_C : T(C(M)) \to C(M)$ is the projection) $\tilde{V}$ is given by

$$\tilde{V}_z = \frac{\partial f^i}{\partial t}(\pi(z), 0) \lambda_i^a(z)\ \frac{\partial}{\partial u^a}\bigg|_z, \quad z \in \pi^{-1}(U).$$

It follows that $\tilde{V} = V^{\uparrow}$. Also, if $\theta_i = \theta(\partial/\partial \tilde{x}^i)$, then we may differentiate $\theta(X_t) = 0$ to get $(\partial f^i/\partial t)\theta_i = 0$ i.e., $V \in H(M)$. Thus $F_\theta(\tilde{V}, S) = 0$ and then (using (7.28) and integrating along the fibre, i.e., applying the identity $\int_{C(M)} (f \circ \pi) d\,\mathrm{vol}(F_\theta) = 2\pi \int_M f\,\omega$, for any $f \in C^\infty(M)$)

$$0 = \frac{d}{dt}\{\mathbb{E}(X_t^{\uparrow})\}_{t=0}$$
$$= 2\pi \int_M g_\theta(V, \Delta_b X - 2\tau JX - 4\nabla_T JX - 6\phi JX)\,\omega$$

which together with the constraint $g_\theta(V, X) = 0$ leads to

$$-\Delta_b X + 4\nabla_T JX + 2\tau JX + 6\phi JX = \lambda(X) X \tag{7.32}$$

where

$$\lambda(X) = -\|\pi_H \nabla X\|^2 + 4 g_\theta(\nabla_T JX, X) + 2 g_\theta(\tau JX, X) + 6 g_\theta(\phi JX, X). \tag{7.33}$$

Here $\pi_H \nabla X$ is the restriction of ∇X to $H(M)$, hence

$$\|\pi_H \nabla X\|^2 = \sum_{a=1}^{2n} g_\theta(\nabla_{E_a} X, \nabla_{E_a} X)$$

on U.

Proposition 7.29 *The operator Δ_b acting on $\Gamma^\infty(H(M))$ is subelliptic of order* $1/2$.

Let us recall

Definition 7.30 A formally self adjoint second order differential operator $L: C^\infty(M) \to C^\infty(M)$ is *subelliptic* of order ϵ $(0 < \epsilon \leq 1)$ at a point $x \in M$ if there is an open neighborhood U of x such that $\|u\|_\epsilon^2 \leq C(|\langle Lu, u\rangle| + \|u\|^2)$ for any $u \in C_0^\infty(U)$, where $\|u\|_\epsilon$ is the Sobolev norm of order ϵ and $\|u\|^2 = \langle u, u\rangle$, while $\langle\ ,\ \rangle$ is the L^2 inner product on functions, i.e., $\langle u, v\rangle = \int_M u\bar{v}\,\omega$. L is *subelliptic* if it is subelliptic at each point $x \in M$. ∎

A typical example in CR geometry is indicated in the following

Definition 7.31 The *sublaplacian* $H: C^\infty(M) \to C^\infty(M)$ given by $Hu = -\mathrm{div}(\nabla^H u)$, where $\nabla^H u$ is the *horizontal gradient* of u, i.e., locally $\nabla^H u = \sum_{a=1}^{2n} E_a(u) E_a$. ∎

By a well-known lemma of E.V. Radkevic, [259], H is subelliptic of order $1/2$. The meaning of Proposition 7.29 is that locally $\Delta_b X$ is given by a subelliptic operator acting on the coefficients of $X = f^a E_a$, plus lower order terms. Precisely, there is a matrix Q of first order differential operators such that

$$(\Delta_b X)^a = Hf^a - Q^a_c f^c. \tag{7.34}$$

Indeed (7.29) may be locally written in the form (7.34) with $Q = A + B$, where

$$A = \left[\sum_{c=1}^{2n} \Gamma^a_{cb} E_c \right]_{1 \leq a,b \leq 2n},$$

$$B = \left[\sum_{c=1}^{2n} \{ E_c(\Gamma^a_{cb}) + \Gamma^a_{cd}\Gamma^d_{cb} - \Gamma^d_{cc}\Gamma^a_{db} \} \right]_{1 \leq a,b \leq 2n},$$

and $\nabla_{E_a} E_b = \Gamma^c_{ab} E_c$.

Equation (7.32) admits the following variational interpretation.

Theorem 7.32 *Let M be a compact strictly pseudoconvex CR manifold, of CR dimension n, and θ a contact form on M. Let $X \in H(M)$ be a unit vector field and let us set*

$$\mathcal{B}_\theta(X) = -\frac{1}{2} \int_M \lambda(X)\, \theta \wedge (d\theta)^n,$$

where $\lambda(X)$ is given by (7.33). Then (7.32) is the Euler-Lagrange equation corresponding to $\{d\mathcal{B}_\theta(X_t)/dt\}_{t=0} = 0$, for any smooth 1-parameter variation $\{X_t\}_{|t|<\delta}$ of X through unit vector fields $X_t \in H(M)$, $|t| < \delta$.

Clearly Theorem 7.32 follows from the identity

$$\mathbb{E}(X^\uparrow) = (n+1)2^{n+1}\pi\, n!\, \mathrm{Vol}(M) + 2\pi \mathcal{B}_\theta(X). \tag{7.35}$$

To establish (7.35) we start from

$$\mathbb{E}(X^\uparrow) = (n+1)\mathrm{Vol}(C(M)) + \frac{1}{2} \int_{C(M)} F_\theta(\nabla^{C(M)} X^\uparrow, \nabla^{C(M)} X^\uparrow)\, d\mathrm{vol}(F_\theta).$$

By a result in [26] $\omega = 2^n n!\, d\mathrm{vol}(g_\theta)$, hence

$$\mathrm{Vol}(C(M)) = 2^{n+1}\pi\, n!\, \mathrm{Vol}(M).$$

Moreover (by (7.22)–(7.27))

$$F_\theta\left(\nabla^{C(M)} X^\uparrow, \nabla^{C(M)} X^\uparrow\right) = \sum_{a=1}^{2n} F_\theta\left(\nabla^{C(M)}_{E_a^\uparrow} X^\uparrow, \nabla^{C(M)}_{E_a^\uparrow} X^\uparrow\right)$$

$$+ 4F_\theta\left(\nabla^{C(M)}_{T^\uparrow} X^\uparrow, \nabla^{C(M)}_{S} X^\uparrow\right) = \sum_a \{g_\theta(\nabla_{E_a} X, \nabla_{E_a} X)$$

$$+ 2(d\theta)(E_a, X)[A(E_a, X) + g_\theta(\phi E_a, X)]F_\theta(T^\uparrow, S)\}$$

$$+ 4g_\theta(\nabla_T X + \phi X, JX) = \|\pi_H \nabla X\|^2$$

$$- 2g_\theta(JX, \tau X - \phi X) + 4g_\theta(\nabla_T X + \phi X, JX)$$

that is

$$F_\theta(\nabla^{C(M)} X^\uparrow, \nabla^{C(M)} X^\uparrow) = -\lambda(X) \circ \pi.$$

7.4. BOUNDARY VALUES OF BERGMAN-HARMONIC MAPS

In their seminal 1998 paper J. Jost & C-J. Xu studied (cf. [180]) the existence and regularity of weak solutions $\phi : \overline{\Omega} \to (S, h)$ to the nonlinear subelliptic system

$$H\phi^i + \sum_{a=1}^{m} \left(\Gamma^i_{jk} \circ \phi\right) X_a(\phi^j) X_a(\phi^k) = 0, \quad 1 \leq i \leq \nu, \tag{7.36}$$

where $H = \sum_{a=1}^m X_a^* X_a$ is the Hörmander operator associated to a system $X = \{X_1, \ldots, X_m\}$ of smooth vector fields on a open set $\Omega \subseteq \mathbb{R}^n$, verifying the Hörmander condition on Ω, (S, h) is a Riemannian manifold and Γ^i_{jk} are the Christoffel symbols associated to the Riemannian metric h. Their study is part of a larger program aiming to the study of hypoelliptic nonlinear systems of variational origin similar to the harmonic maps system, but degenerate elliptic. Indeed, if $X = b^A_a(x)\,\partial/\partial x^A$ then $X_a^* f = -\partial(b^A_a(x) f)/\partial x^A$ for any $f \in C^1_0(\Omega)$ hence

$$Hu = -\sum_{A,B} \frac{\partial}{\partial x^A}\left(a^{AB}(x) \frac{\partial u}{\partial x^B}\right)$$

where $a^{AB}(x) = \sum_{a=1}^{m} b_a^A(x) b_a^B(x)$ so that in general $[a^{AB}]$ is only semi-positive definite. Hence H is degenerate elliptic (in the sense of J.M. Bony, [58]). As successively observed (cf. E. Barletta et al., [25]) solutions of systems of the form (7.36) may be built within CR geometry as S^1-invariant harmonic maps $\Phi : C(M) \to S$ where $S^1 \to C(M) \xrightarrow{\pi} M$ is the canonical circle bundle over a strictly pseudoconvex CR manifold M and harmonicity is meant with respect to the Fefferman metric F_θ (associated to a choice of contact form θ on M, cf. J.M. Lee, [203]). Base maps $\phi : M \to S$ corresponding (i.e., $\Phi = \phi \circ \pi$) to such Φ were termed *pseudoharmonic maps* and shown to satisfy

$$-\Delta_b \phi^i + \sum_{a=1}^{2n} \left(\Gamma^i_{jk} \circ \phi \right) X_a(\phi^j) X_a(\phi^k) = 0, \quad 1 \leq i \leq \nu, \tag{7.37}$$

where Δ_b is the sublaplacian associated to (M, θ) and $\{X_a : 1 \leq a \leq 2n\}$ is a local orthonormal frame of the maximal complex, or Levi, distribution $H(M)$ of M. The sublaplacian may be locally written $\Delta_b = \sum_{a=1}^{2n} X_a^* X_a$ hence the similarity among the systems (7.36) and (7.37). The formal adjoint X_a^* of X_a is however meant with respect to the L^2 inner product $(u, v) = \int u \bar{v} \theta \wedge (d\theta)^n$, while in [180] one integrates with respect to the Lebesgue measure on Ω (the precise quantitative relationship among the two notions is explained in the next section).

The derivation of (7.36) by analogy to the harmonic map system (replacing the Laplace-Beltrami operator with the Hörmander operator) is nevertheless rather formal. Indeed CR manifolds appear mainly as boundaries of smooth domains Ω in $\mathbb{C}^n$ and it was not known previous to the work in [103] whether boundary values of harmonic maps from Ω extending smoothly up to $\partial\Omega$ were pseudoharmonic. We may state

Theorem 7.33 (Y. Kamishima et al., [103]) *Let $\Omega \subset \mathbb{C}^n$ $(n \geq 2)$ be a smoothly bounded strictly pseudoconvex domain and g the Bergman metric on Ω. Let S be a complete ν-dimensional $(\nu \geq 2)$ Riemannian manifold of sectional curvature* $\mathrm{Sect}(S) \leq \kappa^2$ *for some $\kappa > 0$. Assume that S may be covered by one coordinate chart $\chi = (y^1, \ldots, y^\nu) : S \to \mathbb{R}^\nu$. Let $f \in W^{1,2}(\Omega, S) \cap C^0(\overline{\Omega}, S)$ be a map such that $f(\overline{\Omega}) \subset B(p, \mu)$ for some $p \in S$ and some $0 < \mu < \min\{\pi/(2\kappa), i(p)\}$ where $i(p)$ is the injectivity radius of p. Let $\phi = \phi_f : \overline{\Omega} \to S$ be the solution to the Dirichlet problem*

$$\tau_g(\phi) = 0 \quad \text{in } \Omega, \quad \phi = f \quad \text{on } \partial\Omega. \tag{7.38}$$

Then

$$N(f^i) = -\frac{1}{2(n-1)}\left(H_b f\right)^i, \quad 1 \le i \le \nu, \tag{7.39}$$

for any local coordinate system (ω, y^i) *on* S *such that* $\phi(\overline{\Omega}) \cap \omega \neq \emptyset$ $(f^i = y^i \circ f)$. *Also* $N = -JT$ *and* T *is the characteristic direction of* $\partial\Omega$. *In particular if* $f \in C^\infty(\partial\Omega, S)$ *and* $N(f^i) = 0$ *then* $f : \partial\Omega \to S$ *is a pseudoharmonic map.*

The proof (similar to that of Theorem 2.35 in Chpater 2 of this monograph) is beyond the scope of this book (which is confined to the study of harmonic vector fields and their generalizations). Let us mention however that the key idea in the proof of Theorem 7.33 is (as first observed by A. Korányi & H.M. Reimann, [195]) that the Kählerian geometry of the interior of Ω and the contact geometry of the boundary $\partial\Omega$ may be effectively related through the use of the Bergman kernel $K(z,\zeta)$ of Ω. The main technical ingredient in the proof is then the existence of an ambient linear connection ∇ (the *Graham-Lee connection*, cf. R. Graham et al., [151], or Appendix A in [26]) defined on a neighborhood of $\partial\Omega$ in Ω and inducing the Tanaka-Webster connection (cf. [279], [307]) on each level set of $\varphi(z) = -K(z,z)^{-1/(n+1)}$ $(z \in \Omega)$. See also Section 5.3 in [24], p. 87–95.

7.5. PSEUDOHARMONIC MAPS

Let $(M, T_{1,0}(M))$ be a $(2n+1)$-dimensional orientable CR manifold, of CR dimension n. We assume throughout that M is strictly pseudoconvex i.e., the Levi form G_θ (given by $G_\theta(X,Y) = (d\theta)(X,JY)$ for any $X, Y \in H(M)$) is positive definite for some pseudohermitian structure θ. Let θ be a contact form on M such that G_θ is positive definite and let T be the characteristic direction of $d\theta$ i.e., the globally defined nowhere zero tangent vector field on M, everywhere transverse to $H(M)$, determined by $\theta(T) = 1$ and $T \rfloor d\theta = 0$. As previously demonstrated, a strictly pseudoconvex CR manifold comes equipped with a natural second order differential operator Δ_b (similar to the Laplace-Beltrami operator on a Riemannian manifold) given by

$$\Delta_b u = -\mathrm{div}\left(\nabla^H u\right), \quad u \in C^2(M), \tag{7.40}$$

the sublaplacian of (M,θ). Here div is the divergence with respect to $\Psi = \theta \wedge (d\theta)^n$ i.e., $\mathcal{L}_X \Psi = \mathrm{div}(X)\Psi$ where $\mathcal{L}_X$ is the Lie derivative, and

$\nabla^H u = \pi_H \nabla u$ (the horizontal gradient of u). Also ∇u is the gradient of u with respect to the Webster metric g_θ i.e., the Riemannian metric

$$g_\theta(X, Y) = G_\theta(X, Y), \quad g_\theta(X, T) = 0, \quad g_\theta(T, T) = 1, \tag{7.41}$$

for any $X, Y \in H(M)$, and $\pi_H : T(M) \to H(M)$ is the projection relative to the decomposition $T(M) = H(M) \oplus \mathbb{R}T$. The sublaplacian is degenerate elliptic (in the sense of J.M. Bony, [58]) and subelliptic of order $1/2$ (cf. G.B. Folland, [118]) hence hypoelliptic (cf. L. Hörmander, [171]). Let us assume that M is compact and consider the energy functional

$$E(\phi) = \frac{1}{2} \int_M \mathrm{trace}_{G_\theta}(\pi_H \phi^* h)\ \Psi \tag{7.42}$$

where $\pi_H B$ denotes the restriction to $H(M)$ of the bilinear form B. Here E is defined on the set of all C^∞ maps $\phi : M \to S$ from M into a ν-dimensional Riemannian manifold (S, h).

Definition 7.34 A *pseudoharmonic* map is a C^∞ map $\phi : M \to S$ such that $\{dE(\phi_t)/dt\}_{t=0} = 0$ for any smooth 1-parameter variation $\phi_t : M \to S$ of ϕ i.e., $\phi_0 = \phi$. ∎

Let us set

$$(H_b\phi)^i \equiv -\Delta_b \phi^i + \sum_{a=1}^{2n} \left(\Gamma^i_{jk} \circ \phi\right) X_a(\phi^j) X_a(\phi^k), \quad 1 \le i \le \nu.$$

The Euler-Lagrange equations of the variational principle $\delta E(\phi) = 0$ are $H_b(\phi) = 0$ (cf. [25]). Let $\phi : M \to S$ be a pseudoharmonic map. Let (U, x^A) and (V, y^i) be a local coordinate systems on M and S such that $\phi(U) \subseteq V$. Let $\{X_a : 1 \le a \le 2n\}$ be local G_θ-orthonormal frame in $H(M)$ defined on the open set U. As a consequence of the nondegeneracy of M the vector fields $\{(d\varphi)X_a : 1 \le a \le 2n\}$ form a Hörmander system on $\Omega = \varphi(U) \subseteq \mathbb{R}^{2n+1}$ where $\varphi = (x^1, \ldots, x^{2n+1})$. As the formal adjoint of $X_a = b^A_a \partial/\partial x^A$ with respect to Ψ is given by $X^*_a u = -\partial\left(b^A_a u\right)/\partial x^A - b^B_a \Gamma^A_{AB} u$ one may conclude that $f \equiv \phi \circ \varphi^{-1} : \Omega \to S$ is a subelliptic harmonic map if and only if $Lf^i = 0$ in Ω where L is the (purely local) first order differential operator $Lu = \sum_{a=1}^{2n} b^B_a \Gamma^A_{AB} X_a u$ and Γ^A_{BC} are the local coefficients of the Tanaka-Webster connection of (M, θ) with respect to (U, x^A). If for instance $M = \mathbb{H}_n$ (the *Heisenberg group*, cf. e.g., [110], p. 11-12) then $L \equiv 0$ and the two notions coincide.

To demonstrate a class of pseudoharmonic maps, we look at $\overline{\partial}_b$-pluriharmonic maps of a nondegenerate CR manifold into a Riemannian manifold. We need a few additional notions of pseudohermitian geometry (cf. e.g., [110], Chapter 1). The tangential Cauchy-Riemann operator is the first order differential operator

$$\overline{\partial}_b : C^\infty(M) \to \Gamma^\infty(T_{0,1}(M)^*)$$

defined by $(\overline{\partial}_b f)\overline{Z} = \overline{Z}(f)$ for any C^∞ function $f : M \to \mathbb{C}$ and any $Z \in T_{1,0}(M)$. A $(0,1)$-form is a $\mathbb{C}$-valued differential 1-form η on M such that $T_{1,0}(M) \rfloor \eta = 0$ and $T \rfloor \eta = 0$. See also the preparation of CR geometry earlier in this chapter.

Definition 7.35 A $(1,1)$-*form* is a $\mathbb{C}$-valued differential 2-form ω on M such that

$$\omega(Z, W) = \omega(\overline{Z}, \overline{W}) = 0, \quad T \rfloor \omega = 0,$$

for any $Z, W \in T_{1,0}(M)$. ■

Definition 7.36 Let $\Lambda^{0,1}(M) \to M$ and $\Lambda^{1,1}(M) \to M$ be the corresponding vector bundles. Besides from $\overline{\partial}_b$ we define the differential operator

$$\partial_b : \Gamma^\infty(\Lambda^{0,1}(M)) \to \Gamma^\infty(\Lambda^{1,1}(M))$$

as follows. Let η be a $(0,1)$-form. Then $\partial_b \eta$ is the unique $(1,1)$-form on M coinciding with $d\eta$ on $T_{1,0}(M) \otimes T_{0,1}(M)$. ■

Definition 7.37 A C^2 function $u : M \to \mathbb{R}$ is said to be $\overline{\partial}_b$-*pluriharmonic* if $\partial_b \overline{\partial}_b u = 0$. ■

Cf. [102] or Section 5.6 in [24], p. 112. The notion of a $\overline{\partial}_b$-pluriharmonic function admits a natural generalization to smooth maps $\phi : M \to S$ with values in a Riemannian manifold.

Definition 7.38 (R. Petit, [255]) The *second fundamental form* of ϕ is given by

$$\beta_\phi(X, Y) = (\phi^{-1}\nabla^h)_X \phi_* Y - \phi_* \nabla_X Y, \quad X, Y \in \mathcal{X}(M). \tag{7.43}$$

■

As to the notations adopted in (7.43), ∇^h is the Levi-Civita connection of (S,h), ∇ is the Tanaka-Webster connection of (M,θ), and $\phi_* X$ is the cross-section in the pullback bundle $\phi^{-1}TS \to M$ given by $(\phi_* X)_x = (d_x\phi)X_x$ for any $x \in M$. Also $\phi^{-1}\nabla^h$ is the connection in $\phi^{-1}TS \to M$ induced by ∇^h i.e., locally

$$(\phi^{-1}\nabla^h)_{\partial_A} X_k = \frac{\partial \phi^j}{\partial x^A}\left(\Gamma^i_{jk} \circ \phi\right) X_i .$$

Here (U, x^A) and (ω, y^i) are local coordinate systems on M and S respectively (with $\phi(U) \subseteq \omega$), ∂_A is short for $\partial/\partial x^A$, $\phi^i = y^i \circ \phi$, and X_i is the natural lift of $\partial/\partial y^i$ i.e., $X_i(x) = (\partial/\partial y^i)_{\phi(x)}$ (so that $\{X_i : 1 \leq i \leq \nu\}$ is a local frame in $\phi^{-1}TS \to M$ defined on the open set $\phi^{-1}(V)$).

Definition 7.39 We say $\phi : M \to S$ is *$\overline{\partial}_b$-pluriharmonic* if

$$\beta_\phi(X,Y) + \beta_\phi(JX, JY) = 0 \tag{7.44}$$

for any $X, Y \in H(M)$. ∎

Equivalently $\phi : M \to S$ is $\overline{\partial}_b$-pluriharmonic if and only if $\beta_\phi(Z, \overline{W}) = 0$ for any $Z, W \in T_{1,0}(M)$. This may be locally written

$$(\partial_b \overline{\partial}_b \phi^i)(Z, \overline{W}) + Z(\phi^j)\overline{W}(\phi^k)\, \Gamma^i_{jk} \circ \phi = 0$$

hence if $S = \mathbb{R}^\nu$ then $\partial_b \overline{\partial}_b \phi^i = 0$ i.e., each ϕ^i is a $\overline{\partial}_b$-pluriharmonic function. We may state

Proposition 7.40 (Y. Kamishima et al., [103]) *Let M be a strictly pseudoconvex CR manifold and S a Riemannian manifold. Then every $\overline{\partial}_b$-pluriharmonic map $\phi : M \to S$ is a pseudoharmonic map.*

For a proof one may see [103].

As largely reported on in this monograph, the theory of harmonic vector fields on a Riemannian manifold M was started by G. Wiegmink, [309], and C.M. Wood, [316], starting from the observation that the total space $T(M)$ of the tangent bundle over a Riemannian manifold (M,g) carries a Riemannian metric G_s naturally associated to g (the Sasaki metric). Then one could consider the ordinary Dirichlet energy functional $E(X) = \frac{1}{2}\int_M \operatorname{trace}_g(X^* G_s)\, d\operatorname{vol}(g)$ defined on $C^\infty(M, T(M))$. As it turned out, a vector field $X : M \to T(M)$ is a harmonic map, i.e., a critical point of E for *arbitrary* smooth 1-parameter variations of X if and only if X is absolutely parallel and the same result is achieved when looking for critical points of E restricted to the space of all smooth vector fields $\mathfrak{X}(M)$

(cf. Chapter 2 of this monograph). A new and wider notion of harmonicity was however obtained by looking at *unit* vector fields X and by restricting oneself to variations of X through unit vector fields. A rather different theory (of harmonic vector fields) was seen to arise, aspects of which (e.g., stability of Hopf vector fields on spheres, the interplay with contact geometry) were subsequently investigated by many authors such as F.C. Brito, [71], D-S. Han et al., [157], A. Higuchi et al., [164], C. Oniciuc, [230], D. Perrone, [237]-[247], and A. Yampolsky, [323]. A similar approach also led to the more general theory of harmonic sections in vector bundles (cf. K. Hasegawa, [159], J.J. Konderak, [193]). Their main results were reported on in Chapters 2 to 5 of this book.

Inspired by the geometric interpretation of subelliptic harmonic maps (in terms of the Fefferman metric, cf. [25]) together with the extension of the harmonic vector field theory to semi-Riemannian geometry (cf. O. Gil-Medrano et al., [130]) D. Perrone et al. studied a subelliptic analog to harmonic vector fields, cf. [107]. There one considered vector fields $X \in H(M)$ on a strictly pseudoconvex CR manifold M endowed with a contact form θ (with G_θ positive definite) such that $G_\theta(X,X) = 1$ and the horizontal lift $X^\uparrow : C(M) \to T(C(M))$ (with respect to the connection 1-form $\sigma \in \Gamma^\infty(T^*(C(M)) \otimes L(S^1))$ in C.R. Graham, [150]) is harmonic with respect to the Fefferman metric F_θ (which is a Lorentzian metric on $C(M)$, cf. [203]). By a result in [107] any such X was seen to satisfy

$$-\Delta_b X + 4\nabla_T JX + 2\tau JX + 6\phi JX = \lambda(X)\, X \tag{7.45}$$

where

$$\lambda(X) = -\|\pi_H \nabla X\|^2 + 4g_\theta(\nabla_T JX, X) + 2g_\theta(\tau JX, X) + 6g_\theta(\phi JX, X).$$

See also (7.32) in this chapter. This is a nonlinear subelliptic system of variational origin (actually (7.45) are the Euler-Lagrange equations associated to the functional $\mathcal{B}_\theta(X) = -\frac{1}{2}\int_M \lambda(X)\,\theta \wedge (d\theta)^n$, cf. Theorem 7.32 in this chapter) yet formally rather dissimilar from the harmonic vector fields system in Riemannian geometry. In the present chapter we build (following [103]) another subelliptic analog to the theory of harmonic vector fields, starting from the functional (7.42) restricted to the space of all unit vector fields (with respect to the Webster metric g_θ) and allowing only for variations through unit vector fields.

7.6. THE PSEUDOHERMITIAN BIEGUNG

7.6.1. Total Bending

Related to (7.40), we consider the sublaplacian on vector fields. Let M be a strictly pseudoconvex CR manifold and θ a contact form on M such that G_θ is positive definite. If X is a C^2 vector field on M then $\Delta_b X$ is the vector field locally given by

$$(\Delta_b X)^i = \Delta_b X^i - 2a^{jk}\Gamma^i_{j\ell}\frac{\partial X^\ell}{\partial x^k} - a^{jk}\left(\frac{\partial \Gamma^i_{j\ell}}{\partial x^k} - \Gamma^s_{j\ell}\Gamma^i_{ks} + \Gamma^s_{jk}\Gamma^i_{s\ell}\right)X^\ell \quad (7.46)$$

where $X = X^i\,\partial/\partial x^i$ and Γ^i_{jk} are the coefficients of the Tanaka-Webster connection ∇ of (M,θ) with respect to the local coordinate system (U, x^i) on M. Also $a^{ij} = g^{ij} - T^i T^j$ and $[g^{ij}] = [g_{ij}]^{-1}$ where $g_{ij} = g_\theta(\partial_i, \partial_j)$, $\partial_i = \partial/\partial x^i$. Let

$$\mathcal{U}(M,\theta) = S(M, g_\theta) = \{X \in \mathcal{X}^\infty(M) : g_\theta(X,X) = 1\}$$

be the set of all C^∞ unit vector fields on (M, g_θ).

Definition 7.41 The pseudohermitian *biegung* (or *total bending*) is the functional $\mathcal{B}_H : \mathcal{U}(M,\theta) \to [0, +\infty)$ given by

$$\mathcal{B}_H(X) = \frac{1}{2}\int_M \|\nabla^H X\|^2 \,\Psi, \quad X \in \mathcal{U}(M,\theta). \quad (7.47)$$

Here $\nabla^H X \in \Gamma^\infty(H(M)^* \otimes T(M))$ is the restriction of ∇X to $H(M)$. ■

The biegung (7.47) is a pseudohermitian analog to R. Wiegmink's total bending (cf. [309]) of a vector field on a Riemannian manifold (and $\mathcal{B}_H(X)$ measures the failure of X to satisfy $\nabla_Y X = 0$ for any $Y \in H(M)$). We adopt the following definition.

Definition 7.42 A *subelliptic harmonic vector field* is a C^∞ unit vector field $X \in \mathcal{U}(M,\theta)$ which is a critical point of $\mathcal{B}_H$ with respect to 1-parameter variations of X through unit vector fields. ■

For simplicity, we assume that M is compact (otherwise we may modify the definition (7.47) by integrating over a relatively compact domain $\Omega \subset M$ and consider only variations supported in Ω). Subelliptic harmonic vector fields will be shown to satisfy the nonlinear subelliptic system

$$\Delta_b X - \|\nabla^H X\|^2 X = 0, \quad (7.48)$$

(the Euler-Lagrange equations of the constrained variational principle associated to (7.47)). The pseudohermitian biegung (7.47) is related to the functional (7.42). To see this we need the CR analog to the Sasaki metric.

7.6.2. Tangent Bundles over CR Manifolds

Let $\pi^{-1}TM \to T(M)$ be the pullback of the tangent bundle, where $\pi : T(M) \to M$ is the projection. If X is a vector field on M, then $\hat{X} = X \circ \pi$ is as usual the natural lift of X. Let θ be a contact form on M with G_θ positive definite. The Tanaka-Webster connection ∇ of (M, θ) induces a connection $\hat{\nabla}$ in $\pi^{-1}TM \to T(M)$ which one describes as customary in local coordinates. Let $(U, \tilde{x}^i)$ be a local coordinate system on M and $(\pi^{-1}(U), x^i, y^i)$ the naturally induced local coordinates on $T(M)$. Let X_i be the natural lifts of $\partial/\partial\tilde{x}^i$. Let Γ^i_{jk} be the local coefficients of ∇ with respect to $(U, \tilde{x}^i)$. Then $\hat{\nabla}$ is locally given by

$$\hat{\nabla}_{\partial_j} X_k = (\Gamma^i_{jk} \circ \pi) X_i, \quad \hat{\nabla}_{\dot{\partial}_j} X_k = 0, \tag{7.49}$$

where $\partial_i = \partial/\partial x^i$ and $\dot{\partial}_i = \partial/\partial y^i$ for simplicity. The following constructions (leading to the CR analog S_θ of the Sasaki metric) are formally similar to those in Chapter 1 (and reported here for the convenience of the reader). Let $\mathcal{L} = y^i X_i$ be the Liouville vector. A tangent vector field $\mathcal{X}$ on $T(M)$ is *horizontal* if $\hat{\nabla}_{\mathcal{X}} \mathcal{L} = 0$. A calculation based on (7.49) shows that $\mathcal{X} = \mathcal{X}^i \partial_i + \mathcal{X}^{i+m} \dot{\partial}_i$ is horizontal if and only if $\mathcal{X}^{i+m} = -N^i_j \mathcal{X}^j$. Here $N^i_j = \Gamma^i_{jk} y^k$ and $m = 2n+1$. Let

$$\mathcal{H}_u = \{\mathcal{X}_u : \mathcal{X} \text{ horizontal}\}, \quad u \in T(M).$$

Then

$$\delta_i = \partial_i - N^i_j \dot{\partial}_j, \quad 1 \le i \le m, \tag{7.50}$$

is a local frame of $\mathcal{H} \to T(M)$ on $\pi^{-1}(U)$ hence $\mathcal{H}$ is a C^∞ distribution of rank m on $T(M)$ and

$$T(T(M)) = \mathcal{H} \oplus \operatorname{Ker}(d\pi). \tag{7.51}$$

Thus the restriction to $\mathcal{H}$ of

$$L : T(T(M)) \to \pi^{-1}TM, \quad L_u \mathcal{X} = (d_u \pi) \mathcal{X},$$
$$\mathcal{X} \in T_u(T(M)), \quad u \in T(M),$$

is a bundle isomorphism whose inverse is denoted by $\beta : \pi^{-1}TM \to \mathcal{H}$ (the *horizontal lift* associated to ∇). Let $\gamma : \pi^{-1}TM \to \text{Ker}(d\pi)$ be the *vertical lift* i.e., locally $\gamma X_i = \dot{\partial}_i$. The *Dombrowski map* is the bundle morphism

$$K : T(T(M)) \to \pi^{-1}TM, \quad K = \gamma^{-1} \circ Q,$$

where $Q : T(T(M)) \to \text{Ker}(d\pi)$ is the projection associated to the decomposition (7.51). The given data induces a Riemannian metric S_θ on $T(M)$ given by

$$S_\theta(\mathcal{X},\mathcal{Y}) = g_\theta(L\mathcal{X},L\mathcal{Y}) + g_\theta(K\mathcal{X},K\mathcal{Y}), \quad \mathcal{X},\mathcal{Y} \in T(T(M)).$$

As well as in Riemannian geometry (cf. D.E. Blair, [42]), S_θ is referred to as the *Sasaki metric* of (M,θ). The total space of the tangent bundle of a strictly pseudoconvex CR manifold possesses a rich geometric structure whose investigation is (as opposed to the Riemannian case, cf. [42] and references therein) far from being complete. For instance, note that the Riemannian manifold $(T(M), S_\theta)$ carries the compatible almost complex structure

$$J(\beta X) = \gamma X, \quad J(\gamma X) = -\beta X, \quad X \in \pi^{-1}TM.$$

A simple calculation shows that the Nijenhuis tensor field of J is given by

$$N_J(\beta X, \beta Y) = \gamma R(X,Y)\mathcal{L} + \beta T(X,Y),$$
$$N_J(\gamma X, \beta Y) = \beta R(X,Y)\mathcal{L} - \gamma T(X.Y),$$
$$N_J(\gamma X, \gamma Y) = -\gamma R(X,Y)\mathcal{L} - \beta T(X.Y),$$

for any $X, Y \in \pi^{-1}TM$. Here

$$R(X,Y)Z = R^{\hat{\nabla}}(\beta X, \beta Y)Z, \quad T(X,Y) = T^{\hat{\nabla}}(\beta X, \beta Y).$$

Also $R^{\hat{\nabla}}$ is the curvature tensor field of $\hat{\nabla}$ and $T^{\hat{\nabla}}$ is defined by

$$T^{\hat{\nabla}}(\mathcal{X},\mathcal{Y}) = \hat{\nabla}_{\mathcal{X}} L\mathcal{Y} - \hat{\nabla}_{\mathcal{Y}} L\mathcal{X} - L[\mathcal{X},\mathcal{Y}]$$

for any tangent vector fields $\mathcal{X},\mathcal{Y}$ on $T(M)$. As the Tanaka-Webster connection has torsion, J is never integrable.

7.6.3. The First Variation Formula

Let us consider the functional $E : C^\infty(M, T(M)) \to \mathbb{R}$ given by

$$E(\phi) = \frac{1}{2} \int_M \text{trace}_{G_\theta} \left(\pi_H \phi^* S_\theta \right) \Psi$$

where $\pi_H \phi^* S_\theta$ denotes the restriction of the bilinear form $\phi^* S_\theta$ to $H(M) \otimes H(M)$. We shall show that

Theorem 7.43 *Let M be a compact strictly pseudoconvex CR manifold and θ a contact form with G_θ positive definite. Let X be a smooth vector field on M. Then*

$$E(X) = n\,\mathrm{Vol}(M,\theta) + \mathcal{B}_H(X), \tag{7.52}$$

where $\mathrm{Vol}(M,\theta) = \int_M \Psi$. *Consequently* i) $E(X) \geq n\,\mathrm{Vol}(M,\theta)$ *with equality if and only if* $\nabla^H X = 0$. *Also* ii) $X : (M,\theta) \to (T(M), S_\theta)$ *is a pseudoharmonic map if and only if* $\nabla^H X = 0$. *Let us assume additionally that* $X \in \mathcal{U}(M,\theta)$ *and let* $\mathcal{X} : M \times (-\delta,\delta) \to \mathcal{U}(M,\theta)$ *be a smooth* 1*-parameter variation of X through unit vector fields* ($\mathcal{X}(x,0) = X(x)$, $x \in M$) *and let us set* $V = (\partial X_t/\partial t)_{t=0}$ *where* $X_t(x) = \mathcal{X}(x,t)$, $x \in M$, $|t| < \delta$. *Then* iii) $g_\theta(V,X) = 0$ *and*

$$\frac{d}{dt}\{E(X_t)\}_{t=0} = \int_M g_\theta(V, \Delta_b X)\,\Psi. \tag{7.53}$$

Consequently iv) *a C^∞ unit vector field X on M is a subelliptic harmonic vector field if and only if X is a C^∞ solution to* (7.48).

Statement (ii) extends a result by T. Ishihara, [176], and O. Nouhaud, [223], to the subelliptic case.

Proof of Theorem 7.43. Let $x \in M$ and $\{E_a : 1 \leq a \leq 2n\}$ be a local orthonormal (with respect to G_θ) frame of $H(M)$, defined on an open neighborhood of x. If $E_a = \lambda_a^j \partial/\partial \tilde{x}^j$ then

$$(d_x X) E_{a,x} = \lambda_a^j(x)\{\delta_j + [(\nabla_j X^i) \circ \pi]\dot{\partial}_i\}_{X(x)}$$

where δ_j are given by (7.50) and $\nabla_j X^i = \partial X^i/\partial \tilde{x}^j + \Gamma^i_{jk} X^k$. Then

$$\left(\mathrm{trace}_{G_\theta} \pi_H X^* S_\theta\right)_x = \sum_{a=1}^{2n} (X^* S_\theta)(E_a, E_a)_x$$

$$= \sum_a S_{\theta, X(x)}((d_x X) E_{a,x}, (d_x X) E_{a,x})$$

$$= \sum_a \lambda_a^j(x) \lambda_a^k(x) \{S_\theta(\delta_j, \delta_k) + (\nabla_j X^r)(\nabla_k X^s) S_\theta(\dot{\partial}_r, \dot{\partial}_s)\}_{X(x)}$$

$$= \sum_a \{g_\theta(E_a, E_a) + \lambda_a^j \lambda_a^k (\nabla_j X^r)(\nabla_k X^s) g_{rs}\}_x.$$

Let $T = T^i \partial/\partial\tilde{x}^i$. As $\sum_{a=1}^{2n} \lambda_a^i \lambda_a^j = g^{ij} - T^i T^j$ it follows that

$$\mathrm{trace}_{G_\theta}(\pi_H X^* S_\theta) = 2n + \|\nabla X\|^2 - \|\nabla_T X\|^2$$

where $\|\nabla X\|^2 = g^{ij}(\nabla_i X^k)(\nabla_j X^\ell) g_{k\ell}$. As $\{E_j : 0 \le j \le 2n\}$ (with $E_0 = T$) is a local g_θ-orthonormal frame of $T(M)$ one also has $\|\nabla X\|^2 = \sum_{j=0}^{2n} g_\theta(\nabla_{E_j} X, \nabla_{E_j} X)$ hence (7.52). Clearly (7.52) yields statement (i) in Theorem 7.43.

Let us prove (ii). If $\nabla^H X = 0$ then X is a pseudoharmonic map and an absolute minimum for E in $\Gamma^\infty(M, T(M))$. Vice versa, let us assume X is a pseudoharmonic map of M into the Riemannian manifold $(T(M), S_\theta)$. Thus $\{dE(X_t)/dt\}_{t=0} = 0$ for any smooth 1-parameter variation $X_t : M \to T(M)$ of X ($X_0 = X$). In particular for the variation $X_t(x) = (1-t)X_x$, $x \in M$, $|t| < \epsilon$ (by (7.52))

$$0 = \left.\frac{dE(X_t)}{dt}\right|_{t=0} = \frac{d}{dt}\{n\,\mathrm{Vol}(M,\theta) + \mathcal{B}_H(X_t)\}_{t=0}$$

$$= \frac{d}{dt}\left\{\frac{(1-t)^2}{2}\int_M \|\nabla^H X\|^2 \Psi\right\}_{t=0} = -\int_M \|\nabla^H X\|^2 \Psi.$$

Let $X \in \mathcal{U}(M,\theta)$. To prove the first variation formula (7.53), we need some preparation. Let $N = M \times (-\delta, \delta)$ and let $p : N \to M$ be the projection. Let $p^{-1}TM \to N$ be the pullback of the tangent bundle $T(M) \to M$ by p. Then $\mathcal{X}$ may be thought of as a C^∞ section in $p^{-1}TM \to N$. If Y is a tangent vector field on M we set $\hat{Y} = Y \circ p$. The Webster metric g_θ induces a bundle metric $\hat{g}_\theta$ in $p^{-1}TM \to N$ uniquely determined by $\hat{g}_\theta(\hat{Y}, \hat{Z}) = g_\theta(Y, Z) \circ p$. Also let D be the connection in $p^{-1}TM \to N$ induced by the Tanaka-Webster connection ∇. Precisely let $\tilde{Y}$ be the tangent vector field on $T(M)$ given by

$$\tilde{Y}_{(x,t)} = (d_x i_t) Y_x, \quad x \in M, \quad |t| < \delta,$$

where $i_t : M \to N$, $i_t(x) = (x,t)$. Then D is determined by

$$D_{\tilde{Y}} \hat{Z} = \widehat{\nabla_Y Z}, \quad D_{\partial/\partial t} \hat{Z} = 0, \quad Y, Z \in T(M).$$

Moreover a simple calculation shows that $D\hat{g}_\theta = 0$ and

$$(D_{\tilde{Y}} \mathcal{X})_{(x,t)} = (\nabla_Y X_t)_x, \quad (x,t) \in N.$$

Then

$$\mathcal{B}_H(X_t) = \frac{1}{2}\int_M \sum_{a=1}^{2n} \hat{g}_\theta(\nabla_{E_a} X_t, \nabla_{E_a} X_t)_x \Psi(x)$$
$$= \frac{1}{2}\int_M \sum_a \hat{g}_\theta(D_{\tilde{E}_a}\mathcal{X}, D_{\tilde{E}_a}\mathcal{X})_{(x,t)} \Psi(x)$$

hence

$$\frac{d}{dt}\mathcal{B}_H(X_t) = \int_M \sum_a \hat{g}_\theta(D_{\partial/\partial t} D_{\tilde{E}_a}\mathcal{X}, D_{\tilde{E}_a}\mathcal{X})_{(x,t)} \Psi(x)$$
$$= \int_M \sum_a \hat{g}_\theta(D_{\tilde{E}_a} D_{\partial/\partial t}\mathcal{X}, D_{\tilde{E}_a}\mathcal{X})_{(x,t)} \Psi(x)$$

as $R^D(\partial/\partial t, \tilde{E}_a)\mathcal{X} = 0$ and $[\partial/\partial t, \tilde{E}_a] = 0$. Moreover (by $D\hat{g}_\theta = 0$)

$$\frac{d}{dt}\mathcal{B}_H(X_t) = \int_M \sum_a \{\tilde{E}_a(\hat{g}_\theta(D_{\partial/\partial t}\mathcal{X}, D_{\tilde{E}_a}\mathcal{X}))$$
$$- \hat{g}_\theta(D_{\partial/\partial t}\mathcal{X}, D_{\tilde{E}_a} D_{\tilde{E}_a}\mathcal{X})\}_{(x,t)} \Psi(x).$$

For each fixed $|t| < \delta$ we define $Y_t \in H(M)$ by setting

$$G_\theta(Y_t, Y)_x = \hat{g}_\theta(D_{\partial/\partial t}\mathcal{X}, D_{\tilde{Y}}\mathcal{X})_{(x,t)}$$

for any $Y \in H(M)$ and any $x \in M$. Then (by $\nabla g_\theta = 0$)

$$\tilde{E}_a(\hat{g}_\theta(D_{\partial/\partial t}\mathcal{X}, D_{\tilde{E}_a}\mathcal{X})) = E_a(g_\theta(Y_t, E_a))$$
$$= g_\theta(\nabla_{E_a} Y_t, E_a) + g_\theta(Y_t, \nabla_{E_a} E_a).$$

As $\nabla\Psi = 0$ the divergence operator (see Section 2) is also given by

$$\operatorname{div}(Y) = \operatorname{trace}\{Z \mapsto \nabla_Z Y\} = \sum_{j=0}^{2n} g_\theta(\nabla_{E_j} Y, E_j).$$

Finally (by Green's lemma)

$$\frac{d}{dt}\{E(X_t)\}_{t=0} = -\int_M \hat{g}_\theta(D_{\partial/\partial t}\mathcal{X}, \sum_a \{D_{\tilde{E}_a} D_{\tilde{E}_a}\mathcal{X} - D_{\widetilde{\nabla_{E_a} E_a}}\mathcal{X}\})_{(x,0)}\Psi(x)$$

$$= \int_M g_\theta(V, \Delta_b X)_x \Psi(x)$$

and (7.53) is proved. If X is a critical point i.e., $\{dE(X_t)/dt\}_{t=0} = 0$ then (7.53) together with the constraint $g_\theta(V, X) = 0$ (obtained by differentiating $g_\theta(X_t, X_t) = 1$ at $t = 0$) imply that $\Delta_b X = -\lambda X$ for some $\lambda \in C^\infty(M)$ and taking the inner product with X shows that $\lambda = g_\theta(\Delta_b X, X) = \|\nabla^H X\|^2$. Theorem 7.43 is proved.

See also B. Franchi & E. Serra, [120], who studied subelliptic harmonic vector fields on domains in $\mathbb{R}^2$. Precisely [120] considers the Dirichlet problem $-HU = |XU|^2 U$ in Ω, $|U| = 1$ in Ω, and $U = U_0$ on $\partial\Omega$, where $\Omega \subset \mathbb{R}^2$ is a bounded open set whose boundary $\partial\Omega$ is a smooth simple closed regular curve, $H = X_1^2 + X_2^2$ is the Hörmander operator (sum of squares of vector fields) associated to the Hörmander system $X = \{X_1, X_2\}$ given by $X_1 = \partial/\partial x^1$, $X_2 = x^1 \partial/\partial x^2$, and $U_0 : \partial\Omega \to S^1$ is a smooth function. The authors establish uniqueness and smoothness of the solution U to the above problem (by a lifting argument relying essentially on their assumption $\deg(U_0, \partial\Omega) = 0$). ■

7.6.4. Unboundedness of the Energy Functional

Under the assumptions of Theorem 7.43, we may prove the following

Corollary 7.44 *The characteristic direction T of $d\theta$ is a subelliptic harmonic vector field and an absolute minimum of the energy functional $E : \mathcal{U}(M, \theta) \to [0, +\infty)$. Moreover, for any nonempty open subset $\Omega \subseteq M$ and any unit vector field X on M such that $X \in H(M)$ there is a sequence $\{Y_\nu\}_{\nu \geq 1}$ of unit vector fields such that each Y_ν coincides with X outside Ω and $E(Y_\nu) \to \infty$ for $\nu \to \infty$. In particular, the energy functional E is unbounded from above.*

Proof. The first statement in Corollary 7.44 follows from $\nabla T = 0$ (and then $E(T) = \inf_{X \in \mathcal{U}(M,\theta)} E(X) = n\mathrm{Vol}(M, \theta)$). To prove the second statement let $h = (x^1, \ldots, x^m) : U \to \mathbb{R}^m$ be a local coordinate system on M such that

$U \subseteq \Omega$, $h(U) \supset [-2\pi, 2\pi]^m$ and $X = \partial/\partial x^1$ on U (cf. the proof of the classical Frobenius theorem, e.g., [235], p. 91-92). Moreover, let $\varphi \in C_0^\infty(M)$ be a test function such that i) $0 \le \varphi(x) \le 1$ for any $x \in M$, ii) $\varphi = 1$ in a neighborhood V of the compact set $K = h^{-1}([-\pi, \pi]^m)$ such that $\overline{V} \subset U$, and iii) $\varphi = 0$ outside $h^{-1}([-2\pi, 2\pi]^m)$. For each $\nu \in \mathbb{Z}$, $\nu \ge 1$, let f_ν be the C^∞ extension to M of the function $\sin(\nu x^1)$ (thought of as defined on the closed set $\overline{V}$) and let us set $\alpha_\nu = \varphi f_\nu$. Next let us consider the C^∞ vector field

$$Y_\nu = (\cos \alpha_\nu) X + (\sin \alpha_\nu) T, \quad \nu \ge 1.$$

Then Y_ν is a unit vector field coinciding with X outside Ω. As we may complete X to a local frame of $H(M)$ (and $\nabla T = 0$, $\theta(\nabla_X X) = 0$)

$$\|\nabla^H Y_\nu\|^2 \ge g_\theta(\nabla_X Y_\nu, \nabla_X Y_\nu) = X(\alpha_\nu)^2 + (\cos^2 \alpha_\nu)\|\nabla_X X\|^2 \ge X(\alpha_\nu)^2.$$

On the other hand $X(\alpha_\nu) = X(\varphi) f_\nu + \varphi \nu (\cos \nu x^1)$ on U so that $X(\alpha_\nu) = \nu \cos \nu x^1$ on $V \supset K$. Hence

$$2E(Y_\nu) \ge \int_K \|\nabla^H Y_\nu\|^2 \Psi \ge \int_K X(\alpha_\nu)^2 \Psi = \nu^2 \int_K \cos^2(\nu x^1)\, \Psi.$$

If $d\,\mathrm{vol}(g_\theta) = \sqrt{G(x)}\, dx^1 \wedge \cdots \wedge dx^m$ is the Riemannian volume form of (M, g_θ) (with $G(x) = \det[g_{ij}(x)]$) there is a constant $c_n > 0$ such that $\Psi = c_n\, d\,\mathrm{vol}(g_\theta)$ (and $c_n = 2^n n!$, cf. [293]). Let us set $a = \inf_{x \in K} \sqrt{G(x)}$. Then $a > 0$ and

$$\int_K \cos^2(\nu x^1) \Psi \ge a c_n \int_{[-\pi,\pi]^m} \cos^2(\nu t^1)\, dt^1 \cdots dt^m = a 2^{n-1} n!\, (2\pi)^m.$$

Hence $E(Y_\nu) \ge a 2^{n-2} n!\, (2\pi)^m \nu \to \infty$ for $\nu \to \infty$. ■

7.7. THE SECOND VARIATION FORMULA

Let $X \in \mathcal{U}(M, \theta)$ and let us consider a smooth 2-parameter variation of X

$$\mathcal{Y} : M \times I_\delta^2 \to T(M), \quad I_\delta = (-\delta, \delta), \quad \delta > 0,$$

$$X_{t,s} = \mathcal{Y} \circ i_{t,s}, \quad t, s \in I_\delta, \quad X_{0,0} = X.$$

Here we set $N = M \times I_\delta^2$ and $i_{t,s} : M \to N$, $i_{t,s}(x) = (x, t, s)$ for any $x \in M$. We shall prove the following

Theorem 7.45 *Let* $V = (\partial X_{t,s}/\partial t)_{t=s=0}$ *and* $W = (\partial X_{t,s}/\partial s)_{t=s=0}$. *Let us assume that* $X_{t,s} \in \mathcal{U}(M,\theta)$ *for any* $t,s \in I_\delta$. *If* $X \in \mathcal{U}(M,\theta)$ *is a smooth subelliptic harmonic vector field then*

$$\frac{\partial^2}{\partial t \partial s}\left\{\mathcal{B}_H(X_{t,s})\right\}_{t=s=0} = \int_M g_\theta(V, \Delta_b W - \|\nabla^H X\|^2 W)\Psi. \tag{7.54}$$

In particular for any smooth 1*-parameter variation of* X

$$\frac{d^2}{dt^2}\{\mathcal{B}_H(X_t)\}_{t=0} = \int_M \{\|\nabla^H V\|^2 - \|\nabla^H X\|^2 \|V\|^2\}\Psi. \tag{7.55}$$

The identity (7.54) is the *second variation formula* (of the pseudohermitian biegung). To prove Theorem 7.45, let $p : N \to M$ be the projection and $p^{-1}TM \to N$ the pullback of $T(M)$ by p. Then $\mathcal{Y}$ is a C^∞ section in $\pi^{-1}TM \to N$. Let $\hat{g}_\theta$ and D be respectively the Riemannian bundle metric induced by g_θ and the connection induced by the Tanaka-Webster connection ∇ in $\pi^{-1}TM \to N$. Similar to the conventions adopted in the proof of Theorem 7.43 we set

$$\tilde{Y}_{(x,t,s)} = (d_x i_{t,s})Y_x, \quad x \in M, \quad t,s \in I_\delta.$$

For simplicity we set $\mathbb{T} = \partial/\partial t$ and $\mathbb{S} = \partial/\partial s$ ($\mathbb{T},\mathbb{S} \in \mathcal{X}^\infty(N)$). Then (by $D\hat{g}_\theta = 0$)

$$\frac{\partial}{\partial t}\mathcal{B}_H(X_{t,s}) = \int_M \sum_{a=1}^{2n} \hat{g}_\theta(D_{\mathbb{T}} D_{\tilde{E}_a}\mathcal{Y}, D_{\tilde{E}_a}\mathcal{Y})\Psi = \int_M \sum_a \hat{g}_\theta(D_{\tilde{E}_a} D_{\mathbb{T}}\mathcal{Y}, D_{\tilde{E}_a}\mathcal{Y})$$

due to

$$[\mathbb{T}, \tilde{E}_a] = 0, \quad R^D(\mathbb{T}, \tilde{E}_a)\mathcal{Y} = 0.$$

Then

$$\frac{\partial^2}{\partial s \partial t}\mathcal{B}_H(X_{t,s}) = \int_M \sum_a \frac{\partial}{\partial s}\hat{g}_\theta(D_{\tilde{E}_a} D_{\mathbb{T}}\mathcal{Y}, D_{\tilde{E}_a}\mathcal{Y})\Psi$$

$$= \int_M \sum_a \left\{\hat{g}_\theta(D_{\mathbb{S}} D_{\tilde{E}_a} D_{\mathbb{T}}\mathcal{Y}, D_{\tilde{E}_a}\mathcal{Y}) + \hat{g}_\theta(D_{\tilde{E}_a} D_{\mathbb{T}}\mathcal{Y}, D_{\mathbb{S}} D_{\tilde{E}_a}\mathcal{Y})\right\}\Psi \tag{7.56}$$

(as $[\mathbb{S}, \tilde{E}_a] = 0$ and $R^D(\mathbb{S}, \tilde{E}_a)\mathcal{Y} = 0$)

$$= \int_M \sum_a \left\{ \hat{g}_\theta(D_{\tilde{E}_a} D_{\mathbb{S}} D_{\mathbb{T}} \mathcal{Y}, D_{\tilde{E}_a}\mathcal{Y}) + \hat{g}_\theta(D_{\tilde{E}_a} D_{\mathbb{T}}\mathcal{Y}, D_{\tilde{E}_a} D_{\mathbb{S}}\mathcal{Y}) \right\} \Psi$$

$$= \int_M \sum_a \Big\{ \tilde{E}_a(\hat{g}_\theta(D_{\mathbb{S}} D_{\mathbb{T}}\mathcal{Y}, D_{\tilde{E}_a}\mathcal{Y})) - \hat{g}_\theta(D_{\mathbb{S}} D_{\mathbb{T}}\mathcal{Y}, D_{\tilde{E}_a} D_{\tilde{E}_a}\mathcal{Y})$$

$$+ \tilde{E}_a(\hat{g}_\theta(D_{\mathbb{T}}\mathcal{Y}, D_{\tilde{E}_a} D_{\mathbb{S}}\mathcal{Y})) - \hat{g}_\theta(D_{\mathbb{T}}\mathcal{Y}, D_{\tilde{E}_a} D_{\tilde{E}_a} D_{\mathbb{S}}\mathcal{Y}) \Big\} \Psi.$$

For each fixed $(t,s) \in I_\delta^2$ we define $Y_{t,s} \in H(M)$ by

$$G_\theta(Y_{t,s}, Z)_x = \hat{g}_\theta(D_{\mathbb{S}} D_{\mathbb{T}}\mathcal{Y}, D_{\tilde{Z}}\mathcal{Y})_{(x,t,s)}, \quad Z \in H(M).$$

Then

$$\sum_a \tilde{E}_a(\hat{g}_\theta(D_{\mathbb{S}} D_{\mathbb{T}}\mathcal{Y}, D_{\tilde{E}_a}\mathcal{Y})) = \sum_a E_a(g_\theta(Y_{t,s}, E_a)) \circ p$$

$$= \sum_a \left\{ g_\theta(\nabla_{E_a} Y_{t,s}, E_a) + g_\theta(Y_{t,s}, \nabla_{E_a} E_a) \right\} \circ p$$

$$= \operatorname{div}(Y_{t,s}) \circ p + \hat{g}_\theta\left(D_{\mathbb{S}} D_{\mathbb{T}}\mathcal{Y}, \sum_a D_{\widetilde{\nabla_{E_a} E_a}}\mathcal{Y} \right).$$

Similarly, given $Z_{t,s} \in H(M)$ determined by

$$G_\theta(Z_{t,s}, Z)_x = \hat{g}_\theta(D_{\mathbb{T}}\mathcal{Y}, D_{\tilde{Z}} D_{\mathbb{S}}\mathcal{Y})_{(x,t,s)}$$

one has

$$\sum_a \tilde{E}_a\left(\hat{g}_\theta\left(D_{\mathbb{T}}\mathcal{Y}, D_{\tilde{E}_a} D_{\mathcal{S}}\mathcal{Y} \right) \right)$$

$$= \operatorname{div}(Z_{t,s}) \circ p + \hat{g}_\theta\left(D_{\mathbb{T}}\mathcal{Y}, \sum_a D_{\widetilde{\nabla_{E_a} E_a}} D_{\mathbb{S}}\mathcal{Y} \right).$$

Going back to (7.56) one has (by Green's lemma)

$$\frac{\partial^2}{\partial s \partial t}\left\{\mathcal{B}_H(X_{t,s})\right\}_{t=s=0}$$

$$= \int_M \Big\{ \hat{g}_\theta \Big(D_{\mathbb{S}} D_{\mathbb{T}} \mathcal{Y}, \sum_a \{ D_{\widetilde{\nabla_{E_a} E_a}} \mathcal{Y} - D_{\tilde{E}_a} D_{\tilde{E}_a} \mathcal{Y} \} \Big)$$

$$+ \hat{g}_\theta \Big(D_{\mathbb{T}} \mathcal{Y}, \sum_a \{ D_{\widetilde{\nabla_{E_a} E_a}} D_{\mathbb{S}} \mathcal{Y} - D_{\tilde{E}_a} D_{\tilde{E}_a} D_{\mathbb{S}} \mathcal{Y} \} \Big) \Big\}_{t=s=0} \Psi$$

$$= \int_M \{ g_\theta(U, \Delta_b X) + g_\theta(V, \Delta_b W) \} \Psi$$

where we have set $U = (\partial^2 X_{t,s}/\partial t \partial s)_{t=s=0}$. Moreover (by differentiating $\hat{g}_\theta(\mathcal{Y}, \mathcal{Y}) = 1$)

$$g_\theta(U, X)_x = \hat{g}_\theta(D_{\mathbb{S}} D_{\mathbb{T}} \mathcal{Y}, \mathcal{Y})_{(x,0,0)}$$
$$= \{ \mathbb{S}(\hat{g}_\theta(D_{\mathbb{T}}\mathcal{Y}, \mathcal{Y})) - \hat{g}_\theta(D_{\mathbb{T}}\mathcal{Y}, D_{\mathbb{S}}\mathcal{Y}) \}_{(x,0,0)} = -g_\theta(V, W)_x$$

and (as X is subelliptic harmonic i.e., a smooth solution to (7.48))

$$\frac{\partial^2}{\partial t \partial s}\left\{\mathcal{B}_H(X_{t,s})\right\}_{t=s=0} = \int_M \{ \|\nabla^H X\|^2 g_\theta(U, X) + g_\theta(V, \Delta_b W) \} \Psi$$
$$= \int_M g_\theta(V, \Delta_b W - \|\nabla^H X\|^2 W) \Psi$$

and (7.54) is proved. Finally given an arbitrary smooth 1-parameter variation $\mathcal{X} : M \times I_\delta \to T(M)$ of X through unit vector fields the identity (7.55) follows from (7.54) for the particular 2-parameter variation $\mathcal{Y} : M \times I_{\delta/2} \to T(M)$ given by $\mathcal{Y}(x, t, s) = \mathcal{X}(x, t + s)$ for any $x \in M$ and any $t, s \in I_{\delta/2}$. Indeed

$$\frac{d^2}{dt^2}\{\mathcal{B}_H(X_t)\}_{t=0} = \int_M g_\theta(V, \Delta_b V - \|\nabla^H X\|^2 V) \Psi. \qquad (7.57)$$

On the other hand, for any smooth vector field V on M

$$-\Delta_b\|V\|^2 = \sum_{a=1}^{2n}\{E_aE_a\|V\|^2 - (\nabla_{E_a}E_a)\|V\|^2\}$$

$$= 2\sum_a\{E_a(g_\theta(\nabla_{E_a}V, V)) - g_\theta(\nabla_{\nabla_{E_a}E_a}V, V)\}$$

$$= 2\sum_a\{g_\theta(\nabla_{E_a}\nabla_{E_a}V, V) + g_\theta(\nabla_{E_a}V, \nabla_{E_a}V) - g_\theta(\nabla_{\nabla_{E_a}E_a}V, V)\}$$

hence

$$\Delta_b\|V\|^2 = 2\{g_\theta(\Delta_b V, V) - \|\nabla^H V\|^2\}. \tag{7.58}$$

The identity (7.58) is an obvious pseudohermitian ananlog to (2.23) in Lemma 2.15 (cf. Chapter 2 of this monograph). Now (7.55) follows from (7.57)–(7.58) and Green's lemma.

CHAPTER EIGHT

Lorentz Geometry and Harmonic Vector Fields

Contents

This chapter is devoted to the study of harmonic vector fields on semi-Riemannian manifolds with an emphasis on the Lorentzian case. There are quite a few books on semi-Riemannian geometry. We follow mainly the conventions and notations in B. O'Neill, [229], and J.K. Beem & P.E. Ehrlich, [31]. For the general relativity notions one may consult R. Adler et al., [10], M. Göckeler & T. Schücker, [136], and Chapter IV in S. Ianuş, [174]. See also E. Barletta et al., [24], p. 37–43, for a foliation theoretic approach. The results we report on are mainly due to O. Gil-Medrano & A. Hurtado, [130].

8.1. A FEW NOTIONS OF LORENTZ GEOMETRY

Let (M,g) be an n-dimensional semi-Riemannian manifold of index $1 \leq \nu \leq n-1$. A tangent vector $v \in T_x(M) \setminus \{0_x\}$ is *timelike* (respectively *nonspacelike*, *null*, or *spacelike*) if $g_x(v,v) < 0$ (respectively if $g_x(v,v) \leq 0$, $g_x(v,v) = 0$, or $g_x(v,v) > 0$). A continuous vector field X on M is *timelike* if $g_x(X_x, X_x) < 0$ for any $x \in M$.

Let (M,g) be a Lorentzian manifold i.e., a semi-Riemannian manifold of index $\nu = 1$. In general (M,g) may fail to admit a globally defined timelike vector field. If (M,g) does admit a (globally defined) timelike vector field X then (M,g) is said to be *time oriented* by X. A time-oriented Lorentzian manifold is a *space-time*. In the spirit of semi-Riemannian geometry and in order to develop a theory suitable for applications in general relativity, a

© 2012 Elsevier Inc. All rights reserved.

space-time M is customarily assumed to be 4-dimensional and noncompact. In any case if M is compact then it cannot be simply connected. Indeed if M were compact and $\pi_1(M) = 0$ then the first Betti number would vanish ($H^1(M, \mathbb{R}) = 0$). By the Poincaré duality the third Betti number would vanish as well ($H^3(M, \mathbb{R}) = 0$). Consequently the Euler-Poincaré characteristic would be positive $\chi(M) > 0$ in contradiction with the fact that M carries a nowhere zero globally defined vector field.

On a space-time (M, g, X) nonspacelike tangent vectors are divided in two classes, as follows. Let $v \in T_x(M)$ be a nonspacelike vector i.e., $v \neq 0$ and $g_x(v, v) \leq 0$. Then v is *future directed* (respectively *past directed*) if $g_x(X_x, v) < 0$ (respectively if $g_x(X_x, v) > 0$).

The space of all Lorentz metrics on M is denoted by Lor(M). It is well known that the space of all Riemannian metrics on a compact manifold (thought of as carrying the compact-open topology) is contractible, as is the space of all Riemannian metrics of given total volume. Also (by the Gram-Schmidt process) it may be shown that if the manifold is parallelizable, then for any Riemannian metric there is a globally defined orthonormal frame. The same properties translated into the Lorentzian realm decidedly fail to hold. For instance on a compact manifold M the space Lor(M) has in general many connected components $\mathcal{C}$ with $\pi_1(\mathcal{C}) \neq 0$ (cf. P. Mounoud, [217]).

A smooth curve $C : (a, b) \to M$ in a semi-Riemannian manifold (M, g) is *timelike* (respectively *nonspacelike, null*, or *spacelike*) if its tangent vector $\dot{C}(t)$ is timelike (respectively nonspacelike, null, or spacelike) for any $a < t < b$. Let ∇ be the Levi-Civita connection of (M, g). Let $C : (a, b) \to M$ be a geodesic of ∇. Then

$$\frac{d}{dt}\left\{g_{C(t)}(\dot{C}(t), \dot{C}(t))\right\} = 2g_{C(t)}\left((\nabla_{\dot{C}}\dot{C})_{C(t)}, \dot{C}(t)\right) = 0.$$

Consequently if $\dot{C}(t_0)$ is timelike (respectively null or spacelike) for some $a < t_0 < b$ then $\dot{C}(t)$ is timelike (respectively null or spacelike) for any other value of the parameter $a < t < b$.

Let (M, g, X) be a space-time and $C : (a, b) \to M$ a nonspacelike curve (i.e., $g_{C(t)}(\dot{C}(t), \dot{C}(t)) \leq 0$ for any $a < t < b$). We say C is *future-directed* (respectively *past-directed*) if the tangent vector $\dot{C}(t)$ is future-directed (respectively past-directed) for any $a < t < b$.

Definition 8.1 Let (M, g) be a space-time and $x, y \in M$. We write $x \ll y$ if there is a smooth future-directed timelike curve joining x and y. Also we write $x \leq y$ if either $x = y$ or there is a smooth future-directed nonspacelike curve from x to y. The *chronological future* and the *chronological past* of x are respectively the sets

$$I^+(x) = \{y \in M : x \ll y\}, \quad I^-(x) = \{y \in M : y \ll x\}.$$

The *causal future* and *causal past* of x are respectively the sets

$$J^+(x) = \{y \in M : x \leq y\}, \quad J^-(x) = \{y \in M : y \leq x\}.$$

■

The relations $<<$ and $\leq$ on M are transitive. Also

$$x \ll y \text{ and } y \leq z \Longrightarrow x \ll z,$$

$$x \leq y \text{ and } y \ll z \Longrightarrow x \ll z,$$

cf. e.g., [31], p. 22. If there is a future-directed timelike curve joining x and y then there is an open neighborhood $U \subseteq M$ of y such that any point of U may be reached from x by a future-directed timelike curve (cf. [31], p. 22). Consequently, for any point $x \in M$ the sets $I^{\pm}(x)$ are open in M. As it may be shown by examples, in general the sets $J^{\pm}(x)$ are neither open nor closed.

When $x \in I^+(x)$ there is a timelike loop at x and the space-time (M, g) is said to have a *causality violation*.

Example 8.2 If the cylinder $M = S^1 \times \mathbb{R}$ is endowed with the Lorentzian metric $g = -d\theta \otimes d\theta + dt \otimes dt$ then the circles $t =$ constant are closed timelike curves in (M, g) hence $I^+(x) = M$ for any $x \in M$.

Definition 8.3 Let (M, g) be a space-time. We say (M, g) is *chronological* if there are no closed timelike curves in M i.e., $x \notin I^+(x)$ for any $x \in M$. Also (M, g) is referred to as *causal* if it contains no closed nonspacelike curves. ■

Example 8.4 Let $M = S^1 \times \mathbb{R}$ carry the Lorentzian metric $g = d\theta \odot dt = \frac{1}{2}(d\theta \otimes dt + dt \otimes d\theta)$. Then (M, g) is a chronological space-time which is not causal.

By a well-known result (cf. e.g., Proposition 2.6 in [31], p. 23), compact space-times do contain closed timelike curves and hence fail to be chronological.

Definition 8.5 Let (M, g) be a space-time. We say (M, g) is *distinguishing* if given $x, y \in M$

$$I^+(x) = I^+(y) \text{ or } I^-(x) = I^-(y) \Longrightarrow x = y.$$

■

Therefore in a distinguishing space-time, distinct points have distinct chronological futures and distinct chronological pasts.

8.2. ENERGY FUNCTIONALS AND TENSION FIELDS

We follow the exposition by O. Gil-Medrano & A. Hurtado, [130]. Let (M,g) and (N,h) be two semi-Riemannian manifolds and $\phi : M \to N$ a smooth map. Let $L_\phi \in \Gamma^\infty(T^*(M) \otimes T(M))$ be determined by

$$g(L_\phi X, Y) = (\phi^* h)(X,Y), \quad X, Y \in \mathfrak{X}(M).$$

Let then $e(\phi) \in C^\infty(M)$ be defined by $e(\phi) = (1/2)\,\text{trace}(L_\phi)$. If $\{E_i :\leq i \leq n\}$ is a local g-orthonormal frame defined on an open set $U \subseteq M$ then

$$e(\phi) = \frac{1}{2}\sum_{i=1}^{n} \epsilon_i\,(\phi^* h)(E_i, E_i)$$

on U where $\epsilon_i = g(E_i, E_i) \in \{\pm 1\}$. Let us assume that M is oriented. Given a relatively compact domain $\Omega \subset M$ we define the energy functional

$$E_\Omega(\phi) = \int_\Omega e(\phi)\, d\,\text{vol}(g).$$

For a local coordinate system (U, x^i) on M one of course has $d\,\text{vol}(g) = \sqrt{|a|}\, dx^1 \wedge \cdots \wedge dx^n$ on U where $a = \det[g_{ij}]$. The Euler-Lagrange equations corresponding to the variational principle $\delta\, E_\Omega(\phi) = 0$ are $\tau_g(\phi) = 0$ where $\tau_g(\phi) \in \Omega^0(\phi^{-1}TN)$ is locally given by

$$\tau_g(\phi) = \sum_{i=1}^{n} \epsilon_i \left\{ (\phi^{-1}\nabla^h)_{E_i} \phi_* E_i - \phi_* \nabla_{E_i} E_i \right\}$$

on U. Here ∇ and ∇^h are respectively the Levi-Civita connections of the semi-Riemannian manifolds (M,g) and (N,h) while $\phi^{-1}\nabla^h$ is the connection induced by ∇^h in the pullback bundle $\phi^{-1}TN \to M$.

As in Section 7.1 of this monograph we may consider the Sasaki metric G_s on $T(M)$ associated to the semi-Riemannian metric g. Then G_s is a semi-Riemannian metric of index 2ν. This is usually referred to as the *canonical metric* (in [107]) or the *Kaluza-Klein metric* (in [130]) on $T(M)$. Let $V : M \to T(M)$ be a smooth tangent vector field, thought of as a map of the semi-Riemannian manifolds (M,g) and $(T(M), G_s)$. It may be shown (by following the arguments in the positive definite case, cf. Proposition 2.3 in this monograph) that

$$(V^* G_s)(X,Y) = g(X,Y) + g(\nabla_X V, \nabla_Y V), \quad X,Y \in \mathfrak{X}(M).$$

Consequently

$$L_V = I + (\nabla V)^t \circ (\nabla V)$$

where the transpose is meant with respect to the pointwise inner product g (similar to (2.5) in Chapter 2). It follows that

$$E_\Omega(V) = \frac{n}{2}\mathrm{Vol}(\Omega) + \frac{1}{2}\int_\Omega g^*(\nabla V, \nabla V)\,\mathrm{vol}(g).$$

See also Section 7.3 of this book. By a result in [130] the tension field of V may be locally written

$$\tau_g(V) = \sum_{i=1}^n \left\{\epsilon_i \left(R(\nabla_{E_i} V, V)E_i\right)^H - \left(\Delta_g V\right)^V\right\} \circ V \tag{8.1}$$

on U where Δ_g is the rough Laplacian i.e., locally

$$\Delta_g V = -\sum_{i=1}^n \epsilon_i \left(\nabla_{E_i} \nabla V\right) E_i = -\sum_{i=1}^n \epsilon_i \{\nabla_{E_i} \nabla_{E_i} V - \nabla_{\nabla_{E_i} E_i} V\}$$

on U. Cf. again Section 6.3. Also, if $X \in \mathfrak{X}(M)$ then $X^H, X^V \in \mathfrak{X}(T(M))$ are the horizontal and vertical lifts of X. The proof of (8.1) is formally similar to that of (2.21) in Proposition 2.12 and the reader will encounter no difficulty in supplying the details. As in Chapter 2, the identity (8.1) may be used to show that for a critical point $V \in \mathfrak{X}(M)$ of the energy functional $E_\Omega : C^\infty(M, T(M)) \to \mathbb{R}$ one must have

$$\int_\Omega g^*(\nabla V, \nabla V)\, d\,\mathrm{vol}(g) = 0.$$

As g is not positive definite, one may not conclude at this point that V is parallel (and a case by case discussion based on causal characters will be necessary).

8.3. THE SPACELIKE ENERGY

Let (M,g) be an n-dimensional space-time and $Z \in \Gamma^\infty(S_{-1}(M,g))$ a *reference frame* (i.e., a unit timelike vector field on M). We set

$$A_Z = -\nabla Z, \quad P_Z(X) = X + g(X,Z)Z, \quad X \in \mathfrak{X}(M).$$

Also let us set $A'_Z = A_Z \circ P_Z$ for further use.

Definition 8.6 The function

$$\tilde{b}(Z) = \frac{1}{2} g^*(A'_Z, A'_Z) \in C^\infty(M)$$

is called the *spacelike energy density* of Z while for each relatively compact domain $\Omega \subset M$

$$\tilde{\mathcal{B}}_\Omega(Z) = \int_\Omega \tilde{b}(Z)\, d\,\mathrm{vol}(g)$$

is referred to as the *spacelike energy* of Z. If M is compact and $\Omega = M$, we write simply $\tilde{\mathcal{B}}_M(Z) = \tilde{\mathcal{B}}(Z)$. ■

Let $(\mathbb{R}Z)^\perp \subset T(M)$ be the orthogonal complement of $\mathbb{R}Z$. As Z is timelike, the distribution $\mathbb{R}Z$ is nondegenerate, hence $T(M) = \mathbb{R}Z \oplus (\mathbb{R}Z)^\perp$ and the restriction of g to $(\mathbb{R}Z)^\perp$ is positive definite. Let $\{E_\alpha : 1 \leq \alpha \leq n-1\}$ be a local g-orthonormal frame of $(\mathbb{R}Z)^\perp$ defined on the open set $U \subseteq M$. Then $\{E_i : 1 \leq i \leq n\} = \{E_1, \ldots, E_{n-1}, Z\}$ is a local orthonormal frame of $T(M)$. As

$$A'_Z E_i = -\nabla_{E_i} Z - g(E_i, Z)\nabla_Z Z$$

the spacelike energy density of Z is locally given by

$$\tilde{b}(Z) = \frac{1}{2} \sum_{\alpha=1}^{n-1} g(\nabla_{E_\alpha} Z, \nabla_{E_\alpha} Z) \tag{8.2}$$

on U. As $\epsilon_n = g(Z,Z) = -1$ it follows that

$$0 = E_\alpha(g(Z,Z)) = 2g(\nabla_{E_\alpha} Z, Z)$$

hence $\nabla_{E_\alpha} Z \in (\mathbb{R}Z)^\perp$. Using (8.2) and the fact that $g|_{(\mathbb{R}Z)^\perp}$ is positive definite it follows that

Proposition 8.7 *Let (M,g) be a space-time. For any reference frame $Z \in \Gamma^\infty(S_{-1}(M,g))$ one has $\tilde{b}(Z) \geq 0$. Also $\tilde{B}_\Omega(Z) \geq 0$ for any relatively compact domain $\Omega \subset M$ and $\tilde{B}_\Omega(Z) = 0$ if and only if $A'_Z = 0$ on Ω.*

A simple formula relates the ordinary energy and spacelike energy functionals

$$e(Z) = \frac{1}{2}\,\mathrm{trace}_g\,[g + g(\nabla Z, \nabla Z)] = \frac{n}{2} + \frac{1}{2}\sum_{i=1}^{n} \epsilon_i g(\nabla_{E_i} Z, \nabla_{E_i} Z)$$

(by $\epsilon_1 = \cdots = \epsilon_{n-1} = 1 = -\epsilon_n$ and (8.2))

$$= \frac{n}{2} + \tilde{b}(Z) - \frac{1}{2} g(\nabla_Z Z, \nabla_Z Z)$$

hence

$$E_\Omega(Z) = \frac{n}{2}\mathrm{Vol}(\Omega) + \tilde{\mathcal{B}}_\Omega(Z) - \frac{1}{2}\int_\Omega g(A_Z Z, A_Z Z)\, d\,\mathrm{vol}(g). \tag{8.3}$$

Our next task is to derive the first variation formula for the spacelike energy functional $\tilde{\mathcal{B}}_\Omega$. We follow the presentation in [130].

Lemma 8.8 *For any reference frame* $Z \in \Gamma^\infty(S_{-1}(M,g))$

$$\tilde{b}(Z) = \frac{1}{2}\left\{\mathrm{trace}\left(A_Z^t \circ A_Z\right) + g(\nabla_Z Z, \nabla_Z Z)\right\}. \tag{8.4}$$

Proof. Let $\{E_i : 1 \le i \le n\} = \{E_\alpha : 1 \le \alpha \le n-1\} \cup \{Z\}$ be a local orthonormal frame of $T(M)$. Then

$$\mathrm{trace}\left(A_Z^t \circ A_Z\right) = \sum_{i=1}^{n} \epsilon_i g((A_Z^t \circ A_Z) E_i, E_i)$$

$$= \sum_{\alpha=1}^{n-1} g\left(\left(A_Z^t \circ A_Z\right) E_\alpha, E_\alpha\right) - g\left(\left(A_Z^t \circ A_Z\right) Z, Z\right)$$

$$= \sum_\alpha g(\nabla_{E_\alpha} Z, \nabla_{E_\alpha} Z) - g(\nabla_Z Z, \nabla_Z Z)$$

(by (8.2))

$$= 2\tilde{b}(Z) - g(\nabla_Z Z, \nabla_Z Z).$$

■

Let $\mathcal{Z} : M \times (-\delta, \delta) \to S_{-1}(M,g)$ be a smooth 1-parameter variation of Z such that $\mathcal{Z}(x,0) = Z(x)$ for any $x \in M$. Let us set $Z_s(x) = \mathcal{Z}(x,s)$ for any $x \in M$ and any $|s| < \delta$. Locally

$$Z_s(x) = f^i(x,s)\left.\frac{\partial}{\partial x^i}\right|_x, \quad x \in M.$$

If $X = \{dZ_s/ds\}_{s=0}$ then

$$X_x = \frac{\partial f^i}{\partial s}(x,0)\left.\frac{\partial}{\partial x^i}\right|_x .$$

We also adopt the notation $Z'_s = (dZ_s/ds)(s)$ so that $Z'_0 = X$.

Lemma 8.9 *Let Z be a reference frame on (M,g). Then*

$$\frac{d}{ds}\left\{\operatorname{trace}\left(A^t_{Z_s} \circ A_{Z_s}\right)\right\} = 2\operatorname{trace}\left(A^t_{Z_s} \circ A_{Z'_s}\right)$$

for any $|s| < \delta$.

Proof. Let $\{E_i : 1 \le i \le n\}$ be a local orthonormal frame of $T(M)$. Let us set

$$\nabla_{E_i}\frac{\partial}{\partial x^j} = \Gamma^k_{ij}\frac{\partial}{\partial x^k}$$

so that

$$\nabla_{E_i} Z_s = \left\{E_i(f^j) + \Gamma^j_{ik} f^k\right\}\frac{\partial}{\partial x^j},$$

$$g(\nabla_{E_i} Z_s, \nabla_{E_i} Z_s)$$
$$= E_i(f^j)E_i(f^k)g_{jk} + 2E_i(f^j)f^k\Gamma^\ell_{ik}\, g_{j\ell} + f^j f^k \Gamma^p_{ij}\Gamma^q_{ik}\, g_{pq},$$

hence (by $[\partial/\partial s, E_i] = 0$)

$$\frac{d}{ds}\left\{\operatorname{trace}\left(A^t_{Z_s} \circ A_{Z_s}\right)\right\} = \frac{d}{ds}\sum_{i=1}^n \epsilon_i g(\nabla_{E_i} Z_s, \nabla_{E_i} Z_s)$$

$$= 2\sum_i \epsilon_i \left\{E_i\left(\frac{\partial f^j}{\partial s}\right)E_i(f^k)g_{jk} + \frac{\partial f^j}{\partial s} f^k \Gamma^p_{ij}\Gamma^q_{ik}\, g_{pq}\right.$$

$$\left. + \left[E_i\left(\frac{\partial f^j}{\partial s}\right)f^k + E_i(f^j)\frac{\partial f^k}{\partial s}\right]\Gamma^\ell_{ik} g_{j\ell}\right\}$$

$$= 2\sum_i \epsilon_i g(\nabla_{E_i} Z'_s, \nabla_{E_i} Z_s) = 2\sum_i \epsilon_i g((A^t_{Z_s} \circ A_{Z'_s})E_i, E_i)$$

and Lemma 8.9 is proved. ■

Lemma 8.10 *Let $Z \in \Gamma(S_{-1}(M,g))$ be a reference frame. Then*

$$\frac{d}{ds}\left\{g(\nabla_{Z_s} Z_s, \nabla_{Z_s} Z_s)\right\} = 2g(\nabla_{Z'_s} Z_s + \nabla_{Z_s} Z'_s, \nabla_{Z_s} Z_s)$$

for any $|s| < \delta$.

Proof. Let us set $\partial_j = \partial/\partial x^j$ for simplicity. One has

$$g(\nabla_{Z_s} Z_s, \nabla_{Z_s} Z_s) = f^j f^k g(\nabla_{\partial_j} Z_s, \nabla_{\partial_k} Z_s)$$

hence

$$\begin{aligned}
&\frac{d}{ds}\{g(\nabla_{Z_s} Z_s, \nabla_{Z_s} Z_s)\} \\
&= 2\left\{f^j f^k g(\nabla_{\partial_j} Z_s', \nabla_{\partial_k} Z_s) + \frac{\partial f^j}{\partial s} f^k g(\nabla_{\partial_j} Z_s, \nabla_{\partial_k} Z_s)\right\} \\
&= 2\{g(\nabla_{Z_s} Z_s', \nabla_{Z_s} Z_s) + g(\nabla_{Z_s'} Z_s, \nabla_{Z_s} Z_s)\}.
\end{aligned}$$

∎

Using Lemmas 8.8 to 8.10

$$\frac{d}{ds}\left\{\tilde{b}(Z_s)\right\} = \operatorname{trace}\left(A^t_{Z_s} \circ A_{Z_s'}\right) + g(\nabla_{Z_s'} Z_s + \nabla_{Z_s} Z_s', \nabla_{Z_s} Z_s)$$

for any $|s| < \delta$. Therefore we obtain the following first variation formula

Proposition 8.11 (O. Gil-Medrano & A. Hurtado, [130]) *Let* (M, g) *be a space-time and* Z *a reference frame on* M. *Let* $\{Z_s\}_{|s|<\delta}$ *be a smooth 1-parameter variation of* Z *such that* $Z_0 = Z$ *and let us set* $X = Z_0' = \{dZ_s/ds\}_{s=0}$. *Then*

$$\begin{aligned}
&\frac{d}{ds}\left\{\tilde{\mathcal{B}}_\Omega(Z_s)\right\}_{s=0} \\
&= \int_\Omega \left\{\operatorname{trace}\left(A^t_Z \circ A_X\right) + g(\nabla_X Z + \nabla_Z X, \nabla_Z Z)\right\} d\operatorname{vol}(g) \qquad (8.5)
\end{aligned}$$

provided that $\operatorname{supp}(X) \subset \Omega$.

Let us consider a local orthonormal frame $\{E_\alpha : 1 \le \alpha \le n-1\}$ of $(\mathbb{R}Z)^\perp$. Then

$$\begin{aligned}
&\operatorname{trace}\left(A^t_Z \circ A_X\right) \\
&= \sum_{\alpha=1}^{n-1} g((A^t_Z \circ A_X)E_\alpha, E_\alpha) - g((A^t_Z \circ A_X)Z, Z)
\end{aligned}$$

$$= \sum_\alpha g(A_X E_\alpha, A_Z E_\alpha) - g(A_X Z, A_Z Z)$$

$$= \sum_\alpha g(\nabla_{E_\alpha} X, \nabla_{E_\alpha} Z) - g(\nabla_Z X, \nabla_Z Z)$$

hence the integrand in (8.5) may be locally written as

$$\sum_\alpha g(\nabla_{E_\alpha} X, \nabla_{E_\alpha} Z) + g(\nabla_X Z, \nabla_Z Z).$$

To proceed further we need a few conventions of tensor calculus. Given a manifold M there is a unique $C^\infty(M)$-linear map $C : \Gamma^\infty(\mathcal{T}^{1,1}(M)) \to C^\infty(M)$ such that $C(X \otimes \omega) = \omega(X)$ for any $X \in \mathfrak{X}(M)$ and any $\omega \in \Omega^1(M)$. Moreover, if $1 \leq i \leq r$ and $1 \leq j \leq s$ we define a $C^\infty(M)$-linear map

$$C^i_j : \Gamma^\infty(\mathcal{T}^{r,s}(M)) \to \Gamma^\infty(\mathcal{T}^{r-1,s-1}(M))$$

as follows. Let $A \in \Gamma^\infty(\mathcal{T}^{r,s}(M))$. We set

$$(C^i_j A)(\omega^1, \ldots, \omega^{r-1}, X_1, \ldots, X_{s-1})$$
$$= C(A(\omega^1, \ldots, \omega^{i-1}, \cdot, \omega^i, \ldots, \omega^{r-1}, X_1, \ldots, X_{j-1}, \cdot, X_j, \ldots, X_{r-1}))$$

for any $\omega^1, \ldots, \omega^{r-1} \in \Omega^1(M)$ and any $X_1, \ldots, X_{s-1} \in \mathfrak{X}(M)$.

Lemma 8.12 *Let $K \in \Gamma^\infty(\mathcal{T}^{1,1}(M))$. Then*

$$\left(C^1_1 \nabla K\right) X = -\text{trace}(K \circ \nabla X) - \delta\alpha \tag{8.6}$$

where $\alpha \in \Omega^1(M)$ is given by $\alpha(Y) = g(K(X), Y)$ for any $Y \in \mathfrak{X}(M)$ and $\delta\alpha = -\text{div}(\alpha^\sharp)$.

Here $\sharp$ denotes raising of indices by g i.e., $g(\alpha^\sharp, Y) = \alpha(Y)$. To prove Lemma 8.12 we consider a local orthonormal frame $\{E_1, \ldots, E_n\} \subset \mathfrak{X}(U)$ and set $\omega^i(Y) = g(E_i, Y)$ for any $Y \in \mathfrak{X}(U)$. Then $\nabla K \in \Gamma^\infty(\mathcal{T}^{1,2}(M))$ may be locally written

$$\nabla K = (\nabla K)^i_{jk} E_i \otimes \omega^j \otimes \omega^k$$

for some $(\nabla K)^i_{jk} \in C^\infty(U)$. It follows that

$$(\nabla K)(\cdot, \cdot, X) = \epsilon_k X^k (\nabla K)^i_{jk} E_i \otimes \omega^j$$

hence

$$(C^1_1 \nabla K) X = C((\nabla K)(\cdot, \cdot, X)) = \epsilon_i \epsilon_k X^k (\nabla K)^i_{ik}. \tag{8.7}$$

If $X = X^i E_i$ for some $X^i \in C^\infty(U)$ then $\omega^j(X) = \epsilon_j X^j$ (no summation). Next

$$\begin{aligned}\operatorname{trace}(K \circ \nabla X) &= \sum_i \epsilon_i g(K(\nabla X)E_i, E_i) \\ &= \sum_i \epsilon_i (\nabla_i X^j)\epsilon_j K_j^k g(E_k, E_i)\end{aligned}$$

so that

$$\operatorname{trace}(K \circ \nabla X) = \sum_i (\nabla_i X^j) K_j^i \epsilon_j . \tag{8.8}$$

Moreover

$$\begin{aligned}\delta\alpha = -\operatorname{div}(\alpha^\sharp) &= -\sum_i \epsilon_i g(\nabla_{E_i}\alpha^\sharp, E_i) \\ &= -\sum_i \epsilon_i \{E_i(\alpha(E_i)) - \alpha\left(\nabla_{E_i}E_i\right)\} \\ &= -\sum_i \epsilon_i \{E_i(g(K(X), E_i)) - g(K(X), \nabla_{E_i}E_i)\} \\ &= -\sum_i \epsilon_i g(\nabla_{E_i}KX, E_i) \\ &= -\sum_i \epsilon_i \{g((\nabla_{E_i}K)X, E_i) + g(K\nabla_{E_i}X, E_i)\}\end{aligned}$$

that is

$$\delta\alpha = -\operatorname{trace}(K \circ \nabla X) - \sum_i \epsilon_i g((\nabla_{E_i}K)X, E_i). \tag{8.9}$$

We wish to compute the last term in (8.9). We have

$$(\nabla_{E_i}K)X = \epsilon_i \epsilon_j X^j (\nabla K)_{ij}^k E_k$$

hence (by (8.7))

$$\sum_i \epsilon_i g((\nabla_{E_i}K)X, E_i) = \epsilon_i \epsilon_j X^j (\nabla K)_{ij}^i = \left(C_1^1 \nabla K\right) X$$

and then (6.9) implies (8.6).

We wish to establish the following

Theorem 8.13 (O. Gil-Medrano & A. Hurtado, [130]) *For any reference frame $Z \in \Gamma^\infty(S_{-1}(M,g))$ and any smooth 1-parameter variation $\{Z_s\}_{|s|<\epsilon}$*

of Z such that $Z_s \in \Gamma(S_{-1}(M,g))$ for any $|s| < \epsilon$ and $Z_0 = Z$, the following first variation formula for the spacelike energy holds

$$\frac{d}{ds}\left\{\tilde{\mathcal{B}}_\Omega(Z_s)\right\}_{s=0} = \int_\Omega \tilde{\omega}(X)\, d\,\mathrm{vol}(g) \tag{8.10}$$

provided that $\{Z_s\}_{|s|<\epsilon}$ is supported in Ω i.e., $\mathrm{Supp}(X) \subset \Omega$ *where $X = Z_0' = \{dZ_s/ds\}_{s=0}$ and*

$$\tilde{\omega}_Z = -C_1^1 \nabla \tilde{K} + \left(\tilde{K}\nabla_Z Z\right)^\flat, \quad \tilde{K} = ((\nabla Z)\circ P_Z)^t. \tag{8.11}$$

Consequently, a reference frame Z is a critical point of $\tilde{\mathcal{B}}_\Omega$ if and only if $\tilde{\omega}_Z$ vanishes on $(\mathbb{R}Z)^\perp$.

Cf. also Corollary 3.3 and Proposition 3.4 in [130], p. 88. Here $\flat$ denotes lowering of indices by g i.e., for each $V \in \mathfrak{X}(M)$ the 1-form $V^\flat \in \Omega^1(M)$ is given by $V^\flat(Y) = g(Y,V)$ for any $Y \in \mathfrak{X}(M)$.

Proof of Theorem 8.13. We start from the first variation formula (8.5) in Proposition 8.11. First we compute the term $\mathrm{trace}\left(A_Z^t \circ A_X\right)$. Let $\{E_\alpha : 1 \le \alpha \le n-1\}$ be an orthonormal frame of $(\mathbb{R}Z)^\perp$. Then

$$\begin{aligned}\mathrm{trace}\left(A_Z^t \circ A_X\right) &= \sum_{\alpha=1}^{n-1} g(A_Z^t A_X E_\alpha, E_\alpha) - g(A_Z^t A_X Z, Z)\\ &= \sum_\alpha g(\nabla_{E_\alpha} X, \nabla_{E_\alpha} Z) - g(\nabla_Z X, \nabla_Z Z).\end{aligned}$$

On the other hand

$$\sum_\alpha g(\nabla_{E_\alpha} X, \nabla_{E_\alpha} Z) = \sum_\alpha g((\nabla Z)^t \circ (\nabla X)E_\alpha, E_\alpha)$$

as $P_Z : T(M) \to (\mathbb{R}Z)^\perp$ is the natural projection associated to the decomposition $T(M) = (\mathbb{R}Z) \oplus (\mathbb{R}Z)^\perp$ (i.e., $P_Z E_\alpha = E_\alpha$ and $P_Z Z = 0$)

$$= \sum_\alpha g((\nabla Z)^t \circ (\nabla X)E_\alpha, P_Z E_\alpha) = \sum_{i=1}^n \epsilon_i g((\nabla Z)^t (\nabla X)E_i, P_Z E_i)$$

(where $E_n = Z$ and $\epsilon_\alpha = -\epsilon_n = 1$, $1 \le \alpha \le n-1$)

$$= \sum_i \epsilon_i g(P_Z^t (\nabla Z)^t (\nabla X)E_i, E_i) = \mathrm{trace}\{\tilde{K} \circ (\nabla X)\}$$

where $\tilde{K}$ is given by the second formula in (8.11). Hence (by (8.5))

$$\frac{d}{ds}\left\{\tilde{\mathcal{B}}_\Omega(Z_s)\right\}_{s=0} = \int_\Omega \left\{\operatorname{trace}\left(\tilde{K}\circ\nabla X\right) + g(\nabla_X Z, \nabla_Z Z)\right\} d\operatorname{vol}(g). \quad (8.12)$$

Next (by (8.12) and Lemma 8.12)

$$\frac{d}{ds}\left\{\tilde{\mathcal{B}}_\Omega(Z_s)\right\}_{s=0} = \int_\Omega \{-(C_1^1\nabla\tilde{K})X - \delta\alpha + g(\nabla_X Z, \nabla_Z Z)\}$$

where

$$\begin{aligned}\alpha(Y) &= g(\tilde{K}X, Y) = g(((\nabla Z)\circ P_Z)^t X, Y)\\ &= g(X, (\nabla Z)PZ_Y) = g(X, \nabla_{P_Z Y} Z)\\ &= g(X, \nabla_Y Z) + g(Y, Z)g(X, \nabla_Z Z)\end{aligned}$$

hence

$$\alpha^\sharp = g(X, \nabla Z)^\sharp + g(X, \nabla_Z Z)Z. \quad (8.13)$$

Let us set $V = g(X, \nabla Z)^\sharp \in \mathfrak{X}(M)$. Let $x \in \operatorname{Supp}(\alpha^\sharp)$ so that $x = \lim_{\nu\to\infty} x_\nu$ for some sequence $x_\nu \in M$ such that $\alpha^\sharp(x_\nu) \neq 0$ for any $\nu \geq 1$. Let (U, x^i) be a local coordinate neighborhood in M such that $x \in U$. Thus $x_\nu \in U$ for any $\nu \geq \nu_0$ and some $\nu_0 \geq 1$. Let us assume that $X(x_\mu) = 0$ for some $\mu \geq \nu_0$. Then (by (8.13))

$$0 \neq \alpha^\sharp(x_\mu) = V(x_\mu) = g^{ji}(x_\mu) g(X, \nabla_{\partial_i} Z)_{x_\mu} \left.\frac{\partial}{\partial x^j}\right|_{x_\mu} = 0,$$

a contradiction. Therefore $X(x_\nu) \neq 0$ for any $\nu \geq \nu_0$ so that $x \in \operatorname{Supp}(X)$. We have shown that $\operatorname{Supp}(\alpha^\sharp) \subseteq \operatorname{Supp}(X) \subset \Omega$ hence (by Green's lemma) $\int_\Omega \delta\alpha \, d\operatorname{vol}(g) = 0$ and the first variation formula becomes

$$\frac{d}{ds}\{\tilde{B}_\Omega(Z_s)\}_{s=0} = \int_\Omega \{-(C_1^1\nabla\tilde{K})X + g(\nabla_X Z, \nabla_Z Z)\} d\operatorname{vol}(g). \quad (8.14)$$

We ought to compute the term $g(\nabla_X Z, \nabla_Z Z)$. As $g(Z_s, Z_s) = -1$ one has (by taking the derivative with respect to s at $s = 0$) $g(X, Z) = 0$. Thus

$$\begin{aligned}g(\nabla_X Z, \nabla_Z Z) &= g(X, (\nabla Z)^t \nabla_Z Z)\\ &= g(P_Z X, (\nabla Z)^t \nabla_Z Z) = g(X, P_Z^t (\nabla Z)^t \nabla_Z Z)\\ &= g(X, ((\nabla Z)\circ P_Z)^t \nabla_Z Z) = g(X, \tilde{K}\nabla_Z Z) = \left(\tilde{K}\nabla_Z Z\right)^\flat X\end{aligned}$$

and (8.14) yields (8.10). As a consequence of (8.10) $Z \in \mathrm{Crit}(\tilde{\mathcal{B}}_\Omega)$ if and only if

$$\int_\Omega \tilde{\omega}_Z(X) d\,\mathrm{vol}(g) = 0 \tag{8.15}$$

for any $X \in \Gamma^\infty((\mathbb{R}Z)^\perp)$. We may decompose $\tilde{\omega}_Z$ as $\tilde{\omega}_Z^\sharp = \lambda Z + Y$ for some $\lambda \in C^\infty(M)$ and $Y \in \Gamma^\infty((\mathbb{R}Z)^\perp)$. Then (8.15) may be written as $\int_\Omega g(X,Y) d\,\mathrm{vol}(g) = 0$. In particular for $X = Y$, we obtain (as $(\mathbb{R}Z)^\perp$ is space-like) $Y = 0$ and Theorem 8.13 is completely proved. ■

Definition 8.14 A reference frame $Z \in \Gamma^\infty(S_{-1}(M,g))$ is said to be *spatially harmonic* if for any relatively compact domain $\Omega \subset\subset M$ the restriction of Z to Ω is a critical point of $\tilde{\mathcal{B}}_\Omega$. ■

Let us consider the second order elliptic operator $\tilde{D}$ given by

$$\tilde{D}X = \Delta_g X - \nabla_X \nabla_X X - \mathrm{div}(X)\nabla_X X + ((\nabla X) \circ P_X)^t \nabla_X X$$

for any $X \in \mathfrak{X}(M)$. We may state the following

Corollary 8.15 *Let $Z \in \Gamma^\infty(S_{-1}(M,g))$ be a reference frame. Then Z is spatially harmonic if and only if $\tilde{D}Z$ is collinear to Z.*

Proof. By the very definition of $\tilde{\omega}_Z$

$$\begin{aligned}\tilde{\omega}_Z(X) &= -\left(C_1^1 \nabla \tilde{K}\right)X + g(X, \tilde{K}\nabla_Z Z) \\ &= \mathrm{trace}\left(\tilde{K} \circ \nabla X\right) - \mathrm{div}(\alpha^\sharp) + g(X, \tilde{K}\nabla_Z Z)\end{aligned}$$

for any $X \in \mathfrak{X}(M)$. We shall compute separately each term in the expression of $\tilde{\omega}_Z(X)$. Let $\{E_i : 1 \le i \le n\}$ be a local orthonormal frame of $T(M)$ where $\{E_\alpha : 1 \le \alpha \le n-1\}$ is a local orthonormal frame of $(\mathbb{R}Z)^\perp$ and $E_n = Z$. First

$$\begin{aligned}\mathrm{trace}\left(\tilde{K} \circ \nabla X\right) &= \sum_{i=1}^n \epsilon_i g(\tilde{K}(\nabla X)E_i, E_i) \\ &= \sum_i \epsilon_i g(\nabla_{E_i} X, (\nabla Z) P_Z E_i)\end{aligned}$$

i.e.,

$$\mathrm{trace}\left(\tilde{K} \circ \nabla X\right) = \sum_{\alpha=1}^n g(\nabla_{E_\alpha} X, \nabla_{E_\alpha} Z). \tag{8.16}$$

Second

$$\operatorname{div}(\alpha^\sharp) = \sum_i \epsilon_i g(\nabla_{E_i}\alpha^\sharp, E_i) = \sum_i \epsilon_i \{E_i(\alpha(E_i)) - \alpha(\nabla_{E_i}E_i)\}$$
$$= \sum_i \epsilon_i \{E_i(g(X, \nabla_{E_i}Z)) + E_i\big(g(E_i, Z)g(X, \nabla_Z Z)\big)$$
$$- g(X, \nabla_{\nabla_{E_i}E_i}Z) - g(\nabla_{E_i}E_i, Z)g(X, \nabla_Z Z)\}$$
$$= \sum_i \epsilon_i \{g(\nabla_{E_i}X, \nabla_{E_i}Z) + g(X, \nabla_{E_i}\nabla_{E_i}Z)\} + Z(g(X, \nabla_Z Z))$$
$$- \sum_i \epsilon_i \{g(X, \nabla_{\nabla_{E_i}E_i}Z) + g(\nabla_{E_i}E_i, Z)g(X, \nabla_Z Z)\}$$
$$= \sum_i \epsilon_i g(X, \nabla_{E_i}\nabla_{E_i}Z - \nabla_{\nabla_{E_i}E_i}Z) + Z(g(X, \nabla_Z Z))$$
$$+ \sum_i \epsilon_i \{g(\nabla_{E_i}X, \nabla_{E_i}Z) - g(\nabla_{E_i}E_i, Z)g(X, \nabla_Z Z)\}.$$

On the other hand

$$\sum_i \epsilon_i g(\nabla_{E_i}E_i, Z) = \sum_i \epsilon_i \{E_i(g(E_i, Z)) - g(E_i, \nabla_{E_i}Z)\} = -\operatorname{div}(Z)$$

hence

$$\operatorname{div}(\alpha^\sharp) = -g(X, \Delta_g Z) + Z(g(X, \nabla_Z Z))$$
$$+ \sum_i \epsilon_i g(\nabla_{E_i}X, \nabla_{E_i}Z) + g(X, \nabla_Z Z)\operatorname{div}(Z). \tag{8.17}$$

Therefore (by (8.16)–(8.17))

$$\operatorname{trace}(\tilde{K} \circ \nabla X) - \operatorname{div}(\alpha^\sharp) = g(\Delta_g Z, X)$$
$$- g(X, \nabla_Z \nabla_Z Z) - g(X, \nabla_Z Z)\operatorname{div}(Z) \tag{8.18}$$

for any $X \in \mathfrak{X}(M)$. Finally

$$\tilde{\omega}_Z(X) = g(\Delta_g Z, X) - g(X, \nabla_Z \nabla_Z Z)$$
$$- g(X, \nabla_Z Z)\operatorname{div}(Z) + g(X, \tilde{K}\nabla_Z Z)$$

so that $\tilde{\omega}_Z^\sharp \in \mathfrak{X}(M)$ is given by

$$\tilde{\omega}_Z^\sharp = \tilde{D}Z. \tag{8.19}$$

An equivalent formulation of Theorem 8.13 is that Z is spatially harmonic if and only if $\tilde{\omega}_Z^\sharp$ is collinear to Z, hence (8.19) yields the conclusion in Corollary 8.15. ∎

Corollary 8.16 *Let Z be a reference frame. If Z is geodesic (i.e., $\nabla_Z Z = 0$) then $Z \in \mathrm{Crit}(\tilde{B}_\Omega)$ if and only if $Z \in \mathrm{Crit}(E_\Omega)$.*

If $\nabla_Z Z = 0$ then $\tilde{D}Z = \Delta_g Z$ and Corollary 8.16 holds true.

A class of spatially harmonic vector fields $Z \in \Gamma^\infty(S_{-1}(M,g))$ are of course those for which the spacelike energy vanishes i.e., $\tilde{\mathcal{B}}_\Omega(Z) = 0$ for any relatively compact domain $\Omega \subset\subset M$. We wish to give an elementary physical interpretation of this situation. Let Z be a reference frame and let S_{symm} and S_{skew} be the symmetric and skew-symmetric parts of $-A'_Z$ i.e.,

$$g(S_{\text{symm}}X, Y) = -\frac{1}{2}\{g(A'_Z X, Y) + g(X, A'_Z Y)\},$$

$$g(S_{\text{skew}}X, Y) = -\frac{1}{2}\{g(A'_Z X, Y) - g(X, A'_Z Y)\},$$

for any $X, Y \in \mathfrak{X}(M)$. In particular

$$-A'_Z = S_{\text{symm}} + S_{\text{skew}}.$$

We also set

$$\Theta = \mathrm{trace}(S_{\text{symm}}), \quad \sigma = S_{\text{symm}} - \frac{\Theta}{n-1} P_Z,$$

so that $\mathrm{trace}(\sigma) = 0$. Summing up, one has the decomposition

$$-A'_Z = S_{\text{skew}} + \sigma + \frac{\Theta}{n-1} P_Z. \tag{8.20}$$

We adopt the following

Definition 8.17 S_{symm} and S_{skew} are respectively called the *deformation* and *rotation* of the reference frame Z. Also Θ and σ are said to be the *expansion* and *shear* of Z, respectively. If $S_{\text{symm}} = 0$ then Z is *rigid*. If $S_{\text{skew}} = 0$ then Z is *irrotational*. ∎

Let $\{E_i : 1 \le i \le n\}$ be a local orthonormal frame of $T(M)$ defined on the open subset $U \subseteq M$ such that $\{E_\alpha : 1 \le \alpha \le n-1\}$ is a local orthonormal frame of $(\mathbb{R}Z)^\perp$ on U and $E_n = Z$. One has $-A'_Z E_\alpha = A^i_\alpha E_i$ for some $A^i_\alpha \in C^\infty(U)$. Then

$$A^n_\alpha = -g(\nabla_{E_\alpha} Z, Z) = -\frac{1}{2} E_\alpha(g(Z,Z)) = 0$$

hence A'_Z maps $(\mathbb{R}Z)^\perp$ into itself. It is fairly easy to check that so do S_{symm}, S_{skew}, σ and P_Z. Summing up, we shall work with the bundle morphisms

$$A'_Z,\ S_{\text{symm}},\ S_{\text{skew}},\ \sigma,\ P_Z : (\mathbb{R}Z)^\perp \to (\mathbb{R}Z)^\perp$$

and the restriction of g to $(\mathbb{R}Z)^\perp$ is positive definite. If we set

$$S_{\text{symm}} E_\alpha = S_\alpha^\beta E_\beta, \quad S_{\text{skew}} E_\alpha = \Omega_\alpha^\beta E_\beta,$$

then $S_\alpha^\beta = S_\beta^\alpha$ and $\Omega_\alpha^\beta = -\Omega_\beta^\alpha$. Consequently

$$\sum_\alpha g(S_{\text{skew}} E_\alpha, S_{\text{symm}} E_\alpha) = 0.$$

Also

$$\sum_\alpha g(\sigma E_\alpha, E_\alpha) = \text{trace}(\sigma) = 0.$$

In the end

$$\tilde{\mathcal{B}}_\Omega(Z) = \int_\Omega \tilde{b}(Z)\, d\,\text{vol}(g)$$

$$= \frac{1}{2}\int_\Omega \{\|S_{\text{skew}}\|^2 + \|\sigma\|^2 + \frac{1}{n-1}\Theta^2\}\, d\,\text{vol}(g).$$

Consequently, $\tilde{\mathcal{B}}_\Omega(Z) = 0$ for any $\Omega \subset\subset M$ if and only if the reference frame Z is both rigid and irrotational.

Example 8.18 (Continuation of Example 8.2) Let $S^1 \times \mathbb{R}$ carry the Lorentz metric $g = -d\theta \otimes d\theta + dt \otimes dt$. We set $E_1 = \partial/\partial t$ and $E_2 = \partial/\partial\theta$ so that $\{E_1, E_2\}$ is a local orthonormal frame of $T(S^1 \times \mathbb{R})$ defined on the open set $U = (0, 2\pi) \times \mathbb{R}$. A reference frame $Z \in \Gamma^\infty(S_{-1}(S^1 \times \mathbb{R}, g))$ is locally given by $Z = \lambda E_1 + \mu E_2$ hence

$$\tilde{b}(Z) = \frac{\epsilon_1}{2}\|\nabla_{E_1} Z\|^2 = \frac{1}{2}\left\{\left(\frac{\partial\lambda}{\partial t}\right)^2 - \left(\frac{\partial\mu}{\partial t}\right)^2\right\}.$$

Therefore $\partial/\partial\theta$ is a rigid irrotational spatially harmonic reference frame on U. As $\partial/\partial\theta$ is geodesic it is also a harmonic vector field. As $g(Z,Z) = -1$ one has $\mu^2 - \lambda^2 = 1$ hence

$$\tilde{b}(Z) = \frac{1}{2}\frac{\lambda_t^2}{\lambda^2 + 1}, \quad \lambda_t = \frac{\partial\lambda}{\partial t}$$

so that $\tilde{\mathcal{B}}_\Omega(Z) = 0$ for any relatively compact domain $\Omega \subset\subset U$ if and only if $\lambda_t = 0$. Consequently the rigid irrotational reference frames are precisely the spherically symmetric ones. The relationship (if any) among the theory of harmonic vector fields on $M = (S^1 \times \mathbb{R}, g)$ and the fact that M has a causality violation is unclear.

Example 8.19 (Continuation of Example 8.4) Let $S^1 \times \mathbb{R}$ be endowed with the Lorentz metric $g = d\theta \odot dt$. We set

$$E_1 = \frac{\partial}{\partial t} + \frac{\partial}{\partial \theta}, \quad E_2 = \frac{\partial}{\partial t} - \frac{\partial}{\partial \theta},$$

so that $\{E_1, E_2\}$ is a local orthonormal frame of $T(S^1 \times \mathbb{R})$ on $U = (0, 2\pi) \times \mathbb{R}$. Let

$$Z = \lambda \frac{\partial}{\partial t} + \mu \frac{\partial}{\partial \theta} \in \Gamma^\infty(U, S_{-1}(S^1 \times \mathbb{R}, g))$$

be an arbitrary reference frame i.e., $\lambda\mu + 1 = 0$ on U. Then

$$\tilde{b}(Z) = \frac{1}{2} \frac{(\lambda_t + \lambda_\theta)^2}{\lambda^2}$$

hence $\tilde{b}(Z) = 0$ (and then Z is a spatially harmonic vector field) if and only if $\lambda_t + \lambda_\theta = 0$. In particular

$$Ce^{a(t-\theta)} \frac{\partial}{\partial t} - \frac{1}{C} e^{-a(t-\theta)} \frac{\partial}{\partial \theta}, \quad C \in \mathbb{R} \setminus \{0\}, \; a \in \mathbb{R},$$

is a family of rigid irrotational spatially harmonic reference frames on (U, g). Moreover

$$\operatorname{div}(Z) = \lambda_t + \lambda_\theta, \tag{8.21}$$

$$\nabla_Z Z = \left(\lambda_t - \lambda^{-2} \lambda_\theta\right) \left(\lambda \frac{\partial}{\partial t} - \mu \frac{\partial}{\partial \theta}\right), \tag{8.22}$$

on U. The identity (8.21) implies the following global statement: *a reference frame on $(S^1 \times \mathbb{R}, g)$ is rigid and irrotational if and only if it is divergence free.* Also (as a consequence of (8.22)) $Z = \lambda\, \partial/\partial t + \mu\, \partial/\partial\theta$, $\lambda\mu + 1 = 0$, satisfies both $\tilde{b}(Z) = 0$ and $\nabla_Z Z = 0$ if and only if $\lambda_t = \lambda_\theta = 0$ on U. Therefore

$$\lambda \frac{\partial}{\partial t} - \frac{1}{\lambda} \frac{\partial}{\partial \theta}, \quad \lambda \in \mathbb{R} \setminus \{0\},$$

are all the rigid irrotational geodesic (and then harmonic) reference frames on U. It is an open problem to relate the behavior of (spatially) harmonic vector fields on $M = (S^1 \times \mathbb{R}, d\theta \odot dt)$ to the facts that M is chronological yet not causal.

Example 8.20 (Projective vector fields) Let (M,g) be a Lorentz manifold. We recall that a vector field $Z \in \mathfrak{X}(M)$ is said to be *projective* if there is a differential 1-form $\omega \in \Omega^1(M)$ such that

$$(\mathcal{L}_Z \nabla)(X, Y) = \omega(X)Y + \omega(Y)X,$$

for any $X, Y \in \mathfrak{X}(M)$. An *affine* vector field is a projective vector field Z with $\omega = 0$. Let Z be a projective vector field. The identities

$$(\mathcal{L}_Z \nabla)(X, Y) = [Z, \nabla_X Y] - \nabla_{[Z,X]} Y - \nabla_X [Z, Y],$$
$$\nabla_X \nabla_Z Y = -R(X, Z)Y + \nabla_Z \nabla_X Y + \nabla_{[X,Z]} Y,$$

lead to

$$(\mathcal{L}_Z \nabla)(X, Y) = \nabla_X \nabla_Y Z - \nabla_{\nabla_X Y} Z + R(X, Z)Y.$$

Therefore Z is a projective vector field if and only if

$$\nabla_X \nabla_Y Z - \nabla_{\nabla_X Y} Z = \omega(X)Y + \omega(Y)X - R(X, Z)Y \tag{8.23}$$

for some $\omega \in \Omega^1(M)$ and any $X, Y \in \mathfrak{X}(M)$. Let $Z \in \Gamma^\infty(S_{-1}(M,g))$ and let us assume that Z is projective. We set $X = Y = Z$ in (8.23) to obtain

$$g(\nabla_Z Z, \nabla_Z Z) = 2\omega(Z). \tag{8.24}$$

Similarly let us replace (X, Y) by (Z, X) where $X \in (\mathbb{R}Z)^\perp$. We derive

$$\omega(X) = g(\nabla_X Z, \nabla_Z Z). \tag{8.25}$$

A curvature calculation based on (8.24)–(8.25) shows that (8.19) may be written as

$$\begin{aligned} \tilde{\omega}_Z(X) = &- \mathrm{Ric}(Z, X) \\ &- g(X, \nabla_{\nabla_Z Z} Z) - \omega(X) - g(X, \nabla_Z Z)\mathrm{div}(Z) \end{aligned} \tag{8.26}$$

for any $X \in (\mathbb{R}Z)^\perp$. Consequently

Proposition 8.21 (O. Gil-Medrano & A. Hurtado, [130]) *Let (M,g) be a compact Lorentz manifold. Let Z be a projective reference frame. Then* i) *Z is spatially harmonic if and only if*

$$g(X, \nabla_{\nabla_Z Z} Z) + \mathrm{Ric}(Z, X) + \omega(X) + g(X, \nabla_Z Z)\mathrm{div}(Z) = 0$$

for any $X \in (\mathbb{R}Z)^\perp$. ii) *If* Z *is affine then* Z *is a critical point of the ordinary energy functional (and then* Z *is spatially harmonic) if and only if* $\mathrm{Ric}(X,Z) = 0$ *for any* $X \in (\mathbb{R}Z)^\perp$. *Let us additionally assume that* (M,g) *is an Einstein manifold i.e.,* $\mathrm{Ric} = \lambda g$ *for some* $\lambda \le 0$. *Let* Z *be an affine reference frame on* M. *Then* iii) Z *is a critical point of the ordinary energy functional. Also,* Z *is geodesic, hence it is spatially harmonic.*

The proof of (8.26) is rather involved yet a mere adaptation of calculations in Chapters 2 and 3 to the Lorentzian case. See [130], p. 91, for details. It is well known that when (M,g) is an Einstein manifold with $\lambda > 0$ there are no unit timelike projective vector fields on M.

Example 8.22 (A rigid irrotational reference frame on a Fefferman space) Let $(M, T_{1,0}(M))$ be a strictly pseudoconvex CR manifold, of CR dimension n. Let θ be a contact form on M such that the Levi form G_θ is positive definite. Let T be the characteristic direction of $d\theta$. Let $S^1 \to C(M) \to M$ be the canonical circle bundle and let F_θ be the Fefferman metric on $C(M)$ corresponding to the chosen contact form θ. Cf. [110] or Chapter 6 of this book. The Lorentzian manifold $(C(M), F_\theta)$ is referred to as a *Fefferman space*. Let $S = [(n+2)/2]\, \partial/\partial\gamma$ where γ is a local fibre coordinate on $C(M)$. Let σ be the connection form on $C(M)$ given by

$$\sigma = \frac{1}{n+2}\left\{ d\gamma + \pi^*\left[i\omega_\alpha^\alpha - \frac{i}{2} g^{\alpha\bar\beta} dg_{\alpha\bar\beta} - \frac{\rho}{4(n+1)}\theta \right]\right\}.$$

Then $\sigma(S) = 1/2$ so that $F_\theta(T^\uparrow - S, T^\uparrow - S) = -1$ i.e., $T^\uparrow - S$ is timelike. Hence $T^\uparrow - S$ is a time orientation and $C(M)$ is a space-time. If M is compact then $C(M)$ is compact as well, and therefore non-chronological. Let $\{E_a : 1 \le a \le 2n\}$ be a local orthonormal $(G_\theta(E_a, E_b) = \delta_{ab},\ 1 \le a, b \le 2n)$ frame of the Levi distribution $H(M)$ such that $JE_\alpha = E_{\alpha+n}$ for any $1 \le \alpha \le n$. Then $\{E_a^\uparrow, T^\uparrow \pm S : 1 \le a \le 2n\}$ is a local orthonormal frame of $(T(C(M)), F_\theta)$. We shall establish the following

Proposition 8.23 *Let us assume that* θ *is a pseudo-Einstein contact form. Then* $Z = T^\uparrow - S$ *is a rigid irrotational reference frame on the Fefferman space* $(C(M), F_\theta)$ *if and only if* ρ *is constant and*

$$\rho_1 \le \rho \le \rho_2, \quad \|\tau\|^2 = -2n + \frac{\rho}{n+1} - \frac{\rho^2}{16n(n+1)^2},$$

where $\rho_{1,2} = 4n(n+1)(2 \pm \sqrt{2})$ *and* ρ, τ *are respectively the pseudohermitian scalar curvature and torsion of* (M,θ).

To prove Proposition 8.23 we shall compute

$$\tilde{b}(Z) = \frac{1}{2}\left\{\sum_{a=1}^{2n} \|\nabla^{C(M)}_{E_a^\uparrow} Z\|^2 + \|\nabla^{C(M)}_{T^\uparrow+S} Z\|^2\right\}$$

where $\nabla^{C(M)}$ is the Levi-Civita connection of $(C(M), F_\theta)$. By a result of E. Barletta et al. (cf. Lemma 2 in [26], p. 27) for any $X, Y \in H(M)$

$$\nabla^{C(M)}_{X^\uparrow} Y^\uparrow = (\nabla_X Y)^\uparrow - (d\theta)(X,Y)T^\uparrow - [A(X,Y) + (d\sigma)(X^\uparrow, Y^\uparrow)]S, \quad (8.27)$$

$$\nabla^{C(M)}_{X^\uparrow} T^\uparrow = (\tau X + \phi X)^\uparrow, \quad (8.28)$$

$$\nabla^{C(M)}_{T^\uparrow} X^\uparrow = (\nabla_T X \phi X)^\uparrow + 2(d\sigma)(X^\uparrow, T^\uparrow)S, \quad (8.29)$$

$$\nabla^{C(M)}_{X^\uparrow} S = \nabla^{C(M)}_{S} X^\uparrow = (JX)^\uparrow, \quad (8.30)$$

$$\nabla^{C(M)}_{T^\uparrow} T^\uparrow = V^\uparrow, \quad \nabla^{C(M)}_{S} S = \nabla^{C(M)}_{S} T^\uparrow = \nabla^{C(M)}_{T^\uparrow} S = 0, \quad (8.31)$$

where $\phi : H(M) \to H(M)$ and $V \in H(M)$ are respectively given by $G_\theta(\phi X, Y) = (d\sigma)(X^\uparrow, Y^\uparrow)$ and $G_\theta(V, Y) = 2(d\sigma)(T^\uparrow, Y^\uparrow)$. By (8.31) one obtains

$$\nabla^{C(M)}_{T^\uparrow+S} Z = V^\uparrow$$

hence

$$F_\theta\left(D_{T^\uparrow+S}Z, D_{T^\uparrow+S}Z\right) = G_\theta(V,V) = \|V\|^2. \quad (8.32)$$

Similarly (by (8.28) and (8.30))

$$\nabla^{C(M)}_{E_a^\uparrow} Z = \left(\tau E_a + \phi E_a - JE_a\right)^\uparrow$$

hence

$$F_\theta\left(\nabla^{C(M)}_{E_a^\uparrow} Z, \nabla^{C(M)}_{E_a^\uparrow} Z\right) = \|\tau E_a + \phi E_a - JE_a\|^2. \quad (8.33)$$

Finally (by (8.32)–(8.33))

$$\tilde{b}(Z) = \operatorname{trace}(\tau^2) + 2\operatorname{trace}(\tau\phi) - 2\operatorname{trace}(\tau J) + 2\operatorname{trace}(J\phi) + 2n + \sum_{a=1}^{2n} G_\theta(\phi E_a, \phi E_a) + G_\theta(V,V).$$

Let $\{W_\alpha : 1 \le \alpha \le n\}$ be a local orthonormal (i.e., $G_\theta(W_\alpha, W_{\overline{\beta}}) = \delta_{\alpha\beta}$) frame of $T_{1,0}(M)$. Then

$$\tau^2(W_\alpha) = A_\alpha^{\overline{\beta}} \tau(W_{\overline{\beta}}) = A_\alpha^{\overline{\beta}} A_{\overline{\beta}}^{\gamma} W_\gamma$$

so that

$$\tau^2 : \begin{pmatrix} A_\alpha^{\overline{\beta}} A_{\overline{\beta}}^{\gamma} & 0 \\ 0 & A_{\overline{\alpha}}^{\beta} A_\beta^{\overline{\gamma}} \end{pmatrix}$$

hence (as the trace of an endomorphism of a real linear space coincides with the trace of its extension by $\mathbb{C}$-linearity)

$$\operatorname{trace}(\tau^2) = 2\sum_{\alpha,\beta} \left| A_\alpha^{\overline{\beta}} \right|^2 = \|\tau\|^2. \tag{8.34}$$

Next

$$\tau\phi W_\alpha = \phi_\alpha{}^\beta \tau(W_\beta) = \phi_\alpha{}^\beta A_\beta^{\overline{\gamma}} W_{\overline{\gamma}}$$

so that

$$\tau\phi : \begin{pmatrix} 0 & \phi_\alpha{}^\beta A_\beta^{\overline{\gamma}} \\ \phi_{\overline{\alpha}}{}^{\overline{\beta}} A_{\overline{\beta}}^{\gamma} & 0 \end{pmatrix}$$

and then

$$\operatorname{trace}(\tau\phi) = 0. \tag{8.35}$$

Similarly

$$\tau J : \begin{pmatrix} 0 & iA_\alpha^{\overline{\beta}} \\ -iA_{\overline{\alpha}}^{\beta} & 0 \end{pmatrix}, \quad J\phi : \begin{pmatrix} i\phi_\alpha{}^\beta & 0 \\ 0 & -i\phi_{\overline{\alpha}}{}^{\overline{\beta}} \end{pmatrix},$$

so that

$$\operatorname{trace}(\tau J) = 0, \tag{8.36}$$

$$\operatorname{trace}(J\phi) = i\left(\phi_\alpha{}^\alpha - \phi_{\overline{\alpha}}{}^{\overline{\alpha}}\right). \tag{8.37}$$

By a result in [26], p. 28, if $\phi W_\alpha = \phi_\alpha{}^\beta W_\beta + \phi_\alpha{}^{\overline{\beta}} W_{\overline{\beta}}$ then

$$\phi^{\alpha\beta} = 0, \quad \phi^{\overline{\alpha}\beta} = \frac{i}{2(n+2)}\left[R^{\overline{\alpha}\beta} - \frac{\rho}{2(n+1)} g^{\overline{\alpha}\beta}\right],$$

where $\phi^{\alpha\beta} = g^{\alpha\overline{\gamma}}\phi_{\overline{\gamma}}{}^{\beta}$, etc. This may be also written as

$$\phi_\alpha{}^\beta = \frac{i}{2(n+2)}\left[R_\alpha{}^\beta - \frac{\rho}{2(n+1)}\delta_\alpha^\beta\right], \quad \phi_{\overline{\alpha}}{}^\beta = 0. \tag{8.38}$$

Then (by (8.37)–(8.38))

$$\phi_\alpha{}^\alpha = \frac{i\rho}{4(n+1)}, \quad \text{trace}(J\phi) = -\frac{\rho}{2(n+1)}. \tag{8.39}$$

Let us compute the term $\sum_{a=1}^{2n}\|\phi E_a\|^2$. To this end, we consider the local frame $\{E_a : 1 \leq a \leq 2n\}$ given by

$$E_\alpha = \frac{1}{\sqrt{2}}(W_\alpha + W_{\overline{\alpha}}), \quad E_{\alpha+n} = \frac{i}{\sqrt{2}}(W_\alpha - W_{\overline{\alpha}}),$$

for any $1 \leq \alpha \leq n$. Then

$$\sum_{a=1}^{2n} G_\theta(\phi E_a, \phi E_a) = 2\sum_{\alpha,\beta}\left|\phi_\alpha{}^\beta\right|^2. \tag{8.40}$$

Summing up the information in (8.34)–(8.36) and (6.46)–(6.48), we may conclude that

$$\sum_{a=1}^{2n} F_\theta\left(\nabla^{C(M)}_{E_a^\uparrow} Z, \nabla^{C(M)}_{E_a^\uparrow} Z\right) = \sum_a \|\nabla^{C(M)}_{E_a^\uparrow} Z\|^2$$
$$= \|\tau\|^2 - \frac{\rho}{n+1} + 2n + \sum_{\alpha,\beta}\left|\phi_\alpha{}^\beta\right|^2. \tag{8.41}$$

Finally we ought to compute V. To this end we differentiate σ

$$(n+2)d\sigma = \pi^*\left[i\, d\omega_\alpha{}^\alpha - \frac{i}{2}dg^{\alpha\overline{\beta}} \wedge dg_{\alpha\overline{\beta}} - \frac{\rho}{4(n+1)}d(\rho\theta)\right]$$

and observe that

$$dg^{\alpha\overline{\beta}} \wedge dg_{\alpha\overline{\beta}} = 0.$$

Let $X \in H(M)$. Then

$$(n+2)(d\sigma)(T^\uparrow, X^\uparrow) = i(d\omega_\alpha{}^\alpha)(T,X) - \frac{1}{4(n+1)}d(\rho\theta)(T,X)$$

and on one hand

$$d(\rho\theta)(T,X) = (d\rho \wedge \theta + \rho\, d\theta)(T,X) = -\frac{1}{2}X(\rho)$$

as $T \rfloor d\theta = 0$. On the other hand

$$d\omega_\alpha{}^\alpha = R_{\lambda\overline{\mu}}\theta^\lambda \wedge \theta^{\overline{\mu}} + \left(W^\alpha_{\alpha\lambda}\theta^\lambda - W^\alpha_{\alpha\overline{\mu}}\theta^{\overline{\mu}}\right) \wedge \theta$$

(cf. e.g., (5.20) in [110], p. 295) hence

$$i(d\omega_\alpha{}^\alpha)(T, T_\beta) = -\frac{i}{2}\left(W^\alpha_{\alpha\lambda}\theta^\lambda - W^\alpha_{\alpha\overline{\mu}}\theta^{\overline{\mu}}\right)(T_\beta) = -\frac{i}{2}\,W^\alpha_{\alpha\beta}\,.$$

Finally

$$(n+2)G_\theta(V, T_\beta) = 2(n+2)(d\sigma)(T^\uparrow, T^\uparrow_\beta) = -iW^\alpha_{\alpha\beta} + \frac{1}{4(n+1)}\,\rho_\beta$$

where $\rho_\beta = T_\beta(\rho)$. Therefore $V = W + \overline{W}$ where

$$W = \frac{1}{n+2}g^{\alpha\overline{\beta}}\left[\frac{1}{4(n+1)}\,\rho_{\overline{\beta}} + iW^{\overline{\gamma}}_{\overline{\gamma}\overline{\beta}}\right]T_\alpha\,. \tag{8.42}$$

Let us assume from now on that θ is a pseudo-Einstein contact form on M i.e., $R_{\alpha\overline{\beta}} = (\rho/n)g_{\alpha\overline{\beta}}$. Throughout $g_{\alpha\overline{\beta}} = \delta_{\alpha\beta}$. For an up-to-date presentation of the main results on pseudo-Einstein structures on CR manifolds the reader may see Chapter 5 in [110]. As a consequence of the pseudo-Einstein condition, the functions $W^\alpha_{\alpha\beta}$ may be determined. Indeed (cf. [110], p. 298)

$$W^\alpha_{\alpha\beta} = -\frac{i}{2n}\,\rho_\beta\,.$$

Consequently

$$V = -\frac{1}{4n(n+1)}\left(\rho^\alpha\,T_\alpha + \rho^{\overline{\alpha}}\,T_{\overline{\alpha}}\right),$$

$$\phi_\alpha{}^\beta = \frac{i\rho}{4n(n+1)}\,\delta^\beta_\alpha\,, \quad \sum_{\alpha,\beta}|\phi_\alpha{}^\beta|^2 = \frac{\rho^2}{16n(n+1)^2}\,.$$

Then $\tilde{b}(Z) = 0$ if and only if $W = 0$ and

$$\|\tau\|^2 = -2n + \frac{\rho}{n+1} - \frac{\rho^2}{16n(n+1)^2}.$$

By the first condition ρ is a real valued CR function. Yet on a nondegenerate CR manifold, the only real valued CR functions are the constants.

8.4. THE SECOND VARIATION OF THE SPACELIKE ENERGY

The purpose of this section is to establish the following

Theorem 8.24 (O. Gil-Medrano & A. Hurtado, [130]) *Let (M,g) be an n-dimensional Lorentzian manifold. Let $Z \in \Gamma^\infty(S_{-1}(M,g))$ be a spatially harmonic reference frame M. Then*

$$\frac{d^2}{ds^2}\{\tilde{B}_\Omega(Z_s)\}_{s=0} = \int_\Omega \{\text{trace}\left((\nabla X)^t \circ (\nabla X)\right)$$
$$+ 2g(\nabla_X X, \nabla_Z Z) + g(\nabla_X Z + \nabla_Z X, \nabla_X Z + \nabla_Z X)$$
$$+ \|X\|^2 \left[g(\nabla_Z Z, \nabla_Z Z) - \left(C_1^1 \nabla \tilde{K}\right) Z\right]\} d\,\text{vol}(g) \tag{8.43}$$

for any smooth 1-parameter variation $\{Z_s\}_{|s|<\epsilon} \subset \Gamma^\infty(S_{-1}(M,g))$ of Z such that $Z_0 = Z$ and $X = Z_0' = (dZ_s/ds)_{s=0}$ is supported in Ω.

Proof. We recall that

$$\frac{d}{ds}\left\{\tilde{b}(Z_s)\right\} = \text{trace}\left(A^t_{Z_s} \circ A_{Z_s'}\right) + g(\nabla_{Z_s'} Z_s + \nabla_{Z_s} Z_s', \nabla_{Z_s} Z_s)$$

where as customary $Z_s' = dZ_s/ds$. We wish to compute the second derivative of $\tilde{b}(Z_s)$ at $s = 0$. First

$$\frac{d}{ds}\left\{\text{trace}\left(A^t_{Z_s} \circ A_{Z_s'}\right)\right\}_{s=0} = \text{trace}\left(A^t_X \circ A_X\right) + \text{trace}\left(A^t_Z \circ A_{Z_0''}\right)$$

where $Z_0'' = (d^2 Z_s/ds^2)_{s=0}$. Second and last

$$\frac{d}{ds}\left\{g(\nabla_{Z_s'} Z_s + \nabla_{Z_s} Z_s', \nabla_{Z_s} Z_s)\right\}_{s=0}$$
$$= g(\nabla_{Z_0''} Z + 2\nabla_X X + \nabla_Z Z_0'', \nabla_Z Z)$$
$$+ g(\nabla_X Z + \nabla_Z X, \nabla_X Z + \nabla_Z X).$$

It follows that

$$\frac{d^2}{ds^2}\left\{\tilde{b}(Z_s)\right\}_{s=0} = \text{trace}\left(A^t_X \circ A_X + A^t_Z \circ A_{Z_0''}\right)$$
$$+ g(\nabla_{Z_0''} Z + 2\nabla_X X + \nabla_Z Z_0'', \nabla_Z Z)$$
$$+ g(\nabla_X Z + \nabla_Z X, \nabla_X Z + \nabla_Z X). \tag{8.44}$$

Next

$$\begin{aligned}
&\operatorname{trace}\left(A_Z^t \circ A_{Z_0''}\right) + g(\nabla_Z Z_0'', \nabla_Z Z) \\
&= \sum_i \epsilon_i g(A_{Z_0''} E_i, A_Z E_i) + g(\nabla_Z Z_0'', \nabla_Z Z) \\
&= \sum_\alpha g(\nabla_{E_\alpha} Z_0'', \nabla_{E_\alpha} Z)
\end{aligned}$$

(as $P_Z Z = 0$ and $P_Z E_\alpha = E_\alpha$)

$$\begin{aligned}
&= \sum_i \epsilon_i\, g((\nabla Z_0'')E_i, (\nabla Z)P_Z E_i) \\
&= \sum_i \epsilon_i\, g(\tilde{K}(\nabla Z_0'')E_i, E_i) = \operatorname{trace}\left(\tilde{K} \circ \nabla Z_0''\right) \\
&= -\left(C_1^1 \nabla \tilde{K}\right) Z_0'' - \delta\beta
\end{aligned}$$

where $\beta \in \Omega^1(M)$ is given by

$$\beta(Y) = g(\tilde{K} Z_0'', Y), \quad Y \in \mathfrak{X}(M).$$

Therefore (8.44) becomes

$$\begin{aligned}
&\frac{d^2}{ds^2}\left\{\tilde{b}(Z_s)\right\}_{s=0} = \operatorname{trace}\left(A_X^t \circ A_X\right) \\
&+ 2g(\nabla_X X, \nabla_Z Z) + g(\nabla_{Z_0''} Z, \nabla_Z Z) \\
&+ g(\nabla_X Z + \nabla_Z X, \nabla_X Z + \nabla_Z X) - \left(C_1^1 \nabla \tilde{K}\right) Z_0'' - \delta\beta. \qquad (8.45)
\end{aligned}$$

As $\partial/\partial s$ is a differential operator, it follows that $\operatorname{Supp}(Z_0'') \subseteq \operatorname{Supp}(X) \subset \Omega$. On the other hand $\operatorname{Supp}(\beta^\sharp) \subseteq \operatorname{Supp}(Z_0'')$ hence we may integrate in (8.45) over Ω and apply Green's lemma to obtain

$$\begin{aligned}
&\frac{d^2}{ds^2}\left\{\tilde{\mathcal{B}}_\Omega(Z_s)\right\}_{s=0} = \int_\Omega \Big\{\operatorname{trace}\left((\nabla X)^t \circ (\nabla X)\right) \\
&+ 2g(\nabla_X X, \nabla_Z Z) + g(\nabla_X Z + \nabla_Z X, \nabla_X Z + \nabla_Z X) \\
&+ g(\nabla_{Z_0''} Z, \nabla_Z Z) - \left(C_1^1 \nabla \tilde{K}\right) Z_0''\Big\}\, d\operatorname{vol}(g). \qquad (8.46)
\end{aligned}$$

Since

$$g(Z_0'', Z) = -g(X, X)$$

it follows that

$$P_Z Z_0'' = Z_0'' + g(Z_0'', Z)Z = Z_0'' - g(X,X)Z$$

hence we obtain the decomposition

$$Z_0'' = P_Z Z_0'' + \|X\|^2 Z. \tag{8.47}$$

We may of course write $\|X\|^2$ instead of $g(X,X)$ as $X \in (\mathbb{R}Z)^\perp$ and $(\mathbb{R}Z)_x^\perp$ is a positive definite subspace of $(T_x(M), g_x)$ for any $x \in M$. As Z is a critical point of $\tilde{\mathcal{B}}_\Omega$ the 1-form $\tilde{\omega}_Z$ vanishes on $(\mathbb{R}Z)^\perp$ (cf. Theorem 8.13 above). Then (as P_Z is $(\mathbb{R}Z)^\perp$-valued) the decomposition (8.47) implies

$$\tilde{\omega}_Z(Z_0'') = \|X\|^2 \tilde{\omega}_Z(Z). \tag{8.48}$$

Finally (by (8.48))

$$\begin{aligned}
&g(\nabla_{Z_0''} Z, \nabla_Z Z) - (C_1^1 \nabla \tilde{K}) Z_0'' \\
&= g(\nabla_{Z_0''} Z, \nabla_Z Z) + \tilde{\omega}_Z(Z_0'') - g(Z_0'', \tilde{K} \nabla_Z Z) \\
&= g(\nabla_{Z_0''} Z, \nabla_Z Z) + \|X\|^2 \tilde{\omega}_Z(Z) - g(Z_0'', ((\nabla Z) \circ P_Z)^t \nabla_Z Z) \\
&= g(\nabla_{Z_0''} Z, \nabla_Z Z) + \|X\|^2 \left[-\left(C_1^1 \nabla \tilde{K}\right) Z + g(Z, \tilde{K} \nabla_Z Z) \right] \\
&\quad - g(\nabla_{P_Z Z_0''} Z, \nabla_Z Z) = \|X\|^2 \left\{ g(\nabla_Z Z, \nabla_Z Z) - \left(C_1^1 \nabla \tilde{K}\right) Z \right\}
\end{aligned}$$

so that (8.46) may be written as

$$\begin{aligned}
\frac{d^2}{ds^2} \left\{ \tilde{\mathcal{B}}_\Omega(Z_s) \right\}_{s=0} = \int_\Omega &\left\{ \operatorname{trace}\left((\nabla X)^t \circ (\nabla X) \right) \right. \\
&+ 2g(\nabla_X X, \nabla_Z Z) + g(\nabla_X Z + \nabla_Z X, \nabla_X Z + \nabla_Z X) \\
&\left. + \|X\|^2 \left[\|\nabla_Z Z\|^2 - \left(C_1^1 \nabla \tilde{K}\right) Z \right] \right\} d\,\mathrm{vol}(g).
\end{aligned} \tag{8.49}$$

Here $\nabla_Z Z = -A_Z Z \in (\mathbb{R}Z)^\perp$ and g is positive definite on $(\mathbb{R}Z)^\perp$ so we wrote $\|\nabla_Z Z\|^2$ instead of $g(\nabla_Z Z, \nabla_Z Z)$. The identity (8.49) is precisely the announced second variation formula (8.43). ■

Let $\left(\mathrm{Hess}\, \tilde{\mathcal{B}}_\Omega\right)_Z(X,X)$ denote the right hand member of (8.49) for any $X \in (\mathbb{R}Z)^\perp$.

Definition 8.25 A spatially harmonic reference frame Z on M is said to be *stable* if $\left(\mathrm{Hess}\, \tilde{\mathcal{B}}_\Omega\right)_Z(X,X) \geq 0$ for any relatively compact domain $\Omega \subset\subset M$ and any $X \in (\mathbb{R}Z)^\perp$. ■

Hopf vector fields on Lorentzian Berger spheres are stable critical points of the spacelike energy (cf. O. Gil-Medrano & A. Hurtado, [130]).

8.5. CONFORMAL VECTOR FIELDS

We start with the following

Definition 8.26 Let (M,g) be a Lorentzian manifold. A tangent vector field $X \in \mathfrak{X}(M)$ is said to be *closed* and *conformal* if there is $\phi \in C^\infty(M)$ such that $\nabla_V X = \phi V$ for any $V \in \mathfrak{X}(M)$. ■

Let (M,g) be a space-time carrying a timelike vector field $X \in \mathfrak{X}(M)$ which is closed and conformal. O. Gil-Medrano & A. Hurtado, [130], looked at the harmonicity of the reference frame

$$\nu = \frac{1}{\sqrt{-g(X,X)}} X.$$

Their approach relies on the following result of S. Montiel, [215]

Theorem 8.27 *Let (M,g) be an n-dimensional ($n \geq 2$) Lorentzian manifold carrying a timelike vector field $X \in \mathfrak{X}(M)$ which is closed and conformal. Then* i) *the distribution $(\mathbb{R}X)^\perp$ is integrable.* ii) *If $\mathcal{F}$ is the foliation tangent to $(\mathbb{R}X)^\perp$ then the functions $g(X,X)$, $\operatorname{div}(X)$ and $X(\phi)$ are constant along the leaves of $\mathcal{F}$.* iii) *The reference frame $\nu = [-g(X,X)]^{-1/2} X$ satisfies*

$$\nabla_\nu \nu = 0, \quad \nabla_Y \nu = \frac{\phi}{\sqrt{-g(X,X)}} Y, \tag{8.50}$$

for any $Y \in \mathfrak{X}(M)$ such that $g(Y,\nu) = 0$.

Using (8.50) one may easily compute the rough Laplacian of ν. Precisely let $\{E_\alpha : 1 \leq \alpha \leq n-1\}$ be a local orthonormal frame of $(\mathbb{R}\nu)^\perp$. Then

$$\begin{aligned}
\Delta_g \nu &= -\sum_{\alpha=1}^{n-1} \left\{ \nabla_{E_\alpha} \nabla_{E_\alpha} \nu - \nabla_{\nabla_{E_\alpha} E_\alpha} \nu \right\} \\
&= -\frac{\phi}{\sqrt{-g(X,X)}} \sum_{\alpha=1}^{n-1} \left(\nabla_{E_\alpha} E_\alpha - P_\nu \nabla_{E_\alpha} E_\alpha \right) = -\frac{\phi^2}{g(X,X)} n\nu.
\end{aligned}$$

Consequently

Theorem 8.28 (O. Gil-Medrano & A. Hurtado, [130]) *Let M be a compact n-dimensional, $n \geq 2$, Lorentzian manifold carrying a timelike closed and conformal vector field $X \in \mathfrak{X}(M)$. Then $v = [-g(X,X)]^{-1/2}X$ is a critical point of the ordinary energy function. Also v is geodesic hence it is spatially harmonic. If additionally M has nonnegative Ricci curvature in the null directions, then v is an absolute minimizer of the spacelike energy.*

For details on the last statement in Theorem 8.28 one may see [130], p. 93.

Example 8.29 A *generalized Robertson-Walker* space-time is a warped product $B \times_f F$ where $B \subseteq \mathbb{R}$ is an open interval equipped with the metric $-dt \otimes dt$ and (F, g_F) is a Riemannian manifold. Also $f : B \to (0, +\infty)$ is a smooth function. The vector field $Z = \partial/\partial t$ is referred to as a *comoving* reference frame on $B \times_f F$. Cf. L.J. Alias et al., [16]. When F is either the unit sphere $S^{n-1}(1)$, or $\mathbb{R}^{n-1}$, or the hyperbolic space $\mathbb{H}^{n-1}(-1)$, the corresponding warped product is referred to as a *Robertson-Walker* space-time. For an elementary introduction to warped products in semi-Riemannian geometry, one may see B. O'Neill, [229], p. 204–207. The simplest models of neighborhoods of stars and black holes are warped products (cf. Chapter 13 in [229]). By a result of S. Montiel, [215], any generalized Robertson-Walker space-time is locally isometric to a Lorentzian warped product one of whose factors is 1-dimensional and negative definite. As a consequence of Theorem 8.28, the comoving reference frame $\partial/\partial t$ on a generalized Robertson-Walker space-time M is spatially harmonic. Also, if M is compact then $\partial/\partial t$ is an absolute minimizer of the spacelike energy functional. ■

Example 8.30 (Reference frames on the Gödel universe) For each $\alpha > 0$ let us consider the Lorentzian metric

$$g = dx_1 \otimes dx_1 + dx_2 \otimes dx_2 - \frac{1}{2} e^{2\alpha x_1} dy \otimes dy - 2e^{\alpha x_1} dy \odot dt - dt \otimes dt$$

on $\mathbb{R}^4$ with the coordinates (x_1, x_2, y, t). This is an exact solution of the Einstein field equations where matter appears as a rotating pressure-free perfect fluid. Then the reference frame $\partial/\partial t$ is a harmonic vector field. Also, $\partial/\partial t$ is geodesic hence a spatially harmonic vector field, as well. Moreover, the reference frame $Z = \sqrt{2} e^{-\alpha x_1} \partial/\partial y$ is not spatially harmonic yet $\tilde{\mathcal{B}}_\Omega(Z) = \tilde{\mathcal{B}}_\Omega(\partial/\partial t)$ for any relatively compact domain $\Omega \subset\subset \mathbb{R}^4$. It may be shown that $\partial/\partial t$ is unstable. For details on the relevant calculations one may see [130], p. 94–97. ■

A study of (timelike) harmonic vector fields on a Lorentzian torus is performed by M. Soret et al., [108], in the spirit of the work by G. Wiegmink, [309].

APPENDIX A

Twisted Cohomologies

As previously seen (cf. Proposition 1.13 and Theorem 1.17 in Chapter 1) the isotropic almost complex structures $J_{\delta,\sigma}$ on $T(M)$ are rarely integrable. It is therefore natural to look at the twisted Dolbeau cohomology of the almost complex manifold $(T(M), J_{\delta,\beta})$, cf. e.g., I. Vaisman, [297]. Appendix A is devoted to briefly reviewing a few notions and results on the *twisted cohomology* of a pseudocomplex of cochains, cf. e.g., S. Halperin & D. Lehmann, [156], with an application to the Dolbeau pseudocomplex of the almost complex manifold $(T(M), J_{1,0})$. Let $\mathcal{C}$ be a sequence of Abelian groups and group homomorphisms

$$\cdots \to C^{p-1}(\mathcal{C}) \xrightarrow{\delta^{p-1}} C^p(\mathcal{C}) \xrightarrow{\delta^p} C^{p+1}(\mathcal{C}) \to \cdots$$

where in general $\delta^p \circ \delta^{p-1} \neq 0$. Then $\mathcal{C}$ is a *pseudocomplex of cochains* (the terminology is due to I. Vaisman, [296]) and $\mathcal{C}$ is an ordinary cochain complex when $\delta^p \circ \delta^{p-1} = 0$ for any p. One natural way to associate a concept of cohomology to a pseudocomplex $\mathcal{C}$ is to set (cf. S. Halperin & D. Lehmann, [155])

$$H^p(\mathcal{C}) = \frac{\mathrm{Ker}(\delta^p)}{\mathrm{Ker}(\delta^p) \cap \delta^{p-1} C^{p-1}(\mathcal{C})}.$$

There are however several chain complexes whose associated cohomology groups are isomorphic to $H^p(\mathcal{C})$. For instance (cf. again [155]) let $\underline{\mathcal{C}}$ be the sequence

$$\cdots \to C^{p-1}\left(\underline{\mathcal{C}}\right) \xrightarrow{\underline{\delta}^{p-1}} C^p\left(\underline{\mathcal{C}}\right) \xrightarrow{\underline{\delta}^p} C^{p+1}\left(\underline{\mathcal{C}}\right) \to \cdots$$

$$C^p\left(\underline{\mathcal{C}}\right) = \mathrm{Ker}(\delta^{p+1} \circ \delta^p) \subseteq C^p(\mathcal{C}), \quad \underline{\delta}^p = \delta^p\big|_{C^p(\underline{\mathcal{C}})}.$$

If $x \in C^p(\underline{\mathcal{C}})$ then $y \equiv \delta^p(x) \in C^{p+1}(\mathcal{C})$ lies indeed in $C^{p+1}(\underline{\mathcal{C}})$ as

$$(\delta^{p+2} \circ \delta^{p+1})y = \delta^{p+2} \circ (\delta^{p+1} \circ \delta^p)x = 0$$

so that the restriction of δ^p to $C^p(\underline{\mathcal{C}})$ is $C^{p+1}(\underline{\mathcal{C}})$-valued. Moreover for any $x \in C^p(\underline{\mathcal{C}})$

$$(\underline{\delta}^{p+1} \circ \underline{\delta}^p)x = (\delta^{p+1} \circ \delta^p)x = 0$$

© 2012 Elsevier Inc. All rights reserved.

so that $\underline{\mathcal{C}}$ is a cochain complex. Its cohomology

$$H^p(\underline{\mathcal{C}}) = \frac{Z^p(\underline{\mathcal{C}})}{B^p(\underline{\mathcal{C}})}$$

coincides with the cohomology of the pseudocomplex $\mathcal{C}$. Indeed

$$\begin{aligned} Z^p(\underline{\mathcal{C}}) &= \mathrm{Ker}(\underline{\delta}^p) = \{x \in C^p(\underline{\mathcal{C}}) : \underline{\delta}^p x = 0\} \\ &= \{x \in \mathrm{Ker}(\delta^{p+1} \circ \delta^p) : \delta^p x = 0\} = \mathrm{Ker}(\delta^p), \\ B^p(\underline{\mathcal{C}}) &= \underline{\delta}^{p-1} C^{p-1}(\underline{\mathcal{C}}) = \delta^{p-1}\mathrm{Ker}(\delta^p \circ \delta^{p-1}) \\ &= \mathrm{Ker}(\delta^p) \cap \delta^{p-1} C^{p-1}(\mathcal{C}), \end{aligned}$$

hence $H^p(\underline{\mathcal{C}}) = H^p(\mathcal{C})$. Another option is the sequence $\overline{\mathcal{C}}$ given by

$$\cdots \to C^{p-1}(\overline{\mathcal{C}}) \xrightarrow{\overline{\delta}^{p-1}} C^p(\overline{\mathcal{C}}) \xrightarrow{\overline{\delta}^{p}} C^{p+1}(\overline{\mathcal{C}}) \to \cdots$$

$$C^p(\overline{\mathcal{C}}) = \frac{C^p(\mathcal{C})}{\delta^{p-1}\delta^{p-2}C^{p-2}(\mathcal{C})}, \quad \overline{\delta}^p \pi^p x = \pi^{p+1}\delta^p x, \quad x \in C^p(\mathcal{C}),$$

where $\pi^p : C^p(\mathcal{C}) \to C^p(\overline{\mathcal{C}})$ is the canonical projection. First it should be observed that $\pi^p x = \pi^p x'$ yields $x' - x = \delta^{p-1}\delta^{p-2} y$ for some $y \in C^{p-2}(\mathcal{C})$. Then

$$\delta^p x' - \delta^p x = \delta^p \delta^{p-1} \delta^{p-2} y \in \delta^p \delta^{p-1} C^{p-1}$$

hence $\pi^{p+1}\delta^p x = \pi^{p+1}\delta^p x'$ that is the definition of $\overline{\delta}^p \pi^p x$ doesn't depend upon the choice of representative. Moreover

$$\left(\overline{\delta}^p \circ \overline{\delta}^{p-1}\right) \pi^{p-1} x = \pi^{p+1}\left(\delta^p \delta^{p-1}\right) x = 0$$

for any $x \in C^{p-1}(\mathcal{C})$ so that $\overline{\mathcal{C}}$ is a cochain complex. Let us show that the cohomology groups of $\mathcal{C}$ and $\overline{\mathcal{C}}$ are isomorphic. Let us compute first the cycles of $\overline{\mathcal{C}}$

$$\begin{aligned} Z^p(\overline{\mathcal{C}}) &= \left\{\pi^p x : \overline{\delta}^p \pi^p x = 0\right\} = \left\{\pi^p x : \delta^p x \in \delta^p \delta^{p-1} C^{p-1}(\mathcal{C})\right\} \\ &= \left\{\pi^p x : \exists y \in C^{p-1}(\mathcal{C}) \text{ such that } x - \delta^{p-1} y \in \mathrm{Ker}(\delta^p)\right\}. \end{aligned}$$

As to the boundaries of $\overline{\mathcal{C}}$

$$B^p(\overline{\mathcal{C}}) = \overline{\delta}^{p-1} C^{p-1}(\overline{\mathcal{C}}) = \overline{\delta}^{p-1} \pi^{p-1} C^{p-1}(\mathcal{C}) = \pi^p \delta^{p-1} C^{p-1}(\mathcal{C}).$$

Next we consider the map $\Phi : H^p(\overline{\mathcal{C}}) \to H^p(\mathcal{C})$ given by

$$\begin{aligned}\Phi : [\pi^p x] &\equiv \pi^p x + B^p(\overline{\mathcal{C}}) \\ &\mapsto [x - \delta^{p-1} y] \equiv x - \delta^{p-1} y + \mathrm{Ker}(\delta^p) \cap \delta^{p-1} C^{p-1}(\mathcal{C}).\end{aligned}$$

If $y' \in C^{p-1}(\mathcal{C})$ is another element such that $x - \delta^{p-1} y \in \mathrm{Ker}(\delta^p)$ then

$$\mathrm{Ker}(\delta^p) \ni (x - \delta^{p-1} y) - (x - \delta^{p-1} y') = \delta^{p-1}(y' - y) \in \delta^{p-1} C^{p-1}(\mathcal{C})$$

hence $(x - \delta^{p-1} y) - (x - \delta^{p-1} y') \in \mathrm{Ker}(\delta^p) \cap \delta^{p-1} C^{p-1}(\mathcal{C})$ i.e. $[x - \delta^{p-1} y] = [x - \delta^{p-1} y'] \in H^p(\mathcal{C})$ so that the definition of $\Phi([\pi^p x])$ doesn't depend upon the choice of $y \in C^{p-1}(\mathcal{C})$. Also if $[\pi^p x] = [\pi^p x']$ then $\pi^p(x' - x) \in B^p(\overline{\mathcal{C}}) = \pi^p \delta^{p-1} C^{p-1}(\mathcal{C})$ i.e., $\pi^p(x' - x) = \pi^p \delta^{p-1} z$ for some $z \in C^{p-1}(\mathcal{C})$. Thus $x' - x - \delta^{p-1} z \in \delta^{p-1} \delta^{p-2} C^{p-2}(\mathcal{C})$ that is

$$x' - x - \delta^{p-1} z = \delta^{p-1} \delta^{p-2} w \tag{A.1}$$

for some $w \in C^{p-2}(\mathcal{C})$. As both $\pi^p x$ and $\pi^p x'$ are cycles of $\overline{\mathcal{C}}$ there exist $y, y' \in C^{p-1}(\mathcal{C})$ such that

$$x - \delta^{p-1} y, \ x' - \delta^{p-1} y' \in \mathrm{Ker}(\delta^p).$$

So on one hand $(x' - \delta^{p-1} y') - (x - \delta^{p-1} y) \in \mathrm{Ker}(\delta^p)$. On the other hand (by (A.1))

$$\begin{aligned}(x' - \delta^{p-1} y') - (x - \delta^{p-1} y) &= x' - x - \delta^{p-1}(y' - y) \\ &= \delta^{p-1}(z + \delta^{p-2} w) - \delta^{p-1}(y' - y) \in \delta^{p-1} C^{p-1}(\mathcal{C}).\end{aligned}$$

We may conclude that $x - \delta^{p-1} y$ and $x' - \delta^{p-1} y'$ are equivalent modulo $\mathrm{Ker}(\delta^p) \cap \delta^{p-1} C^{p-1}(\mathcal{C})$ so that $[x - \delta^{p-1} y] = [x' - \delta^{p-1} y'] \in H^p(\mathcal{C})$. Therefore $\Phi([\pi^p x])$ is well defined.

To check that Φ is a monomorphism let us assume that $\Phi([\pi^p x]) = 0$ i.e., $x - \delta^{p-1} y \in \mathrm{Ker}(\delta^p) \cap \delta^{p-1} C^{p-1}(\mathcal{C})$. In particular $x - \delta^{p-1} y \in \delta^{p-1} C^{p-1}(\mathcal{C})$ hence $x \in \delta^{p-1} C^{p-1}(\mathcal{C})$ from which

$$\pi^p x \in \pi^p \delta^{p-1} C^{p-1}(\mathcal{C}) = B^p(\overline{\mathcal{C}})$$

i.e., $[\pi^p x] = 0 \in H^p(\overline{\mathcal{C}})$.

To check that Φ is an epimorphism let $[z] \in H^p(\mathcal{C})$ so that $z \in \mathrm{Ker}(\delta^p)$. Let us choose just any $y \in C^{p-1}(\mathcal{C})$ and set $x \equiv \delta^{p-1} y + z \in C^p(\mathcal{C})$. Then

$$\overline{\delta}^p \pi^p x = \pi^{p+1} \delta^p x = \pi^{p+1}(\delta^p \delta^{p-1} y + \delta^p z) = 0$$

because $\delta^p\delta^{p-1}y \in \delta^p\delta^{p-1}C^{p-1}(\mathcal{C}) = \mathrm{Ker}(\pi^{p+1})$ and $\delta^p z = 0$. We have shown that $\pi^p x$ is a cycle of $\overline{\mathcal{C}}$ so that we may consider the corresponding cohomology class $[\pi^p x] \in H^p(\overline{\mathcal{C}})$ and compute

$$\Phi([\pi^p x]) = [x - \delta^{p-1}y] = [z].$$

Hence Φ is an isomorphism $H^p(\overline{\mathcal{C}}) \approx H^p(\mathcal{C})$.

The third and last (as far as Appendix A is concerned) alternative is to consider the sequence

$$\cdots \to C^{p-1}(\tilde{\mathcal{C}}) \xrightarrow{\tilde{\delta}^{p-1}} C^p(\tilde{\mathcal{C}}) \xrightarrow{\tilde{\delta}^{p}} C^{p+1}(\tilde{\mathcal{C}}) \to \cdots$$

$$C^p(\tilde{\mathcal{C}}) \equiv C^p(\mathcal{C}) \times \left(\delta^p\delta^{p-1}C^{p-1}(\mathcal{C})\right),$$

$$\tilde{\delta}^p(\lambda,\mu) \equiv (\delta^p\lambda - \mu, \delta^{p+1}(\delta^p\lambda - \mu)),$$

for any $\lambda \in C^p(\mathcal{C})$ and any $\mu \in \delta^p\delta^{p-1}C^{p-1}(\mathcal{C})$. Then

$$\tilde{\delta}^{p+1}\tilde{\delta}^p(\lambda,\mu) = \tilde{\delta}^{p+1}(\delta^p\lambda - \mu, \delta^{p+1}(\delta^p\lambda - \mu)) = 0$$

because of $\delta^{p+1}\lambda' - \mu' = 0$ where $\lambda' \equiv \delta^p\lambda\mu$ and $\mu' \equiv \delta^{p+1}(\delta^p\lambda - \mu)$. Therefore $\tilde{\mathcal{C}}$ is a cochain complex. The cycles of $\tilde{\mathcal{C}}$ are

$$Z^p(\tilde{\mathcal{C}}) = \{(\lambda,\mu) \in C^p(\tilde{\mathcal{C}}) : \delta^p\lambda = \mu\}$$

$$= \{(\lambda, \delta^p\delta^{p-1}\sigma) : \lambda \in C^p(\mathcal{C}),\ \sigma \in C^{p-1}(\mathcal{C}),\ \lambda - \delta^{p-1}\sigma \in \mathrm{Ker}(\delta^p)\}$$

while the boundaries are

$$B^p(\tilde{\mathcal{C}}) = \tilde{\delta}^{p-1}C^{p-1}(\tilde{\mathcal{C}})$$

$$= \left\{(\delta^{p-1}\alpha - \beta, \delta^p(\delta^{p-1}\alpha - \beta)) : \alpha \in C^{p-1}(\mathcal{C}),\ \beta \in \delta^{p-1}\delta^{p-2}C^{p-2}(\mathcal{C})\right\}.$$

Let us build the group homomorphism

$$\Psi : Z^p(\tilde{\mathcal{C}}) \to H^p(\mathcal{C}), \quad \Psi : (\lambda, \delta^p\delta^{p-1}\sigma) \mapsto [\lambda - \delta^{p-1}\sigma].$$

Then

$$\begin{aligned}\mathrm{Ker}(\Psi) &= \left\{(\lambda, \delta^p\delta^{p-1}\sigma) \in Z^p(\tilde{\mathcal{C}}) : \lambda - \delta^{p-1}\sigma \in \mathrm{Ker}(\delta^p) \cap \delta^{p-1}C^{p-1}(\mathcal{C})\right\} \\ &= \left\{(\delta^{p-1}\sigma - \gamma, \delta^p(\delta^{p-1}\sigma - \gamma)) : \sigma \in C^{p-1}(\mathcal{C}),\right. \\ &\qquad \left.\gamma \in \mathrm{Ker}(\delta^p) \cap \delta^{p-1}C^{p-1}(\mathcal{C})\right\}\end{aligned}$$

so that $\mathrm{Ker}(\Psi) \subset B^p(\tilde{\mathcal{C}})$. It follows that Ψ gives rise to a well-defined group homomorphism

$$H^p(\tilde{\mathcal{C}}) \to H^p(\mathcal{C}),\ (\lambda, \delta^p\delta^{p-1}\sigma) + B^p(\tilde{\mathcal{C}}) \mapsto [\lambda - \delta^{p-1}\sigma]. \qquad \text{(A.2)}$$

It remains to be shown that (A.2) is a group isomorphism. To this end it suffices to show that (A.2) is an epimorphism. To this end let $[\alpha] \in H^p(\mathcal{C})$ so that $\alpha \in \text{Ker}(\delta^p)$. Let us consider an arbitrary element $\sigma \in C^{p-1}(\mathcal{C})$ and set $\lambda \equiv \alpha + \delta^{p-1}\sigma$. Then $\lambda - \delta^{p-1}\sigma \in \text{Ker}(\delta^p)$ so that $(\lambda, \delta^p\delta^{p-1}\sigma) \in Z(\tilde{\mathcal{C}})$ and the morphism (A.2) maps $[(\lambda, \delta^p\delta^{p-1}\sigma)]$ into $[\lambda - \delta^{p-1}\sigma] = [\alpha]$. ■

The complex $\tilde{\mathcal{C}}$ (associated to the pseudocomplex $\mathcal{C}$) is the piece of homological algebra needed to build an analog to the Dolbeau cohomology for the almost complex manifold $(T(M), J_{\delta,\sigma})$. We set

$$T^{1,0}(T(M), J_{\delta,\sigma}) = \{X - iJ_{\delta,\sigma}X : X \in T(T(M))\} \quad (i = \sqrt{-1}).$$

Definition A.1 Let $f \in C^1(T(M))$ be a complex valued function on $T(M)$. We say f is *holomorphic* if $\overline{Z}(f) = 0$ for any $Z \in T^{1,0}(T(M), J_{\delta,\sigma})$. ■

We examine only the case of the almost complex structure $J = J_{1,0}$ discovered by P. Dombrowski, [99]. Let (U, x^i) be a local coordinate system on M and $(\pi^{-1}(U), x^i, y^i)$ the induced local coordinates on $T(M)$ so that $\{\delta_i, \dot{\partial}_i : 1 \le i \le n\}$ is a local frame of $T(T(M))$ on $\pi^{-1}(U)$ associated to a fixed nonlinear connection $\mathcal{H}$ on $T(M)$. Here $\delta_i = \delta/\delta x^i = \partial_i - N_i^j \dot{\partial}_j$ and $N_j^i \in C^\infty(\pi^{-1}(U))$ are the local coefficients of the nonlinear connection. Any $Z = \lambda^j\delta_j + \mu^j\dot{\partial}_j \in T^{1,0}(T(M), J)$ is an eigenvector of J corresponding to the eigenvalue i hence $\lambda^j = i\mu^j$ i.e., $\{\dot{\partial}_j + i\delta_j : 1 \le j \le n\}$ is a local frame for $T^{1,0}(T(M), J)$ defined on the open set $\pi^{-1}(U)$. Therefore $f \in C^1(T(M))$ is holomorphic on $T(M)$ if

$$\frac{\partial f}{\partial y^j} - i\frac{\delta f}{\delta x^j} = 0 \tag{A.3}$$

in $\pi^{-1}(U)$ for any local coordinate system (U, x^j) on M. If $f = u + iv$ are the real and imaginary parts of f then (A.3) may also be written

$$\frac{\delta u}{\delta x^j} - \frac{\partial v}{\partial y^j} = 0, \quad \frac{\partial u}{\partial y^j} + \frac{\delta v}{\delta x^j} = 0. \tag{A.4}$$

The equations (A.3) (or (A.4)) are referred to as the *Cauchy-Riemann equations* on $T(M)$. In spite of the formal similarity to the ordinary Cauchy-Riemann equations in $\mathbb{C}^n$, few results in classical complex analysis carry over to the case of solutions to the system (A.3) mainly due to the fact that arguments based on power series are unavailable.

Moreover if $Z_j = \dot{\partial}_j + i\delta_j$ then we may determine the dual local frame $\{\omega^j : 1 \le j \le n\}$ in $T^{1,0}(T(M))^*$ i.e., for the complex 1-forms determined by

$$\omega^j(Z_k) = \delta_k^j, \quad \omega^j(Z_{\overline{k}}) = 0, \tag{A.5}$$

where $Z_{\bar{j}} = \overline{Z}_j$. To this end we define the local 1-forms δy^i by setting

$$\delta y^j = dy^j + N^j_k dx^k, \quad 1 \le j \le n,$$

so that

$$\delta y^j(\delta_k) = 0, \quad \delta y^j(\dot{\partial}_k) = \delta^j_k, \quad dx^j(\delta_k) = \delta^j_k, \quad dx^j(\dot{\partial}_k) = 0.$$

If $\omega_j = f^k_j dx^k + g^k_j \delta y^k$ then (A.5) yields

$$g^j_k + if^j_k = \delta^j_k, \quad g^j_k - if^j_k = 0,$$

or $f^j_k = -\frac{i}{2}\delta^j_k$ and $g^j_k = \frac{1}{2}\delta^j_k$ hence

$$\omega^j = \frac{1}{2}\left(\delta y^j - i dx^j\right), \quad 1 \le j \le n.$$

The complex de Rham algebra of $(T(M), J)$ admits the well-known decomposition

$$\Omega^\bullet(T(M)) = \bigoplus_{p,q=1}^{n} \Omega^{p,q}(T(M)).$$

A form $\eta \in \Omega^{p,q}(T(M))$ admits the local representation

$$\eta = f_{j_1 \dots j_p \bar{k}_1 \dots \bar{k}_q}\, \omega^{j_1 \dots j_p \bar{k}_1 \dots \bar{k}_q}$$

where we set for simplicity

$$\omega^{j_1 \dots j_p \bar{k}_1 \dots \bar{k}_q} = \omega^{j_1} \wedge \dots \wedge \omega^{j_p} \wedge \omega^{\bar{k}_1} \wedge \dots \wedge \omega^{\bar{k}_q},$$

for some C^∞ functions $f_{j_1 \dots j_p \bar{k}_1 \dots \bar{k}_q} : \pi^{-1}(U) \to \mathbb{C}$. Here $\omega^{\bar{j}} = \overline{\omega^j}$. Let $d : \Omega^{p,q}(T(M)) \to \Omega^{p+q+1}(T(M))$ be the ordinary exterior differentiation operator. It admits the decomposition

$$d = \partial_{(1,0)} + \overline{\partial}_{(0,1)} + \mathcal{N}_{(2,-1)} + \overline{\mathcal{N}}_{(-1,2)}$$

where the subscripts (to be omitted in the sequel) indicate the type of the operator. For instance $\partial\eta$ is the $(p+1, q)$-component of $d\eta$ for any $\eta \in \Omega^{p,q}(T(M))$. It is an easy exercise that the property $d^2 = 0$ implies

$$\begin{gathered}\partial^2 + \mathcal{N}\overline{\partial} + \overline{\partial}\mathcal{N} = 0, \quad \overline{\partial}^2 + \partial\overline{\mathcal{N}} + \overline{\partial}\mathcal{N} = 0,\\ \mathcal{N}^2 = 0, \quad \overline{\mathcal{N}}^2 = 0, \\ \partial\overline{\partial} + \overline{\partial}\partial + \mathcal{N}\overline{\mathcal{N}} + \overline{\mathcal{N}}\mathcal{N} = 0, \quad \partial\mathcal{N} + \mathcal{N}\partial = \overline{\partial}\,\overline{\mathcal{N}} + \overline{\mathcal{N}}\,\overline{\partial} = 0.\end{gathered} \tag{A.6}$$

We need to compute the exterior differential of δy^j

$$
\begin{aligned}
d\delta y^j &= d\left(dy^j + N^j_k dx^k\right) = dN^j_k \wedge dx^k \\
&= \left(\frac{\partial N^j_k}{\partial x^\ell} dx^\ell + \frac{\partial N^j_k}{\partial y^\ell} dy^\ell\right) \wedge dx^k \\
&= \left(\frac{\partial N^j_k}{\partial x^m} - N^\ell_m \frac{\partial N^j_k}{\partial y^\ell}\right) dx^m \wedge dx^k + \frac{\partial N^j_k}{\partial y^\ell} \delta y^\ell
\end{aligned}
$$

that is

$$
d\delta y^j = \frac{\delta N^j_k}{\delta x^\ell} dx^\ell \wedge dx^k + \frac{\partial N^j_k}{\partial y^\ell} \delta y^\ell \wedge dx^k. \tag{A.7}
$$

Using

$$
\delta y^j = \omega^j + \omega^{\bar{j}}, \quad dx^j = i(\omega^j - \omega^{\bar{j}}),
$$

we get

$$
\begin{aligned}
dx^j \wedge dx^k &= -\omega^j \wedge \omega^k + \omega^j \wedge \omega^{\bar{k}} + \omega^{\bar{j}} \wedge \omega^k - \omega^{\bar{j}} \wedge \omega^{\bar{k}}, \\
\delta y^j \wedge dx^k &= i\left(\omega^j \wedge \omega^k - \omega^j \wedge \omega^{\bar{k}} + \omega^{\bar{j}} \wedge \omega^k - \omega^{\bar{j}} \wedge \omega^{\bar{k}}\right).
\end{aligned}
$$

Let us assume from now on that (M,g) is a Riemannian manifold and $\mathcal{H}$ is the nonlinear connection associated to the regular connection $\hat{\nabla}$ in $\pi^{-1}TM \to T(M)$ (where ∇ is the Levi-Civita connection of (M,g)). Using (A.7) we shall compute $d\omega^j = \frac{1}{2} d\delta y^j$ in terms of the tensor field $R^i_{jk} = \delta_k N^i_j - \delta_j N^i_k$. Indeed

$$
d\delta y^j = -\frac{1}{2} R^j_{k\ell} dx^k \wedge dx^\ell + \Gamma^j_{k\ell} \delta y^\ell \wedge dx^k
$$

hence

$$
4d\omega^j = R^j_{k\ell} \omega^k \wedge \omega^\ell - 2\left(R^j_{k\ell} + 2i\Gamma^j_{k\ell}\right) \omega^k \wedge \omega^{\bar{\ell}} + R^j_{k\ell} \omega^{\bar{k}} \wedge \omega^{\bar{\ell}}.
$$

Consequently

$$
4\,\partial\omega^j = R^j_{k\ell} \omega^k \wedge \omega^\ell, \quad 2\,\overline{\partial}\omega^j = -\left(R^j_{k\ell} + 2i\Gamma^j_{k\ell}\right) \omega^k \wedge \omega^{\bar{\ell}}, \tag{A.8}
$$

$$
\mathcal{N}\omega^j = 0, \quad 4\overline{\mathcal{N}}\omega^j = R^j_{k\ell} \omega^{\bar{k}} \wedge \omega^{\bar{\ell}}. \tag{A.9}
$$

As a byproduct, one obtains a new proof of Proposition 1.13 in Chapter 1 of this book. Precisely, J is integrable if and only if $\overline{\mathcal{N}} = 0$ (cf. also

Proposition 4.1 in [297], p. 363, relating $\overline{\mathcal{N}}$ to the Nijenhuis torsion of the almost complex structure). Yet $\overline{\mathcal{N}} = 0$ is equivalent to $R^i_{jk} = 0$. On the other hand $R^i_{jk} = R^i_{jk\ell} y^\ell$ so that $R^i_{jk} = 0$ if and only if (M,g) is flat. ■

A cohomology of $(T(M), J)$ similar to the Dolbeau cohomology of a complex manifold is obtained by considering the pseudocomplex $\mathcal{C} \equiv (\Omega^{0,\bullet}(T(M)), \overline{\partial})$ and the associated complex $\tilde{\mathcal{C}} \equiv (D^\bullet(T(M)), \overline{D})$ given by

$$D^q(T(M)) = \Omega^{0,q}(T(M)) \times \overline{\partial}^2 \Omega^{0,q-1}(T(M)),$$

$$\overline{D}(\lambda, \mu) \equiv (\overline{\partial}\lambda - \mu, \overline{\partial}(\overline{\partial}\lambda - \mu)),$$

for any $\lambda \in \Omega^{0,q}(T(M))$ and any $\mu \in \overline{\partial}^2 \Omega^{0,q-1}(T(M))$. By the previous preparation of homological algebra the cohomology groups

$$H^q(D^\bullet(T(M)), \overline{D}) = \frac{\operatorname{Ker}(\overline{D} : D^q(T(M)) \to D^{q+1}(T(M)))}{\overline{D}\, D^{q-1}(T(M))}$$

are naturally isomorphic to the cohomology groups of the pseudocomplex $(\Omega^{0,\bullet}(T(M)), \overline{\partial})$.

Definition A.2 The groups $H^q(T(M), J) \equiv H^q(D^\bullet(T(M)), \overline{D})$ are called the *Dolbeau cohomology* groups of the almost complex manifold $(T(M), J)$. ■

There are no known examples of explicit calculation of the Dolbeau cohomology groups, even in simple instances such as $H^q(T(S^2), J)$, $q \in \{1, 2\}$. Little may be said with elementary techniques. For instance

Proposition A.3 *Let (M,g) be a Riemannian manifold. Then $H^1(T(M), J) = 0$ if and only if for any $\lambda \in \Omega^{0,1}(T(M))$ and any $f \in \Omega^{0,0}(T(M))$ such that*

$$2(Z_{\overline{k}}\lambda_{\overline{\ell}} - Z_{\overline{\ell}}\lambda_{\overline{k}}) + R^j_{k\ell}\left(\lambda_{\overline{j}} + Z_j f\right) = 0 \tag{A.10}$$

for an arbitrary local coordinate system (U, x^i) on M, there is $g \in \Omega^{0,0}(T(M))$ such that $\lambda = \overline{\partial} g$ and

$$R^j_{k\ell} Z_j(f - g) = 0 \tag{A.11}$$

for any local coordinate system (U, x^i) on M.

Note that

$$D^1(T(M)) = \Omega^{0,1}(T(M)) \times \overline{\partial}^2 \Omega^{0,0}(T(M)),$$

$$\overline{D}(\lambda, \overline{\partial}^2 f) = (\overline{\partial}\lambda - \overline{\partial}^2 f, \overline{\partial}(\overline{\partial}\lambda - \overline{\partial}^2 f)),$$

$$\lambda \in \Omega^{0,1}(T(M)), \quad f \in \Omega^{0,0}(T(M)).$$

Locally $\lambda = \lambda_{\bar{j}} \omega^{\bar{j}}$ hence (by (A.8))

$$\overline{\partial}\lambda = \left\{ Z_{\bar{k}}(\lambda_{\bar{\ell}}) + \frac{1}{4} R^{j}_{k\ell} \lambda_{\bar{j}} \right\} \omega^{\bar{k}} \wedge \omega^{\bar{\ell}}$$

hence

$$\overline{\partial}\lambda - \overline{\partial}^{2} f = \left\{ Z_{\bar{k}}(\lambda_{\bar{\ell}} - Z_{\bar{\ell}} f) + \frac{1}{4} R^{j}_{k\ell} \left(\lambda_{\bar{j}} - Z_{\bar{j}} f \right) \right\} \omega^{\bar{k}} \wedge \omega^{\bar{\ell}}.$$

It follows that $\overline{D}(\lambda, \overline{\partial}^{2} f) = 0$ if and only if

$$Z_{\bar{k}} \lambda_{\bar{\ell}} - Z_{\bar{\ell}} \lambda_{\bar{k}} + \frac{1}{2} R^{j}_{k\ell} \lambda_{\bar{j}} = \left[Z_{\bar{k}}, Z_{\bar{\ell}} \right] f + \frac{1}{2} R^{j}_{k\ell} Z_{\bar{j}} f. \tag{A.12}$$

On the other hand

$$\begin{aligned} [Z_j, Z_k] &= i[\dot{\partial}_j, \delta_k] - i[\dot{\partial}_k, \delta_j] + [\delta_j, \delta_k] \\ &= i\left(\dot{\partial}_k N^{\ell}_j - \dot{\partial}_j N^{\ell}_k \right) \dot{\partial}_\ell - R^{\ell}_{jk} \dot{\partial}_\ell \end{aligned}$$

hence $\left[Z_j, Z_k \right] = -R^{\ell}_{jk} \dot{\partial}_\ell$. Consequently (A.12) is equivalent to (A.10) i.e., $(\lambda, \overline{\partial}^{2} f) \in \operatorname{Ker}(\overline{D})$ if and only if (A.10) holds.

Proof of Proposition A.3. $H^1(T(M), J) = 0$ if and only if for any $(\lambda, \overline{\partial}^{2} f) \in \operatorname{Ker}(\overline{D})$ there is $g \in \Omega^{0,0}(T(M))$ such that $\lambda = \overline{\partial} g$ and $\overline{\partial}^{2}(f - g) = 0$. Using (A.10) with $\lambda = \overline{\partial} g$ one shows easily that $\overline{\partial}^{2}(f - g) = 0$ if and only if (A.11) holds good. ■

We suggest several applications of twisted cohomologies to the geometry of almost complex and almost CR structures by quoting O. Muskarov, [218]-[220], J. Davidov & O. Muskarov, [93], P. De Bartolomeis, [95], R. Holubowicz & W. Mozgawa, [169], H. Hashimoto & K. Mashimo, [160], and H. Hashimoto & K. Mashimo & K. Sekigawa, [161].

The Stokes Theorem on Complete Manifolds

Let (M,g) be an n-dimensional orientable complete Riemannian manifold and $*: \Omega^\bullet(M) \to \Omega^{n-\bullet}(M)$ the Hodge operator. For each $\omega \in \Omega^\bullet(M)$ we set

$$|\omega| = \sqrt{*(\omega * \omega)}$$

(the pointwise norm of ω). Let $L^p(\Omega^\bullet(M))$ be the space of all differential forms $\omega \in \Omega^\bullet(M)$ such that $|\omega|^p \in L^1(M)$ i.e.,

$$\int_M |\omega|^p * 1 < \infty.$$

The following special form of the Stokes theorem is by now classical.

Theorem B.1 (M.P. Gaffney, [121]) *Let M be an orientable complete Riemannian manifold whose metric tensor is of class C^2. Let ω be a $(n-1)$-form of class C^1 such that $\omega \in L^1(\Omega^{n-1}(M))$ and $d\omega \in L^1(\Omega^n(M))$. Then $\int_M d\omega = 0$.*

The scope of Appendix B is to give a rigorous proof of Theorem B.1. One reason we include M.P. Gaffney's theorem in these notes is to suggest that many of the results we report on may be generalized to hold on complete Riemannian manifolds (although at present this monograph is mostly one of geometry and analysis on compact Riemannian manifolds). Another reason is frankly didactic.[1] We recall (cfr. e.g., [138], p. 70–72) that

$$\alpha \wedge *\beta = \beta \wedge *\alpha, \quad *\delta\alpha = (-1)^{\deg(\alpha)} d * \alpha\,.$$

[1] Indeed it came to some surprise (to the authors of this book) that the original proof by M.P. Gaffney of his Stokes type theorem on complete Riemannian manifolds (a result to be included in any textbook on Riemannian geometry, in the opinion of the authors) lacked the details and accuracy required for an exposition in front of a student public.

© 2012 Elsevier Inc. All rights reserved.

Let X be a vector field of class C^1 on M and ω the dual 1-form i.e., $\omega^\sharp = X$. Then

$$\operatorname{div}(X) * 1 = -(\delta\omega) * 1 = -*(\delta\omega) = d(*\omega)$$

hence (by Theorem B.1)

$$\int_M \operatorname{div}(X)\, d\operatorname{vol}(g) = 0 \tag{B.1}$$

provided that both $\omega = X^\flat$ and $d\omega$ are in L^1. In this form, Theorem B.1 was applied to prove Theorem 2.14 in Chapter 2 of this book.

To prove Theorem B.1 we need some preparation. Let $j(t)$ be a C^∞ function which vanishes outside the interval $[-1,1]$, is positive inside, and satisfies $\int_{-1}^{+1} j(t)\, dt = 1$. We set

$$j_\epsilon(x) = \epsilon^{-n} j\left(\frac{x_1}{\epsilon}\right)\cdots j\left(\frac{x_n}{\epsilon}\right), \quad x = (x_1,\dots,x_n) \in \mathbb{R}^n, \ \epsilon > 0.$$

The function $j_\epsilon(x)$ is the well-known *Friedrichs mollifier.* A fundamental property of $j_\epsilon(x)$ is stated as the following: Let $h(x)$ be a continuous function of compact support and let us set

$$(J_\epsilon h)(x) = (j_\epsilon * h)(x) = \int_{\mathbb{R}^n} j_\epsilon(x-y) h(y)\, dy.$$

Then $J_\epsilon h$ is a C^∞ function and $J_\epsilon h \to h$ uniformly as $\epsilon \to 0$. Moreover

$$\left|(J_\epsilon h)(x)\right| \leq \sup_{y\in Q_\epsilon(x)} |h(y)| \tag{B.2}$$

where $Q_\epsilon(x) \subset \mathbb{R}^n$ is the cube of width 2ϵ and center x. Cf. for instance Lemma 2.18 in [9], p. 29–30. Of course one should not confuse the convolution product $*$ in the definition of $J_\epsilon h$ with the Hodge operator. Another ingredient we need is

$$|\alpha * \beta| \leq |\alpha|\,|\beta| \tag{B.3}$$

for any $\alpha, \beta \in \Omega^p(M)$. This is an immediate consequence of the fact that $|\alpha + t\beta|^2 \geq 0$ for any $t \in \mathbb{R}$.

Let $r : M \to [0, +\infty)$ be the distance from a fixed point $x_0 \in M$ i.e., $r(x) = d(x_0, x)$ for any $x \in M$. The function r is clearly Lipschitz

$$|r(x) - r(y)| \leq d(x,y), \quad x, y \in M. \tag{B.4}$$

In particular, r is continuous yet it is not differentiable, in general. The main idea in the proof of Theorem B.1 is to approximate r on a large compact set by a C^1 function r_ϵ and then use (B.4) to prove that $|dr_\epsilon|$ is bounded. The approximation is accomplished by using the Friedrichs mollifier.

We set $S_t = \{x \in M : r(x) \geq t\}$ for each $t > 0$. Also let $C(S_t) = M \setminus S_t$ be the complement of S_t. Note that $\int_{C(S_k)} |\alpha| * 1 \to 0$ as $k \to \infty$, for any $\alpha \in \Omega^\bullet(M)$.

Let $f(t)$ be a C^1 function such that $f(t) = 0$ for $t \leq 0$, $0 < f(t) < 1$ for $0 < t < 1$, and $f(t) = 1$ for $t \geq 1$. Let us set $B = \sup_{t\in\mathbb{R}} f'(t)$. Let $\omega \in \Omega^{n-1}(M)$ be a $(n-1)$-form as in Theorem B.1. Then for any $\epsilon > 0$ there is $k_\epsilon \geq 1$ such that

$$\int_{C(S_{k-1})} |d\omega| * 1 < \frac{\epsilon}{2}, \quad \int_{C(S_{k-1})} |\omega| * 1 < \frac{\epsilon}{4B},$$

for any $k \geq k_\epsilon$.

By the assumption of completeness, every closed bounded subset of M is compact. This is of course a consequence of the well-known Hopf-Rinow theorem (cf. e.g. Theorem 4.1 in [189], Vol. I, p. 172). Let $k \geq k_\epsilon$. Then S_{k+4} is a compact set. Let $\mathcal{O}$ be an orientation of M such that $\varphi(U) = \mathbb{R}^n$ for any $(U, \varphi) \in \mathcal{O}$. We choose a finite smooth partition of unity $\{(U_i, \phi_i) : 1 \leq i \leq p\}$ on S_{k+4} subordinated to an open covering of S_{k+4} with coordinate neighborhoods in $\mathcal{O}$ so that

$$(U_i, \varphi_i) \in \mathcal{O}, \quad \operatorname{diam}(U_i) \leq \frac{1}{2},$$
$$K_i = \operatorname{supp}(\phi_i) \subset U_i, \quad 1 \leq i \leq p.$$

Of course

$$S_{k+4} \subset \bigcup_{i=1}^{p} U_i, \quad \phi_i \in C^\infty(M), \quad 0 \leq \phi_i \leq 1, \quad \sum_{i=1}^{p} \phi_i = 1.$$

Given $\epsilon > 0$ we define $J^i_\epsilon(\phi_i r) : M \to \mathbb{R}$ by setting

$$J^i_\epsilon(\phi_i r)(x) = \begin{cases} J_\epsilon\left[(\phi_i r) \circ \varphi_i^{-1}\right](\varphi_i(x)), & \text{if } x \in U_i, \\ 0, & \text{if } x \in M \setminus U_i, \end{cases}$$

for any $x \in M$. To show that $J^i_\epsilon(\phi_i r) \in C^\infty(M)$, it suffices to check that $\operatorname{supp}[J^i_\epsilon(\phi_i r)] \subset U_i$. To this end let us recall (cf. [9], p. 30) that for any domain $\Omega \subseteq \mathbb{R}^n$ one has $J_\epsilon u \in C^\infty_0(\Omega)$ provided that $\operatorname{supp}(u) \subset\subset \Omega$ and

$\epsilon < \text{dist}(\text{supp}(u), \partial\Omega)$. As $\text{supp}(\phi_i r) \subseteq K_i$ it follows that $J_\epsilon(u_i) \in C_0^\infty(\mathbb{R}^n)$ where we set $u_i = (\phi_i r) \circ \varphi_i^{-1} : \mathbb{R}^n \to [0, +\infty)$ for simplicity. Finally if $K_{\epsilon,i} = \text{supp}[J_\epsilon(u_i)]$ then the set $\text{supp}[J_\epsilon^i(\phi_i r)] = \varphi_i^{-1}(K_{\epsilon,i})$ is both compact and contained in U_i.

It is an elementary fact of metric geometry that there is $\epsilon_0 > 0$ such that for each $0 < \epsilon \le \epsilon_0$ one has

$$j_\epsilon(\xi - \eta)(\phi_i r)(\varphi_i^{-1}(\eta)) = 0$$

for $\varphi_i^{-1}(\eta)$ near ∂U_i. Precisely let us set

$$Q_\epsilon(0) = \{\eta \in \mathbb{R}^n : |\eta^i| < \epsilon, \ 1 \le i \le n\}$$

and note that

$$\eta \in Q_\epsilon(0) \Longrightarrow j_\epsilon(\eta) = 0.$$

By changing variables

$$J_\epsilon(\phi_i r)(x) = \int_{\mathbb{R}^n} j_\epsilon(\eta) u_i(\xi - \eta)\, d\eta, \quad x \in U_i, \tag{B.5}$$

where $\xi = \varphi_i(x)$. Let $x_i = \varphi_i^{-1}(0) \in U_i$. There is $\rho_i > 0$ such that $B(x_i, 2\rho_i) \subseteq U_i$. We set $\rho = \min\{\rho_i : 1 \le i \le p\} > 0$. As the sets $\{Q_\epsilon(0) : \epsilon > 0\}$ form a fundamental system of open neighborhoods of the origin we may choose $\epsilon_i > 0$ such that

$$Q_{\epsilon_i}(0) \subseteq \varphi_i(B(x_i, \rho)).$$

We set $\epsilon_0 = \min\{\epsilon_i : 1 \le i \le p\} > 0$. We claim that

$$\text{dist}\left(\varphi_i^{-1}(Q_\epsilon(0)), \partial U_i\right) \ge \rho \tag{B.6}$$

for any $1 \le i \le p$ and any $0 < \epsilon \le \epsilon_0$. The proof is by contradiction. If

$$\rho > \text{dist}\left(\varphi_i^{-1}(Q_\epsilon(0)), \partial U_i\right) = \inf\left\{\text{dist}(x, \partial U_i) : x \in \varphi_i^{-1}(Q_\epsilon(0))\right\}$$

for some $i \in \{1, \dots, p\}$ and some $0 < \epsilon \le \epsilon_0$ then

$$\rho > \text{dist}(x, \partial U_i) = \inf\{d(x, y) : y \in \partial U_i\}$$

for some $x \in \varphi_i^{-1}(Q(\epsilon))$ that is $\rho > d(x, y)$ for some $y \in \partial U_i$. Then

$$\varphi_i(x) \in Q_\epsilon(0) \subseteq Q_{\epsilon_0}(0) \subset \varphi_i(B(x_i, \rho))$$

implies that $x \in B(x_i, \rho)$ so that $d(x, x_i) < \rho$. Finally, by the triangle inequality

$$d(x_i, y) \leq d(x_i, x) + d(x, y) < 2\rho$$

that is $y \in B(x_i, 2\rho)$ which of course means that $B(x_i, 2\rho) \cap \partial U_i \neq \emptyset$, a contradiction. Inequality (B.6) is proved. Now inequality (B.6) may be indeed interpreted as announced i.e., the integrand in the right hand side of (B.5)

$$f_i(y) = j_\epsilon(\varphi_i(y))u_i(\xi - \varphi_i(y)), \quad y \in U_i,$$

vanishes near ∂U_i. For instance f_i vanishes on the one-sided neighborhood of the boundary $\{y \in U_i : \text{dist}(y, \partial U_i) \leq \rho/2\}$.

Let $0 < \epsilon \leq \epsilon_0$ with $\epsilon_0 > 0$ built as above. Let us consider the function

$$r_\epsilon : M \to [0, +\infty),$$

$$r_\epsilon(x) = \begin{cases} \sum_{i=1}^{p} J_\epsilon^i(\phi_i r)(x), & x \in S_{k+3}, \\ k+3, & x \in C(S_{k+3}), \end{cases}$$

for any $x \in M$. Let $x \in S_{k+3}$. As

$$J_\epsilon^i(\phi_i r)(x) \to (\phi_i r)(x), \quad \epsilon \to 0,$$

it follows that

$$r_\epsilon(x) \to \sum_{i=1}^{p} \phi_i(x) r(x) = r(x), \quad \epsilon \to 0.$$

Consequently there is $0 < \epsilon_1 \leq \epsilon_0$ such that for any $0 < \epsilon \leq \epsilon_1$ one has

$$|r_\epsilon(x) - r(x)| < \frac{1}{2}, \quad x \in S_{k+3}. \tag{B.7}$$

Let us set $f_k(t) = f(t-k)$ so that $f_k(r_\epsilon(x)) = 1$ for any $x \in C(S_{k+2})$ and any $0 < \epsilon \leq \epsilon_1$. Indeed if $x \in C(S_{k+2})$ then $r(x) > k+2$ hence

$$r_\epsilon(x) > r(x) - \frac{1}{2} > k + \frac{3}{2} \Longrightarrow r_\epsilon(x) - k > 1.$$

Let $0 < \epsilon \leq \epsilon_1$ and $k \geq k_\epsilon$. We may write $d\omega$ as

$$d\omega = d[f_k(r_\epsilon)\omega] + d\big[\big(1 - f_k(r_\epsilon)\big)\,\omega\big]$$

where $f(r_\epsilon) = f \circ r_\epsilon$. Then

$$\int_M d\omega = \int_M d[f_k(r_\epsilon)\omega] + \int_M d\big[\big(1 - f_k(r_\epsilon)\big)\,\omega\big]$$

$$= \int_{C(S_k^\epsilon)} d[f_k(r_\epsilon)\omega] + \int_{S_{k+2}} d\big[\big(1 - f_k(r_\epsilon)\big)\,\omega\big]$$

where $S_t^\epsilon = \{x \in M : r_\epsilon(x) \le t\}$ for each $t > 0$. Indeed $f_k(r_\epsilon)$ is zero on S_k^ϵ and $1 - f_k(r_\epsilon)$ is zero on $C(S_{k+2})$. We claim that $(1 - f_k(r_\epsilon))\omega$ is supported in the interior of S_{k+2} hence

$$\int_{S_{k+2}} d\big[\big(1 - f_k(r_\epsilon)\big)\,\omega\big] = 0$$

by the ordinary Stokes theorem (for forms with compact support). Indeed if $x \in S_{k+2}$ is such that $1 - f_k(r_\epsilon(x)) \neq 0$ then $r_\epsilon(x) < k+1$ yet (by (B.7)) $-1/2 + r(x) < r_\epsilon(x)$ so that $r(x) < k + 3/2$ that is

$$\{x \in S_{k+2} : 1 - f_k(r_\epsilon(x)) \neq 0\} \subset S_{k+3/2}$$

and by passing to closures

$$\operatorname{supp}[1 - f_k(r_\epsilon)] \subseteq S_{k+3/2} \subset \operatorname{Int}(S_{k+2}).$$

It remains that

$$\int_M d\omega = \int_{C(S_k^\epsilon)} d[f_k(r_\epsilon)\omega]$$

$$= \int_{C(S_k^\epsilon)} f_k'(r_\epsilon)\, dr_\epsilon \wedge \omega + \int_{C(S_k^\epsilon)} f_k(r_\epsilon)\, d\omega$$

$$= \int_{S_{k+1}^\epsilon \setminus S_k^\epsilon} f_k'(r_\epsilon)\, dr_\epsilon \wedge \omega + \int_{C(S_k^\epsilon)} f_k(r_\epsilon)\, d\omega$$

as $f_k'(r_\epsilon)$ is zero outside S_{k+1}^ϵ by the very construction of f_k. We wish to estimate the integral

$$\int_{S_{k+1}^\epsilon \setminus S_k^\epsilon} f_k'(r_\epsilon)\, dr_\epsilon \wedge \omega .$$

To this end we need the following

Proposition B.2 *There is a constant* $C > 0$ *and* $0 < \epsilon_2 \leq \epsilon_1$ *such that* $|dr_\epsilon| \leq C$ *on* S_{k+2} *for any* $0 < \epsilon \leq \epsilon_2$. ∎

The proof of Proposition B.2 will be given later on. We claim that $S^\epsilon_{k+1} \subset S_{k+3}$. Indeed if $x \in S^\epsilon_{k+1}$ then $r_\epsilon(x) \leq k+1$ hence x must lie in S_{k+3} (by the very definition of r_ϵ we know that $r_\epsilon = k+3 > k+1$ outside S_{k+3}). In particular inequality (B.7) must hold on $S^\epsilon_{k+1} \setminus S^\epsilon_k$. We claim that

$$S^\epsilon_{k+1} \setminus S^\epsilon_k \subset C(S_{k-1}). \tag{B.8}$$

Indeed if $x \in S^\epsilon_{k+1} \setminus S^\epsilon_k$ then $x \in C(S^\epsilon_k) \subset C(S_{k-1})$. The last inclusion follows from $S^\epsilon_k \supset S_{k-1}$ which may be seen as follows. Let $y \in S_{k-1}$. Then

$$r_\epsilon(y) - \frac{1}{2} < r(y) \leq k-1$$

hence $y \in S_{k-1/2} \subset S^\epsilon_k$. Next (by (B.8) and Proposition B.2)

$$\left| \int_{S^\epsilon_{k+1}\setminus S^\epsilon_k} f'_k(r_\epsilon)\, dr_\epsilon \wedge \omega \right| \leq \int_{S^\epsilon_{k+1}\setminus S^\epsilon_k} \left|f'_k(r_\epsilon)\right| \, |dr_\epsilon| \, |\omega| * 1$$

$$\leq 2B \int_{C(S_{k-1})} |\omega| * 1 < \frac{\epsilon}{2}.$$

Using again $C\left(S^\epsilon_k\right) \subset C(S_{k-1})$ we may also estimate (as $0 \leq f_k(t) \leq 1$)

$$\left| \int_{C(S^\epsilon_k)} f_k(r_\epsilon)\, d\omega \right| \leq \int_{C(S^\epsilon_k)} \left|f_k(r_\epsilon)\right| \, |d\omega| * 1$$

$$\leq \int_{C(S_{k-1})} |d\omega| * 1 < \frac{\epsilon}{2}.$$

Summarizing the information obtained so far, it remains that

$$\left| \int_M d\omega \right| \leq \frac{\epsilon}{2} + \frac{\epsilon}{2} = \epsilon$$

for arbitrary $\epsilon > 0$ that is $\int_M d\omega = 0$ as claimed. Theorem B.1 is proved. Let us prove Proposition B.2. Let $x \in S_{k+2}$ and $0 < \epsilon \leq \epsilon_1$. Then (as $S_{k+2} \subset$

$\text{Int}(S_{k+3}))$

$$\frac{\partial r_\epsilon}{\partial x^j}(x) = \sum_{i=1}^{p} \frac{\partial J^i_\epsilon(\phi_i r)}{\partial x^j}(x) = \sum_{i=1}^{p} \frac{\partial J_\epsilon(u_i)}{\partial e_j}(\varphi_i(x))$$

where $\partial/\partial e_j$ denotes the directional derivative in the direction e_j (here $\{e_1, \dots, e_n\} \subset \mathbb{R}^n$ is the canonical linear basis). Let us set $\xi = \varphi_i(x)$. Then

$$\frac{\partial r_\epsilon}{\partial x^j}(x) = \sum_{i=1}^{p} \lim_{t \to 0} \frac{1}{t} \left\{ J_\epsilon(u_i)(\xi + te_j) - J_\epsilon(u_i)(\xi) \right\}$$
$$= \lim_{t \to 0} \frac{1}{t} \sum_{i=1}^{p} \left\{ \int_{\mathbb{R}^n} j_\epsilon(\xi + te_j - \eta) u_i(\eta)\, d\eta - \int_{\mathbb{R}^n} j_\epsilon(\xi - \eta) u_i(\eta) d\eta \right\}.$$

Let us change variables $\eta' = \eta - te_j$ in the first integral and then drop the accents. We obtain

$$\frac{\partial r_\epsilon}{\partial x^j}(x) = \sum_{i=1}^{p} \lim_{t \to 0} \frac{1}{t} \left\{ \int j_\epsilon(\xi - \eta) u_i(\eta + te_j) d\eta - \int j_\epsilon(\xi - \eta) u_i(\eta) d\eta \right\}$$
$$= \sum_{i} \lim_{t \to 0} \frac{1}{t} \left\{ \int j_\epsilon(\xi - \eta) \underline{\phi}_i(\eta + te_j) \left[\underline{r}(\eta + te_j) - \underline{r}(\eta) \right] d\eta \right.$$
$$\left. + \int j_\epsilon(\xi - \eta) \left[\underline{\phi}_i(\eta + te_j) - \underline{\phi}_i(\eta) \right] \underline{r}(\eta) d\eta \right\}$$

where $\underline{\phi}_i = \phi_i \circ \varphi_i^{-1}$ and $\underline{r} = r \circ \varphi_i^{-1}$. We need to show that there exist $t_0 \geq 0$ and a constant $C_j > 0$ such that for any $0 < t < t_0$

$$\frac{\rho(\eta, \eta + te_j)}{t} \leq \frac{C_j}{2} \tag{B.9}$$

for any $\eta \in \mathbb{R}^n$ such that $\text{dist}(\varphi_i^{-1}(\eta), \partial U_i) \geq \rho/2$. Here $\rho > 0$ is the number appearing in (B.6) and $\rho(\eta, \eta + te_j)$ is short for $d(\varphi_i^{-1}(\eta), \varphi_i^{-1}(\eta + te_j))$. We may assume without loss of generality that $\{(U_i, \varphi_i) : 1 \leq i \leq p\}$ consists of normal coordinate neighborhoods. As t is sufficiently small we may also assume that both $\varphi_i^{-1}(\eta)$ and $\varphi_i^{-1}(\xi + te_j)$ lie in a simple and convex open set Ω_i contained in U_i. Let $\gamma : [0, \ell(t)] \to \Omega_i$ be the unique minimizing geodesic, parametrized by arc length, joining $\varphi_i^{-1}(\eta)$ and $\varphi_i^{-1}(\eta + te_j)$, where $\ell(t) = \rho(\eta, \eta + te_j)$. As $\varphi_i = (x^1, \ldots, x^n)$ are normal coordinates the components $\gamma^k = x^k \circ \gamma$ of γ are given by

$$\gamma^k(s) = \eta^k + \frac{t}{\ell(t)} \delta^k_j s, \quad 0 \le k \le n,$$

for any $0 \le s \le \ell(t)$. Therefore for any $t > 0$ sufficiently small

$$\rho(\eta, \eta + te_j) = \int_0^{\ell(t)} \left[g_{k\ell}(\gamma(s)) \frac{d\gamma^k}{ds}(s) \frac{d\gamma^\ell}{ds}(s) \right]^{1/2} ds$$

$$= \frac{t}{\ell(t)} \int_0^{\ell(t)} g_{jj}(\gamma(s))^{1/2} ds \le t \frac{C_j}{2}$$

where

$$C_j = 2 \sup_{x \in L_i} |g_{jj}(x)|^{1/2}, \quad L_i = \left\{ x \in U_i : \text{dist}(x, \partial U_i) \ge \frac{\rho}{2} \right\}.$$

Statement (B.9) is proved.

As the directional derivative $(\partial r_\epsilon / \partial x^j)(x)$ exists for any $x \in U_i$ one may use a particular sequence $t \to 0$ with $t > 0$ in the above calculations. Using (B.9) we may estimate the integral

$$I_i(t) = \frac{1}{t} \int j_\epsilon(\xi - \eta) \underline{\phi}_i(\eta + te_j) \left[\underline{r}(\eta + te_j) - \underline{r}(\eta) \right] d\eta$$

as follows

$$|I_i(t)| \le \int j_\epsilon(\xi - \eta) \underline{\phi}_i(\eta + te_j) \left| \frac{\underline{r}(\eta + te_j) - \underline{r}(\eta)}{\rho(\eta, \eta + te_j)} \right| \frac{\rho(\eta, \eta + te_j)}{t} d\eta$$

(by (B.9) and the Lipschitz property of r)

$$\le \frac{C_j}{2} \int j_\epsilon(\xi - \eta) \phi_i(\varphi_i^{-1}(\eta + te_j)) \, d\eta \to \frac{C_j}{2} J_\epsilon(\phi_i \circ \varphi_i^{-1})(\xi), \quad t \to 0^+.$$

The reason one may use (B.9) is that the integrand in $I_i(t)$ vanishes outside the compact set L_i. Moreover

$$\sum_{i=1}^p J_\epsilon \left(\phi_i \circ \varphi_i^{-1} \right)(\xi) \to \sum_{i=1}^p \phi_i(x) = 1, \quad \epsilon \to 0.$$

Hence there is $0 < \epsilon_2' \leq \epsilon_1$ such that for any $0 < \epsilon \leq \epsilon_2'$

$$\sum_{i=1}^{p} \left| \lim_{t \to 0+} I_i(t) \right| \leq \frac{C_j}{2}. \tag{B.10}$$

It remains that we estimate the integral

$$J_i(t) = \frac{1}{t} \int j_\epsilon(\xi - \eta) \left[\underline{\phi}_i(\eta + te_j) - \underline{\phi}_i(\eta) \right] \underline{r}(\eta) d\eta.$$

We have

$$\left| J_i(t) \right| \leq \int j_\epsilon(\xi - \eta) \left| \frac{\underline{\phi}_i(\eta + te_j) - \underline{\phi}_i(\eta)}{t} - \frac{\partial \underline{\phi}_i}{\partial e_j}(\eta) \right| \underline{r}(\eta) d\eta$$

$$+ \int j_\epsilon(\xi - \eta) \frac{\partial \underline{\phi}_i}{\partial e_j}(\eta) \underline{r}(\eta)\, d\eta \to J_\epsilon \left(\underline{r} \frac{\partial \underline{\phi}_i}{\partial e_j} \right)(\xi), \quad t \to 0^+.$$

On the other hand

$$\sum_{i=1}^{p} J_\epsilon \left(\underline{r} \frac{\partial \underline{\phi}_i}{\partial e_j} \right)(\xi) \to \sum_{i=1}^{p} r(x) \frac{\partial \phi_i}{\partial x^j}(x) = 0, \quad \epsilon \to 0,$$

hence there is $0 < \epsilon_2'' \leq \epsilon_1$ such that for any $0 < \epsilon \leq \epsilon_2''$

$$\sum_{i=1}^{p} \left| \lim_{t \to 0^+} J_i(t) \right| \leq \frac{C_j}{2}. \tag{B.11}$$

Let us set $\epsilon_2 = \min\{\epsilon_2', \epsilon_2''\}$. Then (by (B.10)–(B.11)) for any $0 < \epsilon \leq \epsilon_2$

$$\left| \frac{\partial r_\epsilon}{\partial x^j}(x) \right| \leq C_j$$

for any $x \in S_{k+2}$ and any $i \in \{1, \ldots, p\}$ such that $x \in U_i$. Finally let $x \in S_{k+2}$ and let $i \in \{1, \ldots, p\}$ such that $x \in U_i$. Then

$$|dr_\epsilon|^2(x) = g^{jk}(x) \frac{\partial r_\epsilon}{\partial x^j}(x) \frac{\partial r_\epsilon}{\partial x^k}(x) \leq C_j C_k a^{jk},$$

$$a^{jk} = \max \left\{ \sup_{y \in S_{k+2} \cap U_i} \left| g^{jk}(y) \right| : 1 \leq i \leq p \right\},$$

and Proposition B.2 is proved.

APPENDIX C

Complex Monge-Ampère Equations

Contents

C.1. INTRODUCTION

Complex Monge-Ampère equations are nonlinear partial differential equations involving the complex Hessian $u_{j\overline{k}} = \partial^2 u/\partial z^j \partial \overline{z}^k$. These are related to the theory of functions in several complex variables and to complex geometry, in particular foliation theoretic aspects. On an arbitrary complex manifold V, of complex dimension n, one uses the first order differential operators ∂ and $\overline{\partial}$, whose definition is independent of the local complex coordinates one may adopt, and considers equations of the form

$$(\partial\overline{\partial}u)^{p+1} = 0, \quad \text{(C.1)}$$

$$\partial\overline{u} \wedge \overline{\partial}u \wedge (\partial\overline{\partial}u)^p = 0, \quad \text{(C.2)}$$

$$\overline{\partial}u \wedge (\partial\overline{\partial}u)^p = 0, \quad \text{(C.3)}$$

where, to start with, the functions $u : V \to \mathbb{C}$ are of class C^2 on V. Of course $\partial\overline{\partial}u$ is a $(1,1)$-form on V and the exponents denote exterior powers e.g., $(\partial\overline{\partial}u)^p = (\partial\overline{\partial}u) \wedge \cdots \wedge (\partial\overline{\partial}u)$ (p terms).

Definition C.1 The equation

$$\left(\partial\overline{\partial}u\right)^n = 0 \quad \text{(C.4)}$$

is called the *(homogeneous) complex Monge-Ampère* equation. ■

© 2012 Elsevier Inc. All rights reserved.

With respect to a local system of complex coordinates $(U, z^1, \ldots, z^n)$ on V the complex Monge-Ampère equation (C.4) may be written

$$\det\left[\frac{\partial^2 u}{\partial z^j \partial \overline{z}^k}\right] = 0. \tag{C.5}$$

Complex Monge-Ampère equations first appeared in the work of H. Bremmermann, [70]. When $n = 1$ the equation (C.5) is the Laplace equation $\Delta u = 0$ and, as argued in [206], (C.4) is the most natural extension of the Laplace equation to higher dimensional complex manifolds. Several properties of the forms in the left-hand side of the equations (C.1)–(C.13) were discussed by S.S. Chern & H. Levine & L. Nirenberg, [87], when u is a bounded, real, plurisubharmonic function. When restricted to a local coordinate neighborhood (U, z^j) the $(1,1)$-form $\partial\overline{\partial}u$ may be identified to $u_{j\overline{k}}$ so that the equations (C.1)–(C.3) are easily seen to be (locally) equivalent to

$$\operatorname{rank}\left[u_{j\overline{k}}\right] \leq p, \tag{C.6}$$

$$\operatorname{rank}\begin{bmatrix} 0 & u_{\overline{1}} & \cdots & u_{\overline{n}} \\ (\overline{u})_1 & u_{1\overline{1}} & \cdots & u_{1\overline{n}} \\ \vdots & \vdots & & \vdots \\ (\overline{u})_n & u_{n\overline{1}} & \cdots & u_{n\overline{n}} \end{bmatrix} \leq p, \tag{C.7}$$

$$\operatorname{rank}\begin{bmatrix} u_{\overline{1}} & \cdots & u_{\overline{n}} \\ u_{1\overline{1}} & \cdots & u_{1\overline{n}} \\ \vdots & & \vdots \\ u_{n\overline{1}} & \cdots & u_{n\overline{n}} \end{bmatrix} \leq p, \tag{C.8}$$

respectively. The substitutions $u = \exp(\beta v)$ and $u = \log w$ in (C.1) lead respectively to the equations

$$(\partial\overline{\partial}v)^p \wedge (\partial\overline{\partial}v + (p+1)\beta\, \partial v \wedge \overline{\partial}v) = 0, \tag{C.9}$$

$$(\partial\overline{\partial}w)^p \wedge (w\,\partial\overline{\partial}w + (p+1)\partial w \wedge \overline{\partial}w) = 0, \tag{C.10}$$

which are locally equivalent to

$$\operatorname{rank}\begin{bmatrix} \dfrac{1}{\beta} & v_{\overline{1}} & \cdots & v_{\overline{n}} \\ v_1 & v_{1\overline{1}} & \cdots & v_{1\overline{n}} \\ \vdots & \vdots & & \vdots \\ v_n & v_{n\overline{1}} & \cdots & v_{n\overline{n}} \end{bmatrix} \leq p, \tag{C.11}$$

$$\operatorname{rank}\begin{bmatrix} w & w_{\bar{1}} & \cdots & w_{\bar{n}} \\ w_1 & w_{1\bar{1}} & \cdots & w_{1\bar{n}} \\ \vdots & \vdots & & \vdots \\ w_n & w_{n\bar{1}} & \cdots & w_{n\bar{n}} \end{bmatrix} \leq p. \tag{C.12}$$

The form embraced by the equations (C.11)–(C.12) allows one to parallel the formal similarities among complex and real Monge-Ampère equations (the real Monge-Ampère equations being well known to involve determinants of the real Hessian). It is the proper place to make the following remark (following the exposition in [30], p. 544–545): Given a nonparametric real hypersurface $x^{n+1} = u(x^1,\ldots,x^n)$ in $\mathbb{R}^{n+1}$ such that the (real) Hessian of u has constant rank

$$\operatorname{rank}\left[\frac{\partial^2 u}{\partial x^j \partial x^k}\right] = p,$$

the graph of u is developable by real $(n-p)$-dimensional hyperplanes (and actually if $p=1$ and $n=2$ the hypersurface is locally a cylinder). The key ingredient in the proof of this result is the (geometric) interpretation of the Hessian $\left[\partial^2 u/\partial x^j \partial x^k\right]$ as the Jacobi matrix of the normal vector field $(u_{x_1},\ldots,u_{x_n},-1)$. See for instance P. Hartman & L. Nirenberg, [158].

Let us look at a complex analog to this situation. To this end, let us consider a domain $D \subset \mathbb{C}^n$ and a smooth function $u: D \to \mathbb{R}$. The Levi form (cf. e.g., [110], p. 5–6) of the real hypersurface

$$M = \left\{(z', z^{n+1}) \in \mathbb{C}^{n+1} : \log|z^{n+1}| + u(z') = 0, \quad z' \in D, \quad z^{n+1} \neq 0\right\}$$

has at least $n-p$ zero eigenvalues if and only if u satisfies (C.6). If the Levi form of M has exactly $n-p$ zero eigenvalues then by a result of F. Sommer, [272] (cf. also [24], p. 47–48, and p. 134–137) M is foliated by complex manifolds of complex dimension $n-p$. On the other hand, a result by K. Abe, [1], shows that the requirements ensuring that M is locally a cylinder should be much stronger than the condition (C.1) alone. The lack of an appropriate geometric interpretation of (C.1) is responsible for the further development of the theory (cf. E. Bedford & M. Kalka, [30], D. Burns, [73], P-M. Wong, [313; 314], T. Duchamp & M. Kalka, [111]) pursuing the foliation in Sommer's result (cf. *op. cit.*) rather than the "locally a cylinder" aspect.

C.2. STRICTLY PARABOLIC MANIFOLDS

Let V be a connected complex n-dimensional manifold and $\tau : V \to [0, +\infty)$ a C^∞ function with $\delta = \sup \sqrt{\tau} \leq \infty$.

Definition C.2 We call $\tau : V \to [0, +\infty)$ a *strictly parabolic exhaustion* of V if i) for any $r \in \mathbb{R}$ with $0 \leq r < \delta$ the set

$$V[r] = \{z \in V : \tau(z) \leq r^2\}$$

is compact, ii) $\tau < \delta^2$ on V, iii) $dd^c\tau > 0$ on V, and iv) the relations $dd^c \log \tau \geq 0$. $\left(dd^c \log \tau\right)^{n-1} \neq 0$ and

$$(dd^c \log \tau)^n = 0 \tag{C.13}$$

hold on $V_* = V \setminus V[0]$, where $d^c = (i/(4\pi))(\overline{\partial} - \partial)$. A pair (V, τ) consisting of a complex manifold V and a strictly parabolic exhaustion $\tau : V \to [0, +\infty)$ is called a *strictly parabolic manifold*. Also δ is the *maximal radius* of the exhaustion τ. ■

The equation (C.13) is the complex Monge-Ampère equation and the relations $dd^c \log \tau \geq 0$ and $\left(dd^c \log \tau\right)^{n-1} \neq 0$ are nondegeneracy conditions meaning that the (1,1)-form $\partial\overline{\partial} \log \tau$ has exactly $n-1$ strictly positive eigenvalues on V_*.

Example C.3 Let $V = \mathbb{C}^n$ and $\tau_0 : \mathbb{C}^n \to [0, +\infty)$ be given by $\tau_0(z) = |z|^2$ for any $z \in \mathbb{C}^n$. Given $0 < r \leq +\infty$ the set $\mathbb{C}^n(r) = \{z \in \mathbb{C}^n : \tau_0(z) < r^2\}$ is the open ball of radius r centered at 0. Then $(\mathbb{C}^n(r), \tau_0)$ is a strictly parabolic manifold of maximal radius r.

It is a by now classical result of W. Stoll, [275], that up to isomorphism the balls $(\mathbb{C}^n(r), \tau_0)$ if $0 < r < \infty$ and $(\mathbb{C}^n, \tau_0)$ if $r = \infty$ are the only strictly parabolic manifolds. Precisely

Theorem C.4 (W. Stoll, [275]) *Let (V, τ) be a strictly parabolic manifold of complex dimension n and maximal radius δ. There is a biholomorphism $F : \mathbb{C}^n(\delta) \to V$ such that $\tau \circ F = \tau_0$.*

Therefore the biholomorphism F furnished by Theorem C.4 preserves the exhaustions (i.e., $\tau \circ F = \tau_0$) and F is an isometry of Kaehler metrics i.e., $F^*(dd^c\tau) = dd^c\tau_0$. Alternative proofs of Theorem C.4 were given by D. Burns, [73]. A local study of this context was performed (by dropping the assumption that τ be an exhaustion function) by P-M. Wong, [313]. Cf. also Theorem A in [314], p. 226.

C.3. FOLIATIONS AND MONGE-AMPÈRE EQUATIONS

Let M be a real $2n$-dimensional manifold. For any ideal $\mathcal{J}$ in the de Rham algebra $\Omega^\bullet(M)$ we define a distribution $\mathrm{Ann}(\mathcal{J})$ on M by setting

$$\mathrm{Ann}(\mathcal{J})_x = \{X \in T_x(M) : X \rfloor \omega_x = 0, \quad \omega \in \mathcal{J}\}, \quad x \in M.$$

If $\mathcal{F}$ is a foliation of M each of whose leaves is a complex manifold, a function f on M is said to be *holomorphic* (respectively *pluriharmonic*) on $\mathcal{F}$ if the restriction $f\big|_L$ of f to each leaf $L \in M/\mathcal{F}$ is holomorphic (respectively pluriharmonic) on L. Let $T^{0,1}(\mathcal{F}) \to M$ be the complex vector bundle whose portion over each $L \in M/\mathcal{F}$ is the anti-holomorphic bundle $T^{0,1}(L) \to L$. We consider the first order differential operator $\overline{\partial}_{\mathcal{F}} : C^1(M,\mathbb{C}) \to \Gamma^0(T^{0,1}(\mathcal{F})^*)$ given by $\left(\overline{\partial}_{\mathcal{F}} f\right)\overline{Z} = \overline{Z}(f)$ for any C^1 function $f : M \to \mathbb{C}$ and any $Z \in \Gamma^\infty(T^{1,0}(\mathcal{F}))$. Here $T^{1,0}(\mathcal{F}) = \overline{T^{0,1}(\mathcal{F})}$. Then $f \in C^1(M,\mathbb{C})$ is holomorphic on $\mathcal{F}$ if $\overline{\partial}_{\mathcal{F}} f = 0$. The space of all functions holomorphic on $\mathcal{F}$ is denoted by $\mathcal{O}_{\mathcal{F}}(M)$.

As the aspects we discuss in the sequel are of local nature we restrict ourselves to a domain $D \subset \mathbb{C}^n$. We follow essentially the exposition in [30]. One may easily check

Lemma C.5 *Let* $\{\omega_1, \ldots, \omega_k\}$ *be a set of differential* q*-forms on* D *and* $\mathcal{J} \subset \Omega^\bullet(D)$ *the ideal generated by* $\{\omega_1, \ldots, \omega_k\}$. *If* $d\mathcal{J} \subseteq \mathcal{J}$ *then* $\mathrm{Ann}(\mathcal{J})$ *is involutive.*

Let $(z^1, \ldots, z^n)$ be the natural complex coordinates on $\mathbb{C}^n$. Let $\omega = (i/2)\omega_{j\overline{k}}\, dz^j \wedge d\overline{z}^k$ be a $(1,1)$-form on D with $\omega_{j\overline{k}} \in C^\infty(D,\mathbb{C})$. Then ω is real (i.e., $\omega = \overline{\omega}$) if and only if $\omega_{j\overline{k}} = \overline{\omega_{k\overline{j}}}$ i.e., $[\omega_{j\overline{k}}]$ is a Hermitian symmetric matrix. Therefore if ω is real and $A = [\omega_{j\overline{k}}]$ then there is a C^∞ map $U : D \to \mathrm{U}(n)$ such that

$$UAU^{-1} = \mathrm{diag}(\lambda_1, \ldots, \lambda_n)$$

for some $\lambda_j \in C^\infty(D,\mathbb{R})$, $1 \le j \le n$. Let us set $U = [U^j_k]$ and

$$\alpha^j = U^j_k\, dz^k \in \Omega^{1,0}(D), \quad 1 \le j \le n.$$

Then

$$\omega = \frac{i}{2}\lambda_j \alpha^j \wedge \alpha^{\overline{j}}, \quad \alpha^{\overline{j}} = \overline{\alpha^j}.$$

Next we set $\beta_j = |\lambda_j|^{1/2}\alpha_j$ so that

$$\omega = \frac{i}{2}\epsilon(\lambda_j)\,\beta^j \wedge \beta^{\bar{j}},$$

where $\epsilon(\lambda_j)(z)$ is the sign of the real number $\lambda_j(z)$ for any $z \in D$. Let $\beta^j = b^j - iJ^* b^j$ be the real and imaginary parts of β^j as a $(1,0)$-form on D. Here J^* is given by

$$(J^*\alpha)(X) = -\alpha(JX), \quad \alpha \in \Omega^1(D), \quad X \in \mathfrak{X}(D),$$
$$J^*(\alpha \wedge \beta) = (J^*\alpha) \wedge (J^*\beta),$$

and J is the complex structure on D. Then

$$\omega = -\epsilon(\lambda_j)\,b^j \wedge J^* b^j. \tag{C.14}$$

As a consequence of (C.14)

Lemma C.6 *Let $\mathcal{J} \subset \Omega^\bullet(D)$ be the ideal generated by the real $(1,1)$-forms $\{\omega_1, \ldots, \omega_k\}$ on D. Then $\mathrm{Ann}(\mathcal{J})$ is J-invariant.*

Next we shall establish

Proposition C.7 *Let $\omega = (i/2)\omega_{j\bar{k}}\,dz^j \wedge d\bar{z}^k$ be a real $(1,1)$-form on D. Then the distribution $\mathrm{Ann}(\omega) = \{X \in T(D) : X \rfloor \omega = 0\}$ has rank $2(n-p)$, i.e., $\dim_{\mathbb{R}} \mathrm{Ann}(\omega)_z = 2(n-p)$ for any $z \in D$, if and only if*

$$\omega^p \neq 0, \quad \omega^{p+1} = 0, \tag{C.15}$$

everywhere in D.

Proof. Let $Z = Z^j \partial/\partial z^j$ be a vector field of type $(1,0)$ on D. Then

$$Z \rfloor \omega = \frac{i}{4} Z^j \omega_{j\bar{k}}\,d\bar{z}^k. \tag{C.16}$$

Let us set

$$\mathrm{Ann}(\omega)^{1,0} = \{Z \in T^{1,0}(D) : Z \rfloor \omega = 0\}, \quad \mathrm{Ann}(\omega)^{0,1} = \overline{\mathrm{Ann}(\omega)^{1,0}}.$$

As a consequence of (C.16) and the fact that $[\omega_{j\bar{k}}]$ is Hermitian symmetric it follows that

$$\mathrm{Ann}(\omega) \otimes \mathbb{C} = \mathrm{Ann}(\omega)^{1,0} \oplus \mathrm{Ann}(\omega)^{0,1}.$$

Hence $X = Z + \overline{Z} \in \mathrm{Ann}(\omega)$ with $Z \in T^{1,0}(D)$ if and only if $Z \in \mathrm{Ann}(\omega)^{1,0}$ i.e., if and only if $Z^j \omega_{j\bar{k}} = 0$. It follows that

$$\dim_{\mathbb{R}} \mathrm{Ann}(\omega)_z = 2\dim_{\mathbb{C}} \mathrm{Ann}(\omega)^{1,0}_z = n - \mathrm{rank}[\omega_{j\bar{k}}(z)].$$

Finally it may be seen from

$$\omega^p = 2^{-p} i^{p^2} \omega_{j_1\overline{k}_1} \cdots \omega_{j_p\overline{k}_p} dz^{j_1} \wedge \cdots \wedge dz^{j_p} \wedge d\overline{z}^{k_1} \wedge \cdots \wedge d\overline{z}^{k_p}$$

that $\text{rank}[\omega_{j\overline{k}}] = p$ if and only if ω satisfies (C.15). ∎

The main purpose of this section is to establish the following

Theorem C.8 (E. Bedford & M. Kalka, [30]) *Let $D \subset \mathbb{C}^n$ be a domain and let $f \in C^3(D)$ satisfy $(dd^c f)^{p+1} = 0$ and $(dd^c f)^p \neq 0$ everywhere in D. Let $f = u + iv$ be the real and imaginary parts of f. If v is plurisubharmonic then there exists a foliation $\mathcal{F}_p$ of D by complex submanifolds of complex codimension p such that* i) *u, v are pluriharmonic on $\mathcal{F}_p$ and* ii) *$\partial u/\partial z^j$, $\partial v/\partial z^j \in \mathcal{O}_{\mathcal{F}_p}(D)$ for any $1 \leq j \leq n$.*

Proof. Let $\mathcal{J}$ be the ideal in $\Omega^\bullet(D)$ spanned by $dd^c u$ and $dd^c v$. Then the distribution $\text{Ann}(\mathcal{J}) \subset T(D)$ is integrable. Indeed $\text{Ann}(\mathcal{J})$ is involutive (by Lemma C.5) and has constant rank

$$\dim_{\mathbb{R}} \text{Ann}(\mathcal{J})_z = 2(n-p), \qquad z \in D,$$

∎

as a consequence of

Lemma C.9 *Let ω_1 and ω_2 be real $(1,1)$-forms on D such that $\omega_2(Z, \overline{Z}) \geq 0$ for any $Z \in T^{1,0}(D)$. Let λ_j^a, $b_a^j \in C^\infty(D)$, $a \in \{1,2\}$ as in the proof of Lemma* (C.6) *i.e.,*

$$\omega_a = -\sum_{j=1}^{n} \epsilon\left(\lambda_j^a\right) b_a^j \wedge J^* b_a^j, \quad a \in \{1,2\}.$$

Let $\mathcal{J}$ be the ideal in $\Omega^\bullet(D)$ spanned by $\{b_a^j, J^* b_a^j : 1 \leq j \leq n,\ a \in \{1,2\}\}$. Then $\dim_{\mathbb{R}} \text{Ann}(\mathcal{J})_z = 2(n-p)$ for any $z \in D$ if and only if $(\omega_1 + i\omega_2)^{p+1} = 0$ and $(\omega_1 + i\omega_2)^p \neq 0$ everywhere in D.

For a proof of Lemma C.9 the reader may see [30], p. 548–549. By Lemma C.6 the distribution $\text{Ann}(\mathcal{J})$ is J-invariant hence each leaf of the corresponding foliation $\mathcal{F}_p$ (such that $T(\mathcal{F}_p) = \text{Ann}(\mathcal{J})$) is a complex submanifold of D. Let $L \in D/\mathcal{F}_p$ be a leaf of $\mathcal{F}_p$ and let $\iota : L \to D$ be the inclusion. Let ∂_L and $\overline{\partial}_L$ be the ∂ and $\overline{\partial}$ operators relative to L (as a complex manifold). Let $\iota^* u = u|_L$ be the restriction of u to L. As $\text{Ann}(dd^c u) \supseteq T(L)$ it follows that

$$\partial_L \overline{\partial}_L (u|_L) = \partial_L \overline{\partial}_L \iota^* u = \iota^* \partial\overline{\partial} u = 0$$

hence $u|_L$ is pluriharmonic. A similar proof shows that v is also pluriharmonic along the leaves of $\mathcal{F}_p$. To end the proof of Theorem C.8 let us check for instance that $u_j = \partial u/\partial z^j$ is holomorphic along the leaves of $\mathcal{F}_p$. Let L be a leaf and $Z = Z^j \partial/\partial z^j \in T^{1,0}(L)$ a complex vector field of type $(1,0)$ on L. As

$$T(L) \subseteq \mathrm{Ann}(dd^c u), \quad J\,T(L) \subseteq \mathrm{Ann}(dd^c u),$$

it follows that

$$0 = Z \rfloor dd^c u = Z^j u_{j\bar{k}}\, d\bar{z}^k$$

hence

$$\overline{Z}(u_j) = \overline{Z^j}\, u_{\bar{j}k} = 0.$$

C.4. ADAPTED COMPLEX STRUCTURES

By a result of H. Grauert, [152], any real analytic manifold M may be embedded in a complex manifold as a maximal totally real submanifold. This may be seen by complexifying the transition functions defining M. Grauert's complexification is however not unique. The works by V. Guillemin & M. Stenzel, [154], and L. Lempert & R. Szöke, [206], represent an effort of formulating additional conditions on the ambient complex structure to make the complexification canonical for any given C^ω Riemannian manifold M. More precisely, in [154] and [206] one builds a complex structure on some open subset $\mathcal{A}^* \subseteq T^*(M)$ which is compatible with the canonical symplectic structure on $T^*(M)$. Equivalently there is a unique complex structure J on some open subset $\mathcal{A} \subseteq T(M)$ (referred to as an *adapted* complex structure) such that the leaves of the Riemann foliation of $T(M)$ are holomorphic curves. The set $T^r(M) = \{X \in T(M) : \|X\|_g < r\}$ equipped with J is a *Grauert tube*. As the adapted complex structure is derived from the given metric g on M in a canonical way, it turns out that dF is an automorphism of $(T^r(M), J)$ for any $F \in \mathrm{Isom}(M, g)$. If all automorphisms of $T^r(M)$ are obtained this way, the Grauert tube $T^r(M)$ is said to be *rigid*. Rigidity of Grauert tubes was investigated by D. Burns, [74], and D. Burns & R. Hind, [75], when M is compact, and by S-J. Kan, [181], for (not necessarily compact) C^ω homogeneous Riemannian spaces. Adapted complex structures and the corresponding compatible metrics on the tangent bundle are the reason to be of Appendix C, based on the belief that a better understanding of the rich geometric structure of the tangent bundle of a Riemannian manifold will lead to a further development of the analysis of harmonic vector fields and their generalizations.

Let (M,g) be a Riemannian manifold. As seen in Chapter 1 of this book, the decomposition $T_v(T(M)) = \mathcal{H}_v \oplus \mathcal{V}_v$, $v \in T(M)$, arising from g already produces an almost complex structure $J_{1,0}$ on $T(M)$, discovered in [99], yet integrable if and only if g is flat. On the other hand if $X = X^i\,\partial/\partial x^i$ is a unit vector field on an open subset $U \subseteq \mathbb{R}^n$ then X is a harmonic vector field if and only if the map

$$\phi : U \to S^{n-1}, \quad \phi(x) = \left(X^1(x), \ldots, X^n(x)\right), \quad x \in U,$$

is harmonic. Loosely speaking, there is nothing new about the theory of harmonic vector fields on a locally Euclidean manifold.

Another example is due to A. Morimoto & T. Nagano, [216], who built a canonical complex structure on the tangent bundle over a compact, simply connected, Riemannian symmetric space of rank 1. If for instance $M = S^n \subset \mathbb{R}^{n+1}$ then we may identify $T(S^n)$ with

$$\left\{(x,v) \in \mathbb{R}^{n+1} \times \mathbb{R}^{n+1} : \|x\| = 1, \quad \langle x, v\rangle = 0\right\}.$$

Next let $Q_n \subset \mathbb{C}^{n+1}$ be given by

$$Q_n = \left\{ (z^1, \ldots, z^{n+1}) \in \mathbb{C}^{n+1} : \sum_{j=1}^{n+1} (z^j)^2 = 1 \right\}$$
$$= \left\{\xi + i\eta : \langle \xi, \xi\rangle - \langle \eta, \eta\rangle = 1, \quad \langle \xi, \eta\rangle = 0, \quad \xi, \eta \in \mathbb{R}^{n+1}\right\}.$$

Let us consider the map $f : T(S^n) \to Q_n$ defined by

$$f(x,v) = \cosh(\|v\|)x + i\frac{\sinh(\|v\|)}{\|v\|}\,v, \quad (x,v) \in T(S^n).$$

Then f is an $\mathrm{SO}(n+1)$-equivariant diffeomorphism.[1] A complex structure on $T(S^n)$ is obtained by pulling back via f the complex structure of the hyperquadric Q_n. G. Patrizio & P-M. Wong, [236], investigated the relationship among the complex structures discovered by A. Morimoto &

[1] Given $(x,v) \in T(S^n)$ let us consider $\zeta = f(x,v) \in Q_n$. Then $\|v\| = \log\left[\frac{\sqrt{2}}{2}\left(\sqrt{|\zeta|^2+1} + \sqrt{|\zeta|^2-1}\right)\right]$. Consequently the image of the tube $T^\epsilon(S^n)$ under the diffeomorphism $f : T(S^n) \to Q_n$ is $f(T^\epsilon(S^n)) = Q_n \cap B(\sqrt{\cosh(2\epsilon)})$ where $B(r) = \{z \in \mathbb{C}^{n+1} : |z| < r\}$. Also $f(\partial\, T^\epsilon(S^n)) = Q_n \cap S^{2n+1}(\sqrt{\cosh(2\epsilon)})$ where $S^{2n+1}(r) = \partial\, B(r)$.

T. Nagano (cf. *op. cit.*) and the global solutions to the homogeneous complex Monge-Ampère equation.

Yet another example of complex structure was built by L. Lempert, [204], on a tube around the zero section in the tangent bundle over a compact hyperbolic manifold (i.e., a compact quotient of the unit ball in $\mathbb{R}^n$ endowed with the Cayley-Klein metric).

The three examples mentioned above are particular instances of *adapted complex structures* (according to the terminology adopted in [206] and [277]). The remainder of this section is devoted to briefly discussing the construction of adapted complex structures in a sufficiently small neighborhood of $\sigma_0(M) \subset T(M)$ for any C^ω Riemannian manifold M. Moreover we state (following [277]) necessary and sufficient conditions under which an adapted complex structure is defined on the whole of $T(M)$. This happens, for instance, when M is a compact symmetric space. It should be mentioned that the canonical complex structures (on the total space of the cotangent bundle) built by V. Guillemin & M. Stenzel, [154], are actually equivalent to those of R. Szöke, [277].

Let (M,g) be a Riemannian manifold. As in Chapter 1 let $E : T(M) \to [0,+\infty)$ be given by $E(v) = \frac{1}{2}g_{\pi(v)}(v,v)$ for any $v \in T(M)$. We set

$$T^r(M) = \left\{v \in T(M) : 2E(v) < r^2\right\}, \qquad r \geq 0.$$

Let $N_\gamma : T(M) \to T(M)$ denote fibrewise multiplication by $\gamma \in \mathbb{R}$. From now on we assume that (M,g) is complete, so that each geodesic of (M,g) can be continued to the whole $\mathbb{R}$. Let $\gamma : \mathbb{R} \to M$ be a geodesic and let us consider the immersion $\psi_\gamma : \mathbb{C} \to T(M)$ defined by

$$\psi_\gamma(x+iy) = N_y \frac{d\gamma}{dt}(x), \qquad x+iy \in \mathbb{C}.$$

Let $\mathcal{G}(\mathbb{R},M)$ be the set of all maximal geodesics in M. Then

$$\{\psi_\gamma(\mathbb{C} \setminus \mathbb{R}) : \gamma \in \mathcal{G}(\mathbb{R},M)\}$$

are the leaves of a smooth foliation by real surfaces of $T(M) \setminus \sigma_0(M)$.

Definition C.10 The foliation $\mathcal{F}$ of $T(M) \setminus \sigma_0(M)$ defined by $[T(M) \setminus \sigma_0(M)]/\mathcal{F} = \{\psi_\gamma(\mathbb{C} \setminus \mathbb{R}) : \gamma \in \mathcal{G}(\mathbb{R},M)\}$ is called the *Riemann foliation* of $T(M) \setminus \sigma_0(M)$. ■

Each leaf $\psi_\gamma(\mathbb{C} \setminus \mathbb{R})$ of the Riemann foliation carries the natural complex structure got by pushing forward via ψ_γ the complex structure on $\mathbb{C}$.

Definition C.11 Let (M,g) be a complete Riemannian manifold and $0 < r \leq \infty$. A complex structure J on $T^r(M)$ is said to be *adapted* if any leaf of the Riemann foliation is a complex submanifold of $(T^r(M),J)$. ■

By a result of [206], if an adapted complex structure on $T^r(M)$ exists then it is unique. As to the existence question one may state the following

Theorem C.12 (R. Szöke, [277]) *Let M be a compact real analytic manifold endowed with a real analytic Riemannian metric g. Then there is $\epsilon > 0$ such that $T^\epsilon(M)$ carries an adapted complex structure.*

Given a C^∞ complete Riemannian manifold (M,g), it may be shown that whenever $T^r(M)$ admits an adapted complex structure both the manifold M and the metric g are C^ω. Hence the conditions in Theorem C.12 are both necessary and sufficient (for the existence and uniqueness of adapted complex structures on tubes $T^r(M)$). Cf. e.g., Theorem 1.5 in [205], p. 237.

Let $x \in M$ and let $\gamma : \mathbb{R} \to M$ be a geodesic issuing at x (i.e., $\gamma(0) = x$) parametrized by arc length. Let $\{v_1, \ldots, v_{n-1}\} \subset T_x(M)$ be tangent vectors such that $\{\dot\gamma(0), v_1, \ldots, v_{n-1}\}$ is an orthonormal basis of $(T_x(M), g_x)$. Let $\{X_1, \ldots, X_{n-1}, Y_1, \ldots, Y_{n-1}\} \subset J_\gamma$ be the Jacobi fields determined by the initial conditions

$$X_i(x) = v_i, \quad \left(\nabla_{\dot\gamma} X_i\right)(x) = 0,$$
$$Y_i(x) = 0, \quad \left(\nabla_{\dot\gamma} Y_i\right)(x) = v_i,$$

for any $1 \leq i \leq n-1$. Cf. [189], Vol. II, p. 63 for the relevant material (on Jacobi fields along a geodesic in a Riemannian manifold). Then $\{X_1, \ldots, X_{n-1}\}$ are pointwise linearly independent on $\gamma(\mathbb{R} \setminus S)$ where $S \subset \mathbb{R}$ is a discrete subset such that $\gamma(S)$ is precisely the set of points on γ which are conjugate to x. On the other hand, the vector fields $\{X_i, Y_i : 1 \leq i \leq n-1\}$ are orthogonal to $\dot\gamma$. Hence there exist functions $\varphi_{jk} : \mathbb{R} \setminus S \to \mathbb{R}$ such that

$$Y_j \circ \gamma = \sum_{k=1}^{n-1} \varphi_{jk} X_k \circ \gamma, \quad 1 \leq j \leq n-1.$$

The following result gives a precise description of the size of the tube $T^r(M)$ on which an adapted complex structure may be defined.

Theorem C.13 (R. Szöke, [277]) *Let (M,g) be a complete C^ω Riemannian manifold and let $0 < r \leq \infty$. The following statements are equivalent:*

1. *There is an adapted complex structure on $T^r(M)$.*

2. *For any $x \in M$ and any geodesic γ issuing at x there is a matrix $[f_{jk}]$ of meromorphic functions defined on $D^r = \{x + iy \in \mathbb{C} : |y| < r\}$ such that*

  i. *all poles of f_{jk} lie on the real line $\mathbb{R}$,*

  ii. $f_{jk}\big|_{\mathbb{R}} = \varphi_{jk}$, $1 \leq j,k \leq n-1$, *and*

  iii. *the matrix $[\mathrm{Im}(f_{jk})]$ is pointwise invertible on $D^r \setminus \mathbb{R}$.*

Theorem C.13 implies the existence of globally defined adapted complex structures on the tangent bundle over an arbitrary compact symmetric space (not only for the rank one spaces).

Theorem C.14 (R. Szöke, [277]) *Let (M,g) be a complete locally symmetric space.*

a. *If M has nonnegative sectional curvature (in particular if M is a compact symmetric space) then there is an adapted complex structure defined on the whole of $T(M)$.*

b. *If the sectional curvature of M is $\geq \lambda$ for some $\lambda < 0$ then there is an adapted complex structure on $T^{\pi/(2\sqrt{-\lambda})}(M)$.*

By Theorem 4.3 in [206], p. 697, if M is compact then part (b) in Theorem C.14 above is sharp and the complex structure on $T^{\pi/(2\sqrt{-\lambda})}(M)$ cannot be extended any further. Precisely

Theorem C.15 (L. Lempert & R. Szöke, [206]) *Let M be a compact Riemannian manifold such that there is an adapted complex structure on $T^r(M)$ for some $0 < r \leq \infty$. Then the sectional curvature of M is $\geq -\pi^2/(4r^2)$.*

C.5. CR SUBMANIFOLDS OF GRAUERT TUBES

Let (M,g) be a compact connected C^ω Riemannian manifold and $T^r(M)$ a Grauert tube. Let us consider the map $\rho : T^r(M) \to \mathbb{R}$ given by $\rho(v) = 2E(v) = g(v,v)$ for any $v \in T^r(M)$. As previously seen (cf. [154], [206]) i) ρ is strictly plurisubharmonic, ii) $M = \rho^{-1}(0)$, iii) the Kählerian metric h whose Kähler form is $i\partial\overline{\partial}\rho$ is compatible to g i.e., $\sigma_0^* h = g$ where $\sigma_0 : M \to T(M)$ is the zero section, and iv) $\left(\partial\overline{\partial}\sqrt{\rho}\right)^n = 0$ on $T^r(M) \setminus \sigma_0(M)$. Let us set $\Omega_\epsilon = \{v \in T^r(M) : \rho(v) < \epsilon^2\}$ and $M_\epsilon = \partial\Omega_\epsilon$. Then M_ϵ is a strictly pseudoconvex CR manifold, whose CR structure is naturally induced by the adapted complex structure on $T^r(M)$. The relationship among the Riemannian geometry of (M,g) and the CR geometry of M_ϵ is insufficiently explored as yet. By a result of S-J. Kan, [182], if $\epsilon_1 \neq \epsilon_2$ then M_{ϵ_1} and M_{ϵ_2} are inequivalent CR manifolds, provided that $\dim(M) = 2$. Also S-J. Kan computed (cf. again [182]) the Burns-Epstein invariant (cf. [85] for the relevant notions) of M_ϵ.

C.5.1. The Szegö Kernel

Let us assume that $\dim(M) = 2$. If $\iota_\epsilon : M_\epsilon \to T^r(M)$ is the inclusion then $\theta_\epsilon = \iota_\epsilon^*(-i\partial\rho)$ is a contact form on M_ϵ. Let S_ϵ be the Szegö kernel with respect to the volume form $\theta_\epsilon \wedge d\theta_\epsilon$. Cf. e.g., R. Ponge, [257], K. Hirachi, [167]. See also [91], [84]. By adapting the results of L. Boutet de Monvel & J. Sjöstrand, [65], and C. Fefferman, [116], to the domains Ω_ϵ one may show that the Szegö kernel of $(M_\epsilon, \theta_\epsilon)$ admits the development

$$S_\epsilon(z,z) = \varphi(z)\rho_\epsilon(z)^{-2} + \psi(z)\log\rho_\epsilon(z)$$

for some $\varphi, \psi \in C^\infty(\overline{\Omega}_\epsilon)$ and some defining function ρ_ϵ of Ω_ϵ with $\rho_\epsilon > 0$ in Ω_ϵ. We may quote

Theorem C.16 (E. Koizumi, [190]) *The boundary value of the logarithmic term coefficient $\psi_0 = \psi|_{M_\epsilon}$ has the asymptotic expansion*

$$\psi_0 \sim \frac{1}{24\pi^2}\sum_{\ell=0}^{\infty} F_\ell^{\psi_0}\epsilon^{2\ell}, \quad \epsilon \to 0^+, \tag{C.17}$$

where $F_\ell^{\psi_0}(\lambda^2 g) = \lambda^{-2\ell-4} F_\ell^{\psi_0}(g)$ for any $\lambda > 0$. Moreover

$$F_0^{\psi_0} = -\frac{1}{10}\Delta\rho - \frac{2}{5}\left(\epsilon^2 T_\epsilon^2 \rho\right)\Big|_{\epsilon=0} \tag{C.18}$$

where ρ and Δ are the scalar curvature and Laplace-Beltrami operator of (M,g) while T_ϵ is the characteristic direction of $d\theta_\epsilon$.

One may regard $\left(\epsilon^2 T_\epsilon^2 \rho\right)\big|_{\epsilon=0}$ as a function (which is not S^1-invariant) on the total space of a circle bundle over M (cf. Lemma 4.5 in [190]). It may also be shown that ψ_0 is a constant multiple of the Q-curvature (cf. [117], [147] and [168]). This may turn useful in understanding the role of Q-curvature in CR geometry.

C.5.2. Chains and the Characteristic Flow

The purpose of this section is to report on the results of M.B. Stenzel, [274]. Let V be a complex $(n+1)$-dimensional manifold and $\phi : V \to \mathbb{R}$ a smooth, positive, strictly plurisubharmonic function. Let h be the Kählerian metric whose Kähler 2-form is $2i\partial\overline{\partial}\phi$ i.e., $h = 2\phi_{j\overline{k}}\, dz^j \odot d\overline{z}^k$ with respect to a local system of complex coordinates $(\Omega, z^1, \dots, z^{n+1})$ on V. Let $\mathbb{C} \to K_V \to V$ be the canonical line bundle over V i.e., $K_V = \Lambda^{n+1,0}(V)$ (the bundle of $(n+1,0)$-forms on V). We consider the inner product $(\cdot,\cdot)$ on K_V determined by

$$i^{n(n+1)}(n+1)!\, \nu \wedge \overline{\mu} = (\nu, \mu)\left(\partial\overline{\partial}\phi\right)^{n+1},$$

for any $\nu, \mu \in (K_V)_z$ and any $z \in V$. We set $|\nu| = (\nu, \nu)^{1/2}$ for any $\nu \in K_V$. Let $\{Z_j : 1 \le j \le n+1\}$ be a local frame of $T^{1,0}(V)$, defined on the open set $\Omega \subseteq V$. Let $\{\omega^j : 1 \le j \le n+1\}$ be the complex $(1,0)$-forms on Ω defined by

$$\omega^j(Z_k) = \delta^j_k, \quad \omega^j(\overline{Z}_k) = 0.$$

If $\partial\overline{\partial}\phi = a_{j\overline{k}}\omega^j \wedge \omega^{\overline{k}}$ for some C^∞ functions $a_{j\overline{k}} : \Omega \to \mathbb{C}$ then

$$\left|\omega^1 \wedge \cdots \wedge \omega^{n+1}\right| = \left(\det[a_{j\overline{k}}]\right)^{-1/2}.$$

Here $\theta^{\overline{k}} = \overline{\theta^k}$ for any $1 \le k \le n+1$.

Let us define a complex vector field Z of type $(1,0)$ on V by requiring that

$$2\partial\overline{\partial}\phi\,(Z, \overline{W}) = |\partial\phi|^{-2}\left(\overline{\partial}\phi\right)(\overline{W}), \quad W \in T^{1,0}(V). \tag{C.19}$$

Here $|\partial\phi|^2 = h^*(\partial\phi, \overline{\partial}\phi)$ and h^* is the bundle metric on $T^*(V)$ induced by h. It should be observed that $(\partial\phi)(Z) = 1$.

Lemma C.17 (M.B. Stenzel, [274]) *The following statements are equivalent*

1. $\left(\partial\overline{\partial}\sqrt{\phi}\right)^{n+1} = 0$.

2. $|\partial\phi| = \sqrt{2\phi}$.

3. $[Z, \overline{Z}] = (2\phi)^{-1}(Z - \overline{Z})$.

4. *Let Ξ be the complex vector field on V determined by*

$$\Xi \rfloor i\partial\overline{\partial}\phi = -\mathrm{Im}\,\overline{\partial}\phi.$$

Then $\Xi\phi = 2\phi$.

Cf. [274], p. 387–388, for a proof of Lemma C.17.

Let (M, g) be a real $(n+1)$-dimensional compact, connected, C^ω Riemannian manifold and $T^r(M)$ a Grauert tube. Let $S^1 \to C(M_\epsilon) \xrightarrow{\pi_\epsilon} M_\epsilon$ be the canonical circle bundle over $M_\epsilon = \partial T^\epsilon(M)$ (cf. e.g., [110],

p. 119). Let $\theta = -\text{Im}\left(\overline{\partial}\phi\right)$. Then $\theta_\epsilon = \iota_\epsilon^*\theta$ is a contact form on M_ϵ, where $\iota_\epsilon : M_\epsilon \to T^r(M)$ is the inclusion. Let F_{θ_ϵ} be the Fefferman metric on $C(M_\epsilon)$ i.e.,

$$F_{\theta_\epsilon} = \pi_\epsilon^* \tilde{G}_{\theta_\epsilon} + 2(\pi_\epsilon^*\theta_\epsilon) \odot \sigma_\epsilon$$

where

$$\sigma_\epsilon = \frac{1}{n+2}\left\{d\gamma + \pi_\epsilon^*\left(i\omega_\alpha^\alpha - \frac{i}{2}g^{\alpha\overline{\beta}}dg_{\alpha\overline{\beta}} - \frac{\rho}{4(n+1)}\theta_\epsilon\right)\right\},$$

cf. e.g., (2.31) in [110], p. 129. Here γ is a local fibre coordinate on $C(M_\epsilon)$. Also ω_β^α are the local 1-forms of the Tanaka-Webster connection ∇^ϵ of $(M_\epsilon, \theta_\epsilon)$ (cf. Theorem 1.3 in [110], p. 25, for the existence and uniqueness of ∇^ϵ) with respect to a local frame $\{T_\alpha : 1 \le \alpha \le n\}$ of $T_{1,0}(M_\epsilon)$. Moreover $g_{\alpha\overline{\beta}} = L_{\theta_\epsilon}(T_\alpha, T_{\overline{\beta}})$ and ρ is the pseudohermitian scalar curvature of $(M_\epsilon, \theta_\epsilon)$ (cf. e.g., [110], p. 50). As well known (cf. Prposition 2.17 in [24], p. 28) σ_ϵ is a connection 1-form in the principal bundle $S^1 \to C(M_\epsilon) \to M_\epsilon$ and $d\sigma_\epsilon$ is projectable i.e., $d\sigma_\epsilon = \pi_\epsilon^*\Omega_\epsilon$ for some 2-form Ω_ϵ on M_ϵ. Let T_ϵ be the characteristic direction of $d\theta_\epsilon$ (cf. [110], p. 8). It may be shown that

Theorem C.18 (M.B. Stenzel, [274]) *The integral curves of T_ϵ are chains if and only if $T_\epsilon \rfloor \Omega_\epsilon = 0$. In particular if (M,g) is a harmonic manifold then the integral curves of T_ϵ are chains on M_ϵ.*

Here by a chain one understands the projection on M_ϵ of a non vertical null geodesic of the Fefferman metric F_{θ_ϵ}. The proof of Theorem C.18 is to relate the Riemannian geometry of (M,g) to the Kählerian geometry of $(T^r(M),h)$ and to the properties of the solutions to the Monge-Ampère equation $\left(\partial\overline{\partial}\sqrt{\phi}\right)^{n+1} = 0$. An important role in the needed calculations (cf. [274], p. 391–393) is played by the Graham-Lee connection (cf. Proposition 1.1 in [151], p. 701) associated to the foliation by level sets of $2E$ whose transverse curvature is (by Lemma C.17 above) $r = (2\phi)^{-1}$.

Exceptional Orbits of Highest Dimension

A smooth *action* of a Lie group G on a n-dimensional C^∞ manifold M is a C^∞ map $\alpha : M \times G \to M$ such that $\alpha(x,e) = x$ and $\alpha(x,ab) = \alpha(\alpha(x,a),b)$ for any $x \in M$ and $a,b \in G$. The *orbit* of a point $x \in M$ for the action α is the subset $G(x) = \mathcal{O}_x(\alpha) = \{\alpha(x,a) \in M : a \in G\}$. The *isotropy group* of $x \in M$ is the subgroup $G_x = G_x(\alpha) = \{a \in G : \alpha(x,a) = x\}$. Clearly G_x is closed in G. The map $j_x : G \to M$ given by $j_x(a) = \alpha(x,a) = xa$ induces the map $\bar{j}_x : G/G_x \to M$ given by $\bar{j}_x(\bar{a}) = j_x(a)$ where $\bar{a} = G_x \cdot a$. Note that $a^{-1}b \in G_x$ if and only if $xa = j_x(a) = j_x(b) = xb$ hence $\bar{j}_x$ is well defined and one-to-one. It may be shown that G/G_x has a natural structure of a differentiable manifold and $\bar{j}_x$ is an injective immersion whose image is $G(x)$. It is customary to call $\alpha : M \times G \to M$ a *foliated* action if for any $x \in M$ the orbit $G(x)$ has fixed dimension r (cf. e.g., C. Camacho & A.L. Neto, p. 29). Also α is *locally free* when $\dim(G) = r$ as well. If α is a foliated action of G on M it may be easily shown that there is a codimension $n-r$ foliation $\mathcal{F}$ of M such that $M/\mathcal{F} = \{G(x) : x \in M\}$ i.e., the leaves of $\mathcal{F}$ are the orbits of the action (cf. Proposition 1 in [79], p. 29).

Let G be a compact Lie group acting on a n-dimensional topological manifold M. Let $x \in M$ and let G_x^0 be the connected component of the identity in G_x. The order of each quotient group G_x/G_x^0 is finite. Let $m(x)$ be the order of G_x/G_x^0. If $p, q \in \mathbb{Z}$ with $p \geq 0$ and $q > 0$ we set

$$M_{p,q} = \{x \in M : \dim G(x) = p, \quad m(x) = q\},$$
$$M_p = \{x \in M : \dim G(x) = p\}.$$

The following facts are rather well known.

Proposition D.1

i. *For each point $x \in M_p$ there is an open neighborhood $U \subset M$ such that for any $y \in U \cap M_p$ its stability group G_y is conjugate to an open subgroup of G_x and then $m(y)$ is a factor of $m(x)$. In particular $m : M_p \to \mathbb{Z}$ is upper semi-continuous. Also for any $r \in \mathbb{Z}$ with $r > 0$ the set $\bigcup_{q=1}^{r} M_{p,q}$ is open in M_p.*

© 2012 Elsevier Inc. All rights reserved.

ii. *If G is connected then every connected component of $M_{p,q}$ is a fibre bundle whose fibres are orbits.*

iii. *The set $\bigcup_{p=0}^{s} M_p$ is closed for any $s \in \mathbb{Z}$ with $s > 0$.*

Statement (i) in Proposition D.1 follows from the fact that every point $x \in M$ has a neighborhood $U \subset M$ such that G_y is conjugate to a subgroup of G_x for any $y \in U$ (cf. S. Bochner, [49]). Statements (ii)–(iii) follow from (i) and a result by A. Gleason, [135].

Let us assume that there is $r \in \mathbb{Z}$ with $0 \leq r \leq n$ such that $\dim G(x) \leq r$ for any $x \in M$. An orbit $G(x)$ is *singular* if $\dim G(x) < r$. Let us set

$$F = \bigcup_{p=0}^{r-1} M_p$$

i.e., F is the set of points on singular orbits. Then

Theorem D.2 (D. Montgomery & H. Samuelson & L. Zippin, [212]) *The points on orbits of highest dimension form a connected open set whose complement has dimension at most $n-2$ i.e., $M \setminus F$ is a connected open set and* $\dim(F) \leq n-2$.

Let us set

$$k = \inf\{m(x) : x \in M_r\}, \quad E = \bigcup_{q=k+1}^{\infty} M_{r,q}.$$

As a corollary of (i) and (iii) in Proposition D.1

Proposition D.3 *$E \cup F$ is a closed set.*

The main result in [213] is

Theorem D.4 (D. Montgomery & H. Samuelson & C.T. Yang, [213]) *Let M be a n-dimensional C^∞ manifold and let G be a compact connected Lie group acting smoothly on M. If $H_{n-1}(M, \mathbb{Z}_2) = 0$ then the closed set $E \cup F$ has dimension $\dim(E \cup F) \leq n-2$ and $M \setminus (E \cup F)$ is a fibre bundle whose fibres are orbits.*

Here $H_{n-1}(M, \mathbb{Z}_2)$ is the $(n-1)^{\text{th}}$ Čech homology group of M (when M is compact) or of its one-point compactification (when M is noncompact) with coefficients in $\mathbb{Z}_2$.

Let $y \in M$ and let us set

$$X = \left\{x \in M : G_x^0 = G_y^0\right\}, \quad G' = \left\{g \in G : g\,G_y^0 = G_y^0 g\right\}.$$

Then X is closed in M_r, hence it is locally compact. Also G'/G_y^0 acts on X as a transformation group and there is a neighborhood W of the identity in G'/G_y^0 such that $h(y) = h'(y)$ with $h, h' \in W$ implies that $h = h'$. The following result follows from arguments in [135].

Lemma D.5 *There exist*

1. *a closed neighborhood $N \subset X$ of y,*
2. *a closed subset $C \subset N$ such that $y \in C$,*
3. *a compact neighborhood W of the identity in G'/G_y^0, such that every element of N may be uniquely written as hz for some $h \in W$ and $z \in C$. Also N may be taken in any preassigned neighborhood of y.*

Using Lemma D.5 one may show (cf. [213], p. 132–135)

Proposition D.6

i. $\dim(E) \leq n-1$ (*and hence* $\dim(E \cup F) \leq n-1$).

ii. *$E \setminus M_{r,2k}$ doesn't separate M locally i.e., each point of M has an arbitrarily small neighborhood $U \subset M$ such that $U \setminus (E \setminus M_{r,2k})$ is connected.*

iii. *The orbit space of $M_{r,k}$ is connected. In particular if G is connected then $M_{r,k}$ is connected.*

It should be mentioned that Lemma D.5 and Proposition D.6 do not require the differentiability M (the results hold for any topological manifold M). More refined results on the exceptional set E (cf. Theorem D.4 above) are available only in the differentiable case and rely on Lemma 3.1 in [213], p. 136 (the main technical tool there). Most arguments in [212] may be significantly simplified when M is a smooth manifold and G acts smoothly on M (cf. [213], p. 135–140).

One may produce examples showing that none of the two assumptions (G is connected and $H_{n-1}(M, \mathbb{Z}_2) = 0$) in Theorem D.4 may be dropped. For instance let $M = S^n \subset \mathbb{R}^{n+1}$ be the unit sphere with $n \geq 2$ and let G consist of the identical transformation of M and the reflection with respect to the hyperplane $x_{n+1} = 0$. That is $G = \{1_{S^n}, a\}$ where $a(x) = (x', -x_{n+1})$ for any $x = (x', x_{n+1}) \in S^n$ (with $x' \in \mathbb{R}^n$). Then

$$G(x) = \{x, \ (x', -x_{n+1})\}, \quad x \in S^n,$$

so that $\dim G(x) = 0$ and hence $r = 0$. Therefore

$$M_p = \begin{cases} S^n & \text{if } p = 0 \\ \emptyset & \text{if } p \geq 1 \end{cases}, \quad p \in \mathbb{Z}, \quad p \geq 0,$$

and then $F = \emptyset$. On the other hand

$$G_x = \begin{cases} \{1_{S^n}\} & \text{if } x_{n+1} \neq 0 \\ G & \text{if } x_{n+1} = 0 \end{cases},$$

$$m(x) = \left|G_x/G_x^0\right| = \begin{cases} 1 & \text{if } x_{n+1} \neq 0 \\ 2 & \text{if } x_{n+1} = 0 \end{cases},$$

$$k = \inf\{m(x) : x \in M_0\} = 1,$$

so that

$$E = \bigcup_{q=2}^{\infty} M_{0,q} = M_{0,2} = S^n \cap \left\{x \in \mathbb{R}^{n+1} : x_{n+1} = 0\right\} = S^{n-1}$$

i.e., $\dim(E \cup F) = n - 1$. Here $H_{n-1}(S^n, \mathbb{Z}_2) = 0$ yet G isn't connected.

Let M be a complete Riemannian manifold of dimension n, and let G be a Lie group of isometries of M which is closed in the full isometry group $\text{Isom}(M,g)$ of M. We say that M is of *cohomogeneity one* under the action of G if G has at least an orbit of codimension one. For the general theory of cohomogeneity one manifolds the reader may see A.V. Alekseevsky & D.V. Alekseevsky, [14; 15], G.E. Bredon, [70], F. Podesta & A. Spiro, [256], and R.S. Palais & C.L. Terng, [234]. We close this appendix by recalling a few facts about cohomogeneity one manifolds. It is known that their orbit space $\Omega = M/G$ is a Hausdorff space homeomorphic to one of the following spaces

$$\mathbb{R}, \quad S^1, \quad [0, +\infty), \quad [0, 1].$$

Let $\pi : M \to \Omega$ be the natural projection. If $x \in M$ the orbit $G(x)$ is *principal* (respectively *singular*) if the corresponding image in the orbit space Ω is an interior (respectively boundary) point. A point $x \in M$ whose orbit is principal (respectively singular) is called a *regular* (respectively *singular*) point. The subset of all regular points is an open and dense subset of M denoted by M_{reg} and the subset of all singular points is denoted by M_{sing}. If $\Omega^0 \subset \Omega$ is homeomorphic to an open interval in $\mathbb{R}$ and $\mathcal{O}$ is a principal orbit then $\Omega^0 \times \mathcal{O}$ is diffeomorphic to $\pi^{-1}(\Omega^0)$. All principal orbits are diffeomorphic to each other and if $M/G = \mathbb{R}$ then M is diffeomorphic to $\mathbb{R} \times \mathcal{O}$ where $\mathcal{O}$ is a principal orbit. Each singular orbit has dimension less than or equal to $n - 1$. A singular orbit of dimension $n - 1$ is called an *exceptional* orbit. No exceptional orbit is simply connected and if $\pi_1(M) = 0$ then no exceptional orbit may exist. If M is orientable and all principal orbits are connected then any exceptional orbit is non-orientable. If $\mathcal{O}$ is the only singular orbit of M (the case $M/G = [0, +\infty)$) then $\pi_1(M) = \pi_1(\mathcal{O})$. We recall

Definition D.7 A geodesic $\gamma : \mathbb{R} \to M$ on a Riemannian manifold of cohomogeneity one is called a *normal geodesic* if it is orthogonal to each orbit that it meets. ■

A geodesic γ is normal if and only if it is orthogonal to $G(\gamma(t))$ for at least one t. If $M/G = S^1$ or $M/G = [0,1]$ then a normal geodesic $\gamma : \mathbb{R} \to M$ intersects each principal orbit $\mathcal{O}$ infinitely many times (i.e., $\gamma(t) \in \mathcal{O}$ for infinitely many $t \in \mathbb{R}$). If $M/G = [0, +\infty)$ then $\gamma : \mathbb{R} \to M$ intersects each principal orbit in two distinct points. If $M/G = \mathbb{R}$ then $\gamma : \mathbb{R} \to M$ intersects a principal orbit exactly once. Also, the following results are of common use in this context.

Theorem D.8 (P.J. Ryan, [263]) *Let $M(c)$ be a space form of (constant) curvature $c \leq 0$ and let M be a real hypersurface in $M(c)$ whose principal curvatures are constant. Then at most two of the principal curvatures are distinct.*

Theorem D.9 (P.J. Ryan, [264]) *Let $\tilde{M}$ be a real space form and M a hypersurface in $\tilde{M}$. If the principal curvatures of M are constant and at most two of them are distinct then M is congruent to an open set of one of the standard examples.*[1]

Theorem D.10 (R.S. Palais & C.L. Terng, [234]) *Let M be a complete Riemannian G-manifold that admits sections and let $\mathcal{O}$ be a principal orbit of M. Then*

i. *$\exp_x[\nu(\mathcal{O})_x]$ is a properly embedded totally geodesic submanifold of M for each $x \in \mathcal{O}$. Here $\nu(\mathcal{O}) \to \mathcal{O}$ is the normal bundle of $\mathcal{O}$ in M.*

ii. *$\nu(\mathcal{O}) \to \mathcal{O}$ is flat and has trivial holonomy. If $N_1, \cdots, N_k$ is a linear basis in $\nu(\mathcal{O})_x$ then the G-invariant normal vector fields $\tilde{N}_i$ given by $\tilde{N}_i(gx) = (d_x g) N_i$ form a globally defined parallel frame in $\nu(\mathcal{O}) \to \mathcal{O}$.*

iii. *The principal curvatures of $\mathcal{O}$ with respect to any parallel normal field are constant.*

Theorem D.11 *Let M be a complete hypersurface in the Euclidean space $\mathbb{R}^{n+1}$ whose principal curvatures are constant. Then M is isometric to one of the following spaces*

$$\mathbb{R}^n, \quad S^k(c) \times \mathbb{R}^{n-k}, \quad 1 \leq k \leq n, \quad c > 0.$$

[1] When $\tilde{M} = \mathbb{R}^{n+1}$ the standard examples are hyperplanes, spheres, and cylinders over spheres (cf. P.J. Ryan, [264], Section 1).

Proof. Theorem D.11 follows from Theorem D.8 and D.9 (because M is complete). ■

Theorem D.12 (J.A. Wolf, [311]) *Let M be an n-dimensional connected homogeneous Riemannian flat manifold. Then M is isometric to the product $\mathbb{R}^m \times T^{n-m}$ of the Euclidean space with a flat Riemannian torus.*

Theorem D.13 *Let M be a simply connected cohomogeneity one Riemannian manifold of non-positive curvature. Then M has at most one singular orbit.*

The proof of Theorem D.13 is similar to that of Proposition 3.3 in [256]. Cf. also R. Mirzaie & S.M.B. Kashani, [211], p. 187. A complete classification of cohomogeneity one Riemannian manifolds which are simply connected, compact, of positive curvature and of dimension ≤ 6 is due to C. Searle, [269]. Topological properties of cohomogeneity one negatively curved Riemannian manifolds were studied by F. Podesta & A. Spiro, [256], while the flat case was investigated by R. Mirzaie & S.M.B. Kashani, [211]. The reader may also see P. Ahmadi & S.M.B. Kashani, J. Berndt & H. Tamaru, [27], [12], K. Grove et al., [153], A. Kollross, [192], and L. Verhóczki, [301].

APPENDIX E

Reilly's Formula

Let M and N be two oriented Riemannian manifolds of dimensions n and $n-1$ respectively where M is allowed to have a boundary i.e., $\partial M \neq \emptyset$. Let $\{E_A : 1 \leq A \leq n\}$ be a smooth local orthonormal frame on M defined on the open set $U \subset M$. Let $\{\omega_A : 1 \leq A \leq n\}$ be the corresponding dual coframe i.e., $\omega_A(E_B) = \delta_{AB}$. Let ∇ be the Levi-Civita connection of M and R its curvature tensor field. Let ω_B^A be the connection 1-forms associated to ∇ and $\{E_A : 1 \leq A \leq n\}$ i.e.,

$$\nabla E_B = \omega_B^A \otimes E_A, \quad 1 \leq B \leq n.$$

Then (Cartan's structure equations)

$$d\omega_A = \omega_A^B \wedge \omega_B, \quad \omega_A^B = -\omega_B^A, \tag{E.1}$$

$$d\omega_A^B = \omega_A^C \wedge \omega_C^B + \Omega_A^B, \tag{E.2}$$

$$\Omega_A^B = \frac{1}{2} \sum_{C,D=1}^{n} R_{ABCD}\, \omega_C \wedge \omega_D,$$

$$R_{ABCD} = g(R(E_C, E_D)E_A, E_B).$$

Here g denotes the Riemannian metric on M. Our convention[1] for the Ricci tensor in this monograph is $\mathrm{Ric}(X,Y) = \mathrm{trace}\{Z \mapsto R(Z,Y)X\}$ hence $R_{AB} = \mathrm{Ric}(E_A, E_B)$ is given by

$$R_{AB} = \sum_{C=1}^{n} R_{CABC}.$$

If T is a tensor field of type $(0,r)$ on M we set

$$T_{A_1\cdots A_r} = T(E_{A_1}, \ldots, E_{A_r}), \quad T_{A_1\cdots A_r, B} = \left(\nabla_{E_B} T\right)(E_{A_1}, \ldots, E_{A_r}).$$

The components $T_{A_1\cdots A_r, B}$ of the covariant derivative of T satisfy

$$\sum_{B=1}^{n} T_{A_1\cdots A_r, B}\, \omega_B = dT_{A_1\cdots A_r} + \sum_{j=1}^{r} \sum_{A=1}^{n} T_{A_1\cdots A_{j-1} A A_{j+1} \cdots A_r}\, \omega_A^{A_j}. \tag{E.3}$$

[1] The Ricci tensor is defined such that the Ricci tensor of the unit sphere S^n is positive definite.

© 2012 Elsevier Inc. All rights reserved.

Let T be a $(0,r)$-tensor field and $S = \nabla T$ its covariant derivative so that

$$S_{A_1 \cdots A_{r+1}} = S(A_1, \ldots, A_{r+1}) = \left(\nabla_{E_{A_1}} T\right)\left(E_{A_2}, \ldots, E_{A_{r+1}}\right)$$

i.e.,

$$S_{A_1 \cdots A_{r+1}} = T_{A_2 \cdots A_{r+1}, A_1} \, .$$

We adopt the notation

$$T_{A_1 \cdots A_r, BC} = S_{BA_1 \cdots A_r, C} = \left(\nabla_{E_C} S\right)\left(E_B, E_{A_1}, \ldots, E_{A_r}\right).$$

The following commutation formula is useful in the sequel

$$T_{A_1 \cdots A_r, BC} - T_{A_1 \cdots A_r, CB} = \sum_{j=1}^{r} \sum_{D=1}^{n} T_{A_1 \cdots A_{j-1} D A_{j+1} \cdots A_r} R_{DA_j CB} \, . \tag{E.4}$$

To help the reader get accustomed with the formalism in this appendix, we give a proof of (E.4). Applying (E.3) to the $(0, r+1)$-tensor field $S = \nabla T$, one has

$$\sum_{C=1}^{n} S_{BA_1 \cdots A_r, C} \, \omega_C = dS_{BA_1 \cdots A_r} + \sum_{k=0}^{r} \sum_{D=1}^{n} S_{A_0 \cdots A_{k-1} D A_{k+1} \cdots A_r} \, \omega_D^{A_k} \tag{E.5}$$

where we set $A_0 = B$. The differential of $S_{BA_1 \cdots A_r} = T_{A_1 \cdots A_r, B}$ (appearing in (E.5)) may be determined by taking the exterior differential of (E.3). Indeed

$$\sum_{B=1}^{n} \left\{ \left(dT_{A_1 \cdots A_r, B}\right) \wedge \omega_B + T_{A_1 \cdots A_r, B} \, d\omega_B \right\} = \sum_{j=1}^{r} \sum_{B=1}^{n} \left\{ \left(dT_{A_1 \cdots A_{j-1} B A_{j+1} \cdots A_r}\right) \wedge \omega_B^{A_j} + T_{A_1 \cdots A_{j-1} B A_{j+1} \cdots A_r} \, d\omega_B^{A_j} \right\}. \tag{E.6}$$

From now on, to simplify writing we omit sums over repeated indices. Moreover we adopt the temporary notation

$$(D_1 \cdots D_r) = (A_1 \cdots A_{j-1} B A_{j+1} \cdots A_r)$$

and replace $d\omega_B$ and $d\omega_B^{A_j}$ from Cartan's structure equations (E.1)–(E.2). Also, $dT_{D_1 \cdots D_r}$ may be expressed in terms of covariant derivatives by using once again (E.3). We obtain

$$\left(dS_{BA_1\cdots A_r}\right)\wedge\omega_B+S_{BA_1\cdots A_r}\,\omega_B^D\wedge\omega_D=\sum_{j=1}^{r}\left\{\left(T_{A_1\cdots A_{j-1}BA_{j+1}\cdots A_r,D}\,\omega_D\right.\right.$$

$$\left.\left.-\sum_{k=1}^{r}T_{D_1\cdots Dk-1ED_{k+1}\cdots D_r}\,\omega_E^{D_k}\right)\wedge\omega_B^{A_j}+T_{D_1\cdots D_r}\left(\omega_B^E\wedge\omega_E^{A_j}+\Omega_B^{A_j}\right)\right\}$$

hence (by applying both members to the pair (E_R,E_S))

$$\begin{aligned}
&E_R\left(S_{SA_1\cdots A_r}\right)-E_S\left(S_{RA_1\cdots A_r}\right)+S_{BA_1\cdots A_r}\left(\Gamma_{RB}^S-\Gamma_{SB}^R\right)\\
&=\sum_{j=1}^{r}\left\{\Gamma_{SB}^{A_j}S_{RD_1\cdots D_r}-\Gamma_{RB}^{A_j}S_{SD_1\cdots D_r}\right.\\
&\quad-\sum_{k=1}^{r}\left(\Gamma_{RE}^{D_k}\Gamma_{SB}^{A_j}-\Gamma_{SE}^{D_k}\Gamma_{RB}^{A_j}\right)T_{D_1\cdots D_{k-1}ED_{k+1}\cdots D_r}\\
&\quad\left.+\left(\Gamma_{RB}^{E}\Gamma_{SE}^{A_j}-\Gamma_{SB}^{E}\Gamma_{RE}^{A_j}+R_{BA_jRS}\right)T_{D_1\cdots D_r}\right\}.
\end{aligned}\tag{E.7}$$

Here $\Gamma_{BC}^A=\omega_C^A(E_B)$. On the other hand (by applying both members of (E.5) to E_C)

$$\begin{aligned}
S_{BA_1\cdots A_r,C}&=E_C\left(S_{BA_1\cdots A_r}\right)+\sum_{k=0}^{r}S_{A_0\cdots A_{k-1}DA_{k+1}\cdots A_r}\Gamma_{CD}^{A_k}\\
&=E_C\left(S_{BA_1\cdots A_r}\right)+S_{DA_1\cdots A_r}\Gamma_{CD}^{B}+\sum_{j=1}^{r}S_{BA_1\cdots A_{j-1}DA_{j+1}\cdots A_r}\Gamma_{CD}^{A_j}
\end{aligned}$$

hence (by (E.7))

$$\begin{aligned}
&S_{SA_1\cdots A_r,R}-S_{RA_1\cdots A_r,S}\\
&=E_R\left(S_{SA_1\cdots A_r}\right)-E_S\left(S_{RA_1\cdots A_r}\right)+\left(\Gamma_{RD}^S-\Gamma_{SD}^R\right)S_{DA_1\cdots A_r}\\
&\quad+\sum_{j=1}^{r}\left\{\Gamma_{RD}^{A_j}S_{SA_1\cdots A_{j-1}DA_{j+1}\cdots A_r}-\Gamma_{SD}^{A_j}S_{RA_1\cdots A_{j-1}DA_{j+1}\cdots A_r}\right\}\\
&=\sum_{j=1}^{r}T_{A_1\cdots A_{j-1}BA_{j+1}\cdots A_r}R_{BA_jRS}-\sigma
\end{aligned}$$

where

$$\sigma = \sum_{j=2}^{r}\sum_{k=1}^{j-1}\left(\Gamma^{A_k}_{RE}\Gamma^{A_j}_{SB} - \Gamma^{A_k}_{SE}\Gamma^{A_j}_{RB}\right)T_{A_1\cdots A_{k-1}EA_{k+1}\cdots A_{j-1}BA_{j+1}\cdots A_r}$$
$$+\sum_{j=1}^{r}\sum_{k=j+1}^{r}\left(\Gamma^{A_k}_{RE}\Gamma^{A_j}_{SB} - \Gamma^{A_k}_{SE}\Gamma^{A_j}_{RB}\right)T_{A_1\cdots A_{j-1}BA_{j+1}\cdots A_{k-1}EA_{k+1}\cdots A_r}.$$

A calculation shows that the terms of the last two double sums cancel in pairs so that $\sigma = 0$ and (E.4) is proved.

The divergence of T is the $(0, r-1)$-tensor field $\mathrm{div}(T)$ locally given by

$$\mathrm{div}(T)(X_1, \ldots, X_{r-1}) = \sum_{B=1}^{n}(\nabla_{E_B}T)(X_1, \ldots, X_{r-1}, E_B)$$

on U, for any $X_1, \ldots X_{r-1} \in \mathfrak{X}(M)$. The local components of $\mathrm{div}(T)$ are

$$(\mathrm{div}(T))_{A_1\cdots A_{r-1}} = \mathrm{div}(T)(E_{A_1}, \ldots, E_{A_{r-1}}) = \sum_{B=1}^{n} T_{A_1\cdots A_{r-1}B,B}.$$

For each smooth function $f \in C^\infty(M)$ its Hessian $\mathrm{Hess}(f)$ is given by

$$\mathrm{Hess}(f) = \nabla\, df.$$

Locally we set

$$\begin{aligned} f_{AB} &= \mathrm{Hess}(f)(E_A, E_B) = \left(\nabla_{E_A} df\right) E_B \\ &= E_A(E_B(f)) - (\nabla_{E_A}E_B)(f) = E_A(E_B(f)) - \omega^C_B(E_A)E_C(f). \end{aligned}$$

It is customary to set $f_A = E_A(f)$ so that $df = \sum_{A=1}^{n} f_A\omega_A$. Hence $f_{AB} = E_A(f_B) - f_C\,\omega^C_B(E_A)$ i.e.,

$$f_{AB}\omega_C = df_B - f_C\,\omega^C_B. \tag{E.8}$$

Using the fact that ∇ is torsion free and (E.4) one obtains

$$f_{AB} = f_{BA}, \quad f_{A,BC} - f_{A,CB} = f_D R_{DACB}. \tag{E.9}$$

Occasionally we look at the Hessian of f as a field of symmetric linear transformations

$$\left(H_f\right)_x : T_x(M) \to T_x(M), \quad x \in M,$$

by setting $g(H_f X, Y) = \mathrm{Hess}_f(X, Y)$ for any $X, Y \in \mathfrak{X}(M)$.

Let $F : N \to M$ be an isometric immersion. We set

$$\varphi_A = F^*\omega_A, \quad \varphi_A^B = F^*\omega_A^B, \quad \Phi_A^B = F^*\Omega_A^B.$$

From now on we assume that the local orthonormal frame $\{E_A : 1 \leq A \leq n\}$ was chosen in such a way that each E_i is tangent to N i.e., there is a local orthonormal frame $\{e_i : 1 \leq i \leq n-1\}$ on N defined on an open set $V \subset N$ such that $F(V) \subset U$ and $(d_xF)e_{i,x} = E_{i,F(x)}$ for any $x \in V$ and any $1 \leq i \leq n-1$. Then

$$\varphi_n(e_i)_x = (F^*\omega_n)_x(e_{i,x}) = \omega_{n,F(x)}\left((d_xF)e_{i,x}\right) = \omega_{n,F(x)}\left(E_{i,F(x)}\right) = 0$$

for any $x \in V$ and any $1 \leq i \leq n-1$. Therefore $\varphi_n = 0$. Differentiation of $F^*\omega_n = 0$ and (E.1) give

$$0 = F^* d\omega_n = F^*\left(\omega_n^B \wedge \omega_B\right) = \varphi_n^B \wedge \varphi_B = \varphi_n^i \wedge \varphi_i.$$

Hence

$$\varphi_n^i = A_j^i \varphi^j, \quad A_j^i = A_i^j. \tag{E.10}$$

On the other hand, E_n is a unit normal field on N hence (the Weingarten formula)

$$\nabla_X E_n = -aX, \quad X \in \mathfrak{X}(N), \tag{E.11}$$

where a is the Weingarten operator. Applying (E.11) for $X = e_i$ gives $A_j^i = -a_j^i$ where $a(e_j) = a_j^i e_i$ for some $a_j^i \in C^\infty(V)$ (the local components of the Weingarten operator). By Cartan's equations on N

$$d\varphi_i = \varphi_i^j \wedge \varphi_j, \quad \varphi_i^j = -\varphi_j^i, \tag{E.12}$$

where φ_i^j are the connection 1-forms of the induced connection $\hat{\nabla}$ on N (the Levi-Civita connection of the induced metric $\hat{g} = F^*g$ on N). Next we need to recall the Gauss-Weingarten equations

$$g(R(X,Y)Z,W) = \hat{g}(\hat{R}(X,Y)Z,W)$$
$$-g(h(X,W),h(Y,Z)) + g(h(Y,W),h(X,Z)), \tag{E.13}$$
$$g(R(X,Y)Z,E_n) = (\nabla_X h)(Y,Z) - (\nabla_Y h)(X,Z), \tag{E.14}$$

for any $X, Y, Z \in \mathfrak{X}(N)$. Here $\hat{R}$ is the curvature tensor field of $\hat{\nabla}$ while h is the second fundamental form of $F : N \to M$ i.e.,

$$g(h(X,Y),E_n) = \hat{g}(aX,Y), \quad X, Y \in \mathfrak{X}(N).$$

Then $h(e_i, e_j) = h_{ij}E_n$ and $h_{ij} = a^i_j = -A^i_j$. Let us apply the Gauss equation (E.13) for $X = e_k$, $Y = e_\ell$, $Z = e_i$ and $W = e_j$. We obtain

$$R_{ijk\ell} = \hat{R}_{ijk\ell} - A^j_k A^i_\ell + A^j_\ell A^i_k. \tag{E.15}$$

Also let us set

$$(\nabla_{e_i} h)(e_j, e_k) = -A_{jk,i} E_n$$

for some (uniquely defined) smooth functions $A_{jk,i} \in C^\infty(V)$. Then the Weingarten equation (E.14) for $X = e_i$, $Y = e_j$ and $Z = e_k$ gives

$$R_{nkij} = A_{jk,i} - A_{ik,j}. \tag{E.16}$$

Of course, besides from (E.13)–(E.14), in arbitrary codimension the immersion satisfies an additional system of PDEs, the Ricci equation. As N is a real hypersurface $\nabla^\perp E_n = 0$ (where $\nabla^\perp$ is the normal connection) hence, as well known, the Ricci equation gives nothing new (it is equivalent to the Weingarten equation).

Let V be a real m-dimensional inner product space and $B : V \to V$ a symmetric linear operator. For each $r \in \mathbb{Z}$ with $0 \le r \le m$ we define the number $S_r(B) \in \mathbb{R}$ implicitly by

$$\det(I + \lambda B) = S_0(B) + S_1(B)\,\lambda + \cdots + S_m(B)\lambda^m.$$

Of course $S_0(B) = 1$.

Definition E.1 The linear operator $T_r(B) : V \to V$ given by

$$T_r(B) = \sum_{j=0}^{r} (-1)^j S_{r-j}(B) B^j$$

is the *Newton operator* associated to B and r. ■

Here $B^0 = I$ (the identical transformation of V). Newton operators are due to K. Voss, [303]. The linear algebra relevant to Newton operators is described in [260]. In [261] one is mainly interested in the cases where I) $m = n - 1$ and $V = T_x(N)$ and $B = -a_x$ with $x \in N$, and II) $m = n$ and $V = T_x(M)$ and $B = \left(H_f\right)_x$ with $f \in C^\infty(M)$ and $x \in M$. We introduce the functions

$$K_r,\ T_r : N \to \mathbb{R},$$
$$K_r(x) = S_r(-a_x),\quad T_r(x) = T_r(-a_x),\quad x \in N,$$
$$S_r(f),\ T_r(f) : M \to \mathbb{R},$$
$$S_r(f)(x) = S_r\left(\left(H_f\right)_x\right),\quad T_r(f)(x) = T_r\left(\left(H_f\right)_x\right),\quad x \in M.$$

Here we follow the presentation in [261], p. 461. Slightly different conventions are adopted in [17], p. 116. Let $\sigma_r : \mathbb{R}^{n-1} \to \mathbb{R}^{n-1}$ be the rth symmetric function i.e.,

$$\sigma_r(x_1, \ldots, x_{n-1}) = \sum_{1 \le i_1 < \cdots < i_r \le n} x_{i_1} \cdots x_{i_r},\quad 1 \le r \le n-1.$$

If a is the Weingarten, or shape, operator of the isometric immersion $F : N \to M$ as above then we denote by

$$\operatorname{Spec}(a_x) = \{\kappa_1(x), \ldots, \kappa_{n-1}(x)\},\quad x \in N,$$

the principal curvatures of the hypersurface N in M and consider the $n-1$ algebraic invariants

$$s_r : N \to \mathbb{R},$$
$$s_r(x) = \sigma_r(\kappa_1(x), \ldots, \kappa_{n-1}(x)),\quad x \in N.$$

It follows easily that

$$\det(tI - a) = \sum_{r=0}^{n-1} (-1)^r s_r t^{n-r}$$

(pointwise) where $s_0 = 1$ by definition. It is then common to adopt the following

Definition E.2 The function $H_r : N \to \mathbb{R}$ given by

$$\binom{n-1}{r} H_r = s_r,\quad 0 \le r \le n-1,$$

is the rth *mean curvature* of the immersion $F : N \to M$. ■

To relate the approaches[2] in [261] and [17] one may use $\det(I - \lambda a) = \sum_{r=0}^{n-1} S_r(-a)\lambda^r$ with $\lambda = 1/t$ to obtain

$$S_r(-a) = (-1)^r s_r = (-1)^r \binom{n-1}{r} H_r.$$

If $f \in C^\infty(M)$ is a given function then

$$\begin{aligned}
\sum_{r=0}^{n} S_r(f)\lambda^r &= \det\left(I + \lambda H_f\right) \\
&= \sum_{\sigma \in \sigma_n} \epsilon(\sigma)\left(\delta_{1\sigma(1)} + \lambda f_{1\sigma(1)}\right)\cdots\left(\delta_{n\sigma(n)} + \lambda f_{n\sigma(n)}\right) \\
&= \sum_{\sigma \in \sigma_n} \epsilon(\sigma)\Big[\delta_{1\sigma(1)}\cdots\delta_{n\sigma(n)} \\
&\quad + \left(f_{1\sigma(1)}\delta_{2\sigma(2)}\cdots\delta_{n\sigma(n)} + \cdots + \delta_{1\sigma(1)}\cdots\delta_{n-1,\sigma(n-1)}f_{n\sigma(n)}\right)\lambda \\
&\quad + O(\lambda^2)\Big] = 1 + (f_{11} + \cdots + f_{nn})\lambda + O(\lambda^2)
\end{aligned}$$

everywhere in U. Here σ_n is the group of permutations of order $n!$ and $\epsilon(\sigma)$ is the sign of the permutation $\sigma \in \sigma_n$. Hence

$$S_1(f) = \sum_{A=1}^{n} f_{AA} = \sum_{A=1} \{E_A(E_A(f)) - (\nabla_{E_A}E_A)(f)\} = -\Delta f.$$

We shall need the following

Lemma E.3 (E. Reilly, [260])
i. *If $f \in C^\infty(M)$ then*

$$\left(\operatorname{div} T_r(f)\right)_A = \frac{1}{(r-1)!}\delta^{A_1\cdots A_r A}_{B_1\cdots B_r B} f_{A_1B_1}\cdots f_{A_{r-1}B_{r-1}} R_{CB_rBA_r} f_C. \quad \text{(E.17)}$$

ii. *If $F : N \to M$ is an isometric immersion with shape operator a and $A = -a$ then*

$$(\operatorname{div} T_r)_i = \frac{1}{(r-1)!}\delta^{i_1\cdots i_r i}_{j_1\cdots j_r j} A^{j_1}_{i_1}\cdots A^{j_{r-1}}_{i_{r-1}} R_{nj_ri_rj}. \quad \text{(E.18)}$$

[2] Note also that the Weingarten operators in [261] and [17] differ by a sign.

In (E.17)–(E.18) repeated indices indicate summation. Also

$$\delta^{A_1\cdots A_r A}_{B_1\cdots B_r B} = 0, \quad \{A_1, \ldots, A_r, A\} \neq \{B_1, \ldots, B_r, B\},$$

$$\delta^{A_1\cdots A_r A}_{B_1\cdots B_r B} = \epsilon(\sigma), \quad \sigma = \begin{pmatrix} A_1 & \cdots & A_r & A \\ B_1 & \cdots & B_r & B \end{pmatrix} \in \sigma_{r+1}\,.$$

Let us assume from now on that M is a compact oriented Riemannian manifold with boundary $N = \partial M$. Let $\Omega = d\,\mathrm{vol}(g)$ and $\Psi = d\,\mathrm{vol}(\hat{g})$ be the Riemannian volume elements on M and N respectively and ν the exterior unit normal field on N.

Lemma E.4 (E. Reilly, [261]) *If $f : M \to \mathbb{R}$ is a C^∞ function then*

$$\int_M (r+1) S_{r+1}(f)\,\Omega = \int_N g\big(T_r(f)\,\nabla f\,,\,\nu\big)\,\Psi$$

$$-\frac{1}{(r-1)!}\delta^{A_1\cdots A_r A}_{B_1\cdots B_r B}\int_M f_{A_1B_1}\cdots f_{A_{r-1}B_{r-1}}\,R_{CB_rBA_r} f_C f_A\,\Omega. \tag{E.19}$$

Lemma E.5 (E. Reilly, [261])
Let $f \in C^\infty(M)$ and $z = f\big|_N = f \circ F \in C^\infty(N)$. Also let $u = \partial f/\partial\nu = g(\nabla f, \nu) : N \to \mathbb{R}$. Then

$$\int_N g(T_r(f)\nabla f\,,\,\nu)\,\Psi = \int_N \Big\{\tilde{S}_r(z)u - \big(\tilde{T}_{r-1}\big)_{ij} z_i \big(u_j + z_k A_{kj}\big)\Big\}\,\Psi \tag{E.20}$$

where

$$\tilde{S}_r(z) = \frac{1}{r!}\,\delta^{i_1\cdots i_r}_{j_1\cdots j_r}\,\big(z_{i_1j_1} - uA_{i_1j_1}\big)\cdots\big(z_{i_rj_r} - uA_{i_rj_r}\big),$$

$$\big(\tilde{T}_{r-1}\big)_{ij} = \frac{1}{(r-1)!}\,\delta^{i_1\cdots i_{r-1} i}_{j_1\cdots j_{r-1} j}\,\big(z_{i_1j_1} - uA_{i_1j_1}\big)\cdots\big(z_{i_{r-1}j_{r-1}} - uA_{i_{r-1}j_{r-1}}\big).$$

The proof of Lemmas E.3, E.4 and E.5 is omitted (see [261], p. 461–462). Combining Lemmas E.3 up to E.5 leads to the following

Theorem E.6 *Let M be a compact oriented Riemannian manifold with boundary $N = \partial M$ and let ν be the exterior unit normal field on N. Then*

$$(r+1)\int_M S_{r+1}(f)\,\Omega = \int_N \left\{\tilde{S}_r(z)u - \left(\tilde{T}_{r-1}\right)_{ij} z_i \left(u_j + z_k A_{kj}\right)\right\}\Psi$$

$$-\frac{1}{(r-1)!}\delta^{A_1\cdots A_r A}_{B_1\cdots B_r B}\int_M f_{A_1B_1}\cdots f_{A_{r-1}B_{r-1}}\,R_{CB_rBA_r}f_C f_A\,\Omega \qquad \text{(E.21)}$$

for any $f \in C^\infty(M)$.

We shall need the identity (E.21) for $r = 1$. If this is the case then

$$2\int_M S_2(f)\Omega = \int_N \left\{\tilde{S}_1(z)u - \left(\tilde{T}_0\right)_{ij} z_i(u_j + z_k A_{kj})\right\}\Psi$$

$$-\delta^{A_1A}_{B_1B}\int_M R_{CB_1BA_1}f_C f_A \Omega \qquad \text{(E.22)}$$

and

$$\tilde{S}_1(z) = \delta^i_j\left(z_{ij} - uA_{ij}\right) = \sum_{i=1}^{n-1}(z_{ii} - uA_{ii})$$

$$= \sum_i [\mathrm{Hess}(z)(E_i, E_i) + u\,\mathrm{trace}(a)] = -\hat{\Delta}z + uH$$

where $\hat{\Delta}$ is the Laplace-Beltrami operator of the Riemannian manifold $(N, \hat{g})$ and $H = \mathrm{trace}(a)$ is the (non-normalized) mean curvature of $N = \partial M$ in (M, g). Also

$$\left(\tilde{T}_0\right)_{ij} z_i(u_j + z_k A_{kj}) = \delta^i_j z_i(u_j + z_k A_{kj}) = z_i(u_i + z_k A_{ki})$$

$$= \hat{g}\left(\hat{\nabla}z, \hat{\nabla}u\right) - \hat{g}\left(a\hat{\nabla}z, \hat{\nabla}z\right)$$

where $\hat{\nabla}z$ and $\hat{\nabla}u$ are the gradients of z and u with respect to the induced metric $\hat{g}$. Finally we may compute the curvature term

$$\delta^{AB}_{CD}R_{ECDA}f_E f_B = \sum_{A,B}\left\{\delta^{AB}_{AB}R_{EABA}f_E f_B + \delta^{AB}_{BA}R_{EBAA}f_E f_B\right\}$$

$$= (R_{EABA} - R_{EBAA})f_E f_B$$

and $R_{EBAA} = 0$ while $R_{EABA} = -R_{AEBA} = -R_{EB}$ hence

$$\delta^{AB}_{CD} R_{ECDA} f_E f_B = -R_{AB} f_A f_B = -\mathrm{Ric}(\nabla f, \nabla f).$$

Consequently (E.22) may be written

$$\int_M \{2 S_2(f) - \mathrm{Ric}(\nabla f, \nabla f)\} \, \Omega$$

$$= \int_{\partial M} \left\{ \left(-\hat{\Delta} z + uH \right) u - \hat{g}\left(\hat{\nabla} z, \hat{\nabla} u \right) + \hat{g}\left(a \hat{\nabla} z, \hat{\nabla} z \right) \right\} \Psi. \qquad \text{(E.23)}$$

The identity (E.23) is referred to as *Reilly's formula*. One may compare (E.23) and the formula (14) in [261], p. 463 (the sign of A_{ij} in [261] is opposite to that of the usual Weingarten operator). Let us compute $S_2(f)$. We have

$$\sum_{A=0}^{n} S_A(f) \lambda^A = \det\left(I + \lambda H_f \right) = 1 - (\Delta f)\lambda + \sum_{\sigma \in \sigma_n} \epsilon(\sigma) s_\sigma \lambda^2 + (\lambda^3)$$

where s_σ is the function

$$f_{1\sigma(1)} f_{2\sigma(2)} \delta_{3\sigma(3)} \cdots \delta_{n\sigma(n)} + \cdots + \delta_{1\sigma(1)} \cdots \delta_{n-2,\sigma(n-2)} f_{n-1,\sigma(n-1)} f_{n\sigma(n)}.$$

Let $I = \{1, \ldots, n\}$ and let $i, j \in I$ such that $i < j$. Next let us consider the mutually disjoint subsets of σ_n

$$A_{ij} = \{\sigma \in \sigma_n : \sigma(k) = k, \quad \forall\, k \in I \setminus \{i, j\}\}.$$

Clearly

$$\sigma \in \sigma_n \setminus \bigcup_{1 \le i < j \le n} A_{ij} \Longrightarrow s_\sigma = 0.$$

Therefore

$$S_2(f) = \sum_{\sigma \in \sigma_n} \epsilon(\sigma) s_\sigma = \sum_{1 \le i < j \le n} \sum_{\sigma \in A_{ij}} \epsilon(\sigma) s_\sigma = \sum_{i<j} \sum_{\sigma \in A_{ij}} \epsilon(\sigma) f_{i\sigma(i)} f_{j\sigma(j)}.$$

On the other hand $A_{ij} = \{\sigma_0, \sigma_{ij}\}$ where σ_0 is the identical permutation and σ_{ij} interchanges i and j so that $\epsilon(\sigma_{ij}) = -1$. Thus

$$S_2(f) = \sum_{i<j} \left(f_{ii} f_{jj} - f_{ij}^2 \right).$$

Finally $\Delta f = -\sum_{i=1}^{n} f_{ii}$ and

$$\begin{aligned}\|\text{Hess}(f)\|^2 &= \sum_{i,j=1}^{n} f_{ij}^2 = \sum_{i} f_{ii}^2 + 2\sum_{i<j} f_{ij}^2 \\ &= \left(\sum_{i} f_{ii}\right)^2 - 2\sum_{i<j} f_{ii}f_{jj} + 2\sum_{i<j} f_{ij}^2 = (\Delta f)^2 - 2S_2(f)\end{aligned}$$

so that

$$2S_2(f) = (\Delta f)^2 - \|\text{Hess}(f)\|^2. \tag{E.24}$$

Using (E.24) Reilly's formula (E.23) may be written as

$$\begin{aligned}&\int_M \left\{(\Delta f)^2 - \|\text{Hess}(f)\|^2 - \text{Ric}(\nabla f, \nabla f)\right\} \Omega \\ &= \int_{\partial M} \left\{\left(-\hat{\Delta} z + uH\right) u - \hat{g}(\hat{\nabla} z, \hat{\nabla} u) + \hat{g}\left(a\hat{\nabla} z, \hat{\nabla} z\right)\right\} \Psi.\end{aligned} \tag{E.25}$$

Note that (E.25) and the formula in [224], p. 1090–1091, coincide except for the different sign convention for the shape operator.

REFERENCES

[1] K. Abe, *A complex analog of the Hartman-Nirenberg cylinder theorem*, J. Diff. Geom., 7(1972), 453–460.

[2] M.T.K. Abbassi & G. Calvaruso & D. Perrone, *Harmonic sections of tangent bundles equipped with g-natural Riemannian metrics*, Quart. J. Math., (2)62(2011), 259–288.

[3] M.T.K. Abbassi & G. Calvaruso & D. Perrone, *Harmonicity of unit vector fields with respect to Riemannian g-natural metrics*, Differential Geometry and its Applications, 27(2009), 157–169.

[4] M.T.K. Abbassi & G. Calvaruso & D. Perrone, *Harmonic maps defined by the geodesic flow*, Houston J. Math., (1)36(2010), 69–90.

[5] M.T.K. Abbassi & G. Calvaruso & D. Perrone, *Some examples of harmonic maps for g-natural metrics*, Ann. Math. Blaise Pascal, 16(2009), 305–320.

[6] M.T.K. Abbassi & M. Sarih, *On some hereditary properties of Riemannian g-natural metrics on tangent bundles of Riemannian manifolds*, Diff. Geometry and its Applications, (1)22(2005), 19–47.

[7] M.T.K. Abbassi & O. Kowalski, *Naturality of homogeneous metrics on Stiefel manifolds* $SO(m+1)/SO(m-1)$, Differential Geometry and its Applications, 28(2010), 131–139.

[8] K.M.T. Abbassi & G. Calvaruso, *g-natural contact metrics on unit tangent sphere bundles*, Monatsh. Math. 151(2006) 89–109.

[9] R. Adams, *Sobolev spaces*, Academic Press, New York-San Francisco-London, 1975.

[10] R. Adler & M. Bazin & M. Schiffer, *Introduction to general relativity*, McGraw-Hill Book Co., New York-St. Louis-San Francisco-Toronto-London-Sydney, 1965.

[11] R.M. Aguilar, *Isotropic almost complex structures on tangent bundles*, Manuscripta Math., (4)90(1996), 429–436.

[12] P. Ahmadi & S.M.B. Kashani, *Cohomogeneity one anti De Sitter space* H_1^3, Bulletin of the Iranian Mathematical Society, (1)35(2009), 223–235.

[13] H. Akbar-Zadeh & E. Bonan, *Structure presque Kählérienne naturelle sur le fibré tangent à une variété Finsleriénne*, C.R. Acad. Sci. Paris, 258(1964), 5581–5582.

[14] A.V. Alekseevsky & D.V. Alekseevsky, *G-manifolds with one dimensional orbit space*, Adv. Sov. Math., 8(1992), 1–31.

[15] A.V. Alekseevsky & D.V. Alekseevsky, *Riemannian G-manifolds with one dimensional orbit space*, Ann. Global Anal. Geom., 11(1993), 197–211.

[16] L.J. Alías & A. Romero & M. Sánchez, *Uniqueness of complete spacelike hypersurfaces of constant mean curvature in generalized Robertson-Walker space-times*, Gen. Relat. Grav., 27(1995), 71–84.

[17] L.J. Alías & N. Gürbüz, *An extension of Takahashi theorem for the linearized operators of the higher order mean curvatures*, Geometriae Dedicata, 121(2006), 113–127.

[18] T. Aubin, *Nonlinear analysis on manifolds. Monge-Ampère equations*, Grundlehren der mathematischen Wissenscaften 252, Springer-Verlag, New York-Heidelberg-Berlin, 1982.

[19] P. Baird & J.C. Wood, *Harmonic morphisms between Riemannian manifolds*, London Math. Soc. Monographs, New Series, Vol. 29, Clarendon Press, Oxford, 2003.

[20] A. Banyaga, *On characteristics of hypersurfaces in symplectic manifolds*, Proc. Symp. Pure Math., 54(1993), 9–17.

[21] E. Barletta, *On the boundary behavior of the holomorphic sectional curvature of the Bergman metric*, Le Matematiche, (2)LXI(2006), 301–316.

[22] E. Barletta, *On the Dirichlet problem for the harmonic vector fields equation*, Nonlinear Analysis, 67(2007), 1831–1846.

[23] E. Barletta & S. Dragomir, *Differential equations on contact Riemannian manifolds*, Ann. Scuola Norm. Sup. Pisa, Cl. Sci., (4)30(2001), 63–95.

[24] E. Barletta & S. Dragomir & K.L. Duggal, *Foliations in Cauchy-Riemann geometry*, Mathematical Surveys and Monographs, Vol. 140, American Mathematical Society, 2007.

[25] E. Barletta & S. Dragomir & H. Urakawa, *Pseudoharmonic maps from a nondegenerate CR manifold into a Riemannian manifold*, Indiana University Mathematics Journal, (2)50(2001), 719–746.

[26] E. Barletta & S. Dragomir & H. Urakawa, *Yang-Mills fields on CR manifolds*, J. Math. Phys., (8)47(2006), 1–41.

[27] J. Berndt & H. Tamaru, *Cohomogeneity one actions on noncompact symmetric spaces with a totally geodesic singular orbit*, Tohoku Math. J., 56(2004), 163–177.

[28] M. Barros & A. Romero, *Indefinite Kähler manifolds*, Math. Ann., 261(1982), 55–62.

[29] W. Barthel, *Nichtlineare Zusammenhänge und deren Holonomiegruppen*, J. Reine Angew. Math., 212(1963), 120–149.

[30] E. Bedford & M. Kalka, *Foliations and complex Monge-Ampère equations*, Communications on Pure and Applied Mathematics, XXX(1977), 543–571.

[31] J.K. Beem & P.E. Ehrlich, *Global Lorentzian geometry*, Marcel Dekker, Inc., New York-Basel, 1981.

[32] M. Benyounes & E. Loubeau & C.M. Wood, *Harmonic sections of Riemannian vector bundles and metrics of Cheeger-Gromoll type*, Differential Geometry and its Applications, 22(2007), 322–334.

[33] M. Benyounes & E. Loubeau & C.M. Wood, *Harmonic maps and sections on spheres*, ArXiv: mathDG0703060v1.

[34] M. Benyounes & E. Loubeau & L. Todjihounde, *Harmonic maps and Kaluza-Klein metrics on spheres*, ArXiv: mathDG0809.2725v1 16 Sep 2008.

[35] J. Berndt, *Homogeneous hypersurfaces in hyperbolic spaces*, Math. Z., 229(1998), 589–600.

[36] J. Berndt, *Real hypersurfaces in quaternionic space forms*, J. Reine Angew. Math., 419(1991), 9–26.

[37] J. Berndt & Y.J. Suh, *On ruled real hyperurfaces in complex space forms*, Tsukuba J. Math., 17(1993), 311–322.

[38] J. Berndt & Y.J. Suh, *Real hypersurfaces in complex two-plane Grassmannians*, Monatsh. Math., 127(1999), 1–14.

[39] J. Berndt & F. Tricerri & L. Vanhecke, *Generalized Heisenberg groups and Damek-Ricci harmonic spaces*, Lecture Notes in Math., Vol. 1598, Springer-Verlag, Berlin-Heidelberg-New York, 1995.

[40] A.L. Besse, *Einstein manifolds*, Springer-Verlag, Berlin, 1987.

[41] P. Bidal & G. de Rham, *Les formes différentielles harmoniques*, Commentarii Mathematici Helvetici, 19(1946–47), 1–49.

[42] D.E. Blair, *Riemannian geometry of contact and symplectic manifolds*, Progress in Math., Vol. 203, Birkhäuser, Boston-Basel-Berlin, 2002.

[43] D.E. Blair, *Curvature of contact metric manifolds*, in *Complex, Contact and Symmetric Manifolds* (in honour of L. Vanhecke), Ed. by O. Kowalski & E. Musso & D. Perrone, Progress in Math., Vol. 234, Birkhäuser, Boston-Basel-Berlin, 2005.

[44] D.E. Blair & S. Dragomir, *Pseudohermitian geometry on contact Riemannian manifolds*, Rendiconti di Matematica, Roma, 22(2002), 275–341.

[45] D.E. Blair & S.I. Goldberg, *Topology of almost contact manifolds*, J. Differential Geometry, 1(1967), 347–354.

[46] D.E. Blair & T. Koufogiorgos & B.J. Papantoniou, *Contact metric manifolds satisfying a nullity condition*, Israel J. Math., 91(1995), 189–214.
[47] D.E. Blair & D. Perrone, *A variational characterization of contact metric manifolds with vanishing torsion*, Canad. Math. Bull., (4)35(1992), 455–462.
[48] D.E. Blair & A. Turgut Vanli, *Corected energy of distributions for 3-Sasakian and normal complex contact manifolds*, Osaka J. Math., 43(2006), 193–200.
[49] S. Bochner, *Compact groups of differentiable trnasformations*, Ann. of Math., (2)46(1945), 372–381.
[50] E. Boeckx, *A class of locally ϕ-symmetric contact metric spaces*, Arch. Math., 72(1999), 466–472.
[51] E. Boeckx & L. Vanhecke, *Harmonic and minimal radial vector fields*, Acta Mathematica Hungarica, (4)90(2001), 317–331.
[52] E. Boeckx & L. Vanhecke, *Radial vector fields and harmonic manifolds*, Bull. Soc. Sci. Math. Roumanie (N.S.), (91)43(2000), no. 3–4, 181–185.
[53] E. Boeckx & L. Vanhecke, *Characteristic reflections on unit tangent sphere bundle*, Houston J. Math., 23(1997), 427–448.
[54] E. Boeckx & L. Vanhecke, *Harmonic and minimal vector fields on tangent and unit tangent bundles*, Differential Geometry and its Applications, 13(2000), 77–93.
[55] E. Boeckx & D. Perrone & L. Vanhecke, *Unit tangent sphere bundles and two-point homogeneous spaces*, Periodica Mathematica Hungarica, (2–3)36(1998), 79–95.
[56] E. Boeckx & J.C. Gonzales-Davila & L. Vanhecke, *Energy of radial vector fields on compact rank* 1 *symmetric spaces*, Ann. Global Anal. Geom., 23(2003), 29–52.
[57] E. Boeckx & J.C. Gonzales-Davila & L. Vanhecke, *Stability of the geodesic flow for the energy*, Comm. Math. Univ. Carolinae, (2)43(2002), 201–213.
[58] J.M. Bony, *Principe du maximum, inégalité de Harnak et unicité du problème de Cauchy pour les opérateurs elliptiques dégénéré*, Ann. Inst. Fourier, Grenoble, (1)19(1969), 277–304.
[59] W.M. Boothby, *An introduction to differentiable manifolds and Riemannian geometry*, Academic Press, New York-San Francisco-London, 1975.
[60] A. Borel, *Compact Clifford-Klein forms of symetric spaces*, Topology, 2(1963), 111–122.
[61] V. Borrelli, *Stability of the Reeb vector field of a Sasakian manifold*, Soochow J. Math., (3)30(2004), 283–292.
[62] V. Borrelli & F. Brito & O. Gil-Medrano, *The infimum of the energy of unit vector fields on odd-dimensional spheres*, Ann. Global Anal. Geom., 23(2003), 129–140.
[63] V. Borrelli & O. Gil-Medrano, *A critical radius for unit Hopf vector fields on spheres*, Math. Ann., (4)324(2006), 731–751.
[64] R. Bott & H. Shulman & J. Stasheff, *On the de Rham theory of certain classifying spaces*, Advances in Math., 20(1976), 43–56.
[65] L. Boutet de Monvel & J. Sjöstrand, *Sur la singularité de noyau de Bergman et de Szegö*, As térisque, 34–35(1976), 123–164.
[66] C. Boyer & K. Galicki, *Einstein manifold and contact geometry*, Proc. Amer. Math. Soc., 129(2001), 2419–2430.
[67] C. Boyer & K. Galicki, *On Sasakian-Einstein geometry*, Intern. J. Math., 11(2000), 873–909.
[68] C. Boyer & K. Galicki, *3-Sasakian manifolds*, in *Essays on Einstein manifolds*, Surveys in Differential Geometry, VI, Intern. Press, Boston MA, 1999, 123–184.
[69] G.E. Bredon, *Introduction to compact transformation groups*, Academic Press, New York, 1972.
[70] H. Bremmermann, *On a generalized Dirichlet problem for plurisubharmonic functions and pseudoconvex domains. Characterization of Shilov boundaries*, Trans. Amer. Math. Soc., 91(1956), 246–276.
[71] F. Brito, *Total bending of flows with mean curvature correction*, Differential Geometry and its Applications, 23(2000), 157–163.

[72] F. Brito & P.G. Walczak, *On the energy of unit vector fields with isolated singularities*, Ann. Polonici Math., (3)LXXIII(2000), 269–274.
[73] D. Burns, *Curvature of Monge-Ampère foliations and parabolic manifolds*, Annals of Mathematics, 115(1982), 349–373.
[74] D. Burns, *On the uniqueness and characterization of Grauert tubes*, Lect. Notes Pure Appl. Math., 173(1995), 119–133.
[75] D. Burns & R. Hind, *Symplectic geometry and the uniqueness of Grauert tubes*, J. Geom. Funct. Anal., 11(2001), 1–10.
[76] E. Calabi, *An extension of E. Hop's maximum principle wirth an application to Riemannian geometry*, Duke Math. J., 25(1958), 45–46.
[77] G. Calvaruso & D. Perrone, *H-contact unit tangent sphere bundles*, Rocky Mountain J. Math., (5)37(2007), 1435–1457.
[78] G. Calvaruso & D. Perrone & L. Vanhecke, *Homogeneity on three-dimensional contact metric manifolds*, Israel J. Math., 114(1999), 301–321.
[79] C. Camacho & A.L. Neto, *Geometric theory of foliations*, Birkhäuser, Boston-Basel-Stuttgart, 1985.
[80] S. Campanato, *Equazioni ellittiche del secondo ordine e spazi* $\mathcal{L}^{2,\lambda}$, Ann. Mat. Pura Appl., 69(1965), 321–380.
[81] P.B. Chacon & A.M. Naveira & J.M. Weston, *On the energy of distributions with applications to the quaternionic Hopf fibrations*, Monatsh. Math., 133(2001), 281–294.
[82] P.B. Chacon & A.M. Naveira, *Corrected energy of distributions on Riemannian manifolds*, Osaka J. Math., 41(2004), 97–105.
[83] B-Y. Chen, *Geometry of submanifolds and its applications*, Scince Univ. Tokyo, 1981.
[84] S-C. Chen, *Real analytic regularity of the Szegö projection on circular domains*, Pacific J. Math., (2)148(1991), 225–235.
[85] J-H. Cheng & J.M. Lee, *The Burns-Epstein invariant and deformation of CR structures*, Duke Math. J., (1)60(1990), 221–254.
[86] S.S. Chern & R.S. Hamilton, *On Riemannian metrics adapted to three-dimensional contact manifolds*, Lecture Notes in Math., 1111, Springer-Verlag, Berlin-Heidelberg-New York 1985, 279–305.
[87] S.S. Chern & H. Levine & L. Nirenberg, *Intrinsic norms on complex manifolds*, Global Analysis (papers in honor of K. Kodaira), University of Tokyo Press, Tokyo, 1969, pp. 119–139.
[88] Q.S. Chi, *A curvature characterization of certain locally rank* 1 *symmetric spaces*, J. Diff. Geometry, 28(1988), 187–202.
[89] B-Y. Choi & J.W. Yim, *Distributions on Riemannian manifolds which are harmonic maps*, Tohoku Math.J., 55(2003), 175–188.
[90] S.C. Chun & J.H. Park & K. Sekigawa, *H-contact unit tangent sphere bundles of Einstein manifolds*, Quart. J. Math., (1)62(2011), 59–69.
[91] Y-B. Chung & M. Jeong, *The transformation formula for the Szegö kernel*, Rocky Mountain J. Math., (2)29(1999), 463–471.
[92] R. Critenden, *Covariant differentiation*, Quart. J. Math., (2)13(1962), 285–298.
[93] J. Davidov & O. Muskarov, *Existence of holomorphic functions on twistor spaces*, Bull. Soc. Math. Belg., (2)40(1988), 131–151.
[94] P. Dazord, *Tores Finsleriéns sans points conjugués*, Bull. Soc. Math. France, 99(1971), 171–192.
[95] P. De Bartolomeis, *Generalized twistor space and applications*, in *Seminari di Geometria*, Università di Bologna, Dipartimento di Matematica, Conferenza tenuta il 3 Dicembre 1985, pp. 23–32, 1986.
[96] G. De Rham, *Variétés différentiables*, Actualités Scientifiques et Idustrielles, Hermann, Paris, 1960.

[97] G. Dincă, *Metode variaţionale şi aplicaţii*, Editura Tehnică, Bucureşti, 1980.

[98] M.P. Do Carmo, *Riemannian geometry*, Ed. by R.V. Kadison & I.M. Singer, Birkhäuser, Boston-Basel-Berlin, 1992.

[99] P. Dombrowski, *On the geometry of the tangent bundle*, J. Reine Angew. Math., 210(1961), 73–88.

[100] S. Dragomir, *Geometria diferenţială a fibrărilor asociate*, in *Capitole Speciale de Geometrie Diferenţială*, Ed. by I.D. Teodorescu, Universitatea din Bucureşti, Facultatea de Matematică, Seminarul de Geometrie *G. Vrânceanu*, Bucureşti, 1981, pp. 175–203.

[101] S. Dragomir, *Cauchy-Riemann submanifolds of Kaehlerian Finsler spaces*, Collect. Math., (3)40(1989), 225–240.

[102] S. Dragomir, *Pseudohermitian immersions between strictly pseudoconvex CR manifolds*, American J. Math., (1)117(1995), 169–202.

[103] S. Dragomir & Y. Kamishima, *Pseudoharmonic maps and vector fields on CR manifolds*, J. Math. Soc. Japan, (1)62(2010), 1–35.

[104] S. Dragomir & P. Nagy, *Complex Finsler structures on CR-holomorphic vector bundles*, Rendiconti di Matematica, Roma, Serie VII, 19(1999), 427–447.

[105] S. Dragomir & S. Nishikawa, *Foliated CR manifolds*, J. Math. Soc. Japan, (4)56(2004), 1031–1068.

[106] S. Dragomir & L. Ornea, *Locally conformal Kähler geometry*, Progress in Mathematics, Vol. 155, Birkhäuser, Boston-Basel-Berlin, 1998.

[107] S. Dragomir & D. Perrone, *On the geometry of tangent hyperquadric bundles: CR and pseudoharmonic vector fileds*, Ann. Global Anal. Geom., 30(2006), 211–238.

[108] S. Dragomir & M. Soret, *Harmonic vector fields on compact Lorentz surfaces*, Ricerche mat., DOI 1007/s11587-011-0113-1.

[109] S. Dragomir & J.C. Wood, *Sottovarietà minimali ed applicazioni armoniche*, Quaderni dell'Unione Matematica Italiana, Vol. 35, Pitagora Editrice, Bologna, 1989.

[110] S. Dragomir & G. Tomassini, *Differential Geometry and Analysis in CR Manifolds*, Progress in Mathematics, Vol. 246, Birkhäuser, Boston-Basel-Berlin, 2006.

[111] T. Duchamp & M. Kalka, *Invariants of complex foliations and the Monge-Ampère equation*, Michigan Math. J., 35(1988), 91–115.

[112] J. Eells & L. Lemaire, *A report on harmonic maps*, Bull. London Math. Soc., 10(1978), 1–68.

[113] J. Eells & J.H. Sampson, *Harmonic mappings of Riemannian manifolds*, Amer. J. Math., 86(1964), 109–160.

[114] J. Eells & A. Ratto, *Harmonic maps and minimal immersions with symmetries*, Princeton University Press, Princeton, New Jersey, 1983.

[115] H. Federer, *Geometric measure theory*, Springer, New York, 1969.

[116] C. Fefferman, *The Bergman kernel and biholomorphic equivalence of pseudoconvex domains*, Invent. Math., 26(1974), 1–65.

[117] C. Fefferman & K. Hirachi, *Ambient metric construction of Q-curvature in conformal and CR geometries*, preprint.

[118] G.B. Folland, *A fundamental solution for a subelliptic operator*, Bull. A.M.S., (2)79(1973), 373–376.

[119] B. Foreman, *Complex contact manifolds and hyperkaehler geometry*, Kodai Math. J., 23(2000), 12–26.

[120] B. Franchi & E. Serra, *Convergence of a class of degenerate Ginzburg-Landau functionals and regularity for a subelliptic harmonic map equation*, J. Anal. Math., 100(2006), 281322.

[121] M.P. Gaffney, *A special Stokes theorem on complete Riemannian manifolds*, Ann. of Math., (1)60(1954), 140–145.

[122] H. Geiges, *Normal contact structure on 3-manifolds*, Tohoku Math. J., 49(1997), 415–422.

[123] H. Geiges & J. Gonzalo, *Contact geometry and complex surfaces*, Invent. Math., 121(1995), 147–209.

[124] H. Geiges & J. Gonzalo, *Contact circles on 3-manifolds*, J. Diff. Geometry, 46(1997), 236–286.
[125] M. Giaquinta, *Multiple integrals in the calculus of variations and nonlinear elliptic systems*, Ann. of Math. Studies, 105, Princeton Univ. Press, New Jersey, 1983.
[126] O. Gil-Medrano, *Relationship between volume and energy of unit vector fields*, Differential Geometry and its Applications, 15(2001), 137–152.
[127] O. Gil-Medrano, *Unit vector fields that are critical points of the volume and of the energy: characterization and examples*, Proceedings of the International Conference Curvature in Geometry in honour of Prof. L. Vanhecke, Progress in Math., Vol. 234, Birkhäuser, Boston-Basel-Berlin, 2005, 165–186.
[128] O. Gil-Medrano & J.C. Gonzales-Davila & L. Vanhecke, *Harmonicity and minimality of oriented distributions*, Israel J. Math., 143(2004), 253–279.
[129] O. Gil-Medrano & J.C. Gonzales-Davila & L. Vanhecke, *Harmonic and minimal invariant unit vector fields on homogeneous Riemannian manifolds*, Houston Math J., 27(2001), 377–409.
[130] O. Gil-Medrano & A. Hurtado, *Spacelike energy of timelike unit vector fields on a Lorentzian manifold*, J. of Geometry and Physics, 51(2004), 82–100.
[131] O. Gil-Medrano & E. Llinares-Fuster, *Second variation of volume and energy of vector fields. Stability of Hopf vector fields*, Math. Ann., 320(2001), 531–545.
[132] O. Gil-Medrano & E. Llinares-Fuster, *Minimal unit vector fields*, Tohoku Math. J., 54(2002), 71–84.
[133] H. Gluck & W. Ziller, *On the volume of a unit vector field on the three sphere*, Comment. Math. Helv., 61(1986), 177–192.
[134] G. Giraud, *Sur le probleme de Dirichlet généralisé*, Annales de l'École Normale Supérieure, 46(1929), 131–245.
[135] A. Gleason, *Spaces with a compact Lie group of transformations*, Proc. Amer. Math. Soc., 1(1950), 35–43.
[136] M. Göckeler & T. Schücker, *Differential Geometry, Gauge Theories, and Gravity*, Cambridge Monographs on Mathematical Physics, Cambridge University Press, Cambridge-New York-Port Chester-Melbourne-Sydney, 1987.
[137] C. Godbillon, *Géométrie différentielle et mécanique analytique*, Collection Méthodes, Hermann, Paris, 1968.
[138] S.I. Goldberg, *Curvature and homology*, Dover. Publ., Inc., New York, 1982.
[139] S.I. Goldberg, *Nonnegatively curved contact manifolds*, Proc. Amer. Math. Soc., 96(1986), 651–653.
[140] S.I. Goldberg, *Rigidity of positively curved contact manifolds*, J. London Math. Soc., 42(1967), 257–263.
[141] S.I. Goldberg & D. Perrone, *Contact 3-manifolds with positive scalar curvature*, Contem. Math. Amer. Math. Soc., 127(1992), 59–68.
[142] S.I. Goldberg & D. Perrone & G. Toth, *Cuvature of contact three manifolds with critical metrics*, Lecture notes in Math., 1410, Spriger-Verlag, Berlin-Heidelberg-New York, 1989, 212–222.
[143] S.I. Goldberg & K. Yano, *Integrability of almost cosymplectic structures*, Pacific J. Math., 31(1969), 373–381.
[144] J.C. Gonzàles-Dàvila & L. Vanhecke, *Examples of minimal vector fields*, Ann. Global Anal. Geom., 18(2000), 385–404.
[145] J.C. Gonzàles-Dàvila & L. Vanhecke, *Minimal and harmonic characteristic vector fields on three-dimensional contact metric manifolds*, J. of Geometry, 72(2001), 65–76.
[146] J.C. Gonzàles-Dàvila & L. Vanhecke, *Energy and volume of unit vector fields on three-dimensional Riemannian manifolds*, Differential Geometry and its Applications, 16(2002), 225–244.
[147] A.R. Gover & C.R. Graham, *CR invariant powers of the subLaplacian*, preprint.

[148] C.R. Graham, *The Dirichlet problem for the Bergman Laplacian*, I-II, Comm. Partial Differential Equations, 8(1983), 433–476, 563–641.

[149] C.R. Graham, *Compatibility operators for degenerate elliptic equations on the ball and Heisenberg group*, Math. Z., 187(1984), 289–304.

[150] C.R. Graham, *On Sparling's characterization of Fefferman metrics*, American J. Math., 109(1987), 853–874.

[151] C.R. Graham & J.M. Lee, *Smooth solutions of degenerate Laplacians on strictly pseudoconvex domains*, Duke Math. J., (3)57(1988), 697–720.

[152] H. Grauert, *On Levi's problem and the embedding of real analytic manifolds*, Ann. Math., 68(1958), 460–472.

[153] K. Grove & L. Verdiani & B. Wilking & W. Ziller, *Non-negative curvature obstructions in cohomogeneity one and the Kervaire spheres*, preprint.

[154] V. Guillemin & M. Stenzel, *Grauert tubes and the homogeneous Monge-Ampère equation*, J. Diff. Geometry, 34(1991), 561–570.

[155] S. Halperin & D. Lehmann, *Cohomologies et classes caractéristiques des choux de Bruxelles*, in *Differetial Topology and Geometry*, Dijon, 1974, Ed. by G.P. Joubert & R.P. Moussu & R.H. Roussarie, Lecture Notes in Math., 484(1975), 79–120.

[156] S. Halperin & D. Lehmann, *Twisted exotism*, in *Differential Geometry and Relativity* (a volume in honour of A. Lichnerowicz), Ed. by M. Cahen & M. Flato, D. Reidel Co., Dordrecht, 1976, 67–73.

[157] S.D. Han & J.W. Yim, *Unit vector fields on spheres which are harmonic maps*, Math. Z., 227(1998), 83–92.

[158] Hartman & L. Nirenberg, *On spherical image maps whose Jacobians do not change sign*, Amer. J. Math., 81(1959), 901–920.

[159] K. Hasegawa, *Harmonic sections of normal bundles for submanifolds and their stability*, J. of Geometry, 82(2005), 57–64.

[160] H. Hashimoto & K. Mashimo, *On some 3-dimensional CR submanifolds in S^6*, Nagoya Math. J., 156(1999), 171–185.

[161] H. Hashimoto & K. Mashimo & K. Sekigawa, *On 4-dimensional CR-submanifolds of a 6-dimensional sphere*, Advanced Studies in Pure Math., 34(2002), 143–154.

[162] P.A. Hästö, *On the existence of minimizers of the variable exponent Dirichlet energy integral*, Communications on Pure and Applied Analysis, (3)5(2006), 415–422.

[163] S. Helgason, *Differential geometry, Lie groups, and symmetric spaces*, Pure and Aplied Mathematics, Vol 80, Academic Press, Inc., New York-London-Toronto-Sydney-San Francisco, 1978.

[164] A. Higuchi & B.S. Kay & C.M. Wood, *The energy of unit vector fields on the 3-sphere*, J. Geometry and Physics, 37(2001), 137–155.

[165] S. Hildebrand & H. Kaul & K. Widman, *An existence theorem for harmonic mappings of Riemannian manifolds*, Acta. Math., 138(1977), 1–16.

[166] S. Hildebrand & K. Widman, *On the Hölder continuity of weak solutions of quasilinear elliptic systems of second order*, Ann. Sc. Norm. Sup. Pisa, 4(1977), 145–178.

[167] K. Hirachi, *Logarithmic singularity of the Szegö kernel and a global invariant of strictly pseudoconvex domains*, preprint, to appear in Ann. of Math.

[168] K. Hirachi, *Scalar pseudohermitian invariants and the Szegö kernel on three-dimensional CR manifolds*, in *Complex Geometry*, Lecture Notes in Pure and Applied Math., 143(1993), 67–76.

[169] R. Holubowicz & W. Mozgawa, *On compact non-Kählerian manifolds admitting an almost Kähler structure*, Rendiconti del Circolo Matematico di Palermo, 54(1998), 53–57.

[170] L. Hörmander, *L^2-estimates and existence theorems for the $\overline{\partial}$-operator*, Acta Mathematica, 113(1965), 89–152.

[171] L. Hörmander, *Hypoelliptic second-order differential equations*, Acta Math., 119(1967), 147–171.

[172] S-T. Hu, *Homotopy theory*, Academic Press, New York-San Francisco-London, 1959.
[173] S. Ianus, *Sulle varietà di Cauchy-Riemann*, Rend. dell'Accad. Sci. Fis. Mat., Napoli, 39(1972), 191–195.
[174] S. Ianuş, *Geometrie Diferenţială cu Aplicaţii în Teoria Relativităţii*, ditura Acadeiei Republicii Socialiste România, Bucureşti, 1983.
[175] S. Ianuş & A.M. Pastore, *Harmonic maps on contact metric manifolds*, Ann. Math. Blaise Pascal, (2)2(1995), 43–53.
[176] T. Ishihara, *Harmonic sections of tangent bundles*, J. Math. Tokushima Univ., 13(1979), 23–27.
[177] S. Ishihara & M. Konishi, *Real contact 3-structure and complex contact structure*, Southeast Asian Bull. Math., 3(1979), 151–161.
[178] S. Ishihara & M. Konishi, *Complex almost contact manifolds*, Kodai Math. J., 3(1980), 385–396.
[179] J. Jost & S-T. Yau, *A nonlinear elliptic system for maps from Hermitian to Riemannian manifolds and rigidity theorems in Hermitian geometry*, Acta Math., 170(1993), 221–254.
[180] J. Jost & C-J. Xu, *Subelliptic harmonic maps*, Trans. of A.M.S., (11)350(1998), 4633–4649.
[181] S-J. Kan, *On rigidity of Grauert tubes over homogeneous Riemannian manifolds*, J. Reine Angew. Math., 577(2004), 213–233.
[182] S-J. Kan, *The asymptotic expansion of a CR invariant and Grauert tubes*, Math. Ann., 304(1996), 63–92.
[183] T. Kashiwada, *A note on a Riemannian space with Sasakian 3-structure*, Nat. Sci. Rep. Ochanomizu Univ., 22(1971), 1–2.
[184] T. Kashiwada, *On a contact 3-structure*, Math. Z., 238(2001), 829–832.
[185] J.L. Kazdan & F.W. Warner, *Curvature functions for compact 2-manifolds*, Ann. of Math., 99(1974), 14–47.
[186] M. Kimura, *Sectional curvatures of holomorphic planes on real hypersurfaces in* $\mathbb{P}^n(\mathbb{C})$, Math. Ann., 276(1987), 487–497.
[187] M. Kimura & S. Maeda, *On real hypersurfaces of a complex projective space*, Math. Z., 202(1989), 299–311.
[188] D. Kinderlehrer & G. Stampacchia, *An introduction to variational inequalities and their applications*, Academic Press, London, 1980.
[189] S. Kobayashi & K. Nomizu, *Foundations of Differntial Gometry*, Wiley Interscience, New Yok, Vol. I, 1963, Vol. II, 1969.
[190] E. Koizumi, *The logarithmic term of the Szegö kernel on the boundary of two-dimensional Grauert tubes*, preprint.
[191] I. Kollár & P.W. Michor & J. Slovàk, *Natural operators in differential geometry*, Springer-Verlag, Berlin, 1993.
[192] A. Kollross, *A classification of hyperpolar and cohomogeneity one actions*, Trans. Amer. Math. Soc., (2)354(2001), 571–612.
[193] M. Kon, *Pseudo-Einstein real hypersurfaces in a complex space form*, J. Diff. Geometry, 14(1979), 339–354.
[194] J.J. Konderak, *On sections of fibre bundles which are harmonic maps*, Bull. Math. Soc. Sci. Math. Roumanie, (4)90(1999), 341–352.
[195] A. Korányi & H.M. Reimann, *Contact transformations as limits of symplectomorphisms*, C.R. Acad. Sci. Paris, 318(1994), 1119–1124.
[196] B. Kormaz, *Normality of complex contact manifolds*, Rocky Mountain J., 30(2000), 1343–1380.
[197] T. Koufogiorgos & C. Tsichlias, *On the existence of a new class of contact metric manifolds*, Canadian Math. Bull., 43(2000), 440–447.
[198] T. Koufogiorgos & C. Tsichlias, *Generalized* (k, μ)*-contact metric manifolds with* $\|\text{grad}k\| =$ *constant*, J. of Geometry, (1–2)78(2003), 83–91.

[199] T. Koufogiorgos & M. Markellos & V.J. Papantoniou, *The harmonicity of the Reeb vector field on contact metric three-manifolds*, Pacific J. Math., 234(2008), 325–344.

[200] O. Kowalski & M. Sekizawa, *Natural transformations of Riemannian metrics on manifolds to metrics on tangent bundles - a classification*, Bull. Tokyo Gakugei Univ., (4)40(1988), 1–29.

[201] C. LeBrun, *Fano manifolds, contact structures, and quaternionic geometry*, Internat. Math. J., 6(1995), 419–437.

[202] J.M. Lee & R. Melrose, *Boundary behaviour of the complex Monge-Ampère equation*, Acta Mathematica, 148(1982), 159–192.

[203] J.M. Lee, *The Fefferman metric and pseudohermitian invariants*, Trans. A.M.S., (1)296(1986), 411–429.

[204] L. Lempert, *Elliptic and hyperbolic tubes*, Proceedings of the Special Year in Complex Analysis at the Mittag-Leffler Institute, Princeton University Press, 1987–1988.

[205] L. Lempert, *Complex structures on the tangent bundle of Riemannian manifolds*, Complex Analysis and Geometry, Ed. by V. Ancona & A. Silva, Plenum Press, New York, 1993.

[206] L. Lempert & R. Szöke, *Golbal solutions of the homogeneous complex Monge-Ampère equation and complex structures of the tangent bundle of Riemannian manfolds*, Math. Ann., 290(1991), 689–712.

[207] P.F. Leung, *On the stability of harmonic maps*, Lecture Notes in Math., Vol. 949, Springer-Verlag, 1982, 122–129.

[208] M. Lohnherr & H. Reckziegel, *On ruled real hypersurfaces in complex space forms*, Geom. Dedicata, 74(1999), 267–286.

[209] M. Matsumoto, *Foundations of Finsler geometry and special Finsler spaces*, Kyoto University Press, Kyoto, 1980.

[210] J. Milnor, *Curvature of left invariant metrics on Lie groups*, Advances in Math., 21(1976), 293–329.

[211] R. Mirzaie & S.M.B. Kashani, *On cohomogeneity one flat Riemannian manifolds*, Glasgow Math. J., 44(2002), 185–190.

[212] D. Montgomery & H. Samuelson & L. Zippin, *Singular points of a compact transformation group*, Ann. of Math., (2)63(1956), 1–9.

[213] D. Montgomery & H. Samuelson & C.T. Yang, *Exceptional orbits of highest dimension*, Ann. of Math., (1)64(1956), 131–141.

[214] S. Montiel & A. Romero, *On some real hypersurfaces of a complex hyperbolic space*, Geom. Dedicata, 20(1986), 245–261.

[215] S. Montiel, *Uniqueness of spacelike hypersurfaces of constant mean curvature in foliated space-times*, Math. Ann., 314(1999), 529–553.

[216] A. Morimoto & T. Nagano, *On pseudo-conformal transformations of hypersurfaces*, J. Math. Soc. Japan, (3)15(1963), 289–300.

[217] P. Mounoud, *Some topological and metric properties of the space of Lorentz metrics*, Differential Geometry and its Appl., 115(2001), 47–57.

[218] O. Muskarov, *Existence of holomorphic functions on almost complex manifolds*, Math. Z., 192(1986), 283–295.

[219] O. Muskarov, *Some remarks on genralized Grassmann manifolds*, Acta Math. Hung., (1–2)54(1989), 69–78.

[220] O. Muscarov, *Almost Hermitian structures on twistor spaces and their types*, Atti Sem. Mat. Fis. Univ. Modena, XXXVII(1989), 285–297.

[221] T. Nagano, *Stability of harmonic maps between symmetric spaces*, Proc. Tulane, Lecture Notes in Mathem., Vol. 949, p. 130–137, Springer-Verlag, Berlin-New York, 1982.

[222] L. Ni, *Hermitian harmonic maps from complete Hermitian manifolds to complete Riemannian manifolds*, Math. Z., 232(1999), 331–355.

[223] O. Nouhaud, *Applications harmoniques d'une variété Riemannienne dans son fibré tangent*, C.R. Acad. Sci. Paris, 284(1977), 815–818.

[224] G. Nunes & J. Ripoll, *A note on the infimum of energy of unit vector fields on a compact Riemannian manifold*, J. Geom. Anal., 18(2008), 1088–1097.

[225] M. Okumura, *Contact hypersurfaces in certain Kählerian manifolds*, Tohoku Math. J., (1)18(1966), 74–102.

[226] M. Okumura, *On some real hypersurfaces of a complex projective space*, Trans. Amer. Math. Soc., 212(1975), 355–364.

[227] S. Olszak, *Contact metric manifolds*, Tohoku Math. J., 31(1979), 247–253.

[228] S. Olszak, *On almost cosymplectic manifolds*, Kodai Math. J., (2)4(1981), 239–250.

[229] B. O'Neill, *Semi-Riemannian geometry*, Academic Press, New York-London-Paris-San Diego-San Francisco-Sao Paulo-Sydney-Tokyo-Toronto, 1983.

[230] C. Oniciuc, *Harmonic sections in the unitary tangent bundle*, Demonstratio Math., (3)34(2001), 681–692.

[231] V. Oproiu, *A Kähler Einstein structure on the tangent bundle of a space form*, Int. J. Math. Sci., 25(2001), 183–195.

[232] L. Ornea & L. Vanhecke, *Harmonicity and minimality of vector fields and distributions on locally conformal Kähler and hyperkähler manifolds*, Bull Belg. Math. Soc., 12(2005), 543–555.

[233] R.S. Palais, *A global formulation of the Lie theory of transformation groups*, Memoirs of A.M.S., no. 22, 1957.

[234] R.S. Palais & C.L. Terng, *A general theory of canonical forms*, Trans. Amer. Math. Soc., 300(1987), 771–789.

[235] D.I. Papuc, *Geometrie diferenţială*, Editura Didactică şi Pedagogică, Bucureşti, 1982.

[236] G. Patrizio & P-M. Wong, *Stein manifolds with compact symmetric center*, Math. Ann., 289(1991), 355–382.

[237] D. Perrone, *Contact Riemannian manifolds satisfying* $R(X,\xi)R = 0$, Yokohama Math. J., 39(1992), 141–149.

[238] D. Perrone, *Ricci tensor and spectral rigidity of contact Riemannian three manifolds*, Bull. Inst. Acad. Sinnica, 24(1996), 127–138.

[239] D. Perrone, *Homogeneous contact Riemannian three manifolds*, Illinois Math. J., (2)42(1998), 243–256.

[240] D. Perrone, *Weakly* φ*-symmetric contact metric spaces*, Balkan J. Geom. Appl., (2)7(2002), 67–77.

[241] D. Perrone, *Harmonic characteristic vector fields on contact metric three manifolds*, Bull. Austral. Math. Soc., 67(2003), 305–315.

[242] D. Perrone, *Contact metric manifolds whose characteristic vector field is a harmonic vector field*, Differential Geometry and its Applications, 20(2004), 367–378.

[243] D. Perrone, *Stability of the Reeb vector field of* H*-contact manifolds*, Math. Z., 263(2009), 125–147.

[244] D. Perrone, *On the volume of unit vector fields on Riemannian three manifolds*, C.R. Math. Rep. Acad. Sci. Canada, (1)30(2008), 11–21.

[245] D. Perrone, *Contact Riemannian manifolds with* ξ*-parallel torsion*, in *Selected topics in Cauchy-Riemann geometry*, ed. by S. Dragomir, Quaderni di Matematica, Dipartimento di Matematica, II Università di Napoli, Caserta, 9(2001), 308–336.

[246] D. Perrone, *Taut contact circles on* H*-contact three manifolds*, Intern. Math. Forum, (26)1(2006), 1285–1296.

[247] D. Perrone, *The rough Laplacian and harmonicity of Hopf vector fields*, Ann. Global Anal. Geom., 28(2005), 91–106.

[248] D. Perrone, *Hypercontact metric three-manifolds*, C.R. Math. Rep. Acad. Sci. Canada, (3)24(2002), 97–101.

[249] D. Perrone, *Corrected energy of the Reeb distribution of a 3-Sasakian manifold*, Osaka J. Math., 45(2008), 615–627.

[250] D. Perrone, *Unit vector fields on real spaces forms, which are harmonic maps*, Pacific J. Math., (1)239(2009), 89–104.

[251] D. Perrone, *Minimality, harmonicity and CR geometry for Reeb vector fields*, Int. J. of Math., (9)21(2010), 1189–1218.

[252] D. Perrone, *Instability of the geodesic flow for the energy functional*, Pacific J. Math., (2)249(2011), 431–446.

[253] D. Perrone, *Un'introduzione alla geometria riemanniana*, Aracne Editrice, Roma, 2011.

[254] D. Perrone & L. Vergori, *Stability of contact metric manifolds and unit vector fields of minimum energy*, Bull. Austr. Math. Soc., 76(2006), 269–283.

[255] R. Petit, *Harmonic maps and strictly pseudoconvex CR manifolds*, Communications in Analysis and Geometry, (3)10(2002), 575–610.

[256] F. Podesta & A. Spiro, *Some topological properties of cohomogeneity one manifolds with negative curvature*, Ann. Global Anal. Geom., 14(1996), 69–79.

[257] R. Ponge, *Szegö projections and new invariants for CR and contact manifolds*, ArXiv: math.DG/0601370 v1 15 Jan 2006.

[258] W.A. Poor, *Differential geometric structures*, McGraw Hill, New York, 1981.

[259] E.V. Radkevic, *Hypoelliptic operators with multiple characteristics*, Math. USSR Sb., 8(1969), 181–205.

[260] E. Reilly, *On the Hessian of a function and the curvatures of its graph*, Michgan Math. J., 20(1973), 373–383.

[261] E. Reilly, *Applications of the Hessian operator in a Riemannian manifold*, Indiana Univ. Math. J., 26(1977), 459–472.

[262] P. Rukimbira, *Criticality of K-contact vector fields*, J. Geometry and Physics, 40(2002), 209–214.

[263] P.J. Ryan, *Homogeneity and some curvature conditions for hypersurfaces*, Tohoku Math. J., 21(1969), 363–388.

[264] P.J. Ryan, *Hypersurfaces with parallel Ricci tensor*, Osaka J. Math., 8(1971), 251–259.

[265] J.H. Sampson, *Some properties and applications of harmonic mappings*, Ann. Sc. Ec. Norm. Sup., 11(1978), 211–228.

[266] S. Sasaki, *On differential geometry of tangent bundles of Riemannian manifolds*, Tôhoku Math. J., 10(1958), 338–354.

[267] S. Sasaki, *Spherical space forms with normal contact metric 3-structure*, J. Differential Geometry, 6(972), 307–315.

[268] R. Schoen, *The effect of curvature on the behavior of harmonic functions and mappings*, in *Nonlinear partial differential equations in differential geometry*, IAS/Park City Mathematics Series, Ed. by R. Hardt & M. Wolf, American Mathematical Society, Institute for Advanced Study, Vol. 2, 127–184, 1996.

[269] C. Searle, *Cohomogeneity and positive curvature in low dimensions*, Math. Z., 214(1993), 491–498.

[270] N. Shimakura, *Partial differential operators of elliptic type*, Translations of Mathematical Monographs, Vol. 99, American Mathematical Society, Providence, Rhode Island, 1992.

[271] B. Solomon, *Harmonic maps to spheres*, J. Differential Geometry, 21(1985), 151–162.

[272] F. Sommer, *Komplex-analytische Blätterung reeler Hyperplachen im $\mathbb{C}^n$*, Math. Ann., 137(1959), 392–411.

[273] R.T. Smith, *The second variation formula for harmonic mappings*, Proc. Amer. Math. Soc., 47(1975), 229–236.

[274] M.B. Stenzel, *Orbits of the geodesic flow and chains on the boundary of the Grauert tube*, Math. Ann., 322(2002), 383–399.

[275] W. Stoll, *The characterization of strictly parabolic manifolds*, Ann. Sc. Norm. Sup., Pisa, VII(1980), 87–154.

[276] R. Stong, *Contact manifolds*, J. Differential Geometry, 9(1974), 219–238.

[277] R. Szöke, *Complex structures on tangent bundles of Riemannian manifolds*, Math. Ann., 291(1991), 409–428.

[278] S. Tachibana, *On harmonic tensors in compact Sasakian spaces*, Tohoku Math. J., (2)17(1965), 271–284.

[279] N. Tanaka, *A differential geometric study on strongly pseudo-convex manifolds*, Kinokuniya Book Store Co., Ltd., Kyoto, 1975.

[280] S. Tanno, *The topology of contact Riemannian manifolds*, Illinois J. Math., 12(1968), 700–717.

[281] S. Tanno, *Variational problems on contact Riemannian manifolds*, Trans. Amer. Math. Soc., 314(1989), 349–379.

[282] S. Tanno, *The standard CR structure on the unit tangent bundle*, Tôhoku Math. J., 44(1992), 535–543.

[283] S. Tanno, *Sasakian manifolds with constant ϕ-holomorphic sectional curvature*, Tohoku Math. J., 21(1969), 501–507.

[284] S. Tanno, *Harmonic forms and Betti numbers of certain contact Riemannian manifold*, J. Math. Soc. Japan, (3)19(1967), 308–316.

[285] Y. Tashiro, *On contact structures of tangent sphere bundles*, Tohoku Math. J., 21(1969), 117–143.

[286] C.B. Thomas, *A classifying space for the contact pseudogroup*, Mathematika, 25(1978), 191–201.

[287] C.B. Thomas, *Almost regular contact manifolds*, J. Differential Geometry, 9(1974), 219–238.

[288] Ph. Tondeur, *Foliations on Riemannian manifolds*, Springer-Verlag, New York-Berlin-Heidelberg-London-Paris-Tokyo, 1988.

[289] F. Tricerri & L. Vanhecke, *Homogeneous structures on Riemannian manifolds*, London Math. Soc. Lect. Notes Series, 83, Cambridge Univ. Press, Cambridge, 1983.

[290] K. Tsukada & L. Vanhecke, *Minimal and harmonic vector fields in $G_2(\mathbb{C}^{m+2})$ and its dual spaces*, Monatsh. Math., 130(2000), 143–154.

[291] K. Tsukada & L. Vanhecke, *Minimality and harmonicity for Hopf vector fields*, Illinois J. Math., 45(2001), 441–451.

[292] H. Urakawa, *Calculus of variations and harmonic maps*, Transl. Math. Monographs Amer. Math. Soc., 132, 1993.

[293] H. Urakawa, *Yang-Mills connections over compact strongly pseudo-convex CR manifolds*, Math. Z., 216(1994), 541–573.

[294] H. Urakawa, *Stability of harmonic maps and eigenvalues of the Laplacian*, Trans. Amer. Math. Soc., (2)301(1987), 557–589.

[295] H. Urakawa, *The first eigenvalue of the Laplacian for positively curved homogeneous Riemannian manifolds*, Compositio Mathem., 59(1986), 57–71.

[296] I. Vaisman, *Remarques sur la théorie des formes-jet*, C.R. Acad. Sci. Paris, 264-A(1967), 351–354.

[297] I. Vaisman, *New examples of twisted cohomologies*, Bollettino U.M.I., (7)**7-B**(1993), 355–368.

[298] L. Vanhecke, *Geometry in normal and tutbular neighborhoods*, Rend. Sem. Fac. Sci. Univ. Cagliari, 58(1988), 73–176.

[299] E. Vergara-Diaz & C.M. Wood, *Harmonic almost contact structures*, Geometriae Dedicata, (4)123(2006), 131–151.

[300] E. Vergara-Diaz & C.M. Wood, *Harmonic almost contact metric structures, and submersions*, Int. J. of Math., (2)20(2009), 209–225.

[301] L. Verhóczki, *Special cohomogeneity one isometric actions on irreducible symmetric spaces of Types I and II*, Beiträge zur Algebra und Geometrie, Contributions to Algebra and Geometry, (1)44(2003), 57–74.

[302] M.H. Vernon, *Contact hypersurfaces of a complex hyperbolic space*, Tohoku Math. J., 39(1987), 215–222.
[303] K. Voss, *Einige differentialgeometrische Kongruenzsatze für geschlossene Flčhen und Hyperflächen*, Math. Ann., 3(1956), 180–218.
[304] G. Vrânceanu, *O proprietate remarcabilă a torului*, Opera Matematică, Vol. IV, pp. 297–301, Editura Academiei Republicii Socialiste România, Bucureşti, 1977.
[305] F.W. Warner, *Extension of the Rauch comparison theorem to submanifolds*, Trans. Amer. Math. Soc., 122(1966), 341–356.
[306] Y. Watanabe, *Geodesic symmetries in Sassakian ϕ-symmetric spaces*, Kodai Math. J. (1)3(1980), 48–55.
[307] S.M. Webster, *Pseudohermitian structures on a real hypersurface*, J. Diff. Geometry, 13(1978), 25–41.
[308] R.O. Wells, *Differential Analysis on Complex Manifolds*, Graduate Texts in Mathematics, Springer-Verlag, New York-Heidelberg-Berlin, 1980.
[309] G. Wiegmink, *Total bending of vector fields on Riemannian manifolds*, Math. Ann., (2)303(1995), 325–344.
[310] G. Wiegmink, *Total bending of vector fields on the sphere S^3*, Differential Geometry and its Applications, 6(1996), 219–236.
[311] J.A. Wolf, *Spaces of constant curvature*, Publish or Perish, Inc., Wilmington, Delaware, 1984 (fifth edition).
[312] J.A. Wolf, *A contact structure for odd dimensional spherical space forms*, Proc. Amer. Math. Soc., 19(1968), 196.
[313] P-M. Wong, *Geometry of the complex homogeneous Monge-Ampère equations*, Invent. Math., 67(1982), 261–274.
[314] P-M. Wong, *On umbilical hypersurfaces and uniformization of circular domains*, Proceedings of Symposia in Pure Mathematics, 41(1984), 225–252.
[315] C.M. Wood, *An existence theorem for harmonic sections*, Manuscripta Math., 68(1990), 69–75.
[316] C.M. Wood, *On the energy of a unit vector field*, Geom. Dedicata, 64(1997), 319–330.
[317] C.M. Wood, *The energy of Hopf vector fields*, Manuscripta Math., 101(2000), 71–78.
[318] C.M. Wood, *Harmonic sections of homogeneous fibre bundles*, Diff. Geometry and its Applications, (2)19(2003), 193–210.
[319] C.M. Wood, *Bending and stretching unit vector fields in Euclidean and hyperbolic 3-space*, ArXiv: mathDG/0612286.
[320] C.M. Wood, *Harmonic almost complex structures*, Compositio Math., 99(1995), 183–212.
[321] Y.L. Xin, *Some results on stable harmonic maps*, Duke Math. J., (1980), 319–330.
[322] C-J. Xu, *Subelliptic variational problems*, Bull. Soc. Math. France, 118(1990), 147–159.
[323] A. Yampolsky, *A totally geodesic property of Hopf vector fields*, Acta Math. Hungar., (1–2) 101(2003), 93–112.
[324] K. Yano & S. Ishihara, *Tangent and cotangent bundles: differential geometry*, Pure and Applied Mathematics, No. 16. Marcel Dekker, Inc., New York, 1973, 423 pp.
[325] K. Yano & M. Kon, *CR submanifolds of Kaehlerian and Sasakian manifolds*, Progress in Math., Vol. 30, Ed. by J. Coates & S. Helgason, Birkhäuser, Boston-Basel-Stuttgart, 1983.
[326] K. Yano & T. Nagano, *On geodesic vector fields in a compact orientable Riemannian space*, Comm. Math. Helv., 35(1961), 55–64.
[327] Z.R. Zhou, *Uniqueness of subelliptic harmonic maps*, Ann. Global Anal. Geom., 17(1999), 581–594.

INDEX

Printed and bound by CPI Group (UK) Ltd, Croydon, CR0 4YY
08/05/2025
01864880-0001